PATTY'S INDUSTRIAL HYGIENE AND TOXICOLOGY

Fourth Edition

Volume I, Parts A and B
GENERAL PRINCIPLES

Volume II, Parts A, B, C, D, E, and F
TOXICOLOGY

Third Edition

Volume III, Parts A and B
THEORY AND RATIONALE
OF INDUSTRIAL HYGIENE
PRACTICE

PATTY'S INDUSTRIAL HYGIENE AND TOXICOLOGY

Third Edition
Volume III, Part B
Theory and Rationale
of Industrial Hygiene
Practice: Biological Responses

LEWIS J. CRALLEY
LESTER V. CRALLEY
JAMES S. BUS

Editors

CONTRIBUTORS

A. A. Bove
J. S. Bus
A. Cohen
L. J. Cralley
L. V. Cralley
J. E. Gibson
K. A. Grant
N. C. Hawkins
R. D. Heimbach

R. T. Hitchcock
J. P. Hughes
J. L. Mattsson
J. P. J. Maurissen
S. M. Michaelson
W. E. Murray
R. J. Nolan
R. D. Novak
C. N. Park

R. M. Patterson
V. Putz-Anderson
M. H. Smolensky
W. T. Stott
J. A. Swenberg
R. G. Thomas
R. S. Waritz
P. G. Watanabe
P. K. Working

A Wiley-Interscience Publication

JOHN WILEY & SONS, INC.

New York / Chichester / Brisbane / Toronto / Singapore

Copyright © 1995 by John Wiley & Sons, Inc.

All rights reserved. Published simultaneously in Canada.

Reproduction or translation of any part of this work beyond
that permitted by Section 107 or 108 of the 1976 United
States Copyright Act without the permission of the copyright
owner is unlawful. Requests for permission or further
information should be addressed to the Permissions Department,
John Wiley & Sons, Inc., 605 Third Avenue, New York, NY
10158-0012.

Library of Congress Cataloging in Publication Data:

Theory and rationale of industrial hygiene practice.
    At head of title: Patty's Industrial hygiene and
toxicology, volume III.
    Includes bibliographical references and indexes.
    Contents: 3A. The work environment—3B. Biological responses
    1. Industrial hygiene. 2. Industrial toxicology.
I. Cralley, Lewis J., 1911–   . II. Cralley,
Lester V. III. Harris, Robert L., 1924–   .
IV. Patty's Industrial hygiene and toxicology.
    RC957.T48   1993   613.6'2   93-23747
    ISBN 0-471-53066-2 (v. 3, pt. A: acid-free paper)
    ISBN 0-471-53065-4 (v. 3, pt. B: acid-free paper)

Printed in the United States of America

10 9 8 7 6 5 4 3 2 1

Contributors

Alfred A. Bove, M.D., Ph.D., Section of Cardiology, Temple University Hospital, Philadelphia, Pennsylvania

James S. Bus, Ph.D., Dow Chemical Company, Toxicology Research Laboratory, Midland, Michigan

Alexander Cohen, Ph.D., Division of Biomedical and Behavioral Science, NIOSH—Taft Laboratories, Cincinnati, Ohio

Lester V. Cralley, Ph.D., Fallbrook, California

Lewis J. Cralley, Ph.D., Cincinnati, Ohio

James E. Gibson, Ph.D., Dow Elanco and Company, Indianapolis, Indiana

Katharyn A. Grant, Ph.D., Division of Biomedical and Behavioral Science, NIOSH—Taft Laboratories, Cincinnati, Ohio

Neil C. Hawkins, Ph.D., Dow Chemical Company, Midland, Michigan

Richard D. Heimbach, M.D., Ph.D., Methodist Hospital, San Antonio, Texas

R. Timothy Hitchcock, M.S.P.H., CIH, Cary, North Carolina

James P. Hughes, M.D., Piedmont, California

Joel L. Mattsson, D.V.M., Ph.D., Toxicology Research Laboratory, Dow Chemical Company, Midland, Michigan

Jacques P. J. Maurissen, Ph.D., Toxicology Research Laboratory, Dow Chemical Company, Midland, Michigan

Sol M. Michaelson, D.V.M., Deceased

William E. Murray, M.S., Physical Agents Effects Branch, Division of Biomedical and Behavioral Science, NIOSH, Cincinnati, Ohio

Richard J. Nolan, Ph.D., Toxicology Research Laboratory, Dow Chemical Company, Midland, Michigan

R. D. Novak, Ph.D., M.P.H., M.P.A., Department of Medicine, Case Western Reserve University, Cleveland, Ohio

Colin N. Park, Ph.D., Issues Management/Biostatistics, Dow Chemical Company, Midland, Michigan

Robert M. Patterson, Sc.D., CIH, Wayne, Pennsylvania

Vernon Putz-Anderson, Ph.D., Division of Biomedical and Behavioral Science, NIOSH—Taft Laboratories, Cincinnati, Ohio

M. H. Smolensky, Ph.D., School of Public Health, University of Texas, Houston, Texas

William T. Stott, Ph.D., Toxicology Research Laboratory, Dow Chemical Company, Midland, Michigan

James A. Swenberg, D.V.M., Ph.D., Molecular Carcinogenesis and Mutagenesis, University of North Carolina, Chapel Hill, North Carolina

Robert G. Thomas, Ph.D., Gaithersburg, Maryland

Richard S. Waritz, Ph.D., Industrial Toxicology, BioSante International, Inc., Wilmington, Delaware

Philip G. Watanabe, Ph.D., Toxicology Research Laboratory, Dow Chemical Company, Midland, Michigan

Peter K. Working, Ph.D., DABT, Liposome Technologies, Inc., Menlo Park, California

Preface

The events and social impacts that led to the creation of industrial hygiene as a science and profession are equally important today as those that occurred around the turn of this century, which demanded resolution.

A health crisis during that period resulted from a tremendous increase in production to satisfy the rapidly developing industrial revolution taking place without giving forethought to what effect this would have on worker health. As a result, upward of 50 percent of the exposed workers who had been employed for over 20 years in many industries became afflicted with serious and fatal occupational diseases from exposures to harmful environmental agents in the workplace.

This situation was greatly abetted by the nonexistence of a science or profession that could research and address the problem.

A group of dedicated professionals, mainly engineers, chemists, physicists, statisticians, and physicians, concluded that a new multidisciplinary profession was necessary to research, understand, and resolve the ongoing worker health crisis. Their challenge was to develop methods and procedures for identifying harmful agents in the work environment, measure and characterize exposure levels, determine harmful biological responses, and develop adequate preventive and control procedures for protecting worker health.

This effort was successful and created the new science and profession of "industrial hygiene."

It permitted the conduct of epidemiological studies relating worker health and exposures to known harmful agents and stresses in the workplace as well as surveillance programs to detect unsuspected harmful exposures.

The profession has grown in scope, knowledge, and expertise in practice to anticipate and provide worker health protection against harmful environmental exposures in the workplace and at the same time permitting the expansion of industrial technology to meet future demands in serving the public.

As an example, the intensive research and rapid application of technological advances brought on by the burgeoning electronic science is now influencing almost every aspect of industrial, community, and home environments. This has brought into prominence the ergonomic aspects of the profession. Repetitive motion, vibration, abnormal posturing, and restricted motion are but a few of the stresses and insults associated with the operation of word processors, computers, and other electronic equipment.

The public demand to keep abreast of health problems associated with new technological thrusts along with their prevention and control gives increased reliance on the profession as an essential service. Increasingly, the profession is being accepted as the interface protecting health against harmful environmental exposures wherever their occurrence.

Determining the biological response of the body to harmful environmental agents and physical stresses is an essential component of the profession. These provide the parameters needed in the medical management of exposed individuals, the development of safe levels of exposure, and in designing proper preventive and control procedures.

In this revision, ten chapters from the second edition have been updated and expanded to reflect the state of knowledge in their respective coverage. Four new chapters have been added, attesting to increased concerns and need for further coverage in these specific areas. These include: Chapter 5, Reproductive Toxicology; Chapter 6, Neurotoxicology; Chapter 7, Carcinogenesis; and Chapter 8, Cancer Risk Assessment.

LEWIS J. CRALLEY
LESTER V. CRALLEY
JAMES S. BUS

Cincinnati, Ohio
Fallbrook, California
Midland, Michigan
January 1995

Contents

PATTY'S INDUSTRIAL HYGIENE AND TOXICOLOGY

Third Edition

Volume III Part B
THEORY AND RATIONALE
OF INDUSTRIAL HYGIENE
PRACTICE: BIOLOGICAL RESPONSES

CHAPTER ONE

Rationale

**Lewis J. Cralley, Ph.D., Lester V. Cralley, Ph.D., and
James S. Bus, Ph.D.**

1 BACKGROUND

The emergence of industrial hygiene as a science has followed a predictable pattern. Whenever a gap of knowledge exists and an urgent need arises for such knowledge, dedicated people will gain the knowledge.

The harmful effects from exposures to toxic substances in mines and other work places, producing diseases and death among workers, have been known for more than two thousand years. Knowledge on the toxicity of materials, the hazards of physical and biologic agents, and of ergonomic stressors encountered in industry, and means for their evaluation and control were not available during the earlier period of industrial development. With few exceptions, the earliest attention given to worker health was in applying the knowledge at hand, which concerned primarily the recognition and treatment of illnesses associated with a job.

The demand for increased production to meet the needs of an exploding industrial revolution, however, brought about a national health crisis around the turn of the twentieth century. Unless addressed and resolved, this emergency threatened the work force supporting the industrialization of our nation and its social structure.

Emerging medical data showed that upward of 50 percent of the work force employed in many major industries for over 20 years became afflicted with severe occupational diseases associated with harmful environmental exposures in the workplace.

Patty's Industrial Hygiene and Toxicology, Third Edition, Volume 3, Part B, Edited by Lewis J. Cralley,
Lester V. Cralley, and James S. Bus.
ISBN 0-471-53065-4 © 1995 John Wiley & Sons, Inc.

No science or profession existed that could research and resolve this problem. It became evident early on that the talents and dedication of a new multidisciplinary science and profession would be required for this task.

A group of dedicated professionals, mainly engineers, chemists, physicists, toxicologists, statisticians, and physicians combined and dedicated their talents to this end. Their challenge was to develop methods and procedures for identifying harmful agents in the workplace, measure and characterize levels of exposure, determine harmful biological responses, and develop preventive and control procedures for protecting worker health.

Epidemiological studies, made possible through such advances, showed that with few exceptions, worker exposure to harmful environmental agents in the workplace could be reduced to a level that would not adversely affect worker health over a lifetime of employment.

This and subsequent research permitted the establishment of official exposure limits to harmful agents in the workplace for worker health protection as well as permitting advancements on a broad front for industrial production.

These early professional researchers gave the name ''industrial hygiene'' to this new science and profession. The growth and acceptance of this new profession has been phenomenal. The profession has expanded to embrace all aspects in the recognition, characterization and measurement, biological response, and prevention and control of environmental hazards in the workplace and its extension.

The science and profession is supported through ongoing programs in industry, governmental agencies, unions, academia, research foundations, and other institutions requiring the resources provided.

The profession has proven its dedication and contribution to society in protecting workers and the public from excessive harmful environmental exposures. The profession has earned the right and protection to practice wherever the need occurs in service to our society.

The aim of this chapter is not to document or present chronologically the major past contributors to worker health and their relevant works or the events and episodes that gave urgency to the development of industrial hygiene as a science and a profession. Rather, the purpose of the chapter is to place in perspective the many factors involved in relating environmental stresses to health and the rationale upon which the practice of industrial hygiene is based, including the recognition, measurement, evaluation, and control of workplace stresses, the biological responses to these stresses, the body defense mechanisms involved, and their interrelationships.

The individual chapters of this volume and its Part A companion volume cover these aspects in detail.

Similarly, it is not the intent of Parts A and B of this volume to present procedures, instrumental or otherwise, for measuring levels of exposure to chemical, physical, biological, or other stress agents. This aspect of industrial hygiene practice is covered in *Patty's Industrial Hygiene and Toxicology, Volume I, General Principles*. Rather, attention in Parts A and B of this volume is devoted to other aspects of workplace exposures such as representative and adequate sampling and measurement, variations in exposure levels, exposure durations, interpreting results, the rationale for control, and the like.

2 INSEPARABILITY OF ENVIRONMENT AND HEALTH

Knowledge is constantly being developed on the ecological balance that exists between the earth's natural environmental forces and the existing biological species, and how the effects of changes in either may affect the other. In the earth's early biologic history this balance was maintained by the natural interrelationships of stresses and accommodations between the environment and the existing biological species at each particular site. This system related as well to the ecological balance within species, both plant and animal.

Studies of past catastrophic events such as the ice ages have shown the effects that changes in this balance can have on the existing species. The forces that brought on the demise of the dinosaurs that lived during the Mesozoic Era are uncertain. Most probably major geological events were involved. Studies have also shown that in the earth's past history a great many other animal species, as well, have originated and disappeared.

The human species, however, has been an exception to the ecological balance that existed between the natural environment and the evolving biological species in the earth's earlier history. The human ability to think, create, and change the natural environment has brought on changes above and beyond those of the existing natural forces and environment. These changes have had, and continue to have, an ever-increasing impact upon the previous overall ecologic balance.

The capacity of humans to alter the environment to serve their purposes is beyond the bounds of anticipation. In early human history these efforts predictably addressed themselves to better means of survival, that is, to food, shelter, and protection. As these efforts succeeded, humankind was able to devote some of its energy to gaining knowledge concerning factors affecting human health and well-being. Thus evolved the medical sciences, including public health. In some instances these efforts resulted in intervention in the ecological balance in the control of disease. In other situations the environment may have been altered to make desirable resources available, for example, in the damming of streams for flood control and developing hydroelectric power. This type of alteration of the localized natural environment and the associated ecological systems may have an impact by developing additional stresses in readjusting the existing ecological balance.

Of more recent impact on health has been the stress of living brought on by activities associated with personal gratifications such as life-styles as well as those associated with an ever more complex and advancing technology in almost all areas of human endeavor.

The quality of the indoor environment is receiving increasing attention in relation to good health. This applies to the study and control of factors giving rise to psychological stresses associated with living or working in enclosed spaces, as well as air pollution arising from life-styles, building designs, and materials and activities.

The advantages associated with changes in the environment for human benefit and improving the essential quality of living should be assessed for their cost-effectiveness as well as their potential to produce deterioration of the environment and concomitant new stresses. That humans, for optimum health, must exist in harmony with their surrounding 24-hr daily environment and its stresses is self-evident.

For better understanding of the significance of on-the-job environmental health hazards, an overview of the 24-hr daily stress patterns of workers is helpful. This permits a perspective in which the overall component stresses are related to the whole of workers' health.

Our habitat, the earth and its flora and fauna, is in reality a chemical one, that is, an entity that can be described in terms of tens of millions of related elements and compounds. It is the habitat in which the many species have evolved and in which a sort of symbiosis exists that supports the survival of individual species. The intricacy of this relationship is illustrated in the recognition that copper, chromium, fluorine, iodine, molybdenum, manganese, nickel, selenium, silicon, vanadium, and zinc, in trace amounts, are essential to human health and well-being. All, however, are toxic when absorbed in excess, and all are listed in standards relating to permissible exposure limits in working environments. Some compounds of several of these elements are classified as carcinogens. It is most revealing that some trace elements essential for survival are under some circumstances capable of causing our destruction. Thus, the matter of need or hazard is a question of "How much?"

The environment is both friendly and hostile. The friendly milieu provides the components necessary for survival: oxygen, food, and water. On the other hand, the hostile environment constitutes a combination of stresses in which survival is constantly challenged.

Although numerous factors are obviously involved in the optimal health of an individual, stresses arising out of the overall environment, that is, the workplace, life-style, and off-the-job activities, are substantial. The stresses encountered over full 24-hr daily periods have an overall impact on an individual's health. Any activity over the same period of time that can be stress relieving will have a beneficial effect in helping the body to adjust to the remaining insults of the day.

The inseparability of the environment and its relation to health is presented graphically in Figure 1.1.

An environmental health stress may be thought of as any agent in the environment capable of significantly diminishing a sense of well-being, causing severe discomfort, interfering with proper body organ functions, or causing illness or disease. These stresses may be chemical, physical, biologic, psychologic, or ergonomic in nature. They may arise from natural or created sources.

2.1 Macrocosmic Sources

Macrocosmic sources of environmental stress agents, such as the hemispheric or global quality or state of air, soil, or water, are those emanating from the sun or from extensive geographical perturbations and are capable of affecting large geographical areas. Examples of natural stress agents are ultraviolet, thermal, and other radiations from the sun, major volcanic eruptions that release huge quantities of gases and particulate material into the upper atmosphere, the changing of the upper air jetstream and other factors that influence climate, and the movement of the earth's tectonic plates resulting in earthquakes and tidal waves.

Examples of human-created macrocosmic stresses include interference, through release to the atmosphere of some synthetic organic compounds, with the upper

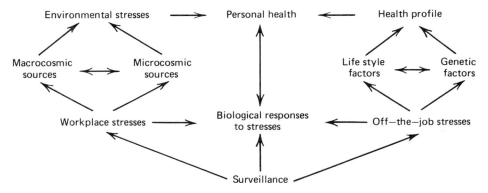

Figure 1.1 Inseparability of environment and health.

atmosphere ozone layer that shields the earth from excessive ultraviolet radiation; the burning of fossil fuels that increases the carbon dioxide level in the atmosphere and alters the earth's heat balance and surface temperature; and the destruction of forests and pollution of oceans that inhibit the biosphere's oxygen-producing capability.

These macrocosmic stresses may act directly upon individuals through such conditions as excessive exposure to ultraviolet and thermal radiation or indirectly by influencing the earth's climate—sunshine, rain, and temperature—thus affecting vegetation and habitability.

2.2 Microcosmic Sources

Microcosmic sources of stress agents are those emanating from localized areas and generally affecting a single region. These are most commonly at the regional or community level and may also include the home and work environments.

Examples of natural sources of microcosmic stress agents are pollen, which gives rise to sensitization, allergy, and hay fever; water pollution from ground sources having high mineral or salt content; and the release of methane, radon, sulfur gases, and other air contaminants from underground and surface areas. It is noteworthy that human evolution has taken place in the presence of natural macrocosmic and microcosmic sources.

Human-created microcosmic sources of health stress agents at the community level include noise from everyday activities such as lawn mowing, motorcycle and truck traffic, and loud music; air pollution from motor vehicle exhaust, release of industrial emissions into the air, emissions from refuse and garbage landfills, toxic waste disposal sites, spraying of crops, and life-style; surface water pollution through the release of contaminants from home, community, agricultural, and industrial activities into streams; and seepage into ground water of contaminants from landfills, agricultural activities, and from industrial and other waste disposal sites.

Stress agents from these sources may cause direct responses, such as the effects of noise on hearing and toxic exposures on health, or indirect responses such as acid smog and rain affecting vegetation and soil quality.

Regarding microcosmic sources, it is noteworthy that segments of industry are taking seriously their obligations and opportunities for protection of the health and well-being of both their employees and the communities in which their plants operate. In late 1988 the Chemical Manufacturers Association (CMA) adopted an initiative identified as Responsible Care: A Public Commitment. Each CMA member company, as a condition of membership, will manage its business according to the following listed principles (CMA, 1988):

To recognize and respond to community concerns about chemicals and our operations.

To develop and produce chemicals that can be manufactured, transported, used, and disposed of safely.

To make health, safety, and environmental considerations a priority in our planning for all existing and new products and processes.

To report promptly to officials, employees, customers, and the public information on chemical related health or environmental hazards and to recommend protective measures.

To counsel customers on the safe use, transportation, and disposal of chemical products.

To operate our plants and facilities in a manner that protects the environment and the health and safety of our employees and the public.

To extend knowledge by conducting or supporting research on the health, safety, and environmental effects of our products, processes, and waste materials.

To work with others to resolve problems created by past handling and disposal of hazardous substances.

To participate with government and others in creating responsible laws, regulations, and standards to safeguard the community, workplace, and environment.

To promote the principles and practices of responsible care by sharing experiences and offering assistance to others who produce, handle, use, transport, or dispose of chemicals.

These principles pledge good practices in controlling microcosmic sources, and, through those principles applying to employees, endorse good industrial hygiene practices. The Responsible Care initiative bodes well for community and worker health, as well as for the long-term financial health of the industry.

2.3 Life-Style Stresses

The life-style of individuals, including habits, nutrition, off-the-job activities, recreation, exercise, and rest, may have beneficial effects as well as stresses that exert a profound influence on health. Extensive knowledge has been developed, and continues to expand, on the influence of habits such as smoking, alcohol consumption, and use of drugs on health. Stresses from such activities may have additive, accumulative, and synergistic actions or may exert superimposed responses on the effects of other exposures arising in places of work. These off-the-job agents can be causes of respiratory, cardiovascular, renal, or other diseases, and may create grave health problems in individuals over and above any effects of exposures encountered in workplaces.

Knowledge is available on the deleterious effects on the health of offspring caused by smoking, alcohol consumption, and drug use by women during pregnancy; effects include malformation and improper functioning of body organs and systems, low birth weight, and so forth.

The benefits to health of good, adequately balanced nutrition, that is, vitamins, minerals, and other essential food intake, are gaining increased attention in relation to general fitness, weight control, prevention of disease, supporting the natural body defense mechanisms, recovery from exposure to environmental stresses, and aging. Conversely, malnutrition and obesity are associated with many diseases or dysfunctions, and may have synergistic effects with exposures to other stresses.

Recreational activities are important aspects of good health practices, releasing tension brought on through both off- and on-the-job stresses. Conversely, many recreational activities may be harmful, such as listening to excessively loud music, which may lead to hearing decrement, frequent engagements in events and schedules that interfere with the body's internal rhythmic functions, failing to observe needed precautions while using toxic agents in hobby activities, and pursuing activities to the point of exhaustion.

Both exercise and rest are important for maintaining good health. Exercise helps in maintaining proper muscular tone as well as weight control. Exercise, however, should be designed for specific purposes, maintained on a regular basis, and structured to accommodate the body physique and health profile of the performing individual. Otherwise more harm than benefit may result. Rest provides time for the body to recuperate from physical and psychological stresses.

The hours between work shifts and during weekends provide time for the body to excrete substantial portions of some chemical agents absorbed during a work shift or a workweek. The significance of these nonexposure recovery periods, of course, depends on the biologic half-time of the chemical agent of interest. Work patterns that disturb this recovery period, as in moonlighting, may have an especially deleterious effect if similar stresses are encountered on the second job. Likewise, smoking, alcohol consumption, and use of drugs may impair the body's proper recuperation from previous stresses.

2.4 Off-the-Job Stresses

Workers may encounter a host of stresses outside places of work. These may be chemical, physical, biologic, or psychologic in nature and usually are encountered during the 8- to 10-hr period between the end of the work period and the beginning of sleep time, and on weekends. Many off-the-job activities, if performed in excess and without regard to necessary precautions, are capable of producing stress and injury. Hobby and recreational activities may account for a substantial portion of this time. Hobby participants may encounter environmental exposures from activities such as soldering, welding, cleaning, gluing, woodworking, grinding, sanding, and painting, which are experienced by workers on the job but that are well controlled on the job. The home hobbyist, however, often does not have available the appropriate protective devices such as local exhaust ventilation, protective clothing, goggles, and respirators. The same hobbyist may neglect other good safety practices; he or she may ignore precautionary labels, take shortcuts to save time, fail to use the basic principles of keeping toxic materials from the skin, neglect thorough washing after skin contact, and may smoke or eat while working with such materials.

The off-the-job gamut of health stresses is wide and formidable. To deal with these stresses satisfactorily requires that a degree of accommodation be reached based on judgment, feasibility, personal options, objectives, and other factors. It is evident that although the components of the total sum of environmental health stresses must be considered on an individual basis, no single component can stand alone and apart from the others.

2.5 Workplace Stresses

Places of work may be the most important sources of health stresses if workplace operations have not been studied thoroughly and the associated health hazards have not been eliminated or controlled. This was evidenced during the time of early industrial development when little information was available on methods for identifying, measuring, evaluating, and controlling work-related stresses. During this period workplace exposures were often severe, leading to high prevalence of diseases and excess deaths.

Beginning about the turn of this century, and especially since the 1950s, management, labor union, government, academic, and other groups have taken substantial interest in worker health and in the control of exposures to stresses in places of work. This, along with the setting of standards of exposure limits, has created a broad support for expanding knowledge and practice to prevent illnesses and diseases associated with on-the-job exposures. This volume focuses on the theory and rationale for recognizing, evaluating, and controlling on-the-job health hazards.

2.6. Biological Response to Environmental Health Stresses

The human body consists of a number of discrete organs and systems derived from the embryonic state, encased in a dermal sheath, and developed to perform specific functions necessary for the overall functioning of the body as an integral unit. These organs are interdependent so that a malfunction in one may affect the functioning of many others. As an example, a malfunction in the alveoli, which hinders the passage of oxygen into the blood transport system, may have a direct effect on other organs through their diminished oxygen supply. Similarly, any hormonal imbalance or enzyme aberration may affect the functioning of many other body organs. Once absorption has occurred, toxic agents may selectively target one or more of the organs. Each organ has its own means for accommodating to, or resisting, stress and adjusting to injury, and its own propensity for repair.

Research has shown that the biological response to even the same harmful environmental agent or stress may vary widely depending on its chemical and physical characteristics, levels and duration and frequency of exposures, route of entry into the body, superimposed infections, repetitive nature of insult, psychological factors involved, and other associated insults.

1. Acute Responses. High and repetitive exposures to hazardous environmental agents even over relatively short durations may cause a severe response and death. This phenomenon was especially observed during the early phase of the industrial revolution. Examples include exposures to lead, mercury, carbon monoxide, and carbon tetrachloride.

2. Chronic Response. When the exposure pattern to even the same agent is changed from that of an acute response to one of a lower level over a prolonged duration, the biological response often becomes significantly different. Examples include the heavy metals, solvents, carbon monoxide, and most hazardous environmental agents.

3. Mode of Entry into Body. The mode of entry of toxic agent into the body, that is, by inhalation, ingestion, or cutaneous absorption, is extremely important since the nature of the response is directly related to the total dose as well as specific selective organ sites affected. It is interesting that many of the essential elements for life are also fatal in high concentrations. Thus, it is the dose, exposure pattern, and metabolic pathway that determines the biological response of an exposure to a chemical.

4. Latency Period. Exposure to some hazardous environmental agents requires a prolonged latency period of 20 or more years before certain clinical symptoms appear. Asbestos exposures and cancer as well as free silica exposures and silicosis are examples.

5. Superimposed Infectious Agents. In some instances a superimposed infectious agent may drastically affect the otherwise usual biological response to a harmful environmental agent. Tubercular infection superimposed on an individual with silicosis has a very progressive and fatal response. This situation was very common in workers exposed to free silica dust around the turn of the twentieth century.

6. Synergism. In multiple exposures to harmful environmental agents, some combinations of exposures have a biological response several times greater than would be predicted from their additive effects alone. Examples include cigarette smoke and alcohol intake in combination with asbestos and solvent exposures.

7. Ergonomic Factors. The human body may show adverse effects when exposed to specific physical environmental stresses and insults. Repetitive motions over a prolonged period of time may cause carpal tunnel syndrome. Abnormal postures, improper lighting, restricted motion, and other insults associated with the operation of data processing, word processors, and other manually operated electronic equipment have been found to exhibit negative biological responses.

8. Physiologically Based Pharmacokinetics (PB-PK). Some chemicals, when absorbed, are broken down into specific metabolites that can serve as markers of the original exposure that may no longer be recognized. As an example, benzene is metabolized by the body into muconic acid and other chemicals and excreted in the urine. The muconic acid is apparently the carcinogenic agent rather than the original benzene exposure. The presence of muconic acid in the urine signifies a previous exposure to benzene (Johnson and Lucier).

Although the organ structures and functions of several experimental animal species have many similarities to those of the human body, with some more similar than those of humans, care has to be taken in extrapolating research data on any one experimental animal species to the human. Similarly, research data obtained in isolated systems such as cell cultures have to be cautiously interpreted when applied to even the same cells in the whole integrated human body organ system.

2.7 Body Protective Mechanisms Against Environmental Stresses

The human body, in coexisting with the hostile stresses of the external environment, has developed a formidable system of protection against many of these stresses. This is accomplished in a remarkable manner by the ectodermal and endodermal barriers resisting absorption of noxious agents through inhalation, skin contact, and ingestion, and supported by the backup mesodermal and biotransformation mechanisms once absorption has occurred. These external and internal protective mechanisms, however, are not absolute and can be overwhelmed by a stress agent to the extent that they are ineffective, with resultant disease and death. Also, these protective mechanisms may become impaired in various degrees from insults associated with lifestyles and other daily activities. In studying the effects of specific stresses on health, it is important to be aware of the body's protective mechanisms. Suitable control of a specific stress should supplement the body's protective response to that stress.

2.8 Coaptation of Health and Environmental Stresses

To survive, the human body must live in balance with the surrounding environment and its concomitant stresses. Since these stresses, singly or combined, are not constant in value even for short periods of time or over limited geographical areas, the body must have built-in mechanisms for adjusting to differing levels of stresses through adaptation, acclimatization, and other accommodating mechanisms. There is a limit, however, to which the body can protect itself against these stresses without a breakdown occurring in its protective systems.

An extremely thought-provoking concept on associations between environmental stresses and health decrements has been presented by Theodore Hatch (Hatch, 1962, 1972). His concept examines associations between stresses and the human body's adjustments, compensations, and finally breakdown and failure, in response to them. The concept is particularly useful in understanding the effects of multiple risk factors when they include those of both occupational and nonoccupational origin.

If humans are to have freedom of geographic movement and of living in highly hostile environments that produce environmental insults greater than can be handled by the body's protective and adaptive mechanisms, some means of protection other than accommodation by the body must be provided. Extreme examples of the need for such protection are living and working in confined spaces where the immediate human environment is under absolute control against outside catastrophic stresses as in the cases of space travel and submarine activities.

More typical of this coaptive relationship is exposure to ultraviolet radiation from the sun. It is obvious that avoidance of all ultraviolet radiation from this source is impractical. In addition to the ozone layer of the upper atmosphere shielding the earth from major levels of ultraviolet radiation emitted by the sun, and the body's own protective mechanisms such as skin pigmentation, further accommodation is reached through the use of special clothing, skin barriers, eye protection, and a managed limitation to exposure.

At high altitudes where the partial pressure of oxygen is diminished from that to which the human body may be accustomed at lower elevations, the body acclimates in time by increasing the number of red blood cells and hemoglobin that carry oxygen to the tissues. When the availability of atmospheric oxygen decreases below the limit of acclimation, further accommodation may be provided externally through the use of supplemental oxygen supply.

The human body has a number of regulating mechanisms to keep its temperature within normal limits when it is exposed to excessively high or low environmental temperatures. This permits living in a limited but wide range of environmental temperatures. Further accommodation to extremes in environmental temperatures may be provided through special clothing, protective equipment, and living and working in climate-controlled structures.

Where excessive exposures to environmental stresses exist in a workplace, emphasis is placed on their elimination or on lowering them to acceptable levels. Stress levels should be lowered through recognized engineering or administrative control procedures to the point that the body defense mechanisms can adequately prevent injury to health. Some situations may not be amenable to engineering or administrative control and may properly require the use of personal protective equipment or other control strategies.

There is a limit to what can be done to alter existing environmental stresses from natural macrocosmic sources. Thus, these become ubiquitous background stresses upon which other exposures from microcosmic sources such as community and industrial pollution, off-the-job activities, life-styles, and on-the-job activities are added. Rationally, then, it is primarily the stresses created from predominantly microcosmic and a few macrocosmic sources that are amenable to control. These must be kept within acceptable limits to permit humans to avoid health damage from the stresses of an increasingly complex technologic age.

3 INDUSTRIAL TECHNOLOGICAL ADVANCES

In the early history of humankind, the many activities associated with living were at the tribal level where emphasis was placed upon survival, that is, on procuring adequate food, protection, and shelter. The tribes, many of whom were nomadic, were undoubtedly aware that they lived in accommodation with their environments.

This would have been evidenced through the appropriate use of clothing, safe use of fire, and observing climatic patterns.

It was inevitable that the nomadic way of life would give way in most instances to a more settled life-style in which cooperative efforts for food, shelter, and protection were more dependable than those based on individual or small group effort and ingenuity. During this transition period accommodations to the natural elements were made easier through more permanent shelter and a more organized pattern of living.

The next advancement in accommodation came through the realization that increased production could be attained through specialization of work pursuits wherein a designated work group devoted its principal effort to the making of a single commodity such as clothing, pottery, or tools, or the growing of foods. Each group shared its commodity in exchange for the products of other groups. This type of trade evolved to cottage-type industries that related primarily to the community level of commerce. Even at that level of production many of the health stresses associated with different pursuits were intensified over those of nomadic living in which every person was a sort of jack-of-all-trades. This was especially true where the operations tended to be restricted to confined and crowded spaces.

As means for communications and transportation improved, trade increased between adjoining communities, and the search continued for ways of producing commodities in increased volume with less manpower. Similar types of production operations tended to expand and to be concentrated within the same housing structure. This led to increases in the health stresses of the whole work force in instances where the stress agents were cumulative in intensity and response.

This trend toward industrialization continued and intensified with the advent of the steam engine. Developments such as the steam engine gave rise to the industrial revolution, which brought about larger factories and new production techniques, along with increased health risks. While in earlier times of cottage industries there were relatively few health risks in any one workplace, the new technology and industrialization led to more complex patterns of exposures.

Since the advent of the industrial revolution technological advances and their application to production have expanded at an ever-increasing pace. In the latter half of the twentieth century the application in industry of knowledge gained through space technology and other such research has rapidly expanded into the current high-technology electronic age.

The advent of this high technology and its application to production is having its effect upon both the nature of employment and the concomitant health stresses. While in the past workers needed only special instructions to perform most job operations— and this will continue for some time—the move into higher technology has created a demand for highly trained employees for many job positions. This trend can be expected to increase dramatically. Computers, word processors, video display terminals, lasers, microwave, and other electronic equipment are becoming commonplace in industry. Robots, which require sophisticated management and control, are taking over many repetitive operations such as in painting and metalworking.

The nature and extent of associated health stresses are becoming more complex with the advent of high-technology industry. At the same time, the health effects of these stresses are becoming more detectable with more sophisticated measurement and diagnostic tools.

The urgent need for knowledge concerning the effects of exposures to health stresses associated with an ever-expanding industrial technology, along with the methodology for their recognition, evaluation, and control, gave rise to the science of industrial hygiene. This science must keep attuned to the ever-changing applications of technology in industry.

4 EMERGENCE OF INDUSTRIAL HYGIENE AS A SCIENCE AND PROFESSION

Science may be defined as an organized body of knowledge and facts established through research, observations, and hypotheses. As such, a science may be basic, as exemplified by the fundamental physical sciences, or it may be applied in the sense that the principles of other sciences are brought to bear in developing facts and knowledge in a specific area of application.

During the early history of industrial development the lack of knowledge on the effects of health stresses associated with industrial operations and how these could be controlled, with the concomitant massive exposures to toxic materials in places of work, led to many serious episodes of illness, disease, and death among workers. An example is the high incidence of silicosis and silicotuberculosis that existed a century ago among workers in hard rock mines, the granite industry, and in tunneling operations, wherever the dust had a high free silica content.

During this early period the major effort on behalf of workers' health was to apply the knowledge at hand, which related primarily to the recognition and treatment of occupational illnesses.

It was not until around the turn of this century that specific attention began to be devoted to the preventive aspects of industrial illnesses. Scientists and practitioners, including engineers, chemists, and physicists, began to apply their knowledge and skills toward the development of methods and procedures for identifying, measuring, and controlling exposures to harmful airborne dusts and chemicals in workplaces. At that time there were no recognized procedures for carrying out these activities.

After trying various potential air-sampling procedures, the impingement method was judged the most adaptable one at that time for collecting many of the airborne contaminants such as particulates, mists, some fumes, and gases. The light-field microscopic dust counting technique was developed for enumerating levels of mineral dust in the air; and conventional wet chemical analytical methods available at the time were adapted for measuring quantities of chemical agents in these samples.

Even during this early period of development of airborne sample collection and analytical procedures, scientists realized that the exposure patterns that existed were more complex and complicated than the instruments for sample collection and analysis could define. These scientists also knew of many of the deficiencies associated with the measurement procedures being developed. They were aware that the data being collected represented only segments of the overall exposure patterns. They believed, however, that these segments could be used as indices that would represent overall exposure patterns so long as the production techniques and other operational factors remained the same. It must be remembered that at that time information was not available on respiratory deposition and dust size. It was imperative to them, and rightly so, that some method, with whatever deficiencies that may have been incumbent, be developed for indexing airborne levels of contamination in workplaces, both for estimating levels of exposure and for use as benchmarks in determining degrees of air quality improvement after controls had been established.

Since the earliest instruments for collecting airborne contaminant samples were nonportable, the collected samples represented general room levels of contamination, and the results depended on where in the workplace the samples were taken.

The procedures described above for airborne sampling and analysis in workrooms, as primitive as they may seem today, served well during that period of time. They accounted for the drastic reduction in massive exposures that existed in many work sites and were the methods and procedures upon which future refinements would be made.

These early scientists showed that it was feasible to lower the massive workplace airborne contamination levels that often existed at that time; and by relating the data to the health profiles of workers, they observed that the lowering of exposure levels also lowered the incidence of the associated diseases. Thus began the first field studies that were to have a profound influence on the development of the earliest permissible exposure limits and in developing the rationale upon which the practice of industrial hygiene is predicated. The insights developed in these field studies were to be further substantiated through laboratory and clinical research.

The development of the hand-operated midget impinger pump during the 1930s was a decided improvement over the standard impinger pump since it was portable and permitted movement about a workplace while airborne samples were being taken. Samples of particulate or gaseous agents could be taken with glass impingers or fritted glass bubblers near workers as they moved about their tasks. This new worker exposure data demonstrated that workers often had higher levels of exposure than those indicated by the general room airborne levels.

Other instruments, such as the electrostatic precipitator and evacuated containers, came into use during this period. In the late 1940s the paper and membrane filter methods for collecting some airborne particulate samples came into use. The filter sampling procedure was found to be superior in many respects to the impingement method. The method did not fracture particles or disperse agglomerates, which often accompanied impingement collection, and could be performed in ways that permitted direct microscopic observation of particles and gravimetric measurement of samples.

Insights also began to emerge on the importance of particle size in relation to deposition and retention of particulate material in the respiratory tract. Electron microscopes and membrane filters made it possible to study particles of submicron sizes.

A surge of improved and sophisticated techniques for quantifying workers' exposures to health stress agents took place in the 1960s. This applies both to sample collection and analytical techniques in which much lower levels of exposure to specific agents could be determined. There also began a dramatic increase in toxicologic and epidemiological studies by government, industry, universities, and foundations, directed to obtaining data upon which to base exposure standards as well as good industrial hygiene practices.

Another major advancement in developing better methods for studying occupational diseases occurred at midcentury. Toxicologic and other studies had revealed that lowering the level of exposure and extending the exposure duration changed the dose-response pattern of many toxic agents. As an example, a high level of exposure to airborne lead produces an acute response over a relatively short period of time. In contrast, lowering the exposure level of this agent and extending the exposure time shows a different dose-response pattern, a chronic form of lead poisoning. Thus, in studying the effects of exposure to health stress agents it is important to obtain relevant dose-response data over an extended period of exposure time.

One method of obtaining relevant health profile data on workers is through study of their medical records. Another method is through the study of causes of death among worker populations using death certificates located through Social Security, management, retirement system, and labor union records. Such studies have revealed that a lifetime of work exposure to an agent, or an extended observation period of 20 or more years from time of initial exposure, may be necessary to fully define the wide range of dose-response relationships. This may be especially true for carcinogenic and other long latency types of exposure response.

A more recent advancement relates to chronobiology, the study relating to the body's internal biological rhythms and their effects on organ functions, and so forth. The workweek schedule can have a direct effect on these rhythmic patterns and health. Also, there is some evidence that the rate of absorption of toxic materials and reactions to stress may relate in some way to an individual's chronobiology.

The establishment of professional associations to support the interests and growth of the profession has played an important role in developing industrial hygiene as a science. In the United States in the 1930s the American Public Health Association had a section on industrial hygiene that supported the early growth of the profession. The American Conference of Governmental Industrial Hygienists was organized in 1938. The American Industrial Hygiene Association was organized in 1939. The American Board of Industrial Hygiene was created and held its first meeting in 1960. This board certifies qualified industrial hygienists in the comprehensive practice of industrial hygiene and in the past has certified in six additional industrial hygiene specialties as well. Industrial hygienists certified by the board have the status of Diplomates and as such are eligible for membership in the American Academy of Industrial Hygiene.

The American Academy of Industrial Hygiene has developed and adopted a 15-point Code of Ethics for the professional practice of industrial hygiene. The code addresses industrial hygienists' responsibilities to the profession, to workers, to employers and clients, and to the public. It is noteworthy that the Code of Ethics specifies that the primary responsibility of an industrial hygienist is to protect the health of employees (i.e., workers) and that responsibility to any employer or client is subservient to that to workers.

The American Industrial Hygiene Association administers a laboratory accreditation program with the objective of assisting those laboratories engaged in analyses of industrial hygiene samples in achieving and maintaining performance levels within acceptable ranges.

The American Industrial Hygiene Foundation was established in 1979 under the auspices of the American Industrial Hygiene Association. The functions of the foundation are carried out by an independent Board of Trustees. The foundation provides fellowships to worthy industrial hygiene graduate students, encourages qualified science students to enter the industrial hygiene profession, and stimulates major universities to establish and maintain industrial hygiene graduate programs.

In the United States a number of occupational health regulations were established in the early 1900s with emphasis, in several, on listing limits of exposures to a relatively few agents. These regulations were effective at the local, state, and federal levels depending on governmental jurisdiction.

The Social Security Act of 1935 and the Walsh–Healy Act of 1936, had an immense impact in giving increased stability, incentive, and expanded concepts in the practice of industrial hygiene. These acts stimulated industry to incorporate industrial hygiene programs as an integral part of management. They also stimulated broad-base programs in industry, foundations, educational institutions, insurance carriers, labor unions, and government that address the causes, recognition, and control of occupational diseases. These acts established the philosophy that the worker had a right to earn a living without endangerment to health and were the forerunners for the passage of the Occupational Safety and Health Act of 1970.

Passage of the Occupational Safety and Health Act of 1970, which has the purpose of assuring ''so far as possible every man and woman in the nation safe and healthful working conditions'' had a very broad bearing on the further development and practice of the industrial hygiene profession. These enabling acts, and the regulations deriving from them, have been substantial factors in the broad recognition of industrial hygiene as a science and a profession. It has been necessary to expand the profession in all of its concepts and technical aspects to meet its expanded responsibilities.

Other industrialized countries have had similar experiences in the professional recognition and growth of the science relating to the recognition, measurement, evaluation, and control of work-related health stresses.

4.1 Definition of Industrial Hygiene

The American Industrial Hygiene Association defines industrial hygiene as ''that science and art devoted to the recognition, evaluation, and control of those environmental factors or stresses, arising in or from the workplace, which may cause sickness, impaired health and well-being, or significant discomfort and inefficiency among workers or among the citizens of a community.'' Because the science and practice of industrial hygiene continues to evolve, the association is reviewing this definition for possible revision.

By any definition, however, industrial hygiene is an applied science encompassing the application of knowledge from a multidisciplinary profession including the sciences and professions of chemistry, engineering, biology, mathematics, medicine, physics, toxicology, and other specialties. Industrial hygiene meets the criteria for the definition as a science since it brings together in context and practice an organized body of knowledge necessary for the recognition, evaluation, and control of health stresses in the work environment.

In the early 1900s the major thrust in the control of workplace health stresses was directed toward those areas in industry having massive exposures to highly toxic materials. The professional talents of engineers, chemists, physicians, physicists, and statisticians were those largely used in these programs. As industrial technology advanced, the complexity of workers' exposures also increased, along with an increase in the professional knowledge and skills needed to study the new health effects and to develop the methods for recognition, evaluation, and control of the new environmental stresses. The need for new knowledge and skills continues now with the advent of high technology in the electronic and allied industries. Factors such as improper lighting and contrast, glare, posture, fatigue, need for intense concentration, tension, and many other stresses arise in the operation of computers, word processors, video display terminals, and laser, microwave, and ionizing radiation equipment, which are becoming commonplace in industry. Thus, concerns for the health of employees above and beyond that of toxicity response arise. The study and control of these nonchemical stress agents point to the need for the occupational health nurse, psychologist, human factors engineer, ergonomist, and others to join the professional team in studying the effects and control of the ever-widening list of health stress agents in places of work.

In the early practice of occupational health nursing, emphasis was placed on such activities as the emergency treatment of traumatic injuries stipulated in written orders of a physician and in maintaining records and information relating to physical examinations and the like. With the current advanced training of occupational health nurses, this limited role has been found to be wasteful of professional talent and resources. Occupational health nurses are often the first interface between workers and pending health problems and are in a position to gain information on situations and health stresses both on and off the job that may, unless addressed, lead to more serious responses. Occupational health nurses have increasingly become members of the multidisciplinary team needed in the recognition of job-associated health stresses. Similarly, psychologists, in the study of effects of strain, tension, and similar stresses, and human factors engineers and ergonomists in designing machines, tools, and equipment compatible with physical and morphologic limitations of workers, are examples of other professionals joining the multidisciplinary team studying the effects and control of the increasingly complex health stresses associated with advanced industrial technology.

The complexity of the multifaceted professional effort needed for carrying out the responsibilities of professional practice in the protection of worker health is further illustrated in the 28 technical committees of the American Industrial Hygiene Association and the 7 different specialty areas of certification that have been used in the past by the American Board of Industrial Hygiene.

4.2   Rationale of Industrial Hygiene Practice

The practice of industrial hygiene is based on the following observations, experiences, and rationales:

1. Environmental health stresses in the workplace can be quantitatively measured and expressed in terms that relate to the degree of stress.
2. Stresses in the workplace, in the main, show a dose–response relationship. The dose can be expressed as a value integrating the concentration or intensity, and the time duration, of the exposure to the stress agent. In general, as the dose increases the severity of the response also increases. As the dose decreases the biological response decreases and may at some time and dose value exhibit a different kind of response, chronic versus acute, depending on the time duration of the stress, even though the total stress expressed as a dose–response value may be the same.
3. The human body has an intricate mechanism of protection, both in preventing the invasion of hostile stresses into the body and in dealing with stress agents once invasion has occurred. For most stress agents there is some point above zero level of exposure that the body can tolerate over a working lifetime without injury to health.
4. Levels of exposure of workers to specific stress agents should always be kept within prescribed safe limits. Regardless of their type, all exposures in the workplace should be kept at lower than prescribed levels as are reasonably attainable through good industrial hygiene and work practices.
5. Some stress agents may cause serious biological responses among a few workers at such low levels that exposures should be controlled to levels as low as reasonably achievable regardless of any higher regulatory limit. An example of such an agent is one having genotoxic properties.
6. The elimination of health hazards through process design and/or the use of nonhazardous substitute materials should be the first objective in maintaining a safe workplace. When this is not feasible, recognized engineering or administrative controls should be used to keep exposures within acceptable limits. In some cases, when feasible engineering and administrative controls are insufficient, supplemental programs such as the use of personal protective equipment or other control practices have application.
7. Surveillance of both the work environment and workers should be maintained to assure a healthful workplace.

4.3   Elements of an Industrial Hygiene Program

The purpose of an industrial hygiene program is to assure a healthful workplace for employees. It should include all the functions needed in the recognition, evaluation, and control of occupational health hazards associated with production, office, and other work. This requires a comprehensive program designed around the nature of the operations, documented to preserve a sound retrospective record, and executed in a professional manner.

The basic components of a comprehensive program include the following:

1. Coordinated technical activities capable of detecting occupational health stresses in any part or process of the establishment.
2. Capability to conduct or obtain measurement and evaluation activities for the assessment of occupational health stresses anywhere in the establishment.
3. Capability to determine the need for and to obtain and maintain effective engineering and administrative controls necessary for safe and healthful workplaces throughout the establishment.
4. Participation in the periodic review of worker exposure and health records to detect the emergence of insufficiently controlled health stresses in the workplace.
5. Participation in research, including toxicological and epidemiological studies designed to generate data useful in establishing safe levels of exposure.
6. Maintaining a data storage system that permits appropriate retrieval of information necessary for the study of long-term effects of occupational exposures.
7. Assuring the relevancy of the data being collected.

An integrated program is capable of responding to the need for the establishment of appropriate exposure controls, both for current needs and for those that may result from technological advances and associated process changes.

The almost universal availability of high-capacity and powerful desk-top computers that has taken place over the past decade has greatly facilitated the conduct of industrial hygiene programs. A great amount of industrial-hygiene-related software for record keeping, technical reference (e.g., regulations, safety data sheets, etc.), sampling data analysis, exhaust ventilation design, and other such industrial hygiene functions has become available from commercial sources or through professional journals, professional associations, and individual industrial hygienists. The American Industrial Hygiene Association has a Computer Applications Committee whose mission is to provide a forum for advancing the use of computer applications by occupational and environmental health professionals. Among other activities this committee reviews and reports on available software.

The industrial hygienist at the corporate or equivalent level should report to top management. His or her responsibility involves appropriate input whenever product, technological, operational, or process changes, or other corporate considerations, may have an influence on the nature and extent of associated health hazards. When new plants or processes are planned, the corporate industrial hygienist should assure that adequate controls are incorporated at the design stage.

5 HEALTH HAZARD RECOGNITION

An important aspect of a responsive industrial hygiene program is that it is capable of recognizing potential health hazards or, when new materials and operations are encountered, to exercise expert judgment in maintaining an adequate surveillance program until any associated health hazards have been defined. This should not be a problem in cases involving operations, procedures, or materials for which adequate knowledge is available. In such cases it is primarily a matter of application of available knowledge and techniques. In operations and procedures involving a new substance or material for which relevant information is limited or unavailable, it may be necessary to extrapolate information from other kindred sources, to use professional judgment in setting up a control program with a reasonable factor of safety, and to incorporate an ongoing surveillance program to further define health hazards that may emerge. In some instances it may be necessary to undertake toxicological research prior to the production stage to define parameters needed in setting up the control and surveillance program.

One of the basic concepts of industrial hygiene is that the environmental health stresses of the workplace can be quantitatively measured and recorded in terms that relate to the degree of stress.

The recognition of potential health hazards is dependent on such relevant basic information as:

1. Detailed knowledge of the industrial process and any resultant emissions that may be harmful
2. The toxicological, chemical, and physical properties of these emissions
3. An awareness of the sites in the process that may involve exposure of workers
4. Job work patterns with energy requirements (i.e., metabolic levels) of workers
5. Other coexisting stresses that may be important

This information may be expressed in a number of ways depending on its ultimate use. A very effective form is a material process flowchart that lists each step in the process along with the appropriate information just noted. This permits the pinpointing of areas of special concern. The effort in whatever form it may take, however, remains only a tool for the use of the industrial hygienist in the actual identification of the stresses in the workplace. In the quantification itself, many approaches may be taken depending on the information sought, its intended use, the required sensitivity of measurement, the level of effort and instruments available, and the practicality of the procedures.

Aside from the production workplace with the attendant toxicological, physical, and other related health stresses, a new area of concern is rapidly gaining special attention where employees may be subjected to a high degree of stress from tension, physical and mental strain, fatigue, excessive concentration, and distraction such as may exist for operators of computers, word processors, video display terminals, and other operator-intensive equipment. Off-the-job stresses, life-style factors, and the immediate room environment may become increasingly important for such workers. The recognition of associated health stresses and their evaluation require a special battery of psychological and physiological body reaction and response tests to define and measure factors of fatigue, tension, eyestrain, deficits in the ability to concentrate, and the like.

6 EXPOSURE MEASUREMENTS

Both direct and indirect methods may be used to measure exposures to stress agents. Table 1.1 illustrates direct and indirect measurements of chemical agents.

6.1 Direct Measurements

To measure directly the quantity of a chemical agent actually absorbed by the body, fluids, tissues, expired air, excreta, and so on must be analyzed for the agent per se or for a biotransformation product. The quantification of a body burden of the agent requires information regarding the biological half-time of the agent or its metabolite as well. Such procedures may be quite involved, since the evaluation of the data at times depends on previous information gathered through epidemiological studies and animal research. Studies on animals, moreover, may have used indirect methods for measuring exposure to the stress agent, necessitating appropriate extrapolation in the use of such values. The current adopted list of Biological Exposure Indices published by the American Conference of Governmental Industrial Hygienists lists 61 determinants (the agent or a metabolite) for 33 compounds and gives notice of intent to adopt 5 more determinants for 5 compounds. Fourteen additional compounds are under study by the committee for establishment of biological exposure indices.

One decided advantage of biological monitoring coupled with information on the biological half-time of an agent or its metabolite is that exposure can be integrated on a time-weighted basis. Such integration is difficult to estimate through ambient air sampling when the exposure is highly intermittent or involves peak exposures of

Table 1.1 Methods for Measuring Worker Exposure to Absorbed Chemical Stress Agents

Direct	Indirect
Body dosage	Environment
Tissues	Ambient air
Fluids	Interface of body and stress
Blood	Physiological response
Serum	Sensory
Excreta	Pulse rate and recovery pattern
Urine	Body temperature and recovery pattern
Feces	Voice masking, etc.
Sweat	
Saliva[a]	
Hair[a]	
Nails[a]	
Mother's milk[a]	
Alveolar air	

[a] Not usually considered to be excreta; see Volume 3B, Chapter 3.

varying duration. Conversely, biological monitoring may fail to reflect adequately the influence of peak concentrations per se that may have special meaning for acute effects. Urine analysis may also provide valuable data on body burden in addition to current exposures when the samples are collected at specific time intervals after exposure. In general, quantitative body burden interpretation of analytic values for a biological specimen requires knowledge of biologic half-time for the agent of interest and an appropriate exposure-sampling time sequence and schedule.

Sampling of alveolar air may be an appropriate procedure for monitoring exposures to organic vapors and gases. An acceleration of research in this area can be anticipated because of the ease with which samples can be collected.

The use of biologic specimens, particularly for purposes of research, ordinarily requires the informed consent of each study subject who provides a sample. This is clearly necessary for an invasive procedure, such as blood sampling, and may apply even to the collection of excreta and exhaled air. Invasion of privacy may be at issue, for example, detection of alcohol consumption, in the analysis of exhaled breath or other excreta.

6.2 Indirect Measurement

The most widely used technique for the evaluation of occupational health hazards is indirect in that the measurement is made at the interface of the body and the stress agent, for example, the breathing zone or skin surface. In this approach the stress level actually measured may differ appreciably from the actual body dose. For example, all the particulates of an inhaled dust are not deposited in the lower respiratory tract. Some are exhaled and others are entrapped in the mucous lining of the upper respiratory tract and eventually are expectorated or swallowed. The same is true of gases and vapors of low water solubility. Thus, the target site for inhaled chemicals is scattered along the entire respiratory tract, depending on their chemical and physical properties. Another example is skin absorption of a toxic material. Many factors, such as the source and concentration of the contaminant, that is, airborne or direct contact, and its characteristics, body skin location, and skin physiology, relate to the amount of the contaminant that reacts with or is absorbed through the skin. For chemical agents the sampling and analytical procedures must relate appropriately to the chemical and physical properties, such as particle size, solubility, and limit of sensitivity of analytical procedures, for the agent being assessed. Other factors of importance are weighted average values, peak exposures, and the job energy requirements, which are directly related to respiratory volume and pulmonary deposition characteristics.

Exposures to physical agents such as noise and ionizing radiation are almost always measured by indirect methods such as dosimetry or assessment of work area intensity levels.

The indirect method of health hazard assessment is, nevertheless, a valid one when the techniques used are the same or equivalent to those relied on in the studies that established the standards.

7 ENVIRONMENTAL EXPOSURE QUANTIFICATION

Procedures for measuring airborne exposure levels of a stress agent depend to a great degree on the reasons for making the measurements. Some of these are: (1) obtaining worker exposure levels over a long period of time on which to base permissible exposure limits, (2) compliance with standards, and (3) performance of process equipment and controls. It is essential that the data be valid regardless of the purpose for which they were collected and that they be capable of duplication. This is a key factor in establishing exposure limits to be used in standards and in fact-finding related to compliance. Since judgment and action will in some way be passed on the data, validity is paramount if the data are to be used as a bona fide basis for action.

7.1 Long-Term Exposure Studies

In epidemiologic studies in which the relationship between a stress agent and the body response is sought, ideally the stress factor would be characterized in great detail. This could require massive volumes of data suitable for statistical analysis and a comprehensive data-collecting procedure so that a complete exposure picture may be accurately constructed. The sampling procedures and strategy should be fully documented, including number and length of time of samples, their locations, their types, that is, personal or area samples, and should be adequate to cover the full work shift activities of the worker. Any departure from normal activities should be noted. These are important since the data may be used at a later date for a purpose not anticipated at the time of sample collection.

This ideal situation is seldom the case in epidemiologic studies. In research on long-term health effects the typical epidemiologic study involves use of surrogates for exposures or efforts to retrospectively reconstruct exposures. The collection of valid retrospective data may be extremely difficult. If available at all, actual sampling data may be scanty; the sample collection and analytical procedures used in the past may not have been well documented as to precise methodology, and may have been less sensitive and efficient, or may have measured different parameters, than do current procedures; sampling locations and types may not be well defined; and the job activities of the workers may have changed considerably even though the job designations may be the same. Other factors that need to be considered in securing retrospective data relate to contrasting past and current plant operations, including changes in technology and raw materials, effectiveness of exposure control procedures and their surveillance, and housekeeping and maintenance practices. In many instances an attempt to accommodate these differences has been made through broad assumptions and extrapolations with an unknown degree of validity and without expressing the limitations of such derived data.

The effect of national emergencies may significantly change the nature and extent of workers' exposures to stress agents. The experience during World War II is an example. The number of worker hours per week were increased substantially in many industries. Control equipment was allocated to specified industries and denied to others. Local exhaust ventilation systems at times became ineffective or completely inoperable due to lack of maintenance and replacement parts. Less attention was given to plant maintenance, housekeeping, and monitoring procedures. Substitute or lower quality raw materials had to be used in many instances.

Although the major impact of World War II upon levels of exposure to harmful agents occurred from around 1940 to the early 1950s, the effects of many of these exposures may not have shown up among members of that work force and its retirees until decades later.

Thus, expressing exposure levels in the past for more than a few years may be only extrapolated guesses unless factors such as the above can be clearly examined and the data validity established.

7.2 Compliance with Standards

In contrast to the collection of data for epidemiological studies, data collected for the purpose of compliance with standards, as they are now interpreted, may require relatively few samples if the values are clearly above or below the designated value for the agent of interest. If the values are borderline, evaluation may call for a more comprehensive sampling exercise and may be a matter for legal interpretation. The nature and type of samples taken should meet the criteria upon which the standards were based. Scientifically, though, the data should be adequate to establish a clear pattern with no one single value being given undue weight and should meet data analysis requirements.

7.3 Spot Sampling

The exposure of a worker may arise from a number of sources, including the ambient levels of the agent in the general room air, which in turn may be influenced by ambient levels of the agent in the community air, leaks from improperly maintained operating equipment such as from pump seals and flanges, the inadequate performance of engineering control equipment, and the care workers take in performing job operations. Spot sampling can easily detect the effects of any one of these factors on the overall worker exposure level and point the direction for further exposure control action.

8 DATA EVALUATION

The evaluation of the intensity of a physical agent or of airborne levels of a chemical agent to determine compliance with a standard or to determine specific sources of the stress agent is generally uncomplicated and straightforward. The evaluation of environmental exposure data that serve as a basis for determining whether a health hazard exists is more complicated and requires a denominator that characterizes a satisfactory workplace. Similarly, the use of environmental exposure data for establishing safe levels of exposure or a permissible exposure level, as in epidemiologic studies, requires their correlation with other parameters such as the health profile of the workforce.

As pointed out earlier, the early field studies of the 1920s and 1930s showed that when the very high exposures of workers were lowered, there was a corresponding lowering of the related disease incidence in workers. These and other studies gave support to the dose–response rationale upon which the practice of industrial hygiene is primarily based, that is, there is a dose–response relationship between the extent of exposure and severity of biological response to most stress agents and in which the response is negligible at some point above zero level.

There is great difficulty, however, in determining lower levels of exposure to a specific agent and its effect on the health of workers over a working lifetime. Often this is done through extrapolation of other data or by trying to estimate past exposures. In the lower range of the dose–response region, the incidence of disease from exposure to an agent may be so low that it approaches the level for that disease in the community outside the industry under study. This results in part from exposure of the general population to stress agents such as smoking, alcohol consumption, drug use, hobby activities, community and in-house pollution, and the like. These incidental stresses may be similar in magnitude to those on the job or may be additive to, accumulative, or synergistic with stress from on-the-job exposures. For various reasons, often including limited study population size, even well-controlled studies may not be sensitive enough to give data that can be reliably extrapolated to the lowest dose–response region for lifetime exposure.

The problems of estimating dose–response of human populations at low levels of exposure to hazards has given rise in recent years to a new scientific field of endeavor called risk analysis. There is not yet unanimity of opinion among scientists in the field on the most appropriate models and estimating procedures for all types of risk situations, including those involving lifetime exposures to low levels of health stressors.

It is known that the body has protective mechanisms to guard against the effects of low levels of exposure to many environmental agents. For a great number of agents encountered in the industrial environment, data on dose–response relationships support the industrial hygiene rationale that the level of exposure does not have to be zero over a lifetime of work to avoid injury to workers' health. Some agents, however, such as those having genotoxic properties (i.e., being able to directly damage genetic material in cells), and perhaps some associated with hypersensitivity, may not have a threshold of biologic response, and the lowest achievable level of exposure for workers should be required.

Evaluation of data from exposure stresses relating to tension, fatigue, annoyances, irritation, decrements in ability to concentrate and discern, and the like are often subjective and may also involve the personal background, traits, habits, and so forth of those being stressed. Such stresses may require that attention be given to personal behavior for proper definition and control. Evaluation of such situations generally must be done by specialists other than an industrial hygienist.

9 ENVIRONMENTAL CONTROL

The cornerstone of an effective industrial hygiene program can be described as:

1. Proper identification of on-the-job health hazards
2. The measurement of levels of exposure to such hazards
3. Evaluation of all exposure data in context with work schedules and job demands
4. Environmental control

In essence, the success of the entire program depends on the success of the control effort. The technical aspects of the program must encompass sound practices and must be related both to workers and to the medical preventive program.

The heart of a control program must rest with process and/or engineering controls properly designed and properly operated to protect workers' health. The most effective and economic control is that which has been incorporated at the stage of process design and production planning, and which has been made part of the process. With new processes this can be accomplished by bringing industrial hygiene input into the bench level, pilot plant, and final stages of process development. It is neither good industrial hygiene practice nor sound economics to neglect exposure control in process design with the intention of adding supplemental control hardware piecemeal as indicated by future production or to comply with regulations.

Although engineering and administrative means should be used wherever feasible to achieve control of exposures, the need may exist for the judicious use of personal protective equipment under unique circumstances, for example, during breakdowns, spills, accidental releases, and some repair jobs. Personal protective equipment should be used only sparingly and under appropriate circumstances, and never as a substitute for more reliable and effective engineering or administrative controls.

Increasingly, engineering controls are being supported with automatic alarm systems to give an alert when controls are malfunctioning and excessive air contamination or physical agent intensity is occurring.

The control of stresses associated with high technology in the operation of equipment such as computers, word processors, video display terminals, microwave equipment, and the like requires a different engineering approach from that used in the control of toxic stresses. Providing optimal lighting, adjusting equipment to the operator's stature, and maintaining an overall general room compatibility with tasks are required when such stresses are encountered. Additional considerations including special rest periods and designated exercises may be appropriate.

10 EDUCATIONAL INVOLVEMENT

In the late 1930s very few universities in the United States offered programs leading to degrees in industrial hygiene at either the undergraduate or graduate level. In 1994 more than 30 colleges and universities offer programs leading to undergraduate or graduate degrees in industrial hygiene. This reflects the enormous growth in the profession that has taken place over the past 50 years. The passage of the Occupational Safety and Health Act of 1970 had a major impact on this growth.

The American Industrial Hygiene Association has increased in membership from 160 in 1940 to more than 10,000 in 1994. In the 1982–1994 decade the number of Diplomates in the American Academy of Industrial Hygiene more than tripled from 1865 to 5500; in the spring of 1994 there were over 900 industrial hygienists in training.

Professional organizations such as the American Industrial Hygiene Association, the American Conference of Governmental Industrial Hygienists, and the American Academy of Industrial Hygiene offer excellent opportunities for the interchange of professional knowledge and the continuing education of industrial hygienists. These professional organizations invite participation through technical publications, lectures, committee activities, seminars, and refresher courses. As an example, the American Industrial Hygiene Conference and Exhibition of 1994 listed 469 technical papers covering a wide range of subjects. The same conference offered 174 professional development courses and other aids for the purpose of increasing knowledge and skills in the practice of industrial hygiene. This participation by experts in the many facets of the profession enhances the overall performance of practitioners and permits industrial hygienists to keep abreast of new technology in the recognition, measurement, and control of workplace stresses. These associations and the American Academy of Industrial Hygiene support the profession in its fullest concept. In return, practicing industrial hygienists are obliged to keep involved in the educational and knowledge sharing process by making professional information available to others who have interest in, and responsibility for, the health and well-being of workers.

Industrial hygienists should take active roles in educating management concerning environmental stresses in places of work and the means for their control. Alert management can bring pending processes and production changes to the attention of industrial hygienists for study and follow-up, thus avoiding inadvertent occurrences of health problems.

The educational involvement of workers is extremely important. Workers have the right to know the status of their working environments, the stresses that may be deleterious to their health if excessive exposures occur, and of the control systems that have been instituted for their protection. Knowledgeable workers are in a position to enhance their own protection through the proper use of control equipment, the proper response to administrative controls, and, where it is needed, to the proper use of personal protective equipment. When a control system malfunctions, a worker is often the first to observe it and can call it to the attention of management. In cases of spills and leaks, or equipment breakdown, a worker who is knowledgeable of the hazardous nature of the materials involved can better follow prescribed emergency procedures. Industrial hygienists are in an excellent position to participate in worker protection education programs.

11 SUMMARY

Gigantic strides have been made during the past four or five decades in characterizing and controlling environmental health hazards in places of work. In many industrial plants where comprehensive industrial hygiene programs are in effect and exposures to work hazards are well controlled, the off-the-job stresses such as smoking, alcohol consumption, drug use, and hobby activities may have greater effect on workers' health than do their on-the-job stresses.

Industrial technology changes rapidly. As technology changes and new technology is applied, new and different on-the-job health stresses emerge. Industrial hygienists must stay abreast of these changes and with the procedures for their recognition, evaluation, and control. It is vital that the techniques used in measuring occupational health stresses cover all the relevant components of each stress and that these are incorporated into any resulting control program. The practice of industrial hygiene rests on having valid data, on proper judgment in evaluating these data, and on effective follow through.

The science and profession of industrial hygiene has a vital role in industry. A well-implemented comprehensive industrial hygiene program leads to a healthful workplace.

The following chapters are devoted to the theoretical basis and rationale for the science and profession of industrial hygiene.

REFERENCES

CMA (1988). *CMA News*, Vol. 16, Number 8, November 1988, Chemical Manufacturers Association, Washington, D.C.

Hatch, T. (1962). "Changing Objectives in Occupational Health," *AIHAJ*, **23**, 1–7.

Hatch, T. (1972). "The Role of Permissible Limits for Hazardous Airborne Substances in the Working Environment in the Prevention of Occupational Disease," *Bull. World Health Organization*, Geneva, **47**, 151–159.

Johnson, Eric S. and George Lucier, (1992). "Perspectives on Risk Assessment Impact of Recent Reports on Benzene," *Am. J. Indust. Med.*, **21**, 749–757.

CHAPTER TWO

Toxicologic Data in Chemical Safety Evaluation

Richard J. Nolan Ph.D., William T. Stott Ph.D., and Philip G. Watanabe Ph.D.

1 INTRODUCTION

The ultimate objective of toxicological research on chemicals is to obtain information that will form a sound basis for recommendation of "safe" levels of exposure for humans contacting these substances during manufacture, use and disposal. The information generally sought in such research is the dose–response function that characterizes the untoward effects produced by selected doses of a chemical. Subsequently, the safe level of exposure has been selected traditionally by judgment to be 1/10 to 1/5000 of the highest dose that produced no discernible effect (Weil, 1972; Zapp, 1977). Safety factors are selected in accordance with the seriousness and persistence of the adverse effects and the shape of the dose–response curve.

Safe as used here is a relative term. No matter what adverse effect is of concern, absolute safety to everyone regardless of conditions can never be assured for any chemical or for that matter any activity in which humans participate.

In general, data depicting the adverse response of humans to selected doses of a chemical are unavailable. In some instances data from human experimentation are available; however, the adverse effects selected for study are generally transient and mild, such as skin irritation or slight depression of central nervous system (CNS) function. Rarely, epidemiological studies of people exposed during the manufacture and use of a chemical provide dose–response information for more severe effects.

Patty's Industrial Hygiene and Toxicology, Third Edition, Volume 3, Part B, Edited by Lewis J. Cralley, Lester V. Cralley, and James S. Bus.
ISBN 0-471-53065-4 © 1995 John Wiley & Sons, Inc.

Such information, when available, is most important. Because human experimentation is not feasible for elucidation of serious manifestations of toxicity, it is necessary to use experiments in animals to characterize such responses. However, the use of animals to predict the response of humans creates a dilemma. It is not possible to assure absolutely that people will not be more or less sensitive than the animal populations; therefore, the range of doses producing a given effect in humans may be much larger than in the animal species selected for experimentation. Indeed, even experimental results from studies on small numbers of humans are not totally reliable for predicting the response of large population groups for the same reasons.

Therefore, to accomplish the ultimate objective, data collected in studies using animals must be extrapolated to predict the response in humans. This chapter provides insight into such extrapolation. The basic premise is that toxicity is manifest as a result of the presence of the toxic agent at a specified concentration at the target site. Species difference in response to given exposures to a chemical occur because of differences in how the chemical is absorbed into the body, distributed and biotransformed once in the body, and subsequently excreted from the body. Species differences can also be a result of a response due to a species-specific receptor.

2 DOSE–RESPONSE FUNCTION

Fundamental to the extrapolation process, whether intraspecies or interspecies, is the dose–response function. This concept was enunciated in the sixteenth century by the physician–alchemist Philippus Aureolus Theophrastus Bombast von Hohenheim, better known as Paracelsus. In a Third Defense of the Principle written in 1538, he said: "Was ist das nit gift? Alle ding sind gift und nichts ohn gift. Allenin die dosis macht eng ding ken gift ist." ("What is it that is not poison? All things are poison and none without poison. Only the dose determines that a thing is poison.")

Figure 2.1 simulates, for a normally distributed population, the percentage of individuals responding to a range of doses for two chemicals. For chemical I, the range of doses producing a discernible response in the entire population is considerably smaller than that for chemical II. Curves for both chemicals indicate that at some selected doses on the low side, only a few in a population will respond. At the other extreme are those individuals requiring large doses to elicit the response. Dose X for both chemicals will cause 50 percent of the population to respond, since the areas under the two curves to the left of this point constitute one-half of the total areas under the respective curves. The dose causing 50 percent of the population to respond is the median effective dose, or ED_{50}; specifically for lethality, the dose is termed LD_{50}.

The primary reason for showing simulated curves for two chemicals in Figure 2.1 is to emphasize the necessity of using judgment to select a safe level of exposure. A sharply defined curve such as that of chemical I allows greater assurance that one-tenth of the dose not producing a response will be safe. Because of the wide range of doses producing a response to chemical II, a larger safety factor is needed. For this reason the use of a single value such as ED_{50} or LD_{50} for assessment of the relative hazard of a chemical can be misleading.

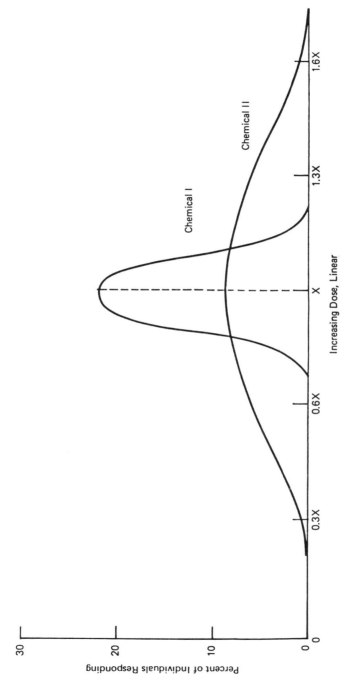

Figure 2.1. Illustrative dose–response curves for two fictitious chemicals.

If the data in Figure 2.1 are plotted as the cumulative percentage of the responding population, a sigmoid curve will result, going from 0 to 100 percent. For greater sharpness of the sigmoid curve, the dose–response may be plotted frequently as the percentage responding versus the logarithm of the dose (see Fig. 2.7). Since straight lines rather than sigmoid curves are preferred, a more desirable representation of the data is the use of a logarithmic probability (probit) display (Litchfield and Wilcoxon, 1949).

Utilizing a logarithmic probability display of data for the hepatotoxicity incurred from single doses of various chlorinated solvents to mice, Plaa et al. (1958) published the results appearing in Figure 2.2. Obviously the ED_{50} values for the various compounds provide some indication of the relative hepatotoxicity. However, with respect to judgment as to what dose will not produce hepatotoxicity to any of the population, a much more conservative safety factor must be used for carbon tetrachloride and chloroform than for the other materials because of the shallow slope for these compounds.

Another important criterion for selection of a safety factor for a chemical lies in a comparison of the doses producing a subtle unnoticed effect with those producing an effect that can be readily discerned. To illustrate, refer to Figures 2.3–2.5, depicting the dose responses for hepatotoxicity, narcosis, and death for carbon tetrachloride, chloroform, and 1,1,1-trichloroethane (Gehring, 1968).

For carbon tetrachloride and chloroform, hepatotoxicity can occur with exposure levels less than those that will produce narcosis. For 1,1,1-trichloroethane, however, injury to the liver occurs only at exposure levels in excess of those needed to produce

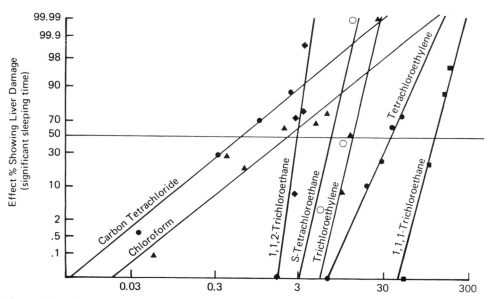

Figure 2.2. Dose–response curves for the effect of each of seven halogenated hydrocarbons on prolongations of pentobarbital sleeping time in mice.

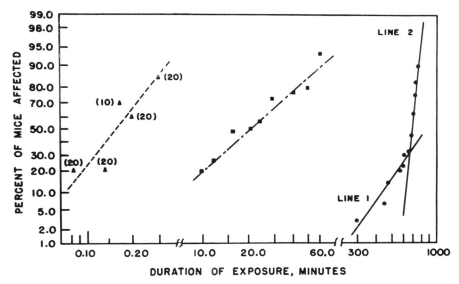

Figure 2.3. Carbon tetrachloride vapor, 8500 ppm. Percentage plotted on probability scale of mice anesthetized (squares) dead (circles), or having a significant serum glutamic acid–pyruvic acid–transaminase (SGPT) elevation (triangles), as a function of the $\log_{10}$ duration of exposure. Each point for anesthesia and lethality was obtained by using a single group of 30 mice; the number in each group used to obtain the points for SGPT activity is given in parentheses.

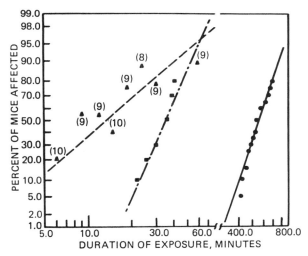

Figure 2.4. Chloroform vapor, 4500 ppm. Percentage plotted on probability scale of mice anesthetized (squares) dead (circles), or having a significant SGPT elevation (triangles), as a function of the $\log_{10}$ duration of exposure. Each point for anesthesia was obtained by using a single group of 10 mice, and each point for lethality was obtained with a single group of 20 mice; group size used for determining SGPT activity is indicated in parentheses.

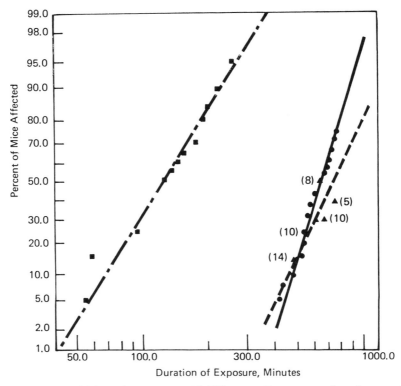

Figure 2.5. 1,1,1-Trichloroethane vapor, 13,500 ppm. Percentage plotted on probability scale of mice anesthetized (squares), dead (circles), or having a significant SGPT elevation (triangles), as a function of the $\log_{10}$ duration of exposure. Each experimental point for anesthesia and lethality was calculated by using composite groups of 20–135 mice. Individual group sizes used to obtain SGPT activity are indicated in parentheses.

narcosis. Indeed, liver injury occurs only when the exposure levels are sufficient to cause death to some individuals. For 1,1,1-trichloroethane, these results provide considerable confidence that prevention of narcosis, a noticeable effect, will preclude development of liver disease.

For carbon tetrachloride and chloroform, more restrictive judgments for tolerable exposures are needed to prevent the hazard of hepatotoxicity. The shallow curve depicting the hepatotoxicity incurred from various exposures to chloroform indicates greater variability in the population; therefore, a more conservative safety factor is justifiable for chloroform than for carbon tetrachloride. Stated another way, flat dose–response curves indicate that susceptibility to the chemical is highly variable in the population, and there is a greater probability that a small number of individuals may be adversely affected even though most will be unaffected. Diethylene glycol is another substance that gives rise to a flat dose–response in most species. In humans this has been reflected in human poisoning cases in which some individuals have died from ingestion of small doses, whereas others have survived relatively large doses (Geiling and Cannon, 1938). In contrast, Conyoulex toxin, a substance occurring occasionally in clams and mussels, has a steep dose–response curve and one-fourth the LD_{50} can be ingested without measurable risk (Dewberry, 1950).

3. ROUTE OF EXPOSURE

The toxic effects of a chemical depend on the absorption of a toxic substance into the organism. Under conditions of occupational exposure, a toxic substance can gain entry into the human body through the respiratory tract, the skin, and the gastrointestinal tract. The entry of a chemical through these routes is dependent on the chemical-physical properties of the agent. Whether a toxic effect occurs depends on the concentration of the substance at the target site and the susceptibility of the biological system.

3.1 Inhalation

In cases of occupational exposure, inhalation frequently represents an important mode of entry of toxic substances into the body. In almost every manufacturing process the atmosphere is likely to contain some amount of vapors, dust particles, mists, and other substances. The absorption of substances by way of the respiratory tract depends on its physical and chemical properties.

Gaseous or vaporous substances generally pass into the body mainly by way of the alveoli and are distributed throughout the body by the bloodstream. However, some materials are effectively absorbed in the upper respiratory tract. The degree of retention of such substances in the upper air passages depends on their water solubility: the more soluble the substance, the greater the degree of its absorption through the upper air passages, and vice versa. Some are quickly broken down on moist mucosal surfaces.

The retention of inhaled particles (aerosols) occurs throughout the respiratory tract, from the nasal cavity downward depending on the particle size. As the particle size increases, an increasing number of particles tend to be retained in the upper airways. Particles about 1 μm in size settle mainly in the alveoli, whereas those over 5 μm tend to deposit in the upper airways. The site and degree of retention also depend on the density, shape, and electric charge of the particles.

3.2 Cutaneous Absorption

Absorption through the skin is also an important route of entry for many substances into the body. Too often skin absorption is given inadequate emphasis in assessment of exposure potential. The skin may be considered as a multilayered protective cover for the organism. This barrier, however, is not continuous; it is perforated by hair roots and follicles that penetrate deep into the dermis and subcutaneous tissue. The ability of a substance to penetrate the skin depends mainly on its lipid and water solubilities. Lipid-soluble substances are capable of moving through the fatty layers of the skin, but their further absorption will be hindered if their hydrophobic properties retard their dissolution in the blood. Skin absorption is also affected by such factors as temperature, contact surface area, and duration of contact.

3.3 Ingestion

Ingestion is not usually considered as a form of occupational exposure (except if it occurs accidentally) if the basic rules of occupational hygiene are followed, and overexposure due to ingestion following contamination of hands, food, beverage, cigarettes, and other materials at the workplace rarely occurs. Accidentally ingested materials are absorbed into the body at one or more sites in the gastrointestinal tract.

Gastric absorption depends on the nature and quantity of gastric contents. Gastric and intestinal secretions may considerably alter the chemical nature of a substance and increase its solubility. Chemicals may also be transformed by intestinal bacteria. For example, aromatic nitro compounds are reduced to amines by bacteria in the intestine.

After absorption from the stomach and intestine, chemicals are carried by the blood to the liver, where they are frequently biotransformed or excreted in the bile. Biotransformation processes in the liver generally oxidize and/or conjugate a foreign substance in preparation for excretion from the body. Materials can be metabolically detoxified or activated in this process.

4 ACUTE TOXICITY

Acute toxicity of chemicals in animals is evaluated by administration of single doses by various routes of exposure. Although oral administration is uncommon for exposure to most industrial chemicals, it is commonly used for a first assessment of the potential of a chemical to cause toxicity. Too frequently, lethality is considered to be the most important adverse effect, and other signs of adverse effects such as the physical condition of the animals and damage to particular organs are overlooked. Rigorous evaluation of these parameters frequently reveals the organ system affected by the chemical and occasionally provides insight as to the mechanism of toxicity.

For determination of acute oral toxicity, strict attention should be given to the persistence of the effects. For example, delayed deaths occurring 2 or more days following administration, or depression of body weight gain 1 or 2 weeks after administration, suggests either persistence of the chemical in the body or slow repair of the damage incurred. In either case more conservative handling of the chemical to preclude adverse effects on health is warranted. It may also be anticipated that repeated exposure to a chemical eliciting persistent manifestations of toxicity subsequent to a single exposure will constitute a greater hazard than exposure to a chemical that does not.

Exposure to industrial chemicals occurs most frequently by way of contamination of the skin or inhalation of the vapor or dust. Contamination of skin may result in local damage and in some instances absorption into the body and systemic toxicity.

Adverse local reactions of either skin or eyes are generally extrapolated directly to humans, and appropriate measures are instituted to preclude contamination of skin and eyes with injurious concentrations of chemicals.

Some chemicals penetrate the skin readily. An indication of rapid penetration is an equivalency or near equivalency of acutely toxic doses, whether given orally or by skin application. Evidence of significant penetration of a chemical through the skin warrants institution of precautions to minimize skin contamination. The rigor of these precautions will be dictated by the probability of dermal exposure and the severity of potential effects as revealed by acute, subchronic, and chronic toxicity, usually revealed in studies relying on nondermal routes for administration.

Exposure to chemicals by way of inhalation is a major concern in an industrial environment. Frequently the acute toxicity of a chemical by way of inhalation is initially assessed by exposing animals to a concentrated atmosphere of the chemical for 4–6 hr. If injury or death occurs, the duration of exposure and the exposure concentrations are decreased progressively until the adverse effects disappear. The objective of such experimentation is to characterize the toxicity as a function of duration and concentration of exposure. Frequently, this function can be visualized by plotting the log of the concentration versus the log of the exposure needed to produce a given effect or lack of effect.

To illustrate, consider the data of Adams et al. (1952) depicting the acute toxicity of carbon tetrachloride (Fig. 2.6). For rats, line *CD* represents the most severe exposures not resulting in death, and *EF* represents exposures causing no discernible effects. The large difference between exposures represented by these lines carries the same interpretive significance as discussed previously for the data in Figure 2.3.

In the absence of human experience, tolerable single exposure to carbon tetra-chloride would be set at least 10-fold less than those represented by line EF. Data for additional species—dog, monkey, and rat—indicating lesser or equivalent susceptibility, increase confidence that a 10-fold margin of safety is adequate.

For many chemicals such as carbon tetrachloride, human experience will augment the judgment. When Adams et al. (1952) reported their results, sufficient human experience was available to support the conclusion that single exposures such as those represented by line *EF* did not produce discernible adverse health effects in humans.

5 SUBCHRONIC AND CHRONIC TOXICITY

Only initial judgments are possible from acute toxicity tests. If carefully done, they provide insight into the potential toxicity on repeated exposure to a chemical, but repeated exposure studies are needed to provide a more definitive assessment of subchronic and chronic toxicity.

In studies to characterize potential subchronic and chronic toxicity, chemicals are administered to animals daily for 5–7 days per week for an entire lifetime or a fraction thereof. When administered orally, the chemicals are mixed with the diet when feasible. Inhalation exposure is conducted in chambers with a regimen usually of 6–8 hr per day, 5 days per week. Sometimes other routes of exposure are used (e.g.,

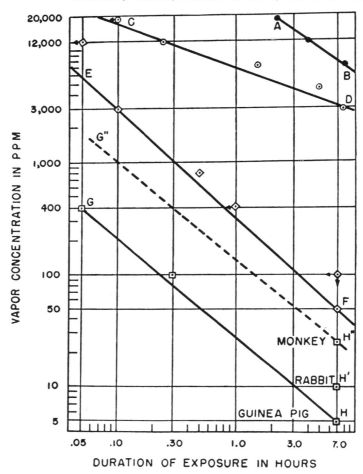

Figure 2.6. Vapor toxicity of carbon tetrachloride. Line *AB* represents the least severe single exposures causing death of all rats tested and line *CD* the most severe single exposures, permitting survival of all rats tested. Line *EF* represents the most severe single exposures without detectable adverse effect and line *GH* the most severe repeated exposures without detectable adverse effect in rats. Points, *H*, *H'*, and *H"* represent the most severe repeated exposures without detectable adverse effect in the guinea pig, the rabbit, and the monkey, respectively. Line *G"H"* represents the most severe repeated exposures with little probability of adverse effect in human beings.

daily application to the skin or injection). Extensive antemortem and postmortem evaluations are generally made with the use of the most recent clinical, chemical, and pathological methods. Typically, more than 50–100 different parameters are assessed.

Adams et al. (1952) also reported the subchronic toxicity of carbon tetrachloride. Line *GH* in Figure 2.6 represents daily exposures that did not produce discernible adverse effects in rats over a duration of 6 months. Also shown as single points are no-effect exposures for guinea pigs, rabbits, and monkeys exposed 7 hr per day, 5 days per week, for 6 months. In these studies rats and guinea pigs appeared to be most susceptible; rabbits and monkeys were intermediate and least susceptible, respectively. The liver was the target organ. Since human experience indicated that humans were less susceptible than rats, a line $G''H''$ parallel to GH was drawn to represent exposures judged to be safe in humans. As a result, 5 ppm was recommended as the standard for the work environment for 7-hr daily exposures.

In the absence of human experience, exposures represented by line GH or possibly even lower exposures may have been selected. To preclude subchronic toxicity, a workroom standard of 5 ppm would have been justified in that this would have protected the most sensitive species, rats and guinea pigs. The volume of air breathed by these species per unit body weight is greater than that for humans. Thus the dose received by humans is less, providing a built-in safety factor.

A key question in subchronic and chronic toxicology studies is, how long should the animals be exposed? Ideally, chronic toxicology studies should involve treatment of the animals for a significant portion of their lifetime, considered to be 2 years for rats and 18 months to 2 years for mice. When the chemical is administered by routes other than in the diet, administration for a lifetime requires an extensive effort. For example, daily generation of atmospheres containing the vapors of a chemical and chemical analyses to assure the desired concentration are time-consuming and costly, as is handling of the animals. In addition, physical facilities for performing inhalation studies are costly and limited in availability. In the past, chronic inhalation studies have been conducted by using daily exposures for a fraction of the lifetime (e.g., 6–18 months) and subsequently maintaining the animals for the duration of the lifetime. This protocol duplicates, reasonably well, exposures incurred by workers. Also, the track record of this protocol for revealing carcinogenic effects of a chemical is acceptable, although few definitive studies comparing the result of daily administration for a lifetime versus a fraction of a lifetime are available. Chronic inhalation studies currently are designed to expose rats or mice 6 hr per day, 5 days per week for 24 to 18–24 months, respectively. Following exposure, the test species are euthanized and undergo a complete clinical and histopathologic examination.

A number of authors have demonstrated that toxicology studies exceeding 3 months in duration are usually unnecessary to reveal the potential chronic toxicity of chemicals (Barnes and Denz, 1954; Paget, 1963; Bein, 1963; Frawley, 1967; Peck, 1968; Hayes, 1972). Weil and McCollister (1963) compared the minimum dose causing a toxic effect and a maximum dose causing no effect for 33 chemicals in short-term (29–200 days) versus long-term studies. The ratios of the indicated dosage levels for short-term versus 2-year feeding studies were 2.0 or less for 50 percent of the chemicals. A ratio greater than 5 occurred for only three chemicals; the largest ratio for a maximum no-effect level was 12.0. These results suggest that in lieu of chronic toxicity data, selection of one-tenth the no-effect level in a subchronic study as a no-effect for chronic toxicity may be appropriate until a definitive study is conducted. Again judgment should be tempered by human experience when available. As indicated earlier, subchronic toxicity studies often adequately reveal the potential chronic toxicity of a chemical. However, as more chemicals are studied, exceptions are becoming too common to permit over-reliance on this generality. Therefore, subchronic toxicity studies should be considered only interim assessments of potential chronic toxicity.

As indicated previously, chronic studies by way of inhalation exposure constitute a major and costly undertaking. For many chemicals, only studies utilizing exposure by way of ingestion may be available. To estimate the maximum dose that will not cause an effect by way of inhalation, one may assume that a worker will inhale 10 m^3 of air per 8-hr workday (Stokinger and Woodward, 1958). Since not all the chemical inhaled is retained, a retention factor is suggested as follows (McCollister et al., 1951; Stokinger and Woodward, 1958):

	Water Solubility (%)	Retention Factor (%)
Essentially insoluble	<0.1	10
Poorly soluble	0.1–5	30
Moderately soluble	5–50	50
Highly soluble	50–100	80

The estimated dose is calculated in accordance with the equation

$$\text{Estimated dose} = \frac{(10 \text{ m}^3) \ (\text{mg/m}^3) \ (\text{retention factor})}{\text{kg body weight}}$$

Estimation in accordance with the foregoing is frequently useful. However, it is not recommended that such an estimation be used to satisfy the need for more definitive studies (i.e., a chronic study using inhalation exposure or a pharmacokinetic evaluation to characterize the fate of inhaled versus ingested chemicals).

Currently, assessment of the chronic toxicity potential of chemicals is considered the best means of providing a database to set tolerable chronic exposure levels for humans. Typically, in chronic studies, 50–100 animals per sex per species are exposed to each of three or four levels of the agent, together with an equal or greater number of controls. Frequently, two species are studied. Unfortunately, tunnel vision has resulted too often in the use of such studies to study only carcinogenicity. Chronic toxicity studies should be aimed at assessing as best as possible any manifestation of an untoward effect. Furthermore, interpretation of a carcinogenic response without assessment of other manifestations of toxicity is an unscientific exercise in futility, since secondarily induced carcinogenesis occurs (Gehring and Blau, 1977) and since other manifestations of toxicity may be equally serious with respect to worker's health.

6   RELIABILITY OF ANIMAL STUDIES FOR PREDICTING TOXICITY IN HUMANS

There is no comprehensive, definite means of assessing the reliability of predicting toxicity or lack thereof in humans with the use of toxicity data collected in animals. Although surprising initially, this apparent deficiency is understandable because such an assessment requires dose–response data for humans as well as animals. Doses of chemicals causing frank signs of toxicity cannot be administrated intentionally to people. Even when human studies are conducted, reliance on symptoms such as nausea, headache, anxiety, depression, disorientation, weakness, or irritation as perceived by the subjects precludes, for the most part, a correlation with animal studies in which the parameters are limited to those that can be observed and measured by the investigator: signs versus symptoms.

Correlations of the adverse effects in humans with those in animals exist predominantly for drugs. Review of this extensive literature will cause considerable anxiety because it contains many instances of studies in animals that have revealed both false-positive and false-negative results with respect to humans (Baker et al., 1970; Baker, 1971).

In the study of six unnamed drugs, Litchfield (1961) found a significant relationship between the signs of toxicity in humans, rats, and dogs. However, 23 of 234 signs of toxicity were seen only in humans. Studies on rats and dogs did not predict symptoms unique to humans—headache, loss of libido, and so on. Dogs were found to be somewhat more useful in prediction of drug effects in humans than rats.

Classical examples of drugs that induce false-positive results in animals are fluroxene, an anesthetic that has been used uneventfully in humans but kills dogs, cats, and rabbits, and penicillin, which produced lethality in guinea pigs at doses in the therapeutic range for humans (Wardell, 1973).

Rall (1969) demonstrated an excellent correlation between toxicity data for 18 direct acting anticancer agents in mice and humans when the dose was based on milligrams per square meter of body surface (mg/m^2). The mouse was consistently 12-fold less susceptible than humans, when the data were examined on a milligram per kilogram body weight basis. In the original publication of the correlation of toxic responses in humans and animals for the antineoplastic agents, monkeys, dogs, and rats were also considered (Freireich et al., 1966). Assessments in these and other animal species added little except to further substantiate the existence of a good quantitative correlation when the dose is expressed as milligrams per square meter rather than milligrams per kilogram.

Administration of biologically active chemicals, per se, such as antineoplastic agents frequently induces equivalent responses when the dose is administered proportional to body surface (Pinkel, 1958). This relationship is gaining recognition in estimating the risk incurred by humans from exposure to chemicals in the environment (National Research Council, 1977). In utilizing this relationship, however, one must recognize that the original relationship was developed for biologically active agents. Since metabolism and other physiological processes involved in detoxification are generally more active in smaller animals, the dose of a biologically active chemical per unit mass required to produce a given effect increases as the body mass decreases, whereas the dose per unit surface area remains relatively constant. For a chemical requiring activation to the biologically active toxic form, however, the opposite is likely to be generally true because the rate of activation will be roughly proportional to the body surface area. For a chemical requiring conversion to the active toxicant, therefore, a greater dose of the active form may be received on a milligrams per kilogram body weight by a larger than small species of animal because the latter has a greater surface area : body weight ratio.

In reviewing animal experimental data for assessing the carcinogenicity of chemicals in humans, Wands and Broome (1974) reported that studies in animals have revealed the carcinogenicity of all known human carcinogens except arsenic. Even for arsenic, an increased incidence of leukemia and malignant lymphomas in pregnant Swiss mice has been reported (Osswald and Goertler, 1971). Wands and Broome (1974) also pointed out the existence of positive carcinogenic responses in animals for roughly 1000 compounds having no evidence for a carcinogenic response in humans.

With respect to the foregoing discussion of the apparent lack of reliability of data from animal experiments for predicting toxicity in humans, how can the usefulness of animal experimentation be advocated for this purpose? It must be recognized that exceptions to correlations rather than correlations per se tend to be reported. Existence of toxicity data in animals and application of precautions to avoid toxicity in humans for literally thousands of chemicals preclude visibility of a correlation.

Undoubtedly, the false-positive finding of carcinogenic activity in animals for roughly 1000 chemicals is attributed to a number of factors such as the subject's limited exposure to these agents. For many of these false-positive chemicals, the doses used to elicit a response were sufficient to overwhelm detoxication mechanisms and to cause prominent manifestation of toxicity other than carcinogenicity.

For drugs, it must be recognized that the apparent weakness, not absence, of a correlation for signs of toxicity is frequently qualitative. This is not surprising because administration of therapeutic or near therapeutic doses of biologically active agents to highly integrated biological systems may be expected to produce variable qualitative intra- and interspecies responses. Such differences in the qualitative manifestations of toxicity in response to toxic doses of a chemical should engender less concern than the absence of any manifestations of toxicity in animals given a dose of a chemical later found to be toxic in humans. Although of less concern, qualitative manifestations of toxicity in animal experimentation are important because they provide a basis for selecting parameters to be monitored in an exposed population of people.

For further conceptualization of the problems inherent in extrapolation of toxicity from animal experimentation for prediction of the hazard in humans, the dynamics of toxicity must be realized. The toxicity of a chemical is a function of the absorption of that chemical into the body, distribution, and biotransformation of the chemical. Once the chemical is absorbed in the body, interaction of the chemical per se or a biotransformation product with the biological receptor leads to the ultimate action. Excretion of the chemical itself or its biotransformation products terminates activity. If each of these five processes is an independent variable and each has a correlation coefficient of 0.9 for humans versus animals, the overall correlation coefficient will be 0.6. In spite of the poor correlation predicted for equivalent toxicity in humans and animals, tolerable exposure levels based on animal data have proven effective in precluding manifestations of toxicity in humans. Undoubtedly, the safety factors used in establishing tolerable exposure levels have been largely responsible for this.

Regardless of the success experienced in using toxicity data from animal experimentation as a basis for recommending tolerable exposures for humans, future development of the science of toxicology must be directed at improving the reliability of this extrapolation. Consideration of the five factors influencing the toxicodynamics of chemicals immediately reveals how this can be done. First, it requires quantitative knowledge of the absorption, distribution, biotransformation, biological reaction, and excretion of a chemical or its biotransformation products in humans versus animals. This may be accomplished in pharmacokinetic evaluations of the chemical, which are becoming more common. Second, the reaction of the active chemical with the biological receptor or the mechanism of action must be elucidated. The mechanism of the toxicity of a chemical is not, however, as easily revealed as its pharmacokinetics. Advances in cellular and molecular biology are allowing mechanisms of toxicity to be elucidated for a few chemicals. In the meantime, elucidation of the pharmacokinetic parameters will dramatically improve the extrapolation process.

7 THRESHOLD

In a typical dose–response curve (Fig. 2.7), the solid line represents an observable increase in the percentage of individuals responding to increasing doses. This type of dose–response curve occurs when the response of individuals within a population is distributed normally, which is the situation characteristic for most pharmacological and toxicological responses to chemicals. The concept of threshold has been accepted for most pharmacological and toxicologic responses. Biologically, the term *threshold* has been interpreted to mean that there is a dose below which no response will occur in a population of animals. Hence, when a dose is found that does not produce a toxic response in a reasonable number of subjects (laboratory animals or humans), it is assumed that the dose is subthreshold.

Threshold for an adverse response to a chemical differs with species and is influenced by various physiological factors such as age, sex, diet, and stress. In a population where these variables have been controlled, it is assumed that a large number of animals or people may be exposed to a subthreshold amount of the chemical without a response. Such an assumption renders relatively easy prediction of a safe exposure to a chemical for the population. After a given dose has been found to elicit no response in an experiment, a safety factor is applied to this dose to account for intra- and interspecies differences. Subsequently, no experience of an adverse effect in a few individuals of a population is taken as confirmation of the appropriateness of the safety factor, and the existence of a subthreshold is assumed for the entire population.

A threshold concept is used to deem safe many events or materials in our lives other than chemicals. If a building is constructed to withstand a wind force of 100 mph, it is assumed that construction of millions of duplicates will not result in some that will be devastated by a 10-mph wind. In the absence of such a concept, a judgment of safe for any human endeavor is impossible.

A threshold concept, however, is not universally accepted for chemical carcinogenesis and mutagenesis, toxic responses that are linked mechanistically (see below). Figure 2.7 represents an idealized dose–response curve for the production of heritable mutations or tumors in a test species by a chemical carcinogen. While it is generally accepted that increasing dosages of a mutagen or carcinogen will produce an in-

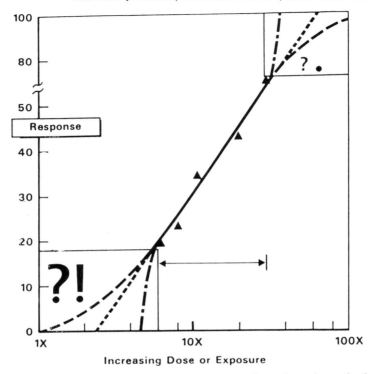

Figure 2.7. Simulated percentage of individuals responding adversely to the logarithm of selected doses. Measurable responses are represented by triangles. The sigmoid (dashed) curve represents a population described by normal distribution; in theory, the percentage responding never reaches zero on the low end or 100 on the high end. The other curves represent a threshold for the response. The boxed-in portions represent regions in which prediction of incidence depends on stochastic, statistical projection.

creased incidence of mutations or tumors in a dosage range accessible to experimental observation, as represented by the solid line, there is disagreement about the response obtained at relatively low dosages, as represented by the lower boxed-in area. The dilemma is whether the incidence of the response drops off rapidly, as would be predicted by a threshold response; continues to baseline as a linear function of dosage, as would be predicted to occur in the absence of a threshold; or is actually higher at lower dosages than predicted by the available data (i.e., has an elevated potency at lower dosages). Unfortunately, the establishment of a clear, statistically supportable threshold at these dosages would require impractical experimental protocols involving thousands or even millions of test subjects.

The fundamental basis of the nonthreshold carcinogenic model is that cancer is an expression of a permanent, replicable defect initiated by reaction between a single molecule of a chemical carcinogen (or mutagen) and a critical molecule such as DNA (i.e., a single "hit"). Support for this argument has primarily come from experimentation on radiation-induced carcinogenesis that failed to establish absolute thresholds within the realm of statistical reliability. However, equating radiation-induced cancer with chemical-induced cancer is tenuous at best, since the latter involves the basic pharmacologic processes of absorption, distribution, biotransformation, and excretion, which control exposure of cells and subcellular organelles to xenobiotics. At the molecular level, numerous repair processes exist (Lindahl, 1982; Setlow, 1983; Hanawalt, 1991) that may eliminate premutagenic lesions arising from a xenobiotic chemical–DNA interaction as well as those arising continuously from normal molecular degradative processes and the myriad of naturally occurring endogenous and exogenous reactive compounds (Saul and Ames, 1986; Ames and Gold, 1990a; 1991; Hoeijmakers and Bootsma, 1990; Harris, 1991; Simic and Bergtold, 1991). Clear evidence of the potential role of these repair mechanisms has been demonstrated experimentally in cultured cell lines and clinically in patients lacking an effective DNA repair mechanism (Maher et al., 1979; Yang et al., 1980; Rudiger et al., 1989; Lindahl et al., 1991). Even in the case of radiation-induced carcinogenesis, a disproportionate decrease in the potency of radium has been demonstrated at lower dosages suggesting repair of genetic lesions (Evans, 1974).

The premise that a single mutagenic lesion in the genome of an animal can result in tumor development is also not consistent with the results of recent mechanistic studies in human tumor tissues. These studies have demonstrated the involvement of a multiplicity of mutagenic and chromosomal changes in cancer development (Fearon and Vogelstein, 1990; Harris, 1991). Indeed, one form of colon cancer appears to require at least five or six distinct gene activation or inactivation steps involving distinct mutagenic events and chromosomal rearrangements for tumor formation to proceed (Vogelstein et al., 1988). In addition, numerous chemicals that have been observed to cause tumors in test animals are not mutagenic at all (see below). Neither these compounds nor their known metabolites are mutagenic in standard mutagenicity assays and they do not bind to DNA. Extensive data suggests that these so-called nongenotoxic carcinogens induce tumors in animals secondary to chronic morphologic and/or enzymatic changes, treatment-related effects that are often readily reversible and have clearly established thresholds.

The resolution of this issue may have a profound impact upon risk assessment and regulation of chemical carcinogens as potential human exposure to many of these compounds occurs only at levels substantially lower than those administered in experimental animal bioassays. In the absence of experimental confirmation of absolute thresholds, a number of nonthreshold-based methodologies have often been employed to fit available tumor data and predict carcinogenic responses at low dosages of suspected carcinogens. In some instances, the more conservative of these extrapolation techniques have been an integral part of the risk assessment process for many chemical carcinogens, for example, the use of the linearized multistage model by the U.S. Environmental Protection Agency (U.S. EPA, 1986). When combined with interspecies conversion factors, safety margins obtained utilizing a nonthreshold-based analysis of tumor data may differ from that of a threshold-based analysis of the same data by several orders of magnitude.

8 PHARMACOKINETICS AS AN AID IN INTER- AND INTRASPECIES EXTRAPOLATIONS OF TOXICITY DATA

As indicated previously, the dynamics of toxicity depend on the absorption of the chemical into the body, its distribution and metabolism in the body, and the excretion of the chemical itself or its metabolite from the body. Toxic manifestations produced by a chemical are a function of the concentration of the toxic entity at the target sites and the duration of exposure of these sites to the toxic entity. For assessment of the hazard of a chemical to humans as well as other species, therefore, it is essential to elucidate the kinetics for its absorption, distribution, biotransformation, and ultimate excretion—that is, its pharmacokinetics. Only with acquisition of such information can interspecies, intraspecies, high-dose–low-dose, and route-to-route extrapolations of potential toxicity be made definitively. Gehring et al. (1976) should be referred to for a detailed discussion of the use of compartmental pharmacokinetics for assessment of the toxicologic and environmental hazards of chemicals. Examples of both compartmental and physiologically based pharmacokinetics (PB-PK) models will follow.

8.1 Dose–Interspecies Response

Figure 2.8 can be used to help envision the complex problem of extrapolating toxicology data across dose and species. In Figure 2.8(a), the chemical dose level (on a body weight basis), the species size (body weight), and the incidence (or severity) of the toxic response all increase in an outward direction from the coordinate intersection. The height of the surface above the dose–species plane in the absence of a chemical dose represents the normal background incidence of a given response. The shaded area in Figure 2.8(a) shows the experimentally observable dose–response relationship in the smaller animal species and high dose levels used in routine toxicology studies. The arrows on the surface of the plane indicate the directions of extrapolation of toxicology data across the toxicity surface in order to estimate the potential risk to humans at realistic exposure levels. When a toxic response is elicited by a chemical, the dose–response curve can be envisioned to lie in the vertical plane for a given species as shown in Figure 2.8(b). As the dose level decreases, the toxic response also diminishes until it virtually vanishes into the background incidence.

At a given dose level (on a body weight basis) the toxic response may either increase or diminish as the species size increases, and is depicted in Figures 2.8(c) and 2.8(d), respectively. Since the basal metabolic rate of different mammalian species is proportional to the body surface area-to-volume ratio, and this ratio increases with decreasing body size, small animals will generally metabolize chemicals more rapidly on a body weight basis than larger animals. Therefore, whether a larger animal species will be more or less sensitive to a chemical than a smaller species is often dependent on whether metabolism of the parent chemical constitutes a detoxification or toxification process. Ultimately, the ratio of the rates of activation and inactivation will determine the concentration of the biologically effective molecule and can be more complex than the simple size relationship described above. The greater susceptibility of the rat versus the mouse to 2-acetylaminofluorene-induced

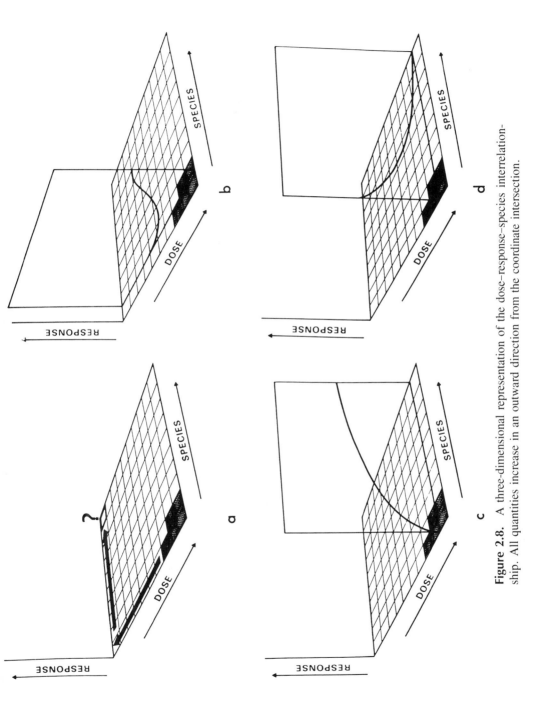

Figure 2.8. A three-dimensional representation of the dose–response–species interrelationship. All quantities increase in an outward direction from the coordinate intersection.

47

hepatocarcinogenesis has been related to higher sulfotransferase activity in the rat, that is, the enzyme that activates 2-acetylaminofluorene (Miller et al., 1964; DeBaun et al., 1968; Miller, 1970). However, as other species have been studied, it has become apparent that the relationship between 2-acetylaminofluorene toxicity and sulfotransferase activity is more complex than initially envisioned (Grantham et al., 1976).

The foregoing considerations imply that the toxicity surface lying above the plane shown in Figure 2.8(a) may easily assume a complex shape between the extremes of dose level and species size and that the shape of this surface is quite likely different for different chemicals. Studies elucidating the predominant mechanisms of toxicity of a given chemical can reveal directional trends across this surface as both species and dose levels change. Likewise, knowledge of the pharmacokinetics of a chemical can provide both qualitative and quantitative information concerning the expected toxic response with changing dose levels in different species. The complex nature of the surface emphasizes the necessity of integrating the different disciplines of toxicology and utilizing all available information in a rational manner in order to obtain the most realistic estimate of the potential risk to humans.

The following examples were selected to illustrate how knowledge of the pharmacokinetics, metabolism, and mechanism of actions of a chemical can influence the assessment of dose-related relationships and interspecies extrapolation.

8.2   2,4,5-Trichlorophenoxyacetic Acid (2,4,5-T): Linear and Nonlinear Pharmacokinetics and Species Differences (Compartmental Model)

2,4,5-Trichlorophenoxyacetic acid is a herbicide that is more toxic in dogs than rats. In order to elucidate the potential hazard of 2,4,5-T in humans, the pharmacokinetics of 2,4,5-T were characterized in several species including man (Piper et al., 1973). Figures 2.9 and 2.10 show the elimination of ^{14}C activity from plasma and in urine following a 5 mg/kg oral dose of ^{14}C-labeled 2,4,5-T in rats and dogs. Plasma concentration–time curves were characterized by a one-compartment open model with a first-order absorption and elimination. The half-lifes ($t_{1/2}$) for the decrease in plasma 2,4,5-T in rats and dogs were 4.7 and 77.0 hr, respectively. The $t_{1/2}$ for the elimination of 2,4,5-T in the urine was 13.6 hr in rats and 86.6 hr in dogs. The more rapid elimination of ^{14}C activity of 2,4,5-T from the plasma than in the urine of rats suggested that the compound was concentrated in the kidney prior to excretion. The slower rate of elimination of 2,4,5-T by dogs than by rats is consistent with the higher toxicity in dogs (Drill and Hiratzka, 1953; Rowe and Hymas, 1954). Generally, the toxicity of an agent to an individual or species is related inversely to the ability of the individual or species to eliminate the compound.

Almost all of the 2,4,5-T excreted by rats was via the urine, whereas approximately 20 percent of that excreted by dogs was through the feces. Measurable amounts of breakdown products of 2,4,5-T were not produced by rats given 5 mg/kg, whereas 10 percent of the ^{14}C activity excreted in the urine of dogs was attributable to breakdown products. This illustrates a pertinent principle: If maximum elimination by one route, in this case urine, is exceeded, a greater fraction will be eliminated by other routes.

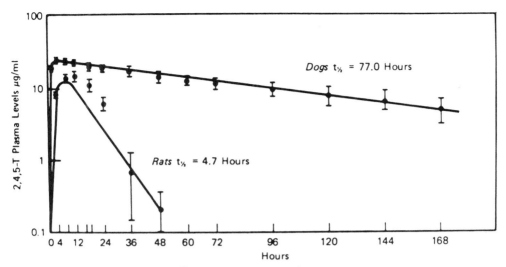

Figure 2.9. Concentration of ^{14}C activity expressed as microgram equivalents 2,4,5-T per milliliter of plasma in dogs and rats following a single oral dose of 5 mg/kg $[^{14}C]$-2,4,5-T. Each point represents a mean and SE.

Since 2,4,5-T is an organic acid, it was hypothesized that the poorer ability of the dog than the rat to eliminate it was associated with the poor organic acid secretory process of the dog kidney. If 2,4,5-T is eliminated primarily by an active secretory process in the kidney, elimination should be nonlinear, and saturable in the rat by administration of higher doses. The data in Figures 2.11 and 2.12 demonstrate that the elimination of 2,4,5-T from both the plasma and the body was shown to be nonlinear after administration of 100 and 200 mg/kg. The previous suggestion that the rapid elimination from the plasma of rats given 5 mg/kg 2,4,5-T was in part due to uptake by the kidney was fortified by the finding that the rates of elimination of 2,4,5-T from the plasma and body were equivalent after doses of 100 or 200 mg/kg. At these higher doses, the active uptake of 2,4,5-T, by the kidney had been saturated.

In addition to the lack of superposition of the plasma 2,4,5-T concentration–time curves with increasing doses, another criterion for nonlinear pharmacokinetics was revealed. A larger percentage of the ^{14}C activity administered as $[^{14}C]$ 2,4,5-T was excreted in the feces as the dose was increased. Furthermore, degradation products of 2,4,5-T were found in the urine of rats given the 100 or 200 mg/kg dose levels but not found in the urine of rats given 5 or 50 mg/kg dose levels.

In order to better characterize the nonlinear pharmacokinetics of 2,4,5-T rats were given intravenous (IV) doses of 5 or 100 mg/kg (Sauerhoff et al., 1976). The clearance of 2,4,5-T from the plasma of rats given 100 mg/kg dose level (Fig. 2.13) was in accordance with the Michaelis–Menten equation. Estimates of the maximum transport rate (V_{max}) and Michaelis constant (K_m) were 16.6 ± 1.8 $\mu g/g/hr$ and $127.6 \pm$

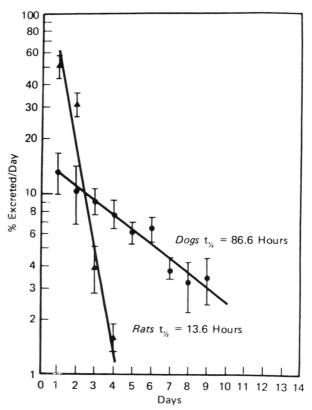

Figure 2.10. Percentage of administered ^{14}C activity excreted by dogs and rats during successive 24-hr intervals following a single oral dose of 5 mg/kg [^{14}C]-2,4,5-T. Each point represents a mean and SE.

μg/g, respectively. During the linear phase of excretion, that is, between 36 and 72 hr after administration of 100 mg/kg, the $t_{1/2}$ was 5.3 $\pm$ 1.2 hr. This half-life is not significantly different from that found for rats given 5 mg/kg.

The apparent volume of distribution (V_d) increased from 190 $\pm$ 8 to 235 $\pm$ 10 mL/kg for rats receiving 5 and 100 mg/kg [^{14}C]-2,4,5-T, respectively. This increase in V_d indicates that as the dose is increased, a disproportionately larger fraction finds its way into tissue and cells. As a consequence the potential for toxic effects to occur would undoubtedly also be expected to increase disproportionately at high dose levels.

As indicated previously, the discrepancy between the elimination rates of 2,4,5-T from the plasma and in the urine suggested that the kidney concentrated the compound prior to excretion. Table 2.1 depicts plasma and kidney concentrations and the kidney–plasma ratios of ^{14}C activity at various times after IV injection of 100 mg/kg [^{14}C] 2,4,5-T. The data show clearly that at low plasma concentrations

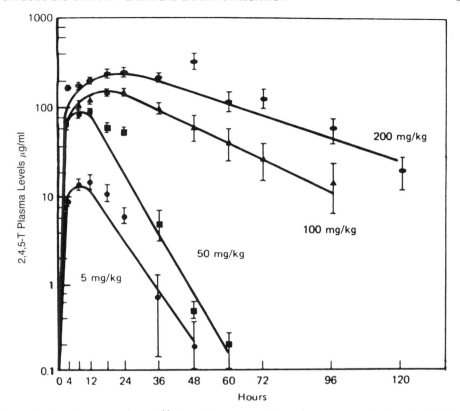

Figure 2.11. Concentration of ^{14}C activity expressed as microgram equivalents 2,4,5-T per milliliter plasma in rats following a single dose of 5, 50, 100, or 200 mg/kg [^{14}C]-2,4,5-T. Each point represents a mean and SE.

the kidney concentrates 2,4,5-T and that the kidney's ability to concentrate 2,4,5-T is saturable.

To further characterize the saturable active transport of 2,4,5-T by rat and dog kidney, in vitro studies were conducted using renal slices (Hook et al., 1974). These studies demonstrated conclusively that (1) 2,4,5-T is actively transported by the organic acid excretory mechanism of the kidney, (2) kidney tissue of dogs and newborn rats have less capacity to transport 2,4,5-T than does that of adult rats, and (3) large concentrations of 2,4,5-T overwhelm the transport system.

The fate of 2,4,5-T following oral doses of 5 mg/kg has also been investigated in humans (Gehring et al., 1973). The elimination of 2,4,5-T from plasma and in the urine followed apparent first-order kinetics with a $t_{1/2}$ of 23.1 hr (Fig. 2.14). Comparison of the elimination rates with those obtained for dogs and rats suggests that the toxicity of 2,4,5-T to humans would be greater than that to rats but less than that to dogs (Fig. 2.15). The higher peak plasma levels were attained in humans

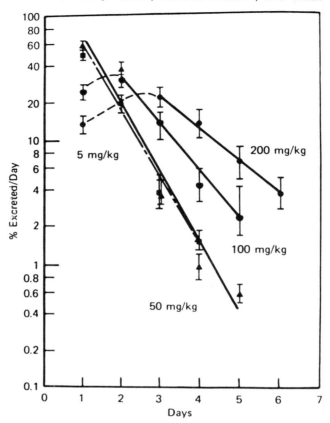

Figure 2.12. Percentage of administered ^{14}C activity excreted by rats during successive 24-hr intervals following a single oral dose of 5, 50, 100, or 200 mg/kg [^{14}C]-2,4,5-T. Each point represents a mean and SE.

with a dose of 5 mg/kg than in either dogs or rats, and they were associated with a greater degree of plasma protein binding in humans. In humans, the volume of distribution was only 80 mL/kg, attesting to the retention of 2,4,5-T in the vascular compartment.

Figure 2.16 illustrates simulated levels of 2,4,5-T in the plasma that would be attained in humans with repeated ingestion. The simulation shows that plasma 2,4, 5-T levels plateau after 3 days and that the maximum plasma concentration of 2,4, 5-T will be approximately $23D_0$ μg/mL. For an individual ingesting 0.1 mg/kg day of 2,4,5-T this equates to a maximum plasma 2,4,5-T concentration of 2.3 μg/mL, which is ~30-fold lower than observed in volunteers given a single 5 mg/kg oral dose of 2,4,5-T. Finally, determination of renal clearance values revealed values exceeding glomerular filtration, as was expected from the results of the studies in animals described previously.

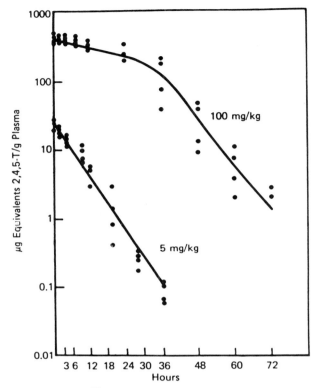

Figure 2.13. Concentration of ^{14}C activity expressed as microgram equivalents 2,4,5-T per gram of plasma of rats following a single intravenous dose of 5 or 100 mg/kg [^{14}C]-2,4,5-T. The curve for rats receiving the latter dose typifies dose-dependent elimination.

Table 2.1. Concentration of ^{14}C Activity Expressed as Microgram Equivalents 2,4,5-T per Gram in Kidney and Plasma after Intravenous Administration of 100 mg/kg [^{14}C]-2,4,5-T

Time Interval (hr)	Concentration of 2,4,5-T (μg/g) in		Kidney: Plasma
	Plasma	Kidney	
4	415 ± 19	297 ± 11	0.71 ± 0.02
8	359 ± 20	281 ± 17	0.78 ± 0.03
16	315 ± 21	297 ± 29	0.94 ± 0.04
32	142 ± 21	198 ± 18	1.40 ± 0.11
64	25 ± 9	122 ± 45	4.95 ± 0.59

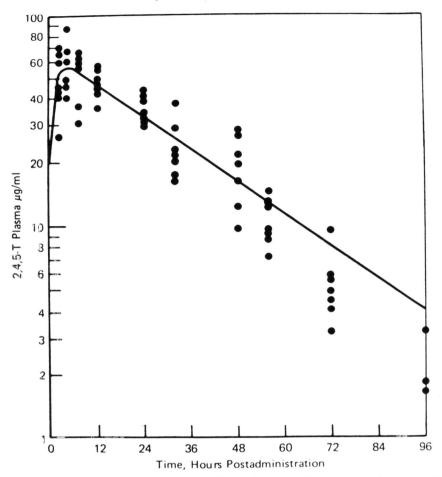

Figure 2.14.   Concentration of 2,4,5-T in blood plasma of humans as a function of time following a single oral dose of 5 mg/kg. The points are values for seven different subjects.

These pharmacokinetic studies of 2,4,5-T in humans and animals, together with the results of toxicological studies in animals, support the conclusion that the hazard from exposure to small amounts of 2,4,5-T encountered during its recommended use is negligible.

8.3   Construction of a Physiologically Based Pharmacokinetic (PB-PK) Model

The concept of incorporating physiological principles into pharmacokinetic modeling was discussed more than 50 years ago by Teorell (1937). Implementing this intuitively logical step has been frustrated, until recently, by the limitations in available computer technology. However, with the advent of powerful computers and user-friendly simulation languages, it has become relatively easy for toxicologists to construct quantitative models of complex biological phenomena. A good example of

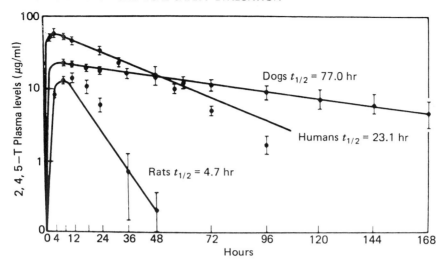

Figure 2.15. Concentration of 2,4,5-T in blood plasma of dogs, humans, and rats following a single oral dose of 5 mg/kg. Each point represents a mean and SD.

such a model is the PB-PK model developed by Andersen et al. (1987) for methylene chloride.

A diagram of the PB-PK model developed by Andersen et al. (1987) is presented in Figure 2.17. The mammalian organism is represented as a collection of compartments, each composed of tissues with similar blood flow rates and partition coeffi-

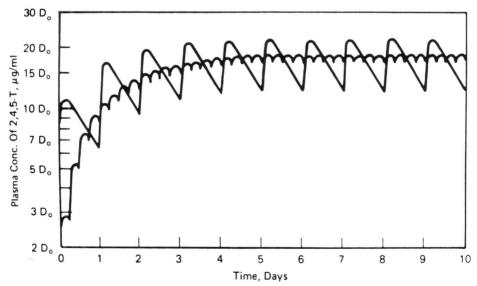

Figure 2.16. Predicted plasma concentrations of 2,4,5-T in humans ingesting a dose of D_0 (mg/kg) every 24 hr (curve with large excursions) or a dose of $\frac{1}{4} D_0$ every 6 hr (curve with small excursions).

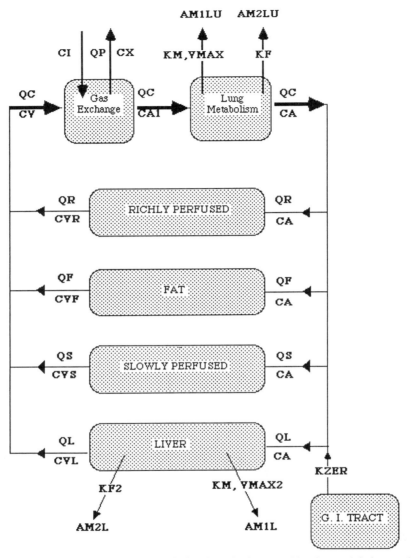

Figure 2.17. Diagram of a physiologically based pharmacokinetic model for methylene chloride.

cients. These compartments are linked together and to the sites chemicals enter the body, in this case the gastrointestinal tract and the lung, by circulating blood. The size of each compartment corresponds to known volumes of the tissues in the human and rodent species. Other physiological variables, such as pulmonary ventilation and blood flow rates, are set to values appropriate for each species. These values are readily available in the published scientific literature.

In this model, chemicals may enter the body of the animal through inhalation, ingestion, or intravenous injection (Fig. 2.17). Other routes of exposure such as dermal absorption of vapors or liquids have been described but are not shown in this model. Once in the body, the chemical is distributed to the tissues by circulating blood. The concentration of the chemical in the blood leaving a tissue is assumed to be in equilibrium with that tissue and its concentration will be determined by the blood–tissue partition coefficient for that chemical. Inhaled and exhaled air also reach equilibrium at concentrations related to the blood–air partition coefficients. Partition coefficients as used in these models are constants that may be determined experimentally for specific chemicals and tissues according to a variety of techniques.

In this model, chemicals may be eliminated from the body in expired air or may be destroyed by metabolic enzymes present in the organism. The concentration-dependent rates of metabolism in the various species can also be determined experimentally through a variety of in vivo and in vitro procedures and then incorporated into the PB-PK model in a quantitative fashion.

Once the physiological, physicochemical, and biochemical constants are known for a particular chemical–species combination, a series of simultaneous differential equations are written describing the rates of change of chemical in each compartment. An excellent discussion of the derivation of the equations for a PB-PK model describing the disposition of styrene in rodents and humans has been provided by Ramsey and Andersen (1984). This general approach, with appropriate modifications, is applicable to almost any chemical, species, or route of exposure.

The system of simultaneous differential equations obtained by this process is complex, and frequently an exact mathematical solution of the equations cannot be obtained. However, this is of little practical consequence because once the differential equations are elaborated, they can be formulated as a computer program and solved to any degree of accuracy desired through the process of successive approximation. Current simulation languages such as SIMUSOLV* contain powerful integration algorithms to accomplish this end.

8.4 Application of a 1,1,1-Trichloroethane PB-PK Model

In order to demonstrate the power and utility of PB-PK modeling techniques, studies with 1,1,1-trichloroethane (methyl chloroform, MC) will be presented. The objective will be to demonstrate the ability of a PB-PK model to deal with dose-dependent metabolic pathways, varied routes of administration, and interspecies extrapolation.

Initially, a PB-PK model was developed to describe the disposition of inhaled MC in the rat. Metabolic rate constants for this model were obtained from studies conducted by Schumann et al. (1982) who exposed rats to 150 and 1500 ppm of ^{14}C-MC for 6 hr. Partition coefficients for MC were determined experimentally, and physiological parameters were taken from the scientific literature. Details of the model have been reported elsewhere (Reitz et al., 1988). As can be seen from Figure 2.18(a), this model accurately described the time course of MC in the venous blood

---

*Trademark of The Dow Chemical Company.

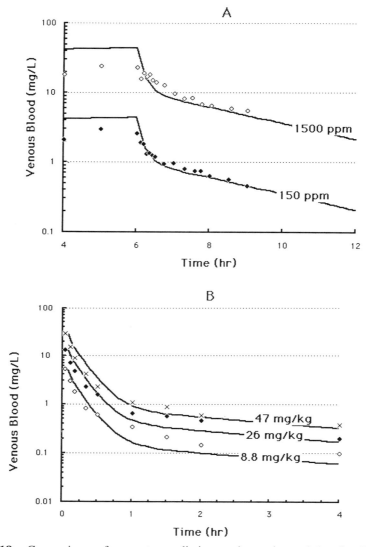

Figure 2.18. Comparisons of computer predictions and experimental data for disposition of methylchloroform in rats: (a) blood levels for inhalation exposure; (b) blood levels following intravenous injection; (c) Rate of elimination in exhaled air for administration in drinking water.

of rats during and following the 6-hr inhalation exposures to 150 and 1500 ppm of MC.

In addition to the blood MC levels, several other types of data were available from the studies of Schumann et al. (1982) to examine the reliability of the PB-PK model. These were: (1) the total body burden of MC and metabolites after 6 hr of

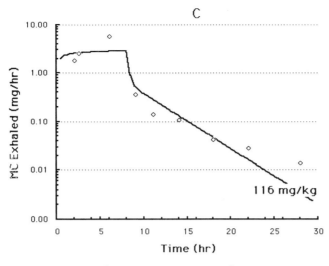

Figure 2.18. (*Continued*)

exposure, (2) the concentration of radioactivity in fat tissue at the end of the exposure, and (3) the concentration of radioactivity in liver tissue at the end of the exposure. Model predictions and measured values for each of these parameters are compared in Table 2.2. Overall, the ratio of predicted to actual data for these two inhalation exposures ranged from a low of 0.73 to a maximum of 1.89 indicating reasonable agreement between the model and the actual experimental data.

The PB-PK model was then used to predict the disposition of MC administered

Table 2.2. Comparison of Predicted and Observed Values for Rats Exposed by Inhalation and via the Drinking Water to MC[a]

	Obser.	Pred.	Ratio (Pred./Obs.)
150 ppm inhalation exposure			
End exposure blood level (mg/L)	2.81	4.43	1.58
Body burden at 6 hr (μmol)	33.0	25.1	0.76
Conc. in fat (μmol/L)	724.0	1304.0	1.80
Conc. in liver (μmol/L)	68.2	49.6	0.73
1500 ppm inhalation exposure			
End exposure blood level (mg/L)	23.7	44.7	1.89
Body burden at 6 hr (μmol)	264.0	241.0	0.91
Conc. in fat (μmol/L)	8403.0	13126	1.56
Conc. in liver (μmol/L)	504	502	1.00
Drinking water (dose, 143 mg/kg)			
Amount metabolized (μmol)	8.19	3.58	0.44

[a]Inhalation data are from Schumann et al. (1982).

by other routes. This was accomplished by modifying the differential equations describing the rate of input to the animal for the different routes of administration, otherwise, no changes were made in the PB-PK model. Figure 2.18(b) shows the agreement between the predicted and observed blood MC levels in the rat given intravenous injection of MC dissolved in rat plasma and administered at dose levels of 8.8, 26, and 47 mg/kg.

In another study, rats were given a saturated solution of ^{14}C-MC in water to drink for an 8-hr period. Samples of excreta and expired air were collected during the 8-hr exposure and for 24 hr post exposure. Predicted and observed rates of elimination of unmetabolized MC in expired air are shown in Figure 2.18(c). Predicted and observed amounts of metabolites are listed in Table 2.2.

This exercise demonstrates that a properly formulated PB-PK model can accommodate data from various routes of administration. It is even possible to describe the disposition of materials administered simultaneously by two or more different routes.

The PB-PK model for MC in the rat was then used for interspecies extrapolation to mice and humans. To accomplish this, the physiological parameters (i.e., tissues volumes, pulmonary ventilation, and blood flow rates) were adjusted to values appropriate for the different species, and the metabolic rate constants were scaled allometrically as outlined by Ramsey and Andersen (1984). No other changes were made in the PB-PK model.

Predicted and observed venous blood MC concentrations in mice exposed to either 150 or 1500 ppm of MC are shown in Figure 2.19. The PB-PK model predicted that MC would be eliminated from the mouse more rapidly than the rat, and this is consistent with the observed data. This is also consistent with the observation that

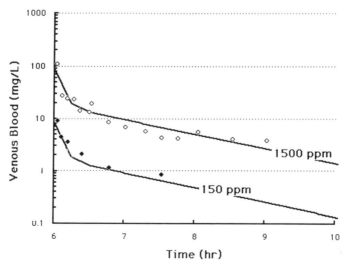

Figure 2.19. Comparisons of computer predictions and experimental data for disposition of methylchloroform in mice: blood levels for inhalation exposure.

elimination half-lives for MC in mice were 5- to 10-fold less than the corresponding half-lives in rats (Schumann et al. 1982). Comparisons of predicted and observed values for other parameters in young mice exposed to MC are summarized in Table 2.3. The ratio of predicted to actual data ranged from a low of 0.60 to a high of 1.19 indicating there was reasonable agreement between the model and all the experimental data.

An important question for any pharmacokinetic model developed with animal data is its applicability to humans. Fortunately, pharmacokinetic data gathered in human volunteers exposed to MC (Nolan et al., 1984) were available to check the validity of extrapolations based on this PB-PK model to humans. Experimentally observed and predicted concentrations of MC in venous blood and expired air taken from humans exposed to 35 or 350 ppm MC for 6 hr are shown in Figures 2.20(a) and 2.20(b), respectively. Additionally, the PB-PK model predicted that 32.2 μmol equivalents of MC would be metabolized during the 240 hr following a 6-hr exposure to 35 ppm MC, and 32.4 μmol equivalents of MC metabolites were actually recovered by Nolan et al. (1984). Similarly, the model predicted that 236 μmol equivalents of MC metabolites would be formed during the 240 hr following a 6-hr exposure to 350 ppm MC, and 246 μmol were actually recovered (Table 2.3).

The good agreement between the model predictions and the experimental data in rats, mice and humans demonstrates that, with appropriate scaling of the critical physiological variables, a properly formulated PB-PK can be used for interspecies extrapolations.

Table 2.3. Comparison of Predicted and Observed Values for Mice and Human Volunteers Exposed by Inhalation to MC[a]

	Obser.	Pred.	Ratio (Pred./Obs.)
Mouse: 150 ppm inhalation exposure			
End exposure blood level (mg/L)	9.27	8.70	0.94
Body burden at 6 hr (μmol)	4.97	2.96	0.60
Amount metabolized (μmol)	0.65	0.62	0.95
Conc. in fat (μmol/L)	1329.0	1575.0	1.19
Conc. in liver (μmol/L)	76.0	51.7	0.68
Mouse: 1500 ppm inhalation exposure			
End exposure blood level (mg/L)	112.0	87.9	0.79
Body burden at 6 hr (μmol)	39.9	25.2	0.63
Amount metabolized (μmol)	1.19	1.23	1.03
Conc. in fat (μmol/L)	16198	15915	0.98
Conc. in liver (μmol/L)	631	525	0.83
Human: 35 and 350 ppm inhalation exposures			
Amount metabolized 35 ppm (μmol)	32.4	32.2	0.99
Amount metabolized 350 ppm (μmol)	246.0	236.0	0.96

[a]Inhalation data for mice are from Schumann et al. (1982) and for human volunteers are from Nolan et al. (1984).

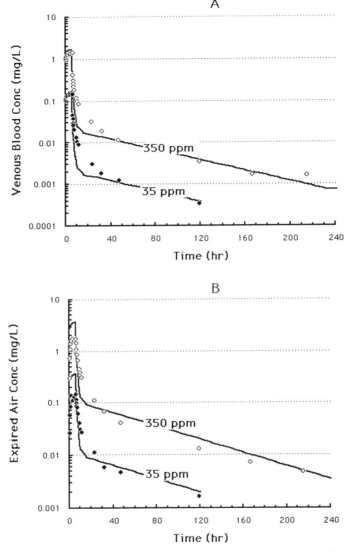

Figure 2.20. Comparisons of computer predictions and experimental data for disposition of methylchloroform in humans: (a) blood levels for inhalation exposure; (b) concentrations in exhaled air for inhalation exposure.

8.5 Relating Animal Studies to Humans

Methyl chloroform has been studied in two long-term inhalation bioassays in which animals were exposed to MC for 6 hr/day, 5 days/week for either 1 (rats; Rampy et al., 1978) or 2 (mice; Quast et al., 1984) years. In both studies toxicity was limited to reversible microscopic alterations in the liver, and exposure to MC was not associated with increases in either benign or malignant tumors. The no observed adverse effect levels (NOAEL) established in these studies were 875 ppm in rats and 1500 ppm in mice.

Assumptions are an implicit part of any interspecies extrapolation. Four assumptions were made in order to integrate the bioassay and pharmacokinetic data and use this information to evaluate the potential for effects to occur in humans. First, based on the type of effects observed in the bioassays, it was assumed that MC per se, and not a metabolite, would be responsible for any effects in humans. The liver was assumed to be the site where adverse effects would occur in humans since the liver is the target organ in both rats and mice. Finally, it was assumed that humans are not more sensitive to MC than rodents and that effects would be proportional to the average concentration of MC in the liver, that is, effects would not occur in humans unless the concentration of MC in liver equaled or exceeded that present in the animal studies. The PB-PK model was used to calculate the average concentration of MC in the liver (ACL, μmol/L) of rats and mice at the NOAEL in the long-term inhalation studies. The ACL for rats was about 55.2 μmol/L during the 1-year exposure period, or about 28 μmol/L averaged over the 2-year life span (rats were exposed for 1 year). The corresponding value for B6C3F1 mice was calculated to be 95 μmol/L over a 2-year exposure period (Table 2.4). The ACL in humans exposed to the maximum permitted concentration of MC in the workplace (350 ppm) during the entire work shift throughout their life would be 21 μmol/L. Thus, the maximum possible ACL in humans is lower than levels that failed to produce any effect in lifetime animal studies, suggesting that there is no significant hazard associated with current workplace standards for MC.

This PB-PK model can also be used to provide perspective for other types of exposure to MC. Lower concentrations of MC (typically 1–10 ppb, highest observed 300 ppb) have been detected in some finished drinking water supplies. No long-term animal studies have been conducted in which MC was administered via the drinking water, and even if there were, the low solubility of MC in water (~ 0.7 g/L) would limit the amount of MC that could be given by this route. However, the ACL associated with human consumption of water containing 10 ppb can be readily cal-

Table 2.4. Average Concentrations of MC (μmol/L) in the Liver of Rats, Mice, and Humans under Specific Exposure Conditions[a]

	Conc. (ppm)	Route	Ave. Conc. in Liver (μmol/L)	Safety Factor Relative to Mouse	to Rat
Rat	875	Inhal.	28		
Mouse	1500	Inhal.	95		
Human	350	Inhal.	21		
Human	0.01	Water	3.4×10^{-4}	$2.7 \times 10^{+5}$	$8.1 \times 10^{+4}$

[a]The concentrations in rats and mice correspond to the NOAEL in long-term inhalation toxicology studies (6 hr/day, 5 days/week). The concentrations in humans correspond to occupational exposure at the threshold limit value (350 ppm) for 6 hr/day, 5 days/week or lifetime consumption of 2 L/day of water containing 10 ppb of MC. Safety factors for the water exposures were calculated by dividing the calculated dose to the animals by the predicted dose to humans.

culated with the PB-PK model and is only 3.4×10^{-4} μmol/L. The ACL in humans consuming such water are much lower than the ACL the animals experienced in the bioassays. For example, the ACL in humans drinking water containing 10 ppb of MC is 81,000-fold lower than the ACL in the rat, and 270,000-fold lower than the ACL in the mouse exposed at the NOAEL. Consequently, there seems little chance that humans drinking water containing small amounts of MC will develop the mild liver effects seen in rodents during inhalation studies.

These are just a few of the ways in which pharmacokinetic data and models can help reduce the uncertainty inherent in estimating human hazard from animal studies. Other promising applications include elucidating mechanisms of toxicity, description of dermal absorption, and predicting the shape of the dose–response curve for carcinogenic and other effects.

9 CARCINOGENESIS

9.1 Animal Cancer Testing and Risk Assessment

Just as animal toxicity testing represents the foundation for interspecies extrapolation of nontumorigenic chemical toxicity, the animal oncogenicity bioassay represents the foundation for predicting the potential carcinogenicity of chemicals in humans. By definition, toxicity testing necessitates the administration of toxic dosages of chemicals to test species in order to characterize their toxic potential. This concept has also been applied to the bioassay of potentially tumorigenic effects of chemicals. The last 25 years have witnessed the advent and widespread use of the so-called maximum tolerated dosage (MTD) rodent oncogenicity bioassay in toxicology, the practice of administering relatively high, often toxic dosages of chemicals to rats and mice over their lifetime without causing an unacceptably high level of mortality (Sontag et al., 1976). The often controversial MTD-based bioassay was initially developed as a protocol of choice by the National Cancer Institute to overcome the necessary statistical limitations of laboratory rodent bioassays. However, critics claim that by the chronic induction of toxic effects the test does not reflect a realistic human exposure scenario and that many times tumor formation occurs secondary to other, nontumorigenic effects of treatment (Stott and Watanabe, 1982; Ames and Gold, 1990b; Cohen and Ellwein, 1990, 1991).

Roughly one-half of all chemicals, either of synthetic or natural origin, tested for oncogenic potential using the MTD protocol has been found to cause an increased incidence of tumors in rats and/or mice (Ames and Gold, 1990). In many of these cases, tumorigenic dosages of chemicals were observed to result in altered metabolism and pharmacokinetics of endogenous and exogenous compounds, physiological adaptations (e.g., hepatic hypertrophy), chronic increases in the rates of cell turnover (e.g., chronic regenerative cell proliferation), and/or secondary effects in multiple tissues. Indeed, significant alterations in the general homeostasis of test animals would appear to be a necessary feature of all MTD bioassays. This association has fueled the debate over whether tumors occurring in these animals, often having a high spontaneous incidence of specific tumors to start with, are relevant to the extrapolation of risk at lower dosages to humans. This issue may ultimately be resolved by understanding the mechanisms by which chemicals cause tumors in animals.

9.2 Mechanisms of Chemical Carcinogenesis

Since the first report of chemically induced carcinogenesis in a controlled laboratory setting by Yamagiwa and Ichikawa (1918), researchers have sought to understand the means by which chemicals cause cancer in animals. Since this time, a number of theories have been proposed to explain the often diverse experimental and epidemiological carcinogenesis database; however, it is generally agreed today that chemicals induce cancer in animals via their ability to cause a number of critical, heritable mutations in nuclear DNA. First outlined by Boveri (1929), the so-called Somatic Mutation Theory of Chemical Carcinogenesis would appear to account for the carcinogenicity of a number of chemical carcinogens that are known to bind nuclear DNA. Indeed, research aimed at the identification of key molecular *targets*, specific protooncogenes and suppressor genes, and the associated changes in gene expression that may, ultimately, result in cellular transformation and tumor development is well underway (Bishop, 1991; Harris, 1991; Kern et al., 1991).

The association of mutagenesis and carcinogenesis has stimulated a large amount of research devoted to the detection of potentially mutagenic chemicals and resulted in the development of a whole battery of relatively rapid or "short-term" mutagenicity assays. Results of these assays, however, have revealed that many compounds found to be tumorigenic in MTD-based bioassays either lack genotoxic potential altogether or are only weakly active despite the great sensitivity of these assays. This difference in the potential of chemical carcinogens to interact with genetic material coupled with the metabolic and physiologic changes and/or signs of toxicity observed in animals at tumorigenic dosages suggests a multiplicity of *nongenotoxic* mechanisms of chemical carcinogenesis. A number of these apparently distinct mechanisms have been proposed and/or elucidated.

The significance of the difference between genotoxic and nongenotoxic mechanisms in terms of safety evaluation of chemical carcinogens, lies in the means by which protooncogene activation or suppressor gene inactivation may occur and in potential effects of the chemical, directly or indirectly, upon subsequent key events following activation (e.g., clonal expansion). The generation of tumors by nongenotoxic chemicals is not necessarily at variance with the Somatic Mutation Theory of Chemical Carcinogenesis. Mutations, including the hypomethylation of DNA—a potentially significant factor in the regulation of gene activity—have been reported to occur in a number of oncogenes in spontaneously occurring tumors and tumors induced by genotoxins and nongenotoxins alike (Fox and Watanabe, 1985; Reynolds, et al., 1987; Harris, 1991; Vorce and Goodman, 1991). A common theme emerging from these studies at the molecular level is the relatively large number of mutagenic and chromosomal changes that must occur for tumor formation to proceed and the involvement of enhanced cell proliferation in this process. Nongenotoxic carcinogens appear to cause these genetic changes primarily via indirect means that are readily reversible and have clearly definable thresholds (i.e., follow accepted physiological and pharmacological principles) (Hildebrandt, 1987; Stott et al., 1989; Ames and Gold, 1990b; 1991; Cohen and Ellwein, 1990). At dosages failing to elicit chronic treatment-related effects (i.e., physiological, pharmacological, and/or toxicological changes), there is little chance of tumor formation. In addition, the basic mechanism by which tumors are initiated in test rodent species may have no human correlate, for example, in the case of α-2u-globulin related renal tumors in male Fischer rats (Borghoff et al., 1990). Thus, chemicals causing tumors in animals via nongenotoxic mechanisms will likely pose a significantly lower carcinogenic threat than those causing tumors via direct interaction with genetic material (i.e., genotoxins).

Figure 2.21 provides a general overview of potential mechanisms by which a chemical may initiate the process of chemical carcinogenesis in animals. Once absorbed, a chemical may interact directly with genetic material (direct genotoxin) or a nonreactive compound may be metabolized to a genotoxic compound(s) (indirect genotoxin). In some instances, competing metabolic pathways may result in the metabolic activation of a chemical only at relatively high dosages when normal, low dosage, metabolism is saturated. Functionally, these latter compounds can be considered nongenotoxic since alternative metabolism and activation to any significant degree occurs only at high dosages. Importantly, the rapid detoxification and/or excretion of any genotoxic chemical or metabolite in vivo may negate any potential mutagenic activity.

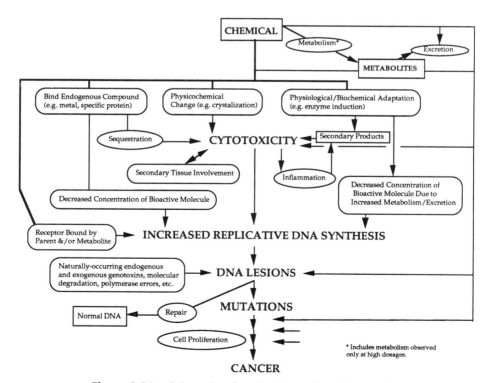

Figure 2.21. Schematic of mechanisms of carcinogenesis.

Increased rates of replicative DNA synthesis by itself may enhance the normal rate at which spontaneous mutations occur in all cells via several means. For example, enhanced rates of replication increase the time that DNA will exist in a particularly susceptible state to alkylation from endogenous and naturally occurring exogenous genotoxins (single-stranded state without enveloping histone proteins). In addition, chemically induced cell proliferation shortens the time available between waves of replicative synthesis when alkylation sites, spontaneous polymerase "errors," and lesions due to spontaneous degradation of the molecule itself (e.g., due to thermodynamic degradation, tautomerization, etc.) are normally repaired (Saul and Ames, 1986; Stott et al., 1989; Ames and Gold, 1990a,b; Cohen and Ellwein, 1990; Butterworth and Slaga, 1992). Genotoxic by-products of physiological adaptations and/or cytotoxicity-stimulated inflammatory processes (e.g., reactive oxygen species) may also pose a mutagenic threat to proliferating cells. Further, these secondary products and the very process of enhanced cell proliferation itself may, in turn, enhance the progress of a previously initiated cell through promotion and progression phases of foci development leading to tumor development.

It is important to note that chronic increases in replicative DNA synthesis in response to cytotoxicity are also expected to greatly enhance the rate of mutation in cells being challenged with cytotoxic dosages of genotoxic compounds. This effect of cytotoxicity has even been modeled for a theoretical genotoxin using a multihit model of carcinogenesis by Reitz (1987) and Conolly et al. (1989). Understandably, the number of mutations expected to occur in animals challenged with cytotoxic dosages of a genotoxic chemical were predicted to far outstrip those expected to occur with a nongenotoxic chemical. However, in both cases, a disproportionate increase in the mutation rate was predicted with increasing dosages in excess of a threshold cytotoxic level. Indeed, a disproportionate increase in tumor incidence has been observed in the large ED01 bioassay of the known genotoxin, acetylaminofluorene (Cohen and Ellwein, 1990).

9.3 Examples of Nongenotoxic Chemical Carcinogens

A number of examples of chemicals that induce tumor formation in animals via nongenotoxic mechanisms of action are reviewed below. These compounds are subdivided based on the principal means by which they initiate the chain of events outlined in Figure 2.21.

9.3.11 Bind Endogenous Compound

The potential tumorigenicity of several compounds has been attributed to their ability to bind endogenous materials such as trace metals or specific proteins. These interactions, through a variety of mechanisms, initiate cytotoxicity resulting in increases in regenerative cell replication.

Nitrilotriacetic acid (NTA) is a nongenotoxic compound that induces renal tubular cortical cell tumors in rats and mice and transitional epithelial cell tumors of the renal pelvis, ureter, and bladder in rats only following high oral doses in chronic bioassays. Anderson and co-workers (Anderson et al., 1985) have shown that administration of high, tumorigenic dosages of the sodium salt or free acid form of NTA results in secondary complexation of endogenous zinc. Tumorigenic dosages of NTA also result in an increased concentration of zinc in renal tubular cells, presumably through renal filtration of the zinc–NTA complex and subsequent extraction of zinc from the complex by renal cortical tubular cells. The elevated cellular zinc is postulated as the cytotoxic mechanism, which stimulates the ultimate appearance of cortical cell tumors. Importantly, lower nontumorigenic dosages of NTA did not elevate renal zinc concentrations.

In contrast to the mechanism of renal tubular cell tumor induction, the bladder transitional epithelial cell tumors observed in NTA-treated animals were shown to closely correlate with the ability of uncomplexed NTA present in the luminal filtrate bathing these cells (present only after high NTA doses) to specifically extract calcium from these cells. The loss of calcium from epithelial cells appeared to be the mechanism underlying this tissue-specific cytotoxicity. Thus, similar to the stimulus for the renal tubule cell tumors, NTA-induced tumors at this site were attributed to chronic cytotoxicity, which was only associated with exposure to high doses of chemical.

Another nongenotoxic mechanism of tumor formation based on the binding of xenobiotics with an endogenous compound has been the interaction of a number of chemicals with the male-rat-specific protein α-2u-globulin leading to formation of specific renal cortical tumors (Borghoff, 1990). α-2u-Globulin is a low-molecular-weight protein synthesized in large quantities in male rat liver, secreted into blood, and because of its low molecular weight, filtered completely by the kidney. Approximately equal portions of the filtered protein load is resorbed by renal proximal tubule cells and excreted into urine. Administration of compounds including 2,2,4-trimethylpentane, isophorone, d-limonene, and p-dichlorobenzene all produce a male rat nephropathy characterized by increased size and number of hyaline droplets in renal tubule cells. The hyaline droplets result from specific accumulation of α-2u-globulin within the lysosomes of the renal tubule cells. Studies have demonstrated that the lysosomal accumulation of α-2u-globulin is likely due to noncovalent binding of the nephrotoxic chemicals and/or their metabolites to this male-rat-specific protein and that this binding subsequently disrupts the ability of the renal cell lysosomes to proteolytically degrade the protein. The lysosomal protein accumulation is thought to result in rupture of the lysosomes, which in turn leads to renal tubule cell cytotoxicity, death, and a compensatory hyperplasia that is maintained during chronic chemical exposure. Consistent with this is the appearance of tumors relatively late in the animal's life. Female rats and mice of either sex do not synthesize α-2u-globulin and also do not exhibit the hyaline droplet nephropathy and associated renal tumors. Most significantly humans also do not synthesize α-2u-globulin, thereby eliminating the relevance of this male rat tumorigenic endpoint for humans.

9.3.2 Physicochemical Change

The potential tumorigenicity of several nonreactive compounds has been attributed to chronic tissue damage/regenerative synthesis caused by their specific physicochemical characteristics. An example of such a compound is terephthalic acid (TPA), which produces epithelial cell ureter and bladder tumors when given to rats at a very high concentration (5%) in the diet over a majority of their lifetime (Chin et al., 1981). The tumorigenic response was specifically associated with the presence of urinary tract stones. Chronic doses of TPA below 1 percent in the diet produced no evidence of toxicity, stone formation, or tumors.

Chin and co-workers (Chin et al., 1981) have demonstrated that stone formation is characterized by a relatively steep dose–response curve and occurred in the urinary bladder lumen of rats only after administration of high dosages of TPA. Stone formation was dependent on the absolute concentration of TPA and calcium in the urine and essentially resulted from the precipitation of complexes of TPA and calcium in the acidic rodent urine. Significantly, the presence of the stones appeared to be a necessary requirement for induction of transitional cell hyperplasia, which was postulated to be the critical element in the tumorigenicity of this nongenotoxic compound. Since stone formation is governed by the solubility limit of the complex in urine, lower doses producing urinary TPA concentrations below the solubility limit (reflecting a more realistic environmental exposure) would appear to have little to no tumorigenic potential.

9.3.3 Metabolism/Pharmacokinetic Properties

The metabolism and kinetics of a chemical in vivo may dictate the means by which the chemical may cause tumors in animals. A particular concern about the use of MTD in bioassays is that dose–dependent alterations in the metabolism/pharmacokinetics of a compound may occur. Examples of carcinogenic chemicals falling into this category are ortho-phenylphenol (OPP) and methylene chloride (MCl).

Ortho-phenylphenol represents a fairly straightforward example of the impact that alternate, high-dose metabolism may have upon the tumorigenic potential of a chemical. A fungicide with wide applications, OPP has been generally negative in a variety of short-term genotoxicity assays, yet caused a high incidence of kidney and urinary bladder tumors when fed to rats at dosages in excess of 500 mg/kg day (Hiraga and Fuji, 1981). Subsequent investigations by Reitz and co-workers (Reitz and Watanabe, 1985) revealed that at dosages of 5 or 50 mg/kg, OPP is excreted primarily via the urine as several relatively nontoxic conjugates. However, at a dosage of 500 mg/kg OPP, significant levels of an additional metabolite, identified as dihydroxybiphenyl, was also excreted. Production of this potentially toxic and reactive semiquinone was found to correlate with histopathologic changes, significant increases in replicative DNA synthesis and disproportionate increases in the macromolecular binding of OPP metabolites in the transitional epithelium of the urinary bladders of treated rats. The tumorigenicity of OPP was thus attributed to the dose–dependent metabolism (Fig. 2.22) of this compound to a reactive metabolite (Pathak and Roy, 1992) and subsequent chronic damage and regeneration of epithelial cells during its excretion.

The widely used solvent MCl represents a more complex example of how metabolism and pharmacokinetics may influence the potential tumorigenicity of a compound. This chemical has been shown to cause an apparent species and, possibly, route-specific tumorigenic response in test animals (Singh et al., 1985). In addition,

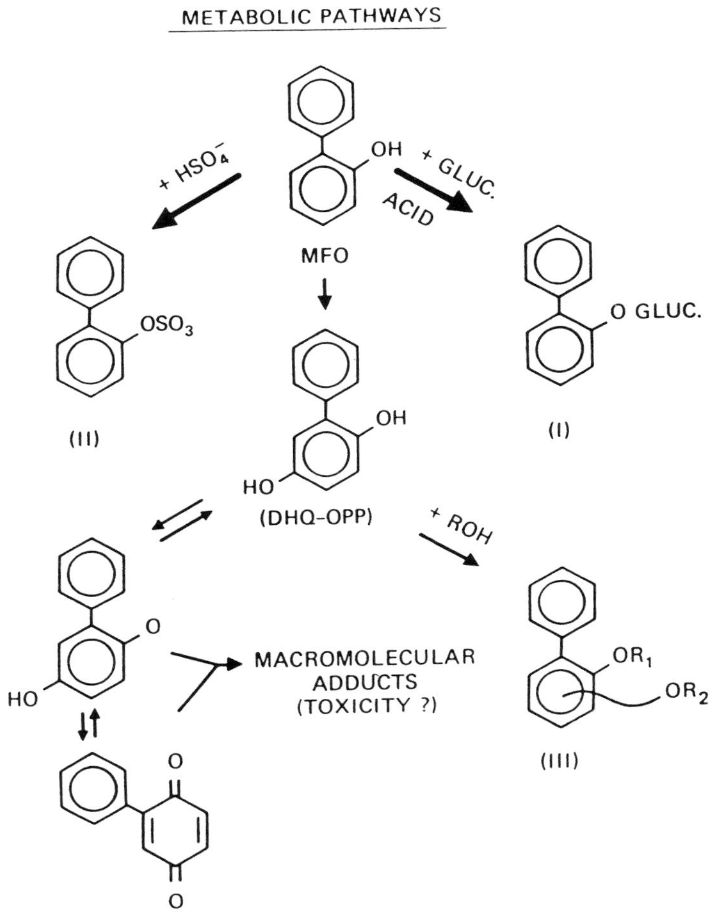

METABOLIC PATHWAYS

Figure 2.22. Hypothetical scheme for the metabolism of OPP in rats, based on identification of metabolic products and dose–response studies in F344 male rats.

MCl has provided variable responses in a number of short-term genotoxicity assays, being generally positive in bacterial mutagenicity assays, yet negative in animal and mammalian cell culture assays. An increased incidence of lung and liver tumors was observed in B6C3F1 mice at relatively high exposure levels in an MTD-based inhalation bioassay. In contrast, no evidence of an increase in lung and liver tumors has been observed in oral mouse and rat bioassays (drinking water), two rat inhalation bioassays, and a hamster inhalation bioassay.

A high-dose, alternate metabolism to a reactive compound has been suggested by Andersen, Reitz, and co-workers (Andersen et al., 1987; Reitz et al., 1989) as an explanation of the apparent contradictory bioassay data for MCl. Methylene chloride appears to be metabolized via two competing pathways in vivo; a high-affinity but low-capacity mixed function oxygenase (MFO) and a low-affinity and high-capacity glutathione transferase. At low dosages, MCl is metabolized primarily via the MFO pathway, which results in its detoxification and excretion. At relatively high dosages, however, the MFO pathway becomes ''saturated'' and more and more MCl is conjugated with glutathione. Subsequent catabolism of the glutathione conjugate results in the production of formaldehyde, a potentially genotoxic compound that causes DNA–protein crosslinking (Casanova et al., 1992). The dosage of formaldehyde to target tissues in the different test species were subsequently predicted by incorporating these dose–dependent metabolic data and species-dependent physiological data into several physiologically based pharmacokinetic models. The calculated dosages correlated well with the observed bioassay results in exposed animals. Only at relatively high exposure concentrations of MCl, such as those utilized in the mouse inhalation bioassay (4000 ppm), were appreciable levels of an active metabolite predicted in hepatic and pulmonary tissues. At lower, nontumorigenic, inhalation or oral dosages of MCl, MFO metabolism predominated and little if any formaldehyde was believed to be generated. By incorporating human metabolic data obtained in vitro, it was also possible to predict the potential activation of MCl in humans. Mice exposed to a tumorigenic 4000 ppm vapors were estimated to have a ''tissue dose'' of more than five orders of magnitude more than humans exposed to 1 ppm MCl vapors.

9.3.4 Biochemical/Physiological Changes

Treatment-related increases in the incidences of tumors in test animals have also often been associated with fairly specific biochemical/physiological changes, especially with the advent of MTD bioassay protocols. These changes may be as straightforward as cytotoxicity in a secondary tissue in response to a greatly increased work load or the loss of a feedback control mechanism for hormonal tissue(s) or as complex as disruptions in normal cellular homeostasis due to massive proliferation of specific subcellular organelles and/or the induction of specific enzyme activities.

A large increase in the metabolic or physical demands placed upon a tissue as a result of primary or secondary effects of a chemical exposure may induce a hypertrophic and/or hyperplastic response in that tissue in order to increase functional tissue mass. This stress may result in the buildup of toxic levels of endogenous compounds in a tissue, in turn, resulting in chronic increases in cell turnover. An example of this latter mechanism is the induction of splenic tumors in rats by chronic administration of aniline in the diet (Bus and Popp, 1987).

Aniline and several structurally related compounds such as o-toluidine, p-chloroaniline, and D&C Red No. 9 dye all share the common characteristics of being acutely toxic to erythrocytes and producing a low incidence of late-developing spleen tumors at high doses in chronic bioassays. Using aniline as a model compound, evaluation of the pathogenesis of the splenic lesions coupled with a characterization of the metabolic disposition in animals suggests that the spleen tumors may be a secondary response resulting from chemically mediated erythrocyte toxicity. Bus and Popp (1987) have proposed that aniline-derived toxicity to erythrocytes results in scavenging of damaged red blood cells by the spleen, which subsequently initiates a series of events critical to the appearance of spleen tumors. These events potentially include a combination of (1) transport of parent compound or toxic metabolite(s) specifically to the spleen by damaged erythrocytes; (2) splenic deposition of erythrocytic debris, particularly iron, which may catalyze tissue-damaging free radical reactions; and (3) induction of a compensatory splenic hyperplasia resulting from scavenging of aniline-injured erythrocytes. The linkage of the splenic tumorigenicity of aniline and its structural congeners to an initial toxic event in erythrocytes suggests that the carcinogenicity of such compounds may be determined by definable thresholds, that is, events leading to carcinogenicity are initiated only at the large doses of compound sufficient to overwhelm the capacity of erythrocytes to absorb the toxic insult.

The enhanced metabolic activity of a tissue in response to an increased metabolic demand or "enzyme inducer" may also indirectly cause a tumorigenic response in a secondary tissue via its elimination of metabolic feedback control mechanisms. A significant example of this is in the hepatic-endocrine axis, which controls systemic levels of hormones produced by endocrine tissues. Enhanced hepatic metabolism of a hormone will result in hypertrophy and hyperplasia of the endocrine tissue as it is stimulated to maintain homeostatic levels of the hormone. As reviewed by McClain (1989), this mechanism likely explains the potential of a number of compounds to cause thyroid tumors in rodent bioassays in which chronic goitrogenic activity has been linked to hepatic enzyme induction.

Homeostatic levels of thyroxine are maintained by a feedback inhibition of the production of thyroid-stimulating hormone (TSH) by the pituitary. This hormone stimulates thyroid gland production of thyroxine, which is subsequently removed from systemic circulation by the action of hepatic thyroxine-glucuronyl transferase. The induction of the activity of this latter enzyme by a number of compounds (e.g., polycyclic hydrocarbons, organochlorine insecticides, barbiturates) and the resultant increase in biliary excretion of the thyroxine glucuronide has been observed to correlate with their goitrogenic activity in treated animals (McClain, 1989). Ultimately, chronic goitrogenic activity and the associated proliferative effect may lead to tumor formation. In the absence of significant alterations in thyroxine metabolism, both anabolic and catabolic pathways, there would be little chance of tumor formation.

In addition to affecting the metabolism of systemic compounds, the induction of enzyme activities by a chemical in a given tissue may result in the intracellular production of cytotoxic and genotoxic compounds. A growing number of nongenotoxic, structurally diverse compounds (hypolipidemic drugs, solvents, plasticizers and agrochemicals) have been observed to induce the proliferation of peroxisomes, a subcellular organelle, concomitant with the production of hepatic tumors in rodents (Stott, 1988; Nemali et al., 1989). Peroxisomes contain a number of amino acid oxidases, catalase, and enzymes involved in the β oxidation of fatty acids. A by-product of peroxisomal oxidases is hydrogen peroxide (H_2O_2), which, under normal circumstances, is degraded by peroxisomal catalase. During the chemical induction of peroxisomal proliferation, however, a disproportionate induction of peroxisomal enzymes may occur resulting in increased intracellular H_2O_2 production. It has been hypothesized that chronic increases in H_2O_2-related oxidative stress is linked to tumor formation in treated animals (Nemali et al., 1989). In this situation, potentially reactive, genotoxic oxidation products would be generated secondary to the enzymologic and morphologic changes associated with peroxisome proliferation. In the absence of chronic peroxisome proliferation, there would be little chance of tumor formation.

While there appears to be a relatively strong qualitative association of peroxisome proliferation with tumor formation, the lack of a quantitative relationship has suggested the involvement of other factors in this process (Conway et al., 1989). The potential of some proliferators to induce chronic increases in replicative synthesis and the ability of proliferators to promote spontaneously transformed cells in vivo have been suggested as alternative mechanisms of tumorigenesis for these compounds (Conway et al., 1989; Kraupp-Grasl et al., 1990). While the potential interrelationship of cell proliferation, promotion and peroxisome proliferation, and their relative involvement in tumor formation remains to be elucidated, it is clear that these compounds induce tumors in animals via a nongenotoxic mechanism of action. It is expected that tumorigenic response will only be observed at dosages that induce a clearly identifiable treatment-related change in hepatic tissues of treated animals. In addition, it has been shown that higher mammalian species are quite resistant to peroxisome proliferation suggesting that, for all practical purposes, peroxisome proliferation is primarily a rodent-specific phenomenon lacking a human correlate (Stott, 1988).

10 CONCLUSIONS

Absolute safety is the goal of society in all endeavors. Absolute safety, however, is never achieved, whether in skyscraper construction or environmental management. An imperfect system is not to be condoned but constantly improved through experience and research as rapidly as possible within limitations placed on the system by society itself.

In the evaluation of the safety of chemicals, as in all other fields of human endeavor, some degree of risk must be considered acceptable to society. The alternative would be the needless prohibition of important benefits. The estimation of the safety (or hazard) of a chemical for a particular use can span an enormous range of complexity. Nevertheless, the objective of all safety testing is to ensure attainment of the desired benefits of use without incurring needless risks. There must, of course, be some balance between the benefit and the cost of assessment, just as there must be a balance between benefit and acceptable risk.

Investigations of "potential hazard" as applied to chemicals include a demonstration of the toxicologic properties of an agent by appropriate test procedures and a determination of an exposure level that does not produce detectable adverse effects by the same procedure. For humans, the margin of safety of an agent is evaluated by relating the predicted human exposure to the maximum exposure level in experimental animals that produces no detectable adverse effects. An agent with a small margin of safety has potential for producing adverse effects; conversely, an agent with a large margin of safety has less potential for producing adverse effects.

For optimum assurance of safety within the existing limitation of capabilities and skills for evaluation, resources should be concentrated on environmental situations in which there is a reasonable expectation that exposure to chemicals may cause real hazards. The evaluation of toxicity data for judging the safety or hazard associated with the material can be made only in light of the anticipated amount and circumstances of human exposure. If the anticipated human exposure is very close to, or perhaps greater than, the maximum dose that produced no observable effect in an adequately sized group of experimental animals, the type and extent of controls to be applied should be considered, in order to reduce human exposure to acceptable levels. The acceptable risk determination includes an evaluation of the benefits to the individual and to society of the proposed exposure and an evaluation of the nature and severity of the anticipated effects. To quote Golberg (1971): "Neither neglect nor panic is the answer (for safety evaluation). Somewhere between these two extremes lies the course of reasonable action appropriate to each chemical-contaminant. Given good science, good judgment, and above all, freedom from extraneous pressures, the right course can be found."

REFERENCES

Adams, E. M., H. C. Spencer, V. K. Rowe, D. D. McCollister, and D. D. Irish (1952). AMA *Arch. Ind. Hyg. Occup. Med.*, **6**, 50–66.

Ames, B. N. and L. S. Gold (1990a). *Proc. Natl. Acad. Sci.*, **87**, 7772–7776.

Ames, B. N. and L. S. Gold (1990b). *Science*, **249**, 970–971.

Ames, B. N. and L. S. Gold (1991). *Mut. Res.*, **250**, 3–16.

Anderson, R. L., W. E. Bishop, and R. L. Campbell (1985). *CRC Crit. Rev. Toxicol.*, **15**, 1–102.

Andersen, M. E., H. J. Clewell, M. L. Gargas, F. A. Smith, and R. H. Reitz (1987). *Toxicol. Appl. Pharmacol.*, **87**, 185–205.

Baker, S. (1971). *Proc. Eur. Soc. Drug Toxicol.*, **12**, 81.

Baker, S., J. Tripod, and J. Jacob (1970). *Proc. Eur. Soc. Drug Toxicol.*, **11**, 9–242.

Barnes, J. M. and I. A. Benz (1954). *Pharmacol. Rev.*, **6**, 191–242.

Bein, H. J. (1963). *Proc. Eur. Soc. Drug Toxicol.*, **2**, 15–25.

Bishop, J. M. (1991). *Cell*, **64**, 235–248.

Borghoff, S. J., B. G. Short, and J. A. Swenberg (1990). *Ann. Rev. Pharmacol. Toxicol.*, **30**, 349–367.

Boveri, T. (1929). *The Origin of Malignant Tumors*, Williams & Wilkins, Baltimore.

Bus, J. S. and J. A. Popp (1987). *Fd. Chem. Toxic.*, **25**, 619–626.

Butterworth, B. E. and T. Slaga, eds. (1992). *Chemically Induced Cell Proliferation: Implications for Risk Assessment*, Wiley-Liss, New York.

Casanova, M., H. d'A. Heck, and D. F. Deyo (1992). *Toxicologist*, **12**, 254 #966.

Chin, T. Y., R. W. Tyl, J. A. Popp, and H. d'A. Heck (1981). *Toxicol. Appl. Pharmacol.*, **58**, 307–321.

Cogliano, V. J., W. H. Farland, P. W. Preuss, J. A. Wiltse, L. R. Rhomberg, C. W. Chen, M. J. Mass, S. Nesnow, and P. D. White (1991). *Science*, **251**, 606–608.

Cohen, S. M. and L. B. Ellwein (1990). *Science*, **249**, 1007–1011.

Cohen, S. M. and L. B. Ellwein (1991). *Can. Res.*, **51**, 6493–6505.

Conolly, R. B., R. H. Reitz, H. J. Clewell, and M. E. Anderson (1989). *Toxicol. Lett.*, **43**, 189–200.

Conway, J. G., R. C. Cattley, J. A. Popp, and B. E. Butterworth (1989). *Drug Metab. Rev.*, **21**, 65–102.

DeBaun, J. R., J. Y. Rowley, E. C. Miller, and J. A. Miller (1968). *Proc. Soc. Exp. Biol. Med.*, **129**, 268–273.

Dewberry, E. B. (1950). *Food Poisoning*, 3rd ed., Leonard Hill Limited, London, pp. 205–217.

Drill, V. A. and T. Hiratzka (1953). *Arch. Ind. Hyg. Occup. Med.*, **7**, 61–67.

Evans, R. D. (1974). *Hlth. Phys.*, **27**, 497–510.

Fearon, E. R. and B. Vogelstein (1990). *Cell*, **61**, 759–767.

Fox, T. R. and P. G. Watanabe (1985). *Science*, **228**, 596–597.

Frawley, J. P. (1967). *Food Cosmet. Toxicol.*, **5**, 293–308.

Freireich, E. J., E. A. Gehan, D. P. Rall, L. H. Schmidt, and H. E. Skipper (1966). *Can. Chemother. Rep.*, **50**, 219–244.

Gehring, P. J. (1968). *Toxicol. Appl. Pharmacol.*, **13**, 287–298.

Gehring, P. J. and G. E. Blau (1977). *J. Environ. Pathol. Toxicol.*, **1**, 163–179.

Gehring, P. J., C. G. Kramer, B. A. Schwetz, J. Q. Rose, and V. K. Rowe (1973). *Toxicol. Appl. Pharmacol.*, **26**, 352–361.

Gehring, P. J., P. G. Watanabe, and G. E. Blau (1976). *Advances in Modern Toxicology— New Concepts in Safety Evaluation*, Halstead Press, New York, pp. 195–270.

Geiling, E. M. K. and P. R. Cannon (1938). *JAMA*, **111**, 919–926.

Golberg, L. (1971). *Food Cosmet. Toxicol.*, **9**, 65–80.

Grantham, P. H., L. C. Mohan, and E. K. Weisburger (1976). *J. Natl. Cancer Inst.*, **56**, 649–651.

Hanawalt, P. C. (1991). *Mut. Res.*, **247**, 203–211.

Harris, C. C. (1991). *Can. Res.*, **51**, 5023s–5044s.

Haseman, J. K. (1985). *Fund. Appl. Toxicol.*, **5**, 66–78.

Hayes, W., Jr. (1972). *Essays in Toxicology*, Vol. 3, Academic Press, New York, pp. 65–77.

Hildebrandt, A. G. (1987). *Arch. Toxicol.*, **60**, 217–223.

Hiraga, K. and T. Fuji (1981). *Food Cosmet. Toxicol.*, **19**, 303–310.

Hoeijmakers, J. H. J. and D. Bootsma (1990). *Can. Cells*, **2**, 311–320.

Hook, J. B., M. D. Bailie, J. T. Johnson, and P. J. Gehring (1974). *Food Cosmet. Toxicol.*, **12**, 209–218.

Kern, S. E., K. W. Kinzler, S. J. Baker, J. M. Nigro, V. Rotter, A. J. Levine, P. Friedman, C. Prives, and B. Vogelstein (1991). *Oncogene*, **6**, 131–136.

Kraupp-Grasl, B., W. Huber, B. Putz, U. Gerbracht, and R. Schulte-Hermann (1990). *Can. Res.*, **50**, 3701–3708.

Lindahl, T. (1982). *Ann. Rev. Biochem.*, **51**, 61–87.

Lindahl, T., R. D. Wood, and P. Karran (1991). In *Origins of Human Cancer*, Cold Spring Harbor Laboratory Press, Cold Spring Harbor, NY, pp. 163–170.

Litchfield, J. T. (1961). *JAMA*, **177**, 34–38.

Litchfield, J. T. and F. Wilcoxon (1949). *J. Pharmacol. Exp. Ther.*, **96**, 99–113.

76   RICHARD J. NOLAN, WILLIAM T. STOTT, AND PHILIP G. WATANABE

Maher, V. M., D. J. Dorney, A. L. Mendrala, B. Konze-Thomas, and J. McCormick (1979). *Mut. Res.*, **62**, 311–323.

McClain, R. M. (1989). *Toxicol. Pathol.*, **17**, 294–306.

McCollister, D. D., W. H. Beaman, G. L. Atchison, and H. C. Spencer (1951). *J. Pharmacol. Exp. Ther.*, **102**, 112–124.

Miller, J. A. (1970). *Can. Res.*, **30**, 559–576.

Miller, E. C., J. A. Miller, and M. Enomoto (1964). *Can. Res.*, **24**, 2018–2026.

National Research Council (1977). Committee on Safe Drinking Water, Report, National Academy of Sciences.

Nemali, M. R., M. K. Reddy, N. Usuda, P. G. Reddy, D. Comeau, M. S. Rao, and J. K. Reddy (1989). *Toxicol. Appl. Pharmacol.*, **97**, 72–87.

Nolan, R. J., N. L. Freshour, D. L. Rick, L. P. McCarty, and J. H. Saunders (1984). *Fund. Appl. Toxicol.*, **4**, 654–662.

Osswald, H. and K. Goertler (1971). *Verh. Deut. Ges. Pathol.*, **55**, 289–293.

Pathak, D. N. and D. Roy (1992). *Toxicologist*, **12**, 252 #956.

Paget, G. E. (1963). *Proc. Eur. Soc. Drug Toxicol.*, **2**, 7–18.

Peck, H. M. (1968). *Importance of Fundamental Principles in Drug Evaluation*, Raven Press, New York, pp. 449–471.

Pinkel, D. (1958). *Can. Res.*, **18**, 853–856.

Piper, W. N., J. Q. Rose, M. L. Leng, and P. J. Gehring (1973). *Toxicol. Appl. Pharmacol.*, **26**, 339–351.

Plaa, G. L., E. A. Evans, and C. H. Hine, (1958). *J. Pharmacol. Exp. Ther.*, **123**, 224–229.

Quast, J. F., L. L. Calhoun, and M. J. McKenna (1984). *Toxicologist*, **5**, 14.

Rall, D. P. (1969). *Environ. Res.*, **2**, 360–367.

Rampy, L. W., J. F. Quast, B. K. J. Leong, and P. J. Gehring (1978). *Proceedings of the 1st International Congress on Toxicology*, G. L. Plaa and W. A. M. Duncan, Eds., Academic Press, New York.

Ramsey, J. R. and M. E. Andersen (1984). *Toxicol. Appl. Pharmacol.*, **73**, 159–175.

Reitz, R. H. (1987). In *Banbury Report 25: Nongenotoxic Mechanisms in Carcinogenesis*, B. Butterworth, and T. Slaga, Eds., Cold Spring Harbor Laboratory Press, Cold Spring Harbor, NY, pp. 107–117.

Reitz, R. H. and P. G. Watanabe (1985). In *Banbury Report 19: Risk Quantitation and Regulatory Policy*, D. G. Hoel and R. A. Merrills, Eds., Cold Spring Harbor Laboratory Press, Cold Spring Harbor, NY, pp. 241–251.

Reitz, R. H., J. N. McDougal, M. W. Himmelstein, R. J. Nolan, and A. M. Schumann (1988). *Toxicol. Appl. Pharmacol.*, **95**, 185–199.

Reitz, R. H., A. L. Mendrala, and F. P. Guengerich (1989). *Toxicol. Appl. Pharmacol.*, **97**, 230–246.

Reynolds, S. H., S. J. Stowers, R. R. Maronpot, S. A. Aaronson, and M. W. Anderson (1987). *Science*, **237**, 1309–1316.

Rowe, V. K. and T. A. Hymas (1954). *Am. J. Vet. Res.*, **16**, 622–629.

Rudiger, H. W., U. Schwartz, E. Serrand, M. Stief, T. Krause, D. Nowak, H. Dorerjer, and G. Lehnert (1989). *Can. Res.*, **49**, 5623–5626.

Sauerhoff, M. W., W. H. Braun, G. E. Blau, and P. J. Gehring (1976). *Toxicol. Appl. Pharmacol.*, **36**, 491–501.

Saul, R. L. and B. N. Ames (1986). In *Mechanisms of DNA Damage and Repair*, M. G. Simic, L. Grossman, and A. C. Upton, Eds., Plenum Press, New York, pp. 529–535.

Setlow, R. B. (1983). In *Human Carcinogenesis*, C. C. Harris and H. Autrup, Eds., Academic Press, NY, pp. 231–254.

Schumann, A. M., T. R. Fox, and P. G. Watanabe (1982). *Toxicol. Appl. Pharmacol.*, **62**, 390–401.

Simic, M. G. and D. S. Bergtold (1991). *Mut. Res.*, **250**, 17–24.

Singh, D. V., H. L. Spitzer, and P. D. White (1985). *Addendum to the Health Assessment Document for Dichloromethane (Methylene Chloride). Updated carcinogenicity assessment of dichloromethane.* EPA/600/8-82/004F.

Sontag, J. M., N. P. Page, and U. Saffiotti (1976). *Guidelines for Carcinogen Bioassay in Small Rodents*, DHHS Publication (NIH 76-801), NCI, Bethesda, MD.

Stokinger, H. E. and R. L. Woodward (1958). *J. Am. Water Works Assoc.*, **50**, 515–529.

Stott, W. T. (1988). *Regul. Toxicol. Pharmacol.*, **8**, 125–159.

Stott, W. T. and P. G. Watanabe (1982). *Drug Metabol. Rev.*, **13**, 853–873.

Stott, W. T., T. R. Fox, R. H. Reitz, and P. G. Watanabe (1989). In *Carcinogenicity and Pesticides*, N. N. Ragsdale and R. E. Menzer, Eds., ACS Books, Washington, D.C., pp. 43–77.

Teorell, T. (1937). *Archs. Im. Pharmacodyn. Ther.*, **57**, 205–240.

U.S. Environmental Protection Agency (1986). *Fed. Reg.*, **51**, 33992–34003.

Vorce, R. L. and J. I. Goodman (1991). *J. Toxicol. Environ. Hlth.*, **34**, 367–384.

Vogelstein, B., E. R. Fearon, S. R. Hamilton, S. E. Dern, A. C. Preisinger, M. Leppert, Y. Nakumura, R. White, A. M. M. Smits, and J. L. Bos (1988). *N. Engl. J. Med.*, **319**, 525–532.

Wands, R. C. and J. H. Broome (1974). In *Proceedings of the Fifth Annual Conference on Environmental Toxicology*, NTIS (National Technical Information Service) AMRL-TR-74-125 Springfield, VA pp. 237–258.

Wardell, W. M. (1973). *J. Anesthesiol.*, **38**, 309–312.

Weil, C. S. (1972). *Toxicol. Appl. Pharmacol.*, **21**, 454–463.

Weil, C. S. and D. D. McCollister (1963). *J. Agric. Food Chem.*, **11**, 486–491.

Yamagiwa, K. and K. Ichikawa (1918). *J. Can. Res.*, **3**, 1–21.

Yang, L. L., V. M. Maher, and J. J. McCormick (1980). *Proc. Natl. Acad. Sci. (USA)*, **77**, 5933.

Zapp, J. A. (1977). *J. Toxicol. Environ. Health*, **2**, 1425–1433.

CHAPTER THREE

Biological Markers of Chemical Exposure, Dosage, and Burden

Richard S. Waritz, Ph.D.

1 INTRODUCTION AND DEFINITIONS OF TERMS

"In analysis of biological specimens for solvents, however, it appears that each solvent represents a different problem, and some knowledge of the peculiarities of the metabolism and excretion of each is necessary, else one may be led astray in one's interpretation of the results" (Elkins, 1954). Elkins's words of caution are as valid today as they were when he wrote them in 1954.

Elkins defined solvent to include any liquid industrial chemical. His observation about metabolic peculiarities applies equally well to solids, gases, vapors, organic chemicals, and inorganic chemicals. As is discussed later, it also is important to know whether the exposure chemical or a metabolite is the toxicologically determining material and to assure that the only source of the chemical being measured is the workplace chemical whose exposure is being monitored.

In the decades since Elkins's observation, knowledge in toxicology, biochemistry, cell biology, pharmacology, enzymology, physiology, and subcellular anatomy and physiology has grown. Knowledge growth in these areas has been accompanied with, and to a large extent made possible by, growth in instrumentation and methodologies capable of making accurate and precise measurements at the subcellular level. This research has made it possible to expand Elkins's industrial hygiene monitoring concept of biological analysis for solvents. It now includes analysis of blood and any excreted material for (1) the workplace chemical of exposure, (2) its metabolites as

Patty's Industrial Hygiene and Toxicology, Third Edition, Volume 3, Part B, Edited by Lewis J. Cralley, Lester V. Cralley, and James S. Bus.
ISBN 0-471-53065-4 © 1995 John Wiley & Sons, Inc.

classically defined, (3) its adducts with macromolecules, (4) the excision products from the biochemical repair of these macromolecular adducts, (5) conjugated derivatives, and (6) second-generation chemicals that are not derived from the workplace chemical but whose concentration in excreted materials is controlled by the body burden of the workplace chemical. Elkins's term *solvent* has been replaced by the term *biological marker* to cover this panoply of direct and indirect biological measurements of the amount of chemical exposure in the workplace and the resulting body burden.

To the extent that the biochemical derivatives of the xenobiotic mentioned in (4) and (5) above and the second-generation chemicals mentioned in (6) are characteristic of the xenobiotic of exposure, they offer measures of exposure to the xenobiotic beyond these envisioned by Elkins. If these adducts and second-generation chemicals have a dose–response relationship with the xenobiotic, they offer another opportunity for quantitating exposure. To the extent that they correlate with dosage of the toxicant in the target organ, they offer a measure of target organ dosage.

If these adducts are excreted by any of the routes to be discussed, they give another measure of exposure, if the analytical method detects bound and unbound forms of the biological marker (see Sections 6.1.4 and 6.7). If they are not excreted, their utility in the workplace will be limited, since an invasive technique will be required to obtain samples and ongoing invasive techniques used on a regular basis generally are not well received by workers (see Sections 2, 6, and 9).

In some cases these adducts could trigger the excretion of another chemical (a second-generation chemical), that qualitatively or quantitatively indicated exposure to the workplace chemical. Without the biological marker research leading to the discovery of this excreted chemical, its presence in the body excretion media probably would not be discovered (see Section 7).

The adducts also can change the concentration of a chemical normally excreted. Without the biological marker research, exposure-related changes in the normally excreted chemical probably would not be detected (see Section 7).

Biological markers also can be important adjuncts to epidemiological studies. Since most workers are exposed to a variety of chemicals, retrospective epidemiology studies often are not able to definitively ascribe a toxic effect to a single chemical. The use of macromolecule–xenobiotic adducts as biological markers offers opportunity for improving the effectiveness and reliability of retrospective epidemiology studies. However, to realize this potential, a macromolecule–xenobiotic adduct biological marker must have a long metabolic half-life and be characteristic of a single xenobiotic or a family of xenobiotics with similar toxic effects. In prospective epidemiological studies, the metabolic half-life is less important.

Biological markers also offer the opportunity to measure important metabolic variations in individual workers prior to employment. The importance of this evaluation is discussed in Section 4.4. Biological markers in categories (4), (5), and (6) above often offer the potential to measure these variations following exposure of the workers to a lower dose of the xenobiotic than would be required for a classical evaluation.

It is likely that the biosynthesis of most biological markers will be controlled by enzymatic reactions rather than by physical factors such as diffusion. The confounding factors affecting biosynthesis are discussed in Section 6.

As more research has been carried out on biological markers, it has become apparent that the quantitation of the exposure chemical or its catabolite in the various bodily excretion media are only two aspects of the universe of measurement of worker exposure in the workplace. As already stated Elkins's term *solvent* has been replaced by the term *biological marker* to cover this panoply of direct and indirect biological measurements to determine the amount of chemical exposure in the workplace and the resulting body burden. However, since quantitation of the exposure chemicals and their metabolites in body excretions, as envisioned by Elkins, has been studied more than any other biological markers, most of the discussion in this chapter will deal with these two types of biological markers. It is likely that the biosynthesis of all biological markers will be controlled by the same biochemical factors that control the body's handling of these two types of biological markers.

This chapter is not an in-depth review of particular biologic evaluations such as lead or mercury in blood or urine. Individual biological analyses usually have an extensive literature devoted to them and should be consulted by those interested in the particular analysis (Piotrowski, 1977; Baselt, 1983; Lauwerys, 1983). Nor are analytical methodologies discussed in detail; they also are readily located in specialized texts and scientific articles. The reader also is referred to relevant specialized texts and scientific papers for detailed guidance in the areas of biochemistry, enzymology, pharmacokinetics, and pharmacology (Neilands and Stumpf, 1958; West and Todd, 1966; LaDu et al., 1971; Gibaldi and Ferrier, 1975).

In this chapter, an overview of the concepts and realities of monitoring biological markers to measure worker exposure in the workplace is presented.

The chapter discusses (1) the general concept of biological analysis as a tool to measure workplace exposure to exogenous chemicals with some biological activity (xenobiotics), (2) possible utilities of the concept in the workplace, (3) the problems to be expected in developing a biological monitoring method for a particular chemical or element, (4) the problems involved in applying a method, and (5) the problems and uncertainties encountered in interpreting the results.

Biochemical reactions, enzyme kinetics, pharmacokinetics, metabolism, fat storage, protein binding of chemicals, route of exposure, concurrent exposure to several chemicals, sex of the worker, and age of the worker are some of the things that must be considered in developing, and interpreting the results of, any biological assay for a biological marker. Examples from the published literature are used to illustrate these points.

The terms *biological analysis* and *biological monitoring*, as used in this chapter, indicate analysis of exhaled air; analysis of some biological fluid, such as urine, blood, tears, or perspiration; or analysis of some body component, such as hair or nails, for evaluation of past exposure to a chemical. The chemical of analytical interest is referred to as the biological marker or biomarker.

The classical biological markers for industrial hygiene purposes have been the exposure chemical or a classical metabolite of it. Since most of the developmental and analytical work on biological markers has been carried out on these two types of chemicals, most of the scientific literature is on them. Consequently, most of the illustrations and discussion in this chapter will be based on the lessons learned from and about them. There is no question that most of the caveats discussed for these two types of biological markers will apply to all of the other classes of biological markers that have been developed and will be developed.

2 RATIONALE FOR BIOLOGICAL EVALUATION OF EXPOSURE

Blood and urine analyses for certain components have long been used by the medical profession to facilitate differentiation between the normal and diseased states (Wallach, 1974). Since most industrial chemicals that might cause systemic effects will be transported by the blood system, metabolized by enzymes, and excreted by one of the excretory systems, it seems a logical extension of the medical groundwork in these areas to attempt to measure exposure to industrial chemicals by biological analysis for the industrial chemical(s) of interest.

The objective of industrial hygiene is to prevent an effect level of an agent from reaching that agent's target organ(s) or tissue(s) in the worker. The agent may be physical or chemical. Physical agents such as sound or electromagnetic radiation or particulate radiation occur in only one form, and control of that one form will control the worker's total exposure to that agent. However, workers may be exposed to chemicals in the form of gases, vapors, liquids, or solids, either singly or combined. Workers may be exposed to chemicals through inhaling them, absorbing them through the skin, or ingesting them. Exposure may be by several routes concurrently.

Within limits, the effect of a chemical on an organ or tissue is directly proportional to the amount of the chemical reaching or reacting with that organ or tissue. Ideally, then, to relate toxicity to dose, one should correlate the amount of chemical in the target organ(s) or tissue(s) with the effect seen. This would first be done in animal experiments and these results then extrapolated to humans. However, it is even more impractical to continually and routinely make such direct measurements on an organ in a worker than in a laboratory test animal. Therefore, it would be desirable to have some secondary or tertiary measurements that could be made easily and routinely on a potentially exposed person—measurements that would reliably bear a known relationship to the appearance or absence of toxic effects. Ideally, the concentration found for the biological marker also would give some idea of what additional burden of the exposure chemical or its metabolites would be acceptable before reaching the threshold of a toxic effect.

Within limits, the amount of the chemical reaching the target organs or tissues actually will vary directly with the amount unbound (see Section 6.1.4) in the blood. Therefore, the amount reaching these target organs or tissues can be reasonably approximated by measuring blood levels. However, as with organ analysis, blood sampling is an invasive technique and is not suited to routine daily monitoring of workers.

In most cases, again within limits that vary for each chemical, the amount of the toxicologically determining chemical reaching the target organ(s) or tissue(s) will vary directly with the dose of the exposure chemical received by any or all routes.

Historically, inhalation has been the major exposure route of concern for the industrial worker. Consequently, industrial hygienists have concentrated on keeping atmospheric levels of chemicals below effect levels. The Chemical Substances TLV Committee of the American Conference of Industrial Hygienists (ACGIH; TLV is a registered trademark of the ACGIH and is an acronym for threshold limit value— the dosage that is on the threshold of causing an effect) has established acceptable industrial atmospheric levels for several hundred industrial chemicals. These are levels that they believe "nearly all workers may be repeatedly exposed (by inhalation) day after day without effects" (Amer. Conf. Governmental Ind. Hygienists, 1991). The TLVs for some of these chemicals also have been incorporated into the Occupational Safety and Health Act of 1970 (PL 91-596).

Examination of the TLV list shows that at least one member of most classes of organic compounds listed has the notation "Skin," indicating that toxicologically significant amounts of that chemical in the liquid form or in solution can be absorbed through the skin. If, in addition to inhaling one of these chemicals, workers spill the liquid form of one of these chemicals on their skin, their shoes, or their clothing, or if the worker's hands are immersed in the liquid, the target organs or tissues will receive the chemical through both inhalation and through absorption through the skin. Obviously, in this case, atmospheric levels of the chemical do not give an accurate indication of the amount of chemical that might reach the target organ(s) or tissue(s). In fact, the amount reaching the target organ(s) or tissue(s) from inhaling the chemical may be only a small part of the total reaching them from both routes of entry (Eldridge, 1924; McCord, 1931; McCord, 1934; Yant and Schrenk, 1937; Hill, 1953; Stewart and Dodd, 1964; Dutkiewicz and Tyras, 1967, 1968; Piotrowski, 1971; Tada et al., 1975). Since the amount of the liquid that might contact the skin in an industrial situation is variable, this concurrent route of exposure cannot be reliably assessed as an adjunct to an inhalation exposure of test animals and cannot be factored into deciding safe inhalation levels of the chemical.

Multiple routes of exposure also are possible when workers are exposed to particulate material. Inhaled dust particles may be cleared from the lungs through the trachea and thus be ingested. In these instances, exposure of workers by inhalation also constitutes an exposure by more than one route. If the absorption coefficient for gastrointestinal absorption is greater than that for pulmonary absorption, the second route of exposure again may be more important toxicologically than inhalation. The ratio of absorption by the two routes also may vary with particle size. However, in this situation, in contrast to the inhalation–skin absorption situation, inhalation exposure of test animals will give some measure of concurrent absorption by ingestion. If the particle size is different in the industrial situation than in the animal exposures, however, the amounts concurrently absorbed by the two routes may be more or less than expected from the inhalation studies in animals. Consequently, the amount of the chemical reaching the target organ(s) or tissue(s) may vary, and the animal inhalation exposures may not be quantitatively predictive of the effect on exposed humans.

Obviously, then, measurement of atmospheric concentrations of a chemical will not always give a reliable measure of the amount of the chemical that may reach target organ(s) or tissue(s) in exposed workers.

As already mentioned, the ideal measure is the measure of the concentration in the target organ(s). This is impractical on a continuing basis as mentioned, although it may be practical for single, isolated analyses. Blood sampling, the next best measure of the amount in the target organ(s) or tissue(s), also is impractical on a continuing basis for two reasons: (1) hazards introduced to the worker by taking samples frequently and (2) worker resistance to the procedure. However, this technique is suited for occasional isolated or infrequently repeated analyses.

The ideal method for measuring all of the xenobiotic absorbed, and thus the total risk to the worker, is a noninvasive technique that accurately reflects the degree of toxic insult of the chemical to the worker and is not disruptive or repugnant to the worker.

3 HISTORY OF BIOLOGICAL MONITORING

As mentioned previously, biological monitoring has been used by the medical professions for decades. The formal application of this concept on a broad scale to industrial hygiene is probably most rightfully attributed to Hervey Elkins (Elkins, 1954), although individual correlations of exposure levels, toxicity, and blood or excretory levels of industrial chemicals precede his broad-scale advocacy by many years (McCord, 1931; Widmark, 1933; Neymark, 1936; Schrenk et al., 1941; Von Oettigen et al., 1942 and Sayers et al., 1942). Since his initial advocacy, Elkins has published several papers on the subject (Elkins, 1961, 1967), Elkins and Pagnatto (1965), and Dutkiewicz (1971) has proposed an integrated index of absorption.

Elkins initially tried to correlate exposure levels, urinary levels, and the onset of toxic effects. Since then, the concept of biological monitoring for xenobiotics has been extended to include monitoring of the biliary-fecal route, exhaled air, perspiration, tears, fingernails and toenails, hair, milk, and saliva.

Although analyses of feces, perspiration, tears, nails, hair, milk, or saliva have been used in special situations to give an estimation of body burden of a chemical, analyses of these substances have not been used for routine frequent monitoring of workers to determine immediate past exposure. Analyses of exhaled air for volatile xenobiotics and/or their metabolites and of urine for water-soluble xenobiotics and/or their metabolites have received the greatest attention as a measure of industrial exposure and body burden.

4 USEFULNESS OF BIOLOGICAL MONITORING

Since biological monitoring usually occurs after exposure, it cannot replace good industrial hygiene practices. It does, however, supplement a good industrial hygiene program and provides additional information of value in the overall worker protection program. Some specific utilities of biological monitoring are described in the following sections.

4.1 Indication of Unsuspected Employee Exposure

The unsuspected exposure may arise from skin absorption, as mentioned previously, from unsuspected equipment leaks, which increase air levels of the exposure chemical, or from off-the-job sources (Cralley et al., 1990).

4.2 Guidance to Physician in Deciding Whether to Administer Risky Therapy

Some antidotes, such as atropine or pyridine-2-carboxaldehyde monoxime (2-PAM) used for treating overexposure to certain organophosphates or carbamates are used at dosages that might themselves be toxic if the patient had not received a prior excessive dosage of the organophosphate or carbamate.

4.3 Documentation of Overexposure or Acceptable Exposure

If a worker has received an excessive dose of an agent, good industrial hygiene practice dictates that the worker be removed from that environment for a period of time. Biological monitoring can document the fact of overexposure or acceptable exposure. In the case of overexposure, it can give the degree of overexposure and indicate how long the worker must remain free from additional exposure.

4.4 Preemployment Screen for Metabolic Variations

Some people are unusually sensitive to the adverse effects of chemicals (Amer. Conf. Governmental Ind. Hygienists, 1991, p. 2). For their own protection, these people should not be exposed to levels of a chemical that may not adversely affect the rest of the exposed population but that may affect them. In some cases this unusual sensitivity is due to atypical metabolic routes or to atypical enzyme kinetics. Some of these cases could be evaluated in a preassignment physical examination by administering controlled, low doses of chemicals that the worker would be exposed to on the job. This would be followed by determining whether the worker's individual biological half-life values for excretion of these chemicals are in the normal range and whether the worker excretes the usual metabolites in the usual ratios.

A subnormal half-life might be acceptable, but a greater than normal half-life could predict bioaccumulation and possible hazard under exposure conditions that would not be harmful to most people (Ikeda and Imamura, 1973). Abnormal metabolic products or ratios might or might not indicate greater hazard.

4.5 Validating TLVs Calculated for Concurrent Exposure to Chemicals Affecting the Same Target Organ(s)

When a worker is concurrently exposed to more than one chemical affecting the same target organ(s), exposure levels that were safe for the individual chemicals of exposure usually are not safe for the combined exposure. The TLV handbook contains a formula for calculating reduced atmospheric exposure levels for this situation

(Amer. Conf. Governmental Ind. Hygienists, 1991, p. 48). Direct biological monitoring for the biological marker will not document the safety of the calculated safe exposure levels. If the organ damage caused by overexposure results in elevated levels of some usual blood or excretory fluid component, however, normal postexposure levels of the component(s) would confirm the validity of the calculation. Conversely, abnormal values would indicate the inapplicability of the correction formula in this situation (see Section 5).

4.6 Validation of Area or Personal Monitoring

Personal monitoring, manual area monitoring, and automated area monitoring are the most common techniques now used to monitor worker exposure to atmospheric chemicals. All three offer opportunities for error. Regular biological monitoring may be useful to validate the atmospheric monitoring. However, biological monitoring also offers opportunities for error as discussed in this chapter, and these must be considered in evaluating the biological monitoring results.

5 TYPES OF BIOLOGICAL MONITORING

Most monitoring of biological fluids (and, in some cases, hair, nails, and expired air) can utilize one of two approaches: direct or indirect.

In what I call direct analysis, the biological marker is the exposure chemical per se or a metabolite. The metabolite could result from (1) catabolism of the xenobiotic, (2) reaction of the xenobiotic or a characteristic metabolite with a body macromolecule, (3) reaction of the xenobiotic or metabolite with a chemical that increases its water solubility or processing by the kidney, or (4) excision repair of the macromolecular reaction product.

In what I call indirect analysis, the biological marker is an effect that resulted from the action of the agent on some body system, organ, or tissue. For example, the biological markers' blood urea nitrogen or urinary glutamic-pyruvic transaminase (GPT) levels could be utilized as measures of effect on kidneys (Cornish, 1971; Wallach, 1974). Changes in blood cell numbers per unit volume and distribution between types could be used as biological markers for effect on the hemopoietic system (Wallach, 1974). Changes in isoenzyme patterns could be considered as biological markers, for example, on liver or heart (Cornish, 1971), and cholinesterase levels in whole blood or plasma could be used as an indirect biological marker for organophosphate exposure (Long, 1975).

Indirect analysis of this type is seldom a desirable approach for routine worker monitoring for at least the following reasons:

1. The measured effect frequently occurs only after damage to the worker has occurred, and the thrust of industrial hygiene should be to prevent deleterious effects on workers from agents in the workplace.

2. Appearance of the measured effect may not occur for some time following exposure to the agent and thus might be difficult to attribute to a specific cause.

Also, in this situation, there could have been protracted overexposure before the secondary effect was seen, and this is undesirable.

3. The effect measured may not vary directly with the amount of exposure, making it difficult to determine the degree of exposure.

4. The effect may not be specific and may have several possible etiologies, not all of which occur in the workplace. This should not be interpreted to mitigate the health significance of the change for the individual worker, however.

5. If blood sampling is required, it usually is not suited for routine monitoring in the workplace on a frequent basis.

6   BIOCHEMICAL PROBLEMS ASSOCIATED WITH BIOLOGICAL MONITORING

As with most techniques, additional studies have disclosed many possible sources of variations in the correlation between biological marker levels and exposure levels, other than those associated with technology. These apply not only to analyses carried out on most of the fluids or body components mentioned previously but also apply to the interpretation of the results.

General factors that must be considered in developing, applying, and interpreting a biological analysis include:

1. Metabolic variations.
2. Changes in the ratio of bound to free chemical in the blood.
3. Special situations where excreted levels of the biological marker do not indicate current exposure levels.
4. Concentration changes due to volume changes in the bioassay material.
5. Nonworkplace progenitors of the biological marker and resulting baseline variation in its concentration in the bioassay material.
6. Normal range of the biological marker to be expected in the bioassay material.
7. Time required for the biological marker to appear in the bioassay material.
8. Analytical methodology.
9. Route of exposure.

6.1   Metabolism

Xenobiotics can be handled by the body in many ways, including:

1. Metabolism to a component of a naturally occurring series of reactions. The component may then be oxidized to CO_2 and H_2O or else anabolized (biochemical synthesis) to a body component.
2. Excretion of the chemical unchanged.
3. Metabolism to a chemical that is more readily excreted.

4. Storage in an organ, bone, body structure, or organized element of some sort, such as body fat or brain.
5. Accumulation without storage (e.g., in the blood or the hepatobiliary-intestinal loop).
6. Reaction with a macromolecular component of the body.

Many xenobiotics follow a combination of routes 2 and 3. For these, all the factors that can affect metabolism become important in determining (1) what fraction of the original exposure chemical is excreted unchanged and by what route and (2) how much of each possible metabolite is excreted and by what route.

Factors that can affect either or both the (1) rate of excretion and (2) ratios of the excreted xenobiotic to the various possible metabolites excreted, obviously can affect the estimate of exposure based on analysis of excreted biological markers.

Anything that can increase or decrease the normal accumulation or the storage of xenobiotics handled by routes 4 or 5 will have the inverse effect on excreted materials. In these situations, biological marker analyses relied on to indicate level of exposure will indicate correspondingly high or low exposure.

Most of these factors apply to organic compounds. There are not as many opportunities for individual variation in biological monitoring of metal or inorganic metal salt exposures. Unlike organic compounds, these materials normally are not metabolized. Nevertheless, one should expect individual variations in route and rate of excretion of these materials. Also, to the extent that storage or excretion of these metals or salts may be hormonally controlled, concurrent exposure to medicinals or industrial chemicals that affect hormonal levels should be considered when evaluating the body burden of metals or metal salts through analysis of excreted biological markers.

Inorganic metal salts normally are not absorbed through intact skin. Thus, monitoring of the workroom air usually will give a good estimation of worker exposure to such salts in contrast to organic chemicals that may be absorbed through intact skin as well as inhaled. Organic salts, organometallic compounds, and organic complexes of metal ions or metals (e.g., butyl chromate, tetraethyl lead, tetramethyl lead, dialkyl mercury, methylcyclopentadienyl manganese tricarbonyl, organic tins, and organic silanes) may be absorbed through the skin. The metabolism of the organic portion of these compounds, and consequently the storage and excretion of the metal component, may be affected by the factors mentioned previously, which are discussed more fully in the following paragraphs.

Some industrial organic chemicals that have low water solubility and no sites that can be easily modified enzymatically are stored by the body. Examples are polychlorinated or polybrominated biphenyls (PCB, PBB) and 2,3,7,8-tetrachlorodibenzo-p-dioxin (Waritz et al., 1977). Some metals, such as lead and mercury, also are stored by the body. However, just as with foodstuffs and pharmaceuticals most industrial chemicals are metabolized and excreted (Williams, 1959; Parke, 1968). Chemicals that are totally catabolized (biochemical breakdown) seldom yield characteristic end products and are not suitable for analysis other than in blood shortly after exposure. However, chemicals that are excreted either without modification or after modification to metabolites not occurring naturally offer excellent opportunities for measuring worker exposure by measuring the amount excreted.

Most excretory systems of the body are aqueous. If the xenobiotic is not catabolized to a component of one of the body's many metabolic sequences, the thrust of its metabolism will be toward increased water solubility. This means that, in general, a hydrocarbon xenobiotic will be hydroxylated or at least one carbon will be oxidized to a keto group. Frequently, alkyl side chains on aromatic compounds are oxidized to a carboxylic acid group or all the way to benzoic acid. An aliphatic alkane may have a terminal carbon oxidized to a carboxylic acid moiety or one or both carbons oxidized to keto group(s) (Perbellini et al., 1980). Thus, the chemical(s) excreted following exposure may not be the same as the exposure chemical. Examples of some of the metabolic routes that have led to metabolites in blood or to excretion products in liquid excreta are shown in Figure 3.1.

As can be seen, several metabolites of one chemical may be excreted in the urine. In addition, glucuronates, sulfates, mercapturic acids, or amides of metabolites may be excreted. These derivatives are called conjugates. Analyses of biological excretory liquids for a biological marker should measure conjugated, bound, and free forms of the biological marker since ratios of the three may change.

Since the direction of metabolism usually is toward increased water solubility, metabolites generally will be less volatile than the chemical of exposure. Hence, metabolites generally will not be excreted in exhaled air. [There are some notable exceptions. For example, some metals, such as selenium and tellurium, are methylated by the body. The methylated compound is more volatile than the metal and is excreted in exhaled air (McConnell and Portman, 1952).] Accordingly, some of the complications introduced by metabolism are correspondingly less important when biological monitoring can be based on direct analysis for the workplace chemical in exhaled air. For this latter monitoring, generally only the exposure chemical need be considered, regardless of the route by which it was absorbed.

Neither the kinetics of a particular metabolic pathway nor the relative kinetics of alternate metabolic pathways are constant in the same person at all times or between individuals. Since a biological monitoring program relies on the rate of appearance or the amount of the biological marker in the bioassay material, anything that affects this rate or amount will affect the estimate of the amount of workplace exposure. Many factors can affect metabolic rates and thus excretion, singly or in combination. These include:

1. Individual variations in enzyme complement and enzyme kinetics.
2. Diet and dietary state.
3. Stimulation or inhibition of enzymes involved in the metabolic sequences.
4. Ratio of bound to free chemical in the blood.
5. Dose of the exposure chemical.
6. Competition for a necessary enzyme.
7. Obesity and fat–muscle ratio.
8. Age of the worker.

Pathways

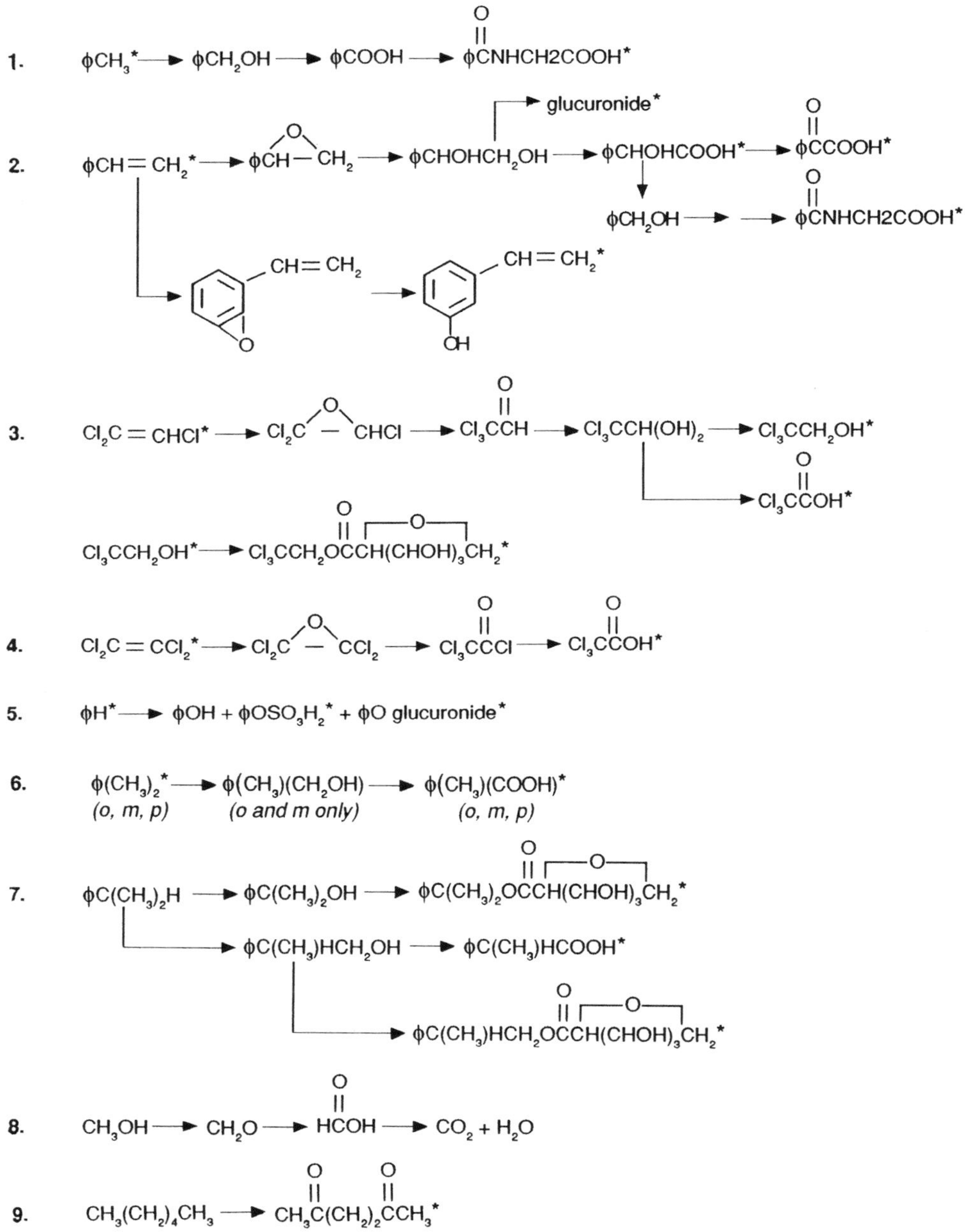

Figure 3.1 Principal metabolic pathways of some industrial chemicals. Other minor metabolites may be formed (Bakke and Scheline, 1970) and excreted, but they are not of value for biological monitoring. Symbols: ϕ = phenyl; * = identified in urine of humans or animals.

9. Disease.
10. Concurrent use of medicinals or other chemicals.
11. Sex of the worker.
12. Pregnancy.
13. Smoking.

Furthermore, any one or combination of several of these factors may be more or less important on any given day.

6.1.1 Individual Variations in Enzyme Complement and Kinetics

Individual variations in metabolic pathways and kinetics have been recognized in the field of biochemistry for years (Hommes, 1973). Hommes indicates that, in many cases, these variations are known to be due to genetic variations in individuals. For years, these variations were regarded as only biochemical curiosities. The importance of these genetic variations in pharmaceutical metabolism was appreciated only comparatively recently as more effort was directed toward studying medicinal toxicity as a consequence of individual variations in metabolism and excretion. The study of genetic-based nonusual metabolic pathways and kinetics has been named *pharmacogenetics* by the pharmacologists. Pharmacogenetic-based perturbations of metabolism of xenobiotics is just as important to biological monitoring and metabolism/excretion of workplace chemicals as it is to metabolism/excretion and net dosage of medicinals.

If a worker is missing an enzyme or has an enzyme with specific activity different from that of most other workers and that enzyme is critical in the metabolism and excretion of a workplace chemical, concentrations of the expected metabolite(s) in the biological medium analyzed will give a false picture of the worker's exposure, if they are compared with values from most workers. As discussed earlier, if, as a result of this aberrant enzyme, the elimination half-time ($t_{1/2}$) of a workplace chemical is increased, the error will be of the worst type; that is, it will indicate less exposure than the worker has received. As discussed earlier, increased $t_{1/2}$ for a chemical to which a worker is exposed could lead to unexpected accumulation in the body with possible toxic consequences.

If metabolism by the worker leads to unexpected products, the fact and degree of the worker's exposure could be dangerously underestimated because of the absence of the expected biological markers.

Individual differences in the toxicity of sulfa drugs and the antitubercular drug

References: Reaction 1—Bakke and Scheline, 1970; Reaction 2—Carpenter et al., 1944, Danishefsky and Willhite, 1954, El Masri et al., 1958, Stewart et al., 1968, Ohtsuji and Ikeda, 1971, Slob, 1973, Pfäffli et al., 1981; Reaction 3—Butler, 1949, Cooper and Friedman, 1958, Souček and Vlachová, 1960, Byington and Liebman, 1965, Cole et al., 1975; Reaction 4—Yllner, 1961, Daniel, 1963; Reaction 5—Porteous and Williams, 1949; Reaction 6—Bray, et al., 1949, Baake and Scheline, 1970; Reaction 7—Robinson, et al., 1955, Sénczuk and Litewka, 1976; Reaction 9—Perbellini, et al., 1980.

isoniazid were found to be due to genetically controlled individual differences in the rate of acetylation and hence excretion of these medicinals (Evans and White, 1964; Jenne, 1965). It is believed that approximately 50 percent of the U.S. population can be classified as "slow" acetylators (Greim, 1981). Obviously, acetylation-dependent excretion of an industrial chemical would be expected to be slower in one-half of the exposed U.S. workers than it would be in the other half. To further complicate the picture, Kalow (1980) has reported that, of over 2000 Japanese subjects tested, only 12 percent were slow acetylators. This suggests that norms would have to be established or validated for each race for each biological monitoring procedure developed.

Sloan et al. have reported (Sloan et al., 1978) that about 5 percent of a Caucasian population studied had reduced capability to hydroxylate aliphatic hydrocarbons. This reduced capability was of genetic origin. Defective O-demethylation of genetic origin also has been reported (Kitchen et al., 1979).

Greim (1981) has tabulated other genetically controlled enzymatic variations of importance to pharmaceutical metabolism and excretion. He also reported the frequency of these variations in the population. To the extent that these enzymes are important in an individual's metabolism of industrial chemicals, a similar percentage of variations could be expected in workers.

Vessell and Passananti (1971) have reported that the $t_{1/2}$ in plasma for dicumarol varied almost 10-fold in fraternal twins. The $t_{1/2}$ in plasma for the medicinals antipyrine and phenylbutazone varied 3-fold and 6-fold, respectively, in these same twins. Variations at least as large are to be expected in the general population as well as similar variations in the metabolism and excretion of industrial chemicals.

Tang and Friedman (1977) examined the liver microsomal oxidase activity* from 10 humans varying in age from 3 days to 92 years. Activity varied from "undetectable" to "highly active" as measured by mutations induced in *Salmonella typhimurium* TA-100 incubated with various aromatic amines known to require activation by microsomal oxidase for mutagenic activity. Six of Tang and Friedman's cases were between the ages of 22 and 57 years, inclusive. Both sexes were represented. Some had died violent deaths, some had died from disease, and one had died following surgery for an acute condition. Thus, not all had died of disease and not all were aged. Therefore, Tang and Friedman's findings should be representative of what could be expected for a general worker population.

6.1.2 Diet and Dietary State

Many studies have been carried out showing the effects of diet on enzymatic activity and consequently on the metabolism of chemicals. For example, excessive (3–9 g/day) levels of niacinamide have been reported to cause elevation of serum glutamic-oxalic transaminase (SGOT) and alkaline phosphatase (APase) in humans (Winter

---

*If a xenobiotic dose not fit a specific enzyme, its metabolism generally will be mediated by the nonspecific microsomal enzymes. These enzymes do not seem to require the enzymatic "fit" that characterizes most enzymatically catalyzed reactions. They are active in varying degrees on a variety of structures. They are known to carry out epoxidations, demethylations, reductions, and hydrolyses. Findings should be representative of what could be expected for a general worker population.

and Boyer, 1973). Protein deficiency has been reported to cause reduced hepatic metabolism in rats (Newberne, 1975) and could be expected to have the same effect in humans. High protein diet, on the other hand, has been shown to enhance the metabolism of chemicals in humans (Alvares et al., 1976). A charcoal-broiled beef diet has been shown to increase microsomal oxidase [benzo(a)pyrene hydroxylase] activity in the liver of pregnant rats (Harrison and West, 1971), suggesting a stimulation of the nonspecific microsomal oxidase systems. However, when humans were placed on a diet containing charcoal-broiled beef, there was no change in the plasma half-life of the test chemical used to monitor changes in microsomal oxidase activity (Conney et al., 1977).

Brussels sprouts, cabbage, and cauliflower have been shown to contain chemicals that, when ingested by rats, increase liver microsomal oxidase activity as measured by oxidation of benzo[a]pyrene. Furthermore, specific activity was found to vary between cultivars of brussels sprouts and cabbage (Loub et al., 1975).

Diets high in corn oil and low in starch or sugar also have been reported to increase liver microsomal enzyme activity in rats (Yaffe et al., 1980). High-fiber diets have been reported to cause a negative calcium balance (Anon., 1977).

It is thus probable that diet, particularly a diet unbalanced with regard to any component, could lead to changes in enzymatic activity followed by changes in metabolism and in excretion of workplace chemicals.

6.1.3 Stimulation or Inhibition of Enzymes That Metabolize the Chemical

As mentioned earlier, many xenobiotics are metabolized through the nonspecific microsomal enzymes that can catalyze at least four types of reactions: oxidation (epoxidation), demethylation, reduction, and hydrolysis. More than 200 chemicals are known to stimulate some or all of the known microsomal enzymes and thus increase the metabolism of many xenobiotics (Conney, 1967). This could result in an elevated concentration of the metabolites of workplace chemicals in excreted fluids of workers. If the evaluation of worker exposure is based on the normal concentration of excreted metabolites formed by means of the microsomal enzymes, this increased microsomal enzyme activity could result in an erroneously high estimate of worker exposure. Conversely, if the analysis is based on unmetabolized xenobiotic, an erroneously low estimate of worker exposure would result.

A wide variety of chemical structures have been shown to stimulate the microsomal enzyme systems, including structures having medicinal, pesticide, or industrial uses. Chemicals shown to stimulate these enzyme systems include phenobarbital [one of the most potent stimulators (Conney, 1967)], spironolactone (Gillette, 1971), condensed polynuclear aromatic hydrocarbons, and polychlorinated or polybrominated biphenyls].

Cigarette smoke also has been found to stimulate the activity of liver microsomal enzymes in pregnant rats (Conney et al., 1977).

Ingestion of ethanol has been shown to both increase and decrease the blood and urine levels of xenobiotics, depending on the time of alcohol ingestion relative to the time of exposure (Sato and Nakajima, 1987; Sato et al., 1991). Ethanol has been reported to retard the formation of hippuric acid from toluene in humans (Wallén et al., 1984). Ethanol (0.8 g/kg) drunk before or after exposure of humans to 200 ppm of methyl ethyl ketone (MEK) for 4 hr retarded the formation of the biological marker 2,3-butanediol (Liira et al., 1990).

Pretreatment with phenobarbital has been shown to stimulate the oxidation of styrene in rats (Ohtsuji and Ikeda, 1971) and the oxidative step in the metabolism of 1,1,1-trichloroethylene (T_3CE) and 1,1,2,2-tetrachloroethylene (T_4CE) in rats and hamsters (Ikeda and Imamura, 1973). In both cases the oxidation rate was almost doubled. Other chemicals [e.g., morphine (Gillette, 1971)] are known to inhibit the microsomal enzyme systems. Styrene and benzene epoxidation by microsomal enzymes also has been reported to be hindered in the presence of toluene (Ikeda et al., 1972b). Obviously, these altered enzymatic rates could lead to increased or decreased appearance of metabolites in the excretory fluid analyzed. This in turn could lead to a high or low estimate, respectively, of workplace exposure if metabolite concentration in the excretory fluid were the basis of estimation of workplace exposure.

Regardless of whether the effect is stimulation or inhibition of the microsomal enzymes, the effect could alter the rate of appearance of metabolites in the excretory fluids only if the microsomal enzymes catalyzed rate-limiting steps in the metabolism of the xenobiotic.

It has been known for many years that continued exposure to a chemical can stimulate metabolism of that chemical (Knox and Mehler, 1951; Neilands and Stumpf, 1958). Ikeda and Imumura (1973) have shown that pretreatment with T_3CE can either stimulate or depress metabolism of subsequently administered T_3CE, depending on the pretreatment time and dose. Pretreatment with T_4CE, on the other hand, only stimulates the metabolism of subsequently administered T_4CE.

Ikeda et al. (1971) reported a subject who was addicted to sniffing T_3CE and whose $t_{1/2}$ for urinary excretion of metabolites was twice the usual $t_{1/2}$. It was not known whether the prolonged $t_{1/2}$ preceded or followed the addiction. Thus, repeated exposure to a chemical may not always stimulate its metabolism.

6.1.4 Ratio of Bound to Free Chemical in the Blood

Many medicinal chemicals have been found to be transported in blood in the unbound state (free) as well as bound to proteins. Generally, the binding is to albumin in the plasma, but it also may be to other plasma proteins and/or to proteins that are structural parts of formed elements in the blood (Gillette, 1971). Since industrial chemicals basically differ from medicinal chemicals only in their application and in the relative magnitude of their biological effects, there is no a priori reason why they should differ from medicinal chemicals in their ability to bind to blood components. Therefore, as expected, industrial chemicals also are transported in the blood in both the bound and free forms. Also, as with medicinal chemicals, the ratio of bound to unbound would be expected to vary from chemical to chemical.

The binding phenomenon in most cases probably is a simple complex formation, subject to all the mass action mathematics and dynamics of inorganic or other organic complexes. The stability constants also probably vary over several orders of magnitude, just as do those for classical complexes. Thus, bound and unbound fractions would be in dynamic equilibrium. This has important ramifications in biological monitoring, as discussed in the paragraphs that follow.

Total blood concentration (bound plus free) of the xenobiotic and its metabolites should reflect workplace exposure of chemicals that are absorbed into the worker's system. However, the amount excreted per unit time generally reflects only the unbound concentration in the blood. As already mentioned, systemic toxicity generally will vary directly with the concentration of the unbound chemical. Situations can occur in which the normally bound fraction of a workplace chemical (the fraction that was bound when the correlations between dose, toxicity, and amount of biological marker excreted were developed) is decreased. In these situations, the amount of biological marker excreted still will represent the unbound concentration in the blood and thus the hazard to the worker (Reidenburg, 1974). However, it will not be as closely related to workplace exposure and will suggest recent exposures higher than actually occurred.

Decreases in the ratio of bound to unbound chemical in the blood generally are caused by a decreased number of available binding sites. This usually is due either to a deficiency of binding protein or to concurrent administration of a chemical that is more tightly bound. Disease also has been shown to reduce binding of chemicals (Reidenburg, 1974).

In some workers a decreased amount of binding protein may be due to genetic factors. However, it most frequently is due to disease. Chronic disease conditions accompanied by lowered plasma levels of protein include chronic liver disease, rheumatoid arthritis, diabetes, and essential hypertension. Nephritis and gastrointestinal diseases such as peptic ulcer and colitis also can be accompanied by lowered plasma protein. Other ways in which disease can affect the ratio of bound to unbound chemical are discussed in Section 6.1.9.

A decrease in the number of available binding sites associated with a normal level of transport protein usually is caused by the presence of other chemicals that are more tightly bound to the protein. Phenylbutazone, a medicinal occasionally prescribed for arthritis and other aches and pains, is very tightly bound to plasma proteins. It is bound so tightly that it will displace most other bound chemicals from the transport proteins in blood. If present in sufficient concentration in the blood, it could displace significant amounts of a bound workplace chemical. This would result in elevated blood levels of the free form of the latter and thus could result in an increased rate of metabolism and excretion. If workplace exposure were being estimated from excreted levels of the exposure chemicals and/or their metabolites, this could result in an erroneously high estimate of exposure. However, the excreted level would still reflect the hazard to the worker, since that would usually vary with the concentration of unbound chemical in the blood.

Another common medicinal that also is strongly bound is salicylic acid (Reidenburg, 1974). Less common medicinals that could displace bound workplace chemicals include clofibrate, a cholesterol-lowering medicinal (Reidenburg, 1974).

Since workplace chemicals also are bound in varying degrees to blood proteins, as already mentioned, concurrent high exposure to two or more industrial chemicals could decrease the ratio of bound to unbound for the one with the weaker bond.

Among the diseases, cirrhosis has been shown to result in decreased binding of medicinals, with accompanying increased blood levels of the unbound form (Reidenburg, 1974). There undoubtedly are other diseases that also reduce binding.

In some cases the xenobiotic or one of its metabolites reacts with an intracellular or extracellular macromolecule to form an actual covalent bond. This new compound potentially has a half-life equal to that of the original macromolecule. As discussed in Section 9, this new macromolecule, or a fraction of it containing the reactant from the xenobiotic, can be a biological marker in its own right.

6.1.5 Effect of Dose of Exposure Chemical on Metabolism

It has been known for many years that the metabolism of chemicals can change with the dose administered, as mentioned previously. This obviously can affect the dose–excretion curves for metabolites used to measure exposure. Quick (1931), for example, found that when high doses of benzoic acid were administered to humans, the glucuronide became an important excretory product, as well as hippuric acid. Von Oettingen et al. (1942) found that after about 8 hr of exposure to approximately 300 ppm of toluene, benzoic acid glucuronide also was excreted in addition to hippuric acid.

Sabourin et al. (1989) found that at low doses of benzene, muconic acid was the predominant metabolite excreted in the urine of rats instead of phenol. Phenol predominates at higher levels. Bechtold et al. (1991) reported this dichotomous metabolism also occurs in humans.

Tanaka and Ikeda (1968) found the usual metabolic pathway for trichloroacetic acid (TCA) formation in humans from T_3CE became saturated after about 8 hr of inhalation exposure at 150 ppm. The excretion rate of TCA reached a plateau at that concentration.

More recently, Ikeda et al. (1972b) reported that the usual metabolic pathway for T_4CE was saturated after exposure at 50 ppm for several hours. They also reported that, in the rat, formation of phenylglyoxylic and mandelic acids plateau at styrene exposure levels ≥ 100 ppm for 8 hr.

Götell et al. (1972) reported that in humans the urinary concentration of phenylglyoxylic acid not only reached a peak after exposure to styrene for approximately 8 hr at 150 ppm but actually decreased at higher atmospheric levels.

Watanabe et al. (1976) also have reported that the usual metabolic pathway for vinyl chloride monomer (VCM) is saturable in the rat. At low VCM levels, it is principally metabolized and the metabolites are excreted in the urine. Only a small percentage is excreted unchanged in exhaled air. At high VCM levels, the reverse is true and most of the VCM is excreted unchanged in exhaled air.

The metabolic route of T_3CE in rats also has been reported to be saturable (Dallas et al., 1991). However, saturation occurred somewhere between the Occupational Safety and Health Administration (OSHA) permissible exposure limits (PELs) (50 ppm, 8-hr time-weighted average, in 1991) and 10 times that concentration. Thus, metabolic saturation would not be expected to confound biological monitoring of T_3CE in the workplace.

These studies all indicate the importance of not extrapolating biological monitoring equations or dose–response curves beyond the conditions used to derive the equations or reference curves.

6.1.6 Competition for a Necessary Enzyme

If a worker is concurrently exposed to two or more chemicals that require the same enzyme(s) for metabolism, the resulting ratios of metabolites and parent compound in blood and excreted fluids can be dramatically altered. The chemical that is preferentially metabolized by the enzyme(s) will have the most nearly normal pattern. This is elegantly illustrated by concurrent exposure to ethanol and methanol.

The first step in the metabolism of ethanol and methanol is a dehydrogenation catalyzed by the liver enzyme alcohol dehydrogenase (ADH) as shown in reactions 1 and 2:

$$CH_3CH_2OH \xrightarrow{ADH} CH_3\overset{\overset{\displaystyle O}{\|}}{C}H \tag{1}$$

$$CH_3OH \xrightarrow{ADH} H\overset{\overset{\displaystyle O}{\|}}{C}H \tag{2}$$

Alcohol dehydrogenase acts preferentially on ethanol. Therefore, in the presence of ethanol, only a small fraction of the methanol is dehydrogenated, and most of the methanol circulates unmetabolized in the blood. Consequently, blood, urinary, and exhaled methanol levels are elevated over what they would be if the same dose of methanol were not competing with ethanol for the available ADH (Leaf and Zatman, 1952). Under these conditions, biological monitoring would suggest higher workplace exposure to methanol than occurred. [In this situation the elevated methanol levels reflect not increased hazard of eye damage but rather decreased hazard, since the metabolite causing the eye damage appears to be formic acid, a metabolite (Tephly, 1977).]

Enzyme competition also may explain the observation by Stewart et al. (1970) that the urinary excretion of 1,1,1-trichloroethanol (TCE) was greater than expected and that of TCA was less than expected after T_3CE exposure, if the subject concurrently consumed alcohol. If T_3CE exposure were being measured by urinary TCE excretion, this also would result in higher estimates of exposure than occurred.

Ingestion of ethanol before, or closely after, inhalation exposure to MEK also alters the oxidative metabolism kinetics of MEK (Liira et al., 1990).

From a toxicological standpoint, the importance of all these possible metabolism variations depends on whether the most important toxicant is a metabolite of the exposure chemical or the exposure chemical itself. Obviously, anything that decreases the concentration of a toxicant decreases the hazard, and anything that increases the concentration of a toxicant increases the hazard.

For biological monitoring purposes, these causes of variation indicate that the range of concentrations of a workplace chemical or its metabolites in blood, exhaled air, excretory fluids, or excretory solids following a given exposure normally will be broad. This finding also would be expected if one examines the range of blood and urine concentrations reported for chemicals normally occurring in these fluids in humans (Wallach, 1974). Two to threefold variations between the upper and lower "normal" values are usual. Furthermore, recognizing this normal range, physicians seldom rely solely on urine or blood levels of a chemical to diagnose a disease condition.

From a worker monitoring standpoint, one must be particularly careful about interpreting blood or excretory fluid levels of a chemical that is always present in these fluids and that also may arise from workplace exposure. In this situation the concentration range to be expected from a particular exposure level may vary by four- to sixfold (see Sections 7 and 9).

6.1.7 Obesity and Fat–Muscle Ratio

Obesity can decrease the rate of elimination of both water- and fat-soluble chemicals. A 50–100 percent increase in elimination half-life has been reported for both classes of medicinals in obese persons (Abernathy et al., 1981). For the obese person and the fat-soluble medicinal, this was due to the increased amount of fat available for distribution and storage of the chemical. For the obese person and the water-soluble medicinal, this was due to an increased volume of distribution. Both of these factors tend to decrease blood concentrations of the exposure chemical and its metabolites and reduce their elimination rate.

If the fat–muscle ratio increases, the rate of elimination of a fat-soluble chemical would be expected to decrease for the reason already mentioned. The fat–muscle ratio usually is higher in older workers than it is in younger workers. The ratio also is usually higher in women (see also Section 6.1.11).

A similar relationship has been predicted for T_3CE using a physiologically based pharmacokinetic model (Sato et al., 1991).

6.1.8 Age of the Worker

Blood level and excretion of medicinals following a given dose have been found to change with age. Since metabolism and excretion of industrial chemicals also utilizes enzymes as well as the same excretion mechanisms as medicinals, similar changes could be expected for industrial chemicals. For example, in disease-free and pathology-free elderly subjects, Triggs et al. (1975) found a 50–100 percent increase in the elimination half-life of two medicinals when compared with disease-free and pathology-free young subjects. Absorption of the medicinal by these groups did not differ significantly.

Glomerular filtration rate has been reported to decrease by about 35 percent in the elderly (Rowe et al., 1976). This could result in delayed appearance of the biological marker in the urine, as well as a reduced slope of the exposure–urinary concentration curve if this curve originally had been determined in young workers.

Older workers also frequently have a higher fat–muscle ratio or weigh more than younger workers of the same height. Both of these factors may cause decreased excretion rate of a biological marker (see Section 6.1.7).

Hepatic blood flow has been reported to decrease as much as 40–45 percent with age (Geokas and Haverback, 1969). This could result in a decrease in the rate of formation of the biological marker, if it is formed in the liver from the exposure chemical.

At the other end of the age spectrum, metabolic variations in biological marker concentrations in neonates are not as important from an industrial hygiene standpoint since neonates are not usually in the industrial workplace. However, they may be present there in newly industrializing countries, and industrial hygienists practicing in such countries should be cognizant of many metabolic differences between a neonate and an adult. These differences may affect the relationship between workplace exposure and biological marker concentration in the bioassay material. In any country a neonate may be exposed to an industrial chemical if the mother is exposed to xenobiotics in the workplace and is nursing the child (see Section 11).

In addition to neonatal exposure, fetal exposure almost certainly occurs in the workplace. The umbilical chord presents no more of a barrier to industrial chemicals than it does to any other organic or inorganic chemical. Unfortunately, no noninvasive method for obtaining biological markers from fetuses is available, and invasive sampling carries great risk to the fetus. Although many studies have been carried out showing the teratogenicity of industrial chemicals to animals (Rawat et al., 1986; Johannsen et al., 1987; and Tyl et al., 1988), none of these investigators tried to develop a dose–response curve for a biological marker.

6.1.9 Disease

In addition to the effect of some diseases on the amount of binding proteins, diseases can affect the ratio of bound chemical to free chemical in other ways. This, in turn, may affect the rate of appearance of the biological marker in the biological assay material.

Any disease that affects metabolism and/or excretion also could be expected to alter the rate of appearance of the biological marker in the bioassay material. Decreased renal clearance, for example, could result in higher blood levels and lower urinary levels of the biological marker than would be expected from studies on subjects with normal renal clearance (Reidenburg, 1974). As already mentioned, binding of xenobiotics to serum protein is decreased with cirrhosis. A biological monitoring program based on urinary levels of the biological marker would suggest lower exposure levels than occurred in this situation.

If the concentrations of the biological marker in urine are normalized by calculating the ratio to creatinine (see section 6.4) and creatinine excretion also is reduced, the error should be self-correcting. If the biological monitoring is based on blood, tears, perspiration, or saliva analyses, decreased renal clearance would result in a higher than normal concentration of the biological marker in these fluids. This would result in higher estimates of workplace exposure than actually occurred.

6.1.10 Concurrent Use of Medicinals

In addition to the previous specific examples of the effects of concurrent administration of medicinals on any biological monitoring procedures, medicinals can cause erroneous results in other ways. For example, the elimination half-life of diazepam was increased in patients taking oral contraceptives with low estrogen content (Abernathy et al., 1982). Cimetidine has been reported to decrease liver blood flow and the elimination of propanolol (Feely et al., 1981). The possible effects of reduced blood flow on biological monitoring was discussed in Section 6.1.8.

These results indicate that the effect of concurrently administered common medicinals should be investigated for any biological monitoring procedure being developed. Such medicinals might include analgesics such as aspirin, acetaminophen, and ibuprofen; tranquilizers; phenytoin; barbiturates; antacids, and oral contraceptives. Alcohol, a common medicinal excipient, also affects the rate of alkane oxidations (see Section 6.1.3).

6.1.11 Sex of the Worker

Chemicals are known to have different lethality in the two sexes. For example, many of the organophosphate insecticides are three to four times more acutely lethal to one sex than to the other (Hayes, 1963). The more sensitive sex varies with the chemical. This difference may be due, for example, to differences in absorption, hormonal controls, metabolic route, or metabolic kinetics. Obviously, different exposure correlation factors would have to be used for each sex when metabolism and metabolic kinetics are different, since the same level of biological marker in each sex of worker could indicate different exposure. Since these same factors would affect the appearance of other toxic signs or symptoms, it should not be surprising to find differences in these latter effects between sexes.

Daniel and Gage (1965) found that, after dosing for up to 50 days, female rats had stored about 50 percent more orally administered butylated hydroxy toluene (BHT) in their fat than males, at the same dietary levels. After single oral doses, females excreted about 50 percent more in the urine than did the males.

Ikeda and Imamura (1973) found the urinary half-life of total trichloro compounds (TTC) from T_3CE or T_4CE exposure to be almost twice as great in female workers as in male workers. They reported similar findings for other biological markers used to assess exposure to these two compounds. Conversely, Nomiyama and Nomiyama (1971) reported a significantly greater ($p < 0.05$) urinary excretion of TCA in females than males in the first 24 hr following inhalation exposure to T_3CE. For the first 12 hr following exposure, males excreted twice as much TCE as females. In accordance with these findings, Nomiyama (1971) found that it was necessary to use a larger factor for females than males when correlating exposure concentration with exhaled air concentration of T_3CE.

Nomiyama and Nomiyama (1969) also found that female subjects absorbed approximately twice as much (41.6%) of a given inhalation dose of benzene as did males (20.3%), as mentioned previously.

Ikeda and Ohtsuji (1969a) found that the normal urinary concentration of hippuric acid in about 30 female Japanese students was approximately twice that of about 36 male Japanese students (specific gravity and creatinine corrected). Conversely, the normal urinary level of TCA was twice as high in males as in females.

Sexual differences in lethality and toxicity of up to 1000 percent have been reported in animals and humans (Calabrese, 1986). A comprehensive review of these differences has been published and the reader is referred to this text for a more complete discussion of sexual differences in the pharmacokinetics and metabolism of organic chemicals (Calabrese, 1985). As would be expected, these sexual differences exist regardless of the end-use of the chemical: industrial, pharmaceutical, agricultural, recreational, or avocational, since metabolic processes are insensitive to man's intended use of any chemical that becomes systemic.

6.1.12 Pregnancy

It has been reported that the concentration of unbound medicinals in blood increased in pregnant women versus nonpregnant controls or the same women after delivery. This was true for weakly acidic medicinals, weakly basic medicinals, and a neutral steroid (Dean et al., 1980). The importance of decreased binding with regard to biological monitoring and worker hazard is discussed in Section 6.1.4. Decreased binding can lead to a misleading evaluation of the exposure, but probably not of the hazard.

For discussion of biological markers and the fetus, see Section 6.1.8.

6.1.13 Smoking

Cigarette smoking has been reported to both stimulate (Jick, 1974; Conney et al., 1977; Jusko, 1979; Hjelm et al., 1988) and have no effect (Desmond et al., 1979) on xenobiotic metabolism. Metabolism of the medicinal chlordiazepoxide has been reported to be both stimulated by smoking and to be unaffected by smoking (Jick, 1974; Desmond et al., 1979). This suggests that some unknown factor(s) also may have been affecting these studies.

Ethylene oxide in cigarette smoke also has been reported to interfere with evaluation of ethylene oxide exposure in the workplace utilizing the ethylene oxide adduct of hemoglobin as the biological marker (Törnqvist et al., 1986b; see Section 9).

Thus, it would be prudent to evaluate the effect of smoking on any biological assay where the biological marker is a metabolite or derivative of the exposure chemical. This also would be prudent when the biological marker is a secondary, chemically unrelated, chemical whose concentration is due to the xenobiotic, a metabolite, or a derivative.

6.2 Special Situations Where Excreted Levels of the Biological Marker Would Not Indicate Current Exposure Levels

Situations also exist where all of the biological marker in the biological monitoring material may not arise from recent exposure or even from workplace exposure. Conversely, the biological marker may not be excreted in the monitoring material at the rate expected, as previously discussed. Obviously, one should ensure that such factors are not operating in any particular assay on an individual.

It should be stressed that, to the extent that the concentration of the biological marker in the biological material being monitored reflects the hazard to the worker, steps should be taken to reduce the hazard, if necessary, regardless of the etiology of the toxicant. If the biological marker did not arise largely from the workplace, however, trying to reduce workplace exposure would not reduce the hazard to the worker (see also Section 6.4).

Competition for the same enzyme between concurrently administered methanol and ethanol is an example that already has been discussed. Leaf and Zatman (1952) found that urinary methanol levels following methanol exposure were up to 100 percent higher in the presence of ethanol than in its absence. Obviously, correlations of urinary methanol and exposure, made in the absence of concurrently administered ethanol, would be invalid in this situation. Use of such inapplicable correlations would cause large overestimates of the workplace exposure and the hazard.

Drug therapy for mobilizing stored chemicals or for initiating excretion of recycling chemicals can lead to misleading levels of a chemical in the biological monitoring materials. For example, the use of penicillamine to treat Wilson's disease (Walshe, 1956) or schizophrenia (Nicholson et al., 1966) or the use of tetraethylthiuram disulfide for alcoholism therapy would lead to elevated copper excretion and it would be totally unrelated to current workplace exposure and the hazard.

Any medicinal that could release a stored, recycling, or protein-bound biological marker or its progenitors could cause such misleading results, regardless of the reason the medicinal was being given. As mentioned previously, the numbers probably would reflect hazard to the worker but not workplace exposure (see also Section 6.7).

In a procedure using urinary measurement of the biological marker, decreased renal clearance, active lactation, excessive sweating, or excessive salivation all could cause decreases in urinary levels that would result in decreased estimates of workplace exposure.

Absorption of chemicals through the skin or by the lungs also can vary between individuals. If a person absorbed significantly more or less of the administered dose than did the subjects used to make the correlations of exposure and excretion, the individual analysis would indicate correspondingly greater or lower exposure than occurred. The hazard estimate probably again would be correct, however.

Individual absorption may vary by as much as 100 percent. In a study of oral absorption of ampicillin (MacLeod et al., 1974), the amount absorbed varied from 32 to 64 percent in nine healthy men aged 20–40 years. The absorption of benzene by inhalation also has been reported to vary by 100 percent between males and females, with females absorbing 41.6 percent of the dose and males 20.3 percent (Nomiyama and Nomiyama, 1969).

Assuming no metabolic differences, absorption differences between individuals thus could lead to a 100 percent difference in the level of biological marker in the biological monitoring material from two individuals with the same exposure.

6.3 Concentration Changes Due to Volume Changes in Biological Assay Liquids

The rate of appearance of the biological marker at the excretory organ generally will be independent of the rate of excretion of the body material being used for biological monitoring. Thus the concentration of the biological marker in the body material will change with the volume of excretion of the latter. This is not a serious problem for hair or nail analyses. For exhaled air analyses, the subject generally is resting when samples are taken and will not be doing anything to greatly increase respiratory volume per unit time. Therefore, this also will not be an important factor in exhaled air analyses. However, the excretion rate of liquids can vary greatly. The rate of perspiration excretion during and closely following exertion versus resting, or the rate on a hot day versus a cool day, obviously is different. Similarly, the rate of excretion of tears in subjects who are crying is obviously different from the rate in subjects who are not crying. Salivation rate also can be altered by emotional, physical, and chemical factors. Since these biological fluids seldom are used for industrial biological monitoring, excretion rate changes of these fluids seldom are important. Excretion rate, however, is a variable that must be considered in any assay utilizing these fluids.

Urine volume normally varies by as much as 300 percent (600–2500 mL/24 hr) between individuals and in the same individual at different times (Wallach, 1974). Factors that can cause volume variations include fluid intake, emotional state, medicines, pregnancy, disease, menstruation, ambient temperature, and body temperature.

Urinary concentrations of the biological marker also may be diluted by urine already in the bladder when the biological marker reaches the kidneys. This will be most important in situations where (1) end-of-shift "spot" urine samples (see Section 7.3.4) are analyzed, (2) the biological marker has a relatively long excretion half-life, and (3) several hours elapse between the penultimate urine sample and the sample taken for analysis. Voiding at a set time prior to taking the urine sample for analysis might reduce the spot urine analysis variability due to this factor. This technique has been utilized by several investigators (Piotrowski, 1971; Ikeda et al., 1972b; Ikeda and Ohtsuji, 1969a,b).

If the content of the biological marker in the total 24-hr volume of urine is being measured, the dilution effect will be minimized. However, if only spot samples of urine are being analyzed, dilution factors may be extremely important.

Recognizing that the same kinetics of presentation at the excretion organ apply to normal constituents of urine as well as to xenobiotics and their metabolites, investigators have used various properties or components of urine to normalize analytical values and thus compensate for concentration changes due to volume differences.

As mentioned previously, all other things being equal, the same ultimate insertion factors that control the concentration of the biological marker in urine generally will control the concentration of normal constituents in the urine. Therefore, if it were possible to measure a colligative property of urine or the concentration of a usual urinary constitutent that has a constant rate of presentation to the urine and rarely arises from a xenobiotic, it should be possible to normalize urinary concentrations and thus correct for volume differences.

Specific gravity and osmolality are colligative properties of urine that are easy to measure. Specific gravity normally varies by only about 10-fold, whereas osmolality varies by over 20-fold. Although the concentration of xenobiotic or its metabolites will make some contribution to specific gravity, the contribution generally will be small compared to the effect due to normal constituents. The xenobiotic or its metabolites probably would make a greater contribution to changes in osmolality. For whatever reasons, specific gravity has been used widely to normalize urine values and osmolality has not. Creatinine has been used most frequently for normalizing analytical results for xenobiotics in urine. Creatinine excretion normally varies less than twofold [1.1–1.6 g/24 hr or 15–25 mg/kg body weight/24 hr (Wallach, 1974)].

Other chemicals normally present in urine could be used instead of creatinine content for normalizing analytical results for the biological markers. Elkins normalized on the basis of sulfate excretion (Elkins, 1954), using the same rationale discussed previously. However, this normalization has not been used extensively.

Ogata et al. (1970) found that the "rate" of excretion of hippuric acid from 23 young adult human males following inhalation exposure to toluene is a more consistent measure than normalizing by the use of specific gravity. To determine rate, they collected urine samples every 3 or 4 hr during a 3- or 7-hr exposure. The total amount of hippuric acid in the sample was then divided by the minutes of exposure. The resulting quotient was the rate in milligrams per minute. The rates for the 0- to 4-hr urine samples or the 4- to 8-hr samples were then plotted against the airborne toluene concentration. The standard deviations were calculated, using both the specific gravity and rate-normalizing methods. At an airborne toluene concentration of 100 ppm, they were not greatly different for either time period as shown in Table 3.1. At toluene concentrations of 200 ppm, the standard deviations in the rates were less than the standard deviations of specific-gravity-corrected concentrations.

As can be seen from Table 3.1, a direct dose–response relationship was found with both methods of normalizing. However, one standard deviation from the hippuric acid concentration value at 100 ppm overlapped one standard deviation at 200 ppm by either method of normalizing. Shorter sampling periods would be more desirable for validating the concept. The approach would not be too practical in an industrial situation since it would require that the worker void urine a specific number of hours prior to voiding the urine sample to be analyzed. Levine and Fahy (1945) found that calculating rates did not result in reduced variability of lead analyses in the urine of exposed workers (Engström et al., 1976).

There has been some controversy as to whether 1.016 or 1.024 more nearly represents the average urinary specific gravity (Van Haaften and Sie, 1965; Rainsford and Davies, 1965; Fishbeck et al., 1975).

Table 3.1 Effect of Two Normalizing Treatments on Urine Analysis for Hippuric Acid Following Toluene Exposure

Exposure sampling period (hr):	0–4		4–8	
Toluene exposure (ppm):	100	200	100	200
Hippuric acid conc. in urine, uncorrected (mg/L)				
MEAN	2.95	3.74	3.09	8.19
SD[a]	0.83	0.59	0.70	2.62
Specific gravity corrected (mg/L)[b]				
MEAN	2.58	3.71	2.81	5.85
SD	0.40	0.67	0.66	1.24
Rate (mg/min)				
MEAN	2.09	3.13	3.10	4.61
SD	0.35	0.36	0.84	0.80

[a]SD = standard deviation.
[b]Corrected to 1.024.
Source: Ogata et al. (1970)

Normalizing will have the same effect on variability regardless of which of the two numbers is used because of the nature of the correction equation, as shown in Eqs. (3) and (4). Levine and Fahy (1945) used Eq. (3) to correct to specific gravity 1.024. Equation (4) is used by the National Institute for Occupational Safety and Health (NIOSH) (U.S. Dept. Health, Educ., Welfare, 1974) to correct to specific gravity 1.024.

$$\text{Corrected value} = (\text{observed value}) \times \frac{24}{(\text{specific gravity} - 1) \times 10^3} \qquad (3)$$

$$= (\text{observed value}) \times \frac{24}{\text{last two digits of specific gravity}} \qquad (4)$$

It can be seen that (3) reduces to Eq. (4) in practice, since (specific gravity − 1) for a specific gravity of, say, 1.021 = 0.021. When this number is multiplied by 10^3, it becomes 21. These are the last two digits of the specific gravity value. Obviously, Eq. (3) or (4) could be used to normalize to any other specific gravity by substituting the last two digits of that specific gravity for 24 in the numerator of Eq. (3) or (4).

The question of which specific gravity to use can be avoided by normalizing the urinary analyses on the basis of creatinine content, for the reasons already discussed. In the field of industrial hygiene, creatinine concentration was initially used by Hill (1953) to normalize analyses of aniline in urine. Its use as a basis for normalizing has increased since then.

Mixed results have been reported from normalizing with either specific gravity or creatinine. As can be seen in the work of Ikeda or Ogata reported in Table 3.2, neither creatinine nor specific gravity corrections consistently gave significantly smaller standard deviations in comparison to each other or to uncorrected values. This also has been the experience of other investigators. Seki et al. (1975) found that specific gravity correction slightly reduced the scatter, whereas creatinine correction did not. Conversely, Pagnotto and Liebermann (1967) found that creatinine correction reduced scatter over that found in uncorrected values. Ellis (1963) found that neither correction significantly reduced scatter over that found in uncorrected samples.

Creatinine clearance also has been reported to decrease with age, especially in subjects over 60 years of age (Rowe et al., 1976). No such change has been reported for specific gravity.

In practice, since both specific gravity and creatinine are usual urinary determinations, many authors now report the uncorrected value as well as values corrected by both methods. The data presently available do not indicate that either method of normalizing analytical results reduces the standard deviation over that found in uncorrected values.

Reporting the uncorrected value of the biological marker as well as the specific gravity and creatinine concentration is recommended. This permits the readers to make their own corrections for comparison with other data or for other purposes.

Table 3.2 Effect of Corrections on Variations in Urine Analysis[a]

Exposure Chemical		Biological Marker Name	Biological Marker Concentration in Urine[b]			Number of Subjects	References
Name	Exposure Concentration (ppm)		Uncorrected	Corrected to Specific Gravity 1.016	Corrected for Creatinine (g/g)		
None	0	Phenylglyoxylic acid	0.017 (59–159)	0.011 (64–155)	0.013 (69–138)	35	Ikeda et al. (1974)
None	0	Mandelic acid	0.057 (61–163)	0.036 (64–158)	0.043 (63–160)	35	Ikeda et al. (1974)
None	0	Hippuric acid	0.35 (57–177)	0.29 (55–175)	0.24 (58–171)	31	Ikeda and Ohtsuji (1969b)
Toluene	20	Hippuric acid	1.84 (66–152)[c]	1.18 (72–139)[c]	1.06 (76–132)[c]	10	Ikeda and Ohtsuji (1969b)
	60	Hippuric acid	2.27 (59–171)[c]	1.21 (58–174)[c]	1.14 (66–151)[c]	10	Ikeda and Ohtsuji (1969b)
None	—	Hippuric acid	0.30 (25–420)	0.29 (21–480)	0.23 (20–512)	36	Ikeda and Ohtsuji (1969a)
None	—	Phenol	0.026 (32–312)	0.023 (31–316)	0.019 (32–318)	36	Ikeda and Ohtsuji (1969a)
T_3CE	3	TTC	0.039 (67–149)	0.035 (75–133)	0.041 (74–135)	9	Ikeda et al. (1972b)
T_3CE	45	TTC	0.339 (72–138)	0.253 (77–130)	0.338 (79–126)	5	Ikeda et al. (1972b)
T_3CE	120	TTC	0.915 (85–118)	0.481 (86–116)	0.519 (68–146)	4	Ikeda et al. (1972b)

[a]Figures in parentheses are standard deviation ranges as a percentage of the geometric mean.
[b]Spot samples (see Section 7.3.4).
[c]Exposure values uncorrected for background level.

6.4   Nonworkplace Progenitors of the Biological Marker and Resulting Baseline Variation in Its Concentration in the Biological Assay Material

If the biological marker also has a background concentration range in the biological assay material being analyzed, any biological monitoring procedure must determine (a) the normal background range, (b) common nonoccupational factors that affect the normal range, and (c) whether or not a small fraction of the maximum safe workplace exposure to the biological marker or its workplace progenitor causes a significant elevation above the normal background concentration of the chemical. If one has a series of analyses on a worker over a period of time, interindividual variations become less important. However, if there is an elevation in one analysis, possible nonworkplace causes still should be investigated.

For example, extremely varied normal levels of various metals have been reported for human scalp hair. Topical contamination of the hair from such causes as cosmetics or hair dyes, which do not represent workplace exposure, may be the cause. Because of this great variation, hair levels that might be assumed to represent industrial exposure must be set quite high, unless one is regularly monitoring individual workers.

Examples of a similar situation with organic compounds include the use of (a) urinary hippuric acid levels for monitoring styrene, toluene, or certain N-alkyl benzene exposures or (b) urinary phenol levels as a measure of benzene or phenol exposure.

As mentioned earlier, styrene, toluene, and certain n-alkylbenzene compounds are metabolized to benzoic acid, which usually is condensed with glycine to form the excretory product: the biological marker hippuric acid. Hippuric acid occurs normally in the urine, generally as a result of ingestion of benzoic acid, sodium benzoate, or quinic acid, a metabolic precursor of benzoic acid. Sodium benzoate is used as a food preservative. Coffee beans, prunes, plums, and cranberries are known to contain benzoic acid or quinic acid.

With widespread distribution of benzoic acid and its progenitors in foodstuffs, it is not surprising that its background level in urine should vary. As shown in Table 3.3, the normal urinary level of hippuric acid fluctuates 3-fold in U.S. subjects, up to 13-fold in Japanese subjects, and almost 40-fold in Finnish subjects.

This variation essentially negates the use of hippuric acid as a biological marker for measuring styrene exposure to airborne concentrations up to four times as high as the TLV for several reasons. If one assumes for humans (a) 10 percent conversion of styrene to urinary hippuric acid (Ikeda et al., 1974), (b) a pulmonary absorption coefficient of 0.8 (Lorimer et al., 1976), (c) a 24-hr urine volume of 1.2 L, and (d) an 8-hr exposure during which $10 m^3$ of air was breathed, an 8-hr exposure to 200 ppm of styrene would be necessary before the average level of urinary hippuric acid reported by Stewart et al. (1968) would be raised to the upper normal level they reported. In fact Stewart et al. found that exposure to 100 ppm of styrene for 7 hr did not cause significantly elevated urinary hippuric acid concentration up to 48 hr postexposure.

In partial agreement with Stewart's findings, Ikeda et al. (1974) found significantly elevated urinary hippuric acid levels in most subjects exposed to styrene for up to

Table 3.3 Normal Urinary Levels and Ranges (mg/liter) of Hippuric Acid as Reported by Various Authors[a]

| | Corrected Value Based on | | | | |
Uncorrected	Specific Gravity	Creatinine	Subjects: No. Sex, Country	Analytical Procedure	References
1583[a] (833–2583)	ND[b]	ND	9, M, U.S.	UV absorption	Stewart et al. (1968)
350[c]	290[c,d]	240[c]	31, M, Japan	PC	Ikeda and Ohtsuji (1969) Ogata et al. (1962)
(199–616) 301[c]	(160–510) 290[c,d]	229[c]	36, M, Japan		
398[c]	570[c,d]	449[c]	30, F, Japan		
NR	1037[e] (126–4844)	739 (86–2340)	39, M, Finland	GC	Engstrom et al. (1976)
800 (400–1400)	ND	ND	NR, M, U.S.	UV absorption	Pagnatto and Lieberman (1967)
NR	184[f] (35–444)	ND	NR, M, Japan	PC	Ogata et al. (1962)
335	ND	ND	6M, U.S.	Titration	Von Oettingen et al. (1942)

[a]Calculated by R. S. Waritz from data in reference. Assumed 1.21 urine volume for 24 hr.
[b]ND = not determined; NR = not reported; M = male; F = female; PC = paper chromatography; GC = gas chromatography.
[c]Geometric mean.
[d]Corrected to specific gravity of 1.016.
[e]Corrected to specific gravity of 1.018.
[f]Corrected to specific gravity of 1.024.

160 min at concentrations of up to 200 ppm but no significant increase at concentrations of ≤ 60 ppm for 2 hr or less. Twenty-four urine samples were collected and specific gravity was corrected to 1.016. Spot urine samples taken by these same investigators from these same workers after the workers had been exposed for approximately 6 hr to styrene levels up to 30 ppm did not show significant elevations of hippuric acid (Ohtsuji and Ikeda, 1970).

The significant urinary elevation of hippuric acid following these latter exposures might be explained by the delayed appearance of hippuric acid in the urine following styrene exposure. Ikeda et al. (1974) found that hippuric acid did not appear in workers until about 24 hr postexposure. This would not explain Stewart's results since he collected total urine samples for over 24 hr postexposure.

Thus, the combination of small percentage conversion and high, variable, natural levels must be assumed to mitigate against the use of urinary hippuric acid as a biological marker for styrene exposure.

Hippuric acid in urine has been used successfully as a marker for toluene exposure. Since toluene is almost 70 percent metabolized to hippuric acid (Ogata et al., 1970), the high background levels of hippuric acid become less important. Any exposure in excess of 50 ppm for 8 hr would be expected to elevate an average background hippuric acid level beyond Stewart's upper normal (Stewart et al., 1968). This assumes 79 percent absorption (Ogata et al., 1970), 70 percent metabolism to hippuric acid, 10 m^3 of air respired, and a 24-hr urine volume of 1.2 L.

In agreement with this, Engström et al. (1976) found very poor correlation between blood levels of free toluene and urinary hippuric acid (creatinine corrected) in painters concurrently exposed for 8 hr to airborne levels of toluene less than 50 ppm and unspecified low concentrations of m- and p-xylene. This is in contrast to the findings of Tardif et al. (1991) who found that concurrent exposure of human males for 7 hr to a mixture of 50 ppm of toluene and 40 ppm of xylene (isomer mixture unspecified) did not alter the blood levels of these compounds from those seen in exposures to each individually. They also found no changes in urinary levels of hippuric acid or methyl hippuric acid (corrected on a creatinine basis) compared to exposures to the individual compounds.

The use of urinary phenol levels as a measure of benzene exposure also could be frustrated at low benzene levels by the presence of phenol in urine from nonbenzene sources. For many of these sources, urinary phenol levels would correspond to much lower worker hazard than would be indicated if the phenol had arisen from benzene.

Phenol as well as metabolic precursors of urinary phenol, such as the amino acid tyrosine and the essential amino acid phenylalanine, occur in the diet. Whereas the normal urinary output usually is considered to be about 10 mg/day (Parke, 1968), the normal range can be much greater than this as shown in Table 3.4. In three groups of Japanese subjects (in Japan) totaling 97 subjects, the geometric means of the three groups ranged from 18.2 to 34.8 mg/L. The 95 percent fiduciary limits varied from 7.3 to 123.8 mg/L.

These values were for urine corrected to specific gravity 1.016. If the values were corrected to specific gravity 1.024 as recommended by NIOSH (U.S. Dept. Health, Education and Welfare, 1974), they would be 50 percent higher. The use of geometric means by Ikeda and Ohtsuji (1969a,b) also suggests a distribution skewed

Table 3.4 Normal Urinary Levels and Ranges (mg/L) of Phenol as Reported by Various Authors

Uncorrected[b]	Corrected[a] Specific Gravity Adjustment	Creatinine Adjustment	Subjects: No., Sex, Country	Analytical Procedure	References
NR[b]	5.5[c]	NR	20, M/F, U.S.	GC	Van Haaften and Sie (1965)
26.1[c]	23.3[c,d]	18.9[c]	36, M, Japan	Colorimetric (Gibbs, 1927)	Ikeda and Ohtsuji (1969a)
(8.3–81.5)[e]	(7.3–73.7)[e]	6.0–60.1)[e]			
25.2[c]	34.8[c,d]	28.5[c]	30, F, Japan	Colorimetric (Gibbs, 1927)	Ikeda and Ohtsuji (1969a)
(8.4–75.5)[e]	(9.8–123.8)[e]	(11.0–71.1)[e]			
22.8[c]	8.2[c,d]	14.5[c]	31, F, Japan	Colorimetric (Gibbs, 1927)	Ikeda and Ohtsuji (1969a)
(8.9–58.3)[e]	(8.0–41.5)[e]	(7.1–29.9)[e]			
(1–80)		NR	52, U. S.	GC	Roush and Ott (1977)
(4–5.5)	(5–11.0)[f]	NR	1, U.S.	GC	Fishbeck et al. (1975)
NR	(12–144)[d]		109, Gr. Brit.	Colorimetric (Gibbs, 1927)	Rainsford and Davies (1965)

[a]Only values obtained using the Gibbs colorimetric method (Gibbs, 1927) or a GC method are reported. Since the other common colorimetric method, Thies and Benedict (Theis and Benedict, 1924) is known to also give positive results with p-cresol (Van Haaften and Sie, 1965). The p-cresol content in control urine may be 10 times greater than the phenol content (Parke, 1968; Van Haaften and Sie, 1965). Both Gibbs and the Theis and Benedict methods give reactions with o- and m-cresol, but these are usually present in urine in much lower concentrations than p-cresol.

[b]Abbreviations: NR = not reported; M = male; F = female; GC = gas chromatography.

[c]Corrected to specific gravity of 1.016.

[d]Geometric mean.

[e]95 percent fiducial limit.

[f]Corrected to specific gravity of 1.024.

110

toward the high values. Roush and Ott (1977) also found preexposure urinary phenol values of up to 80 mg/L (corrected to specific gravity 1.024). Fishbeck et al. (1975) reported that the urinary phenol concentration of one person could vary by 100 percent (from 5.0 to 11.0 mg/L) in a 6-week period. Because of such variations, Roush and Ott (1977) have suggested that urinary phenol measurements do not reliably indicate benzene exposure if the exposure is less than 8 hr at 5 ppm. Van Haaften and Sie's data (1965) support this contention. Thus, urinary phenol measurements would not be suitable for routine monitoring of employee exposure to benzene under the current OSHA standard (U.S. Dept. Labor, 1991c) or the proposed ACGIH TLV (0.1 ppm) for atmospheric benzene levels in the workplace (Amer. Conf. Governmental Ind. Hygienists, 1991).

Another factor that also mitigates against the unqualified and sole use of some arbitrary urinary phenol level as an indicator of excessive benzene or phenol exposure is the presence of urinary phenol progenitors in many medicinals. Phenylsalicylate, phenol, and sodium phenate are active ingredients in common over-the-counter medicinals and prescription medicinals. Phenylsalicylate is used as an enteric coating in other common medicinals. Fishbeck et al. (1975) for example, found that Pepto-Bismol® (a registered trademark of Procter & Gamble Co., Cincinnati, OH) or Chloraseptic® (a registered trademark of Procter & Gamble Co., Cincinnati, OH) cold lozenges (both of which contain phenylsalicylate) when taken as directed, led to peak urinary phenol levels of 480 and 498 mg/L, respectively (corrected to specific gravity of 1.024). Common medicinals containing phenol include "P&S" liquid or ointment, a shampoo suggested for use in cases of psoriasis and seborrheic dermatitis; Oraderm® (a registered trademark of R. Schattner Co., Washington, D.C.), a lip balm; and Campho-Phenique® (a registered trademark of Sterling Drug, Inc., New York), a formulation suggested for cuts, burns, cold sores, and fever blisters.

Thus, before concluding that particular urinary phenol levels indicate particular benzene or phenol exposures, one should determine that the subject is not taking any over-the-counter or prescription pharmaceuticals that contain metabolic precursors of urinary phenol. It also would be desirable to have background levels of urinary phenol for the subject in addition to determining that there have been no changes in dietary habits that could increase urinary phenol levels.

In some cases the confounding effect of this normal, variable background "noise" on the biological marker concentration can be circumvented by utilizing for biological analysis another metabolite that does not arise from nonworkplace sources. In the case of benzene, muconic acid has been shown to be a characteristic biological marker for humans at atmospheric levels around 5 ppm (see Section 6.1.5 and Bechtold et al., 1991).

6.5 Defining Normal Range of Biological Marker Expected in the Biological Assay Material

In Section 6.4 the wide concentration range of biological marker that can occur in the biological assay material if the marker also can arise in the assay material from nonwork sources was discussed. All the factors discussed in the preceding five subsections suggest that when the biological marker is a metabolite of the exposure chemical, a broad range of concentrations of the biological marker in the biological assay material is to be expected, even if the marker does not occur naturally. This range also would be expected from medical experience with blood and urine analyses. If measurements are repeatedly made on the same group of workers over a period of time, the ranges should be narrower. Then an excursion could be seen more readily and quantitated.

It has been reported that the amount of a nonspecific metabolite excreted in urine decreased after 1 month on the job even though worker exposure did not decrease. By 4 months, the concentration had increased to preexposure levels, and by 5 months the amount excreted exceeded preemployment values. Presumably the kinetics of excretion also varied. Measurements were made on 177 workers, so the results cannot be explained by the influence of one worker (Kilpikari and Savolainen, 1982). This finding cannot be explained on the basis of adaptive enzyme formation. It presents another possible variable that should be examined in developing any biological monitoring procedure.

In breath analyses the biological marker usually is the exposure chemical not a metabolite. Therefore, the concentration range of exhaled chemical for a particular exposure concentration would be expected to be narrower. Nevertheless, even here a range is to be expected.

6.6 Time Required for the Biological Marker to Appear in the Biological Assay Material

The excretion half-life of chemicals varies from a few hours to days (see Section 7.3.4). This means that some xenobiotics will appear in biological assay materials in a few minutes, and others will not appear in significant concentrations for days following exposure. Obviously, in the latter situation, it would be useless to monitor at the end of the shift to measure that day's exposure to the chemical. There also may be differences in rate of appearance of biological markers in the various possible biological assay materials. For example, many inorganic cations will appear reasonably promptly in urine, tears, perspiration, or saliva. They will not appear in the external part of the hair shaft or nails for weeks because of the slow generation of these assay materials.

Hippuric acid from styrene exposure, for example, does not appear in urine until about 24 hr after exposure (Tanaka and Ikeda, 1968). Similarly, TCA does not appear in significant amounts in the urine of exposed subjects until 24–36 hr after exposure to T_3CE (Ogata et al., 1971).

6.7 Analytical Methodology

Another factor that must be considered in biological monitoring is the analytical procedure used. The analytical procedure potentially introduces two principal sources of variation: (1) conjugated versus free biological marker and (2) the analytical procedure itself.

As shown in Figure 3.1, many biological markers are excreted, in urine at least, in both the conjugated and the free form. Also, as discussed in Section 6.1.4, chemicals may be transported in the blood in either the free or protein-bound form. Obviously, analyses of only the free biological marker should not be compared with the analyses of the total chemical, that is, free plus conjugated or bound. The data of Slob (1973) shown in Table 3.5 illustrate this very well. The three subjects were exposed to various unspecified concentrations of styrene in the workplace. The exposure increased from subject 1 to subject 3. The analyses were for mandelic acid. As can be seen, the apparent urinary concentration of mandelic acid almost doubled following hydrolysis of the mandelic acid conjugate that also was excreted.

The measuring methodology also can be an important source of variation. In general, two types of instrument methodology have been used: visible or ultraviolet (UV) spectrophotometry and other instruments.

Many chemicals present in biological assay materials may have visible or UV absorption spectra similar to the biological marker, thus potentially introducing large and uncontrolled error. Light absorption values of colored derivatives measured at a particular wavelength also can be misleading, since most chromophoric reagents are nonspecific and may react with other chemicals having structures similar to the biological marker. For example, the Theis and Benedict reagent for phenol (Theis and Benedict, 1924) also detects *p*-cresol. The concentration of *p*-cresol normally present in the urine from unexposed subjects may be 10 times the concentration of phenol (Van Haaften and Sie, 1965). The analytical procedure of Ikeda and Ohtsuji (1972) for urinary TTC also detects dichloro compounds.

This lack of specificity can be overcome to some extent by removing interfering materials or extracting the biological marker. However, the extraction procedure may not be complete. For example, Pagnotto and Lieberman (1967) and Walkley et al. (1961) attempted to purify their urinary phenol by steam distillation, not knowing what interfering materials they were trying to remove. Unfortunately, *p*-cresol, the major analytical interference, should steam distill almost as well as phenol (Stull, 1947; Glasstone, 1949). Therefore, their purification procedure did not remove the principal impurity that gives false-positive results with their colorimetric procedure. For this reason, the data this group obtained with this method has qualitative value but is of uncertain quantitative value. Either removal of interfering materials or isolation of the biological marker frequently leads to losses of the biological marker, which creates further error if the losses are not regular. Even if the losses are regular, this error will cause discrepancies in results between methods that require cleanup

Table 3.5 Effect of Conjugate Hydrolysis on Apparent Urinary Mandelic Acid Concentration in Humans

	Urinary Mandelic Acid Concentration (mg/L)		
Subject	Without Hydrolysis	With Hydrolysis	Increase Following Hydrolysis (%)
1	12	16	33
2	61	118	93
3	349	598	71

Source: Slob (1973)

and other methods that do not. In addition, the removal of interfering substances may be so time-consuming that it is impractical for routine analysis of biological materials from workers.

Gas chromatography, size exclusion chromatography, and high-performance liquid chromatography (HPLC) are three instrumental techniques that usually are readily amenable to separation of the biological marker without loss and give results that are characteristic and unique for the biological marker.

Additional analytical techniques that have been used in the research laboratory include postlabeling, enzyme-linked immunosorbent assays (ELISA), radioisotope dilution, X-ray fluorescence, and radioimmune assay and other antigen–antibody reactions. These techniques can be very specific. Not all of these techniques, however, are commonly available in the clinical or industrial hygiene laboratory.

6.8 Route of Exposure

As mentioned previously, one of the goals of biological monitoring is to measure total exposure to a chemical regardless of the route by which the chemical is received. Achievement of this goal must be shown to be true for each method developed. In at least two reported cases, this goal may not have been reached.

Dutkiewicz and Tyras (1967) found that about 5 percent of ethylbenzene absorbed through the skin is converted to urinary mandelic acid, whereas Bardodej and Bardodejova (1961) found 60 percent of inhaled ethylbenzene excreted as urinary mandelic acid. Dutkiewicz and Tyras concluded that urinary monitoring of mandelic acid as a measure of total ethylbenzene exposure would not be acceptable.

Yant and Schrenk (1937) also reported different relative methanol levels in several organs and tissues depending on whether the methanol was administered by inhalation or by subcutaneous administration. This suggests that blood levels of unbound methanol vary depending on the administration route. If that is the case, excretion levels would be expected to vary with the two routes of administration.

Sabourin et al. (1989) found that the ratios of multiple metabolites to each other varied with the same total dose delivered orally compared to delivery by inhalation.

Urinary levels of mandelic acid from styrene, in contrast, have been reported to reflect total exposure by both skin absorption and inhalation in humans. The kinetics of the appearance of mandelic acid in the urine do not seem to be affected by the route of exposure (Bowman et al., 1990).

7 URINE ANALYSIS TO MEASURE INDUSTRIAL EXPOSURE

7.1 Correlation of Exposure with Biological Marker Concentrations in Urine

Urine analysis to measure exposure usually relies on analysis for a metabolite. As already discussed, a great number of factors affect the concentration of the unbound exposure chemical and its metabolites in the blood and thus the amount excreted per unit of urine per unit time. Thus, it should not be surprising that there are great variations reported between authors for the translation of urinary levels of the biological marker to exposure levels of the workplace chemical.

Despite the large uncertainty factor in translating urinary levels to body burden or exposure, there is no question but that for many (and probably all) chemicals a relationship does exist. There also is no question, as stated in Section 1, that we do not yet know all the factors that contribute to variability and thus do not know how to correct for all of them.

At the present stage of development, urine analyses for most organic compounds probably are best suited to (1) preemployment monitoring to detect obvious metabolic variations that could increase individual hazard and (2) regular monitoring of an employee to develop a continuing baseline against which an overexposure would be obvious and could lead to remedial measures in the workplace. At the present stage of development for most industrial organic chemicals, a single urine analysis could be correlated only within a broad range of body burden or prior exposure.

The problem is compounded when the biological marker can arise from dietary or other nonoccupational sources, as has been discussed previously for the use of urinary phenol to evaluate phenol or benzene exposure and for urinary hippuric acid to evaluate styrene or toluene exposure. In the latter three instances, natural levels are so high and so variable that they invalidate the technique for exposure levels of interest for routine worker monitoring. For example, as previously discussed, urinary levels of phenol from subjects without any workplace exposure to benzene or phenol can exceed 80 mg/L (Roush and Ott, 1977). Roush and Ott, and also Van Haaften and Sie (1965), have published data suggesting that urinary phenol levels are not statistically valid measures of individual benzene exposure at levels of 5 ppm or below because of normal variability. For these same reasons, Ikeda and Ohtsuji (1969a) suggested that urinary hippuric acid was not statistically valid for estimating toluene exposures below 130 ppm. As already mentioned, the data of Ogata et al. (1970) suggest that it may not be valid below 200 ppm. Several publications suggest that urinary hippuric acid is not a usable index of styrene concentration at atmospheric levels of workplace interest because of this variability and high background level (Ohtsuji and Ikeda, 1970; Ikeda et al., 1974).

Except for the situations mentioned earlier, where urinary phenol is elevated because of medication, it probably is a good index of workplace phenol exposure. The work of Ohtsuji and Ikeda (1972) indicates that an 8-hr exposure to 5 ppm in the air would be expected to result in about 400 mg of phenol/L of urine in a spot sample of urine taken at the end of the workday.

In addition to the classical biological markers just discussed, adducts of xenobiotics with macromolecules and excision products from these adducts also may be excreted in urine, where they may be quantitated and correlated with workplace exposure. From a workplace monitoring standpoint, this would be beneficial in that it would yield another biological marker that could be obtained by a noninvasive procedure. Lewalter and Korallus (1985) reported the presence of an excision product from a xenobiotic–hemoglobin adduct in urine. Unfortunately, the derivative in the urine was an acetylated derivative and, only slow acetylators (see Section 6.1.1) carried out the reaction. Thus, even with macromolecular biological markers, before exposure can be correlated with urinary concentration of acetylated excision products, it will be necessary to determine whether the worker is a slow or fast acetylator (see Section 6.1.1).

Albumin adducts should appear in the urine within a day after exposure and may continue to appear up to a week after a single exposure. Thus, they offer the potential for measuring an integrated exposure over a workweek. However, urinary levels would not drop to zero during 2 days not at work, and there would be carry-over in the urine into the next workweek (Henderson et al., 1989). Presumably this could be handled mathematically once the kinetics were determined.

The 4-aminobiphenyl–hemoglobin adduct also has been reported to be cleaved in humans and a characteristic excision product excreted in urine (Skipper et al., 1986).

Deoxyribonucleic acid (DNA) repair leading to excision and urinary excretion of alkylated purine and pyrimidine bases occurs much more rapidly than excision of hemoglobin adducts (see Section 9) and offers potential for workplace monitoring of exposure to alkylating agents (Henderson et al., 1989). Urinary measurements of these modified bases also offer the potential of comparing the DNA repair efficiency of individual workers. Alkylated purine and pyrimidine bases are examples of biological markers that might not be looked for, if the metabolic handling of an alkylating workplace chemical were not known.

3-Methyladenine has been reported in the urine of humans exposed to methylating agents (Shuker et al., 1987). 8,9-dihydro-(7′-guanyl)-9-hydroxyaflatoxin B_1 has been detected in the urine of humans consuming foodstuffs contaminated with aflatoxin B_1 (Autrup et al., 1985). There appeared to be rough correlation between the guanine derivative concentration in the subjects' urine and their intake of aflatoxin B_1, but it was not possible to quantitate the intake.

Many other adducts of putative carcinogens or their metabolites with the DNA bases in animals have been reported, but no attempts were made to locate the base–xenobiotic adduct in the animal's urine (Singer, 1976; Westra et al., 1976; Ashurst and Cohen, 1981).

Urinary levels of 2-thiothiazolidine-4-carboxylic acid have been shown to correlate with carbon disulfide exposure in humans (Campbell et al., 1985), but no correlation with neurotoxicity has been made.

Urinary levels of hydroxypyroline and hydroxylysine consequent to increased destruction of lung collagen by pulmonary toxicants have been investigated as early biological markers for processes that could lead to fibrous tissue formation. Hydroxyproline release, however, may also be indicative of collagen synthesis as well as collagen degradation and thus may not be suitable as a biological marker for this process (Evans et al., 1989).

Evans et al. (1989) studied the urinary concentrations of hydroxylysine after single exposures and multiple exposures of rats to nitrogen dioxide (NO_2). The 1992 OSHA PEL for NO_2 is (C)5 ppm (U.S. Dept. Labor, 1991a). They found that two 6-hr exposures to 1.3 or 7.5 ppm NO_2 did not increase 24-hr urinary levels of hydroxylysine significantly over control values ($p > 0.05$). Significance ($p < 0.05$) was not achieved until the rats were exposed to 15 ppm. Urinary hydroxyproline levels were not investigated in these experiments.

When rats were exposed 6 hr per day, 5 days per week for 4 weeks, to 1 ppm of NO_2, a urinary increase in hydroxylysine varying directly with days of exposure was seen, but its absolute concentration did not become significant ($p < 0.05$) versus the controls until after the fourth week of exposure. Four weeks after the last exposure, values had dropped to normal. Urine samples (24-hr) were collected "periodically." In another experiment rats were exposed to 0.5 ppm NO_2. All other exposure parameters were the same as before. Urine samples (24-hr) were collected "at frequent intervals." In this experiment a significant ($p < 0.05$) increase in urinary hydroxylysine concentration was seen after three exposures. The concentration reached a plateau in the second week and remained there throughout the exposure, returning to normal 10 days after the last exposure.

Urinary concentrations of hydroxyproline appear to have promise as a biological marker for collagen destruction, a possible early marker for fibrous tissue formation. However, urine sampling schedule appears to be critical (see Section 7.3.4 for further discussion of the importance of sampling schedule). In addition, background levels can vary directly with fish and meat intake (Evans et al., 1989). Thus, diet can confound the measurement and again illustrates the importance of diet in the interpretation of biological marker analytical results (see also Section 6.1.2).

7.2 Biologic Threshold Limit Values

Instead of using excretion levels of a biological marker to quantitate the level of exposure, it has been proposed that excretion levels of a biological marker be used simply to indicate the presence or absence of overexposure. The term *biologic threshold limit value* was proposed by Elkins (1967). The concept is not restricted to the workplace chemical, its classical metabolites, or to urine analyses but would apply to any biological marker. It would apply to blood, to exhaled air, to any medium of excretion, and to any liquid, solid, or gaseous biological excretion. It would apply to the xenobiotic, its classical metabolites, its adducts to macromolecules, the excision-repair fragments from macromolecular adducts, and to chemically unrelated biological markers whose biosynthesis was related to the presence of the xenobiotic, as long as there was a dose–response relationship.

Nothing presently known about metabolism and excretion and their kinetics invalidates the concept. In general, the normal variation in excretion of chemicals, particularly metabolites, would have to be considered in setting such discriminator levels. The application of this concept is further complicated where the biological marker occurs naturally in urine and thus has a high and variable background level. In this situation two courses are possible: (1) the discriminator number could either be set high enough to eliminate all "normal" values or (2) it could be set at a lower fiducial or otherwise derived limit that could include many people that had not been overexposed to the workplace chemical.

If the first alternative is used, the discrimination level might never be reached for some overexposed individuals. The second alternative would require some alternate confirmation that the elevated levels really were due to workplace exposure. Urinary hippuric acid as a biological marker of toluene or styrene exposure already has been discussed in this regard. The U.S. Department of Labor (U.S. Dept. of Labor, 1978) proposed that urinary phenol levels of > 75 mg/L, adjusted to specific gravity of 1.024, be considered indication of overexposure to benzene, leading to certain subsequent blood measurements. Yet, as discussed in Section 6.4, Roush and Ott (1977) have shown that urinary phenol levels in unexposed individuals may exceed this value. Fishbeck et al. (1975) also have shown that the recommended dosage of a common over-the-counter medicinal can lead to urinary phenol levels over five times this great. Even prescribed use of a common lozenge recommended for sore throats can cause a five- to sixfold elevation of urinary phenol. As discussed earlier, other common medicinal uses of metabolic precursors of urinary phenol also could cause levels of urinary phenol that could be interpreted as industrial overexposure to benzene.

In terms of worker hazard, the second course would be the safest, but it easily could be abused in situations where the biological marker has nonworkplace progenitors. However, as mentioned before, if urinary levels of the biological marker truly reflect systemic hazard to the worker, the source of the chemical is irrelevant. Every effort should be made to find the cause of the elevated concentration and reduce the exposure. However, if the source is not the workplace, attempts to reduce workplace exposure will not result in reduced hazard to the worker.

For reasons discussed in the following paragraphs, the discriminator number for organic compounds in urine also must be based on a urine sample collected a certain number of hours postexposure or erroneously high or low values relevant to the discriminator number conditions will be obtained.

Some of the workplace exposure chemicals that have shown a urinary dose–response relationship are listed in Table 3.6. Also shown are the biological markers, the exposure–excretion (dose–response) equations, the type of analysis, and the type of urine sample. In some cases the investigators did not calculate the dose–response equation, and this was done by the present author. These instances have been noted. If the equation was based on analyses corrected by any of the methods discussed in Section 6.3, this also is noted. Expected metabolite concentrations based on the excretion equations also are shown for a particular atmospheric concentration of the workplace chemical.

Table 3.6 illustrates many of the problems of estimating total workplace exposure from the analysis of a biological marker in excretions. These are discussed in the following sections, using the data in Table 3.6 for illustration.

Where more than one metabolite is excreted, as in the cases of T_3CE and styrene, it has been suggested that the total concentration of all of them may correlate better with exposure than the concentration of any particular one (Elia et al., 1980).

Also, depending on the shape of the excretion curve, the equation giving the best fit may require use of the log of x and the log of y; for example, $\log y = a \log x + b$ (Elia et al., 1980).

7.3 Factors to be Considered in Using the Methodology

7.3.1 Slope of the Dose–Response Curve

Methylchloroform metabolite concentrations in urine follow a normal dose–response relationship, at least to 50 ppm in air, the upper airborne exposure level studied. However, the standard deviation is approximately 40 percent of the exposure value, and two standard deviations would be 80 percent. When this is coupled with the shallow slope of the curve (0.073–0.28, depending on the metabolite measured), it is doubtful that two standard deviations below the measured urinary metabolite level resulting from exposure to 350 ppm (1992 TLV) would be greater than two standard deviations above the zero exposure level of metabolites. The problem of the importance of the slope of the dose–response curve is treated in greater detail by Imamura and Ikeda (1973). In general, because of the variability of metabolism and excretion, the dose–response curve should have a fairly steep slope. This is particularly important when the biological marker has nonworkplace progenitors.

7.3.2 Nonlinear Response

The equations found by Seki et al. (1975) that relate urinary metabolite levels to inhalation exposure of methyl chloroform are straight lines, at least up to atmospheric concentrations of 50 ppm for 8 hr. However, they did not extend their observations to 350 ppm exposures, the present TLV of methyl chloroform.

In some cases, as already discussed, the urinary metabolite concentration plateaus (Ikeda et al., 1974) or may even decrease (Götell et al., 1972) with increasing concentration. Thus one cannot extrapolate from lower to higher concentrations but must actually carry out measurements at the exposure levels of interest.

Similarly, you cannot extrapolate to lower doses from higher doses, since the relative importance of a metabolic pathway may change (Sabourin et al., 1989; Bechtold et al., 1991).

7.3.3 Different Analytical Responses for Conjugated and Unconjugated Forms of the Biological Marker

The importance of hydrolysis of conjugates prior to analysis already has been discussed in Section 6.8. As shown in Table 3.6, analytical procedures should include a procedure for conjugate hydrolysis.

7.3.4 Spot Urine Samples versus 24-hr Sample

In their analyses, Seki et al. (1975) used pseudo spot samples of urine. These are samples taken at a particular time relative to exposure, in contrast to 24-hr samples in which all urine voided over a 24-hr period or longer is collected. Elkins (1965) has raised the question of whether spot samples or 24-hr samples are better. The present author believes that this is not a matter of real concern; rather, it is more important that:

1. The sample be taken after there has been sufficient time for adequate metabolism and excretion to minimize variation due to analysis of small amounts and to reach a more or less steady state of excretion versus intake. This will vary with the urinary elimination half-life, which may vary from less than 4 hr for phenol to almost a week for T_4CE. Representative elimination half-life values for some industrial chemicals are shown in Table 3.7.

2. The sample always be taken at the same time relative to exposure. This applies not only to the original investigator but to others who want either to compare results or to use the original investigator's results. This is necessary because the kinetics of

Table 3.6 Analytical Details for Analyses of Urine from Males Exposed to

Exposure Chemical	Biological Marker	Equation	Correction	Analytical Method Hydrolyzes Conjugates
Methyl chloroform (1,1,1-trichloroethane)	TTC	$Y = 0.27X + 0.54$	None	Yes
		$= 0.18X + 0.29$	Specific gravity $= 1.016$	
		$= 0.28X + 0.16$	mg/g creatinine	
	TCE	$= 0.19X + 0.27$	None	
	TCA	$= 0.073X + 0.21$	None	
T_3CE	TTC	$Y = 8.37X + 17.12$	Specific gravity $= 1.024$	Yes
	TCE	$= 5.19X + 12.28$	None	
	TCA	$= 3.17X + 4.84$	None	
T_3CE	TTC	$Y = 7.25X + 5.5$	None	Yes
		$= 4.97X + 1.9$	Specific gravity $= 1.016$	
		$= 5.50X + 6.2$	mg/g creatinine	
	TCE	$= 5.57X + 4.4$	None	
	TCA	$= 2.74X + 0.7$	None	
Benzene	Phenol	$Y = 19X + 30^b$	Specific gravity $= 1.016$	Yes
Phenol	Phenol	$Y = 108X + 28^b$	None	Yes
		$= 80X + 40^b$	Specific gravity $= 1.016$	
		$= 70X + 45^b$	mg/g creatinine	
Toluene	Hippuric acid	$Y = 23X + 800^b$	None	—
		$= 16X + 600^b$	Specific gravity $= 1.016$	
		$= 18X + 350^b$	mg/g creatinine	
Toluene	Hippuric acid	$Y = 40X + 550^b$	Specific gravity $= 1.024$	—
		$= 24X + 500^b$	mg/g creatinine	
Toluene	Hippuric acid	$Y = 9.8X + 400^b$	None	—
Toluene	Hippuric acid	$Y = 31X + 400^b$	None	—
		$= 25X + 300^b$	Specific gravity $= 1.024$	
m-Xylene	m-Methylhippuric acid	$Y = 27X^b$	None	—
		$= 26X^b$	Specific gravity $= 1.024$	
p-Xylene	p-Methyl-hippuric acid	Not calculable	None	—
		Not calculable	Specific gravity $= 1.024$	
Styrene	Mandelic acid	$Y = 18.4X + 149^d$	Specific gravity $= 1.024$	No
	Mandelic acid	$= -14.0X + 6125^e$	Specific gravity $= 1.024$	
	Phenylglyoxylic acid	$= 2.7X + 79^d$	Specific gravity $= 1.024$	
	Phenylglyoxylic acid	$= -1.0X + 540^e$	Specific gravity $= 1.024$	

[a]Abbreviations: U = unknown; Y = concentration of biological marker in urine in mg/L of urine or mg/g of creatinine; X = concentration of exposure chemical in ppm; NA = not available.
[b]Calculated by RSW from investigator's data.
[c]Calculated by investigators.
[d]Exposures $\leq$ 150 ppm.

Various Chemicals[a]

Maximum Atmospheric Concentration	Calculated Urinary Concentration (mg/L) at Given Exposure ppm	Standard Deviation (ppm) at Given Exposure ppm	Urine Sample Type	Analytical Method	References
50 ppm × 8 hr × 3 days	14 @ 50 9 @ 50 14 @ 50 10 @ 50 4 @ 50	~20 @ 50[b] ~15 @ 50[b] ~15 @ 50[b] ~20 @ 50[b] ~25 @ 50[b]	Spot; after 3 days of exposure	Colorimetric (Tanaka and Ikeda, 1968)	Seki et al. (1975)
40 ppm × 8 hr × 2.5 days	854 @ 100[c] 531 @ 100[c] 322 @ 100[c]	NA	Spot; after 2.5 days of exposure	Colorimetric (Tanaka and Ikeda, 1968)	Ogata et al. (1971)
175 ppm × 8 hr × 3 days	730 @ 100 500 @ 100 550 @ 100 575 @ 100 210 @ 100	~40 @ 100[b] ~20 @ 100[b] ~40 @ 100[b] ~40 @ 100[b] —	Spot; after 3 days of exposure	Colorimetric (Tanaka and Ikeda, 1968)	Ikeda et al. (1972b)
150 ppm × 8 hr	120 @ 50[b]	NA	Spot; at end of workday	Colorimetric (Gibbs, 1927)	Rainsford and Davies (1965)
3.5 ppm × 8 hr	568 @ 5[b] 440 @ 5[b] 395 @ 5[b]	~1 @ 3.5[b] ~1 @ 3.5[b] ~1 @ 3.5[b]	Spot; at end of workday	Colorimetric (Ikeda, 1964)	Ohtsuji and Ikeda (1972)
240 ppm × 6 hr	3100 @ 100[b] 2200 @ 100[b] 2150 @ 100[b]	~45 @ 100[b] ~45 @ 100[b] ~45 @ 100[b]	Spot; near end of workday	Colorimetric (Ikeda and Ohtsuji, 1969a)	Ikeda and Ohtsuji (1969b)
170 ppm × 8 hr	4550 @ 100[b] 2900 @ 100[b]	NA NA	Spot, at end of workday	Spectrophotometric[f]	Pagnotto and Lieberman (1967)[g]
600 ppm × 8 hr 800 ppm × 6 hr	1380 @ 100[b]	NA NA	24-hr and end of day	Titration with 0.5N NaOH	Von Oettingen et al. (1942)
200 ppm × 7 hr	3100 @ 100[b] 2500 @ 100[b]	~10 @ 100[b] ~15 @ 100[b]	Spot; at end of exposure	Colorimetric (Ogata et al., 1969)	Ogata et al. (1970)
200 ppm × 7 hr	2700 @ 100[b] 2600 @ 100[b]	~55 @ 100[b] ~30 @ 100[b]	Spot; at end of exposure	Colorimetric (Ogata et al., 1969)	Ogata et al. (1970)
100 ppm × 7 hr	1420 @ 100[b] 3090 @ 100[b]	NA NA	Spot; at end of exposure	Colorimetric (Ogata et al., 1969)	Ogata et al. (1970)
290 ppm × 8 hr with occasional excursions of 1500 ppm × 10 min	1000 @ 50[c] 2000 @ 100[c] NA NA	NA	Spot; at end of workday	Colorimetric (Ohtsuji and Ikeda, 1970)	Götell et al. (1972)

[c]Exposure > 150 ppm.
[f]Toluic acid interferes.
[g]Includes female workers, but results for females were not separated by investigators.
[h]Observed values.

Table 3.7 Human Urinary Excretion Half-Life Values for Some Workplace Chemicals

Exposure Chemical	Biological Marker	Excretion Half-Life (hr)	Urine Samples	References
Styrene	Phenylglyoxylic acid	8.5	24 hr	Ikeda et al. (1974)
	Mandelic acid	7.8	24 hr	
Styrene	Phenylglyoxylic acid	10	Not given	Wolffe et al. (1978)
	Mandelic acid	10		
Toluene	Hippuric acid	~4 (α phase)[a]	24 hr	Ogata et al. (1970)
	Hippuric acid	~12 (β phase)[a]	24 hr	
Toluene	Hippuric acid	6.3[b]	24 hr	Ogata et al. (1970)
T$_3$CE	TTC	41	Spot	Ikeda and Imamura (1973)
T$_3$CE	Trichloroethanol	~7 (α phase)[a]	Spot	Kilpikari and Savolainen (1982)
		~76 (β phase)[a]		
Phenol	Phenol	3.4[b]	Every 2 hr	Piotrowski (1971)
Xylene	Methyl hippuric acid	3.8[b]	24 hr	Ogata et al. (1970)
Xylene	Methyl hippuric acid	~1 (α phase)	24 hr	Riihimäki et al. (1979);
		~20 (β phase)	24 hr	Engström et al. (1977)
T$_4$CE	TTC	144[b]	24 hr	Ikeda and Imamura (1973)
1,1,1-Trichloroethane	TTC	8.7	Spot	Seki et al. (1975)

[a]Estimated by R. S. Waritz from graphs in publication; α phase is the initial, rapid decay; β phase is the later, slower decay.
[b]Calculated by Ikeda and Imamura (Ikeda and Imamura, 1973) from data in publication.

absorption, distribution, and excretion may give different dose–response equations for metabolite levels at different times after exposure or after a single exposure versus multiple exposures. This is particularly true if the chemical has a long elimination half-life.

7.3.5 Equivalency of Analytical Methods

Seki et al. (1975) used a colorimetric method of analysis. The problems associated with various methods of analysis already have been discussed. If comparisons are to be made between investigators, all should either use the same analytical procedure or the equivalency of the different procedures should be demonstrated. If a discriminator number (biological TLV) to indicate overexposure is to be used, the same requirements should be met.

When the same analytical procedure and comparable urine sampling strategy are used, comparable results are possible as shown by the equations developed by Ogata et al. (1971) and by Ikeda et al. (1972b) for urinary TTC, TCE, and TCA following T_3CE exposure. Accordingly, the urinary concentrations of TTC, TCE, and TCA reported by both groups following T_3CE exposure to 100 ppm are very close (Table 3.6).

Despite the similarity of the equations developed by Ogata et al. and Ikeda et al., when the urinary TCA and TCE data reported by Stewart et al. (1974) are substituted in these equations, the exposures reported by Stewart et al. are underestimated by approximately one-third on the basis of TCE content and by approximately one-half on the basis of TCA content. Stewart et al. analyzed 24-hr urine samples after three 7.5-hr exposures to 100 ppm of T_3CE, in contrast to Ogata et al. and Ikeda et al., who analyzed spot urine samples after two $\frac{1}{2}$-hr or three 8-hr exposures to maximum concentrations of 40 or 175 ppm, respectively, of T_3CE. This probably is not a major procedural difference. The major difference appears to be in analytical procedures. Ogata et al. and Ikeda et al. used the same colorimetric procedures for TTC, TCE, and TCA. Stewart et al. used a gas chromatographic (GC) procedure. The data suggest that the slope of the standard curve for the colorimetric procedure is greater than the slope of the standard curve for the GC procedure.

The deviations to be expected with different analytical procedures also are illustrated in Table 3.6 by the equations developed by various investigators for urinary hippuric acid levels following toluene exposure. In this case the situation is further complicated by the high variable background level of hippuric acid, as already mentioned. However, the investigators carried out the exposure at airborne levels that would minimize this interference. Thus these differences probably also principally reflect analytical differences, for the most part. Von Oettingen (1942), for instance, whose equation indicates one of the lowest background levels of hippuric acid, precipitated the hippuric acid from urine and then titrated it. The precipitation almost certainly was not quantitative. Ogata et al. (1970), whose equation indicates the next lowest background level of hippuric acid, extracted the hippuric acid before forming a derivative that had a characteristic light absorption spectrum. It is possible that the initial extraction was not complete. The method of Pagnotto and Lieberman (1967) also depends on an extraction, followed by determination of light absorption at a specific wavelength. Ikeda and Ohtsuji (1969b) separated the hippuric acid by paper chromatography followed by reaction to form a colored derivative, extraction of the derivative from the paper, and determination of its light absorption at a characteristic wavelength. Ogata et al. (1969) later showed that the color depended on the filter paper used.

7.4 Analysis for Inorganic Ions

Urine analysis also can be used to measure exposure to metallic ions. Some of the metal ions excreted in urine include beryllium, cadmium, copper, iron, lead, lithium, magnesium, manganese, mercury, selenium, and zinc. The factors to be considered in developing and applying the methodology, and interpreting the results do not differ from those already discussed. Dose–response relationships exist and can be utilized. As already mentioned, it is unlikely that metabolic factors will be as important for inorganic salt excretion as they are for organic chemical excretion. Therefore, the normal range of excretion accompanying a given exposure may be narrower.

Metals are unlike organic chemicals in that they rarely have nonworkplace precursors that are less toxic than the industrial exposure chemical. Consequently, metal ion excretion levels will represent the true hazard to the worker, except in the few instances to be discussed. Thus, biologic TLVs for metals can be set with greater assurance that excursions above the biologic TLV represent greater hazard to the worker. However, it should be remembered that exposure to metals also can occur outside the workplace. Regardless of where they occur, excretion levels generally indicate the true hazard and exposure should be reduced.

Elkins (1967) has reviewed urine levels for several metals. On the basis of his experience, he feels that 0.2 mg lead/L of urine (corrected to specific gravity of 1.024) represents significant absorption but not necessarily a toxic level. Similarly, he reported that 0.25 mg mercury/L would represent significant absorption of mercury but not necessarily a toxic effect. The 1972 NIOSH review (U.S. Dept. Health, Education and Welfare, 1972) of the data correlating urinary lead concentration with biological effects concurs with Elkin's assessment.

Urine analyses also have been used to measure exposure to inorganic anions. Extensive work has been reported, for example, on the correlation of urinary fluoride and exposure to inorganic fluoride in the aluminum industry. NIOSH (U.S. Dept. of Health, Education and Welfare, 1975) has suggested that the work of Kaltreider et al. (1972) and Derryberry et al. (1963) indicates that postshift urinary values of ≤7 mg fluoride/L urine (corrected to specific gravity of 1.024) indicate exposure to inorganic fluoride levels that would not be expected to cause osteofluorosis.

However, because of the ubiquity of fluorine ions, for example, inorganic fluoride in drinking water, tea, and cereal grains, and organic fluoride in, for example, refrigerants and degreasing solvents, urinary fluoride levels may not always reflect solely occupational exposure to inorganic fluoride. This urinary level also cannot be taken to indicate safe exposure to inorganic fluorides such as oxygen difluoride, nitrogren trifluoride, sulfur pentafluoride, sulfur tetrafluoride, tellurium hexafluoride, or any other inorganic fluoride with innate toxicity significantly greater than the inorganic fluorides used to set this standard.

Organic fluoride levels in urine may indicate more or less hazard than corresponding inorganic fluoride levels, depending on the comparative toxicity of the organo-fluorine compound and any of its metabolites.

7.5 Optimum Conditions for Using Urine Analyses

The studies to date indicate that urine analyses for a particular organic biological marker will be most reliable if:

1. The biological marker has no nonworkplace progenitors.
2. The slope of the dose–response curve is fairly steep (≥ 0.5).
3. The elimination half-life is no greater than 8 hr and preferably no greater than 4 hr.
4. The analytical method is specific for the biological marker. This requirement tends to eliminate colorimetric procedures. If GC procedures are used, the peak identity should be confirmed with a second column with different retention characteristics, or a GC/mass spectrometer combination should be used in developing the method.
5. The method is validated in humans at the highest and lowest exposure levels of interest. The dose–response curves should not be extrapolated beyond the extremes experimentally validated.
6. Urine collection times are consistent and consider excretion half-life.
7. Urine samples are analyzed shortly after collection. If they cannot be analyzed promptly, they should be frozen or at least refrigerated (Rainsford and Davies, 1965; Elkins, 1967).
8. The method and equations are first validated for the group of workers of interest before routinely applying the correlation equations.
9. The worker is not on a diet, suffering from a disease, or taking a medicine that could alter the relevant kinetics of the reactions of interest or any normalizing procedures used.
10. The worker is not being exposed off the job to the biological marker or another progenitor of the biological marker.
11. The urinary level of the biological marker is relatable to the amount of exposure chemical absorbed by all routes.
12. The dose–response equations have been shown to apply to both men and women, or separate equations are developed for (and applied to) each sex.
13. The workday of the group of interest and of the group used to develop the equations and/or discriminator number (biological TLV) are the same, if the half-life of elimination is much greater than 8 hr.

8   EXHALED AIR ANALYSIS TO MEASURE INDUSTRIAL EXPOSURE

8.1   Background and Advantages of Methodology

It has been known for decades that many industrial chemicals that are inhaled and enter the vascular system of the human body are later excreted to some degree in exhaled air. Metabolites of some also may be exhaled. A representative listing is shown in Table 3.8.

It has been observed that the concentration of many of these chemicals in exhaled air decreases regularly with time (Stewart et al., 1961). It also has been observed that some chemicals absorbed through the skin are excreted, unchanged, in exhaled air. The concentration of these chemicals in the exhaled air also has been found to decrease regularly with time after exposure (Stewart and Dodd, 1964). If the concentration of the exposure chemical in exhaled air varies in some regular fashion with body burden, regardless of the route of adsorption, a very desirable method for measuring industrial exposure would be possible. It would have at least the following advantages over various other methods:

1. Metabolism usually would not be involved, so all the metabolic factors that can affect the rate of appearance of the biological marker would not be involved. In most cases physical or physical-chemical factors would predominate, and these should be fairly constant between individuals.
2. The biological marker appears rapidly in the exhaled air. It is not necessary to wait hours or weeks for the biological marker to appear in the biological assay material, as is necessary in many other biological assays.
3. The analysis would be amenable to GC techniques. These can be made quite specific, thus eliminating analytical interference from nonbiological markers. Gas chromatographic techniques are able to quantitate small amounts and can be used for simultaneous analysis of several chemicals. Thus, concurrent exposure to several chemicals could be quantitated fairly rapidly and inexpensively.
4. Several samples can be taken in rapid succession, so the assessment of ex-

Table 3.8 Some Workplace Chemicals Detected in Human Exhaled Air Following Exposure

Chemical	References
Benzene	Derryberry et al. (1963)
Carbon tetrachloride	Stewart and Dodd (1964); Stewart et al. (1961)
Diethyl ether	Haggard (1924)
Methanol	Elkins (1954)
Methyl acetate	Elkins (1954)
Methylene chloride	Stewart and Dodd (1964); DiVincenzo et al. (1972)
Styrene	Stewart et al. (1968); Götell et al. (1972)
Toluene	Von Oettingen et al. (1942)
1,1,1-Trichloroethane	Stewart and Dodd (1964)
T_3CE	Stewart and Dodd (1964); Nomiyama and Nomiyama (1971)
T_4CE	Stewart et al. (1974a); Guberan and Fernandez (1974); Fernandez et al. (1976)
Vinyl chloride monomer	Baretta et al. (1969)
Xylene	Riihimäki et al. (1979)

posure can be based on either a kinetic analysis or substitution in an already derived equation.

5. In many cases the subject could be observed while providing the sample to assure that collection instructions are being followed.

6. Very few nonworkplace progenitors exist for the biological marker. Thus, the excreted material more likely represents workplace exposure than in many analyses of other materials excreted.

7. The technique is noninvasive.

8. The technique measures individual exposure without the bother of a personal monitor or the uncertainty of area monitoring.

8.2   Factors Affecting Exhaled Air Levels of Biological Markers

Assuming that transport through the lungs into the vascular system and back into the lungs is simple diffusion rather than "active transport" (Piotrowski, 1977, p. 113), several factors can be proposed that possibly will affect the postexposure concentrations of biological marker in exhaled air.

1. *Concentration of Biological Marker in Inhaled Air.* If movement of the chemical into the bloodstream is a simple diffusion process, the amount entering the blood will vary with the rate of diffusion through the alveolar wall and the partial pressure (concentration) of the chemical in the workplace atmosphere. It should not vary with the duration of exposure, once equilibrium between blood and the atmosphere has been established.

2. *Blood Concentration.* If movement in and out of the blood is a simple diffusion process, the concentration of the biological marker in exhaled air will vary directly with the concentration of that chemical in the blood. There will, of course, be some lag due to the rate of diffusion. The important blood concentration usually will be the concentration of unbound chemical. (Factors affecting binding already have been discussed in Section 6.1.4.) The rate of metabolism of the chemical also will affect the unbound concentrations in the blood. (Factors affecting this were discussed in Section 6.1.)

In addition to the general effects of disease (discussed in Section 6.1.9), emphysema can uniquely affect the interpretation of the results. If the effective lung surface is decreased, the total diffusion rate of the biological marker from the blood into the atmospheric side of the lungs will be decreased, resulting in a lowered concentration in the exhaled air in comparison to the disease-free person. This would decrease the calculated exposure. Since the absorption coefficient of this emphysematous person also would be decreased in comparison to the normal person, the exhaled air concentration of biological marker still would be a measure of that person's body burden, but the equation relating the exhaled air concentration to exposure concentration would be different from that applicable to the normal person.

3. *Solubility in Tissues or Fat and Binding to Them.* Some chemicals will be more soluble in, or more strongly bound to, these materials than will others. Except in unusual situations where there is a high degree of solubility or binding, this should

not have a significant effect on exhaled air concentrations of the biological marker at the concentrations and in the time frame of interest.

4. *Dilution in Exhaled Air.* This is the pulmonary counterpart of urinary dilution discussed in Section 6.4. Every postexposure breath dilutes the biological marker in the alveoli with air that does not contain the biological marker. This lowers the concentration of biological marker in the exhaled air, which in turn usually will decrease the precision of the analysis. This dilution can be overcome to some extent by analyzing *end-tidal*, or *alveolar*, air. To do this, the subject inhales and exhales normally through the collecting apparatus two or three times. At the end of the last breath, the final few milliliters are either collected in the sampler for later analysis or diverted directly into the analytical instrument (Fernandez et al., 1976). According to DiVincenzo et al. (1972), the alveolar air may be 50 percent richer in biological marker than the usual exhaled air. This technique is possible because instruments such as the gas chromatograph can routinely analyze a few hundredths of a milliliter of a gas with high precision and accuracy.

Factors that have been shown to affect the concentration of the biological marker in exhaled air include:

1. *Nonworkplace Progenitors of the Biological Marker.* Although the instrumentation generally used for exhaled air sampling is quite specific, it cannot indicate whether the chemical it is detecting appeared in the breath sample from the bloodstream or the mouth. It cannot tell if the chemical got into the bloodstream from prior inhalation or from ingestion. For example, phenol from a lozenge for sore throats (see Section 6.4) will invalidate exhaled air analyses for phenol. Acetone from severe untreated diabetes will similarly confound exhaled air analyses for acetone absorbed from the workplace. Depending on the chemical of interest, lozenges, candy, chewing gum, tobacco, mouthwash, and toothpaste also could be sources of interference.

2. *Respiratory Rate.* Until the blood and extracellular fluid compartment are saturated, respiratory rate can affect the rate of uptake. It similarly can affect the rate of desorption as shown by DiVincenzo et al. (1972). This suggests that correlation equations from subjects at rest should not be used on subjects that have been exerting themselves, and vice versa.

3. *Sex.* As already discussed (Section 6.1.11), the absorption coefficients for some vapors have been reported to vary for the two sexes (Nomiyama and Nomiyama, 1969; Piotrowski, 1971). This will affect both the time to saturation and, probably, the rate of desorption. Nomiyama (1971) also has reported an absolute difference between sexes in the concentration of T_3CE in the exhaled air in the β phase (see footnote a, Table 3.7) of respiratory elimination. The female concentration is lower. However, the slopes of the β-phase curves are parallel. The difference is due to a prolonged α phase in women. Stewart et al. (1974), on the other hand, found no sexual difference in the decay curves for T_3CE in the exhaled air of men and women following concurrent exposure of both sexes to T_3CE.

If there are sexual differences, the use of equations derived for one sex on analyses of exhaled air from the other sex would give erroneous estimates of workplace exposure. Another factor that will affect the rate of saturation of the blood and extracellular fluid compartment is the difference in blood volume and, probably, total extracellular fluid. For instance, the male blood volume is 75 mL/kg body weight, whereas the female volume is only 90 percent of that, or 67 mL/kg body weight (Wallach, 1974). Since there easily can be a 100 percent difference in body weights between the sexes, there could be large differences in the volume of this compartment and thus the amount of the biological marker in the compartment after a given exposure. Again, this could result in the use of an inappropriate correlation equation to estimate workplace exposure.

Obesity and fat–muscle ratio will alter the amount of fat-soluble exposure chemicals that are stored and, consequently, the rate of excretion, as already discussed in Section 6.1.7. Women generally have a higher fat–muscle ratio as discussed.

4. *Skin Absorption of the Exposure Chemical.* In addition to these factors that could lead to erroneous correlations between exposure levels and biological marker concentration in exhaled air, Stewart and Dodd (1964) have shown that the alveolar decay kinetics for some solvents may be different for skin absorption exposure than for inhalation exposure. If this is true for other solvents, exhaled air concentration may not be a reliable medium for evaluating total body burden when the burden arises from various routes of exposure.

Stewart and Dodd (1964) studied the skin absorption of carbon tetrachloride, methylene chloride, T_3CE, T_4CE, and methylchloroform (1,1,1-trichloroethane). They found that, for thumb immersion or nonoccluded topical exposure, the postexposure alveolar air decay was linear, whereas for total hand immersion the decay was exponential. Thus, the breath excretion kinetics may vary with the size of the dose for noninhalation administration. This may simply reflect saturation of the metabolic pathway that predominates at lower doses. If so, it emphasizes the importance of preparing standard curves for all expected exposure concentrations. The decay curves developed to date for excretion in breath following inhalation administration all have been exponential decays. As the technique is further studied, additional possible causes of variation undoubtedly will be found.

Obviously, just as with urine analyses (Section 7), it will have to be demonstrated for each chemical that the kinetics of its excretion in breath are the same no matter what the dose or route of exposure. If the kinetics change, multiple routes of exposure will invalidate the methodology for measuring body burden and exposure.

8.3 Shortcomings of the Methodology

Studies to date indicate that the technique has some shortcomings that must be recognized:

1. It does not appear to be widely usable for samples taken within, variously, 0–2 hr postexposure. The decay curves for various concentrations of xenobiotic in exhaled air versus exposure times frequently are indistinguishable in this time period.

If a chemical is excreted so rapidly in exhaled air that none remains the next morning, samples would have to be taken off the job by the worker and returned for analysis the next morning. DiVincenzo et al. (1972) have found that workers may not follow instructions exactly for collection of alveolar air samples.

2. Although the body tends to integrate the exposure, thus giving exhaled breath concentrations representative of the average exposure, some data suggest that samples taken shortly after exposure will principally reflect the latest exposure level (Stewart et al., 1974).

8.4 Experimental Studies

For inhalation administration, the results published to date have been very encouraging for all the compounds studied. The methodology has given good correlation between concentration decrease of the biological marker in postexposure exhaled air and prior exposure, at least for the first 3–5 hr postexposure.

As mentioned previously, the decay curve is exponential and generally fits Eq. (5):

$$C_t/C_0 = K_A e^{-\alpha t} + K_B e^{-\beta t} + \cdots K_N e^{-nt}$$

where C_t = concentration of the biological marker in exhaled air at time t

C_0 = exposure concentration

α, β, and n = rate constants for respective decay periods

K_A, K_B, K_N = zero time coefficient for that curve segment

Except for protracted studies over many hours or for chemicals where the decay curve seems to have more than two segments, all terms beyond the first two usually are unnecessary. K_A, α, K_B, β, K_N, and n all can be obtained graphically from the semilogarithmic plot of C_t/C_0 versus t (Gibaldi and Ferrier, 1975). C_t is determined experimentally at time t. The equation can then be applied to unknown exposure situations and C_0 can be calculated.

The data of Fernandez et al. (1976) suggest that the time of exposure, at least for T_4CE, determines the decay curve. They studied alveolar air concentrations up to 4 hr postexposure. They found that the same equation described the postexposure decays after 2 hr of exposure to 100 or 200 ppm, if the ratio C_t/C_0 was plotted against postexposure time. A similar result was found for 4-hr exposures to 100, 150, or 200 ppm of T_4CE and for 8-hr exposures to these same three concentrations. The curves for the three time periods were different. Their equations are shown in Table 3.9. Baretta et al. (1969) found that the decay curves for VCM in human alveolar air following 8-hr exposures to 50, 100, 250, or 500 ppm also were a family for the decay period studied. Calculation of C_t/C_0 for all curves at various common reported postexposure times yielded a common curve, within experimental limits, for the 20-hr decay period reported (see footnote d in Table 3.9). Because neither the y intercept(s) for these curves nor the data points for the first hour postexposure

Table 3.9 Postexposure Decay Equations for Xenobiotics in Human Exhaled Air

Chemical	Exposure Period (hr)	Exposure Concentration (ppm)	Postexposure Study Period (hr)	Air Sample	Subjects (No., Sex)	K_A	K_B	α	β	References
						Coefficients and Exponents[a] $(Y = K_A e^{-\alpha t} + K_B e^{-\beta t})$				
Methylene chloride	2	100, 200	3	E[b]	11, M[c]	1.8×10^{-2}	10^{-1}	1.1×10^{-1}	10^{-2}	DiVincenzo et al. (1972)
Styrene	8	25, 115, 260	5	E	15, M	1.5×10^{-2}	5×10^{-3}	1.01	1.3×10^{-1d}	Götell et al. (1972)
	7	99	6	A	6, M	1.5×10^{-2}	5×10^{-3}	1.01	1.3×10^{-1d}	Engström et al. (1976)
T_4CE	7	101	110	A	16, M	3.5×10^{-2}	3.5×10^{-2}	8.3×10^{-2}	8.8×10^{-3d}	Stewart et al. (1970)
	2	100, 150, 200	4	A	23, M; 1,F	2.9×10^{-1}	1.6×10^{-1}	1.2×10^{-2}	8.0×10^{-3}	Fernandez et al. (1976)
	4		4			2.1×10^{-1}	2.0×10^{-1}	9.9×10^{-2}	7.5×10^{-3}	
	8		4			10^{-1}	2.2×10^{-1}	4.9×10^{-2}	4.4×10^{-3}	
T_3CE	1	20, 100, 20	22	A	3–9 (M, F)	4.5×10^{-2}	4.8×10^{-2}	4.6×10^{-1}	5.1×10^{-2d}	Stewart et al. (1974a)
						5×10^{-2}	1.3×10^{-2}	1.6	8.1×10^{-2d}	
	$7\frac{1}{2}$					9×10^{-2}	2.6×10^{-2}	1.8	6.9×10^{-2d}	

[a] Y = ratio of concentration in exhaled air at time t (hr) to exposure concentration (i.e., C_t/C_0).

[b] E = normal exhaled air; A = alveolar air.

[c] M = male; F = female.

[d] These calculations were made by R. S. Waritz, using data obtained from graphs presented in the author's original scientific paper and without access to the original raw data. Therefore, they are subject to errors introduced by the original translation to graph form and by printing. They should be considered illustrative only.

131

are shown in the paper by Baretta et al., it is not possible to determine whether the first term in the equation describing the decay curves will be the same for all. Nevertheless, the data are strongly suggestive, and the time period reported would be suitable for calculating exposure levels. Decay curves for shorter exposure times were not reported, so it is not possible to tell if those curves would be different from those for the 8-hr exposure.

Within experimental limits, Stewart et al. (1974) also found that human alveolar air decay curves for the first 20 hr following 7.5 hr of exposure to 20, 100, or 200 ppm T_3CE also were a family. For this family, as with the T_4CE decay curves of Fernandez et al. (1976), the curves expressing the ratio C_t/C_0 versus postexposure time were coincident within experimental limits and could be represented by the one equation shown in Table 3.9. Likewise, the curves for alveolar decay ratios following 1- or 3-hr exposures at 100 and 200 ppm were identical for both exposures at each time period. The coefficients and exponents for the curves from all three exposure times (1, 3, and 7.5 hr) were not identical. The decays following the 1- and 3-hr exposures at 20 ppm were not followed by Stewart et al. for a long enough time to provide enough data points for comparisons.

The data of DiVincenzo et al. (1972) also suggest that the equations for the decay curves of the C_t/C_0 ratio following 2-hr human exposures to either 100 or 200 ppm methylene chloride also will be the same for both concentrations. This equation is shown in Table 3.9.

Both Stewart et al. (1968) and Götell et al. (1972) studied the decay of styrene in exhaled air following 8-hr exposures of humans. Stewart's subjects were exposed in a chamber to 99 ppm of styrene. Götell's subjects were workers exposed to a time-weighted average of 89–139 ppm in the workplace. Stewart collected alveolar air samples for analysis. Götell did not specify the samples collected.

Despite these inconsistencies, the agreement between the two decay curves was remarkable. Both decay curves when transformed to C_t/C_0 versus time (in hours) could be expressed as the same equation over the time period studied. This is shown in Table 3.9, and this is in contrast to the findings of Fernandez et al. (1976) and Stewart et al. (Stewart et al., 1970) for T_4CE breath decay. In this case the former group found that the β phase did not start until about 1.5–2 hr following an 8-hr exposure to 100 ppm. Stewart's group did not find the β phase starting until approximately 45 hr after a 7-hr exposure to 101 ppm.

There seems to be no question that in order to calculate exposure from exhaled air decay curves, it will be necessary to know the exposure time.

Although this aspect of the use of exhaled air decay curves to calculate exposure is very encouraging, the variability reported for some solvents is discouraging. For example, DiVincenzo et al. (1972) found that with 11 subjects the breath decay curves for 100 and 200 ppm of methylene chloride were within two standard deviations of each other. Similarly, the data of Götell et al. for styrene (Götell et al., 1972) indicate his decay curves would not reliably distinguish between 25 and 115 ppm or 115 and 260 ppm.

Conversely, Baretta et al. (1969) found that exhaled air decay curves for VCM readily could distinguish between 50 and 100 ppm or 100 and 250 ppm with only four to six subjects. Unfortunately, not enough studies have been published with such comparisons to indicate the probable general situation. The technique appeared to demonstrate worker accumulation of T_4CE after four and five 7-hr exposures to about 100 ppm (Stewart et al., 1970).

Overall, exhaled air analyses seem to hold good promise as a way of determining previous exposure to certain industrial chemicals. As can be seen in Table 3.9, not all chemicals are excreted at the same rate, and calibration curves will have to be developed for each chemical. It is probable that curves also will have to be developed for each group of workers, but not enough data have yet been developed to judge. The data strongly suggest that in order to apply the technique quantitatively, the duration of exposure must be known. However, Stewart et al. (1974) have published data suggesting that, within certain concentration and time limits, the decay curve is determined by the product of concentration and time.

For optimum utilization of the technique, it may be necessary to develop individual decay data over a period of time. The technique, with or without concurrent urine analysis, should be usable to screen a worker for unusual metabolism and excretion of many workplace chemicals prior to assignment to an area where the worker may be exposed to these chemicals.

9   BLOOD ANALYSIS AS A MEASURE OF INDUSTRIAL EXPOSURE

Blood analysis is an invasive technique that carries some resultant risk to the worker. Most workers also find it objectionable and would probably object strongly to daily or even weekly samples being drawn. Generally, the order of worker preference for sampling for biological monitoring would be expected to be exhaled air, urine and, last of all, blood.

Blood analysis has not been used extensively in industrial hygiene for measuring exposure, and comparatively few papers have been published in this area.

Blood analyses are used extensively in pharmacology and in clinical trials for new medicinals and are a valuable tool in these areas. They have provided great insight into the metabolism and excretion of xenobiotics and into the individual variation in these body processes. Pharmaceutical chemists and biochemists have been responsible for the greatest developments in these areas. The reader is referred to texts in this specialized field for in-depth information on its utility and drawbacks (Gillette, 1971; LaDu et al., 1971; Greim, 1981).

There is no question that total blood levels of xenobiotics and their metabolites reflect dosage. There also is no question that the factors discussed in Sections 6.1, 6.4, and 6.5 play a great role in individual variations. Thus all the caveats presented in that section must be considered in interpreting individual analyses for organic and inorganic industrial chemicals.

The correlations of inorganic lead exposure, biological effects, and blood levels of inorganic lead probably have been more extensively studied than for any other industrial compound. The early work by Kehoe and others has been reviewed by Kehoe (1963). This work and later studies also have been reviewed by NIOSH (U.S. Dept. Health, Education, and Welfare, 1972). The reader is referred to these reviews for excellent summaries and discussions of the work that led to the present standards for biological monitoring of inorganic lead exposure.

Kehoe (1963) suggested that the maximum acceptable blood level for inorganic lead in adult workers is < 80 $\mu g/100$ g blood. The National Institute for Occupational Safety and Health concurred (U.S. Dept. Health, Education, and Welfare, 1972). The Occupational Safety and Health Administration (U.S. Dept. of Labor, 1975) initially proposed a level of > 60 $\mu g/100$ g as the level that would dictate worker removal from exposure areas. The OSHA lead standard that finally issued has an action level of 40 $\mu g/100$ g of whole blood. It also has a sliding scale of maximum blood levels, initially coupled with maximum airborne exposure levels of lead. Both decreased with the age of the standard. Allowable blood levels started at 80 μg lead/ 100 g of whole blood and decreased to 50 $\mu g/100$ g of whole blood in the fifth year.

In its proposed National Ambient Air Quality Guide, the EPA (U.S. Environmental Protection Agency, 1977) has suggested that mean blood lead levels in excess of 15 $\mu g/100$ mL of blood in children aged 1–5 years could be accompanied by adverse biological effects.

Blood analysis also has been used successfully by Osterman-Golkar and her colleagues (Osterman-Golkar et al., 1976) to quantitate workplace exposure to ethylene oxide. In a series of papers, they showed that ethylene oxide reacted with several sites on the hemoglobin molecule and that the hemoglobin adduct–DNA adduct ratio was constant in rats over a range of ethylene oxide exposure concentrations. They further showed that there was a dose–response relationship between exposure and adduct formation (Ehrenberg and Osterman-Golkar, 1980; Osterman-Golkar et al., 1983). The application of this biological marker to monitoring human exposure to ethylene oxide in the workplace has been reported by Osterman-Golkar and her colleagues (Farmer et al., 1986; Törnqvist et al., 1986a) and by Van Sittert et al. (1985).

Shortly after Osterman-Golkar's first publication on the ethylene oxide–hemoglobin biological marker, Calleman et al. (1978) published a similar correlation between exposure to ethylene oxide, hemoglobin adduct formation, and DNA adduct formation. This was then correlated with the radiation dosage having an equivalent effect on DNA, and the leukemia risk was predicted. An epidemiological follow-up study several years later found an actual incidence very close to that predicted (Hogstedt, 1986). Mayer et al. (1990) also have reported a dose–response relationship in humans between ethylene oxide exposure and ethylene oxide–hemoglobin adducts. The relationship was valid in the region of the OSHA PEL of 1 ppm (v/v).

This technique of measuring hemoglobin adducts should be applicable to any workplace epoxide or unsaturated chemical that can be metabolized to an epoxide. Segerback has shown that ethylene forms an adduct with hemoglobin (Segerbäck, 1983). Brenner et al. (1991) have shown that styrene forms an adduct with hemoglobin. While there did seem to be a dose–response relationship between exposure and adduct formation in this latter study, it was not significant ($p = 0.127$) using the Wilcoxon rank-sum test.

Sun et al. (1990) reported that benzene also formed an adduct in rats and mice with hemoglobin. A linear dose–response was observed for doses up to 500 $\mu mol/kg$. Both species also gave a linear dose–response for up to three daily doses, the maximum studied. Studies have not yet been carried out in humans. Bond et al. (1991) found that inhaled isoprene formed an adduct with hemoglobin. Adduct formation was linear between atmospheric concentrations of 20 and 200 ppm and between 200 and 2000 ppm, but the slope of the latter line was only one-fifth of the former.

Skipper and Tannenbaum (1990) have reviewed the literature on hemoglobin adducts of carcinogens and the reader is referred to this review for further examples.

All body systems are in a dynamic state of synthesis and degradation. Half-lives vary. Fortunately, red blood cells have a comparatively long life time, and the hemoglobin adducts do not appear to be excised rapidly by body repair mechanisms. Thus, hemoglobin adducts offer the opportunity of detection of current exposure and an integrated exposure over a period of days or weeks. Unfortunately, obtaining hemoglobin samples requires invasive techniques and may not be acceptable to workers as a frequent, regular procedure.

Serum levels of bone-Gla-protein have been reported to correlate well with cadmium exposure (Kido et al., 1991). However, the serum concentration of this protein varies inversely with glomerular filtration rate (GFR) and age (Rowe et al., 1976), as well as any agents or disease that decrease GFR would increase serum levels of this protein. Thus, while its level in serum correlates with cadmium exposure, it is not specific for cadmium.

Acrylamide also has been shown to bind to hemoglobin (Hashimoto and Aldridge, 1970), but no correlations with neurotoxicity and the amount bound have been made.

The immune system also offers the potential for biological markers that can be used to assay worker exposure to xenobiotics in the workplace. As discussed above and in Section 6.1.4, xenobiotics can form adducts with various blood proteins. If these stable adducts trigger the formation of antibodies, then the antibody presence can be used at least as a qualitative indicator of exposure. Alarie (Karol et al., 1978a, b) has combined various chemicals with proteins, administered the complex to laboratory animals by inhalation, and induced respiratory sensitization. Others have reported finding human antibodies to formaldehyde (Grammer et al., 1990; Thrasher et al., 1990; Madison et al., 1991), isocyanates (Karol et al., 1978c; Cartier et al., 1989), and trimellitic anhydride (Bernstein et al., 1983). All of these chemicals are known to cause respiratory sensitization. Transient antibodies to morphine were reported in workers making morphine (Biagini et al., 1990). In addition, hundreds of industrial chemicals are known to cause skin sensitization. Thus, in theory at least, the immune system offers the potential of specific biological markers for quantitating industrial exposure to many chemicals. However, the quantitative methodology specific to a chemical presently is not available.

The use of the immune system as a source of biological markers has been reviewed briefly by Vogt (1991) and in greater depth by the Subcommittee on Immunotoxicology of the Committee on Biologic Markers, National Research Council (1992). The reader is referred to these reviews for in-depth discussions in this area.

As discussed in Section 5, blood analyses also are routinely used in the medical field to detect organ damage by measuring transfer enzyme levels, blood urea nitrogen, and levels of various chemicals always present in blood. However, changes are not specific for any particular causative agent. Anything damaging the organ could cause an increase in the blood level of these enzymes. The causative agent could be disease, medicinals, or industrial chemicals. These measurements are nonspecific biological markers of organ damage; therefore, from an industrial standpoint, this approach is deficient in two respects:

1. It is not specific to a particular industrial chemical or even to industrial chemicals.
2. It measures an effect that occurs after injury, instead of measuring a leading effect that could be used to forestall injury, which is the goal of industrial hygiene.

In summary, there is no question that blood analyses for biological markers can be developed to measure exposure to workplace chemicals. There also is no question that blood levels of these biological markers can be used to set acceptable exposure levels for industrial chemicals. However, they will show the same individual variability already seen in the medical profession for naturally occurring blood chemicals and medicinals. The causes of the expected variability already have been discussed in Sections 6 and 7. Because of worker objections to the technique and the slight risk to the worker, blood analyses probably will not be used if biological markers in urine or exhaled air can provide equivalent reliability in measurements of worker exposure.

10 HAIR ANALYSIS TO MEASURE INDUSTRIAL EXPOSURE

It has been known for many years that hair contains metals. As discussed earlier, hair may be considered an excretory mechanism for these metals, since the metals appear to have no functional role in the hair protein and, for at least a few of the metals, their content in hair seems to vary with exposure to the metal (Jawarowski, 1965; El-Dakhakhny and El-Sadik, 1972; Hammer et al., 1972; Yoakum, 1976; Henley et al., 1977; Bhat et al., 1982).

Some of the metals reported in hair to date are aluminum, arsenic, beryllium, cadmium, calcium, copper, iron, mercury, lead, manganese, molybdenum, potassium, selenium, silicon, thallium, titanium, and zinc (Haggard, 1924; Bate and Dyer, 1965; Schroeder and Nason, 1969; Anke et al., 1971; Petering et al., 1971; El-Dakhakhny and El-Sadik, 1972; Nishiyama and Nordberg, 1972; Renshaw et al., 1972; Suzuki, et al., 1973; Hopps, 1974; Hurlburt, 1976; Henley et al., 1977).

In addition, the nonmetals, chlorine and phosphorus, also have been reported (Henley et al., 1977). Historically, hair has not been considered suitable for measuring exposure to organic compounds. Since hair is protein with a small amount of colorant, it was not believed capable of carrying organic compounds from systemic circulation. In 1974 Harrison et al. (1974) reported the presence of ^{14}C in the hair of guinea pigs that had been dosed intraperitoneally with ^{14}C-labeled amphetamine. Since then, heroin and morphine metabolites have been reported in human scalp hair of admitted users of these drugs (Baumgartner et al., 1979). The amount and location of the drug along the hair shaft correlated directly with the subject's use history of the drug. The analytical procedure used was radioimmune assay. Analyses were confirmed by urine analysis. Using the same methodology, these authors later reported the presence of phencyclidine (Baumgartner et al., 1981) and cocaine (Baumgartner et al., 1982) in scalp hair of admitted drug users. Similar correlations of drug use periods and location of the drug on the hair shaft were found. Again, radioimmune assay was used and drug use was confirmed by urine analysis.

The presence of organomercurials in the scalp hair of humans after eating fish from lakes contaminated with organomercurials and inorganic mercury also has been reported (Phelps et al., 1980).

Hair is not suitable as a dynamic system for evaluating immediate past exposure to chemicals since the individual hair shafts do not have access throughout their length to any fluid transport system. Thus, the xenobiotic content of the hair reflects that available at the time any particular portion of the shaft was being synthesized. Since hair grows at the rate of about 1 cm in 30 days, clippings would be expected to reflect a historical exposure at best. How many months back in history would be indicated by the length of the clipping and the length of the remaining proximal hair shaft, as already mentioned.

However, if the range of biological marker content of hair normally is sufficiently narrow in a population and the hair concentration varies regularly with blood concentration, suitably timed postexposure analyses could be used to confirm or refute suspected overexposure or continuing acceptable exposure. Unfortunately, as with most evolving areas of science, there are conflicting reports in the literature on the utility of hair analyses. Also, in addition to some of the already discussed factors that can lead to aberrant exposure estimates based on urinary concentration of chemicals, hair analyses have unique problems that may lead to aberrant conclusions. These must be considered in developing and applying any procedure for correlating hair levels with exposure levels.

One of the additional unique problems of hair analysis is that of suitable cleansing of the hair to remove adsorbed contaminants prior to analysis without removing endogenous materials. Hair normally develops an oily coating, and this oil can trap exogenous metals or organic materials. Since this adsorbed material does not represent body burden, it must be removed prior to analysis.

Several preanalysis hair cleaning procedures have been reported. The simplest is washing with detergent followed by distilled deionized water rinses and oven drying (Yoakum, 1976; Nishiyama and Nordberg, 1972; Baumgartner et al., 1979; Oleru, 1976). Additional washings have included acetone (Petering et al., 1971, 1973), methanol (Baumgartner et al., 1979), nitric acid (El-Dakhakhny and El-Sidek, 1972; and Nishiyama and Nordberg, 1972), ether (Petering et al., 1971; Renshaw et al., 1972) and trisodium ethylenediamine tetraacetic acid (Nishiyama and Nordberg, 1972). Most investigators assume that detergent washes, followed by (1) distilled water washes to remove detergent and (2) organic solvent washes, remove all adsorbed materials. In some cases the initial wash has been with the organic solvent (Anon., 1977). Nishiyama and Nordberg (1972) found that washing procedures that removed all adsorbed cadmium also removed the endogenously derived cadmium in the hair. Experimental results similar to those of Nishiyama and Nordberg have been reported for hair analyses of other metals (Bate and Dyer, 1965).

Petering et al. (1971) suggested that ionic detergents were more appropriate than nonionic ones for washing surface metals from hair because the former could complex the metal ion or form salts with it, thus aiding its removal from the exterior of the hairshaft. Obviously, if the detergent charge could be a factor in metal ion removal, anionic detergents would be more suitable than cationic detergents, unless the metal is present as a negatively charged complex or radical.

Renshaw et al. (1972) reported that the concentration of lead in single hairs from one woman increased with the distance from the scalp. Since they had cleansed the hair by diethyl ether reflux in a Soxhlet extractor prior to analysis, they assumed that all adsorbed external lead had been removed, and this distal increase represented lead that had deposited on the hair from external sources and then diffused into the body of the hairs. No details of the work history, residence, or cosmetics use of the woman were given. The increased lead content at the distal portion also could be surface lead that had been incompletely removed by the ether wash. Certainly, the mass of the evidence suggests that hair concentrations of metals, and apparently organic compounds, do bear a relationship to body burden (Suzuki et al., 1958; El-Dakhakhny and El-Sidek, 1972; Commerce Clearing House, 1978) if the hair is adequately cleansed. The uncertainty regarding contamination by airborne material presumably could be removed by using body, pubic, or axillary hair instead of scalp hair. However, hair from these sites may grow at different rates from scalp (vertex) hair. Axillary hair grows at about two-thirds the rate of scalp hair and pubic hair at about one-half the rate. Body hair grows at approximately the same rate (Hopps, 1974). Rate differences obviously would affect time correlations.

The same factors discussed in Sections 6 and 7 can control the appearance of the biological marker in hair. For example, many of the metals (metal salts) are transported in blood predominantly in the bound form or are stored in the body. Anything that caused their release, with a resulting increase of the free metal/metal salt in liquid transport systems accessing the hair root, could result in a short or extended shaft section with elevated concentration. This could be interpreted as a short- or long-term overexposure. Medicinals or industrial chemicals that were more tightly bound to transport or storage proteins could replace, and thus cause the release of, bound metals/metal salts. This would be followed by increased hair uptake of the released metal/metal salt. Wasting diseases that liberate stored metals/metal salts could have a similar effect.

Therefore, interpretation of isolated hair analyses for estimating workplace exposure should be coupled with a careful and complete medical history to assure that elevated local concentrations along the hair shaft due to nonwork causes are not attributed to work exposure.

Diet also can be expected to affect the level of metals found in hair. Green plants are notorious scavengers of metals from the ground in which they are grown. Thus, hair levels of metals/metal salts could be due not only to the direct ingestion of fruits, vegetables, and cereals but also to the ingestion of meat from animals grazed or fed hay or cereal grains. Conversely, high-fiber diets apparently lower uptake of metals/metal salts from the intestine (Anon., 1977) and would be expected eventually to result in lowered baseline metal content of hair.

Cosmetics and hair dyes also may contain various metal salts or complexes and may contribute to hair levels of metals/metal salts either through (1) absorption of the metal salt or complex followed by uptake by the hair root or (2) by adsorption on the hair shaft. Various salts or complexes of metal ions are permitted in the coloring agents of hair dyes, dye formulations, and cosmetics. These include aluminum, arsenic, barium, chromium (+3), cobalt, copper, iron, lead, mercury, titanium, and zinc (Commerce Clearing House, 1978).

Obviously, the cosmetic and hair-dye use history of the worker also must be determined before attempting to correlate metal content of a worker's hair with workplace exposure to a metal.

In order to use metal content of hair as a biological marker of occupational exposure, sample preparation procedures (e.g., washing) and the analytical method must be validated. In addition, a normal baseline must be established. From the medical experience with urinalyses, and the industrial experience with urinalyses mentioned earlier, this baseline would be expected to be a range rather than a line. Variations with sex also might be expected. The literature data shown in Tables 3.10 and 3.11 indicate this to be the case. For some metals there also were variations apparently due to age. Therefore, before making occupational exposure judgments based on isolated hair levels of metals/metal salts, one must judge the hair analyses in these cases against not only medical and dietary background but also cosmetic and hair-dye use. They also must be matched against control ranges for sex, age, geographic location, and, if possible, prior analyses of the same employee's hair.

As can be seen from Table 3.10, in general, women with no known industrial exposure have higher hair levels of lead than do their male counterparts in the studies. The exception was the group of middle-class urban white females from Cincinnati, Ohio, studied by Petering et al. (1971). Petering et al. also found that the lead level in male scalp hair decreased with age from 2 to 88 years. In this study the lead level was found to increase rapidly in women from age 14 to 30 and then decrease rapidly from age 30 to 84. For both males and females, several values of about 35–40 $\mu g/g$ were observed. Other studies also tried to correlate lead content of hair with age. This would appear from the data of Petering et al. to be necessary, although the slope of the line for men was not great and the upper 95 percent fiducial limit at 60 years was within the fiducial limits at 20 years. The changes with age were dramatic for the women in Petering's study, and age matching for women of working ages definitely would appear to be necessary. If individual historical controls were used, allowance would have to be made for the changes that accompany age.

Klevay (1973), however, reported an age relationship with hair lead content only for males in Panama. In his population, this relationship appeared to reach a nadir between 11 and 20 years. He reported ranges of 1.1–51.2 μg lead per gram of hair for males and 8.7–78.7 μg of lead per gram of hair for females.

It also is obvious that the normal control adult average lead concentration in scalp hair varies by a factor of about 2 for males and about 1.5 for females. Furthermore, values varying by factors of 10–50 between control individuals have been reported (Oleru, 1976).

Suzuki et al. (1958) suggested that 30 μg of lead per gram of hair be considered the upper normal level. El-Dakhakhny and El-Sadik (1972), on the basis of correlation of hair lead levels and clinical signs and/or symptoms, suggested that hair

Table 3.10 Hair Lead Concentrations in Control Populations[a]

Arithmetic Mean Value (μg/g) and [Range]			Significantly Different?	Analytical Method	References
Male	Female	Sex Unknown			
14.7 (A)	19.2 (A)		Yes; $p < 0.001$	Atomic absorption	Oleru (1976); Kraut and Weber (1944)
9.9 (A)	14.6 (A)		ND	Unknown	Suzuki et al. (1958)
17.8 (A)	19.0 (A)		No; $p > 0.05$	Atomic absorption	Schroeder and Nason (1969)
	12 (U), [4–25]		—	Atomic absorption	Renshaw et al. (1972)
		9.4 (U), [3–26]	—	Dithizone (149)	El-Dakhakhny and El-Sadik (1972)
6.1 (T)			—	Atomic absorption	Hammer et al. (1972)
24.5 (AA), [1.1–52.1]	34.6 (AA), [8.7–78.7]		Yes; $p < 0.001$	Atomic absorption	Klevay (1973)
22 (P)	—		ND	Atomic absorption	Petering et al. (1973)
17 (T)	6 (T)				
14 (W)	11 (W)				
11 (R)	10 (R)				
		14.5 (P)	—	Atomic absorption	Yoakum (1976)
4.1 (W)			—	Unknown	Hopps (1974)

[a]Symbols are as follows: A = adult; AA = all ages, urban and rural; ND = not determined; P = preschool; R = 60 years; T = 6–20 years; U = age unknown; W = 20–60 years.

Table 3.11 Hair Content of Various Metals in Control Populations[a]

Metal	Range of Concentrations (µg/g)			Significantly Different?	Analytical Method	References
	Male	Female	Sex Unknown			
Aluminum	1.6–7.8 (T) 1.2–9.2 (T, P) 4.4–5.5 (W)		2–9 (AA)	ND	Neutron Activation	Bate and Dyer (1965)
Antimony	0.1–1.4 (T) 0–4.4 (T)		0.5–4 (AA)	— ND	Unknown Neutron activation	Hopps (1974) Bate and Dyer (1965)
Arsenic	0.07–0.2 (W) 0.4–7.9 (T) 0.7–5.3 (T)		0.18 (P)	— — —	Unknown Atomic absorption Neutron activation	Hopps (1974) Yoakum (1976) Bate and Dyer (1965)
Cadmium	0.08 (mean; A)	0.15 (mean; A)		Yes	Neutron activation	Arunachalam et al. (1979)
	1–2 (P)	1–1.3 (T)		ND	Atomic absorption	Petering et al. (1973)
	2 (T)	1.3–2 (W)	30–530 (AA)	—	Atomic absorption	Oleru (1976)
	1.5–2 (W, R)	2–1.5 (R)	1.06 (P)	—	Atomic absorption	Yoakum (1976)
			> 1000 (U)	No	Unknown	Nishiyama and Nordberg (1972)
	2.76 (AA)	1.77 (AA)		($p > 0.05$)	Atomic absorption	Schroeder and Nason (1969)
Chromium	0.47 (W)			—	Unknown	Hopps (1974)
	0.46 (mean; A)	0.34 (mean; A)		No	Neutron activation	Arunachalam et al. (1979)
Copper	13–30 (P)	20–25 (AA)		ND	Atomic absorption	Petering et al. (1971)
	30 (T) 30–15 (W) 15–10 (R) 7–93 (T) 8–150 (T)		7.8–234 (AA)	—	Neutron activation	Bate and Dyer (1965)
	16.1 (AA)	55.6 (AA)		$p < 0.001$	Atomic absorption	Schroeder and Nason (1969)
	15–17 (W)			—	Unknown	Hopps (1974)
	15.7 (mean; A)	31 (mean; A)		Yes	Neutron activation	Arunachalam et al. (1979)

Table 3.11 (Continued)

Metal	Range of Concentrations (µg/g)			Significantly Different?	Analytical Method	References
	Male	Female	Sex Unknown			
Mercury	0.3–34 (T)	—	0.1–33 (AA)	—	Neutron activation	Bate and Dyer (1965)
	0.5–53 (T)					
	1.7–1.9 (W)					
Selenium	1.30 (mean; A)	1.10 (mean; A)		No	Unknown	Hopps (1974)
Zinc	100–110 (P)	200 (P)		No	Neutron activation	Arunachalam et al. (1979)
	110–140 (T)	200–180 (T)		No	Atomic absorption	Petering et al. (1971)
	140–125 (W, R)	180–150 (W)	51–602 (AA)	—	Neutron activation	Bate and Dyer (1965)
	101–186 (T)					
	85–166 (T)					
	167 (AA)	172 (AA)		No ($p > 0.05$)	Atomic absorption	Schroeder and Nason (1969)
	150–190 (W)			—	Unknown	Hopps (1974)
	140 (mean; A)	160 (mean; A)		Yes	Neutron activation	Arunachalam et al. (1979)

[a]Symbols are as follows: A = adult; AA = all ages, urban and rural; ND = not determined; P = preschool; R = > 60 years; T = 6–20 years; U = age unknown; W = 20–60 years.

lead content greater than 30 $\mu g/g$ be considered indicative of excessive lead exposure. They found that 30 μg lead per gram of hair corresponded to approximately 90 μg lead/100 g of blood.

However, the blood and hair analyses were carried out concurrently, and no mention was made of the residual hair length or the length of the hair sample. Therefore, unless the work exposure to lead had not changed over the number of months represented by the distance of the hair sample from the scalp, the comparison is not valid.

Klevay's data (Klevay, 1973) indicate that 30 $\mu g/g$ may be too low, at least for women. He proposed an upper normal value of 35–40 $\mu g/g$, and some of his hair samples from presumably unexposed persons even exceeded this value, as mentioned previously. However, some of these high values came from urban, nonindustrial areas and may reflect unique urban exposures. In Klevay's study values greater than 35 $\mu g/g$ were seen in male scalp hair only from subjects less than 10 years of age. They were seen in female subjects of all ages. Petering et al. (1973) also reported urban male and female values of approximately 40 $\mu g/g$ from subjects with no apparent industrial exposures. Neither Petering nor Klevay reported any clinical signs or symptoms of plumbism in their subjects, but there is no indication that they looked for them. Thus, the data suggest that the upper normal level for lead content in hair may be even greater than 40 $\mu g/g$.

The data shown in Table 3.11 show age and sex variations reported for normal hair content of several metals, including cadmium, copper, and zinc. In some cases the variations appear to be significant and in others not. For instance, the data of Petering et al. (1971) show an age-related decrease in zinc content for both males and females, but the slopes of the lines are so shallow and the equations of the lines so similar that very little allowance need be made for age or sex. Copper content of hair peaked at about 10 years for males in the study of Petering et al. but showed only a gradual increase with age for females. The 95 percent fiducial limits for the two sexes overlapped for about 20 years. Thus, this data of Petering et al. suggest that hair analyses for copper need to be evaluated against controls matched for age and sex.

Arunachalam et al. (1979) reported significant sex differences for arsenic and copper but not for chromium and selenium.

The available data suggest that because of the variability of individual hair content of metals, it would be difficult to set trigger concentrations or action concentrations (biologic TLVs) for most metals based on hair content and that, taken by themselves, would indicate exposure on the threshold of an effect. Also, since appearance of the metal in the external hair shaft would follow exposure by possibly a month or so, such analyses would not provide as early a warning of overexposure as would urine analyses.

Because of the individual variations reported, hair analyses would be of most value if individual histories of metal content could be developed over a period of years. As with urine and exhaled air analyses, comparisons between populations is difficult, particularly if different washing procedures and analytical techniques are used by the different investigators.

Also, as with other biological assays, high metal content of the hair can come from nonworkplace sources. It also may not represent recent past workplace exposure for the other reasons discussed. Obviously in these cases workplace exposure may be trivial, and trying to reduce exposure will not reduce the worker's hazard. If an undesirably high level of a metal or metals is found in the hair, the source should be conscientiously sought so that possible hazard to the worker may be reduced.

11 MILK ANALYSIS TO MEASURE INDUSTRIAL EXPOSURE

Historically, human breast milk analysis has not been used to monitor industrial exposure to chemicals. It has been realized for over 100 years that milk could be used by the body as an excretory mechanism and that even a mouse mammary cancer virus could be transmitted via a mother mouse's milk. However, it is only within the past two decades that the importance of human milk as an excretory mechanism has been appreciated. Because of the relatively small number of lactating women in the workplace historically, milk analysis has attracted very little interest for monitoring workplace exposure of the mother.

Nursing infants could have body weights as small as one-twentieth the body weight of the mother. Thus, milk levels and corresponding blood or tissue levels that would not be an effect level for the mother could be an effect level, or could accumulate to an effect level, in the infant. In addition, the neonatal enzyme systems for metabolizing the xenobiotic could be incomplete, leading to toxic accumulations.

These considerations, rather than industrial exposure monitoring, have been principally responsible for the interest in monitoring mother's milk. It is unlikely that such analyses will be used extensively for industrial monitoring, but a nursing industrial employee could seek assurance that she did not have a possible effect level of workplace chemicals in her milk.

Since the concern has been principally with nonindustrial exposure, very few studies have related human milk levels of industrial chemicals to blood or plasma levels, to body storage levels, or to workplace exposure levels.

Pesticide levels reported in mother's milk have been summarized (Polishuk et al., 1977). Among the pesticides reported in mother's milk are β-benzene hexachloride, benzenehexachloride, DDT and its metabolites, dieldrin, heptachlor epoxide, and hexachlorobenzene. Strassman and Kutz (1977) also reported finding oxychlordane and *trans*-nonachlor (metabolites of chlordane and heptachlor), respectively, in mother's milk. Curley et al. (1969) also have reported finding α-benzene hexachloride, endrin, aldrin, and mirex in mother's milk. Kepone also has been reported in mother's milk (Giacoia and Catz, 1979).

Mercury and lead (Knowles, 1974), molybdenum (Anke et al., 1971), and iron (Lanzkowsky, 1978) also have been reported in mother's milk.

Many medicinals also are excreted in mother's milk. These have been reviewed in several papers (Stowe and Plaa, 1956; Savage, 1976; Giacoia and Catz, 1979; Lien, 1979; White and White, 1980; Reinhardt et al., 1982; Wolff, 1983; Menella and Beauchamp, 1991). Medicinals reported in mother's milk include barbiturates, sulfonamides, some hormones, oral contraceptives, lithium salts, narcotics, ergotamine, some hypoglycemic agents, acetylsalicylic acid (aspirin), methadone, ethanol, opiates, caffeine, theobromine, theophylline, sulfanilamide, penicillin, erythromycin and other antibiotics, and digoxin.

Subsequent to exposure to styrene, alkyl benzenes, toluene, or trichloroethylene, their metabolites were found in mother's milk (Stowe and Plaa, 1956). Halothane has been reported in the milk of a nursing anesthesiologist (Cote et al., 1976).

Giroux et al. (1992) have tabulated reports of the occurrence of another 130 additional medicinals, chemicals from foodstuffs and household products, home remedies, vitamins, pesticides, and industrial chemicals in mother's milk.

Although few analyses for industrial chemicals have been carried out on mother's milk following industrial exposure, their appearance should be expected since the factors governing the appearance of pesticides or medicinals in mother's milk also would operate for industrial chemicals. Although few studies have been carried out to elucidate the factors, there appear to be many. Thus, a simple spot analysis of an individual mother's milk should be expected to give a result that is indicative of a range of possible plasma concentrations rather than a specific concentration.

The chemicals found in mother's milk were grouped above by their function rather than by their chemical structure and/or physical properties. This was done to make the point that any chemical, whatever its functional role in the world, appears to be a candidate to become a constituent of mother's milk, once it becomes systemic. If one looks at the chemical structures and solubilities of the chemicals that have been reported in mother's milk, it is obvious that representatives of a multitude of inorganic and organic chemical classes are represented. Fat-soluble as well as water-soluble classes also are represented. There does not appear to be a very effective blood–milk barrier. Thus, mother's milk could be expected to be an excellent source of biological markers for measuring workplace exposure to xenobiotics. Unfortunately, it is not widely or consistently available in the work force.

Because milk has both an aqueous and a lipid phase, concentrations of excreted chemicals might be expected to be related to the unbound blood concentration of the chemical (Stowe and Plaa, 1956) and also to levels stored in fatty depots. This is in contrast to urine, saliva, perspiration, and tears that have no, or a much smaller, lipid component. The pH of mother's milk is slightly lower (ca. 7.0) than that of plasma (ca. 7.4). Therefore, weak, unbound bases might be expected to partition preferentially to milk. Polychlorinated biphenyls appear to partition into mother's milk (Polishuk et al., 1977), as do polybrominated biphenyls (PBBs) (Eyster et al, 1983). The milk levels of the latter were reported to be 100 times the maternal serum levels, which would lead to a very unusual standard curve in a biological assay for PBBs in mother's milk.

Other factors that appear to affect the concentration of xenobiotics in mother's milk include age, weight, number of previous pregnancies (Polishuk et al., 1977), sampling time of day, time since last sample, and fat content of the milk (Mes and Davies, 1978). These latter authors also reported that the fat content was higher in mother's milk at the end of a nursing period than at the beginning. Thus, the level of fat-soluble materials in the milk would be expected to be higher at the end of a nursing period than at the beginning.

The factors discussed in Section 6 also would affect the level of a biological marker in mother's milk (Shepherd et al., 1983). Mechanisms, factors, and kinetics of excretion of xenobiotics in mother's milk have been reviewed (Wilson et al., 1980).

Impairment of an alternate excretory mechanism also would be expected to increase levels of a biological marker in milk. Conversely, lactation could be expected to alter the usual urine and blood ratios and concentrations for a nonbound, fat-soluble or water-soluble xenobiotic and its metabolites. This would lead to erroneous calculated exposure levels of the xenobiotics in the workplace if only the urine were analyzed and the fact of lactation were ignored. Furthermore, the error would be one of underestimation of workplace exposure. Similar considerations would hold for analyses of saliva, perspiration, tears, hair, and nails from lactating women.

Increased photoperiod has been shown to increase milk output 15 percent in Holstein cattle (Peters et al., 1978). The effect on the concentrations of chemical components was not studied. It is not known whether photoperiod would affect human milk output or the concentrations of biological markers. In any case it is likely that the fiducial limits of the normal value would be considerably greater than ± 15 percent. Any effect of photoperiod, therefore, would be insignificant in determining whether the measured concentration reflected a safe or unsafe exposure level for the mother.

Xenobiotics and their metabolites would be expected to appear in milk at least as rapidly as in urine. Thus, milk analysis would have an advantage over hair and nail analyses in that it would give a measure of current or immediate past exposures rather than historical exposures.

In summary, mother's milk is a possible excretory fluid for assessing workplace exposure to chemicals. However, because of the limited number of lactating women in the workplace, development of correlations with plasma levels and toxic effect levels in the mother is difficult to justify in preference to similar analytical development for biological markers in urine and exhaled air.

12 SALIVA, TEARS, PERSPIRATION, AND NAIL ANALYSES AS MEASURES OF INDUSTRIAL EXPOSURE

Saliva, tears, and perspiration, although not usually considered excretory fluids, do contain chemicals transferred from blood. Xenobiotics can partition between these fluids and blood. Saliva and perspiration levels of chemicals seem to reflect unbound concentrations of the chemicals in blood (Allen et al., 1975; Schmidt and Kupferberg, 1976; Paxton et al., 1976). It is likely that tear levels also will reflect blood concentrations.

Very little work has been reported on the use of these excreted fluids for biological monitoring. It is probable that all the factors affecting xenobiotic levels discussed in Sections 6 and 7 will apply to these fluids. There also are obvious problems collecting samples of tears and perspiration.

Saliva analysis has demonstrated some utility for following blood levels of drugs such as the antiepileptic drug phenytoin (Schmidt and Kupferberg, 1976 and Paxton et al., 1976).

Metals or metallic salts reported in saliva following industrial exposure include cadmium (Dreizen et al., 1970; Gervais et al., 1981) and mercury (Joselow et al., 1968). Like PBB in mother's milk discussed previously, cadmium concentrations were higher in saliva than in plasma (Gervais et al., 1981).

Saliva analyses might overcome one of the problems of urine analysis: variable dilution of the xenobiotic or its metabolites by urine already in the bladder. Samples could be collected easily at the end of the workday and could provide a viable alternative to urine or blood analyses for nonvolatile workplace chemicals.

Salicylic acid and urea are among the industrially important organic chemicals reported in saliva and tears, respectively (Lanzkowsky, 1978).

Some of the chemicals of industrial importance that have been reported in perspiration include lead (Paxton et al., 1976), arsenic, mercury, iron, manganese, zinc, magnesium, copper, ethanol, benzoic acid, salicylic acid, urea, and phenol (Lanzkowsky, 1978). Arsenic appears in perspiration very quickly after administration (Hopps, 1974).

Although metals have been detected in human fingernails or toenails (Goldblum et al., 1953; Barnett, 1972; Hopps, 1974), analysis of human nails has not been used as a technique for monitoring industrial exposure. Some of the metals reported in human nails are arsenic (Agahian et al., 1990), copper (Barnett et al., 1972), and zinc (Goldblum et al., 1953). The same considerations previously discussed in Section 10 relative to hair analysis apply to nail analysis. For example, Harrison and Clemena (1972) found that it was necessary to wash the nails to remove surface contamination. They also reported that the metal content of a nail varied from nail to nail in the same subject.

Olguin et al. (1983) found elevated arsenic levels in fingernails from 100 subjects who, for 10–60 years, had been drinking water contaminated with arsenic. The arsenic levels in the nail clippings from exposed subjects were significantly higher ($p < 0.001$) than in nail clippings from unexposed controls. The difference in arsenic levels between the group of subjects showing clinical signs of arsenic poisoning and the group that did not was not significant ($p > 0.05$).

Aghanian et al. (1990) studied the arsenic content of fingernails from miners working in a strip mining, crushing, and smelting operation. The ore contained gold and the arsenic mineral, realgar (AsS). Nail clippings from all 10 fingers were pooled for each subject. The subjects were grouped according to whether their exposure to arsenic-containing dust or fumes was high, low, or in between. The maximum airborne arsenic concentration was 24 $\mu g/m^3$. Aghanian et al. were able to derive a simple equation correlating airborne arsenic levels with fingernail content of arsenic:

$$\text{Air level } (\mu g/m^3) = 1.79 \times (\text{nail level in } \mu g/g) - 5.9$$

In relating nail levels of a metal to previous exposure, it should be noted that human fingernails grow approximately 100 $\mu m/day$ and that human toenails grow only approximately 25 $\mu m/day$ (Hopps, 1974). It also has been reported that the nails of each finger grow at a different rate and that the growth rate is different for the fingers on each hand (Dawber, 1970). Approximately 160 days are required to form the nail matrix.

It is likely that nail growth rate will vary with age, nutritional status, and health status.

13 SUMMARY

Toxicological research since 1954 has shown that Elkins's twin concepts of biological monitoring and biological TLVs are viable concepts. The ACGIH has established Biological Exposure Indices for 26 workplace chemicals and announced the intent to set 8 more (American Conference of Governmental Industrial Hygienists, 1991–1992). Toxicologists have expanded on the use of biological monitoring to include not only the exposure chemical and its primary metabolites but also to include secondary metabolic derivatives as well as chemicals normally present in the body, or body excretions, whose concentration is altered by exposure to the xenobiotic of interest. The latter is an extension of the biological monitoring that has been used by the medical profession for scores of years to diagnose disease states. Toxicologists have coined the word *biological marker* or *biomarker* to describe these chemicals that are characteristic not only of workplace chemicals, but are characteristic of any xenobiotic that has entered the body.

Stable biological markers also have been found to be valuable in retrospective epidemiological studies on health effects that may have had a workplace etiology. Workers in the chemical industry historically have been exposed to a multitude of chemicals over their careers, and it has been difficult to ascribe to a particular chemical an effect that occurred many years later. Where the chemical has caused the formation of a long-lived characteristic biological marker, the presence of the marker documents the exposure to the chemical of interest. If there is a quantitative relationship, the biological marker also gives a measure of the magnitude of the exposure.

As more research is carried out, it has become apparent that the task of using biological markers is validating the folk wisdom that every person is different. Groups of people have characteristic enzyme complements and their enzymes have characteristic kinetics. These are different for the various groups. This makes the task of ascribing excretion levels and body burdens of the biological markers to a particular level of exposure difficult and impacts adversely on the setting of biological TLVs. Fortunately, in the case of the worker, it is possible to determine these variabilities in advance of exposure and to modify, for each individual, the quantitative aspects of the relationship between exposure level and levels of the biological marker in body excretion materials and in the body.

Current research is directed toward not only discovering new biological markers characteristic of the workplace xenobiotic but to biological markers that are characteristic of early, reversible stages of development of an adverse health effect.

As more toxicological research is reported, it is becoming apparent that the types of biological markers that may be sought are limited almost solely by the biochemical and toxicological imagination of the investigators and the availability of suitable instrumentation.

REFERENCES

Abernethy, D. R., D. J. Greenblatt, M. Divoll, J. S. Harmatz, and R. I. Shader (1981). *J. Pharmacol. Exp. Ther.*, **217**, 681–685.

Abernethy, D. R., D. J. Greenblatt, M. Divoll, R. Arendt, H. R. Ochs, and R. I. Shader (1982). *N. Engl. J. Med.*, **306**, 791–792.

Agahian, B., J. S. Lee, J. H. Nelson, and R. E. Johns (1990). *Am. Ind. Hyg. Assoc. J.*, **51**, 646–651.

Allen, B. R., M. R. Moore, and J. A. A. Hunter (1975). *Br. J. Dermatol.*, **92**, 715–719.

Alvares, A. P., K. E. Anderson, A. H. Conney, and A. Kappas (1976). *Proc. Natl. Acad. Sci. U.S.A.*, **73**, 2501–2504.

American Conference of Governmental Industrial Hygienists (1991). *1991–1992 Threshold Limit Values for Chemical Substances and Physical Agents and Biological Exposure Indices*, American Conference of Governmental Industrial Hygienists, Cincinnati, OH.

Anke, M., A. Hennig, M. Diettrich, G. Hoffman, G. Wicke, and D. Pflug (1971). *Arch. Tierernähr*, **21**, 505–513.

Anon. (Aug. 13, 1977). *Lancet*, 337–338.

Arunachalam, J., S. Gangadharan, and S. Yegnasubramanian (1979). *Nucl. Act. Tech. Life Sci., Proc. Int. Symp.*, 499–513.

Ashurst, S. W. and G. M. Cohen (1981). *Int. J. Can.*, **27**, 357–364.

Autrup, H., J. Wakhisi, K. Vahakangas, A. Wasunna, and C. C. Harris (1985). *Environ. Hlth. Perspect.*, **62**, 105–108.

Bakke, O. M. and R. R. Scheline (1970). *Toxicol. Appl. Pharmacol.*, **16**, 691–700.

Bardodej, Z. and E. Bardodejova (1966). *Cesk. Hyg.*, **11**, 226–235; *Chem. Abstr.*, **65**, 6086g (1966).

Baretta, E. D., R. D. Stewart, and J. E. Mutchler (1969). *J. Am. Ind. Hyg. Assoc.*, **30**, 537–544.

Barnett, W. B., S. Slavin, and F. J. Fernandez (1972). *Clin. Chem.*, **18**, 716–723.

Baselt, R. C. (1983). *Biological Monitoring Methods for Industrial Chemicals*, Biomedical, Davis, CA.

Bate, L. C. and F. F. Dyer (1965). *Nucleonics*, **23**, 74–81.

Baumgartner, A. M., P. F. Jones, W. A. Baumgartner, and C. T. Black (1979). *J. Nucl. Med.*, **20**, 748–752.

Baumgartner, A. M., P. F. Jones, and C. T. Black (1981). *J. Forensic Sci.*, **26**, 576–581.

Baumgartner, W. A., C. T. Black, P. F. Jones, and W. H. Blahd (1982). *J. Nucl. Med.*, **23**, 790–792.

Bechtold, W. E., G. Lucier, L. S. Birnbaum, S. N. Yin, G. L. Li, and R. F. Henderson (1991). *Am. Ind. Hyg. Assoc. J.*, **52**, 473–478.

Bernstein, D. I., C. R. Zeiss, P. Wolkonsky, D. Levitz, M. Roberts, and R. Patterson (1983). *J. Allergy Clin. Immunol.*, **72**, 714–719.

Bhat, K. R., J. Arunachalam, S. Yegnasubramanian, and S. Gangadharan (1982). *Sci. Total Environ.*, **22**, 169–178.

Biagini, R. E., S. L. Klincewicz, G. M. Henningsen, B. A. MacKenzie, J. S. Gallagher, D. I. Bernstein, and I. L. Bernstein (1990). *Life Sci.*, **47**, 897–910.

Bond, J. A., W. E. Bechtold, L. S. Birnbaum, A. R. Dahl, M. A. Medinsky, J. D. Sun, and R. F. Henderson (1991). *Toxicol. Appl. Pharmacol.*, **107**, 494–503.

Bowman, J. D., J. L. Held, and D. R. Factor (1990). *Appl. Occup. Environ. Hyg.*, **5**, 526–535.

Bray, H. G., B. G. Humphris, and W. V. Thorpe (1949). *Biochem. J.*, **45**, 241–44.

Brenner, D. D., A. M. Jeffrey, L. Latriano, L. Wazneh, D. Warburton, M. Toor, R. W. Pero, L. R. Andrews, S. Walles, and F. P. Perera (1991). *Mutat. Res.*, **261**, 225–236.

Butler, T. C. (1949). *J. Pharmacol. Exp. Ther.*, **97**, 84–92.

Byington, K. H. and K. C. Liebman (1965). *Molec. Pharmacol.*, **1**, 247.

Calabrese, E. J. (1985). *Toxic Susceptibility: Male/Female Differences*, Wiley-Interscience, New York.

Calabrese, E. J. (1986). *Br. J. Ind. Med.*, **43**, 577–579.

Calleman, C. J., L. Ehrenberg, B. Jansson, S. Osterman-Golkar, D. Segerback, K. Svensson, and C. A. Wachtmeister (1978). *J. Environ. Pathol. Toxicol.*, **2**, 427–442.

Campbell, L., A. H. Jones, and H. K. Wilson (1985). *Am. J. Ind. Med.*, **8**, 143–154.

Carpenter, C. P., C. B. Shaffer, C. S. Weil, and H. F. Smyth Jr. (1944). *J. Ind. Hyg. Toxicol.*, **26**, 69–78.

Cartier, A., L. Grammer, J-L. Malo, F. Lagier, H. Ghezzo, K. Harris, and R. Patterson (1989). *J. Allergy Clin. Immunol.*, **84**, 507–514.

Cole, W. J., R. G. Mitchell, and R. F. Salamonsen (1975). *J. Pharm. Pharmacol.*, **27**, 167–171.

Commerce Clearing House, Inc. (1978). *Food Drug Cosmetic Law Reporter*, Chicago, 1978.

Conney, A. H. (1967). *Pharmacol. Rev.*, **19**, 317–366.

Conney, A. H., E. J. Pantuck, K. C. Hsiao, R. Kuntzman, A. P. Alvares, and A. Kappas (1977). *Fed. Proc., Fed. Am. Soc. Exp. Biol.*, **36**, 1647–1652.

Cooper, J. R. and P. J. Friedman (1958). *Biochem. Pharmacol.*, **1**, 76–82.

Cornish, H. H. (1971). *Crit. Rev. Toxicol.*, **1**, 1–32.

Cote, C. J., N. B. Kennep, S. B. Reed, and G. E. Strobel (1976). *Br. J. Anaesth.*, **48**, 541–543.

Cralley, U. U., U. J. Cralley, and W. C. Cooper (1990). *Health and Safety Beyond the Workplace*, Wiley, New York.

Curley, A., F. Copeland, and R. D. Kimbrough (1969). *Arch. Environ. Hlth.*, **19**, 628–632.

Dallas, C. E., J. M. Gallo, R. Ramanathan, S. Muralidhara, and J. V. Bruckner (1991). *Toxicol. Appl. Pharmacol.*, **110**, 303–314.

Daniel, J. W. (1963). *Biochem. Pharmacol.*, **12**, 795–802.

Daniel, J. W. and J. C. Gage (1965). *Food Cosmet. Toxicol.*, **3**, 405–415.

Danishefsky, I. and M. Willhite (1954). *J. Biol. Chem.*, **211**, 549–553.

Dawber, R. (1970). *Br. J. Dermatol.* **82**, 454–457.

Dean, M., B. Stock, R. J. Patterson, and G. Levy (1980). *Clin. Pharmacol. Ther.*, **28**, 253–261.

Derryberry, O. M., M. D. Bartholomew, and R. B. L. Fleming (1963). *Arch. Environ. Hlth.*, **6**, 503–511.

Desmond, P. V., R. K. Roberts, G. R. Wilkinson, and S. Schenker (1979). *N. Engl. J. Med.*, **300**, 199–200.

DiVincenzo, G. D., P. F. Yanno, and B. D. Astill (1972). *J. Am. Ind. Hyg. Assoc.*, **33**, 125–135.

Dreizen, S., B. M. Levy, W. Niedermeier, and J. H. Griggs (1970). *Arch. Oral Biol.*, **15**, 179–188.

Dutkiewicz, T. (1971). *Bromatol. Chem. Toksykol.*, **4**, 39–44; *Chem. Abstr.* (1972), **76**, 10786u.

Dutkiewicz, T. and H. Tyras (1967). *Br. J. Ind. Med.*, **24**, 330–332.

Dutkiewicz T. and H. Tyras (1968). *Br. J. Ind. Med.*, **25**, 243.

Ehrenberg, L. and S. Osterman-Golkar (1980). *Teratogen., Carcinogen., Mutagen.*, **1**, 105–128.

El-Dakhakhny, A-A and Y. M. El-Sadik (1972). *J. Am. Industr. Hyg. Assoc.*, **33**, 31–34.

Eldridge, W. A. (1924). Report 29, *Chemical Warfare Service*; B. R. Allen, M. R. Moore, and J. A. A. Hunter (1975). *Br. J. Dermatol.*, **92**, 715–719.

Elia, V. J., L. A. Anderson, T. J. MacDonald, A. Carson, C. R. Buncher, and S. M. Brooks (1980). *J. Am. Ind. Hyg. Assoc.*, **41**, 922–926.

Elkins, H. B. (1954). *AMA Arch. Ind. Hyg. Occup. Med.*, **9**, 212–221.

Elkins, H. B. (1961). *Pure Appl. Chem.*, **3**, 269–273.

Elkins, H. B., and L. D. Pagnatto (1965). *J. Am. Ind. Hyg. Assoc.*, **26**, 456–460.

Elkins, H. B. (1967). *J. Am. Ind. Hyg. Assoc.*, **28**, 305–314.

Ellis, R. W. (1966). *Br. J. Ind. Med.*, **23**, 263–281.

El Masri, A. M., J. N. Smith, and R. T. Williams (1958). *Biochem. J.*, **68**, 199–204.

Engström, K., K. Husman, and J. Rantanen (1976). *Int. Arch. Occup. Environ. Hlth.*, **36**, 153–160.

Evans, D. A. P. and T. A. White (1964). *J. Lab. Clin. Med.*, **63**, 394–403.

Evans, J. N., D. R. Hemenway, and J. Kelly (1989). *Res. Rep. Hlth. Eff., Inst.*, **29**, 1–17.

Eyster, J. T., H. E. B. Humphrey, and R. D. Kimbrough (1983). *Arch. Environ. Hlth.*, **38**, 47–53.

Farmer, P. B., E. Bailey, S. M. Gorf, M. Törnqvist, S. Osterman-Golkar, A. Kautiainen, and D. P. Lewis-Enright (1986). *Carcinogenesis (Lond.)*, **7**, 637–640.

Feely, J., G. R. Wilkinson, and A. J. J. Wood (1981). *N. Eng. J. Med.*, **304**, 692–695.

Fernandez, J., E. Guberan, and J. Caperos (1976). *J. Am. Ind. Hyg. Assoc.*, **37**, 143–150.

Fishbeck, W. A., R. R. Langner, and R. J. Kociba (1975). *J. Am. Ind. Hyg. Assoc.*, **36**, 820–824.

Geokas, M. C. and B. J. Haverback (1969). *Am. J. Surg.*, **117**, 881–892.

Gervais, L., Y. Lacasse, J. Brodeur, and A. P'an (1981). *Toxicol. Lett. (Amst.)*, **8**, 63–66.

Giacoia G. P. and C. S. Catz (1979). *Clin. Perinatol.*, **6**, 181–196.

Gibaldi M. and D. Ferrier (1975). *Pharmacokinetics*, 1st ed., Marcel Dekker, New York, pp. 284–287.

Gibbs, H. D. (1927). *J. Biol. Chem.*, **72**, 649–664.

Gillette, J. R. (1971). *Ann. NY Acad. Sci.*, **179**, 43–66.

Giroux, D., G. Lapointe, and M. Baril (1992). *Am. Ind. Hyg. Assoc. J.*, **53**, 471–474.

Glasstone, S. (1949). *Elements of Physical Chemistry*, 1st ed., Van Nostrand, New York, pp. 362–363.

Goldblum, R. W., S. Derby, and A. B. Lerner (1953). *J. Invest. Dermatol.*, **20**, 13–18.

Götell, P., O. Axelson, and B. Lindelöf (1972). *Work Environ. Hlth.*, **9**, 76–83.

Grammer, L. C., K. E. Harris, M. A. Shaughnessy, P. Sparks, G. H. Ayars, L. C. Altman, and R. Patterson (1990). *J. Allergy Clin. Immunol.*, **86**, 177–181.

Greim H. A. (1981). In *Drugs and Pharmaceutical Sciences*, P. Jenner and B. Testa, Eds., Vol. 10, *Concepts in Drug Metabolism, Part B*, Marcel Dekker, New York, p. 270.

Guberan E. and J. Fernandez (1974). *Br. J. Ind. Med.*, **31**, 159–167.

Haggard, H. W. (1924). *J. Biol. Chem.*, **59**, 737–751.

Hammer, D. I., J. F. Finklea, R. H. Hendricks, C. M. Shy, and R. J. N. Norton (1972). *Air Pollut. Contr. Off. (U.S.) Publ.*, **AP 91**, 125–134.

Harrison, W. H., R. M. Gray, and L. M. Solomon (1974). *Br. J. Dermatol.*, **91**, 415–418.

Harrison, W. W. and G. G. Clemena (1972). *Clin. Chim. Acta*, **36**, 485–492.

Harrison, Y. E. and W. L. West (1971). *Biochem. Pharmacol.*, **20**, 2105–2108.

Hashimoto, K. and W. N. Aldridge (1970). *Biochem. Pharmacol.*, **19**, 2591–2604.

Hayes, W. J., Jr., (1963) *Clinical Handbook on Economic Poisons*, Government Printing Office, Washington, D.C. p. 13.

Henderson, R. F., W. E. Bechtold, J. A. Bond, and J. D. Sun (1989). *Critical Rev. Toxicol.*, **20**, 65–79.

Henley, E. C., M. E. Kassouny, and J. W. Nelson (1977). *Science*, **197**, 277–278.

Hill, D. L. (1953). *AMA Arch. Ind. Hyg. Occup. Med.*, **8**, 347–349.

Hjelm, E. W., P. H. Näslund, and M. Wallén (1988). *J. Toxicol. Environ. Hlth.*, **25**, 155–164.

Hogstedt, C. (1986). In *Progress in Clinical and Biological Research* (M. Sorsa and H. Norppa, Eds.), Vol. 207, Alan R. Liss, New York, pp. 231–244.

Hommes, F. A., Ed. (1973). *Inborn Errors of Metabolism*, Academic Press, New York.

Hopps, H. C. (1974). *Trace Subst. Environ. Hlth.*, **8**, 59–73.

Hurlbut, J. A. (1976). Natl. Tech. Inform. Serv. Report TID-4500-R64.

Ikeda, M. (1964). *J. Biochem. (Tokyo)*, **55**, 231–243; *Chem. Abstr.*, **60**, 16407b (1966).

Ikeda, M. and T. Imamura (1973). *Int. Arch. Arbeitsmed.*, **31**, 209–224.

Ikeda, M. and H. Ohtsuji (1969a). *Br. J. Ind. Med.*, **26**, 162–164.

Ikeda, M. and H. Ohtsuji (1969b). *Br. J. Ind. Med.*, **26**, 244–246.

Ikeda, M. and H. Ohtsuji (1972). *Br. J. Ind. Med.*, **29**, 99–104.

Ikeda, M., H. Ohtsuji, H. Kawai, and M. Kuniyoshi (1971). *Br. J. Ind. Med.*, **28**, 203–206.

Ikeda, M., H. Ohtsuji, and T. Imamura (1972a). *Xenobiotica*, **2**, 101–106.

Ikeda, M., H. Ohtsuji, T. Imamura, and Y. Komoike (1972b). *Br. J. Ind. Med.*, **29**, 328–333.

Ikeda, M., T. Imamura, M. Hayashi, T. Tabuchi, and I. Hara (1974). *Int. Arch. Arbeitsmed.*, **32**, 93–101.

Imamura, T. and M. Ikeda (1973). *Br. J. Ind. Med.*, **30**, 289–292.

Jaworowski, Z. S. (1965), *Atompraxis*, **11**, 271–273; Klevay, L. M. (1973), *Arch. Environ. Health*, **26**, 169–172.

Jenne, J. W. (1965). *J. Clin. Invest.*, **44**, 1992–2002.

Jick, H. (1974). *Med. Clin. North. Am.*, **58**, 1143–1149.

Johannsen, F. R., G. J. Levinskas, and J. L. Schardein (1987). *Fundam. Appl. Toxicol.*, **9**, 550–556.

Joselow, M. M., R. Ruiz, and L. J. Goldwater (1968). *Arch. Environ. Hlth.*, **17**, 35–38.

Jusko, W. J. (1979). *Drug Metabol. Rev.*, **9**, 221–236.

Kalow, W. (1980). *Trends Pharmacol. Sci.* **1**, 403–405.

Kaltreider, N. L., M. J. Elder, L. V. Cralley, and M. O. Colwell (1972). *J. Occup. Med.*, **14**, 531–541.

Karol, M. H., E. J. Riley, H. H. Ioset, and Y. C. Alarie (1978a). *Toxicol. Appl. Pharmacol.*, **45**, 302.

Karol, M. H., H. H. Ioset, E. J. Riley, and Y. C. Alarie (1978b). *Am. Ind Hyg. Assoc. J.*, **39**, 546–556.

Karol, M. H., H. H. Ioset, and Y. C. Alarie (1978c). *Am. Ind. Hyg. Assoc. J.*, **39**, 454–458.

Kehoe R. A. (1963). In *Industrial Hygiene and Toxicology*, 2nd rev. ed., Vol. II, F. A. Patty, Ed, Wiley, New York, p. 941.

Kido, T., R. Honda, I. Tsuritani, M. Ishizaki, Y. Yamada, H. Nakagawa, K. Nogawa, and Y. Dohi (1991). *Arch. Environ. Hlth.*, **46**, 43–49.

Kilpikari, I. and H. Savolainen (1982). *Br. J. Ind. Med.*, **39**, 401–403.

Kitchen, I., J. Tremblay, J. Andres, L. G. Dring, J. R. Idle, R. L. Smith, and R. T. Williams. *Xenobiotica*, **9**, 397–404.

Klevay, L. M. (1973). *Arch. Environ. Hlth.*, **26**, 169–172.

Knowles, J. A. (1974). *Clin. Toxicol.*, **7**, 69–82.

Knox, W. E. and A. H. Mehler (1951). *Science*, **113**, 237–238.

Kraut, H. and M. Weber (1944). *Biochem. Z.*, **317**, 133–148; Klevay, L. M. (1973). *Arch. Environ. Hlth.*, **26**, 169–172.

LaDu, B. N., H. G. Mandel, and E. L. May (1971). *Fundamentals of Drug Metabolism and Drug Disposition*, 1st ed., Williams and Wilkins, Baltimore.

Lanzkowsky, P. (1978). *N. Engl. J. Med.*, **298**, 343.

Lauwerys (1983). *Industrial Chemical Exposure: Guidelines for Biological Monitoring*, Biomedical, Davis, CA.

Leaf, G. and L. J. Zatman (1952). *Br. J. Ind. Med.*, **9**, 19–31.

Levine, L. and J. P. Fahy (1945). *J. Ind. Hyg. Toxicol.*, **27**, 217–223.

Lewalter, J. and U. Korallus (1985). *Int. Arch. Occup. Environ. Hlth.*, **56**, 179–196.

Lien, E. J. (1979). *J. Clin. Pharmacol.*, **4**, 133–144.

Liira, J., V. Riihimäki, and K. Engström (1990). *Br. J. Ind. Med.*, **47**, 325–330.

Long, K. R. (1975). *Int. Arch. Occup. Environ. Hlth.*, **36**, 75–86; *Chem. Abstr.*, 85, 67416p (1976).

Lorimer, W. V., R. Lilis, W. J. Nicholson, H. Anderson, A. Fischbein, S. Daum, W. Rom, C. Rice, and I. J. Selikoff (1976). *Environ. Hlth. Perspect.*, **17**, 171–181.

Loub, W. D., L. W. Wattenberg, and D. W. Davis (1975). *J. Natl. Cancer Inst.*, **54**, 985–988.

MacLeod, C., H. Rabin, R. Ogilvie, J. Ruedy, M. Caron, D. Zarowny, and R. O. Davies (1974). *Can. Med. Assoc. J.*, **111**, 341–346.

Madison, R. E., E. Broughton, and J. D. Thrasher (1991). *Environ. Hlth. Perspect.*, **94**, 219–223.

Mayer, J., D. Warburton, A. M. Jeffrey, R. Pero, S. Walles, L. Andrews, M. Toor, L. Latriano, L. Wazneh, D. Tang, W-Y. Tsai, M. Kuroda, and F. Perera (1990). *Mutat. Res.*, **248**, 163–176.

McConnell K. P. and O. W. Portman (1952). *J. Biol. Chem.*, **195**, 277–282.

McCord, C. P. (1931) *Ind. Eng. Chem.*, **23**, 931–936.

McCord, C. P. (1934). *Am. J. Pub. Hlth.*, **24**, 677–680.

Mennella, J. A. and G. K. Beauchamp (1991). *N. Engl. J. Med.*, **325**, 981–985.

Mes, J. and D. J. Davies (1976). *Chemosphere*, **7**, 699–706.

Neilands, J. B. and P. K. Stumpf (1958). *Outlines of Enzyme Chemistry*, 2nd ed., Wiley, New York, pp. 379–381.

Newberne, P. M. (1975). *Lab. Animal*, **4**, 20–24.

Neymark, M. (1936). *Skand. Arch. Physiol.*, **73**, 227–236; *Chem. Abstr.*, **30**, 49302 (1936).

Nicolson, G. A., A. C. Greiner, W. J. McFarlane, and R. A. Baker (1966). *Lancet*, **433** (February 12), 344–347.

Nishiyama, K. and G. F. Nordberg (1972), *Arch. Environ. Hlth.*, **25**, 92–96.

Nomiyama, K. (1971). *Int. Arch. Arbeitsmed.*, **27**, 281–292.

Nomiyama, K. and H. Nomiyama (1969). *Ind. Health (Kawasaki, Jap.)*, **7**, 86–87; *Chem. Abstr.*, **72**, 82685a (1970).

Nomiyama, K. and H. Nomiyama (1971). *Int. Arch. Arbeitsmed.*, **28**, 37–48.

Ogata, M., K. Sugiyama, and H. Moriyasu (1962). *Acta Med. Okayama*, **16**, 283–292; see Ikeda and Ohtsuji (1969a) and Pagnatto and Lieberman (1967).

Ogata, M., K. Tomokuni, and Y. Takatsuka (1969). *Br. J. Ind. Med.* **26**, 330–334.

Ogata, M., K. Tomokuni, and Y. Takatsuka (1970). *Br. J. Ind. Med.*, **27**, 43–50.

Ogata, M., Y. Takatsuka, and K. Tomokuni (1971). *Br. J. Ind. Med.*, **28**, 386–391.

Ohtsuji, H. and M. Ikeda (1970). *Br. J. Ind. Med.*, **27**, 150–154.

Ohtusji, H. and M. Ikeda (1971). *Toxicol. Appl. Pharmacol.*, **18**, 321–328.

Ohtsuji, H. and M. Ikeda (1972). *Br. J. Ind. Med.*, **29**, 70–73.

Oleru, U. G. (1976). *J. Am. Ind. Hyg. Assoc.*, **37**, 617–621.

Olguin, A., P. Jauge, M. Cebrian, and A. Albores (1983). *Proc. West. Pharmacol. Soc.*, **26**, 175–177.

Osterman-Golkar, S., L. Ehrenberg, D. Segerbäck, and I. Hällstrom (1976). *Mutat. Res.*, **34**, 1–10.

Osterman-Golkar, S., P. B. Farmer, D. Segerbäck, E. Bailey, C. J. Calleman, K. Svensson, and L. Ehrenberg (1983). *Teratogen., Carcinogen., Mutagen.*, **3**, 395–406.

Pagnotto, L. D. and L. M. Lieberman (1967). *J. Am. Ind. Hyg. Assoc.*, **28**, 129–134.

Parke, D. V. (1968). *The Biochemistry of Foreign Compounds*, 1st ed., Pergamon Press, New York, p. 146.

Paxton, J. W., B. Whiting, F. J. Rowell, J. G. Ratcliffe, and K. W. Stephen (1976). *Lancet*, Sept. 18, 639–640.

Perbellini, L., F. Brugnone, and I. Pavan (1980). *Toxicol. Appl. Pharmacol.*, **53**, 220–229.

Petering, H. G., D. W. Yeager, and S. O. Witherup (1971). *Arch. Environ. Hlth.*, **23**, 202–207.

Petering, H. G., D. W. Yeager, and S. O. Witherup (1973). *Arch. Environ. Hlth.*, **27**, 327–330.

Peters, R. R., L. T. Chapin, K. B. Leining, and H. A. Tucker (1978). *Science*, **199**, 911–912.

Pfäffli, P., A. Hesso, H. Vainio, and M. Hyvönen (1981). *Toxicol. Appl. Pharmacol.*, **60**, 85–90.

Phelps, R. W., T. G. Kershaw, T. W. Clarkson, and B. Wheatley (1980). *Arch. Environ. Hlth.*, **35**, 161–168.

Piotrowski, J. K. (1971). *Br. J. Ind. Med.*, **28**, 172–178.

Piotrowski, J. K. (1977). *Exposure Tests for Organic Compounds in Industrial Toxicology*, National Technical Information Service Publication PB-274 767.

Polishuk, Z. W., M. Ron, M. Wassermann, S. Cucos, D. Wassermann, and C. Lemesch (1977). *Pestic. Monit. J.*, **10**, 121–133.

Porteous, J. W. and R. T. Williams (1949). *Biochem. J.*, **44**, 46–55.

Quick, A. J. (1931). *J. Biol. Chem.*, **92**, 65–85.

Rainsford, S. G. and T. A. L. Davies (1965). *Br. J. Ind. Med.*, **22**, 21–26.

Rawat, A., N. Sethi, and K. Srivastava (1986). *IRCS Med. Sci.*, **14**, 717–718.

Reidenberg, M. M. (1974). *Med. Clin. North Am.*, **58**, 1103–1109.

Reinhardt, D., O. Richter, T. Genz, and S. Potthoff (1982). *Eur. J. Pediatr.*, **138**, 49–52.

Renshaw, G. D., C. A. Pounds, and E. F. Pearson (1972). *Nature*, **238**, 162–163.

Riihimäki, V., P. Pfäffli, K. Savolainen, and K. Pekari (1979). *Scand. J. Work Environ. Hlth.*, **5**, 217–231.

Robinson, D., J. N. Smith, and R. T. Williams (1955). *Biochem. J.*, **59**, 153–159.

Roush, G. J. and M. G. Ott (1977). *J. Am. Ind. Hyg. Assoc.*, **38**, 67–75.

Rowe, J. W., R. Andres, J. D. Tobin, A. H. Norris, and N. W. Shock (1976). *Ann. Intern. Med.*, **84**, 567–569.

Sabourin, P. J., W. E. Bechtold, W. C. Griffith, L. S. Birnbaum, G. Lucier, and R. F. Henderson (1989). *Toxicol. Appl. Pharmacol.*, **99**, 421–444.

Sato, A. and T. Nakajima (1987). *Scand. J. Work Environ. Hlth.*, **13**, 81–93.

Sato, A., K. Endoh, T. Kaneko, and G. Johanson (1991). *Br. J. Ind. Med.*, **48**, 548–556.

Savage, R. L. (1976). *Adverse Drug Reaction Bulletin*, pp. 212–215.

Sayers, R. R., W. P. Yant, H. H. Schrenk, J. Chornyak, S. J. Pearce, F. A. Patty, and J. G. Linn (1942). *Methanol Poisoning: I. Exposure of Dogs to 450–500 ppm. Methanol Vapor in Air, R.I. 3617*, U.S. Department of the Interior, Bureau of Mines; *Chem. Abstr.*, **36**, 45962 (1942).

Schmidt, D. and H. J. Kupferberg (1975). *Epilepsia*, **16**, 735–741.

Schrenk, H. H., W. P. Yant, S. J. Pearce, F. A. Patty, and R. R. Sayers (1941). *J. Industr. Hyg. Toxicol.*, **23**, 20–34.

Schroeder, H. A. and A. P. Nason (1969). *J. Invest. Dermatol.*, **53**, 71–78.

Segerbäck, D. (1983). *Chem. Biol. Interact.*, **45**, 139–152.

Seki, Y., Y. Urashima, H. Aikawa, H. Matsumura, Y. Ichikawa, F. Hiratsuka, Y. Yoshioka, S. Shimbo, and M. Ikeda (1975). *Int. Arch. Arbeitsmed.*, **34**, 39–49.

Sénczuk, W. and B. Litewka (1976). *Br. J. Ind. Med.*, **33**, 100–105.

Shepherd, E. C., L. W. Robertson, N. D. Heidelbaugh, S. H. Safe, and T. D. Phillips (1983). *Toxicologist*, **3**, 6.

Shuker, D. E. G., E. Bailey, A. Parry, J. Lamb, and P. B. Farmer (1987). *Carcinogenesis (Lond.)*, **8**, 959–962.

Singer, B. (1976). *Nature (Lond.)*, **264**, 333–339.

Skipper, P. L. and S. R. Tannenbaum (1990). *Carcinogenesis (Lond.)*, **11**, 507–518.

Skipper, P. L., M. S. Bryant, S. R. Tannenbaum, and J. D. Groopman (1986). *J. Occup. Med.*, **28**, 643–646.

Sloan, T. P., A. Mahgoub, R. Lancaster, J. R. Idle, and R. L. Smith (1978). *Br. Med. J.*, **2**, 655–657.

Slob, A. (1973). *Br. J. Ind. Med.*, **30**, 390–393.

Souček, B. and D. Vlachová (1960). *Br. J. Ind. Med.*, **17**, 60–64.

Stewart, R. D. and H. C. Dodd (1964). *Ind. Hyg. J.*, **25**, 439–446.

Stewart, R. D., H. H. Gay, D. S. Erley, C. L. Hake, and J. E. Peterson (1961). *J. Occup. Med.*, **3**, 586–590.

Stewart, R. D., H. C. Dodd, E. D. Baretta, and A. W. Schaffer (1968). *Arch. Environ. Hlth.*, **16**, 656–662.

Stewart, R. D., E. D. Baretta, H. C. Dodd, and T. R. Torkelson (1970). *Arch. Environ. Hlth.*, **20**, 224–229.

Stewart, R. D., C. L. Hake, and J. E. Peterson (1974a). *Arch. Environ. Hlth.*, **29**, 6–13.

Stewart, R. D., C. L. Hake, A. J. LeBrun, and J. E. Peterson (1974). *Biologic Standards for the Industrial Worker by Breath Analysis: Trichloroethylene*, U.S. Department of Health, Education and Welfare, Cincinnati, p. 96.

Stowe, C. M. and G. L. Plaa (1968). *Ann. Rev. Pharmacol.*, **8**, 337–356.

Strassman, S. C. and F. W. Kutz (1977). *Pest. Monit. J.*, **10**, 130–133.

Stull, D. R. (1947). *Ind. Eng. Chem.*, **39**, 517–518.

Subcommittee on Immunotoxicology, Committee on Biologic Markers, Commission on Life Sciences, National Research Council (1992). *Biologic Markers in Immunotoxicology*, National Academy Press, Washington, D.C.

Sun, J. D., M. A. Medinsky, L. S. Birnbaum, G. Lucier, and R. F. Henderson (1990). *Fundam. Appl. Toxicol.*, **15**, 468–475.

Suzuki, Y., K. Nishiyama, and Y. Matsuka (1958). *Tokushima J. Exp. Med.*, **5**, 111–119; Klevay, L. M. (1973). *Arch. Environ. Hlth.*, **26**, 169–172.

Tada, O., K. Nakaaki, S. Fukabori, and J. Yonemoto (1975). *J. Sci. Labour Part 2 (Rodo Kagaku)*, **51**, 143–153; *Chem. Abstr.*, **83**, 54178w (1973).

Tanaka, S. and M. Ikeda (1968). *Br. J. Ind. Med.*, **25**, 214–219.

Tang, T. and M. A. Friedman (1977). *Mut. Res.*, **46**, 387–394.

Tardif, R., S. Laparé, G. L. Plaa, and J. Brodeur (1991). *Toxicologist*, **11**, 318.

Tephly, T. R. (1977). *Fed. Proc.*, *Fed. Am. Soc. Exp. Biol.*, **36**, 1627–1628.

Theis, R. C. and S. R. Benedict (1924). *J. Biol. Chem.*, **61**, 67–71.

Thrasher, J. D., A. Broughton, and R. Madison (1990). *Arch. Environ. Hlth.*, **45**, 217–223.

Törnqvist, M., J. Mowrer, S. Jensen, and L. Ehrenberg (1986a). *Anal. Biochem.*, **154**, 255–266.

Törnqvist, M., S. Osterman-Golkar, A. Kautiainen, S. Jensen, P. B. Farmer, and L. Ehrenberg (1986b). *Carcinogenesis*, **7**, 1519–1522.

Triggs, E. J., R. L. Nation, A. Long, and J. J. Ashley (1975). *Eur. J. Clin. Pharmacol.*, **8**, 55–62.

Tyl, R. W., I. M. Pritts, K. A. France, L. C. Fisher, and T. R. Tyler (1988). *Fundam. Appl. Toxicol.*, **10**, 20–39.

U.S. Dept. Health, Education and Welfare (1972). *Criteria for a Recommended Standard ... Occupational Exposure to Inorganic Lead*, Cincinnati.

U.S. Dept. Health, Education and Welfare (1974). *Criteria for a Recommended Standard ... Occupational Exposure to Benzene*, Cincinnati, P. 112.

U.S. Dept. Health, Education and Welfare (1975). *Criteria for a Recommended Standard ... Occupational Exposure to Inorganic Fluorides*, Cincinnati.

U.S. Dept. of Labor, Occupational Safety and Health Administration (1975). *Fed. Regis.*, **40**, 45934.

U.S. Dept. of Labor (1978). *Fed. Reg.*, **43**, 5917.

U.S. Department of Labor, Occupational Safety and Health Administration (1991a), 29CFR 1910.1000.

U.S. Department of Labor, Occupational Safety and Health Administration (1991b), 29CFR 1910.1025.

U.S. Department of Labor, Occupational Safety and Health Administration (1991c), 29CFR 1910.1028.

U.S. Environmental Protection Agency (1977). *Fed. Regis.*, **42**, 63076 (1977).

Van Haaften, A. B. and S. T. Sie (1965). *J. Am. Ind. Hyg. Assoc.*, **26**, 52–58.

Van Sittert, N. J., G. deJong, M. G. Clare, R. Davies, B. J. Dean, L. J. Wren, and A. S. Wright (1985). *Br. J. Ind. Med.*, **42**, 19–26.

Vessell, E. S. and G. T. Passananti (1971). *Clin. Chem.*, **17**, 851–866.

Vogt, R. F. (1991). *Environ. Health Perspect.*, **95**, 85–91.

Von Oettingen, W. F., P. A. Neal, and D. D. Donahue (1942). *J. Am. Med. Assoc.*, **118**, 579–584.

Walkley, J. E., L. D. Pagnotto, and H. B. Elkins (1961). *Ind. Hyg. J.*, **22**, 362–367.

Wallach, J. (1974). *Interpretation of Diagnostic Tests*, 2nd ed., Little, Brown, Boston.

Wallén, M., P. H. Näslund, and M. B. Nordqvist (1984). *Toxicol. Appl. Pharmacol.*, **76**, 414–419.

Walshe, J. M. (1956). *Am. J. Med.*, **21**, 487.

Waritz, R. S., J. G. Aftosmis, R. Culik, O. L. Dashiell, M. M. Faunce, F. D. Griffith, C. S. Hornberger, K. P. Lee, H. Sherman, and F. O. Tayfun (1977). *J. Am. Ind. Hyg. Assoc.*, **38**, 307–320.

Watanabe, P. G., G. R. McGowan, and P. J. Gehring (1976). *Toxicol. Appl. Pharmacol.*, **36**, 339–352.

West, E. S. and W. R. Todd (1966). *Textbook of Biochemistry*, 4th ed., Macmillan, New York.

Westra, J. G., E. Kriek, and H. Hittenhausen (1976). *Chem.-Biol. Interact.*, **15**, 149–164.

White, G. J. and M. K. White (1980). *Vet. Hum. Toxicol.*, **22**(Suppl. 1), 1–44.

Widmark, E. M. P. (1933). *Biochem. Z.*, **259**, 285–293.

Williams, R. T. (1959). *Detoxication Mechanisms*, 2nd ed., Wiley, New York.

Wilson, J. T., R. D. Brown, D. R. Cherek, J. W. Dailey, B. Hilman, P. C. Jobe, B. R.

Manno, J. E. Manno, H. M. Redetzki, and J. J. Stewart (1980). *Clin. Pharmacokinet.*, **5**, 1–66.

Winter S. L. and J. L. Boyer (1973). *N. Engl. J. Med.*, **289**, 1180–1182.

Wolff, M. S. (1983). *Am. J. Industr. Med.*, **4**, 259–281.

Wolff, M. S., R. Lilis, W. V. Lorimer, and I. J. Selikoff (1978). *Scand. J. Work Environ. Hlth.*, **4**(Suppl. 2), 114–118.

Yaffe, S., B. Sonawane, H. Lau, P. Coates, and O. Koldovsky (1980). *Fed. Proc.*, *Fed. Am. Soc. Exp. Biol.*, **39**, 751.

Yant, W. P. and H. H. Schrenk (1937). *J. Ind. Hyg. Toxicol.*, **19**, 337–345.

Yllner, S. (1961). *Nature*, **191**, 820.

Yoakum, A. M. (1976). Natl. Tech. Inform. Serv. Report EPA-600/1-76-029.

CHAPTER FOUR

Body Defense Mechanisms to Toxicant Exposure

James S. Bus, Ph.D., and James E. Gibson, Ph.D.

1 INTRODUCTION

At any moment the human body is exposed to an incalculably large number of events capable of causing adverse responses in the body. Although many of these potentially damaging reactions can be attributed to voluntary or involuntary exposure to synthetic substances, it is also true that many of the events are linked to endogenous biological reactions or to agents in the natural environment and hence are totally inescapable. It is obvious, however, that the vast majority of these potential reactions do not result in tissue damage; otherwise the expected lifespan would be short indeed. The reason that all living organisms are able to sustain life in the face of such a continuous onslaught of potentially toxic reactions lies in the ability of the body to defend itself against these reactions by a wide variety of mechanisms.

Although the number of defense mechanisms is exceedingly diverse, they can be divided into three general classifications. Physical or anatomic defenses primarily function as barriers to entry of an agent. Entry must be viewed broadly, including actual entry into the body, entry into organs or tissues once systemic absorption has occurred, and entry into suborgan or subcellular environments. Physiological defenses, which also have as their primary function the prevention of absorption of agents, have the additional important capacity of being able to respond to a variety of toxic insults. The third classification of defense mechanisms, biochemical defenses, act at the cellular level to prevent or repair damaging reactions induced by

Patty's Industrial Hygiene and Toxicology, Third Edition, Volume 3, Part B, Edited by Lewis J. Cralley, Lester V. Cralley, and James S. Bus.
ISBN 0-471-53065-4 © 1995 John Wiley & Sons, Inc.

toxicants. As has been noted by Ames and Gold (1), defense mechanisms developed by animals should be regarded as being of a general type, and thus can be expected to offer protection not only against natural toxic insults but, equally important, also against injury induced by synthetic chemicals.

An important feature of body defense systems is the fact that their diverse nature allows them to effectively function in a multilayered manner to diminish the consequences of toxicant exposure. By way of analogy, biological defense mechanisms can be viewed as functioning much in the same manner as military defenses intended to prevent intrusion of enemy forces. An effective military defense is designed to maximally protect its most vital targets and does so by exacting a progressively greater toll on the attacking force as it approaches those targets. Such defenses increase the probability that vital targets will be attacked only by weakened forces, if at all. Damage sustained from a weakened attacking force is far more likely to be able to be repaired or replaced, and thus maintain the basic viability of the defender.

Figure 4.1 illustrates the multilayered defenses in place in biological systems that prevent both endogenously and exogenously administered compounds from damaging important intracellular macromolecular targets. The first defense barrier is to absorption itself, and if this is breached, additional barriers prevent distribution of the

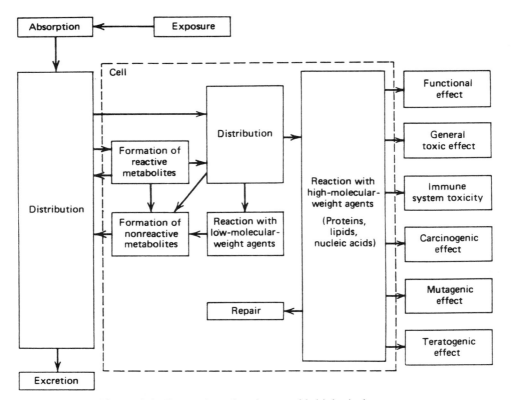

Figure 4.1 Interaction of toxicants with biological systems.

agent to body organs or tissues. Should a toxicant gain entry to the cell, it still may face further physical barriers (such as organelle membranes) to intracellular distribution to critical functional targets. In addition an array of biochemical defenses functioning both inside and outside of cells are active in diminishing the toxic potential of a chemical, and in many cases enhances susceptibility to elimination by excretion. Finally, even if a toxicant should manage to survive these defenses and attack critical cell macromolecules, the potential remains for the damaged macromolecules to be repaired, thus preventing or diminishing expression of a toxic response. If necessary, individual cells irreparably damaged by toxicants may be selectively discarded and replaced, again with the end result of maintaining organism viability.

The purpose of this chapter is to provide a general overview of some of the important defense mechanisms that modulate the toxicity of foreign compounds. It must be pointed out that, for the sake of brevity, not all proposed or demonstrated detoxication mechanisms are discussed.

2 MEMBRANE DEFENSES TO ABSORPTION AND DISTRIBUTION

For a toxicant to exert its activity at the cellular level, it first must pass through a series of membranes. The membranes encountered include those at the site of absorption, primarily skin, lungs, and gastrointestinal tract, and also those at the tissue level such as the cell, mitochondrial, and nuclear membranes. The ability of membranes to physically limit passage of toxicants can be considered a primary line of defense to toxic agents.

The composition of cell membranes, although not constant from tissue to tissue, nonetheless possesses sufficient commonality to permit construction of a unified model describing the ability of chemicals to pass through them. Before consideration of the factors that affect the ability of a chemical to cross cell membranes, however, it is necessary to describe the current understanding of the structure of biological membranes.

2.1 Factors Influencing Membrane Transfer

In general, membranes consist of two layers of lipid molecules, with each layer spatially oriented perpendicular to the membrane surface (Fig. 4.2). The hydrophobic inner core of the membrane is composed primarily of fatty acid hydrocarbon side chains of phospholipids. Varying amounts of cholesterol also are located in the hydrophobic inner core. Each surface of the membrane is composed of the hydrophilic ends of the phospholipid molecules, which in turn are interspersed with various proteins either partially embedded in or completely transversing the membrane (2).

The majority of toxicants cross membranes by the process of simple diffusion. Although low-molecular-weight hydrophilic compounds can diffuse through aqueous channels in the membrane, the passage of higher-molecular-weight compounds (approximately 100 mass units) is determined primarily by their lipid solubility. Because

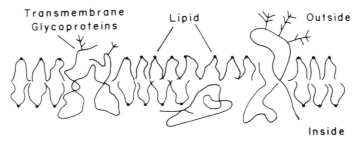

Figure 4.2   Conceptual depiction of a cell membrane. [Redrawn from Bus and Gibson (52).]

of the lipophilic nature of biological membranes, lipid soluble toxicants will diffuse across more readily than toxicants with hydrophilic (water-soluble) properties.

Chemicals may cross membranes by mechanisms other than simple diffusion. These mechanisms, active transport and facilitated diffusion, may not always contribute to the defensive nature of the membrane since they increase the rate of transport across the membrane. Such transport processes can lead to selective accumulation of toxicants in tissues and subsequent overwhelming of the local tissue defense mechanisms (Section 5.4.2). In certain instances, however, transport mechanisms may facilitate elimination of toxicants from the body (Section 4.1), and thus serve as an effective defense mechanism.

Passive diffusion of toxicants through biological membranes is governed by three basic factors: existence of a concentration gradient; lipid solubility, characterized by the oil–water partition coefficient of the toxicant; and degree of ionization (3). Since most toxicants are rapidly distributed away from the inner surface of the membrane on absorption, the latter two factors represent the primary rate-limiting steps for xenobiotic absorption. For many toxicants lipid solubility is reflected by the degree of ionization, which in turn is determined by the pH of the surrounding media. Agents existing in nonionized form will generally have greater lipid solubility than their corresponding ionized form and thus will penetrate membranes more readily.

Ionization of an organic acid or base is a direct function of the pH of the solution in which it is dissolved and the pKa of the agent. The pKa of a chemical is the negative logarithm of the acidic dissociation constant and represents the pH at which a chemical is 50 percent ionized. The degree of ionization of organic acids or bases at any selected pH is given by the Henderson–Hasselbalch equations:

For acids:
$$pKa - pH = \log \frac{\text{nonionized}}{\text{ionized}} \qquad (1)$$

For bases:
$$pKa - pH = \log \frac{\text{ionized}}{\text{nonionized}} \qquad (2)$$

It should be clear from these equations that the extent of ionization of an organic acid, for example, benzoic acid, will decrease with decreasing pH, whereas that for an organic base, such as aniline, will increase with decreasing pH. Thus organic acids will cross membranes more easily when present in a low-pH environment; on the other hand organic bases will cross membranes more easily when in a high-pH solution.

The absorption of toxicants through membranes is governed by the principles outlined in this section. However, each primary site of entry of chemicals into the body and its tissues and organs has unique characteristics that influence the absorption rate from that site.

2.2 Skin

Skin, which probably has the greatest amount of exposure to environmental agents, serves as a relatively impermeable barrier to a large number of potential toxicants. However, skin, like all lipid-containing membranes, will readily permit passage of highly lipid-soluble materials. For example, many cases of human poisoning associated with the agricultural use of the organophosphate insecticides can be directly attributed to the high lipid solubility of these agents and their subsequent ability to readily pass through the skin (4).

For a chemical to move through skin, it must transverse the membranes of a large number of cells. Skin is composed of two multicellular layers, the external epidermis and the underlying dermis (5). The epidermis consists of an outer 8- to 16-cell-thick layer of flattened keratinized cells termed the *stratum corneum*. Beneath that stratum are several layers of living cells that undergo continual differentiation, ultimately ending up as the dead, keratinized cells of the stratum corneum. It is the stratum corneum, however, that offers the primary barrier to passage of chemicals through the epidermis. Removal of this layer of cells by abrasion or other means such as chemical irritation greatly enhances the movement of most chemicals through the skin to the systemic circulation. Once a chemical gains access to the dermis, rapid and complete absorption is usually assured.

Passage of toxicants through the intact epidermis is influenced by several factors (5). The stratum corneum normally contains only small amounts of water, approximately 5–15 percent by weight. If the skin is occluded, which increases hydration by preventing escape of perspiration, the stratum corneum may contain up to 50 percent water. Hydration of the skin can increase permeability up to four- to fivefold. Absorption of chemicals may also be altered by various organic solvents. Dimethyl sulfoxide, for example, has been shown to facilitate movement of chemicals through the skin.

Percutaneous absorption rates may also vary with the anatomical location of the skin. The skin of the palm and heel may be 100–400 times thicker than that of the scrotum. This difference in thickness can significantly alter the amount of chemical absorbed. The ability of nonionized toxicants to pass through various areas of the skin has been estimated as follows: scrotal > forehead > axilla = scalp > back = abdomen > palm (4).

2.3 Gastrointestinal Tract and Lung

The epithelial cells lining the gastrointestinal tract and lung present significantly less of a barrier to absorption than the skin. Unlike skin, there is essentially only a single layer of columnar cells between the lumen of the gastrointestinal tract and the membranes of the systemic capillary beds. In the lung, the cells lining the alveoli are intimately associated with underlying systemic capillaries. Thus, the cells lining the gastrointestinal tract and lung offer little resistance to the penetration of lipid-soluble molecules.

The pH of the gastrointestinal tract varies from 1 to 3 in the stomach to approximately 6 in the intestines (4). As might be expected, this difference in pH significantly affects the absorption of ionizable compounds. Absorption of organic acids occurs more readily from the stomach since the unionized state is preferred in the low-pH environment of the stomach. In contrast, organic bases are more likely to be absorbed from the intestine, where the higher pH favors the nonionized state.

Although the epithelial cells lining the respiratory tract afford little resistance to the systemic absorption of toxicants, the respiratory tract possesses a number of important anatomic and physiological defenses that decrease the potential of inhaled gases, vapors, and particulates to be absorbed. These specialized defense mechanisms are described in Sections 3.1.1–3.1.5.

2.4 Membrane Barriers to Distribution in the Body

The absorption of a toxicant from the external environment to the systemic circulation does not ensure its complete distribution throughout the body. Passage of a toxicant from blood to internal cellular sites requires that additional membranes be crossed. The anatomic environment and physicochemical composition of the capillary membranes serving the various tissues is not uniform, however, and this can affect the ability of a chemical to penetrate various organs. As with other membranes, lipid solubility and ionization state of a toxicant are the primary rate-limiting factors affecting movement across tissue membranes. Thus highly lipophilic agents and ionizable substances that exist primarily in the nonionized state at physiological pH will readily enter tissues.

The distribution of chemicals to the central nervous system, testis, and embryo are examples of nonuniform distribution within an organism. In the central nervous system (CNS) the capillary beds are largely encircled with glial cells, which create additional layers of membranes to be penetrated before access to tissue is obtained (6). A similar situation also exists in the placenta and testis in which multiple cell layers are interposed between the capillary and the ultimate target cells, the embryo and germ cells, respectively (6, 7). Highly water-soluble compounds experience particular difficulty in entering at these sites. Because of the restricted passage of toxicants across these membrane systems, these systems are termed the *blood–brain*, *blood–testis*, and *placental barriers*. It must be pointed out, however, that the term ''barrier'' does not imply an absolute prevention of transfer but only that entry of toxicants to these sites may be more limited when compared to other tissues.

3 NONMEMBRANE DEFENSES TO ABSORPTION AND DISTRIBUTION

3.1 Respiratory Tract

The primary function of the lung is that of gas exchange. It is not surprising, therefore, that the design of the respiratory tract is ideally suited to accomplish this purpose. Inspired air is rapidly humidified and temperature controlled during passage through the upper portions of the respiratory tract. This conditioned air is then delivered to the lung, which, with its vast surface area and only thin separation between the air and the capillary circulation, readily absorbs the oxygen necessary for sustenance of life. Unfortunately, the air that is breathed all too frequently contains numerous potential toxicants, whether in gaseous, vapor, aerosol, or particulate form. In the absence of defense mechanisms, these toxicants would have access to all areas of the respiratory tract, where they not only could damage the delicate membranes of gas exchange but also could undergo systemic absorption.

Fortunately, the respiratory tract has multiple anatomical and physiological defense mechanisms that prevent or moderate the interaction of inhaled toxicants with the exposed tissue. These mechanisms can be divided into four general classes: (a) anatomical barriers, represented by the convoluted pathways that inhaled material must pass through before reaching deep portions of the lung; (b) mucociliary clearance mechanisms, which transport particulate toxicants from the respiratory tract following their deposition there; (c) alveolar macrophages, which remove inhaled particles from the alveoli; and (d) respiratory reflex mechanisms, which reduce the amount of toxicant inhaled. The physicochemical properties of the inhaled toxicant, such as its physical state (aerosol, gas, or vapor) and lipid solubility, and tissue irritant properties determine which, if any, of these defense mechanisms will affect respiratory absorption of the toxicant.

3.1.1 Anatomical Defenses to Particle Deposition

The respiratory tract can be divided into three anatomic regions: the nasopharyngeal, tracheobronchial, and pulmonary regions (8). The nasopharyngeal region begins at the nose and continues to the larynx. This region is a complex series of passages in which the minimum channel width may approach 2 mm in humans (9,10). The tracheobronchial region extends from the larynx to the terminal bronchioles. The trachea bifurcates into the main bronchus to each lung. Each bronchus may have up to 20 divisions before ending in a terminal bronchiole, which has an approximate width of 0.7 mm. The pulmonary region consists of the respiratory bronchioles, alveolar ducts, atria, alveolar sacs, and alveoli. The primary function of this region is gas exchange. The width of a typical alveolus is 0.15 mm.

The complex anatomy of the respiratory tract is a key defense to the deposition of liquid or solid aerosols in the lung. Deposition of particles in various regions of the respiratory tract is controlled by four mechanisms: inertial impaction, gravitational settling, interception, and diffusion (10). Inertial impaction is the process whereby particles carried in an airstream impact into the wall of passage at a bend in the airstream. Since this phenomenon is governed by the inertia of the particle, impaction is a function of particle density, size, and velocity. Gravitational settling, or sedimentation, occurs in areas of low air velocity, primarily the small bronchi, bronchioles, and alveoli. Settling of particles begins when gravitational force on the particles equals the sum of the buoyancy and air resistance, at which time settling will continue at what is described as the terminal settling velocity. Interception is the phenomenon in which particles traveling near the side of a passageway come into contact with the surface. Interception is particularly important for particles of irregular shape such as fibers. Fibers that have small diameters and long lengths may escape impaction in the upper portions of the respiratory tract because of their low density. Fibers encountering the small-diameter openings in the terminal bronchioles and below may not be able to pass, whereas particles of equivalent density but more regular shape pass readily. Diffusion is limited to particles of 0.5 μm or less in size and is due to the impact of gas molecules on the particle. This deposition mechanism is important for particles reaching the small airways.

The net effect of particle deposition mechanisms, acting in concert with the respiratory anatomy, is that the delicate membranes of the alveoli are well protected from deposition of toxic particulates. Relatively large particles of 5–30 μm, which enter the nose at high velocity and rapidly encounter sharp changes in airflow direction, are deposited by impaction in the nasopharyngeal region. Particles of lower density and smaller diameter (1–5 μm) are trapped primarily by sedimentation in the tracheobronchial region, where velocity of the airstream is reduced and less abrupt changes in airflow occur. Only particles of less than 1 μm in size penetrate to the alveoli, where deposition occurs primarily by diffusion (Fig. 4.3).

3.1.2 Mucociliary Clearance

Despite the effectiveness of the anatomical barriers in reducing deposition of particles in the lower and upper respiratory tracts, it should be obvious that significant particle deposition is still likely to occur. Particles deposited in the nasopharyngeal and tracheobronchial regions of the respiratory tract are removed by an additional defense

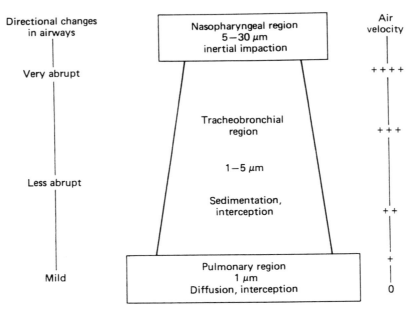

Figure 4.3 Factors influencing deposition of particles in various regions of the respiratory tract.

mechanism, the mucociliary clearance system (8,11), whereas those deposited in the pulmonary region are cleared by alveolar macrophages (8,12). These systems not only prevent accumulation of toxic particulates in the respiratory tract but also diminish or prevent contact of particles with the membranes of cells lining the lumen of the airways.

The nasopharyngeal and tracheobronchial regions of the respiratory tract are lined with both ciliated and mucus-secreting cells (11,13). The two sources of mucus are epithelial goblet cells and submucosal glands. Mucus produced from these sources provides a thin cover for the epithelial cells of the airways. The mucus layer is thought to be biphasic, consisting of a high-viscosity mucus sheet overlaying a layer of low-viscosity serous fluid. The cilia of the epithelial ciliated cells extend primarily into the serous layer of the mucus, with only the cilia tips contacting the high-viscosity outer layer (Fig. 4.4). Cilia beat rapidly (approximately 400–1000 beats/min) in a coordinated pendular fashion, creating a flow of mucus from the tracheobronchial and nasopharyngeal regions that is ultimately swallowed to the stomach. Mucociliary clearance of particles from the respiratory tract varies with location. Mucus flow of 0.4–0.6 mm/min has been reported in the distal areas of the tracheobronchial region and up to 10 mm/min in portions of the trachea (8,13).

3.1.3 Alveolar Macrophages

Particles deposited in the pulmonary region of the respiratory tract are removed by alveolar macrophages (12). These cells, which originate from bone marrow and

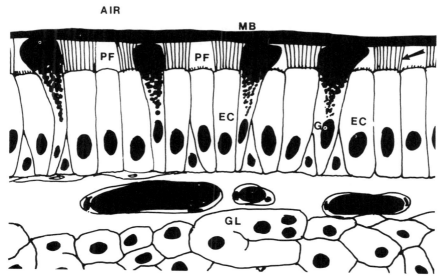

Figure 4.4 Diagram of the components of the mucociliary apparatus. The high-viscosity mucus blanket (MB) floats on the less viscous periciliary fluid (PF) and is propelled by the action of cilia (arrow) extending from the epithelial cells (EC). The mucus is produced by goblet cells (GC) in the epithelium and glands (GL) in the underlying connective tissue.

migrate into the alveolar spaces from the pulmonary interstitium, rapidly phagocytize small particles deposited on the epithelial cells lining the alveoli. Macrophages containing engulfed particles move to the respiratory bronchioles, at which point they are transported out of the respiratory tract by the mucociliary clearance system. Some macrophages also are cleared from the alveoli by migration into the interstitial spaces and lymph systems of the lung.

Ingestion of particles by macrophages does not always result in clearance from the alveoli. Several types of particles, particularly silica and asbestos, cause cell dysfunction and death on ingestion by the macrophages. The toxicity of these types of particles to the macrophages results in the release of toxic constituents, such as lysosomal proteases, into the alveolar spaces. Thus the development of lung toxicity as a result of exposure to silica and asbestos appears to be the direct result of an initial insult to a respiratory defense mechanism, the alveolar macrophage (14).

Lung toxicity may also ensue from an overloading of alveolar macrophages with inhaled particulates that, under low-exposure conditions, would not result in toxicity (15). Macrophage clearance processes are thought to slow when the intracellular particulate volume reaches $25–90$ μm^3 per alveolar macrophage. Macrophage clearance appears to cease completely when particulate volume reaches approximately 600 μm^3 per alveolar macrophage. The progressive accumulation of particulates within macrophages leads not only to diminished mobility but also induces macrophages to release growth factors, chemotactic factors, proteolytic enzymes, and oxidant species. Combinations of these events can lead to secondary toxicity, primarily expressed as fibrosis, in the lung.

3.1.4 Mucus as a Protective Barrier

The mucus layer has an additional protective function above that associated with the mucociliary clearance system. Mucus is composed of approximately 95 percent water, with the remainder consisting of varying amounts of glycoproteins, free proteins, salts, and other materials (16). Consequently, the hydrophilic mucus layer contributes to the upper respiratory tract scrubbing of highly water-soluble toxicants such as sulfur dioxide, chlorine, and formaldehyde, diminishing penetration to the sensitive, unprotected alveolar membranes of the deep lung. Less water-soluble agents such as ozone are not effectively scrubbed during passage through the upper respiratory tract and may cause significant toxic injury to the pulmonary region of the lung (8).

Respiratory tract mucus also serves as a physicochemical barrier to the direct contact of inhaled toxicants with epithelial cells lining the airways. Thus, water-soluble toxicants trapped by mucus do not necessarily diffuse quantitatively through to the epithelial cells, where potential toxic injury may occur. Toxicants absorbed by mucus may react with or otherwise be neutralized by the various components of mucus, effectively reducing the amounts of toxicant delivered to the underlying epithelial cells (16). Since mucus is constantly removed from the respiratory tract and is freshly renewed by mucus-secreting cells, exposure to low concentrations of toxicants may be insufficient to saturate the neutralization capacity of the mucous layer.

There is some evidence that such a protective mechanism may be functioning with exposure to the irritant gas formaldehyde, which produces nasal tumors in rats exposed to high concentrations (15 ppm) of the gas for 2 years (17). Exposure of rats to high concentrations of formaldehyde resulted in both mucostasis and ciliastasis and was associated with injury to the nasal epithelial cells. Exposure to lower concentrations of formaldehyde (0.5 ppm), however, did not alter mucociliary function and was not associated with epithelial cell injury. The ability of formaldehyde to penetrate the mucus barrier, therefore, may have been an important factor in the dose-dependent development of epithelial cell damage following inhalation exposure to formaldehyde.

3.1.5 Respiratory Tract Reflexes

Reflex responses to exposure to inhaled toxicants include (a) coughing or sneezing, which rapidly expel the toxicant; (b) reduction in minute ventilation, which minimizes further uptake of toxicant; and (c) alterations in bronchomotor, cardiovascular, and mucus-secreting activities (13). The precise mechanism of the initiation of the reflexes is not known, although they are thought to be mediated through stimulation of a variety of nerve receptors in the upper respiratory tract (18). Although not a reflex per se, olfactory responses to pungent, acrid, or astringent materials also represents a potential defense by which exposure to noxious agents can be reduced (19). Olfactory detection of noxious odors associated with potentially toxic compounds such a hydrogen sulfide can cause the individual to immediately seek removal from the exposure.

The receptors associated with the trigeminal nerve in the nasal cavity are readily stimulated by a variety of irritant agents. The characteristic response seen when these receptors are stimulated by an inhaled irritant is a rapid decrease in the minute ventilation, which effectively reduces the subsequent dose of the irritant. The reflex inhibition of respiration varies with species, and it is known, for example, that the reflex response of mice to an irritant agent is far greater than that of rats (18). This difference in response appears to have a profound influence on expression of the ultimate toxic effect in the respiratory tract.

A comparison of the differential response of mice and rats to formaldehyde is an example of the role of respiratory reflexes in modulating the toxic effects of exposure to an irritant agent. As mentioned in Section 3.1.4, chronic exposure of rats to 15-ppm formaldehyde produced a 50 percent incidence of tumors in the nasal cavity (17). Mice similarly exposed to formaldehyde exhibited only a 3.3 percent tumor incidence. At a lower (6 ppm) exposure to formaldehyde, a 0 and 1 percent incidence of nasal tumors was found in mice and rats, respectively. An evaluation of respiratory reflex response of mice and rats to acute exposures of formaldehyde suggested that this defense mechanism may explain the dose-dependent and species-specific toxicity of formaldehyde (20). Rats exposed to 15-ppm formaldehyde had only a slight reflex-mediated reduction in minute ventilation, whereas respiration at this concentration in mice was reduced approximately 70 percent [Fig. 4.5(a)]. Since almost all of the inhaled formaldehyde is retained by the nose, the total dose of formaldehyde delivered to the nasal epithelium of each species can be calculated by normalizing the amount of inhaled irritant per unit time over the surface area of the nasal cavity. The amount of inhaled formaldehyde, of course, is dependent on the minute ventilation of the animal, and the surface area of the nasal cavity can be estimated by

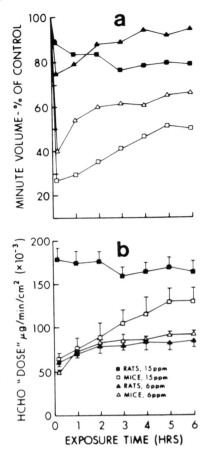

Figure 4.5 (a) Time–response curves for minute volume from rats and mice exposed to 6 or 15 ppm of formaldehyde for 6 hr. (b) Time-weighted averages of the theoretical formaldehyde dose available for deposition on the nasal passages of rats and mice during a 6-hr exposure to 6 or 15 ppm. [Reprinted with permission from Swenberg et al. (17).]

morphometric techniques. When these calculations are compared for mice and rats exposed to 15-ppm formaldehyde, it can be seen that mice received a substantially lower dose of the irritant than did rats [Fig. 4.5(b)]. In fact, because of the marked reduction in minute ventilation in mice exposed to 15-ppm formaldehyde, the actual delivered dose to the nasal cavity at this concentration was comparable to that of rats exposed to 6-ppm formaldehyde. Since these equivalent delivered doses resulted in nonsignificant tumor incidences in both mice and rats, these data suggest that the tumorigenic activity of high doses of formaldehyde in rats versus mice is a direct reflection of the inability of rats to reduce their exposure by reflex inhibition of respiration.

3.2   Protein Binding as a Barrier to Toxicant Distribution

Many toxicants, on absorption into the systemic circulation, bind with varying affinity to plasma proteins. Binding of toxicants to high-molecular-weight proteins may be an important factor in controlling subsequent distribution to critical target macromolecules within cells in that the protein–toxicant complex may have size and charge characteristics that prevent ready passage through cell membranes. Reversible binding of toxicants to proteins consists of hydrogen, van der Waal's, and ionic bonds (4,6) and creates a high-molecular-weight polar complex incapable of crossing cell membranes. Only the fraction of free, unbound toxicant is available for diffusion across membranes, as influenced by the factors described in Section 2.1.

The binding of most toxicants to plasma proteins is nonspecific; consequently, agents with physicochemical characteristics similar to the bound toxicant may displace it from the proteins. Such displacement increases the amount of unbound toxicant available to diffuse into cells, often resulting in enhanced toxicity. An example of this phenomenon is the displacement of unconjugated bilirubin from plasma albumin by sulfonamide drugs. In newborn infants, sulfonamide-induced displacement of bilirubin from plasma protein binding sites may increase the risk of bilirubin encephalopathy (21). Similarly, sulfonamide drugs also have been shown to displace antidiabetic drugs from plasma proteins, resulting in induction of hypoglycemic coma (6).

There are two examples in which binding of toxicants to blood protein have been used as antidotal therapy for the toxicant. The toxicity of both cyanide and hydrogen sulfide is mediated by their complexation with the ferric heme moiety of cytochrome oxidase, resulting in inhibition of oxidative respiration. The toxicity of these agents can be ameliorated, however, by antidotal induction of methemoglobinemia. The mechanism of this protective effect is attributed to the fact that methemoglobin contains a ferric heme group capable of competing effectively with the heme moiety of cytochrome oxidase for cyanide and sulfide binding and thus provides an innocuous binding site for these toxicants (22).

Intracellular protein binding is an important detoxication mechanism for metals such as cadmium and mercury. Many different tissues synthesize a metal binding protein termed *metallothionein*. This unique protein has a high cysteine content, providing abundant thiol binding sites for these metals. Thus the binding of metals to metallothionein prevents their further distribution and binding to other functionally important intracellular sites (23). An interesting aspect of metallothionein function is that its synthesis is rapidly induced following low-level exposure to cadmium. Although the synthesis of additional binding protein offers protection against further exposure to higher doses of cadmium, several investigators have suggested that metallothionein does not function primarily as a defense against cadmium (24,25). Rather, this protein may play an important role in the homeostatic control of tissue zinc and copper concentrations. The protection afforded by metallothionein against cadmium toxicity, therefore, may represent only a fortuitous interaction of a toxic metal with an endogenous control mechanism.

4 EXCRETORY DEFENSE MECHANISMS

Absorbed toxicants are eliminated from the body by a variety of mechanisms. Most toxicants and their metabolites (see Sections 5.1 and 5.2) are excreted by the kidney or liver, although excretion from the lung, gastrointestinal tract, milk, sweat, and saliva can also occur (6).

4.1 Renal Excretion

Passive glomerular filtration represents the predominate mechanism by which toxicants are eliminated by the kidney (6). Compounds with molecular weights of less than 60,000 will readily pass into the urine through pores in the glomerular membrane. Binding of toxicants to plasma proteins will restrict passage into the urine because of the high molecular weight of the toxicant–protein complex. Elimination of toxicants in the urine after filtration through the glomerulus is directly related to lipid solubility of the individual agents. Compounds with high lipid solubility will passively diffuse back into the systemic circulation through the renal tubule cells. Metabolism of toxicants favors elimination in urine, however, since the products of these reactions are generally more water soluble than the corresponding parent compounds (Sections 5.1 and 5.2).

A second important mechanism by which toxicants are excreted into urine is by active transport through renal tubule cells. The active transport process can serve as a defense mechanism by reducing the systemic half-life of toxicants. For example, the herbicide 2,4,5-trichlorophenoxyacetic acid (2,4,5-T) has a higher acute toxicity in dogs (LD_{50} = 100 mg/kg) compared to rats (LD_{50} = 300 mg/kg). A comparison of the pharmacokinetics of 2,4,5-T in these two species has found plasma half-lives of 77 and 4.7 hr for dogs and rats, respectively, offering a plausible explanation for the sensitivity of the dog of 2,4,5-T toxicity. The differential half-lives have been attributed to a diminished renal acid transport capacity in dogs versus rats (26).

There are two types of transport systems, one for organic anions and another for organic cations. The transport systems are not necessarily specific for any given substrate, however, and in certain circumstances elimination of a toxicant may be hindered when multiple substrates are presented to the transport mechanism. For example, the ability of exogenous organic acids to precipitate attacks of gout is due to their ability to compete with endogenous uric acid for active secretion into the urine.

4.2 Liver (Biliary) Excretion

Toxicants (or their metabolites) with molecular weights of approximately 325 or greater may be excreted into the bile (6). Although the mechanisms by which toxicants are excreted into bile are not fully understood, transport mechanisms for organic acids, bases, and neutral compounds have been identified.

The ability of a toxicant to be excreted into bile may profoundly influence its toxicity. In newborn animals the hepatic excretory system is not fully functional. Thus, the 40-fold increased toxicity of ouabain in newborn rats compared to adult animals is associated with an inability of the newborn to eliminate ouabain in the bile (6).

5 BIOCHEMICAL DEFENSES

The absorption of most toxicants into the body is directly influenced by the lipophilicity of the agent; thus the more lipid soluble an agent is, the more likely it is to be absorbed (Section 2.1). This same property, however, decreases the ability of a toxicant to be excreted. This is particularly true for the primary organ of excretion, the kidney, where the aqueous environment of the urine favors elimination of hydrophilic compounds. Fortunately, the body possesses a tremendous capacity to transform foreign compounds into more water-soluble forms, rendering them more susceptible to excretion.

The biotransformation of toxicants to water-soluble metabolites is generally catalyzed by a broad spectrum of cell enzymes. These enzymatic reactions have been divided into two phases: Phase I reactions, consisting of oxidation, reduction, and hydrolysis; and Phase II reactions, mediating conjugation and synthesis (27,28). Phase I reactions involve the insertion of a polar reactive group into the toxicant molecule and usually prepare the toxicant for subsequent Phase II reactions. It must be emphasized, however, that Phase I and II metabolic transformations do not always result in detoxication. In many cases intermediate metabolites having greater biological reactivity than the parent compound are formed. Although generation of reactive intermediates usually results in enhanced toxicity, the ability of such metabolites to attack important biological macromolecules is nonetheless modulated by further Phase I or II reactions (Fig. 4.6).

In addition to formation of reactive metabolites capable of covalently binding to tissue macromolecules, Phase I reductions also may catalyze electron transfer reactions between endogenous electron sources, toxicants, and molecular oxygen. The product of this type of reaction, the superoxide radical (O_2^-), potentially may form extremely toxic oxidant species such as the hydroxyl radical (OH^-) (Figure 4.6). Generation of superoxide radicals does not always result in toxic injury to the cell, however, since several defense mechanisms have been identified that detoxify these reactive oxygen products (Section 5.4).

5.1 Phase I Detoxication Reactions

5.1.1 Microsomal Mixed-Function Monooxygenases

The majority of Phase I transformations of toxicants is carried out by a nonspecific multienzyme system frequently termed the *mixed-function monooxygenase system*. This metabolic system is located in the endoplasmic reticulum (microsomal fraction) of cells and is composed of two component enzymes, NADPH cytochrome P-450 reductase and cytochrome P-450 (28). Both molecular oxygen and reducing equivalents in the form of NADPH are required for catalysis of the oxidation reactions. Multiple forms of cytochrome P-450 have been identified, and each form has some overlap in substrate specificity with the other forms (29,30).

The cytochrome P-450 enzymes catalyze a broad spectrum of oxidation reactions (Table 4.1). The products of the oxidation reactions are more water soluble than the

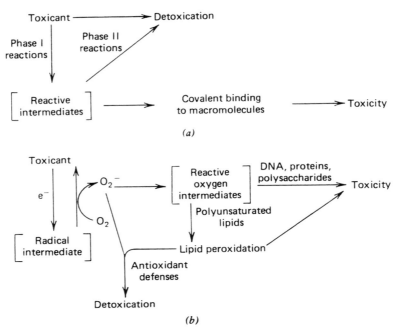

Figure 4.6 Metabolism of toxicants to reactive intermediates capable of (a) covalently binding to cell macromolecules or (b) generating toxic forms of activated oxygen. Because of the intervention of detoxication mechanisms, production of either type of reactive intermediate may not result in toxic responses within the cell.

parent compound and therefore have taken the first of perhaps several steps leading to detoxication. Several products of Phase I metabolism are significantly more reactive than the corresponding parent compounds, however, and formation of such products may be responsible for expression of toxicity. An example of this type of reaction is the formation of epoxides (reaction 4, Table 4.1). Epoxides result from addition of an oxygen atom across an aromatic or olefinic double bond and in many instances are sufficiently electrophilic to form a covalently bound adduct with biological macromolecules. Fortunately, epoxides, along with most other reactive intermediates formed in Phase I reactions, are subject to further Phase II reactions. These reactions generally result in formation of detoxified, readily excretable metabolites. Sections 5.2 and 5.4 describe biochemical defense mechanisms frequently associated with detoxication of products resulting from mixed-function monooxygenase metabolism. It should be noted, however, that if the parent compound contains a reactive structural element similar to that resulting from cytochrome P-450 metabolism, it, too, will be subject to the same detoxication reactions.

In addition to catalyzing oxidation reactions, the mixed function monooxygenase system enzymes also catalyze several reduction reactions (Table 4.1). Single-electron reductions of aromatic nitro groups, bipyridylium compounds, and halogenated hydrocarbons may be carried out by either NADPH cytochrome P-450 reductase or

Table 4.1 Phase I Metabolic Transformations Catalyzed by the Microsomal Mixed-Function Oxidase System Enzymes[a]

Reaction	Example
Oxidation reactions	

1. Aromatic hydroxylation

$$R\text{—}\bigcirc \rightarrow R\text{—}\bigcirc\text{—OH}$$

2. Aliphatic hydroxylation

$$R\text{—}CH_2\text{—}CH_2\text{—}CH_3 \rightarrow R\text{—}CH_2\text{—}CHOH\text{—}CH_3$$

3. N-, O-, or S-dealkylation

$$R\text{—}(N, O, S)\overset{H}{\text{—}}CH_3 \rightarrow R\text{—}(NH_2, OH, SH) + CH_2O$$

4. Epoxidation

$$R\text{—}CH\text{=}CH\text{—}R' \rightarrow R\text{—}\overset{\displaystyle O}{\overset{/\backslash}{CH\text{—}CH}}\text{—}R'$$

5. Desulfuration

$$R_1R_2\overset{S}{\overset{\|}{P}}\text{—}X \rightarrow R_1R_2\overset{O}{\overset{\|}{P}}\text{—}X + S$$

6. Sulfoxidation

$$R\text{—}S\text{—}R \rightarrow R\text{—}\underset{\underset{\displaystyle O}{\|}}{S}\text{—}R'$$

7. N-hydroxylation

$$R\text{—}NH\text{—}\overset{O}{\overset{\|}{C}}\text{—}CH_3 \rightarrow R\text{—}NOH\text{—}\overset{O}{\overset{\|}{C}}\text{—}CH_3$$

Reduction reactions

8. Halogenated hydrocarbon reduction

$$CCl_4 \rightarrow CCl_3{}^{\cdot}$$

9. Nitro reduction

$$R\text{—}NO_2 \rightarrow R\text{—}NO_2{}^{\cdot}$$

10. Bipyridylium reduction

$$CH_3\text{—}{}^{+}N\bigcirc\text{—}\bigcirc N^{+}\text{—}CH_3 \rightarrow$$

$$CH_3\text{—}{}^{+}N\bigcirc\overset{\cdot}{\text{—}}\bigcirc N^{+}\text{—}CH_3$$

cytochrome P-450. The products of these reduction reactions are highly reactive free radicals that cause cell toxicity by either direct reaction with cell macromolecules or reoxidation by molecular oxygen with resultant formation of toxic oxygen species (31,32). The potentially toxic consequences associated with generation of these reactive intermediates within the cell do not go unchecked since several biochemical defenses directed to controlling free-radical-initiated oxidant reactions have been identified (Section 5.4).

5.1.2 Other Microsomal Phase I Metabolism

Although the cytochrome P-450 monooxygenase system is responsible for the majority of Phase I microsomal xenobiotic metabolism, two additional metabolic systems, flavin-containing monooxygenase (33,34) and epoxide hydrolase (35), have been identified in microsomes. Like cytochrome P-450-mediated metabolism, the flavin-containing monooxygenase requires molecular oxygen and NADPH. The substrate specificity of this oxidizing system is primarily focused on oxidation of amine and sulfur-containing compounds, although compounds have also been shown as substrates for this enzyme system.

As noted above, epoxides formed from cytochrome P-450 oxidation of aromatic or olefinic double bonds are potentially reactive with critical cell macromolecules. One mechanism by which reactive epoxides are detoxified is through hydration to dihydrodiols, a reaction catalyzed by epoxide hydrolase. The observation that both epoxide hydrolase and the epoxide generating enzyme, cytochrome P-450, are in close association within the endoplasmic reticulum suggests the importance of this enzyme as an important defense mechanism (35).

5.1.3   Alcohol and Aldehyde Dehydrogenases

The toxicity of a variety of primary alcohols is modulated by the metabolic action of alcohol and aldehyde dehydrogenases (Fig. 4.7). The majority of alcohol dehydrogenase activity is in the cytosolic fraction of tissues, whereas aldehyde dehydrogenase activity is found in cytosol, mitochondria, and microsomal fractions (36,37). The multiple isozymes of both enzymes require NAD^+ as a cofactor. It is important that these two defense enzymes act in sequence since both alcohols and their corresponding aldehydes are potentially toxic; alcohols frequently cause narcotic effects, and aldehydes readily form Schiff base adducts with cell macromolecules (28). Inhibition of aldehyde dehydrogenase by disulfiram, for example, has been used in the treatment of alcoholism. The discomfort caused by accumulation of toxic levels of acetaldehyde when alcohol is consumed by disulfiram-treated individuals acts as a deterrent to further alcohol consumption (37).

An additional aldehyde dehydrogenase, formaldehyde dehydrogenase, appears to have as its primary function the specific detoxication of formaldehyde (38). Formaldehyde is formed not only from the oxidation of methanol but also from N-demethylation of endogenous and exogenous compounds (see reaction 3, Table 4.1). Unlike the aldehyde dehydrogenases described previously, formaldehyde dehydrogenase uses both NAD^+ and reduced glutathione as cofactors.

5.1.4   Hydrolysis Reactions

A wide variety of toxicants contains ester or amide linkages capable of being hydrolyzed by a group of nonspecific esterases and amidases. The products of the

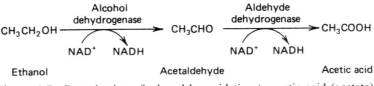

Figure 4.7   Detoxication of ethanol by oxidation to acetic acid (acetate).

hydrolysis reactions, carboxylic acids, alcohols, and free amines, are subject to further Phase I and II detoxication reactions (39).

The importance of enzymatic hydrolysis as a detoxication mechanism is readily appreciated when individuals bearing genetically mediated deficiencies in this enzyme activity are exposed to toxicants deactivated by this mechanism. For example, the duration of action of the neuromuscular blocking agent succinylcholine in humans is usually extremely brief because of its rapid hydrolysis by plasma and liver esterase to less active products. In a few individuals, however, prolonged apnea results from succinylcholine treatment. The aberrant response in these individuals has been attributed to the existence of an atypical pseudocholinesterase incapable of hydrolyzing succinylcholine (40).

Variations among species in the ability to detoxify agents by hydrolysis reactions has been exploited in the development of pesticides. Malathion, a widely used organophosphate insecticide, requires metabolic activation for expression of toxicity in both insects and humans. Although both species rapidly form the toxic intermediate, humans are far more capable than insects of subsequently hydrolyzing the intermediate to nontoxic products. This differential ability to catalyze hydrolysis of reactive intermediates is in part responsible for the species-specific toxicity of malathion and other organophosphate insecticides (41).

5.2 Phase II Detoxication Reactions

5.2.1 Glucuronic Acid Conjugation

Glucuronidation is a major Phase II reaction involved in the detoxication of aliphatic and aromatic alcohols, carboxylic acids, and certain types of amine and thiol-containing compounds (42,43). This reaction is characterized by conjugation of glucuronic acid with the toxicant or its metabolite and is catalyzed by a family of closely related enzymes, the glucuronyl transferases. The majority of the enzyme activity is associated with the endoplasmic reticulum of the liver, although activity also has been found in the kidney, intestine, skin, brain, and spleen.

Conjugation with glucuronic acid does not occur directly but is mediated by formation of an active intermediate, uridine diphosphate glucuronic acid (UDPGA) (43) (Fig. 4.8). Glucuronide conjugates are excreted in both urine and bile.

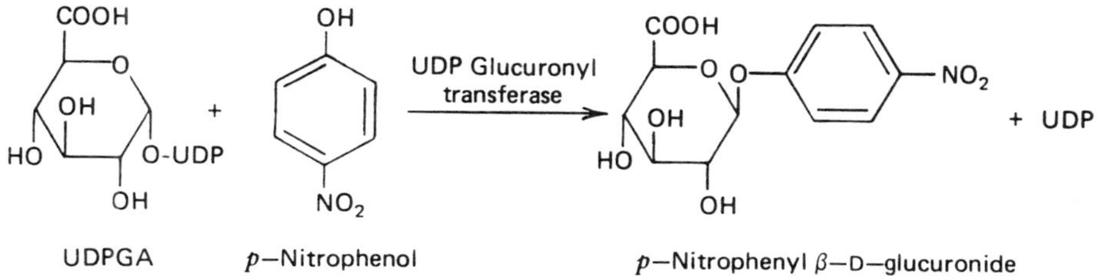

Figure 4.8 Conjugation of *p*-nitrophenol with glucuronic acid.

5.2.2 Sulfate Conjugation

Conjugation of toxicants with sulfate is catalyzed by a group of sulfotransferases found in several tissues (44). The sulfate ester products are highly water soluble and readily excreted by the kidney. Like glucuronic acid conjugation, sulfate conjugation proceeds by means of an activated sulfate intermediate, 3'-phosphoadenosine-5'-phosphosulfate (PAPS). The primary substrates for sulfate conjugation are primary, secondary, and tertiary alcohols, phenols, and to a lesser extent, aromatic amines. In most species, however, conjugation of these substrates with glucuronic acid predominates over sulfate conjugation. The preference for glucuronic acid conjugation versus sulfation has been attributed to the greater total activity of tissue glucuronosyltransferases (28).

5.2.3 Glutathione Conjugation

The toxicity of many compounds is mediated through an electrophilic attack on nucleophilic sites of cellular macromolecules (Fig. 4.6). The resultant covalently bound adducts may cause cell dysfunction or even death. Exposure to electrophilic agents is inescapable, however, since they are ubiquitous in the environment and also are formed in many endogenous metabolic reactions. Conjugation of these reactive agents with glutathione represents a primary mechanism by which the body prevents their attack on critical cellular macromolecules.

Glutathione is a tripeptide consisting of the sequence gamma-glutamate-cysteineglycine and is found in all tissues. Its concentration ranges up to 10 mM in organs such as the liver, which is highly active in forming reactive metabolites from toxicants. The thiol group of the cysteine residue is the functionally important element in detoxication reactions, serving as a readily available nucleophile for electrophilic attack.

Although electrophilic agents spontaneously react with glutathione to varying degrees, this reaction is greatly facilitated in cells by a family of cytosolic enzymes, the glutathione-S-transferases. These enzymes, which have broad substrate specificity, are particularly important in catalyzing the conjugation of hydrophobic, electrophilic toxicants with glutathione (28,45,46).

Conjugation with glutathione enhances elimination of the toxicant from the body in two ways. First, to be excreted into the bile, toxicants must have a molecular weight of 325 or greater (43). Since glutathione has a molecular weight of 307, conjugation of even the simplest electrophilic molecules, when it occurs in the liver, frequently results in extensive biliary excretion. Excretion of glutathione conjugates into bile results in their subsequent elimination in the feces. In certain cases, however, glutathione conjugates excreted into bile may be metabolized by bacteria of the intestinal microflora, resulting in formation of lipophilic toxicants that may be reabsorbed (6). This cyclical process has been termed *enterohepatic circulation* and may result in a significant prolongation of the time required for elimination of certain toxicants from the body.

A second mechanism by which glutathione conjugates are eliminated from the body is by metabolism of the glutathione conjugates to mercapturic acids (43). Mercapturic acids are formed by sequential removal of glutamate and glycine from the glutathione conjugate, followed by acetylation of the amino group of the remaining cysteine conjugate. The enzymes responsible for catalyzing these reactions are found primarily in liver and kidney. Mercapturic acids are very water soluble and consequently are readily excreted by the kidney.

An examination of the dose-dependent hepatotoxicity of acetaminophen provides a good example of the importance of glutathione as a primary detoxication mechanism (47). Acetaminophen, a commonly used analgesic agent, produces severe liver necrosis after high doses in both humans and animals. The liver toxicity has been attributed to microsomal mixed-function monooxygenase metabolism of acetaminophen to an electrophilic intermediate capable of forming covalent adducts with critical cell macromolecules. The data in Figure 4.9, obtained from hamsters treated with increasing doses of acetaminophen, illustrate the protective role of glutathione in preventing arylation of cell macromolecules and associated liver necrosis. As the dose of acetaminophen was increased, the concentration of glutathione in the liver progressively declined as a result of conjugation of the reactive intermediate with hepatic stores of glutathione. It was not until glutathione concentrations were reduced to approximately 30 percent of control values, however, that any marked increase in covalent binding, liver necrosis, and mortality was observed. These data clearly

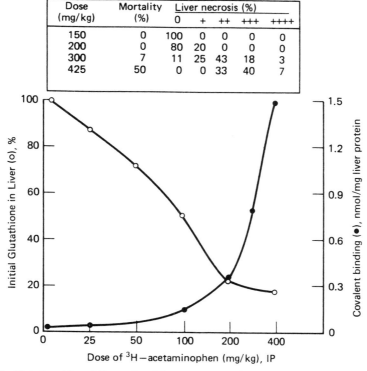

Dose (mg/kg)	Mortality (%)	Liver necrosis (%)				
		0	+	++	+++	++++
150	0	100	0	0	0	0
200	0	80	20	0	0	0
300	7	11	25	43	18	3
425	50	0	0	33	40	7

Figure 4.9 Relationship of liver glutathione concentrations and covalent binding to development of acetaminophen toxicity in hamsters. [Modified from Potter et al. (47).]

suggested that toxicity only resulted when inadequate amounts of glutathione were available to trap the reactive metabolite, thereby freeing it to react with critical cellular macromolecules.

The role of glutathione in modulating acetaminophen toxicity was further demonstrated in a study in which hamsters were treated with diethyl maleate, an agent that reduces hepatic glutathione, prior to administration of acetaminophen (47). In diethyl maleate-pretreated animals, doses of acetaminophen that were minimally hepatotoxic to naive animals now produced extensive liver necrosis. The enhanced toxicity was presumed to be due to decreased availability of glutathione to trap the reactive intermediate of acetaminophen. The results obtained with acetaminophen represent only one example of many in which the toxicity of electrophilic agents is clearly modulated by availability of tissue glutathione.

An important implication from all of these studies is that low doses of an electrophilic toxicant may not result in toxicity; rather, toxicity may result only when the dose is sufficient to overwhelm the glutathione defenses. Thus, drugs such as acetaminophen can be safely used despite their known ability to form a tissue-damaging reactive intermediate.

5.2.4 Acetylation

Aromatic primary amines may be rendered less toxic by formation of N-acetyl conjugates. This conjugation reaction is catalyzed by a group of cytosolic enzymes, the N-acetyltransferases, which utilize acetyl-coenzyme A as the acetyl donor cofactor (28,43,48). Although the N-acetyltransferases catalyze acetylation of hydroxy and thiol groups of endogenous substrates, amines appear to be the only exogenous substrates for these enzymes.

An interesting feature of the N-acetyltransferases is that in humans and rabbits a genetically determined polymorphism exists in the ability to acetylate exogenous amines. Thus one population of individuals is capable of rapidly acetylating toxicants, whereas another conducts this conjugation slowly. This difference in the ability to acetylate makes a profound difference in the susceptibility of these populations to toxicity of amine-containing compounds. For example, individuals who are slow acetylators are more susceptible to the neurotoxic effects of the drug isoniazid, which is eliminated from the body after acetylation. Individuals who are fast acetylators do not accumulate toxic concentrations of isoniazid in their tissues as do the slow acetylators.

5.2.5 Amino Acid Conjugation

Toxicants containing aromatic carboxylic acids may be detoxified by conjugation with glucuronic acid (Section 5.2.1) or amino acids (28,43). Like glucuronic acid conjugation, amino acid conjugation is a multistep process and involves formation of an activated intermediate (Fig. 4.10). The first step of the reaction requires formation of an acyl-S-coenzyme A derivative of the toxicant. This reaction is carried out in mitochondria by enzymes thought to be identical with the fatty acid activating system. The activated intermediate is subject to nucleophilic attack by amino groups of several amino acids, primarily glycine and glutamate, and is catalyzed by mito-

1. Activation of toxicant:

$$\text{CoASH} \quad + \quad \text{RCOOH} \xrightarrow{\text{ATP}} \text{RCOSCoA}$$

| Coenzyme A | Acid-containing toxicant | Acyl-coenzyme A intermediate |

2. Conjugation of acyl-CoA intermediate with amino acid:

$$\text{RCOSCoA} \quad + \quad \text{H}_2\text{N CH}_2\text{COOH} \xrightarrow{N-\text{acyl-transferase}} \text{RCONHCH}_2\text{COOH}$$

| Acyl-coenzyme A intermediate | Glycine | Glycine conjugate |

Figure 4.10 Multistep conjugation of a toxicant containing a carboxyl group with the amino acid glycine.

chondrial and cytosolic N-acyltransferases specific for each amino acid substrate. Conjugation with amino acids not only reduces toxicity of the agent but also enhances rapid renal excretion.

5.2.6 Methylation Reactions

Amine-, phenol-, and thiol-containing toxicants may form N-, O-, and S-methyl conjugates in reactions catalyzed by a variety of methyl transferases. The methyl donor for these reactions is S-adenosylmethionine (43,49). Methylation of N- and O- containing toxicants is seldom regarded as a significant detoxication reaction, however, since the products formed are less water soluble, which may delay elimination of the compound.

S-methylation may represent an important detoxication pathway for hydrogen sulfide produced as a result of bacterial metabolism of dietary protein in the intestines (49). Hydrogen sulfide generated from methionine metabolism in the gut is sequentially metabolized to methanethiol and demethylsulfide by methyltransferases located in the mucosal lining cells of the intestine and liver. Since both methanethiol and particularly demethylsulfide are significantly less toxic than hydrogen sulfide, formation of these conjugates may protect organs such as the brain from the toxic effects of endogenously produced hydrogen sulfide. In fact, coma induced in humans with diminished liver function has been attributed to an accumulation of methanethiol in the blood resulting from an inability to fully methylate this endogenous toxicant. There is no evidence to date, however, that suggests that methylation represents a significant detoxication route for exogenously administered hydrogen sulfide (50).

5.3 Detoxication of Cyanide

The extreme toxicity of cyanide is due to its ability to inhibit oxidative respiration by complexation with cytochrome oxidase. This toxicity can be ameliorated in part by induction of methemoglobinemia, which provides a biologically inert binding site for cyanide (Section 3.2). Although cyanide is tightly bound to methemoglobin, additional steps must be taken to detoxify the small amounts of cyanide that dissociate from the complex. This is accomplished by intravenous administration of thiosulfate, which serves as a sulfur donor for the enzyme rhodanese. Rhodanese converts cyanide to thiocyanate, which is much less toxic than cyanide and is rapidly excreted by the kidney (28).

5.4 Oxidant Defense Mechanisms

Oxygen is essential for survival of all aerobic organisms. Despite its necessity for life, oxygen and its derived products have enormous potential for damaging tissues through oxidant-linked reactions. Consequently, aerobic organisms have developed a sophisticated array of defense to this basic challenge to life (1,51). In addition to oxygen itself, a number of toxicants that cause oxidant injury in tissues have been identified. These toxicants include environmental air pollutants such as ozone and nitrogen dioxide, chlorinated hydrocarbons such as carbon tetrachloride, and agents capable of generating reactive forms of molecular oxygen such as the herbicide paraquat (31,52).

An important mechanism by which oxidants damage tissue is through the membrane-damaging process of lipid peroxidation (52) (Figure 4.11). Lipid peroxidation is a chain reaction process in which lipid radicals are formed by abstraction of allylic hydrogens of polyunsaturated lipids. Once formed, lipid radicals may abstract hydrogens from nearby polyunsaturated lipids, resulting in additional radical formation. Since polyunsaturated lipids are a major component of biological membranes, stim-

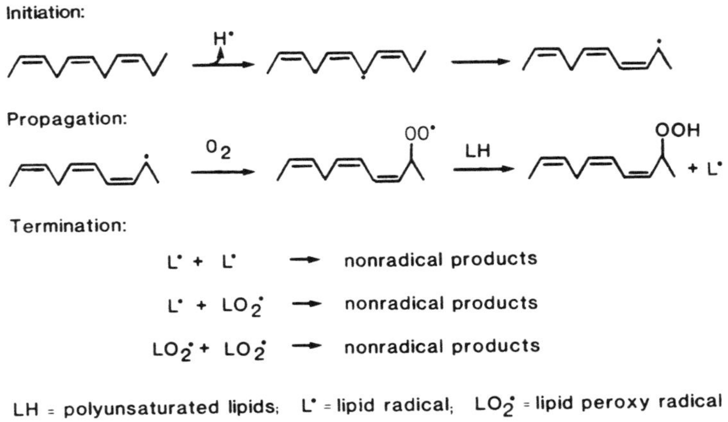

Figure 4.11 Reactions associated with toxicant-initiated lipid peroxidation. Hydrogen abstraction may be mediated by organic radicals, hydroxyl radicals, or other lipid radicals formed in propagation reactions of lipid peroxidation. The chain reaction process of lipid peroxidation is quenched by a series of termination reactions. Antioxidants such as vitamin E inhibit lipid peroxidation by preferentially undergoing lipid radical-mediated hydrogen abstractions, resulting in formation of nonradical products.

ulation of lipid peroxidation may lead to cell dysfunction or even death. Toxicants stimulate lipid peroxidation by a variety of mechanisms (52). Both nitrogen dioxide and ozone react directly with membrane lipids, resulting in lipid radical generation. Carbon tetrachloride is reductively metabolized by microsomal cytochrome P-450 to the trichloromethyl radical, which is capable of directly abstracting hydrogens from unsaturated lipids.

A large number of toxicants appear to stimulate lipid peroxidation through generation of reactive forms of molecular oxygen. The primary mechanism by which toxicants stimulate reactive oxygen formation in tissues is through an enzyme-catalyzed single-electron reduction–oxidation reaction of toxicants (31). In this type of reaction the parent toxicant molecule undergoes a single-electron reduction catalyzed by cellular reductases such as NADPH cytochrome P-450 reductase. The radical products of the reduction reaction are frequently unstable in the oxygen environment of the cell and are immediately reoxidized to the parent compound. The product of the reoxidation reaction is the superoxide radical (O_2^-) (Fig. 4.6). Since O_2^- has weak oxidant activity itself, it is likely that O_2^- per se is not responsible for the oxidant damage associated with O_2^- generation in tissues. Superoxide apparently undergoes an iron-catalyzed Haber–Weiss reaction [Eq. (3)], however, in which O_2^- and hydrogen peroxide (H_2O_2) react to form the extremely potent oxidant, the hydroxyl radical ($OH\cdot$). This radical is readily capable of abstracting hydrogens from unsaturated lipids and consequently stimulating lipid peroxidation reactions. Toxicants

$$O_2^- + H_2O_2 \rightarrow OH\cdot + OH^- + O_2 \tag{3}$$

containing bipyridylium, quinone, or nitro structural elements have been recognized as frequent participants in redox cycling reactions resulting in O_2^- production (31).

Generation of O_2^- has additional toxic consequences beyond the stimulation of lipid peroxidation. In vitro studies have indicated that hydroxyl radicals may produce DNA base and sugar phosphate damage and also may cause DNA strand scission. This type of damage to the genetic material may result in both mutagenic or carcinogenic responses (1,31). Hydroxyl radicals also react with proteins and polysaccharides, resulting in inactivation of other important functional macromolecules.

It should be clear that, if unchecked, oxidant reactions have tremendous potential for producing toxic injury within the cell. In aerobic biological systems O_2^- is a product of many endogenous metabolic reactions (31), mandating an adequate defense against oxidant attack even in the absence of exposure to exogenous toxicants. For example, O_2^- is formed from autoxidation of endogenous substrates such as hydroquinones, catecholamines, and thiols and is also produced as a product of the catalytic activity of enzymes such as xanthine oxidase and aldehyde oxidase. Several oxidants, including O_2^-, also are produced by white blood cells in association with inflammatory responses to tissue injury or disease (1). In fact, it has been estimated that the DNA in every mammalian cell sustains approximately 10^4–10^5 oxidative hits per day as a direct consequence of intrinsic oxidative reactions (1). In aerobic systems, therefore, the ability to combat oxidant stress is essential to survival. In view of the extremely detrimental consequences of O_2^- generation within cells, it is not

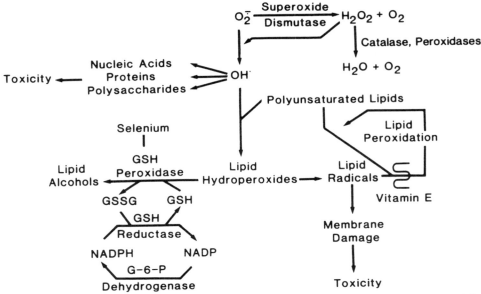

Figure 4.12 Biochemical defense mechanisms that modulate toxic reactions initiated by superoxide radicals. [Modified from Bus and Gibson (26).]

surprising that a multitiered defense system exists to combat oxidant stress originating not only from endogenous metabolic reactions but also from xenobiotic toxicant exposure (Fig. 4.12).

5.4.1 Superoxide Dismutase, Catalase, and Peroxidase System Enzymes

The first line of defense against O_2^- generation is the enzyme superoxide dismutase, which catalyzes the dismutation of O_2^- to hydrogen peroxide and molecular oxygen. Three distinct forms of this metalloenzyme have been identified: iron- or manganese-containing enzymes are commonly found in prokaryotes, whereas a copper–zinc enzyme is found in the cystosol of eukaryotes (53). Mitochondria of eukaryotes, however, contain only the manganoenzyme.

The importance of superoxide dismutase as a biological defense against oxidant toxicity has been elegantly demonstrated by Fridovich and co-workers in a series of experiments in bacteria (53). The ability of *Escherichia coli* to survive in hyberbaric oxygen was found to be directly correlated to the activity of superoxide dismutase, that is, bacteria containing high levels of superoxide dismutase activity tolerated elevated oxygen tensions far better than did those with low enzyme activity. In addition, superoxide dismutase increased in *E. coli* grown under hyperbaric oxygen, which was presumed to represent an adaptive response to oxidant stress. A similar response was observed when bacteria were treated with paraquat, a toxicant that stimulates O_2^- production by redox cycling reactions. An important implication of these findings was that superoxide dismutase does not merely offer a static defense against increased oxidant fluxes but also has the capacity to respond with increased synthesis of the enzyme.

Mammalian systems also have been shown to increase tissue superoxide dismutase activity in response to oxidant exposure. Rats exposed to 85 percent oxygen for 7 days have a marked increase in pulmonary superoxide dismutase activity (54). These rats subsequently exhibited prolonged survival when exposed to an oxidant environment (100% oxygen) that produces complete lethality in nonadapted animals. Return of rats exposed to 85 percent oxygen to room air resulted in a gradual decline of the elevated superoxide dismutase activity to normal levels, which was associated with a corresponding loss of tolerance to 100 percent oxygen exposure.

In order for superoxide dismutase to function effectively as a defense mechanism, the product of the dismutation reaction, H_2O_2, must also be detoxified. Hydrogen peroxide not only has intrinsic oxidant activity but is also a necessary participant with O_2^- in the Haber–Weiss reaction in the production of toxic hydroxyl radicals. Two mechanisms participate in detoxication of H_2O_2: catalase and the glutathione peroxidase system enzymes, which consist of glutathione peroxidase, glutathione reductase, and glucose-6-phosphate dehydrogenase. The three enzymes of the glutathione peroxidase system act in concert, with glucose-6-phosphate supplying reducing equivalents needed for glutathione reductase activity, which in turn maintains adequate concentrations of glutathione required for glutathione peroxidase activity. The consequences of an inadequate ability to detoxify H_2O_2 are evidenced in humans who have a genetically determined deficiency in erythrocyte glucose-6-phosphate dehydrogenase activity. Extensive erythrocyte hemolysis occurs when these individuals are exposed to drugs or toxicants that generate increased concentrations of H_2O_2 within the erythrocyte, presumably because adequate concentrations of reduced glutathione cannot be maintained in the face of diminished glutathione peroxidase system activity (22).

5.4.2 Antioxidant Protection Against Lipid Peroxidation

A second line of defense exists to combat the deleterious effects of lipid peroxidation initiated by oxygen radicals, organic radicals, or gaseous toxicants such as ozone and nitrogen dioxide (Fig. 4.12). The lipid-soluble antioxidant vitamin E serves to terminate the chain reaction process of lipid peroxidation, whereas glutathione peroxidase consumes lipid hydroperoxides that otherwise are a source of additional lipid radicals (52).

The significance of this layer of defense mechanisms in controlling the toxic effects of oxidant agents can be seen from the example of paraquat, a toxicant that stimulates O_2^- generation in cells. Mice with decreased glutathione peroxidase activity or vitamin E concentrations, obtained by respective administration of diets deficient in selenium or vitamin E, were markedly more susceptible to paraquat-induced lethality (31). It is interesting to note that paraquat is primarily a lung toxicant, which is due in large part to the fact that paraquat is preferentially taken up into this organ by an active transport mechanism. However, in mice on selenium deficient diets that had significantly reduced glutathione peroxidase in both liver and lung, paraquat also produced extensive liver toxicity (55). These data suggest that paraquat causes lung toxicity in normal animals because accumulation and retention of toxicant in lung overtaxes the oxidant defense mechanisms in that organ. Toxicity in liver, which does not specifically accumulate paraquat, occurs only when a defense mechanism is compromised.

The glutathione peroxidase system enzymes respond to oxidant stress in the same manner as superoxide dismutase (52). Thus, rats exposed to low concentrations of ozone had significantly increased activity of all three enzymes of this defensive system. This increased oxidant defense capacity appeared to be in part responsible for the development of tolerance to subsequent exposure to high concentrations of ozone.

5.4.3 Other Antioxidant Defenses

In addition to the enzymatic defenses and vitamin E, several other endogenous antioxidants have also been identified (1). These include: ascorbate, a water-soluble plasma antioxidant; uric acid, an important antioxidant in saliva and blood; biliverdin and bilirubin, both degradation products of heme, and potential blood and bile antioxidants; ubiquinol-10, a reduced form of the mitochondrial respiratory chain electron carrier, capable of scavenging lipid membrane radicals in a manner similar to vitamin E; and carnosine, which may serve as an antioxidant in brain and muscle tissue.

5.5 DNA Repair

Cellular DNA, because of its prime role as mediating life, is perhaps the most protected of all cell macromolecules. Nonetheless, it is clear that high doses of toxicants are too sufficient to breach the many layers of defenses protecting DNA from toxicant injury. Toxicants reaching nucleic acids may form lesions that result in mutagenic or carcinogenic responses. Not all toxicant–DNA interactions lead to these responses, however, since DNA is additionally protected from damage by a variety of repair mechanisms. DNA repair is accomplished by two general mechanisms, excision repair and postreplication repair (56).

DNA excision repair in mammalian cells involves several differing processes. DNA damage resulting from exposure to ultraviolet (UV) radiation or electrophilic toxicants with large ring systems (e.g., polycyclic aromatic hydrocarbons) is repaired by what has been termed a ''long-patch'' mechanism. This mechanism involves removal and subsequent replacement of up to 100 adjacent damaged nucleotides. DNA damage caused by exposure to methylating or ethylating agents or by ionizing radiation is repaired by a short-patch mechanism. Short-patch repair involves removal and replacement of short segments of apurinic and apyrimidinic sites in DNA resulting from both nonenzymatic and enzymatic removal of alkylated DNA bases. Although both excision repair processes involve a complex series of enzymatic reactions, excision repair is regarded as a relatively error-free process.

In contrast to the excision repair mechanism, postreplication DNA repair is regarded as being an error-prone process. As its name implies, postreplication repair occurs only during the replication, or S phase, of the cell cycle. DNA damaged by both chemical toxicants or radiation is subject to repair by this mechanism.

The potential importance of DNA repair mechanisms in controlling the onset of cancer can be illustrated from two examples. The human disease xeroderma pigmentosum is characterized by a genetically mediated deficiency in the DNA repair enzymes responsible for removing pyrimidine dimers induced in skin cells as a result of exposure to UV radiation (57). Individuals suffering from this disease exhibit a predisposition for skin cancer as a result of exposure to sunlight. A second example of the importance of DNA repair is seen in animal studies in which the organ-specific carcinogenicity of methylating- or ethylating-type carcinogens was found to directly correlate with a persistence of alkylated DNA bases in the affected tissue (58). Following administration of N-ethyl-N-nitrosourea to rats, O^6-ethylguanine was removed at a far slower rate from brain, which is a target tissue for this carcinogen, than from liver, which is a nontarget tissue. Other studies have shown that transgenic mice expressing human O^6-alkylguanine-DNA alkyltransferase, an enzyme which repairs O^6 DNA alkylation sites, are resistant to cancer induced by the DNA alkylating carcinogen N-methyl-N-nitrosourea (59). These examples illustrate that the susceptibility to environmentally induced carcinogenesis is directly correlated with the ability of the affected tissue to repair promutagenic lesions in DNA.

6 IMMUNOLOGICAL DEFENSE MECHANISMS

The immune system has been recognized in recent years to play a potentially important role in control of the development of cancer (60). The incidence of cancer is markedly higher, for example, in humans chronically treated with immunosuppressive agents or afflicted with primary immunodeficiency diseases. In addition, studies in animals have demonstrated that chemically induced tumors produce strong antitumor immune responses that may ultimately result in regression or elimination of the tumor. These observations suggest the possibility that, even if a toxicant should penetrate the host of defense mechanisms protecting the genetic material of a cell and initiate a malignant transformation, intervention by the immune system may prevent development of a tumor.

7 PROGRAMMED CELL DEATH (APOPTOSIS) AND TUMOR SUPPRESSOR GENES

Massive assault on a cell by a toxicant can obviously overwhelm even the broad array of defenses aligned against the attack. In such circumstances cell function is lost, individual cell death occurs, and focal areas of tissue necrosis results. However, cell death can also be intrinsically programmed within cells and need not necessarily result from overriding toxic insult or cell aging. Programmed cell death, termed apoptosis, may represent a fundamental defense mechanism preventing development of cancer (61).

Apoptosis may play both a positive and negative role in the multistep role of carcinogenesis. As a positive defense mechanism, apoptosis may facilitate the removal of precancerous cells, thus interrupting the ultimate progression to tumors. In contrast, some toxicants may delay or interfere with apoptosis, leading to abnormal accumulation of precancerous cells and associated increased probability of tumor appearance.

The progression of potentially precancerous cells to tumors has also been shown to be inhibited by the protein products of genes, termed tumor suppressor genes (62). Toxicant-induced mutations to tumor suppressor genes, however, may result in expression of altered protein products that have lost ability to properly regulate cell growth, contributing to tumorigenesis. An example of a gene whose mutated form appears to promote the development of cancer is the p53 gene. Studies with the nasal carcinogen formaldehyde have demonstrated that tumors induced by this toxicant contained mutated p53 (63). Other studies have shown that transgenic mice expressing a mutated p53 gene product were unusually prone to the development of a variety of tumors (64). Thus, toxicant-induced alterations in tumor suppressor genes, which represent fundamental defense mechanisms regulating cell growth, facilitate secondary carcinogenic responses.

8 CONCLUSIONS

Appreciation of the role of defense mechanisms in moderating the toxicity of xenobiotics has important practical application to the assessment of potential human hazard associated with exposure to toxicants. The toxicity of any foreign compound is necessarily dependent on the concentration of the toxicant at its critical target site(s) within cells or tissues. Thus, as has been illustrated in this review, intervention of the body defense mechanisms serves as an effective process by which the actual dose of toxicant reaching critical functional targets is dramatically reduced or rendered inconsequential. Toxic responses associated with exposure to many agents exhibit a dose threshold, or a dose below which no response is likely to occur. Since intervention of the body defense mechanisms may be a primary determinant underlying the existence of many thresholds, a comparative evaluation of defense mechanism function in relation to toxicant dose or across species lines may be extremely useful in extrapolating animal toxicity data to humans.

REFERENCES

1. B. N. Ames and L. S. Gold, *Mut. Res.*, **250**, 3 (1991).
2. J. D. Robertson, *Arch. Intern Med.*, **129**, 202 (1972).
3. J. A. Timbrell, *Principles of Biochemical Toxicology*, Taylor and Francis, London, 1982, p. 2147.
4. F. E. Guthrie, "Absorption and Distribution," in *Introduction to Biochemical Toxicology*, E. Hodgson and F. E. Guthrie, Eds., Elsevier, New York, 1980, p.10.
5. B. Idson, *J. Pharm. Sci.*, **64**, 901 (1975).
6. C. D. Klaassen, "Absorption, Distribution, and Excretion of Toxicants," in *Toxicology:*

The Basic Science of Poisons, M. O. Amdur, J. D. Doull, C. D. Klaassen, Eds., 4th ed., Pergamon Press, New York, 1991, p. 50.

7. R. L. Dixon, "Toxic Responses of the Reproductive System," in *Toxicology: The Basic Science of Poisons*, J. D. Doull, C. D. Klaassen, and M. O. Amdur, Eds., 2nd ed., Macmillan, New York, 1980, p. 332.

8. T. Gordon, and M. O. Amdur, "Responses of the Respiratory System to Toxic Agents," in *Toxicology: The Basic Science of Poisons*, M. O. Amdur, J. D. Doull, C. D. Klaassen, Eds., 4th ed., Pergamon Press, New York, 1991, p. 383.

9. D. L. Swift, and D. F. Proctor, "Access of Air to the Respiratory Tract," in *Respiratory Defense Mechanisms, Part 1*, J. D. Brain, D. F. Proctor, and L. M. Reid, Eds., Marcel Dekker, New York, 1977, p. 63.

10. R. F. Hounam and A. Morgan, "Particle Deposition," in *Respiratory Defense Mechanisms*, J. D. Brain, D. F. Proctor, and L. M. Reid, Eds., Part 1, Marcel Dekker, New York, 1977, p. 125.

11. D. F. Proctor, "The Mucociliary Clearance System," in *The Nose: Upper Airway Physiology and the Atmospheric Environment*, D. F. Proctor and I. Andersen, Eds., Elsevier, New York, 1982, p. 245.

12. G. M. Green, G. J. Jakab, R. B. Low, and G. S. Davis, *Am. Rev. Resp. Dis.*, **115**, 479 (1977).

13. J. G. Widdicombe, "Defense Mechanisms of the Respiratory Tract and Lungs," in *Respiratory Physiology II*, J. G. Widdicombe, Ed., Vol. 14, University Park Press, Baltimore, 1977, p. 291.

14. A. C. Allison, "Mechanisms of Macrophage Damage in Relation to the Pathogenesis of Some Lung Diseases," in *Respiratory Defense Mechanisms, Part 2*, J. D. Brain, D. F. Proctor, and L. M. Reid, Eds., Marcel Dekker, New York, 1977, p. 1075.

15. P. E. Morrow, *Toxicol. Appl. Pharmacol.*, **113**, 1, 1992.

16. J. M. Creeth, *Br. Med. Bull.*, **34**, 17 (1978).

17. J. A. Swenberg, C. S. Barrow, C. J. Boreiko, H. d'A. Heck, R. J. Levine, K. T. Morgan, and T. B. Starr, *Carcinogenesis*, **4**, 945 (1983).

18. Y. Alarie, *C. R. C. Crit. Rev. Toxicol.*, **2**, 299 (1973).

19. B. N. Ames, M. Profet, and L. S. Gold, *Proc. Natl. Acad. Sci.*, (USA), **87**, 7772, 1990.

20. J. C. F. Chang, E. A. Gross, J. A. Swenberg, and C. S. Barrow, *Toxicol. Appl. Pharmacol.*, **68**, 161 (1983).

21. L. Z. Benet and L. B. Sheiner, "Pharmacokinetics; the Dynamics of Drug Absorption, Distribution, and Elimination," in *The Pharmacological Basis of Therapeutics*, A. G. Goodman, L. S. Goodman, T. W. Rall, and F. Murad, Eds., 7th ed., Macmillan, New York, 1985, p. 3.

22. R. P. Smith, "Toxic Responses of the Blood," in *Toxicology: The Basic Science of Poisons*, M. O. Amdur, J. Doull, and C. D. Klaassen, Eds., 4th ed., Pergamon, Press, New York, 1991, p. 257.

23. R. A. Goyer, "Toxic Effects of Metals," in *Toxicology: The Basic Science of Poisons*, M. O. Amdur, J. Doull, and C. D. Klaassen, Eds., 4th ed., Pergamon, New York, 1991, p. 623.

24. M. G. Cherian and M. Nordberg, *Toxicology*, **28**, 1 (1983).

25. M. Webb and K. Cain, *Biochem. Pharmacol.*, **31**, 137 (1982).

26. P. J. Gehring, R. G. Watanabe, and G. E. Blau, in *Advances in Modern Toxicology— New Concepts Evaluation*, Halstead Press, New York, 1976, p. 195.

27. R. T. Williams, *Detoxication Mechanisms*, 2nd ed., Chapman and Hall, London, 1959.

28. I. G. Sipes and A. J. Gandolfi, "Biotransformation of Toxicants," in *Toxicology: The Basic Science of Poisons*, M. O. Amdur, J. Doull, and C. D. Klaassen, Eds., 4th ed., Pergamon Press, New York, 1991, p. 88.

29. M. J. Coon and A. V. Persson, "Microsomal Cytochrome P-450: A Central Catalyst in Detoxication Reactions," in *Enzymatic Basis of Detoxication*, W. B. Jakoby, Ed., Vol. 1, Academic Press, New York, 1980, p. 117.

30. J. A. Goldstein and M. B. Folletto, *Environ. Hlth. Perspect.*, **100**, 169 (1993).

31. J. S. Bus and J. E. Gibson, "Role of Activated Oxygen in Chemical Toxicity," in *Drug Metabolism and Drug Toxicity*, J. R. Mitchell and M. G. Horning, Eds., Raven Press, New York, 1984, p. 21.

32. M. W. Anders, "Biotransformation of Halogenated Hydrocarbons," in *Drug Metabolism and Drug Toxicity*, J. R. Mitchell and M. G. Horning, Eds., Raven Press, New York, 1984, p. 55.

33. D. M. Ziegler, "Microsomal Flavin-containing Monooxygenases: Oxygenation of Nucleophilic Nitrogen and Sulfur Compounds," in *Enzymatic Basis of Detoxication*, W. B. Jakoby, Ed., Vol. 1, Academic Press, New York, 1980, p. 201.

34. D. M. Ziegler, *Drug Metabol. Disp.*, **19**, 847 (1991).

35. F. Oesch, "Microsomal Epoxide Hydrolase," in *Enzymatic Basis of Detoxication*, W. B. Jakoby, Ed., Vol. 2, Academic Press, New York, 1980, p. 271.

36. W. F. Borson and T.-K. Li, "Alcohol Dehydrogenase," in *Enzymatic Basis of Detoxication*, W. B. Jakoby, Ed., Vol. 1, Academic Press, New York, 1980, p. 231.

37. J. M. Ritchie, "The Aliphatic Alcohols," in *The Pharmacological Basis of Therapeutics*, 6th ed., A. G. Gilman, L. S. Goodman, and A. Gilman, Eds., Macmillan, New York, 1980, p. 376.

38. H. Weiner, "Aldehyde Oxidizing Enzymes," in *Enzymatic Basis of Detoxication*, W. B. Jakoby, Ed., Vol. 1, Academic Press, New York, 1980, p. 261.

39. E. Heymann, "Carboxylesterases and Amidases," in *Enzymatic Basis of Detoxication*, W. B. Jakoby, Ed., Vol. 2, Academic Press, New York, 1980, p. 291.

40. D. W. Nebert, "Human Genetic Variation in the Enzymes of Detoxication," in *Enzymatic Basis of Detoxication*, W. B. Jakoby, Ed., Vol. 1, Academic Press, New York, 1980, p. 25.

41. A. P. Kulkarni and E. Hodgson, "Comparative Toxicity," in *Introduction to Biochemical Toxicology*, E. Hodgson and F. E. Guthrie, Eds., Elsevier, New York, 1980, p. 106.

42. C. B. Kasper and D. Henton, "Glucuronidation," in *Enzymatic Basis of Detoxication*, W. B. Jakoby, Ed., Vol. 2, Academic Press, New York, 1980, p. 3.

43. W. C. Dauterman, "Metabolism of Toxicants: Phase II Reactions," in *Introduction to Biochemical Toxicology*, E. Hodgson and F. E. Guthrie, Eds., Elsevier, New York, 1980, p. 92.

44. W. B. Jakoby and R. D. Sekura, E. S. Lyon, C. J. Marcus, and J.-L. Wang, "Sulfotransferases," in *Enzymatic Basis of Detoxication*, W. B. Jakoby, Ed., Vol. 2, Academic Press, New York, 1980, p. 199.

45. W. B. Jakoby and W. H. Habig, "Glutathione Transferases," in *Enzymatic Basis of Detoxication*, W. B. Jakoby, Ed., Vol. 2, Academic Press, New York, 1980, p. 63.

46. B. Ketterer, D. J. Meyer and A. G. Clark, "Soluble Glutathione Transferase Isozymes," in *Glutathione Conjugation*, H. Sies and B. Ketterer, Eds., Academic Press, London, 1988, p. 73.

47. W. Z. Potter, S. S. Thorgeirsson, D. J. Jollow, and J. R. Mitchell, *Pharmacology*, **12**, 129 (1974).

48. W. E. Weber and I. B. Glowinski, "Acetylation," in: W. B. Jakoby, Ed., *Enzymatic Basis of Detoxication*, Vol. 2, Academic Press, New York, 1980, p. 169.

49. R. A. Weisiger and W. B. Jakoby, "S-Methylation: Thiol S-Methyltransferase," in *Enzymatic Basis of Detoxication*, W. B. Jakoby, Ed., Vol. 2, Academic Press, New York, 1980, p. 131.

50. R. O. Beauchamp, Jr., J. S. Bus, J. A. Popp, C. J. Boreiko, and D. A. Andjelkovich, C. R. C. *Crit. Rev. Toxicol.*, **13**, 25 (1984).

51. H. Sies, *Am. J. Med.*, **91**, 315 (1991).

52. J. S. Bus and J. E. Gibson, *Rev. Biochem. Toxicol.*, **1**, 125 (1979).

53. I. Fridovich, *Science*, **201**, 875 (1978).

54. J. P. Crapo and D. F. Tierney, *Am. J. Physiol.*, **226**, 1401 (1974).

55. S. Z. Cagen and J. E. Gibson, *Toxicol. Appl. Pharmacol.*, **40**, 193 (1977).

56. D. J. Holbrook, Jr. "Chemical Carcinogenesis," in *Introduction to Biochemical Toxicology*, E. Hodgson and F. E. Guthrie, Eds., Elsevier, New York, 1980, p. 310.

57. J. E. Cleaver, *Nature*, **218**, 652 (1968).

58. T. Kleihues, "The Role of DNA Alkylation and Repair in the Toxic and Carcinogenic Effects of Alkylnitrosoureas," in *Proceedings of the First International Congress on Toxicology: Toxicology as a Predictive Science*, G. L. Plaa and W. A. M. Duncan, Eds., Academic Press, New York, 1978, p. 191.

59. L. L. Dumenco, E. Allay, K. Norton, and S. Gerson, *Science*, **259**, 219 (1993).

60. J. H. Dean and M. J. Murray, "Toxic Responses of the Immune System" in *Toxicology: The Basic Science of Poisons*, M. O. Amdur, J. Doull, and C. D. Klaassen, Eds., 4th ed., Pergamon Press, New York, 1991, p. 282.

61. W. Bursch, F. Oberhammer, and R. Schulte-Hermann, *Trends Pharmacol. Sci.*, **13**, 245, 1992.

62. S. J. Baker and B. Vogelstein, in *Nuclear Processes and Oncogenes*, P. A. Sharp, ed., Academic Press, San Diego, 1992, p. 105.

63. L. Recio, S. Sick, L. Plata, E. Bermudez, E. A. Gross, Z. Chen, K. Morgan, and C. Walker, *Can. Res.*, **52**, 6113 (1992).

64. L. A. Donehower, M. Harvey, B. L. Slagle, M. J. McArthur, C. A. Montgomery, Jr., J. S. Butler, and A. Bradley, *Nature*, **356**, 215 (1992).

CHAPTER FIVE

Reproductive Toxicology

Peter K. Working, Ph.D., DABT

1 INTRODUCTION

Reproductive toxicity is defined as the adverse effects of chemicals on the adult or maturing organism and includes, but is not limited to, deleterious effects on gonadal structure and function, alterations in fertility (e.g., infertility or subfertility), and impaired gamete function (Mattison et al., 1991). Although exposure may occur prior to the time of conception, reproductive toxicity may become evident during fertilization, the embryonic or fetal periods, or even postnatally. A related subject, which will not be discussed here, is developmental toxicity, often called teratology, which is defined as the adverse effects of chemicals on the developing conceptus that are associated with exposure prior to or during pregnancy (U.S. EPA, 1986; Mattison et al., 1989). Developmental toxicity may be manifested during the embryonic or fetal periods, or postnatally to the time of sexual maturation and, potentially, beyond. Excellent and detailed reviews of the subject can be found elsewhere (Manson and King, 1989; Manson and Wise, 1991).

A complete understanding of the potential reproductive toxicity of xenobiotics should be grounded on a basic knowledge of the biology of the reproductive process in males and females. Accordingly, this chapter begins with a brief review of gametogenesis and female and male reproductive biology.

2 GAMETOGENESIS

Gametogenesis is the process by which gametes (sperm and ova) are produced in mammals. The process differs in several important aspects in males and females,

Patty's Industrial Hygiene and Toxicology, Third Edition, Volume 3, Part B, Edited by Lewis J. Cralley, Lester V. Cralley, and James S. Bus.
ISBN 0-471-53065-4 © 1995 John Wiley & Sons, Inc.

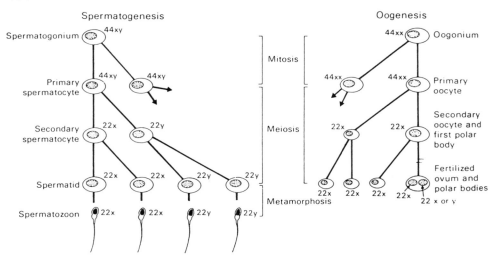

Figure 5.1 The process of meiosis as it occurs during spermatogenesis and oogenesis. (From Dixon, 1986, with permission.)

and these contrasts may account for differential sensitivity to reproductive toxicants sometimes observed in males and females. The central objective of gametogenesis in both sexes is meiosis, a special type of cell division in which the chromosomal count and DNA content of each daughter cell is halved, from the diploid 2N DNA content to the haploid 1N (Fig. 5.1). This is accomplished by the initial replication of the 46 paired chromosomes, as in mitosis, and then two reductive divisions, termed meiosis 1 and meiosis 2, which reduce the chromosomal number to 23, one member of each chromosomal pair.

Gametogenesis starts the same in both sexes. Primordial germ cells migrate into the genital ridges of the fetus during early organogenesis and proliferate into the gonadal complement of stem cells, called oogonia in the female and spermatogonia in the male. From here the processes diverge. The ability to produce spermatogonia is maintained for lifetime in the male, whereas production of oogonia occurs only in early fetal life in females. Stem cells proliferate mitotically until, in a committed division, one daughter cell remains a spermatogonia or oogonia and the other becomes a primary spermatocyte or oocyte. In the male the first reductive division of meiosis occurs shortly thereafter, giving rise to two secondary spermatocytes. Secondary spermatocytes, in turn, undergo a second reductive division, producing two haploid spermatids each, which mature to become a spermatozoa. In contrast, the primary oocytes are arrested in meiotic prophase within the ovary and only undergo the first meiotic division to form a secondary oocyte just prior to ovulation. The second reductive division of meiosis, which forms the haploid ovum, occurs only after sperm penetration of the oocyte in the oviduct.

2.1 Oogenesis

The production of ova is termed oogenesis. In the female the oogonia proliferate mitotically during early fetal life in the first step of oogenesis. Meiosis begins, and the primary oocyte proceeds prenatally only to the diplotene stage of the first meiotic division, which occurs at about week 20 of gestation (Fig. 5.1). The process ceases there, however, with the cells arrested in meiosis as resting, or dictyate, primary oocytes. Once meiosis has begun, the oocytes are no longer able to replicate (i.e., no new ones can be formed). The number of oocytes in the human female reaches a peak of 6–7 million at week 20 of gestation and declines thereafter, being reduced to approximately 2 million at birth (Baker, 1963). The decline continues postnatally as cohorts of oocytes begin to grow, only to degenerate before ovulation for lack of hormonal support in the still immature female. At and after puberty, with oocytes now numbering some 300–400,000, selected oocytes continue on to ovulation, with meiosis reinitiated just prior to ovulation. Just 25,000 primary oocytes remain by 30 years of age. The net result of these events, that is, the cessation of oogonial proliferation after 20 weeks of gestation and the continued recruitment and atresia of oocytes, is the gradual and then precipitous decline in oocyte number in mature females, leading eventually to the near total depletion of oocytes and menopause, when oocyte number may be just a few thousand. Thus, the oocytes produced by gestation week 20 represent the entire lifetime supply of oocytes available to the female. Any event that destroys an oocyte permanently reduces the complement of potential ova and, in theory, can hasten menopause. Only about 400 mature oocytes will be ovulated during a woman's reproductive lifespan.

2.2 Spermatogenesis

The gradual process by which spermatogonial cells divide and differentiate into spermatozoa is known as spermatogenesis. In the fetal male, spermatogonial stem cells are quiescent after the initial brief phase of proliferation. Meiosis is reinitiated at puberty and continues unabated for the lifetime of the male. Because of the continuous replenishment of the stem cells, there is an unceasing supply of spermatocytes, spermatids, and sperm. The continued production of spermatogenic cells is dependent on the existence of stem cell spermatogonia, which have been identified as isolated, undifferentiated A-type spermatogonia (Huckins, 1971). Beginning with the stem cells, the spermatogonia undergo a series of mitotic divisions (six in the rat), the final division yielding preleptotene primary spermatocytes and initiating meiosis. After a prolonged prophase, each tetraploid primary spermatocyte divides to two secondary spermatocytes, which quickly undergo the final meiotic division to yield a total of four haploid spermatids. These spermatids divide no further and, round at formation, are subsequently transformed into the elongated shape typical of mammalian sperm in a process known as spermiogenesis. This process involves extensive nuclear and cytoplasmic reorganization, with condensation of the nuclear material, production of a midpiece, and elaboration of a flagellum.

Each step of spermatogenesis is precisely timed, and complex associations of cells mature in synchrony (Clermont, 1972). As the cycle of the seminiferous epithelium progresses, more mature cells continually become associated with less mature ones, so that in any given area of the tubule, five or six generations of cells appear together, forming cell associations of definite and fixed composition (Leblond and Clermont, 1952; Hess, 1990). Specifically, one or two generations of spermatids are always associated with one or two generations of spermatocytes, as well as with sper-

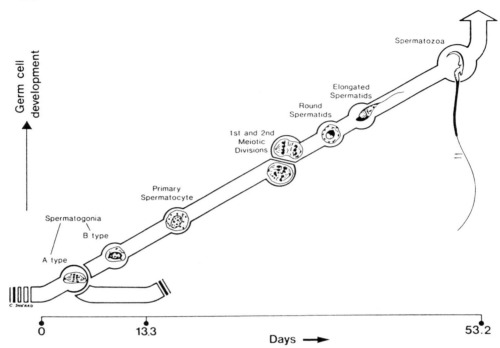

Figure 5.2   Spermatogenesis in the rat. Spermatogonia are continually replenished by mitotic division of stem cells. In the rat differentiated spermatogonia (A type) undergo division to "committed" type B spermatogonia. These cells form preleptotene primary spermatocytes, which undergo a prolonged meiotic prophase and eventually go through two reductive divisions to yield haploid spermatids. Spermatids undergo morphological transformation to become spermatozoa in the process known as spermiogenesis. (From Foster, 1988, with permission.)

matogonia, each in a particular development stage. The cycle of the seminiferous epithelium in the rat has been divided into 14 specific stages (Leblond and Clermont, 1952), each consisting of a specific mixture of cells (Fig. 5.2). Stages differ somewhat in cell population and morphology in the human but generally follow the same pattern. The duration of spermatogenesis corresponds to the time required for a given stem cell to produce mature spermatozoa that are released into the excurrent duct system. In the rat this process takes about 60 days, in the human about 80 days.

3   FEMALE REPRODUCTIVE PHYSIOLOGY

The female reproductive system is a complex system that requires precisely regulated local and systemic circulating hormones for proper functioning. Primary organs of the system are the ovary, the oviducts or fallopian tubes, the uterus, and the vagina (Fig. 5.3). The hormonal regulation of female reproduction is a precisely regulated

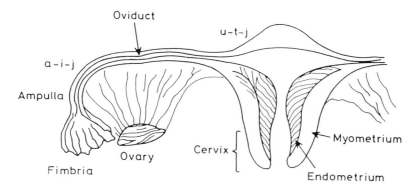

a-i-j: ampullary—isthmic junction
u-t-j: utero—tubal junction

Figure 5.3 Reproductive organs of the human female. (From Gangolli and Phillips, 1988, with permission.)

balance of interactions between the central nervous system, the anterior pituitary and the ovary. A detailed discussion of this neuroendocrine axis can be found in several excellent reviews (Bardin, 1986; Fink, 1988; Matsumoto, 1989).

The adult ovary consists of an outer cortex and an inner medulla, and within the outer cortex are the oocyte-follicle cell complexes (Fig. 5.4). The follicle, composed

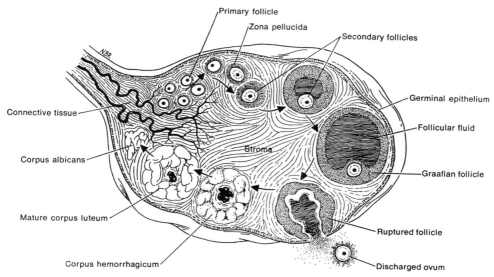

Figure 5.4 Schematic representation of the ovary, seen in sectional view. Arrows indicate the developmental stages of the oocyte as they occur in the ovarian cycle. (From Tortora and Anagnostakos, 1975, with permission.)

of granulosa cells, thecal cells and the oocyte, maintains the hormonal environment necessary to support the growth and maturation of an oocyte. The process, known as folliculogenesis, involves both intra- and extraovarian regulation (Plowchalk et al., 1993). The earliest stage, the primordial follicle, consists of an oocyte (arrested in meiosis I) and a single layer of granulosa cells. It enlarges to the primary oocyte stage in a hormone-independent process. After puberty, groups of primary oocytes are stimulated to become secondary follicles by rising gonadotropin levels. Certain of these continue to develop, and one will become the dominant follicle, destined to be ovulated at the appropriate hormonal signal. The ovarian cycle consists of a cyclic release of pituitary gonadotropins, follicle-stimulating hormone (FSH) and luteinizing hormone (LH), and the secretion of ovarian progesterone and estrogen. Interacting in a delicately balanced cycle, these hormones stimulate the maturation of oocytes and ovulation and prepare the female accessory sex organs for eventual fertilization and implantation.

The oviducts, or fallopian tubes, which lie closely apposed to the ovary, are the passageways for the ovulated oocyte to move to the uterus and the site of fertilization. Transport of gametes and fertilized ova involves the autonomic nervous system, suggesting that agents that affect this system may also have the potential to alter fertility. The uterus, specifically the uterine endometrium, is the site of implantation and, as it is prepared to accept the embryo, reflects the hormonal cycle of the ovary.

Multiple sites in the female reproductive system may be disrupted by xenobiotics. However, few studies have been directed toward, and little data is available for, assessing the actual vulnerability of reproduction in the female to xenobiotics. Carefully designed studies will be required to fully address the potential impact of chemicals on female reproductive function. Several study types that are currently in use are discussed below.

4. MALE REPRODUCTION PHYSIOLOGY

The male reproductive system is also a complex system, sensitive to chemical insult at numerous sites (Fig. 5.5). The reproductive neuroendocrine axis of the male is comprised of the central nervous system, the anterior pituitary gland, and the testes. Input from the central nervous system (CNS) controls the secretion of the pituitary gonadotropins (FSH and LH), which in turn act upon the Leydig and Sertoli cells of the testes to regulate spermatogenesis and hormone production. The male reproductive system, described in more detail below, is comprised of the testes, the accessory sex glands, and the excurrent duct system.

The testicular parenchyma is comprised of two functional compartments, the seminiferous epithelium and the interstitial tissue. The seminiferous epithelium is formed into long convoluted tubules connected at both ends to the beginning of the excurrent duct system, the rete testis (Fig. 5.6). Within the tubules are Sertoli cells, which are somatic cells that extend from the basement membrane of the tubule to its lumen, and germ cells of various stages of development. It is within the tubules that spermatogenesis occurs. The Sertoli cells surround and support the germ cells (Russell, 1980) and tight junctions between the Sertoli cells form the blood–testis barrier. This anatomical barrier effectively divides the tubules into basal and adluminal compart-

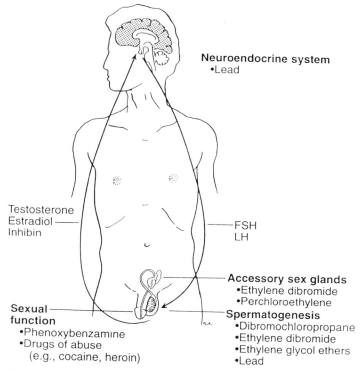

Figure 5.5 The male reproductive system is a complex system, sensitive to chemical insult at numerous sites. Primary targets include the neuroendocrine system, the testes, the accessory sex glands, and sexual function. Examples of toxicants that act at various sites are shown. (From Schrader and Kesner, 1993, with permission.)

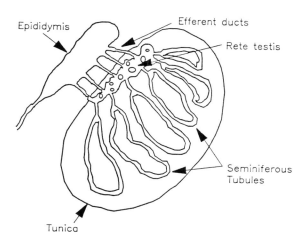

Figure 5.6 Rat testis and excurrent ducts. Sperm are produced in the seminiferous tubules during spermatogenesis. At spermiation, spermatids move out of the testes via the rete testis and efferent ducts into the epididymis. (From Foster, 1989, with permission.)

ments that have differential exposure to blood-borne xenobiotics. In addition, Sertoli cells fulfill a number of other important functions, including secretion of fluid into the seminiferous tubule, coordination of the movement of germ cells from the periphery of the tubule to its lumen, spermiation (the release of mature spermatids at the end of spermatogenesis), and the secretion of a number of important proteins, including androgen-binding protein (Fritz et al., 1976), transferrin (Skinner and Griswold, 1980), and inhibin (Verhoeven and Franchimont, 1983).

The highly vascularized tissue of the interstitial compartment contains the peritubular cells, which are located at the seminiferous tubule periphery, macrophages, smooth muscle cells, and Leydig cells, the primary source of testosterone in the testis (see Fawcett et al., 1973; Connell and Connell, 1977; Ewing, 1983). Toxicants that target the interstitial cells may indirectly affect spermatogenesis via their disturbance of the homeostatic mechanisms of the testis.

The seminiferous tubules, containing the spermatogenic cells, are folded and refolded within the testis, and a cross section of rat testis will reveal several hundred tubules, each identifiable as one of the 14 different spermatogenic stages (Fig. 5.7). Classification of the tubules are based primarily on the morphological appearance and location of the spermatids within the seminiferous epithelium (Leblond and Clermont, 1952). Careful histopathological examination of the testis may permit the determination of the site of action of a toxic agent and can be useful in classifying reproductive toxicants (Chapin, 1989).

After completion of the spermiogenesis phase of spermatogenesis, spermatids move out of the testis into an extensive duct system that is critical to their final maturation into spermatozoa and to their acquisition of fertilizing capacity. This duct system consists of the efferent ducts (ductuli efferentes), epididymis, ductus deferens, and ejaculatory duct (Fig. 5.8). Associated with this duct system are the accessory sex glands, which in humans consist of the seminal vesicles, prostate, and bulbourethral (Cowper's) glands. In rodents, coagulating glands and preputial glands are also present. The duct system and most accessory sex glands in humans and laboratory rodents (rats and mice) are quite similar in structure and also have many common functional properties (Eddy, 1988).

The epididymis is the principal site of sperm maturation and is also a reservoir for the storage of spermatozoa until ejaculation. It consists of three main regions: the head (caput epididymis), the body (corpus epididymis), and the tail (cauda epididymis). Secretory and absorptive functions of the epididymal epithelium are critical for sperm transport, modification of epididymal fluid composition, and changes in sperm morphology, composition, and function that occur during sperm maturation (Bedford, 1975).

The accessory sex glands produce a variety of secretory products that contribute most of the volume of the ejaculate. Fertility is reduced in vivo in the absence of accessory sex gland products, indicating that they contribute to the overall process of reproduction (Eddy, 1988). Sex gland secretions are thought to facilitate transport of spermatozoa within the female reproductive tract after ejaculation and to modify the environment for optimum sperm survival and function. There are few reports of agents causing toxic effects specifically on the accessory sex glands, but it is not clear whether this is due to lack of occurrence or lack of detection. Agents could exert toxic effects on the accessory sex glands either directly or indirectly (secondary to perturbed androgen metabolism). The androgen dependence of the accessory sex glands and the degenerative effects of exogenous hormones such as diethylstilbestrol and estradiol have been studied extensively (Feagans et al., 1961; Cavazos, 1977). In addition, numerous drugs and toxic agents have been found in semen, and these might not only affect the male reproductive tract but also might gain access to the female reproductive tract and possibly conceptus (Hales et al., 1986). Although less is known about the effects of toxic substances on the duct system and accessory sex glands of the male than on the testis, it is clear that both direct and indirect effects are possible. The androgen dependence of the duct system and sex glands makes them susceptible to secondary effects from agents that primarily affect the testis, since testicular damage often results in altered androgen production.

5 XENOBIOTICS AND TOXICITY IN THE MALE AND FEMALE: FUNDAMENTALS

The reproductive process in both sexes involves a complex biological system, vulnerable at multiple sites to the effects of toxic agents. Some reproductive toxicants exhibit a surprising specificity for particular cell types or organs, whereas others have a broader effect, acting at diverse locations. The specificity of many toxicants is a function of both dose level and treatment duration; chemicals that damage only one cell type or reproductive organ often cause more widespread damage if the dose or duration of treatment is increased.

Such dose and duration effects may be one reason for the observation that substances that are toxic to the testis or ovary are frequently also toxic to the epididymis, the accessory sex glands, the oviducts, or the uterus. The similar susceptibility of different parts of the reproductive tract also may be because they are similar both metabolically and physiologically or because the entire reproductive tract is highly dependent on steroid hormones. Thus, agents that directly damage the testis or ovary and alter androgen or progesterone and estrogen production may indirectly affect the remainder of the reproductive tract.

Clearly, to understand the long-term effects of exposure to reproductive toxicants, their mechanisms of action in the reproductive tract must first be understood. To accomplish this goal, it will be necessary to employ more specific approaches to elucidating specific sites and modes of action of reproductive toxicants. Use of such information may permit development of more quantitative ways to monitor changes in the reproductive system in men and women exposed to potentially toxic agents in the environment.

6 MECHANISM OF ACTION OF REPRODUCTIVE TOXICANTS

Toxicants that affect the reproductive system may act directly or indirectly (Table 5.1). Agents that act directly generally do so because of their chemical reactivity (e.g., DNA damage induced by chemotherapeutic agents) or owing to their similarity

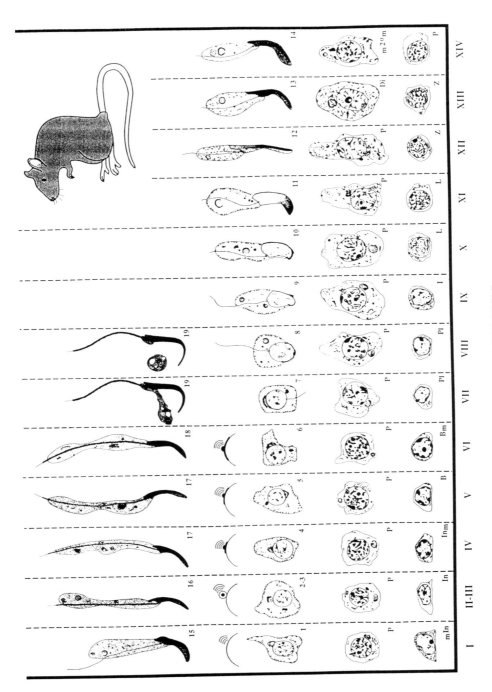

STAGES OF THE CYCLE

202

Figure 5.7 Spermatogenesis in the rat. The vertical columns (Roman numerals) represent cell stages or associations, according to Leblond and Clermont (1952). Abbreviations: In, intermediate spermatogonium; B, type B spermatogonium; Pl, preleptotene primary spermatocyte; L, leptotene primary spermatocyte; Z, zygotene primary spermatocyte; P, pachytene primary spermatocyte; Di, diplotene primary spermatocyte; 2°, secondary spermatocyte. Arabic numbers refer to the steps of spermiogenesis. The subscript m indicates that a mitotic or meiotic division occurs prior to or after the indicated cellular stage. (From Russell et al., 1990, with permission.)

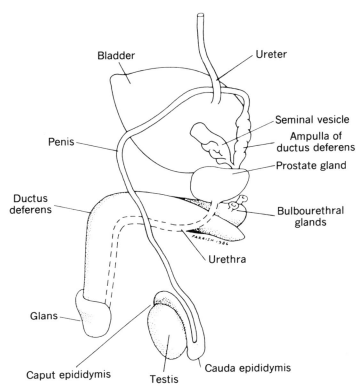

Figure 5.8 Lateral view of the urogenital system of the male human. The duct system includes the epididymis, ductus deferens, and ejaculatory duct. Accessory glands of the male include the seminal vesicles, prostate, and bulbourethral glands. (From Eddy, 1988, with permission.)

to some endogenous compound (e.g., a hormone or hormone antagonist), whereas those that act indirectly may do so via their interference with basic physiological control mechanisms (e.g., induction of enzyme activation systems).

Compounds that are chemically reactive may be nonspecific in their site of action, may act directly on the gonads, or may affect the developing reproductive system. Most are also cytotoxic, mutagenic, or carcinogenic in one or more test systems, although dose levels that disrupt reproduction may be much lower than dose levels that produce tumors in carcinogenicity studies. For example, sterility after cancer chemotherapy is a more common outcome than the induction of a second tumor, despite the fact that most chemotherapy drugs are both mutagenic/carcinogenic and toxic to the reproductive system (Kay and Mattison, 1985). Some direct-acting toxicants, including metals like lead, cadmium, and mercury, are toxic to both the developing and mature reproductive system, often of both sexes (Mattison, 1983).

Direct acting toxicants that act via their structural similarity to endogenous compounds have been described as "biological impostors" (Mattison and Thomford, 1989), in that they mislead biological processes through their interference. Generally,

Table 5.1 Direct and Indirect-Acting Reproductive Toxicants[a]

Mechanism	Compound	Gender
Direct		
Chemical Reactivity	Alkylating agents	M/F
	Cadmium	F
	Boron	M
	Lead	M/F
	Mercury	M/F
	Cyclophosphamide[b]	M/F
	Ethanol[b]	M/F
	Dibromochloropropane[b]	M
Structural Similarity	Steroid hormones	M/F
	Cimetidine	M
	Diethylstilbesterol	M/F
	Nicotine	F
	Azathioprine	M/F
	6-Mercaptopurine	M/F
	Galactose	F
	Halogenated hydrocarbons[b]	M/F
Indirect		
Disrupted homeostasis	Barbiturates	M/F
	Anticonvulsants	M/F
	Halogenated hydrocarbons	M/F
	Polycyclic halogenated hydrocarbons	M/F

[a]Adapted from Mattison and Thomford (1989)
[b]May require metabolic activation

these agents act as hormone agonists or antagonists. Examples include oral contraceptives. Exposure to them during their manufacture represents a well-known example of occupational reproductive toxicity (Harrington et al., 1978). In fact, any xenobiotic that is distributed to the hypothalamus or pituitary and that also acts like a steroid hormone may alter gonadal function. Diethylstilbestrol (DES) is a potent estrogen agonist and can impair reproduction in the female by inhibiting the release of FSH and LH at the level of the hypothalamus and pituitary. Some halogenated hydrocarbons are estrogenic and can alter hypothalamic function and fertility (Kimbrough, 1974; Kupfer, 1975).

Other xenobiotics may act indirectly by disrupting one or more of the complex physiological control mechanisms so important to reproduction. For example, indirect-acting toxicants may induce or inhibit hepatic or gonadal enzyme systems, thereby altering synthesis, secretion, or clearance of the steroid hormones and, subsequently, the hormonal feedback loops required for successful reproduction. Some toxicants may enhance fertility, a class of reproductive toxicity that is often ignored. Xenobiotics that stimulate clearance of the estrogenic components of oral contraceptives, for example, will increase the likelihood of ovulation (Mattison and Thomford, 1989). Metabolic interactions that disrupt hormone interactions are the likely mechanism of action of polycyclic aromatic and halogenated hydrocarbons, such as DDT and polychlorinated and polybrominated hydrocarbons (Welch et al., 1969, 1971; Mattison, 1981).

Both direct-acting and indirect-acting reproductive toxicants, like all toxicants, may require biotransformation to be active (Table 5.1). When metabolic activation must occur, the site of biotransformation can be an important determinant of toxic potential of the xenobiotic, particularly for agents that act directly on the gonads. Highly electrophilic metabolites produced at a distant site from the gonads, like the liver, may not be toxic in the relatively remote gonads owing to the short biological half-lives of these metabolites, that is, they are simply not stable enough to reach the gonads in significant quantity. The large-scale transfer of active metabolites from one organ to another is considered unlikely because of their reactivity (Nelson et al., 1977). Conversely, less reactive metabolic products of hepatic biotransformation may attain significant concentrations in the gonads and, presumably, could be further metabolized to more toxic forms. The gonads contain measurable amounts of cytochrome P450, mixed-function oxidases, epoxide hydrases, aryl hydrocarbon hydrolases and the various transferases necessary for the metabolic activation of many xenobiotics, although these are generally present at only a fraction of their concentration in the liver (Sims and Glover, 1974; Mattison and Thorgeirsson, 1979; Dixon and Lee, 1980; Heinrichs and Juchau, 1980). Nonetheless, these enzymatic activities are present in close proximity to the germ cells and so may be of particular importance in the biotransformation of gonadal toxicants. Most likely, the interaction between gonadal activation and hepatic detoxification of xenobiotics plays a significant role in the modulation of gonadal toxicity. However, gonadal toxicity can also be mediated at a distant site, for example, the hypothalamic–pituitary–gonadal axis, which may be extremely sensitive to toxic insult. Metals, particularly lead, cadmium, and mercury, have been implicated in reproductive dysfunction along the hypothalamic–pituitary–ovarian-uterine axis, with varying impact upon fertility and the female reproductive system (Table 5.2).

Table 5.2 Effects of Metals on the Hypothalamic–Pituitary–Ovarian–Uterine Axis[a]

Compound (route)	Reproductive Effect	Site[b]	Mechanism
Lead acetate (po)	Delayed vaginal opening	H, P	Decreased FSH
Lead acetate	Ovarian atrophy	O, H, P	Decreased FSH, direct toxicity
Triethyl lead chloride (po)	Decreased fertility (blocked implantation)	U, O	Endometrial alteration
Mercuric chloride (sc)	Altered ovarian cycle	O	Direct toxicity
Methyl mercuric chloride	Altered ovarian cycle	O	Direct toxicity
Cadmium chloride (ip)	Follicle necrosis	O	Vascular toxicity
Cadmium	Follicular atresia	O	Vascular toxicity
Cadmium chloride (sc)	Blocked ovulation (PMSG, HCG stimulated)	O	Receptor toxicity (?)

[a]Adapted from Mattison and Thomford (1989)
[b]Site of action: H, hypothalamus; P, pituitary; O, ovary; U, uterus.

7 MUTAGENESIS AND GENOTOXICITY IN GERM CELLS

In many discussions of reproductive toxicity, the potential for the expression of mutagenesis and genotoxic effects (e.g., aneuploidy, clastogenesis, DNA damage) in germ cells is overlooked. There is a good correlation between somatic cell mutagenicity and carcinogenicity (Dunkel, 1985), and the correlation extends to germ cell mutagens: All agents known to be mutagenic in germ cells are also mutagenic in somatic cells or are animal carcinogens. However, although germ cell mutagens are somatic cell mutagens, not all chemicals mutagenic in somatic cells are also mutagenic in gametes, due to a variety of modifying influences. Some germ cell mutagens may require metabolic activation, and, as discussed above, the testis and ovary have significant levels of the required biotransformation hydrolases, reductases, oxidases, and transferases. Although they are present at only a fraction of the levels found in the liver, their proximity to the germ cells renders them a significant factor in modulating the mutagenic activity of genotoxicants in the gonads (Dixon and Lee, 1980).

Other factors may also modulate the action of mutagens in the gonads. In the testes, the potential mutagen must penetrate the blood–testis barrier, whereas there is no evidence of a similar blood–ovary barrier in the female. Hence, the susceptibility of gametes to xenobiotic-induced mutations may be expected to vary between males and females. Differences in gametogenesis may affect the expression of the effects of mutagen exposure in males and females. In females there is a finite supply of oocytes, arrested in the dictyate stage of meiotic prophase I, in the ovary and no DNA synthesis occurs during the remaining stages of oogenesis in the adult female. In the male, in contrast, both mitotic and meiotic DNA synthesis is continually underway in the testes, and the spermatocyte supply (and, hence, the sperm supply) is renewable. As a consequence of these differences in gametogenesis, males and females may have a differential susceptibility to genotoxicants. Males may be more sensitive to agents that require DNA synthesis to express their activity (e.g., certain chemotherapeutic agents), since DNA synthesis is occurring. Females, on the other hand, may be more affected by agents that are lethal to primary oocytes, which are in limited and nonrenewable supply. Thus, from a genetic risk standpoint, the cells of greatest importance in the male are the spermatogonia, the stem cells, since a mutation induced in the stem cells will be carried by all spermatocytes and sperm develop from those stem cells for the lifetime of the male. A mutation induced in a spermatocyte will have a less permanent overall impact, since the affected spermatocytes will in time be gone from the testis. In the female the equivalent oogonial stem cell no longer exists after birth, and instead dictyate primary oocytes comprise the entire germ cell pool. Thus, xenobiotics that kill or cause mutations in oocytes pose the greatest threat to the reproductive genetic integrity of the female. Methods for assessing germ cell mutation and genetic damage will be discussed below.

8 THE USE OF ANIMAL STUDIES TO DEFINE HUMAN RISK

Often, adequate human data are not available to define reproductive risk, so animal studies must be used to protect human reproductive health. It is customary to conduct animal experiments at dosage levels exceeding estimated levels of human exposure, both to increase the likelihood that a weak reproductive toxicant will produce a detectable effect and to compensate for the relatively small numbers of animals used in the test. This results in the necessity for extrapolation of results from relatively high experimental dosage levels to the normally lower levels of human exposure. An important step in characterizing the dose–response relationship in animal studies is to determine the no-observed-effect level (NOEL), that is, the highest exposure level at which no morphological, physiological, or functional modification is detectable under the test conditions. Another widely used concept in toxicology is the no observed adverse effect level (NOAEL), that is, the highest dose level at which no biologically adverse effects occur. In many cases the NOEL and NOAEL both refer to the same exposure level. When the NOEL and NOAEL differ, it is usual for the NOAEL, rather than NOEL, to be used as the basis for establishing permissible levels for human exposure, since it is possible for a substance to have a nonadverse effect at a low dose level and an adverse effect at a higher dose.

Depending on the sensitivity of the endpoint monitored and the test species utilized, different NOAELs may be derived for the same chemical. Generally, if multiple endpoints suggest that the chemical is a reproductive toxicant, then the most sensitive one (that is, the one that occurs at the lowest exposure level) should be used to establish the NOAEL. In the selection of the appropriate NOAEL, the study chosen should use an exposure route that is relevant to the human exposure whenever possible. However, data from other routes should also be evaluated and taken into consideration, especially if supported by pharmacokinetic information. If data from several species/strains are available, the most appropriate species should be used in determining the NOAEL. The most sensitive species should be used in determining the NOAEL, unless there is evidence that data from that species are not relevant to the human. A determination of relevance is based on the effect measured and the existence of comparable anatomical, physiological, pharmacological, pharmacokinetic, metabolic, and pharmacodynamic processes for the effect in the test animal and in humans.

If sufficient data do not exist to determine the NOAEL for an endpoint, then the lowest observed adverse effect level (LOAEL) should be used. Regardless of whether the NOAEL or the LOAEL is used, uncertainty or safety factors are applied to estimate an exposure level for humans at or below which there should be no adverse reproductive or developmental effects (this exposure level is often referred to as the reference dose). The total uncertainty factor usually ranges from 10 to 1000, depending on the number of adjustments needed. Uncertainty factors of 10 each are used (1) when the LOAEL must be used because a NOAEL was not established, (2) to account for differences between species (interspecies extrapolation), and (3) to provide an intraspecies adjustment for variable sensitivity among individuals. In addition, adjustments may be appropriate for length of exposure (e.g., if the animal study is acute and the human exposure is subchronic) and to correct for inadequacy of the NOAEL or LOAEL, which might be concluded after consideration of background variability in the measurements and the sensitivity of the endpoint being utilized. Additional safety or uncertainty factors may also be applied to correct for inadequacies in the design and conduct of the animal studies.

The number of animals per dose group should be determined after consideration of a variety of factors, including the general toxicity of the substance being tested, which will affect the number of animals that survive to provide data, and the expected variation of the endpoints being measured. The likely magnitude of the effect should also be considered, as well as the level of significance desired to establish a positive finding.

9 TESTING FOR REPRODUCTIVE TOXICITY

Alterations in reproductive capacity measured in animals may be sufficient to classify an agent as a hazard to reproduction in humans. Alternatively, less conclusive results may indicate the potential of a chemical to interfere with reproductive processes and suggest the need for further investigation. A variety of reproductive indices and endpoints are commonly measured in breeding studies (Table 5.3), in male animals (Table 5.4), and in female animals (Table 5.5). Some chemicals cause reversible reproductive effects in the adult male and female or developing offspring. This concept is important in the communication of reproductive hazard warnings. Since the effects are reversible and cease after exposure stops and the chemical is cleared from the system, exposures leading to effects in this category are likely to be of lower risk to human reproduction than those that cause permanent damage. However, exposure to even a reversible reproductive toxicant result in a shifting of couples to a smaller completed family size, that is, a temporary effect on fertility may have a permanent effect on family size if that transient effect occurs during attempted reproduction.

9.1 Study Design and Statistical Issues

Before describing the types of studies used to characterize reproductive toxicants in animals, it is useful to consider the effect of study design and the impact of the selection of statistical methods on interpreting their outcome. In the interpretation of data from animal reproductive toxicology studies, the quality of the study, design, conduct, and statistical analyses must be taken into consideration; deficiencies in these factors may lead to the application of additional safety factors. Studies must be of high quality and designed so that the animals are exposed to the test compound by an appropriate route of administration (i.e., relevant to the human route of exposure). Other routes may be relied upon by taking into consideration pharmacokinetic information. Also, exposures should be at the proper time and for the proper duration so as to maximize detection of an effect. Details of study design are presented below, but, in general, the study must include identification of reproductive endpoints suitable for defining an adverse effect (see above).

An important consideration is whether the substance is exerting a selective adverse effect on reproductive function. For a substance to be identified as a reproductive hazard, adverse reproductive effects should occur at doses that do not cause other types of toxicity that could interfere with mating ability or frequency, especially other significant systemic toxicity. When reproductive effects are seen in the presence of systemic toxicity, scientific judgment concerning the probability of reproductive toxicity in the absence of other toxicities (and at lower doses) is needed to determine whether an adverse reproductive effect has occurred.

Another important consideration in evaluating animal data is a determination of the power of the study, which is the probability that the study will demonstrate a true effect. It is dependent on the sample size, as well as the background incidence and variability of the endpoint(s) examined. The apparent lack of an effect may be due to a true lack of activity or the inability of the study to identify an effect because of small sample size. Conversely, some statistically significant effects may arise by chance, especially if a large number of endpoints are analyzed; the use of appropriate historical control data and critical attention to toxicological and reproductive biologic principles may prevent a false assumption of biological relevance in such cases.

Statisticians need also consider that reproductive toxicity studies often involve observations on animals from the same litter, and the presence of litter effects can be recognized as the tendency of litter mates to respond more similarly than animals from different litters (Mantel, 1969; Weil, 1970). It is necessary to take litter effect into account by using the variation between litters rather than the variation within litters as the basis for statistical analysis. In other words the real effects are actually occurring in the female or male that receives the treatment and not the litter; what happens to the litter is biologically independent of what happens to every other litter in the dose group and in the study. Thus, it is well accepted that the appropriate sampling unit (the N) for such studies is the litter, and data are commonly expressed and analyzed on a per litter basis (Gad and Weil, 1989).

Negative findings from animal reproductive toxicology studies deserve special scrutiny regarding study design and conduct. To be useful in risk assessment, such studies must include sufficient numbers of animals to detect an adverse effect, appropriate dose levels and exposure routes must be used, and the data must be evaluated using appropriate statistical methods. Negative studies should also indicate the power to define an adverse effect or the confidence interval on the null hypothesis.

9.2 Animal Breeding Studies

Endpoints that are most commonly determined in animal breeding studies and that may be most useful in determining a potential human reproductive hazard are summarized in Table 5.3. The first eight endpoints listed pertain primarily to indicators of the ability of animals to mate, conceive, or deliver live offspring and, as such, measure overall effects on male and female fertility. These endpoints should be considered collectively when evaluating the results from such studies. The mating index (male or female), which is a measure of libido, is generally a reliable measure in animals. It is not considered a reliable indicator in humans, although there are data on frequency of intercourse that could be a surrogate measure for libido (Working and Mattison, 1993). The survival indices measure pup survival from birth through postnatal day 21, expressed in increments of that time period. The body weights and growth of offspring are sometimes also included in list of useful endpoints but are likely to be insufficient to definitively identify a hazard because of the myriad factors independent of test substance exposure that may contribute to alterations in these growth-related endpoints.

TABLE 5.3 Reproductive Indices Commonly Used in Animal Breeding Studies[a]

Male (Female) Mating Index $= \dfrac{\text{Number of males (females) for which mating was confirmed}}{\text{Number of males (females) used for mating}} \times 100$

Male Fertility Index $= \dfrac{\text{Number of males producing a pregnant female}}{\text{Number of males for which mating was confirmed}} \times 100$

Female Fertility Index $= \dfrac{\text{Number of females confirmed pregnant}}{\text{Number of females for which mating was confirmed}} \times 100$

Gestation Index $= \dfrac{\text{Number of females delivering at least one live offspring}}{\text{Number of females confirmed pregnant}} \times 100$

Number of Implantations per Pregnant Female
Number of Pre- and Post-Implantation Losses
Litter Size at Birth

Live Birth Index $= \dfrac{\text{Mean number of live offspring per litter}}{\text{Mean number of offspring per litter}} \times 100$

Survival Indices $= \dfrac{\text{Number of live offspring on postnatal day 4}}{\text{Number of live offspring born}} \times 100$

$= \dfrac{\text{Number of live offspring on postnatal day 7}}{\text{Number of live offspring on postnatal day 4}} \times 100$

$= \dfrac{\text{Number of live offspring on postnatal day 14}}{\text{Number of live offspring on postnatal day 7}} \times 100$

$= \dfrac{\text{Number of live offspring on postnatal day 21}}{\text{Number of live offspring on postnatal day 14}} \times 100$

Reproductive Capacity of F1 Offspring of Exposed Males and/or Females
(as measured by indices listed above)

[a]From Working and Mattison (1993).

Table 5.4 Indices of Male Fecundity in Laboratory Animals[a]

Disruption of seminiferous epithelium
Alterations in gonadal function causing decreased testicular spermatid number or decreased sperm count in the epididymis, vas deferens, or ejaculate
Decrease in percentage of motile spermatozoa
Significant change in sperm morphology
Alterations in reproductive organ weight (e.g., testes, epididymides, seminal vesicles, or prostate)
Altered concentration or temporal patterns of testosterone, luteinizing hormone (LH) or follicle-stimulating hormone (FSH)

[a]From Working and Mattison (1993).

9.3 Male Reproductive Toxicity Studies

Endpoints that can be used as biomarkers of fecundity in male animals are summarized in Table 5.4. Subjective endpoints, such as those evaluated by traditional histopathological techniques, must be interpreted with caution prior to concluding that a substance is a reproductive hazard, although recent developments in the quantitative evaluation of testicular histology may increase the value of these analyses (Sinha Hakim et al., 1989; Russell, et al., 1990). For endpoints that are more easily quantified, such as testicular spermatid number or the sperm count in the ejaculate, partial alterations are more compelling evidence, provided they are both statistically significant and dose-dependent. A significant increase in the proportion of sperm with morphological abnormalities can also be considered good reason to conclude that a substance is a male reproductive hazard. Endpoints such as changes in reproductive organ weights and hormone profiles, which either are highly variable in humans and laboratory animals or are inconsistent indicators of changes in reproductive potential, should be considered only suggestive of potential reproductive hazard.

9.4 Female Reproductive Toxicity Studies

Endpoints that indicate that a substance may alter female fecundity are listed in Table 5.5. Inhibition of ovulation, inhibition of implantation due to altered uterine histology, delayed puberty, and early reproductive senescence (Greep, 1986) are the most compelling endpoints in defining a potential female reproductive hazard. Changes in other female reproductive endpoints, such as alterations in endocrine and uterine histological patterns that do not inhibit ovulation or implantation and changes in ovarian and/or uterine weight, are suggestive of possible female reproductive hazard but are more difficult to interpret in terms of human female reproductive risk (morphometric data). Recently, quantitative morphometric methods have been used to characterize the ovarian compartments (e.g., oocytes, follicular cells, granulosa cells, etc.) affected by toxicants and to describe the dynamic processes of ovarian toxicity (Plowchalk and Mattison, 1991a, b; Weitzman et al., 1992).

Table 5.5 Indices of Female Fecundity in Laboratory Animals[a]

Estrous cycle disruption resulting in anovulation
Significant reduction in the number of ovarian follicles or oocytes
Altered uterine histology
Altered ovarian histology characterized by reduced corpora lutea or increased number of ovarian cysts
Altered concentration or temporal patterns of testosterone, luteinizing hormone (LH), or follicle-stimulating hormone (FSH)
Alterations in ovarian or uterine weight
Delayed puberty
Premature reproductive senescence

[a]From Working and Mattison (1993).

10 ASSESSING REPRODUCTIVE TOXICITY WITH ANIMAL STUDIES

Several general principles guide the use of animal studies to assess potential human reproductive risk. The data used should be derived from studies of acceptable quality in mammalian species that are predictive of human responses with exposure route(s), level, duration, and frequency that are relevant to human exposure. Data from replicate studies and multiple independent study types should be consistent and reinforcing. When data are discordant, sufficient additional evidence should be available to reconcile the differences. Confidence in a study outcome may be increased by the demonstration of a dose–response relationship (as measured by increased incidence or severity of effect with increasing dose). In all cases endpoints evaluated in such studies should be predictive of reproductive outcomes (Tables 5.3–5.5).

The dose–response distinction between general adult toxicity and reproductive toxicity should be carefully assessed. A substance should be considered a reproductive (male or female) toxicant if it produces its effects at dosage levels lower than those which produce general adult toxicity that is severe enough to interfere with mating ability or frequency.

It is, of course, important to consider the relevance of the animal studies to humans. Evidence of relevancy to humans is based on (a) consistency among animal studies of patterns of exposure, abnormal outcome, and causal associations and (b) evidence indicating biological plausibility of mechanism of action. These factors must be consistent with human biologic principles. When data on human outcome are completely lacking, which will be true for many substances, the decision whether or not to classify a substance as a reproductive toxicant must be based on data from animal studies.

Several different types of protocols are routinely used to test for reproductive toxicity in laboratory animals. Although a detailed discussion of each is beyond the scope of this chapter, a brief description will be provided. However, in addition to these reproductive-specific methods, a substantial amount of information on the effects of chemical exposure on the male and female reproductive system could be obtained from subchronic and chronic toxicity tests if the endpoints were only modestly expanded. Inclusion of high-quality evaluation of testicular histology (Chapin, 1988), the addition of measures of sperm production and quality (e.g., daily sperm production rates and sperm motility measures; Blazak, 1989) and the assessment of estrus cyclicity and vaginal cytology (Morrissey, 1989) could enable these studies to serve as a screen for potential direction reproductive toxicants.

10.1 Single Mating Trials

The single mating trial is a general comprehensive test of reproductive function, which was modified from the Food and Drug Administration (FDA) two-litter test of the 1960s (Palmer, 1981). As its name implies, only a single mating occurs, after a 60- to 70-day exposure period for the males and a 14-day exposure period for the females. The long exposure period in males is obligatory to expose sperm throughout all the stages of spermatogenesis, since the stages are not equally sensitive to toxicants; during this prolonged period, the exposure of the fertilizing sperm during all stages of spermatogenesis is ensured.

The single mating trial can be conducted with just 10 males per treatment group and 20 females (by mating each treated male with two treated females); however, such a practice limits its value in risk assessment because of small sample size (and related statistical problems), as well as because mating of treated males to treated females does not permit the identification of the affected sex. Common modifications that increase its utility include the use of larger group sizes, separate testing of males and females, and the evaluation of reproductive function after a recovery period, which should be equivalent to the length of at least one cycle of spermatogenesis in the test species (Dunnick et al., 1984a,b, 1986; Anderson et al., 1986).

10.2 Serial Mating Trials

This assay is closely related to the single mating trial but utilizes a shorter exposure period (generally about one week in rodents). In order to ensure that all stages of spermatogenesis are tested, exposed males are mated for 8–10 weeks in order to test sperm treated as spermatogonia, as spermatocytes, as spermatids, and as sperm in the epididymis or vas deferens. This assay is applicable only to exposed males but has the advantage of being able to discriminate among the sensitive stages of spermatogenesis. Reduced fertility in the first week, for example, can be attributed to effects on mature spermatozoa in the vas deferens and caudal epididymis (Working et al., 1985), whereas compromised reproductive function in the fifth week would indicate an effect on late-stage primary spermatocytes (Chapin et al., 1985).

A major disadvantage is that this test scheme requires many females, since each male is mated weekly for up to 10 weeks to at least one female per week and sometimes as many as three. Under these conditions, a group of 20 male rats would ultimately use from 200–600 females over the course of a 10-week study. Male mice are typically mated for only 8 weeks because spermatogenesis is shorter, but at least as many females are utilized, since it is recommended that male mice be mated to new females every 4 days, rather than weekly, to reflect the quicker production of sperm (Green et al., 1985).

10.3 Multigeneration Reproduction Tests

Guidelines for the conduct of multigeneration tests have been published by both the FDA and the Environmental Protection Agency (EPA) (Collins, 1978; U.S. EPA, 1982; Lamb, 1985). These references can be consulted for detailed descriptions of this test (see also Lamb, 1988). The test typically involves matings over two or three generations. Although the FDA multigeneration test encompasses three generations, the EPA suggests limiting the test to two generations, since it is unlikely that an adverse effect will be expressed in a third generation and not also seen in an earlier generation (Zenick and Clegg, 1989).

These studies are based on the exposure of males for at least 8 weeks and females for at least 2 weeks prior to mating; treatment of both sexes continues for the remainder of the study. Males are mated to one or two females per mating trial, and each mating trial lasts from 1 to 3 weeks. The more streamlined EPA version utilizes three dose levels and a control, and the parental (P) generation is necropsied after each mating pair produces at least one litter (Fig. 5.9). Exposure of the F_1 generation continues after weaning, and it too is bred to produce at least one litter each prior to necropsy. The F_2 generation is necropsied at weaning. Significant amounts of non-reproductive toxicity data are also collected from all three generations, including

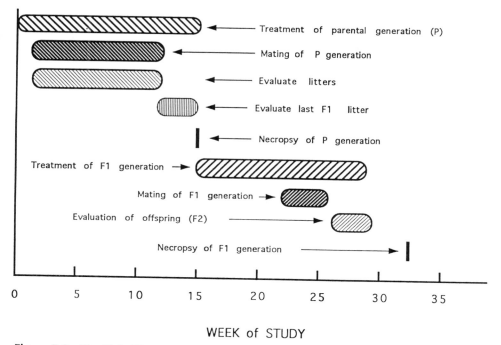

WEEK of STUDY

Figure 5.9 The U.S. EPA two-generation study protocol, a streamlined version of the FDA multigeneration study. Males are treated for at least 8 weeks prior to mating and females for at least 2. Multiple mating trials are employed, with males mated to one or two females per trial. (Modified from Lamb, 1988.)

data on growth, body weight, mortality, and gross and microscopic pathologic changes seen at necropsy. These data can be very valuable in assessing the relevance of the study to human reproductive risk.

As with the single mating study, because treated males are mated to treated females, no discrimination between male-mediated and female-mediated toxic effects is possible. Typically, the question of reversibility is usually also not addressed, although some versions require the production of more than one litter per animal; the second litters can provide animals for developmental toxicity evaluations or a reversibility test, which can add to the value of the study.

The different exposure history of parental and offspring generations must be recognized in evaluating the results of any multigeneration study, be it two generations or three. The animals in all generations but the first have been exposed from before conception throughout life, allowing expression of effects on susceptible developmental stages; the parental generation receives subchronic exposure only during early adult life. If protracted bioaccumulation occurs, or if there are sensitive stages only in the prenatal and prepubertal periods, effects of differing type or severity may be seen in the F_1 or F_2 offspring that are not seen in the parental generation.

10.4 Continuous Breeding Studies

The National Toxicology Program has developed this protocol as an alternative to the multigeneration studies, to both improve the sensitivity of the assay and to distinguish between male and female effects (Lamb, 1985). In this assay, males and females are cohabited one-to-one after just one week of exposure to the test compound; exposure is continued throughout the entire course of the study (Fig. 5.10). The offspring are removed from the cage immediately after delivery, and the females are immediately rebred. The breeding phase lasts 14 weeks, giving each breeding pair time to deliver about five litters. The mating pairs are separated, and the final litters are kept for use in subsequent evaluations, which typically include a single generation mating trial of the second generation and crossover mating of the treated first-generation males and females to untreated animals.

Continued breeding permits observation of the time to onset of any adverse effect, although agents that have cumulative effects could appear increasingly severe with each litter. Nonetheless, the time to first litter correlates to time to pregnancy in humans. Inclusion of the second-generation mating provides data relating to the postnatal development and reproductive capability of exposed offspring, and the use of the crossover mating allowing discrimination between male effects and female effects. However, treatment does not continue during the crossover mating, so reversible effects may present a different picture at necropsy (which occurs after the crossover breeding) than if the necropsy were done immediately at the cessation of treatment.

10.5 Segment I Test

This study type is a variation of the single mating trial. Male rodents are treated for 70 days, a sufficient period of time to expose sperm at all stages of spermatogenesis,

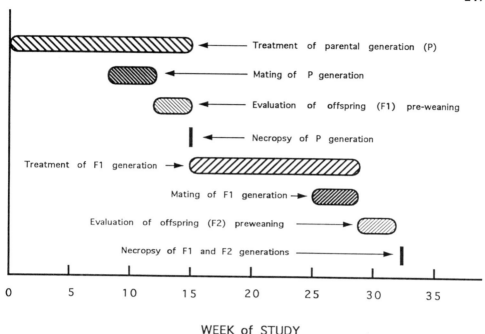

Figure 5.10 Continuous breeding study protocol. Mating begins within a week of the start of treatment and cohabitation continues for up to 14 weeks. Animals are housed as breeding pairs. (Modified from Lamb, 1988.)

and females are exposed for 14 days (approximately 2–3 estrus cycles). Treatment of females is continued during mating, pregnancy, and lactation.

Males are mated to two females each, and one-half of the females are killed at midgestation for examination of uterine contents. Live and dead fetuses are enumerated, fetuses are examined for visceral and skeletal malformations. The remaining females deliver naturally, and the surviving offspring are counted, allowing the calculation of survival indices (Table 5.3). Weanlings may be killed and necropsied for gross and visceral abnormalities. This assay is generally employed as a screening assay for follow-up Segment II and III studies (see below) and provides an overview of fertility, conception rate, pre- and postimplanation survival, parturition, and lactation.

10.6 Segment II Test

This protocol is perhaps the most commonly used assay for assessing the potential developmental toxicity of chemicals and drugs. Studies are typically carried out in two species (a rodent and nonrodent) according to prescribed guidelines (U.S. EPA, 1986) and include three treatment groups and a control group. Dosing should be via the expected human route of exposure, and the high dose should be of sufficient magnitude to cause measurable maternal toxicity. Optimally, the low dose will be a

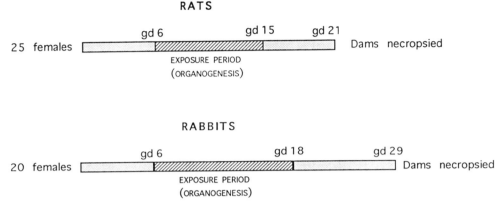

Figure 5.11 Segment II study in rats and rabbits. Exposure to the test agent occurs during the period of organogenesis in each species: gestation days (gd) 6–15 in the rat; gd 6–18 in the rabbit. Necropsy and examination of uterine contents is done on gd 21 in rats and gd 29 in rabbits. Current recommendations are for 25 pregnant rats and 20 pregnant rabbits per exposure group. (Modified from Manson and Kang, 1989.)

NOAEL. The number of animals in each dose group will determine the power of the assay; the less common the endpoint, the fewer the cases above background needed to detect an effect (Manson and Kang, 1989).

Treatment of the females occurs during the period of organogenesis, which varies with the species used (Fig. 5.11); males are not treated. Pregnant females are sacrificed late in gestation (mice on day 18, rats on day 21, and rabbits on day 29 of pregnancy) for examination of uterine contents. Fetal weight, sex, and condition are noted, and fetuses are examined for external malformations. Generally, one-half of the fetuses are processed for detailed examination of viscera, and the other half for skeletal examination. Developmental toxicity is assessed in terms of early or late embryo or fetal deaths, reduced fetal body weight, and the presence of gross, visceral, or skeletal malformations. A common dose–response pattern includes embryo lethality, malformations, and growth retardation of surviving fetuses, with embryo lethality predominating at the high dose and a combination of resorbed, malformed, growth-retarded, and normal fetuses present in the litters at doses within the embryotoxic range. Agents that are toxic, but not teratogenic, may produce a dose–response pattern characterized by growth retardation and embryo lethality without malformations (Neubert et al., 1980).

10.7 Segment III Test

In this test effects on perinatal and postnatal development are assessed. Pregnant females are exposed during the last third of gestation and throughout lactation to weaning. Effects of the test compound on late fetal development, labor and delivery, lactation, neonatal viability, and offspring development and growth are evaluated.

Weanlings may be allowed to survive to adulthood and used to evaluate neurobehavioral deficits, effects on fertility, and the occurrence of perinatally induced cancer.

11 GERM CELL GENOTOXICITY TESTING

Although the primary emphasis of most reproductive toxicity studies is the identification of chemicals that interfere with fertility in some fashion, no discussion of reproductive toxicology would be complete without mention of methods to detect DNA damage or mutation in germ cells. Assay types are limited in number and generally are of two types: those that directly measure heritable damage (Table 5.6) and those that measure effects in germ cells that are, or may be, related to the alteration or damage of their DNA but that may not actually represent heritable genetic injury (Table 5.7).

Table 5.6 Germ Cell Assays That Measure Heritable Genetic Damage

Assay	Reference
Specific Locus Assay	Russell et al. (1981)
Heritable Translocation Test	Generoso et al. (1980)
Dominant Lethal Test	Ehling (1977); Green et al. (1985)
Test for Recessive Lethals	Sheridan (1983)
Nondisjunction Test	Searle and Beechey (1982); Cattanach et al. (1984)
Inversion Assay	Roderick (1971; 1983)
Test for Dominant Mutations	Ehling (1983); Selby (1983)
Sex Chromosome Loss Test	Russell and Matter (1980)
Cytogenetic Analysis (zygote)	Adler and Brewen (1982)

Table 5.7 Germ Cell Assays That May Predict Heritable Genetic Damage

Assay	Reference
DNA Damage and Repair in Oocytes	Pedersen and Brandriff (1980)
DNA Damage and Repair in	Sega (1982)
Spermatogenic Cells	Working and Butterworth (1984) Skare and Schrotel (1985) Sega et al. (1986) Bentley and Working (1988a,b)
Chromosomal aberrations	Adler and Brewen (1982)
Sperm morphology	Wyrobek et al. (1983)
Molecular dosimetry of germ cells	Lee (1978)
DNA adducts in germ cells	Stott and Watanabe (1980)
Sister chromatid exchange in germ cells	Allen and Latt, (1976)

11.1   Assays that Measure Heritable Damage

Of the these assays (Table 5.6), only three (the specific locus assay, the heritable translocation test, and the dominant lethal assay) have been tested with a sufficient number and variety of chemicals to be thought reliable indicators of germ cell mutagenicity (Brusick et al., 1983), and subsequent discussion will be confined to these assays. An older but still timely review of the wide variety of germ cell genotoxicity assays can be found in Russell and Shelby (1985).

11.2   Specific Locus Assay

This assay is the only one of the traditional germ cell assays that directly measures the induction of point mutations in mammals (Russell et al., 1981). It detects intragenic lesions and small deletions at a limited number of marked loci. Most commonly, these effects are measured as visible alterations in the phenotype of F_1 generation offspring of special test stock mice bred to wild-type mice of the same strain that have been exposed to the chemical or drug of interest. The test stock mice are homozygous for recessive traits at seven loci that primarily affect coat color or the morphology of the external ear. Mutation at an appropriate allele in the treated wild-type parent may allow expression of the recessive phenotype in the offspring. Although sensitive (just one mutation in about 1000 offspring defines a positive), the usefulness of this assay is limited by the fact that only a very small portion of the genome is sampled. As a consequence, results cannot be used to estimate the rate of mutation in the total genome and may not be relevant in estimating effects on human health (Russell and Matter, 1980).

Recent assays developed using transgenic mice may prove useful in measuring mutation rates in germ cells. Dose dependent induction of mutations has been observed using bacteriophage shuttle vectors containing a lacI target gene in transgenic mice (Kohler et al., 1991). Mutations are detected using a simple plaque detection assay, and, because of the small size of the target gene, it may be possible to undertake sequence analysis of the mutated genes. Analysis of spleen cells and germ cells after exposure to ethyl nitrosourea (ENU) demonstrated that spontaneous background rates of mutation in the germ cells was approximately threefold lower than in somatic tissues for as yet unknown reasons that may be related to the variation in number of cellular divisions, to the repair capability of the two types of tissue, or even to the presence of the blood–testis barrier. Exposure to ENU resulted in a two- to five-fold increase in mutation rate over background in germ cells. Mutation rate in germ cells was about one-half of that in spleen cells at comparable doses of the chemical. Sequence analysis showed that the predominant class of mutations was base substitutions. The mutation rate measured with this system, however, was significantly lower than those reported in published specific locus studies with ENU (Hitotsumachi et al., 1985), possibly due to differences in the timing of the analysis. Overall, though still unproven, this method may provide a relatively quick method for assessing the potential of chemical or drug to induce mutations in germ cells.

11.3 Heritable Translocation Assay

This assay detects chromosomal rearrangements that are passed on to viable offspring in a process known as translocation heterozygosity (Generoso et al., 1980). Although phenotypically identical to the parental strain, affected offspring can be detected by cytological examination of their chromosomes or, most commonly, by a decrease in their fertility due to the chromosomal translocation. This assay is the only true test of heritable chromosomal effects in germ cells and is quite sensitive because of the low rate of spontaneous translocations (Brusick et al., 1983). Its usefulness is hampered, however, by the apparent lack of heritable translocation in stem cells or primary oocytes (Generoso et al., 1978), the cells of most interest in the male and female, respectively.

11.4 Dominant Lethal Test

This test measures all genetic lesions, small or large, that cause death of the conceptus postfertilization (Fig. 5.12). Although affected offspring do not survive, a

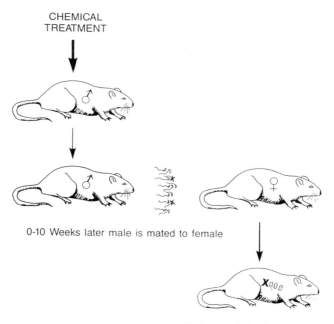

Figure 5.12 Dominant lethal assay in rats. Males are acutely exposed to test agent and subsequently bred to unexposed females for 8–10 weeks after treatment. Chemically induced DNA damage is expressed as increased rates of embryonic and fetal death in the females. (From Working, 1989, with permission.)

positive result in the assay is considered good presumptive evidence for germ cell mutation. There is a good correlation between the specific locus assay and the dominant lethal test, further validating its usefulness (Green et al., 1985). In practice, the assay is performed by exposing the male parent and subsequently mating it to one or more females per week for the duration of spermatogenesis. A decrease in the number of live fetuses counted in the females at necropsy 15–18 days later represents a genotoxic effect.

The dominant lethal test is insensitive to weak mutagens owing to the relatively high background rate of spontaneous intrauterine death in rodents. A more general shortcoming relates to reproductive physiology of the male and the female. The assay is essentially unsuitable for use in treated females since embryonic and fetal death caused by genetic effects cannot be distinguished from those due to general toxicity in the female. An in vivo–in vitro variation of the assay, in which females are treated prior to breeding and the fertilized ova retrieved and cultured through the implantation stage (Mohr and Working, 1990), may permit the enumeration of mutations in females but is quite labor intensive and may not be suited to large scale screening.

11.5 Assays That Predict Heritable DNA Damage

A large number of assays that predict, rather than measure, heritable DNA damage in germ cells have been developed over the years. If sufficiently validated, such assays may serve as useful screens for germ cell mutagens and suggest when further testing is warranted. Generally, when used as screens, positive results are viewed as near-definitive evidence of germ cell mutagenicity, whereas negative results merely indicate the need for further testing (Bentley and Working, 1988a). Positive results in one of these assays can also serve to validate a negative in one of the three assays that directly measure heritable damage, that is, a positive in the second type of assay shows that the chemical arrived at the gonads but was nonmutagenic for one reason or another.

The best validated assays of this type involve the detection of DNA damage, or, conversely, its repair, detection of chromosomal abnormalities in germ cells, or the assessment of abnormal sperm morphology.

Measurement of DNA damage and repair is perhaps the most used assay of this type, and a wide variety of known germ cell mutagens and nonmutagens have been tested using a variety of techniques. The most common methods included the detection of DNA strand breaks or crosslinks using techniques of alkaline elution (Skare and Schrotel, 1985; Sega et al., 1986) and the measurement of DNA repair as a marker of DNA damage, following the incorporation of radiolabeled nucleotides in meiotic and post-meiotic spermatocytes and spermatids (Fig. 5.13). DNA repair is quantitated by liquid scintillation counting of sperm (Sega, 1982) or autoradiographs of cell suspensions (Working and Butterworth, 1984; Bentley and Working, 1988a) or spermatogenic tubule segments (Bentley and Working, 1988b). DNA synthesis of this type is known as excision repair or unscheduled DNA synthesis (UDS) to distinguish it from DNA synthesis that occurs during the S phase of mitosis and meiosis. Far fewer studies have examined chemically induced DNA damage or repair in the oocyte. Fully grown, maturing, and resting-stage oocytes are capable of DNA

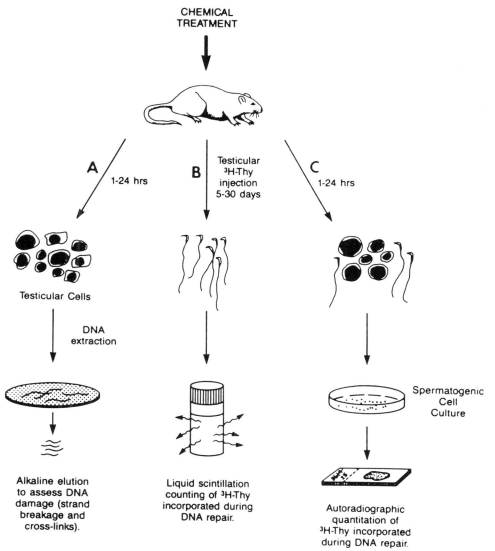

Figure 5.13 Assessment of DNA damage and repair in the male. (A) Alkaline elution. Males are treated and testicular cells are isolated and prepared for alkaline elution techniques to measure DNA strand breaks and crosslinks. (B) Quantitation of unscheduled DNA synthesis (UDS) by liquid scintillation counting. Males are exposed to the test agent and immediately receive an intratesticular injection of radiolabeled thymidine. Five to 30 days later spermatozoa are collected from the epididymides and the amount of radiolabeled base incorporated during DNA repair is quantitated by liquid scintillation techniques. (C) Quantitation of UDS by autoradiography. After chemical exposure, testes are removed up to 24 hr after chemical exposure and spermatogenic cells or tubule segments are cultured in the presence of radiolabeled thymidine. UDS is measured later using autoradiographs prepared from cultured cells or tubule segments. (From Working, 1989, with permission.)

repair after irradiation with ultraviolet (UV) light (Pedersen and Mangia, 1978) and ova can repair chemically induced damage in the fertilizing sperm (Generoso et al., 1979). However, the technical difficulties of this procedure have limited the number of complete and well-controlled studies.

Similar difficulties have limited the number of studies that quantitate chromosomal abnormalities in germ cells, although the assay has long been considered a reliable indicator of potential mutagenic activity (Adler and Brewen, 1982). The majority of chemicals are S-phase-dependent, that is, they induce aberrations only in cells undergoing DNA synthesis at the time of exposure and prior to observation (Adler and Brewen, 1982). Thus, the only cells in the gonads are those in mitosis (the spermatogonia of the testes) and primary and secondary spermatocytes. Postmeiotic spermatids and all oocytes, which do not begin division again until the ovum is penetrated by the sperm, can only be analyzed in the embryo during the first cleavage metaphase.

The use of abnormal sperm morphology as a marker of potential mutagenic effect, once a popular assay because of its ease of application and its apparent usefulness in humans, has fallen into disfavor in recent years owing to our inability to relate changes in sperm shape to a mutagenic cause. Sperm head shape formation and final sperm maturation take place while the sperm is embedded in the Sertoli cell cytoplasm, leading to speculation that alterations in head shape could sometimes be secondary to Sertoli cell toxicity. Increases in the frequency of abnormal sperm morphology can result from causes as diverse as epididymal toxicity (Working et al., 1985) and dietary restriction (Komatsu et al., 1982). Except for the induction of abnormal forms in F_1 males, which are properly considered the induction of dominant mutations, alterations in sperm head morphology cannot be considered a measure of actual genetic damage (Working, 1989).

How predictive are these assays? Results in the DNA repair assay in male rats correlated well with the known mutagenic activity of a number of compounds (Table 5.8). The UDS assay detected 11 of 12 known germ cell mutagens, and DNA damage detected by the use of alkaline elution detected 9 of 10. Neither assay gave a positive response for any known nonmutagen tested. Despite its shortcomings discussed above,

Table 5.8 Correlation of Germ Cell Genotoxicity Assays with Known Germ Cell Mutagenicity[a]

Result	UDS	AE	SpM	ChA
+/+[b]	11/12 (92)	9/10 (90)	8/8 (100)	9/9 (100)
+/−[c]	7/7 (100)	2/2 (100)	4/5 (80)	ND

[a]Abbreviations: ND, not determined; UDS, unscheduled DNA synthesis; AE, alkaline elution; SpM, sperm morphology; ChA, chromosomal abnormalities.

[b]Number of positive responses/number of germ cell mutagens tested. Number in parentheses is percentage "correct" responses. A germ cell mutagen is defined as a compound positive in the specific locus assay, the heritable translocation assay or the dominant lethal test.

[c]Number of negative responses/number of germ cell nonmutagens tested.

Source: Table modified from Working (1989).

the sperm morphology assay was positive with all known mutagens tested and is thus an excellent indicator of gonadal exposure. The detection of chromosomal abnormalities also performed well, but only a few chemicals have been tested in this system, limiting its utility.

Effective assays to measure the rate of mutagenesis in human cells are lacking, but the application of several relevant animal models that are able to detect heritable mutations can clearly be useful in evaluating the mutagenic potential of xenobiotics in humans. Further development of relevant animal models, particularly for the evaluation of mutagenic potential in females, will provide a stronger scientific rationale for human risk assessment in the future.

12 CONCLUSIONS

When attempting to characterize the potential hazard of a drug or chemical exposure on reproduction, whether in the female or in the male, it is necessary to evaluate a hierarchy of information. Only rarely are sufficient human data available. More often, one must depend on animal studies, which have accurately identified most human reproductive hazards (Frankos, 1985; Kimmel et al., 1990), but which have also identified many more toxicants than have been confirmed in humans (Schardein, 1985; Shepherd, 1989). Reproductive risk assessment is clearly not at the point where a single endpoint in animals is indicative of reproductive risk in humans, and the assessment of multiple endpoints in multiple animal models is clearly necessary (Working and Mattison, 1993).

To understand the long-term consequences in humans of exposure to reproductive toxicants, their mechanism(s) of action in the male and the female reproductive tract must first be understood. To this end, more specific approaches must be used to elucidate specific sites and modes of action of reproductive toxicants. Consideration should be given to the redesign of many of the existing reproductive toxicity testing protocols with specific reproductive endpoints in mind. Use of such testing protocols may allow the development of more quantitative assessment of potential human risk based on extrapolation from animal models.

REFERENCES

Adler, I. D. and J. G. Brewen (1982). In *Chemical Mutagens: Principles and Methods for Their Detection*, F. J. de Serres and A. Hollaender, Eds., Vol. 7, Plenum Press, New York, pp. 1–35.

Allen, J. W. and S. A. Latt (1976). *Nature*, **260**, 449–451.

Anderson, J. A., J. A. Petrere, J. E. Fitzgerald, and F. De la Iglesia (1986). *Fundam. Appl. Toxicol.*, **7**, 221–227.

Baker, T. G. (1963). *Proc. Royal Soc. Lond. (Biol.)*, **158**, 417–433.

Bardin, C. W. (1986). In *Reproductive Endocrinology*, S. S. C. Yen and R. B. Jaffe, Eds., W. B. Saunders, Philadelphia, pp. 177–199.

Bedford, J. M. (1975). In *Handbook of Physiology: Endocrinology, Male Reproductive Sys-*

tem, Section 7, D. W. Hamilton and R. O. Greep, Eds., Vol. 5, American Physiological Society, Washington, D.C., pp. 303–317.

Bentley, K. S. and P. K. Working (1988a). *Mutat. Res.*, **203**, 135–142.

Bentley, K. S. and P. K. Working (1988b). *Environ. Mutagen.*, **12**, 285–297.

Blazak, W. F. (1989). In *Toxicology of the Male and Female Reproductive Systems*, P. K. Working, Ed., Hemisphere, New York, pp. 157–172.

Brusick, D., B. J. Kilbey, J. Ashby, H. Bartsch, U. H. Ehling, T. Kada, H. V. Malling, A. T. Natarajan, G. Obe, H. S. Rosenkranz, L. B. Russell, J. Shoneich, A. G. Searle, E. Vogel, J. S. Wassom, and F. K. Zimmerman (1983). *Mutat. Res.*, **114**, 117–177.

Cattanach, B. M., D. Papworth, and M. Kirk (1984). *Mutat. Res.*, **126**, 189–204.

Cavazos, L. F. (1977). In *Frontiers in Reproduction and Fertility Control: A Review of the Reproductive Sciences and Contraceptive Development*, R. O. Greep and M. A. Koblinsky, Eds., MIT Press, Cambridge, pp. 402–410.

Chapin, R. E. (1988). In *Physiology and Toxicology of Male Reproduction*, J. C. Lamb and P. M. D. Foster, Eds., Academic Press, New York, pp. 155–177.

Chapin, R. E. (1989). In *Toxicology of the Male and Female Reproductive Systems*, P. K. Working, Ed., Hemisphere, New York, pp. 179–186.

Chapin, R. E., S. L. Dutton, M. D. Ross, and J. C. Lamb (1985). *Fundam. Appl. Toxicol.*, **5**, 182–189.

Clermont, Y. (1972). *Physiol. Rev.*, **52**, 198–236.

Collins, T. F. X. (1978). In *Handbook of Teratology*, J. G. Wilson and F. C. Fraser, Eds., Plenum Press, New York, pp. 191–214.

Connell, C. J. and G. M. Connell (1977). In *The Testis*, A. D. Johnson and E. R. Groves, Eds., Vol. IV, Academic Press, New York, pp. 333–369.

Dixon, R. L. (1986). In *Toxicology. The Basic Science of Poisons*, C. D. Klaassen, M. O. Amdur and J. Doull, Eds., Pergamon Press, New York, pp. 432–477.

Dixon, R. L. and I. P. Lee (1980). *Fed. Proc.*, **39**, 66–72.

Dunkel, V. C. (1985). In *Advances in Modern Environmental Toxicology*, W. G. Flamm and R.J. Lorentzen, Eds., Vol. 12, Princeton Scientific, Princeton, pp. 61–78.

Dunnick, J. K., B. N. Gupta, M. W. Harris, and J. C. Lamb (1984a). *Toxicol. Appl. Pharmacol.*, **72**, 379–387.

Dunnick, J. K., H. A. Solleveld, M. W. Harris, R. Chapin, and J. C. Lamb (1984b). *Mutat. Res.*, **138**, 213–218.

Dunnick, J. K., M. W. Harris, R. E. Chapin, L. B. Hall, and J. C. Lamb (1986). *Toxicology*, **41**, 305–318.

Eddy, E. M. (1988). In *Physiology and Toxicology of Male Reproduction*, J. C. Lamb and P. M. Foster, Eds., Academic Press, New York, pp. 35–69.

Ehling, U. (1977). *Arch. Toxicol.*, **38**, 1–11.

Ehling, U. (1983). In *Utilization of Mammalian Specific Locus Studies in Hazard Evaluation and Estimation of Genetic Risk*, F. J. de Serres and W. Sheridan, Eds., Plenum Press, New York, pp. 169–190.

Ewing, L. L. (1983). In *Infertility in the Male*, L. I. Lipschultz and S. S. Howards, Eds., Churchill, New York, pp. 43–69.

Fawcett, D. W., W. B. Neaves, and M. N. Flores (1973). *Biol. Reprod.*, **9**, 500–532.

Feagans, W. M., L. F. Cavazos, and A. T. Ewald (1961). *Am. J. Anat.*, **108**, 31–45.

Fink, G. (1988). In *The Physiology of Reproduction*, E. Knobil and J. D. Neill, Eds., Raven Press, New York, pp. 1349–1377.

Foster, P. M. D. (1988). In *Physiology and Toxicology of Male Reproduction*, J. C. Lamb and P. M. D. Foster, Eds., Academic Press, New York, pp. 7–34.

Foster, P. M. D. (1989). In *Toxicology of the Male and Female Reproductive Systems*, P. K. Working, Ed., Hemisphere, New York, pp. 1–14.

Frankos, V. H. (1985). *Fundam. Appl. Toxicol.*, **5**, 615–625.

Fritz, I. B., F. F. G. Rommerts, B. G. Louis, and J. H. Dorrington (1976). *J. Reprod. Fertil.*, **46**, 17–24.

Gad, S. C. and C. S. Weil (1989). In *Principles and Methods of Toxicology*, A. W. Hayes, Ed., Raven Press, New York, pp. 435–483.

Gangolli, S. D. and J. C. Phillips (1988). In *Experimental Toxicology. The Basic Principles*, D. Anderson and D. M. Conning, Eds., Royal Society of Chemistry, London, pp. 348–376.

Generoso, W. M., K. T. Cain, S. W. Huff, and D. G. Gosslee (1978). In *Advances in Modern Toxicology*, W. G. Flamm and M.A. Mehlman, Eds., Vol. 5, Hemisphere, Washington, D.C., pp. 109–129.

Generoso, W. M., K. T. Cain, M. Krishna, and S. W. Huff (1979). *Proc. Natl. Acad. Sci. USA*, **76**, 435–437.

Generoso, W. M., J. B. Bishop, D. G. Gosslee, G. W. Newell, C. Sheu, and E. Von Halle (1980). *Mutat. Res.*, **76**, 191–215.

Greep, R. O. (1986). In *Proceedings of the 1985 Laurentian Hormone Conference*, R. O. Greep, Ed., Academic Press, Orlando, Chapters 8–9.

Green, S., A. Auletta, J. Fabricant, R. Kapp, M. Manandhar, C. Sheu, J. Springer, and B. Whitfield (1985). *Mutat. Res.*, **154**, 39–67.

Hales, B. F., S. Smith, and B. Robaire (1986). *Toxicol. Appl. Pharmacol.*, **84**, 423–430.

Harrington, J. M., G. F. Stein, R. O. Rivera, and A. V. DeMorales (1978). *Arch. Environ. Health*, **33**, 12–15.

Heinrichs, W. L. and M. R. Juchau (1980). In *Extrahepatic Metabolism of Drugs and Other Foreign Compounds*, T. E. Gram, Ed., SP Medical and Scientific Books, New York, pp. 319–332.

Hess, R. A. (1990). *Biol. Reprod.*, **43**, 525–542.

Hitotsumachi, S., D. A. Carpenter, and W. L. Russell (1985). *Proc. Natl. Acad. Sci. USA*, **82**, 6619–6621.

Huckins, C. (1971). *Cell Tissue Kinet.*, **4**, 139–154.

Kay, H. H. and D. R. Mattison (1985). *Contemp. Ob/Gyn.*, **26**, 109–127.

Kimbrough, R. D. (1974). *Crit. Rev. Toxicol.*, **2**, 445–489.

Kimmel, C. A., D. C. Rees, and E. A. Francis (1990). *Neurotoxicol. Teratol.*, **12**, 285–292.

Kohler, S. W., G. S. Provost, A. Fieck, P. L. Kretz, W. O. Bullock, J. A. Sorge, D. L. Putman, and J. L. Short (1991). *Proc. Natl. Acad. Sci. USA*, **88**, 7958–7962.

Komatsu, H., T. Kakizoe, T. Niijima, T. Kawachi, and T. Sugimura (1982). *Mutat. Res.*, **93**, 439–446.

Kupfer, D. (1975). *Crit. Rev. Toxicol.*, **4**, 83–124.

Lamb, J. C. (1985). *J. Amer. Coll. Toxicol.*, **4**, 163–171.

Lamb, J. C. (1988). In *Physiology and Toxicology of Male Reproduction*, J. C. Lamb and P. M. D. Foster, Eds., Academic Press, New York, pp. 137–153.

Leblond, C. P. and Y. Clermont (1952). *Ann. NY Acad. Sci.*, **55**, 548–573.

Lee, W. R. (1978). In *Chemical Mutagens. Principles and Methods for Their Detection*, vol. 5, A. Hollaender, and F. J. de Serres, Eds., Plenum Press, New York, pp. 177–202.

Manson, J. M. and Y. J. Kang (1989). In *Principles and Methods of Toxicology*, A. W. Hayes, Ed., Raven Press, New York, pp. 311–359.

Manson, J. M. and L. D. Wise (1991). In *Toxicology. The Basic Science of Poisons*, M. O. Amdur, J. Doull and C. D. Klaassen, Eds., Pergamon Press, New York, pp. 226–254.

Mantel, N. (1969). In *Progress in Experimental Tumor Research*, Vol. 11, S. Karger, New York, pp. 431–443.

Matsumoto, A. M. (1989). In *The Testis*, 2nd edition, H. Burger and D. deKrestser, Eds., Raven Press, New York, pp. 181–196.

Mattison, D. R. (1981). In *Drug Metabolism in the Immature Human*, L. F. Soyka and G. P. Redmond, Eds., Raven Press, New York, pp. 129–143.

Mattison, D. R. (1983). In *Reproductive and Developmental Toxicity of Metals*, W. T. Clarkson and G. Nordbert, Eds., Plenum, New York, pp. 67–95.

Mattison, D. R. (1991). *Biomed. Environ. Sci.*, **4**, 8–34.

Mattison, D. R. and P. J. Thomford (1989). In *Toxicology of the Male and Female Reproductive Systems*, P. K. Working, Ed., Hemisphere, New York, pp. 101–129.

Mattison, D. R. and S. S. Thorgeirsson (1979). *Cancer Res.*, **39**, 3471–3475.

Mattison, D. R., J. W. Hanson, D. M. Kochhar, and K. S. Rao (1989). *Reprod. Toxicol.*, **3**, 3–12.

Mohr, K. L. and P. K. Working (1990). *Toxicol. In Vitro*, **4**, 115–121.

Morrissey, R. E. (1989). In *Toxicology of the Male and Female Reproductive Systems*, P. K. Working, Ed., Hemisphere, New York, pp. 199–216.

Nelson, S. D., M. R. Boyd, and J. R. Mitchell (1977). In *Drug Metabolism Concepts*, D. M. Jerina, Ed., American Chemical Society, Washington, D.C., pp. 155–185.

Neubert, D., H. J. Barrach, and J. J. Merker (1980). *Curr. Topic Pathol.*, **69**, 242–324.

Palmer, A. K. (1981). In *Developmental Toxicology*, C. Kimmel and J. Buelke-Sam, Eds., Raven Press, New York, pp. 259–287.

Pedersen, R. A. and B. Brandriff (1980). In *DNA Repair and Mutagenesis in Eukaryotes*, W. M. Generoso, M. D. Shelby and F. J. de Serres, Eds., Plenum Press, New York, pp. 389–420.

Pedersen, R. A. and F. Mangia (1978). *Mutat. Res.*, **49**, 425–429.

Plowchalk, D. R. and D. R. Mattison (1991a). In *Growth Factors and The Ovary*, A. N. Hirshfield, Ed., Plenum Press, New York, pp. 427–432.

Plowchalk, D. R. and D. R. Mattison (1991b). *Toxicol. Appl. Pharmacol.*, **107**, 472–481.

Plowchalk, D. R., M. J. Meadows, and D. R. Mattison (1993). In *Occupational and Environmental Reproductive Hazards: A Guide for Clinicians*, M. Paul, Ed., Williams & Wilkins, Baltimore, pp. 18–24.

Roderick, T. H. (1971). *Genetics*, **76**, 109–113.

Roderick, T. H. (1983). In *Utilization of Mammalian Specific Locus Studies in Hazard Evaluation and Estimation of Genetic Risk*, F. J. de Serres and W. Sheridan, Eds., Plenum Press, New York, pp. 135–167.

Russell, L. B. and B. E. Matter (1980). *Mutat. Res.*, **75**, 279–302.

Russell, L. B. and M. D. Shelby (1985). *Mutat. Res.*, **154**, 69–84.

Russell, L. B., P. B. Selby, E. Von Halle, W. Sheridan, and L. Valcovic (1981). *Mutat. Res.*, **86**, 329–354.

Russell, L. D. (1980). *Gamete Res.*, **3**, 99–112.

Russell, L. D., R. A. Ettlin, A. P. Sinha Hikim, and E. D. Clegg (1990). *Histological and Histopathological Evaluation of the Testis*, Cache River Press, Clearwater, FL.

Schardein, J. L. (1985). *Chemically Induced Birth Defects*, Marcel Dekker, New York.

Schrader, S. M. and J. S. Kesner (1993). In *Occupational and Environmental Reproductive Hazards: A Guide for Clinicians*, M. Paul, Ed., Williams & Wilkins, Baltimore, pp. 3–17.

Searle, A. G. and C. V. Beechey (1982). *Cytogenet. Cell Genet.*, **33**, 81–87.

Sega, G. A. (1982). In *Indicators of Genotoxic Exposure, Banbury Report 13*, B. E. Bridges, B. E. Butterworth and I. B. Weinstein, Eds., Cold Spring Harbor Laboratory, Cold Spring Harbor, pp. 503–514.

Sega, G. A., A. E. Sluder, L. S. McCoy, J. G. Owens, and W. M. Generoso (1986). *Mutat. Res.*, **159**, 55–63.

Selby, P. B. (1983). In *Utilization of Mammalian Specific Locus Studies in Hazard Evaluation and Estimation of Genetic Risk*, F. J. de Serres and W. Sheridan, Eds., Plenum Press, New York, pp. 191–210.

Searle, A. G. and C. V. Beechey (1982). *Cytogenet. Cell Genet.*, **33**, 81–87.

Shepherd, T. H. (1989). *Catalog of Teratogenic Agents*, 5th Edition, Johns Hopkins University Press, Baltimore.

Sheridan, W. (1983). In *Utilization of Mammalian Specific Locus Studies in Hazard Evaluation and Estimation of Genetic Risk*, F. J. de Serres and W. Sheridan, Eds., Plenum Press, New York, pp. 125–134.

Sims, P. and P. L. Glover (1974). *Adv. Cancer Res.*, **20**, 165–174.

Sinha Hakim, A. P., A. G. Amador, H. G. Kemcke, A. Bartke and L. D. Russell (1989). *Endocrinology*, **125**, 1829–1843.

Skare, J. A. and K. R. Schrotel (1985). *Environ, Mutagen.*, **7**, 547–561.

Skinner, M. K. and M. D. Griswold (1980). In *J. Biol. Chem.*, **255**, 9253–9525.

Stott, W. T. and P. G. Watanabe (1980), *Toxicol, Appl. Pharmacol.*, **55**, 411–416.

Tortora, G. J. and N. R. Anagnostakos (1975). *Principles of Anatomy and Physiology*, Canfield Press, San Francisco.

U.S. EPA (Environmental Protection Agency) (1982). EPA Publication no. 560/6-82-001. National Technical Information System, Springfield, VA.

U.S. EPA (Environmental Protection Agency) (1986). *Federal Register*, **51**, 34028–34040.

Verhoeven, G. and P. Franchimont (1983). *Acta Endocrinol.*, **102**, 136–143.

Weil, C. S. (1970). *Food Cosmet. Toxicol.*, **8**, 177–182.

Weitzman, G. A., M. M. Miller, S. N. London, and D. R. Mattison (1992). *Reprod. Toxicol.*, **6**, 137–141.

Welch, R. M., W. Levin, and A. H. Conney (1969). *Toxicol. Appl. Pharmacol.*, **14**, 358–367.

Welch, R. M., W. Levin, L. Kuntzmann, M. Jacobson, and A. H. Conney (1971). *Toxicol. Appl. Pharmacol.*, **19**, 234–246.

Working, P. K. (1989). In *Toxicology of the Male and Female Reproductive Systems*, P. K. Working, Ed., Hemisphere, New York, pp. 231–255.

Working, P. K. and B. E. Butterworth (1984). *Environ. Mutagen.*, **6**, 273–286.

Working, P. K. and G. J. Chellman (1989). In *Sperm Measures and Reproductive Success: Institute for Health Policy Analysis Forum on Science, Health and Environmental Risk Assessment*, E. J. Burger, Jr., R. G. Tardiff, A. R. Scialli and H. Zenick, Eds., Alan R. Liss, New York, pp. 211–227.

Working, P. K. and D. R. Mattison (1993). In *Occupational and Environmental Reproductive Hazards: A Guide for Clinicians*, M. Paul, Ed., Williams & Wilkins, Baltimore, pp. 91–99.

Working, P. K., J. S. Bus, and T. E. Hamm, Jr. (1985). *Toxicol. Appl. Pharmacol.*, **77**, 144–157.

Wyrobek, A. J., L. A. Gordon, J. G. Burkhart, M. W. Francis, R. W. Kapp, G. Letz, H. V. Malling, J. C. Topham, and M. D. Whorton (1983). *Mutat. Res.*, **115**, 1–72.

Zenick, H. and E. D. Clegg (1989). In *Principles and Methods of Toxicology*, A. W. Hayes, Ed., Raven Press, New York, pp. 275–309.

CHAPTER SIX

Neurotoxicology: An Orientation

Jacques P. J. Maurissen, Ph.D., and Joel L. Mattsson, D.V.M., Ph.D.

1 INTRODUCTION

Due to an accelerated pace of neurotoxicity testing and research, scientists without great familiarity with the nervous system are going to confront much more neurotoxicity information than ever before. This neurotoxicity information, in material safety data sheets, synopses, and published articles, contains concepts and language that are unfamiliar to many people. The goal of this chapter, therefore, is to provide an orientation for nonneurotoxicologists that include basic concepts, language and applications in neurotoxicology. Thus armed, readers will better understand neurotoxicological information as it relates to their work and interests. Many references are provided for those interested in further insight.

This chapter first places neurotoxicology in a historical perspective and gives a quick overview of some of the activities encountered in this field. It then focuses on the impact of regulations on neurotoxicology and defines neurotoxicity and neurotoxicant in this context. Finally, it cautions the reader against some known pitfalls in the interpretation of neurotoxicological data.

2 HISTORY

The effects of chemicals on behavior were recognized very early in the history of toxicology. As early as 1473, Ulrich Ellenbog wrote one of the first known essays

Patty's Industrial Hygiene and Toxicology, Third Edition, Volume 3, Part B, Edited by Lewis J. Cralley, Lester V. Cralley, and James S. Bus.
ISBN 0-471-53065-4 © 1995 John Wiley & Sons, Inc.

on industrial hygiene and toxicology, which was published posthumously in 1524. In this document he reported headache, visual disturbances, unconsciousness, and paralysis among symptoms of industrial metal poisoning. During the sixteenth century, Paracelsus (1567) described one of the signs of mercury intoxication as "shivering without any frost being felt." At the end of the eighteenth century, John Pearson of London reported mercurial tremor as part of a syndrome for which he proposed the term "mercurial erethism" (Almkvist, 1929). Through the middle of the twentieth century, the field of neurotoxicology was growing slowly as a result of isolated and uncoordinated efforts (Armstrong et al., 1963; Ruffin, 1963).

Actual research in this field started during the latter part of this century, and it was not before the 1970s that neurotoxicology developed into a scientific discipline. Several factors influenced this formalization, such as improved knowledge on specific toxic agents, stricter standards for air pollutants in the Soviet Union due to the use of electrophysiological methods and Pavlovian conditioning, advances in behavioral pharmacology, and the Occupational Safety and Health Act of 1970, which identified psychological factors as a specific concern for workers' health.

In June 1972 Weiss and Laties (1975) organized the first formal meeting entirely devoted to behavioral toxicology in Rochester (New York). Shortly thereafter, in June 1973, a behavioral toxicology workshop was sponsored in part by the National Institute of Occupational Safety and Health (Xintaras et al., 1974). In 1977 the Behavioral Teratology Society was founded, followed by the Behavioral Toxicology Society in 1981. In Europe, the International Neurotoxicology Association was established in 1984 and held its first meeting in 1987. A number of meetings, symposia and conferences have since taken place, and two scientific journals, *Neurotoxicology and Teratology* and *NeuroToxicology*, have been dedicated to publications in the field of neurotoxicology. A number of handbooks have also been published (Roizin et al., 1977; Manzo, 1980; Spencer and Schaumburg, 1980; Mitchell, 1982; Gilioli et al., 1983; Zbinden et al., 1983; O'Donoghue, 1985a; Annau, 1986; Galli et al., 1988; Hartman, 1988; Johnson, 1990; Valciukas, 1991).

In 1976 the U.S. Congress passed the Toxic Substances Control Act (TSCA). In this document behavioral disorders were identified as specific health effect endpoints to be evaluated as part of the toxicological profile of a chemical. As a consequence of the Congress's mandate, the U.S. Environmental Protection Agency (U.S. EPA or EPA) developed a series of neurotoxicity guidelines in 1985 (amended in 1986 and revised in 1991).

During the same period, the International Programme on Chemical Safety (World Health Organization) undertook a study of the principles and methods used to study the effects of chemicals on the nervous system. In 1983 and 1984 members of an International Task Group met in Moscow and in Prague to develop a monograph on this topic (World Health Organization, 1986). More recently, the European Chemical Industry Ecology and Toxicology Center (ECETOC,* 1992, 1993) published a monograph and a special report on the evaluation of the neurotoxic potential of chemicals.

---

*4, Av. E. Van Nieuwenhuyse (Bte 6), B-1160 Brussels, Belgium.

3 SCOPE OF NEUROTOXICOLOGY

The nervous system is a very complex structure that is divided into central and peripheral nervous systems. The central nervous system (CNS) is a group of different organs (nuclei) interconnected by a complex network of feedback inhibitory and excitatory pathways. This multiorgan system is responsible for neuroendocrine, trophic, and autonomic functions, homeostasis, movement, vision, hearing, sensorimotor integration, reproductive behavior, arousal, language, thought, emotions, artistic creation, and so forth (Fig. 6.1). One of the attributes of the brain is its tremendous adaptive reserve capability with its 10–15 billion cells (plasticity). In contradistinction, neurons of the brain of mammals cannot regenerate.

The effects of chemicals on the nervous system have been studied at different levels. *Biochemistry* approaches neurotoxicity from the molecular point of view (e.g., distribution and turnover of various neurotransmitters, receptor binding, enzyme activities, immunohistochemistry, etc.). *Morphology* investigates the structure and ultrastructure of the brain, peripheral nerves and end organs. *Electrophysiology* evaluates the functions of the central and peripheral nervous systems. *Behavioral analysis* describes the effects of chemicals on a particular behavior. *Clinical studies* investigate the effects of chemicals on signs and symptoms of diseases in patients or volunteers. *Epidemiology* attempts to isolate, among various factors, the determinant(s) of the frequency and distribution of diseases in a community, group, or work population.

The following presentation will mainly focus on three broad functional domains of the field of neurotoxicology: sensory functions, motor functions, and higher CNS functions.

3.1 Sensory Functions

Sensory systems provide the interface between the living organism and the external world. In 1860 Gustav Theodor Fechner formalized the psychophysical approach to

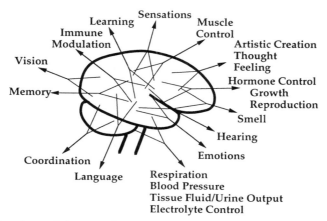

Figure 6.1 Functions of the central nervous system. The forked tails of the arrows depict the integration of specific functions in the central nervous system with all other functions of the brain.

the evaluation of sense organs in his book *Elemente der Psychophysik* (Fechner, 1966). Some of the techniques he described are still being used today. The object of *psychophysics* is the scientific study of the relationship between the physical dimensions of a stimulus and the sensation it elicits as evidenced by the behavioral response it generates. More recently, signal detection theory, which drew upon statistical decision theory and electronic communications, has broadened the scope of the psychophysical approach (Green and Swets, 1974). A wealth of information is now available concerning the scientific study of sensory systems in animals and human beings (Corso, 1967; Stebbins, 1970; Geldard, 1972; Scharf, 1975).

Many approaches have been used to assess the effects of chemicals on sensory systems (*Target Organ Toxicity*, 1982). A few studies will be taken as examples to illustrate some of the experiments performed in sensory neurotoxicology.

3.1.1 Hearing and Toluene

Toluene is a widely used solvent that can cause central nervous depression. In cases of extended periods of toluene abuse, severe nervous system effects have been documented, such as cerebellar encephalopathy and dementia. Hearing loss was also reported in chronic glue sniffers (Ehyai and Freemon, 1983).

Pryor et al. (1983) trained rats to climb or pull a pole suspended from the ceiling of a test chamber to escape or avoid an aversive electrical stimulus delivered to the feet. Each stimulus was preceded by a warning signal (also called a *discriminative stimulus*, e.g., a tone or a change in light intensity). After a number of trials, the animals associated the presence of a warning signal with the imminent occurrence of an aversive stimulus and learned to climb or pull the pole during the warning signal. This behavioral test paradigm is known as the *multisensory conditioned avoidance response* (*CAR*) because the rats were *conditioned* (i.e., trained) to avoid an aversive stimulus by giving a *response* (pole climbing or pulling) to tactile, visual, and auditory stimuli.

Pryor et al. (1983) exposed weanling rats to 1200 and 1400 ppm of toluene for 14 hr a day, 7 days a week for 5 weeks. While the exposed rats were able to acquire the escape/avoidance response to a tactile or a visual stimulus, they were unable to respond to the auditory stimulus. By varying the frequency and the intensity of the tone, it became clear that the hearing of the animals was unimpaired at 4 kHz, slightly impaired at 8 kHz and markedly impaired at 12 kHz.

Following this study, Rebert et al. (1983) performed an electrophysiologic study of the auditory threshold (*auditory brainstem response, ABR*) after exposure to toluene. They used subcutaneous electrodes to record the brain electrical activity in response to a tone of different frequencies and intensity levels. They showed that the thresholds were elevated in the toluene-exposed group for all frequencies of the auditory stimulus, but the effects were greater at the higher frequencies (12 and 16 kHz) compared to the lower frequency (8 kHz). These electrophysiological data confirmed the behavioral ototoxicity findings reported by Pryor et al. (1983) after toluene exposure.

Sullivan et al. (1989) examined the morphological and functional consequences of oral toluene administration on the auditory system of rats (1 mL/kg day in corn oil for 49 consecutive days). The pattern of hair-cell loss in the cochlea corresponded to auditory changes at specific frequencies as reflected by an elevated ABR threshold.

3.1.2 Vibration Sensitivity and Acrylamide

In high enough doses, acrylamide is known to cause central and peripheral nervous system dysfunctions in animals and humans. It induces peripheral nerve axonal degeneration (Fig. 6.2), and causes numbness and sensory loss in the extremities in a glove-and-stocking configuration (O'Donoghue, 1985b).

As blunting of tactile sensitivity is recognized as an early symptom of acrylamide toxicity, Maurissen et al. (1983) set up a computerized system to study the effects of acrylamide on vibration sensitivity in monkeys. Before treatment, the monkeys were conditioned to sit in a chair with the left hand positioned on a vibrating probe. The monkey's task consisted in reporting the presence of an event (i.e., vibration delivered to the fingertip) by releasing a lever (response). Every correct detection was followed by delivery of a squirt of apple juice (a reward, or *positive reinforcement*) into the monkey's mouth.

After a steady baseline was obtained, the animals were dosed orally with 10 mg/kg of acrylamide 5 days a week until the first appearance of toxic signs. The monkeys

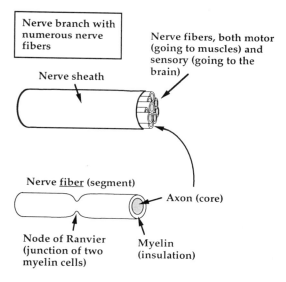

Neuropathy = damage to a nerve
Polyneuropathy = damage to several nerves
Sensory neuropathy = damage to sensory fibers
Motor neuropathy = damage to motor fibers
Axonopathy = axons more severely affected
    than myelin
Myelinopathy = myelin more severely affected
    than axons

Figure 6.2 Basic anatomy of a peripheral nerve.

were tested daily for vibration sensitivity thresholds. Vibration sensitivity decreased during dosing and remained impaired for several months after dosing had ceased, at a time when no other clinical signs of intoxication were apparent.

3.1.3 Neurotoxic Esterase and Orthocresyl Phosphate

Organophosphates have diverse effects on both the peripheral (sensory and motor) and the central nervous systems. The best studied biochemical mechanism of organophosphate intoxication is the inhibition of acetylcholinesterase. Some organophosphates, however, can produce a delayed neuropathy in the peripheral and central nervous systems, also referred to as *organophosphate-induced delayed neuropathy* (*OPIDN*). In humans, the cardinal signs of OPIDN are sensory disturbances and pain in the calf muscles, eventually accompanied with muscle paralysis, and may appear a few weeks after exposure. All animals species are not equally susceptible, and the hen has been the species of choice to demonstrate delayed neurotoxicity. This delayed effect does not appear to be mediated by acetylcholinesterase inhibition but is supposedly linked to *neurotoxic esterase* (*NTE*) inhibition (Davis and Richardson, 1980; Lotti et al., 1993).

In Morocco, Smith and Spalding (1959) reported on an outbreak in humans of pain and tenderness in the calf, accompanied with paresthesias and motor weakness overwhelmingly distal. Thousands of people were affected and developed sensory and motor signs and symptoms of OPIDN. The infective theory was quickly abandoned for a more plausible toxic etiology. The health authorities linked the distribution of cooking oil in the affected area to one wholesale supplier, who had distributed oil contaminated with orthocresyl phosphate.

Lotti et al. (1986) suggested that lymphocytic NTE inhibition could be used as a test to predict the development of OPIDN in humans who have been exposed to organophosphates. Although NTE has clearly been identified as a target for some organophosphates, its physiological role is unknown, and the biochemical mechanisms by which NTE presumably caused delayed neuropathy are still a matter of conjecture.

3.2 Motor Functions

Motor functions are multifarious and involve a wide range of behaviors, from holistic locomotor activity, gait, muscle tone, limb strength, reaction time, posture, steadiness to precision hand movements. Numerous techniques have been developed to quantify motor dysfunctions. A few examples will illustrate this point.

3.2.1 Motor Activity, Auditory Startle, and Pyrethroids

Synthetic pyrethroids are widely used insecticides. They can produce movement disorders. A distinction has been made between two classes of pyrethroids on the basis of structural, neurochemical and electrophysiological properties (Verschoyle and Aldridge, 1980). The Type I pyrethroids (e.g., cismethrin) can cause tremor and seizures in mice, while the Type II (e.g., deltamethrin) can cause unsteady gait and rigidity in rats.

Crofton and Reiter (1984) studied the effects of two pyrethroid insecticides, deltamethrin and cismethrin, on auditory startle and motor activity in male rats. *Auditory startle* was tested in sound-attenuated chambers. Inside each chamber, an animal was placed in a cage mounted on a force transducer. A speaker delivered short and loud tones, and the latency to onset and the peak force of the startle response were measured in response to the acoustic stimuli.

A figure-eight maze was used to test *motor activity* (Fig. 6.3). This maze consists of two continuous alleys that connect in a central area and form a figure eight. Activity was detected by infrared photobeams and sensors located along the alleys. The animals were tested in the figure-eight maze for one hour.

While both deltamethrin and cismethrin produced a dose-related decrease in motor activity, they produced dissimilar effects on acoustic startle response. Deltamethrin decreased force and increased latency in a dose-related manner; cismethrin, on the other hand, increased force and did not change latency. The differential effects of the two pyrethroids were related to their contrasting effects documented with neurochemical and electrophysiological techniques.

The effects of chemicals on motor activity have been extensively studied, and a large variety of devices have been designed to measure it (Evans, 1988), some extremely sophisticated (Kernan et al., 1987).

3.2.2 Electrophysiology, Morphology, and Carbon Disulfide

Carbon disulfide, mainly used in the production of viscose rayon fibers, is a solvent that can induce psychosis and peripheral neuropathy in overexposed workers (Seppäläinen and Haltia, 1980).

Seppäläinen and Linnoila (1976) exposed rats to 750 ppm of carbon disulfide for 6 hr a day, 5 days a week. After 3 weeks of exposure, the rats moved clumsily, and after 6–8 weeks they showed hindlimb weakness and ataxia. The motor nerve conduction velocity, evaluated in the sciatic nerve, decreased two weeks after initiation of exposure and worsened during the next 8–12 weeks. Neuropathologically, Seppäläinen and Haltia (1980) showed that carbon disulfide causes axonal swelling with distal degeneration of the peripheral nerves and spinal cord (i.e., *central-peripheral distal axonopathy*). Figure 6.4 illustrates this concept in humans.

Photocell

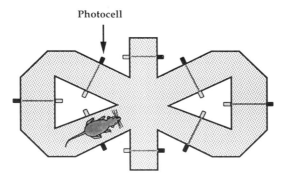

Figure 6.3 Schematic representation of the top view of a figure-eight maze.

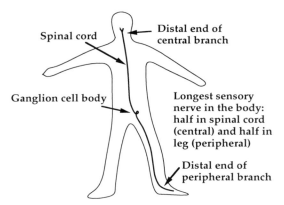

Figure 6.4 Drawing of a long nerve illustrating the distal extremities of the central and peripheral portions of the nerve. Distal axonopathy affects both ends of long nerves in a multifocal pattern and is more accurately referred to as "central-peripheral distal axonopathy."

3.2.3 Reaction Time and Carbon Monoxide

Carbon monoxide is produced by a number of different sources. It can decrease the oxygen transport capability of blood and cause tachycardia, hypoxia, headache, fatigue, nausea, dizziness, unconsciousness, and coma.

Johnson et al. (1974) studied the effects of carbon monoxide on behavioral performance of fare collectors at a toll highway. Alveolar breath samples were collected every 2 hr at the work site during the course of an 8-hr workshift over several days. Ambient air concentration of carbon monoxide, lead, and manganese were also measured. Each day, the subjects were administered a battery of behavioral tests. Among these, an eight-choice *reaction time* was presented.

Eight pushbuttons, within which a miniature red lamp was located, were arranged in a semicircle pattern. At the center of the semicircle, a hold button was located. A cue lamp prompted the subject to press the hold button, and to hold it down. After a random duration following the hold button press, one of the eight pushbuttons was lit, and the subject's task consisted in releasing the hold button and pressing the red illuminated pushbutton as quickly as possible. Total response time was divided into two components: detection time (or decision time) and movement time. *Detection time* was defined as the duration between the onset of the red lamp and the release of the hold pushbutton. *Movement time* was defined as the time between releasing the hold button and pressing the red illuminated button.

Each subject's data were analyzed in terms of changes between preshift and postshift data. Correlation analysis between carboxyhemoglobin levels and reaction time showed that detection time, but not movement time, was positively correlated with increased carboxyhemoglobin levels. In other words there was a tendency for the subjects to have a slow response in selecting the pushbutton, but there was no change in their ability to move quickly from one point to another. Gross motor response was not affected here.

3.2.4 Tremor and Mercury

Wood et al. (1973) studied hand tremor in women who had been occupationally exposed to inorganic mercury vapor during the fabrication of glass pipettes and had developed symptoms of excessive exposure. The patients were instructed to maintain a force on a finger trough (attached to a strain gauge) within some range. Feedback was given to the patient concerning the amount of pressure given. Power spectral analysis performed on the data characterized the frequency and amplitude of the tremor. Over time, tremor severity decreased with decreasing plasma mercury levels. Similarly, Langolf et al. (1978) documented a decrease in tremor frequency and power as a function of urine mercury levels. Methods for the detection and quantitative analysis have also been developed for animals (Shimozaki, 1984).

3.3 Higher Central Nervous System Functions

Higher central nervous system functions are diverse and include memory, learning, attention, vigilance, reasoning, communication, associative processes, intelligence, mood, personality, and so forth. Examples will deal with animal as well as human studies.

3.3.1 Short-Term Memory and Trimethyltin

Organotin compounds have been used as heat stabilizer, catalysts, and biocides, and their effects on the nervous system have been documented. Trimethyltin, however, differs from other organotin compounds in that its primary target in animals is the limbic system, which is known to mediate memory functions, at least partially.

Bushnell (1988) used a delayed response task (*delayed matching to position* or *spatial delayed response*) to study short-term memory in rats. In such a test paradigm, the rats were tested in a quiet experimental chamber (also known as a *Skinner box* or *operant cage*; Fig. 6.5). In this cage there were two retractable levers (serving as discriminative stimuli) and a delivery system for positive reinforcement (some food pellets delivered in a food cup). At the beginning of the trial, one lever was extended inside the chamber. This lever is referred to as the sample lever. The animals were conditioned to press the sample lever, which was then immediately retracted, then to move away from the lever and wait for a variable amount of time (0–20 sec) at a different location. This last step prevented the rats from staying next to the retracted lever. At the end of the delay, both levers were extended into the chamber. A response on the previously extended lever (i.e., sample lever) was

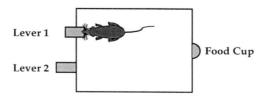

Figure 6.5 Schematic drawing of a Skinner box.

immediately followed by a food pellet delivery (correct response). A response on the nonsample lever was followed by a time-out period during which the houselight was extinguished (incorrect response). The test session comprised a series of discrete trials. In addition, glial fibrillary acidic protein (GFAP), a marker of astrocytes (which perform major homeostatic functions in the central nervous system), was also assessed in the hippocampus at the end of the study.

Sixteen male rats were trained on the spatial delayed response described above, and their body weights were maintained constant by scheduled feeding. Trimethyltin (7 mg/kg) was injected into the tail vein of eight rats; the other eight rats received physiological saline. The number of correct choices (*choice accuracy*) decreased as the delay duration increased in both groups. During the first week following trimethyltin administration, the choice accuracy was significantly decreased in the treated group in comparison with controls. These data show that trimethyltin impairs short-term memory in rats. Increased hippocampal GFAP levels correlated with decreased accuracy in the experimental group.

3.3.2 Attention and 1,1,1-Trichloroethane

1,1,1,-Trichloroethane is mainly used as a solvent, and, in excessive amounts, can depress the central nervous system and cause anesthesia.

Mackay et al. (1987) exposed male volunteers to 0, 175, and 350 ppm of 1,1,1-trichloroethane for 3.5 hr in a balanced design for order effects. Peppermint oil was used to mask the odor of the solvent. After each exposure, the experimenters verified that the subjects were unaware of the exposure conditions. Test performance was assessed before each exposure period and at four separate times during exposure. Both subject and experimenter were blind to treatment conditions.

The Stroop test was used, among other behavioral tests, to assess distractability of attention. In a modified version of this test, the subject was presented with a sheet containing color names, and each name was written in a color different from the name it indicated. For example, the word "RED" was written in blue. Each test item contained one color name on the left and five color names on the right. The subject's task consisted in underlining the word that tells in what color ink the original word was typed. Under the exposure conditions stated above, there was a dose-related improvement in the number of lines of material completed throughout the 3.5-hr exposure session (possibly due to focused attention). The real significance of this enhanced performance is unclear.

4   REGULATORY NEUROTOXICOLOGY

As a mandate from TSCA, EPA developed a series of eight neurotoxicity guidelines in 1985 (U.S. EPA, 1985, amended as U.S. EPA, 1986):

- Functional observational battery (§798.6050)
- Motor activity (§798.6200)
- Neuropathology (§798.6400)

- NTE neurotox assay (§798.6450)
- Schedule-controlled operant behavior (§798.6500)
- Acute delayed neurotoxicity of organophosphorous substances (§798.6540)
- Subchronic delayed neurotoxicity of organophosphorous substances (§798.6560)
- Peripheral nerve function (§798.6850)

The test guidelines were revised in March 1991. The most significant changes of the latest guideline revision (U.S. EPA, 1991) are severalfold:

- The application of the guidelines was extended to neurotoxicity testing under the Federal Insecticide, Fungicide, and Rodenticide Act (FIFRA);
- The functional observational battery, motor activity and neuropathology guidelines were folded into one guideline entitled ''Neurotoxicity Screening Battery.''
- The developmental neurotoxicity screen, previously published as part of the ''Diethylene Glycol Butyl Ether and Diethylene Glycol Butyl Ether Acetate; Test Standards and Requirements'' test rule (U.S. EPA, 1988, §795.250) was revised and included in this revision as ''Developmental Neurotoxicity Study.''
- A new guideline was introduced as an appendix to the ''Neurotoxicity Screening Battery'' under the name ''Guideline for Assaying Glial Fibrillary Acidic Protein.''

Because a great deal of data on new and older chemical substances are going to be generated by the neurotoxicity guidelines on adult animals, the following presentation and discussion will focus on certain requirements of the functional observational battery, motor activity, neuropathology, glial fibrillary acidic protein, and schedule-controlled operant behavior guidelines. The ''Developmental Neurotoxicity Study'' guideline, which is very complex logistically, essentially incorporates most of the elements of the ''Neurotoxicity Screening Battery'' as applied to developing animals and adds measures of developmental landmarks, auditory startle, as well as learning and memory tests.

4.1 Neurotoxicity Screening Battery

This screening battery is designed for rats but may be modified as necessary for other species. The guideline requires four treatment levels (control, low dose, middle dose, and high dose); the high dose must cause significant toxicity or attain a limit dose of 2 g/kg in single-exposure studies or 1 g/kg body weight for repeated dose studies. At least 10 males and 10 females per dose level are given physical examinations and measurements of grip strength, landing foot splay, and motor activity before exposure begins and at spaced intervals throughout the study (e.g., monthly examinations in a 13-week study). At the end of the study, at least 5 males and 5 females per dose are examined for neuropathological changes.

4.1.1 Functional Observational Battery (FOB)

The FOB portion of the screening battery is a careful hand-held and open-field examination for predetermined endpoints such as muscle tone, tremors, abnormal movements, lacrimation, pupillary function, reactivity to sensory stimuli (e.g., handling, sharp noise, tail-pinch, light touch), and measurements of body weight, fore- and hindlimb grip strength, and landing foot splay. Descriptions of other unusual findings also are made. The FOB is performed blind to treatment.

The goal of the FOB is to examine the overall data set for patterns of changes in endpoints, to derive a diagnostic inference from this profile as to the presence of neurotoxicity, and to create a clinical description of the chemical effects (Moser, 1989). A permanent record is therefore made of gradations of response (e.g., scores from 1 to 5) for a large number of variables or for measurements of other variables such as grip strength.

Measurements of *hind- and forelimb grip strength* take advantage of the fact that rats will grasp a bar or wire screen when the paws are placed briskly on the bar or screen. Figure 6.6 shows a device where a screen is attached to a strain gauge or electronic load cell. To measure forelimb strength, the rats are held firmly but gently over the shoulders and by the tail, and then they are set (while still holding the tail) onto the screen. They instinctively grasp the screen and are pulled quickly and smoothly by the tail until the grasp is broken. The strain gauge records the grip strength in grams. Hindlimb strength is measured similarly.

Grip strength measurements contain two elements: The rat must be motivated to grab the screen or bar, and it must have the physical ability to grab the screen (O'Donoghue, 1989). Changes in grip strength, therefore, can reflect treatment-related differences in either motivational or physical domains. Reduced motivation can be due to central nervous system toxicity, to repeated testing, or to secondary effects of systemic toxicity (malaise). Altered physical capabilities can be attributed to peripheral nerve (O'Donoghue, 1989), neuro-muscular junction (Ross and Lawhorn, 1990), or muscle dysfunctions (Squibb et al., 1983).

Landing foot splay is typically measured by putting ink marks on the hind feet, then dropping the rat from a height of about 30 cm onto a piece of paper (Edwards and Parker, 1977). The distance between the ink marks on the paper is a measure of splay. Departure from "normal" splay distances (increases or decreases) have

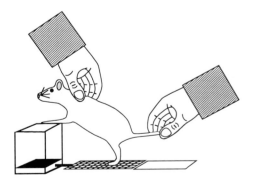

Figure 6.6 Grip strength test apparatus.

been regarded by some as an indication of impaired neuro-muscular integration and/ or equilibrium functions.

4.1.2 Motor Activity

Motor activity is used by EPA in a very specific acception. This test has to fulfill the following basic requirements:

- The device needs to be automated; although the device is not specified, the figure-eight maze is often used and discussed (Figure 6.3).
- The recorded data are motor activity counts, as typically recorded, for example, by the interruption of a photobeam.
- The temporal distribution of activity counts needs to be recorded throughout the motor activity test session (Figure 6.7), because it provides information on the rate of habituation.

After a motor activity system has been put into place, the data are relatively easy to collect, but not to interpret. The differential significance of an increase versus a decrease in motor activity counts remains unclear. As the World Health Organization (WHO) mentions, "if a change in motor activity is observed, additional tests are needed to determine the cause" (WHO, 1986, p. 41). While motor activity has been used because of its *apical* nature (i.e., it requires the integrity of the whole organism) (Reiter and MacPhail, 1979), others have criticized its use for the same reason, that is, it lacks specificity. A change in motor activity can reflect a change in other nonneural organs and does not necessarily reflect neurotoxicity (Gerber and O'Shaughnessy, 1986; Rafales, 1986; Maurissen and Mattsson, 1989). As Dews (1975, p. 445) eloquently remarks, "Neural toxicities [. . .] may not affect behavior such as dancing while nonneural effects may do so, as with diarrhea." Motor disturbance can evidence the direct effect of a chemical on the nervous system, or it can simply express a nonneural effect.

4.1.3 Neuropathology

There are obvious links between nervous system structure and function (Kandel et al., 1991). Many neural structures responsible for sensory, motor, and higher cog-

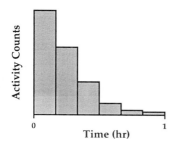

Figure 6.7 Typical pattern of the temporal distribution of motor activity counts throughout a 1-hr test session.

nitive functions are examined in subchronic or chronic toxicity studies. Routinely, the nervous system is histopathologically examined first by removal of brain, spinal cord, and peripheral nerves from the animal at necropsy. The tissues are immersion fixed (preservation). Tissue sections are paraffin embedded and stained to highlight different cell types and then examined by light microscopy.

The neurotoxicity screening battery differs from routine neuropathology by its requirement for *intravascular perfusion* of fixatives into the nervous system and removal of the nervous system postfixation. The peripheral nervous system has to be embedded in plastic instead of paraffin (brain and spinal cord still in paraffin), and a large number of tissue sections have to be examined. If qualitative evaluation of tissues reveals treatment-related changes, then the tissues are to be coded, randomized, and reexamined without awareness of the code (*blind re-evaluation*). An excellent summary of neuropathology methods is found in WHO (1986, pp. 71–100).

Although perfusion fixation is required, it is not without complications and artifacts. The ideal situation is to have some tissues immersion-fixed and some perfusion-fixed, a goal usually accomplished in standard toxicity studies that are combined with neurotoxicity screens. The critical feature of neuropathology lies more in the proper sampling of the nervous system than in the method of fixation. The tissue sampling plan, therefore, must remain flexible and additional sections added depending on FOB findings and information from additional diagnostic tests and review of the open literature.

4.2 Glial Fibrillary Acidic Protein (GFAP)

This test quantifies a type of protein contained in astrocytes. Although astrocytes are not information-processing cells, they perform myriads of normal functions in the central nervous system. "Astrocytes are involved in repair and regeneration processes and in guidance of axons to their proper sites. They are components of the blood-central nervous system barrier, they secrete trophic factors, and they are thought to play a modulatory role in neuronal intercommunication. . . . Their role in the microenvironment again plays a key role." (Fedoroff and Vernadakis, 1986, Vol. 2, p. ix).

Astrocytes react to toxic and physical injury by increasing GFAP. Thus, significant increases in GFAP in particular regions of the central nervous system indicate possible or probable injury to that area, and other research strategies can then focus on those specific areas of the brain. However, because astrocytes have a role in homeostasis of the central nervous system, they respond to physiologic/biochemical challenges as well as to toxic challenges (Fedoroff and Vernadakis, 1986). Thus, one needs to be wary of overinterpretation of small changes in GFAP.

The GFAP guideline requires fresh brain tissues from at least six animals per dose. The test is done by radioimmunoassay, and six regions of the central nervous system are evaluated (cerebellum, cerebral cortex, hippocampus, striatum, thalamus/hypothalamus, and the rest of the brain).

4.3 Schedule-Controlled Operant Behavior

Operant behavior refers to a behavior that is placed under control of a reinforcement, that is, a behavior that is controlled by its consequences (Reynolds, 1968). The relations between stimuli and reinforcement, as defined by the experimenter, are referred to as the *schedule of reinforcement*. In a somewhat broader sense, all temporal and topographical conditions under which a response is followed by a reinforcement, and all relationships involved in the reinforcement of a behavior are called *contingencies of reinforcement*. A large variety of schedules of reinforcement have been designed and used (Ferster and Skinner, 1957). Two schedules of reinforcement are mentioned in the guideline. The *fixed ratio schedule (FR)* requires a fixed number of responses before a reinforcement is given. The *fixed interval schedule (FI)* produces a reinforcement after the first response following a specified delay. The *multiple fixed interval-fixed ratio (mult FI-FR)* schedule alternates the requirements of the FI and the FR schedules, and has been favored by EPA because it produces a steady high *response rate* (number of responses per time unit) during the FR component, and a progressively accelerated response rate (also known as *scallop*, i.e., low rates gradually increasing to higher rates) during the FI component (Fig. 6.8). The "Schedule-Controlled Operant Behavior" guideline is a *performance* guideline in the sense that data are to be evaluated in the context of response rates. Response rates can also be affected by nonspecific effects and are best interpreted when information on sensory and motor functions involved in the operant task is available. However, the principles of operant behavior go far beyond this particular application and can be used to ask specific questions, whether it be about learning, memory or psychophysics, as previously illustrated.

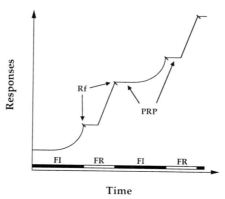

Time

Figure 6.8 Schematic representation of a cumulative record in which the total (cumulated) number of responses is plotted against time. Each occurrence of a response moves the response marking pen one unit toward the top of the paper. The two components of the mult FI-FR alternate: a progressively accelerated rate of responses characterizes the FI, while a consistent high rate typifies the FR performance. Reinforcements (Rf) are indicated by an oblique line and are usually followed by a postreinforcement pause (PRP).

5. DEFINING NEUROTOXICITY AND NEUROTOXICANT

There is much dissension among neurotoxicologists concerning the meaning of neurotoxicity and the definition of a neurotoxicant.

5.1 Neurotoxicity

Neurotoxicity has been defined as "any adverse effect on the structure or function of the nervous system related to exposure to a chemical substance" (U.S. EPA, 1991). The definition mostly hinges upon the interpretation of "adverse."

"Adverse" has been defined in toxicology by Klaassen (p. 15, 1986), in *Casarett and Doull's Toxicology*: "The spectrum of undesired effects of chemicals is broad. Some are deleterious and other are not . . . Some side effects of drugs are never desirable but are deleterious to the well-being of humans. These are referred to as the *adverse, deleterious*, or *toxic* effects of the drug." Thus, *adverse effects are deleterious always*.

In contrast to the "deleterious always" definition, some have interpreted "adverse" as "unwanted." U.S. EPA (1993, p. 41563) writes: "Adverse effects can include both unwanted effects and . . ." This definition of an adverse effect in terms of unwanted effects is social or cultural in nature. Social definitions have virtually no utility because they may even vary as a matter of individual preferences. For example, if one avoids individuals with garlic halitosis, the avoidance behavior garlic elicits would be considered as an adverse effect of garlic according to the preceding definition. If one has already eaten garlic and does not consider the smell objectionable, garlic would not therefore be considered as causing any adverse effect. Note that the toxicologic definition of adverse (Klaassen, 1986) differentiates between undesired effects that are deleterious and those that are not. Thus, the requirement that the effect be always deleterious avoids the quagmire of social definitions.

Not all effects are adverse. Obviously, effects of treatment may be beneficial, indifferent, adverse, or perhaps most commonly, of unknown toxicologic significance. It is important not to label events as adverse simply because the actual consequences to the animal or the individual are poorly understood.

When classifying a change in nervous system *function* as adverse, beneficial, indifferent, or unknown, we propose the following neurofunctional definition of adverse: An adverse effect is a *diminished ability* to perform a function. A diminished ability would be deleterious always. The utility of this definition is that it focuses attention on a specific ability, such as strength, coordinated movement, hearing, attention, or memory. Ability, however, is not measured directly, but is inferred from measures of a behavior.

For example, the intent of grip testing is to measure grip strength, but, in fact, grip tests measure both the *ability* to hold onto a bar or screen and *undefined performance factors* that include the motivation to grip the bar. If a decrease in grip performance is due to the rat learning to avoid the discomfort of being pulled off the apparatus, then the effect might be considered beneficial (accelerated learning). Because of the multitude of possible interpretations, a change in grip performance cannot be classified directly as adverse. Classification requires post hoc evaluation, and since the ability of interest is strength, post hoc tests can be performed to assess the presence of neuropathy (by pathology or electromyography [EMG]), myopathy (by pathology or EMG), neuro-muscular junction dysfunction (by EMG, possibly pathology), or channel dysfunction, such as myotonia (by EMG). If these post hoc tests do not reveal a cause, then the effect falls into the undefined arena, and classification is not possible.

All is not lost when classification is impossible. The index of suspicion of the presence of neurotoxicity can range from "unlikely" to "possibly" to "probably." The index of suspicion will move up or down depending on the pattern of effects discerned on other FOB and classical toxicity measures, and, if the index of suspicion is sufficiently high, subsequent studies may have to be conducted.

Obviously, clinical observations of convulsions, tremor, and ataxia reflect a diminished ability to perform coordinated movements and are adverse. The FOB observation of a diminished startle response to a sharp noise, however, is ambiguous. The ability of interest is hearing, but the observation is performance and includes nonauditory sensorimotor integration factors. A diminished FOB auditory startle response leads to specific questions of auditory function, which can be confirmed by auditory pathway pathology or supplemental tests such as auditory brainstem responses (smaller and slower peak I, acoustic nerve, etc.).

Motor activity is a performance test that does not have defined abilities associated with it. It is an apical test that can change because of any number of undefined reasons, and there are no known consequences for the animal. Hence, the toxicologic significance of motor activity changes is unknown (Maurissen and Mattsson, 1989). Although appreciable changes in motor activity, especially in the absence of classic toxicity, would increase one's index of suspicion of neurotoxicity, motor activity changes do not lead to preplanned post hoc tests nor to refined hypotheses.

The determination of the presence or absence of adverse effects is what differentiates toxicology from biology and psychology. The determination of adverse effects in neurotoxicology is often difficult, but the effort is necessary if neurotoxicology is to be a credible science.

5.2 Neurotoxicant

WHO (1986, p. 69) states that "it is incumbent on the investigator to determine whether an effect on the CNS is due to a primary action of the toxic agent or a secondary one as a result of damage to some other organ, thus altering the afferent input into the CNS or its processing within the CNS." Some, however, have simply defined a neurotoxicant as a chemical that potentially causes neurotoxicity without considering the implications of direct versus indirect effects. As altered behavior can result from the effect of any chemical substance given in sufficient dose, all chemical substances (including food) can be classified as neurotoxicants. For example, unadulterated water would be labeled, according to this simple definition, as a neurotoxicant because, in large doses, it causes electrolyte imbalance (hyponatremia) that itself eventually induces neurotoxic effects, such as gait disturbance, decreased awareness, impaired attention and cognition, delirium, seizures, metabolic encephalopathy, and coma (Labrune et al., 1985; Mattle, 1985; Borowitz and Rocco, 1986; Mediani, 1987).

The previous definition of neurotoxicant is very large and open-ended, and, consequently, has no classificatory value. Another approach incorporates the concept of target organ and requires that a chemical *directly* affect the nervous system in an *adverse* manner before it is classified as a neurotoxicant (ECETOC, 1992, 1993). According to the latter definition, unadulterated water would not be classified as a neurotoxicant because it causes neurotoxicity in an indirect manner, secondarily to a metabolic disturbance.

6   CAUTIONS IN DATA INTERPRETATION

Because of the complexity of the nervous system, considerable knowledge and experience in various aspects of the neural system are necessary. It should be recognized that even the simplest studies are subject to systematic errors that can affect data collection and/or interpretation. Obliviousness to the engineering aspects of a study in neurotoxicology can have profound repercussions on the mere nature of the collected data. Neurotoxicology is a complex field of investigation, and the experienced research neurotoxicologist is in the best position to recognize and avoid confounders and/or to incorporate appropriate methods and experimental designs to deal with them.

6.1   *Post Hoc, ergo Propter Hoc?*

In human neurotoxicology a diagnosis has often simply been made by exclusion of other etiologies. The mere presence of a neurotoxic chemical in an environment is not enough to indict the chemical as a causative agent if some neurological signs or symptoms are found and have not been attributed to a known disease (Valciukas, 1991, p. 556). As Valciukas (1991, p. 297) reminds us, "group differences should never be taken by themselves as the only basis for a judgment of neurotoxicity."

It is often tempting to conclude to a causal relationship between two events solely from their sequential order, especially if this temporal sequence is consistent with a preconceived hypothesis. However, the complexity of the subject of scientific investigation should not, in any way, relax the standards of the scientific proof.

In the field of microbiology, Koch developed a series of premises (known as Koch's postulates) that are required to prove that a microorganism really caused a disease (Jawetz et al., 1972). These postulates are still used to establish the etiologic relationship between a microorganism and a disease.

In the field of neurotoxicology, Spencer and Schaumburg (1985) postulated that three questions must be affirmatively answered before a solvent can be accepted as a neurotoxicant (these questions, however, equally apply to other chemicals):

1. Does the substance or mixture produce a consistent pattern of neurological dysfunction in humans?
2. Can this entity be induced in animals under comparable exposure condition?

3. Are there reproducible lesions in the nervous system or special sense organs of exposed humans and/or animals, and do these abnormalities satisfactorily account for the neurobehavioral dysfunction?

For example, *n*-hexane has been clearly demonstrated to be neurotoxic following prolonged exposure. It produces a consistent pattern of neurological syndromes (paresthesias and weakness of the hands and legs, sensory disturbances, unsteady gait) in humans (Rizzuto et al., 1977; Schaumburg and Spencer, 1979). In rats, *n*-hexane causes hindlimb weakness with foot-drop, weakness of the forelimbs, and unsteady and waddling gait (Schaumburg and Spencer, 1976) at doses roughly comparable to those recorded in humans. The lesions observed in the central and peripheral nervous system are reproducible in humans and animals, and they account for the reported clinical signs and symptoms of neuropathy (Spencer et al., 1980).

6.2 Behavior as an Expression of the Whole Organism

Behavioral toxicity studies should not be performed and interpreted in isolation because behavior is a function of the whole organism, rather than of the brain. Effects on other organs, from the liver to the kidney, should always be kept in mind.

For example, the effects of a chemical on a test of memory should not necessarily be ascribed to a deficit in this function but are often better attributed to changes in other neural or nonneural functions and environmental conditions (Figure 6.9). Only careful analysis and experimental design can tease apart the affected process(es).

6.3 Confounders and Biases in Human Studies

A series of factors have been identified that can confound data interpretation in human studies, such as age, gender, educational level, sampling bias, classification bias, participation bias, occupational link fallacy, diet, alcohol consumption, home environment, job satisfaction, and so forth (Maroni et al., 1977; van Vliet et al., 1990; Valciukas, 1991).

An apparent decrease in neuropsychological performance may not necessarily reflect an impairment of neural functions. Some impairments can precede exposure to a neurotoxic environment, some can be easily simulated (malingering), and results from subjective tests should always be interpreted cautiously in the context of a litigation (Hartman, 1988, pp. 269–271).

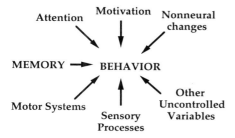

Figure 6.9 Factors possibly affecting the outcome of a memory test.

Some human studies have been performed without a control group (Olson et al., 1981). Others had an inappropriate control group (Lindström, 1980). The importance of a relevant control group has been emphasized by Errebo-Knudsen and Olsen (1986, 1987).

6.4 False Positives and False Negatives

The complexity of the field of neurotoxicology and the requirements of the neurotoxicity guidelines have rendered this area of investigation particularly susceptible to false positives (or Type I error). The number of the statistical p values derived in a simple subchronic adult neurotoxicity study performed under EPA guidelines has gone over 1000 in some cases. Many studies are still inappropriately using multiple t tests, when a univariate or multivariate analysis of variance should have been utilized (Seppäläinen et al., 1980). The larger the number of p values, the greater the probability of deriving a p value statistically significant by chance. Multiplicity of tests is a factor that must be considered. In the context of multiple tests of significance, Gill (1985) gives some examples of how many statistically "significant" p values are needed before it can be concluded that a treatment had any effect at all. If 29 to 40 p values have been derived, it can be calculated according to Gill's approach that a minimum number of 4 significant p values (< 0.05) are required to assert with 95 percent confidence that a true difference exists for one or more of these four "positive" tests.

On the other hand, the absence of a statistically significant difference does not necessarily suggest the absence of an effect if the data variability or the sample size are such that the difference of interest could not be detected (Type II error).

Too many exploratory studies are followed by too few (almost no) confirmatory studies. Nothing can really take the place of a well-designed efficient experiment with a simple hypothesis and a clear statement of purpose (Muller et al., 1984).

REFERENCES

Almkvist, J. (1929). *Acta Med. Scand.*, **70**, 464–476.

Annau, Z. (1986). *Neurobehavioral Toxicology*, Johns Hopkins University Press, Baltimore.

Armstrong, R. D., L. J. Leach, P. R. Belluscio, E. A. Maynard, H. C. Hodge, and J. K. Scott (1963). *Am. Ind. Hyg. Assoc. J.*, **24**, 366–375.

Borowitz, S. M. and M. Rocco (1986). *South. Med. J.*, **79**, 1156–1158.

Bushnell, P. J. (1988). *Neurotoxicol. Teratol.*, **10**, 237–244.

Corso, J. F. (1967). *The Experimental Psychology of Sensory Behavior*, Holt, Rinehart and Winston, New York.

Crofton, K. M. and L. W. Reiter (1984). *Toxicol. Appl. Pharmacol.*, **75**, 318–328.

Davis, C. S. and R. J. Richardson (1980). In *Experimental and Clinical Neurotoxicology*, P. S. Spencer and H. H. Schaumburg, Eds. Williams & Wilkins, Baltimore, pp. 527–544.

Dews, P. B. (1975). In *Behavioral Toxicology*, B. Weiss and V. G. Laties, Eds., Plenum Press, New York, pp. 439–445.

Edwards, P. M. and V. H. Parker (1977). *Toxicol. Appl. Pharmacol.*, **40**, 589–591.

Ehyai, A. and F. R. Freemon (1983). *J. Neurol. Neurosurg. Psychiat.*, **46**, 349–351.

Ellenbog, U. (1524). *Von den gifftingen besen Tempffen und Reuchen*, M. Ramminger, Augsburg.

Errebo-Knudsen, E. O. and F. Olsen (1986). *Sci. Tot. Environ.*, **48**, 45–67.

Errebo-Knudsen, E. O. and F. Olsen (1987). *Br. J. Indust. Med.*, **44**, 71–72.

European Chemical Industry Ecology and Toxicology Centre (1992). *Evaluation of the Neurotoxic Effect of Chemicals*, Monograph No. 18, Brussels, Belgium.

European Chemical Industry Ecology and Toxicology Centre (1993). *Interpretation and Evaluation of the Neurotoxic Effect of Chemicals in Animals*, Special Report No. 6, Brussels, Belgium.

Evans, H. L. (1988). *Toxicology Lett.*, **43**, 345–359.

Fechner, G. (1966). *Elements of Psychophysics*, H. E. Adler (Trans.), D. H. Howes and E. G. Boring, Eds., Holt, Rinehart and Winston, New York.

Fedoroff, S. and A. Vernadakis, Eds. (1986). *Astrocytes*, 3 Vols., Academic Press, New York.

Ferster, C. B. and B. F. Skinner (1957). *Schedules of Reinforcement*, Appleton-Century-Crofts, New York.

Galli, C. L., L. Manzo, and P. S. Spencer (1988). *Recent Advances in Nervous System Toxicology*, Plenum Press, New York.

Geldard, F. A. (1972). *The Human Senses*, Wiley, New York.

Gerber, G. J. and D. O'Shaughnessy (1986). *Neurobehav. Toxicol. Teratol.*, **8**, 703–710.

Gill, J. L. (1985). *J. Animal Sci.*, **60**, 867–870.

Gilioli, R., M. G. Cassitto, and V. Foà (1983). *Neurobehavioral Methods in Occupational Health*, Pergamon Press, Oxford.

Green, D. M. and J. A. Swets (1974). *Signal Detection Theory and Psychophysics*, Robert E. Krieger Publishing, Huntington, NY.

Hartman, D. E. (1988). *Neuropsychological Toxicology: Identification and Assessment of Human Neurotoxic Syndromes*, Pergamon Press, New York.

Jawetz, E., J. L. Melnick, and E. A. Adelberg (1972). *Review of Medical Microbiology*, Lange Medical Publications, Los Altos, CA, pp. 129–130.

Johnson, B. L. (1990). *Advances in Neurobehavioral Toxicology: Applications in Environmental and Occupational Health*, Lewis Publishers, Chelsea, MI.

Johnson, B. L., H. H. Cohen, R. Struble, J. V. Setzer, K. W. Anger, B. G. Gutnik, T. McDonough, and P. Hauser (1974). In *Behavioral Toxicology, Early Detection of Occupational Hazards*, C. Xintaras, B. L. Johnson, and I. de Groot, Eds., U.S. Department of Health, Education, and Welfare, Public Health Service, Center for Disease Control, National Institute for Occupational Safety and Health, HEW Publication (NIOSH) 74-126, Washington, D.C., pp. 306–328.

Kandel, E. R., J. H. Schwartz, and T. M. Jessell, Eds. (1991). *Principles of Neural Science*, Elsevier, New York.

Kernan, W. J., P. J. Mullenix, and D. L. Hopper (1987). *Pharmacol. Biochem. Behav.*, **27**, 559–564.

Klaassen, C. D. (1986). "*Principles of Toxicology*," in *Casarett and Doull's Toxicology*, 3rd ed., C. D. Klaassen, M. O. Amdur, and J. Doull, Eds. Macmillan Publishing New York, p. 15.

Labrune, P., B. Bader, P. Lebras, C. Wood, D. Devictor and G. Huault (1985). Arch. Fr. Pediat., 42, 863–865.

Langolf, G. D., R. Chaffin, and H. P. Whittle (1978). Am. Indust. Hyg. Assoc. J., 39, 976–984.

Lindström, K. (1980). Am. J. Indust. Med., 1, 69–84.

Lotti, M., A. Moretto, R. Zoppellari, R. Dainese, N. Rizzuto, and G. Barusco (1986). Arch. Toxicol., 59, 176–179.

Lotti, M., A. Moretto, E. Capodicasa, M. Bertolazzi, M. Peraica, and M. L. Scapellato (1993). Toxicol. Appl. Pharmacol., 122, 165–171.

Mackay, C. J., L. Campbell, A. M. Samuel, K. J. Alderman, C. Idzikowski, H. K. Wilson, and D. Gompertz (1987). Am. J. Indust. Med., 11, 223–239.

Manzo, L. (1980). Advances in Neurotoxicology, Pergamon Press, Oxford.

Maroni, M., C. Bulgheroni, M. G. Cassito, F. Merluzzi, R. Gilioli, and V. Foà (1977). Scand. J. Work Environ. Hlth., 3, 16–22.

Mattle, H. (1985). Schweiz. Med. Wschr., 115, 882–889.

Maurissen, J. P. J. and J. L. Mattsson (1989). Toxicol. Ind. Hlth., 5, 195–202.

Maurissen, J. P. J., B. Weiss, and H. T. Davis (1983). Toxicol. Appl. Pharmacol., 71, 266–279.

Mediani C. R. (1987). South. Med. J., 80, 421–425.

Mitchell, C. L. (1982). Nervous System Toxicology, Raven Press, New York.

Moser, V. C. (1989). J. Am. Coll. Toxicol., 8, 85–93.

Muller, K. E., C. N. Barton, and V. A. Benignus (1984). Neurotoxicol., 5(2), 113–125.

O'Donoghue, J. L. (1985a). Neurotoxicity of Industrial and Commercial Chemicals, 2 vol., CRC Press, Boca Raton, FL.

O'Donoghue, J. L. (1985b). In Neurotoxicity of Industrial and Commercial Chemicals, J. L. O'Donoghue, Ed., Vol. 2, CRC Press, Boca Raton, FL, pp. 169–177.

O'Donoghue, J. L. (1989). J. Am. Coll. Toxicol., 8, 97–115.

Olson, B. A., F. Gamberale, and B. Grönqvist (1981). Int. Arch. Occup. Environ. Hlth., 48, 211–218.

Paracelsus (1567). Four Treatises of Theophrastus von Hohenheim called Paracelsus, On the Miners' Sickness and other Miners' Diseases, G. Rosen (Trans.), H. G. Sigerist, Ed. (1941), Johns Hopkins Press, Baltimore, p. 115.

Pryor, G. T., J. Dickinson, R. Howd, and C. S. Rebert (1983). Neurobehav. Toxicol. Teratol., 5, 53–57.

Rafales, L. S. (1986). In Neurobehavioral Toxicology, Z. Annau, Ed., Johns Hopkins University Press, Baltimore, pp. 54–68.

Rebert, C. S., S. S. Sorenson, R. A. Howd, and G. T. Pryor (1983). Neurobehav. Toxicol. Teratol., 5, 59–62.

Reiter, L. W. and R. C. MacPhail (1979). Neurobehav. Toxicol., 1, (suppl. 1), 53–66.

Reynolds, G. S. (1968). A Primer of Operant Conditioning, Scott, Foresman, Glenview, IL.

Rizzuto, N., H. Terzian, and S. Galaiazzo-Rizzuto (1977). J. Neurol. Sci., 31, 343–354.

Roizin, L., H. Shiraki, and N. Grcevic (1977). Neurotoxicology, Raven Press, New York.

Ross, J. F. and G. T. Lawhorn (1990). Neurotoxicol. Teratol., 12, 153–159.

Ruffin, J. B. (1963). J. Occup. Med., 5, 117–121.

Scharf, B. (1975). Experimental Sensory Psychology, Scott, Foresman, Glenview, IL.

Schaumburg, H. H. and P. S. Spencer (1976). *Brain,* **99,** 183–192.

Schaumburg, H. H. and P. S. Spencer (1979). *Ann. N.Y. Acad. Sci.,* **329,** 14–29.

Seppäläinen, A. M. and M. Haltia (1980). In *Experimental and Clinical Neurotoxicology,* P. S. Spencer and H. H. Schaumburg, Eds., Williams & Wilkins, Baltimore, pp. 356–371.

Seppäläinen, A. M. and I. Linnoila (1976). *Neuropathol. Appl. Neurobiol.,* **2,** 209–216.

Seppäläinen, A. M., K. Lindström, and T. Martelin (1980). *Am. J. Indust. Med.,* **1,** 31–42.

Shimozaki, H. (1984). *Neurosci. Res.,* **2,** 63–76.

Smith, H. V. and J. M. K. Spalding (1959). *Lancet,* **2,** 1019–1021.

Spencer, P. S. and H. H. Schaumburg (1980). *Experimental and Clinical Neurotoxicology,* Williams & Wilkins, Baltimore.

Spencer, P. S. and H. H. Schaumburg (1985). *Scand. J. Work Environ., Hlth.,* **11,** Suppl. 1, 53–60.

Spencer, P. S., H. H. Schaumburg, M. I. Sabri, and B. Veronesi (1980). *Crit. Rev. Toxicol.,* **7,** 279–356.

Squibb, R. E., H. A. Tilson, and C. L. Mitchell (1983). *Neurobehav. Toxicol. Teratol.,* **5**(3), 331–335.

Stebbins, W. C. (1970). *Animal Psychophysics: The Designs and Conduct of Sensory Experiments,* Appleton-Century-Crofts, New York.

Sullivan, M. J., K. E. Rarey, and R. B. Conolly (1989). *Neurotoxicol. Teratol.,* **10,** 525–530.

Target Organ Toxicity: Eye, Ear and Other Special Senses (1982). Environ. Health Perspect., 44, 1–127.

Toxic Substances Control Act, Public Law 94-469, October 11, 1976.

U.S. Environmental Protection Agency (1985). *Toxic Substances Control Act Test Guidelines, Health Effects Testing Guidelines, Subpart G, Neurotoxicity,* Federal Register, Vol. 50, No. 188, pp. 39458–39470.

U.S. Environmental Protection Agency (1986). *Revision of TSCA Test Guidelines, Federal Register,* Vol. 51, No. 9, p. 1542.

U.S. Environmental Protection Agency (1988). *Diethylene Glycol Butyl Ether and Diethylene Glycol Butyl Ether Acetate; Test Standards and Requirements, Federal Register,* Vol. 53, No. 38, pp. 5932–5953.

U.S. Environmental Protection Agency (1991). *Pesticide Assessment Guidelines, Subdivision F, Hazard Evaluation: Human and Domestic Animals, Addendum 10, Neurotoxicity, Series 81, 82, and 83,* Health Effects Division, Office of Pesticides Programs, Publication PB 91-154617.

U.S. Environmental Protection Agency (1993). *Draft Report: Principles of Neurotoxicity Risk Assessment, Federal Register,* Vol. 58, No. 149, pp. 41556–41599.

Valciukas, J. A. (1991). *Foundations of Environmental and Occupational Neurotoxicology,* Van Nostrand Reinhold, New York.

van Vliet, C., G. M. H. Swaen, A. Volovics, M. Tweehuysen, J. M. M. Meijers, T. de Boorder, and F. Sturmans (1990). *Int. Arch. Occup. Environ. Hlth.,* **62,** 127–132.

Verschoyle, R. D. and W. N. Aldridge (1980). *Arch. Toxicol.,* **45,** 325–329.

Weiss, B. and V. G. Laties (1975). *Behavioral Toxicology,* Plenum Press, New York and London.

Wood, R. W., A. B. Weiss, and B. Weiss (1973). *Arch. Environm. Hlth.,* **26,** 249–252.

World Health Organization (1986). *Principles and Methods for the Assessment of Neurotoxicity Associated with Exposure to Chemicals*, Environmental Health Criteria 60, Office of Publications, World Health Organization, Geneva.

Xintaras, C., B. L. Johnson, and I. de Groot (1974). *Behavioral Toxicology, Early Detection of Occupational Hazards*, U.S. Department of Health, Education, and Welfare, Public Health Service, Center for Disease Control, National Institute for Occupational Safety and Health, HEW Publication (NIOSH) 74–126, Washington, D.C.

Zbinden, G., V. Cuomo, G. Racagni, and B. Weiss (1983). *Application of Behavioral Pharmacology in Toxicology*, Raven Press, New York.

CHAPTER SEVEN

Carcinogenesis

James A. Swenberg, D.V.M., Ph.D.

1 INTRODUCTION

The field of chemical carcinogenesis has been studied for over two centuries in humans, experimental animals, and cell culture. It has gone from being poorly understood with respect to cause-and-effect relationships to a relatively well-understood complex disease process that is comprised of multiple stages and events. This chapter will review important features of the biologic events associated with the induction of cancer in humans and experimental animals. Many of these events are similar across species, although clear examples exist demonstrating species specificity (Swenberg et al., 1992). Major areas of concern exist, however, for quantitative extrapolation of risk for human cancer from high-dose animal studies. Promising new approaches should make it possible to begin incorporating scientific data into this process, rather than relying on mathematical extrapolation alone. For some chemicals, biomarkers of exposure and/or effect are available (Perera and Weinstein, 1982). These may make it possible to quantify the amount of chemical exposure that is actually absorbed and metabolized in order to more clearly demonstrate cause-and-effect relationships and interindividual differences in biotransformation and response.

Historically, the induction of cancer by chemicals was first recognized by astute clinicians who observed associations between occupational or life-style exposures and the occurrence of specific types of cancer. As early as the eighteenth century, Sir Percival Pott recognized a high incidence of scrotal cancer in chimney sweeps, and Hill noted the relationship between nasal cancer and snuff dipping (reviewed in

Patty's Industrial Hygiene and Toxicology, Third Edition, Volume 3, Part B, Edited by Lewis J. Cralley, Lester V. Cralley, and James S. Bus.
ISBN 0-471-53065-4 © 1995 John Wiley & Sons, Inc.

Ruddon, 1987). It was not until more than 150 years later that the carcinogenic potential of chemicals was demonstrated in studies using experimental animals. Today, more than 50 chemicals, mixtures, or processes have been identified as causing cancer in humans (IARC, 1987). In addition, hundreds of chemicals have been shown to cause cancer in experimental animals, including nearly all of the known human carcinogens.

Cancer is caused by both endogenous and exogenous agents and events. It is well established that environmental chemicals, diet, radiation, viruses, and endogenous metabolic processes of our cells can lead to the induction of neoplasia. Many of these events can be controlled to reduce the likelihood of inducing cancer. Clearly, the greatest cause of human cancer is smoking, a totally preventable cause.

Most chemicals known to be human carcinogens cause genetic alterations of one type or another (Shelby, 1988; Shelby and Zeiger, 1990; Ashby and Tennant, 1991). The critical role of genetic alteration in carcinogenesis was first suggested by Boveri in 1914 (Boveri, 1914). Recent advances in molecular genetics have greatly enhanced our knowledge of the critical role of genetic alterations in human cancer. The fact that nearly all known human carcinogens were also carcinogenic in laboratory animals and that genetic alterations were involved in the process led to the paradigm that all chemicals that cause cancer in laboratory animals were likely to cause cancer in humans and that the risk would be proportional to dose (reviewed in Hoel et al., 1985).

Since large numbers of chemicals had not been adequately evaluated for carcinogenic potential, many chemicals underwent carcinogenicity testing during the 1970–1990s. Several hundred chemicals were shown to increase the incidence of benign or malignant tumors in rodents exposed to maximum tolerated doses (MTD) (Huff and Haseman, 1991). Some of these agents were carcinogenic at multiple sites, in both sexes, and in multiple species, while others induced neoplasia at a single site, in one species, and one sex (Ashby and Tennant, 1991). Furthermore, whereas nearly all of the known human carcinogens were genotoxic, nearly half of the rodent carcinogens were not. The dose–response for tumor induction in rodents was highly variable. Some chemicals were capable of inducing cancer using doses that varied over 1000-fold, while other chemicals increased the incidence of tumors at the MTD but not even at one-half that dose (Cohen and Ellwein, 1991; Haseman and Seilkop, 1992). This great diversity in carcinogenic potency and genotoxicity has called into question the blanket assumptions of the previous paradigm (Cohen and Ellwein, 1991).

Industrial hygienists and individuals with associated interests need to have a working knowledge of the principles involved in carcinogenesis to understand current regulations and future changes that are likely to occur. At the present time, most regulations are based on human evidence of carcinogenesis or on extrapolation of the results of animal carcinogenicity bioassays. As is reviewed in much greater depth elsewhere in this book, most regulations setting exposure levels are based on mathematical extrapolation of high-dose animal data. As our knowledge of mechanisms and dose–response relationships improves, however, it is likely that scientific, rather than straight mathematical, extrapolation of risk will assume a greater and greater role. A recent Working Group of the International Agency for Research on Cancer (IARC, 1991) concluded that knowledge of mechanism could either increase or decrease the predicted risk associated with chemical exposure. Recent decisions by regulatory agencies have also begun to incorporate mechanistic data into the estimation of risk and resulting regulation (U.S. EPA, 1991). For this reason this chapter will review the basic principles of carcinogenesis, what is known about important mechanisms that are likely to influence the extrapolation of risk, and future directions of research that might be applicable to industrial hygiene and biomonitoring of exposed individuals.

2 THE NOMENCLATURE OF NEOPLASIA

In discussing carcinogenesis, it is important to understand the terminology used to describe different types of cancer and their associated biology. Most tumor types are described by their cell of origin and their behavior (Table 7.1). For example, ma-

Table 7.1 Classification and Nomenclature of Neoplasia

Tissue	Benign Neoplasm	Malignant Neoplasm
Epithelium		
Squamous	Squamous cell papilloma	Squamous cell carcinoma
Transitional	Transitional cell papilloma	Transitional cell carcinoma
Glandular	Adenoma	Adenocarcinoma
		Carcinoma
Connective tissue		
Adult fibrous	Fibroma	Fibrosarcoma
Embryonic fibrous	Myxoma	Myxosarcoma
Cartilage	Chondroma	Chondrosarcoma
Bone	Osteoma	Osteosarcoma
Fat	Lipoma	Liposarcoma
Muscle		
Smooth muscle	Leiomyoma	Leiomyosarcoma
Skeletal muscle	Rhabdomyoma	Rhabdomyosarcoma
Endothelium		
Lymph vessels	Lymphangioma	Lymphangiosarcoma
Blood vessels	Hemangioma	Hemangiosarcoma
Lymphoreticular		
Lymph nodes	Not recognized	Lymphosarcoma
Hematopoietic		
Bone marrow	Not recognized	Leukemia
Nervous system		
Nerve sheath	Neurilemmoma	Neurogenic sarcoma
Glial cells	Glioma (may give cell type)	Malignant glioma
		Glioblastoma
Embryonic cells	Not recognized	Neuroblastoma

lignant tumors of epithelial cells are called carcinomas, while malignant neoplasms of mesenchymal tissues are called sarcomas. Cancer refers to malignant neoplasms. Benign neoplasms of epithelial tissue are referred to as adenomas or papillomas, while benign mesenchymal lesions are referred to as fibromas, osteomas, gliomas, and so forth. Benign tumors are not cancer, but many benign neoplasms can undergo additional change to become malignant. A malignant neoplasm is more aggressive than a benign lesion, causing local destruction and invasion into adjacent tissues. Malignant tumors can also undergo progressive changes, including the ability to invade blood vessels and lymphatics and to spread to distant sites, a process called metastasis.

Many of the regulations regarding carcinogens require that malignant tumors be induced. Other regulations do not distinguish between the induction of benign and malignant tumors, since benign neoplasms can progress to become malignant. Thus, it is important for the reader to understand the distinction between benign and malignant neoplasms.

3 THE MULTISTAGE CONCEPT OF CARCINOGENESIS

It has long been evident from the development of clinical neoplasia itself that human carcinogenesis involves multiple stages. Clinical and experimental studies have greatly expanded our understanding of the biochemical, genetic, and cellular events involved, from first initiating this complex process to gaining the ability to invade tissues and spread to distant sites (reviewed in Harris, 1991). The simplest multistage model involves two steps called initiation and promotion. Early research supporting the two-stage model was primarily conducted using mouse skin tumors (Berenblum and Shubik, 1947). Since then, two-stage models of carcinogenesis have been demonstrated in many tissues including liver, lung, bladder, thyroid, pancreas, kidney, intestine, stomach, and mammary gland. With the advent of modern molecular biology, this simple model has been extended to incorporate multiple events in experimental animals and humans. For example, eight different events have been identified in human colon carcinogenesis to date, with the possibility that more exist (Vogelstein et al., 1988). It is unlikely that all of these events are required or that an exact sequence must be followed. A general concept that is gaining support is that genomic instability increases with increasing neoplastic progression. This leads to greater amounts and diversity of genetic change in advancing lesions. Recognizing then that carcinogenesis is a multistage process, it is still useful to describe the type of changes that are involved using the simplest models.

3.1 Initiation

Initiation is usually described as a heritable change in a cell that is the result of exposure to an initiating agent. As such, initiation is an irreversible event for the life of that cell and its progeny. Initiating agents or their metabolites can damage the DNA of a cell. When such DNA damage is present during DNA synthesis, it increases the probability of a mutation being induced (Figure 7.1). Such mutations

SOURCES OF MUTATIONS

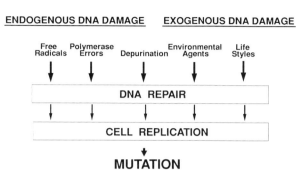

Figure 7.1 Schematic illustrating endogenous and exogenous factors involved in the induction of mutations. DNA damage and/or errors arise from many sources. Some of these are corrected by DNA repair. Those lesions that remain in the DNA at the time of DNA synthesis can result in the induction of mutations. (Modified from Loeb, 1989.)

can be in the form of point mutations, chromosomal aberrations, insertions, and deletions. Whereas the initial DNA damage may be repairable by one or more pathways, the mutational event represents a heritable change that will be passed on to subsequent generations of the mutated cell. Initiation is usually considered an additive process that may exhibit linear or nonlinear dose–responses, but that does not have an absolute threshold.

For many years these mutational events were thought to occur randomly throughout the genome. It is now known that this is not the case. Mutations induced by chemical, physical, and microbial agents preferentially occur at specific locations within individual genes. When these mutations occur at specific sites in genes involved in cell growth and differentiation, that cell can develop a selective growth advantage over its neighbors. Important classes of such genes include oncogenes and tumor suppressor genes (Harris, 1991).

Genotoxic chemical carcinogens are usually potent initiating agents. The amount of initiation induced by these agents can be related to the molecular dose of each type of DNA adduct present at the time DNA synthesis takes place and that adduct's efficiency for causing mutational events (Swenberg et al., 1990; Singer and Essigmann, 1991). In addition to this direct chemically induced DNA damage, spontaneous or background DNA damage is also present in all cells (Loeb, 1989; U.S. NRC, 1989; Harris, 1991). Sources of endogenous DNA damage include depurination, deamination of 5-methylcytosine, and oxy radicals. Estimated rates of spontaneous DNA damage range from 50,000–250,000 events per cell per day (Loeb, 1989; U.S. NRC, 1989). Most of this damage is rapidly repaired and of no genetic consequence. However, if cell replication takes place prior to repair of these lesions, some of them will be converted into mutations (Fig. 7.1). In addition, DNA polymerases have a low rate of error associated with transcription that can result in the insertion of a wrong base. Thus, spontaneously initiated cells are always present.

3.2 Promotion

The seminal feature of tumor promotion is clonal expansion of initiated cell populations (Fig. 7.2). Clonal expansion can occur through increased proliferation or survival of initiated cells relative to normal cells. By increasing the number of initiated cells, the probability of additional genetic alterations occurring is enhanced. Usually, promoters are nongenotoxic agents, that if administered by themselves are not carcinogenic. The temporal sequence of events is important in classifying an agent as a promoter (Fig. 7.3). Classically, if one administers a small dose of an initiator only, no tumors develop. If this same dose of initiator is followed by exposure to a promoting agent, an increased incidence of tumors is observed. Since initiation is an irreversible event, exposure to promoters can increase the incidence and/or multiplicity of tumors whether exposure to the promoter occurs shortly after initiation or months later. If a promoter is given before the initiator or if no initiator is administered, no tumors develop. If an agent causes an increase in tumors when it is given prior to the initiator, it is termed a co-carcinogen rather than a promoter. Nonetheless, promotion is considered to be a reversible, nonadditive process that usually exhibits measurable threshold doses and maximal response doses. Tumor promotion can result from enhanced cell proliferation that is secondary to cytotoxicity or hormonal imbalance (Dietrich and Swenberg, 1991).

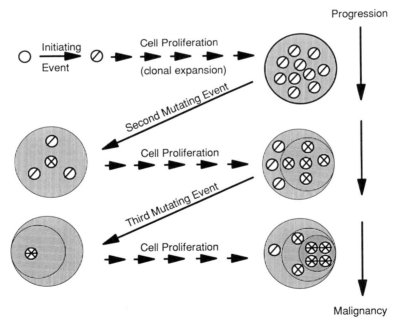

Figure 7.2 Multiple events are required for the progression from a single initiated cell to malignancy. This depicts the critical role of cell proliferation in the fixation of mutational events and in clonally expanding cell populations so that additional genetic changes are probable.

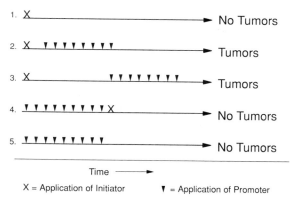

Figure 7.3 Initiation and promotion represent the simplest model of multistage carcinogenesis. The temporal sequence of events and dose are important in defining initiating and promoting agents.

As discussed earlier in this chapter, spontaneous initiation is an infrequent but ongoing process in experimental animals and humans. Thus, significant exposure to a promoting agent alone can result in increased numbers of tumors (Dietrich and Swenberg, 1991). Vigorous debates have been held over whether or not such increases define promoting agents as complete carcinogens (Ames and Gold, 1990; Cohen and Ellwein, 1990; Weinstein, 1991). Clearly, extrapolating risk from high to low doses for such agents is associated with much greater uncertainty than exists for chemicals that have initiating activities at low doses and are complete carcinogens at higher doses.

3.3 Progression

Tumor progression is the process whereby a benign neoplasm becomes malignant, and malignancies become more invasive and eventually metastatic. Increasing evidence suggests that these changes are associated with additional genetic alteration (Fig. 7.2). The clonal origin of tumors has been recognized for many years (Nowell, 1976). With advancing malignancy, tumors develop subpopulations that are genetically heterogeneous. The origin of these genetic changes may be related to further exposure to exogenous or endogenous genotoxic agents, coupled with greater genomic instability in neoplastic cells. Genomic instability may arise from the loss of critical regulator proteins that prevent cells with damaged DNA from entering DNA synthesis (Kastan et al., 1991), the induction of a mutator phenotype (Loeb, 1991), or decreased fidelity of DNA synthesis or repair. Hallmarks of this process are the presence of multiple mutations, chromosomal alterations, and gene amplification in advanced malignancies. The development of resistance to chemotherapy and radiation may be the result of similar alterations.

4 FACTORS AFFECTING DOSE–RESPONSE AND QUANTITATIVE EXTRAPOLATION

The complex nature of chemical carcinogenesis is further complicated by the need to extrapolate from high-dose studies in experimental animals to human exposure scenarios that are often orders of magnitude lower. Pathways and efficiency of biotransformation and the extent of chemically induced oxidative stress and cytotoxicity are factors that frequently differ between high and low exposure and species that are expected to alter the dose–response relationship for tumor induction. Some of the major issues that have been identified are discussed below, along with examples of well-studied carcinogens.

4.1 The Concept of Molecular Dose

The actual amount of a chemical carcinogen that reacts with the target site, such as DNA, is extremely small relative to the total exposure of the individual via the air, water, diet, and so forth. Species differences in biotransformation and DNA repair can have dramatic effects on the dose of a toxicant that reaches or remains at the target site (Swenberg et al., 1990). Likewise, interindividual differences in biotransformation and DNA repair are likely to be much greater in humans than in inbred laboratory rodents. The amount of the chemical that is absorbed, distribution of the chemical within the body, extent of metabolic activation and detoxication in different tissues and cell types, and DNA repair are all factors that affect the *molecular dose* that reaches the genetic material thought to be the target site or that can be measured as surrogate biomarkers (Fig. 7.4). It is well known that many of these processes are enzymatic and can be saturable and/or inducible. When a dose-dependent saturation occurs, a nonlinearity in the amount of a chemical that binds to and/or remains on the DNA as an adduct will exist. Such nonlinearities can either increase or decrease the effect per unit dose of the carcinogen. For example, when saturation of metabolic activation occurs, the amount of the electrophile capable of reacting with DNA to form DNA adducts decreases in proportion to the dose. This is referred to as a supralinear response and is illustrated in curve *b* of Fig. 7.5. In contrast, when detoxication becomes saturated, the amount of the electrophile available to form DNA adducts increases (curve *c*, Fig. 7.5). A similar increase in DNA adducts per unit dose occurs when DNA repair becomes saturated. In both cases it results in a sublinear dose–response curve. At low exposures, where none of these processes are saturated, these reactions proceed according to first-order kinetics and the molecular dose of DNA adducts will be linear with respect to dose (curve *a*, Fig. 7.5). Thus, compared to monitoring external exposure and assuming proportional dose-responses, molecular dosimetry offers the distinct advantage of being capable of integrating dose-dependent differences in absorption, distribution, biotransformation, and DNA repair so that carcinogen exposure can be determined more accurately over a wide range of doses.

While DNA adducts in target organs represent biomarkers that are likely to be causally related to carcinogenesis, it is usually difficult to obtain tissue samples of

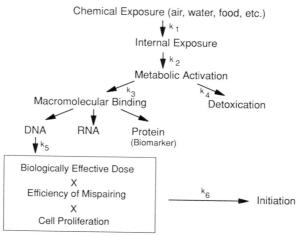

Figure 7.4 Numerous events and pathways are involved in determining the dose–response relationship for macromolecular adducts and initiation. The amount of a chemical that is absorbed (k_1), distributed to various tissues (k_2), metabolized to an electrophile (k_3), and detoxified (k_4) determine the extent of macromolecular binding to DNA and proteins. Such protein adducts may serve as biomarkers of exposure. DNA adducts may be reduced by DNA repair (k_5). Those adducts that persist to the time of scheduled DNA synthesis constitute the biologically effective dose of the carcinogen. Each type of DNA adduct has its own efficiency for causing mutations. When these mutations occur at critical sites in the genome (k_6), initiation results. (Modified from Swenberg et al., 1990.)

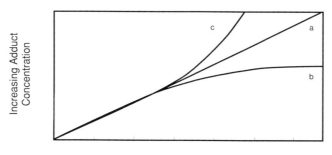

Increasing External Exposure

Figure 7.5 The shape of the dose–response curve will be linear (curve *a*) when none of the biotransformation or DNA repair pathways are saturated, such as occurs at low doses of carcinogens. A supralinear dose–response (curve *b*) results when metabolic activation becomes saturated, yielding a lower molecular dose per unit of exposure at high doses. In contrast, a sublinear dose–response (curve *c*) occurs if detoxication or DNA repair becomes saturated at high doses, yielding a higher molecular dose per unit of exposure. (Modified from Swenberg et al., 1990.)

the organ of concern. Alternative molecular dosimeters include white blood cell (WBC) DNA and protein adducts (Skipper and Tannenbaum, 1990). Such surrogate molecular dosimeters may or may not reflect what is happening at the target site. White blood cell DNA provides information on covalent binding to DNA of electrophilic agents and metabolites that circulate in the blood and, to a lesser extent, that are present in tissues that the WBCs pass through. They may or may not exhibit the same DNA repair activity as the target tissue. Repair of some DNA adducts is similar across tissue and cell types, while others are more/less active in different tissues (Belinsky et al., 1986). Knowledge of these differences is important whenever one is extrapolating from WBC molecular dosimetry.

The situation is even more complex when protein adducts are used. Obvious factors that must be considered include the stability of the adducts in protein versus DNA, the longevity of the protein, and whether or not the comparable DNA adduct is repaired (Walker et al., 1992a,b, 1993). The lack of repair of protein adducts may make these biomarkers ideal measures of exposure, since protein adducts accumulate with continued exposure. Because of differences in persistence, the relationships between DNA and protein adducts will change between single, continuous, and intermittent exposure scenarios (Fennell et al., 1992). Great care is needed in extrapolating protein data to DNA. Nevertheless, surrogate biomarkers offer the potential to more accurately predict risks associated with chemical exposures than do simple mathematical extrapolation.

Many examples of molecular dosimetry data are now available to illustrate the factors that drive the dose responses shown in Figure 7.5. Saturation of metabolic activation of procarcinogens has been well established for the known human carcinogen, vinyl chloride (reviewed in Purchase et al., 1987), and for 4-(Methylnitrosamino)-1-(3-pyridyl)-1-butanone, (NNK), a tobacco-specific nitrosamine (Belinsky et al., 1987). Figure 7.6 illustrates the dose–response in rat lung for O^6-methylguanine (O^6MG), the major promutagenic DNA adduct induced by NNK. One can readily appreciate the greater slope of the molecular dose–response at low versus high doses. This is the result of saturation at higher doses of a more efficient low K_m pathway for NNK activation. The net result is a decrease in the amount of O^6MG formed per unit dose at higher exposures. Assuming a linear dose–response for the induction of lung cancer in rats receiving high doses of NNK would clearly underestimate the risk of low exposures. In contrast, when O^6MG is used as the measure of exposure, the relationship between tumor induction and molecular dose is linear (Belinsky et al., 1990). The low K_m pathway for NNK metabolism can be further localized in the Clara cell population of the rat lung, where it is 38-fold more efficient at low versus high doses (Belinsky et al., 1987). In mice, both the Clara cells and the Type II cells contain this activation pathway (Belinsky et al., 1992).

Detoxication of reactive chemicals is a critical pathway for cell survival in that it reduces the amount of the toxic agent that can react with critical cellular macromolecules. As such, detoxication represents an important host defense that can become saturated. Inhalation exposure to formaldehyde causes the formation of DNA–protein crosslinks (Heck and Casanova, 1990). Oxidation of formaldehyde via formaldehyde dehydrogenase, a glutathione-dependent pathway for detoxication, becomes saturated at exposure concentrations above 4 ppm (Casanova and Heck, 1987).

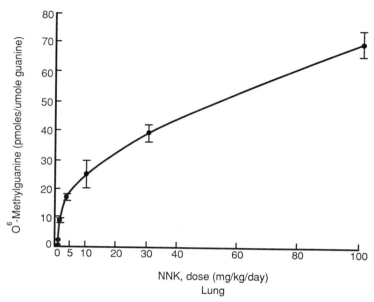

Figure 7.6 Saturation of the metabolic activation of NNK in rat lung results in a supralinear dose–response curve for the promutagenic DNA adduct, O^6-methylguanine. (Modified from Belinsky et al., 1987.)

This results in a sublinear exposure–response curve for the induction of DNA–protein crosslinks (Fig. 7.7) in both rats and monkeys (Heck et al., 1989). To further illustrate the importance of detoxication, Kensler et al. (1986) demonstrated that dietary administration of ethoxyquin, an agent that increases glutathione-S-transferase detoxication, reduced the molecular dose of aflatoxin that reached the DNA and reduced the induction of preneoplastic liver foci by approximately 90 percent.

DNA repair can dramatically alter the molecular dose of a carcinogen that resides on the DNA and the risk for inducing nonrepairable mutations. Different carcinogens induce DNA adducts of differing types and positions within a gene. Each of these adducts carries a different efficiency for the induction of mutations. Thus, some DNA adducts will induce base-pair substitutions one-fourth to one-half of the time when present on template DNA during replication, while others may only cause mutations at frequencies less than 1/1000 (Singer and Essigmann, 1991). For chemicals that induce more than one type of DNA adduct, it is important to have knowledge of the amount and promutagenic efficiency of each adduct formed. Since repair of individual types of DNA adducts utilizes different pathways, the relative molecular dose of these adducts can be very different from what is initially formed. One of the best examples of this is the amount of N-7-methylguanine (7MG) and O^6MG that remains in the DNA after single doses of dimethylnitrosamine (DMN) covering more than four orders of magnitude (Pegg and Hui, 1978). 7MG is formed chemically in the DNA at a ratio of 10:1, relative to O^6MG. O^6MG-DNA–methyltransferase is a specific DNA repair protein that removes the methyl group from O^6MG, but not 7MG, and can be saturated. The effect of saturation is shown in Figure 7.8. The

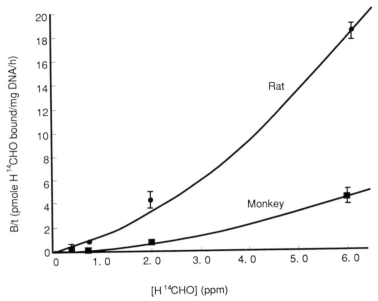

Figure 7.7 Saturation of detoxication results in a sublinear dose–response curve for DNA–protein crosslinks in the nasal mucosa of rats and monkeys exposed to formaldehyde vapor. (Modified from Heck et al., 1989.)

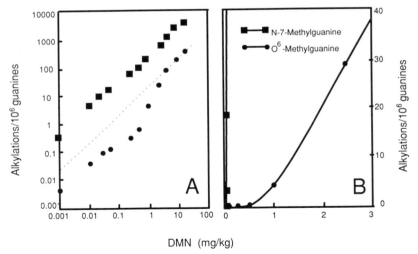

Figure 7.8 Saturation of the DNA repair pathway for (●) O^6-methylguanine results in a sublinear dose–response curve for that adduct, but not for (■) N-7-methylguanine in rats exposed to a single dose of dimethylnitrosamine (DMN). The dashed line in panel A represents the theoretical amount of O^6-methylguanine formed by DMN, whereas the filled circles are the measured amounts (log scale). Panel B illustrates the change in slope as a linear function. (Modified from Pegg and Hui, 1978.)

amount of 7MG remains linear over the entire dose range, while O^6MG increases at a much greater slope at doses above 0.5 mg/kg DMN.

Saturation of DNA repair is even more critical during chronic exposure to carcinogens. If a DNA adduct is not repaired (similar to protein adducts), it will accumulate linearly with increasing length of exposure. However, if repair occurs, the adduct will reach steady-state concentrations when the amount formed each day equals the amount that is removed. In the case of rapidly repaired DNA adducts, the amount present in a cell will be the same after one day or one year of exposure. For many DNA adducts, steady-state concentrations will be achieved after 2–4 weeks of exposure (Poirier et al., 1984; Boucheron et al., 1987).

The net effect of DNA repair is well illustrated by the molecular dosimetry of DNA adducts following 4 weeks exposure of rats to drinking water containing Diethylnitrosamine (DEN) (Fig. 7.9). Several promutagenic DNA adducts are formed following DEN exposure, with O^6-ethylguanine (O^6EG) initially being present in slightly greater amounts than O^2-ethylthymidine (O^2ET), which is present at approximately four-fold greater amounts than O^4-ethylthymidine (O^4ET). Daily ex-

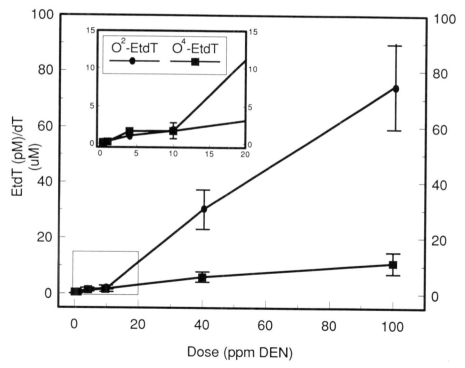

Figure 7.9 The molecular dose of O^4-ethylthymidine is linear in liver DNA from rats exposed for 4 weeks to drinking water containing 0.4–100 ppm diethylnitrosamine. In contrast, O^2-ethylthymidine exhibits a sublinear dose–response due to saturation of DNA repair. (From Nakamura and Swenberg, 1992.)

posures to drinking water containing 10 ppm DEN or less result in molar ratios where $O^2ET = O^4ET \gg O^6EG$, while higher exposures result in molar ratios where $O^2ET > O^4ET \gg O^6EG$ (Nakamura and Swenberg, 1992). Since each of these adducts causes different types of mutations, the change in their molecular dose is likely to alter the mutational spectrum in a dose-dependent manner. This in turn could influence initiation and progression of DEN-induced carcinogenesis.

4.2 Cell Proliferation

The role of cell proliferation in initiation and promotion was discussed earlier in this chapter but will be briefly reviewed in context to dose–response and quantitative extrapolation of carcinogenicity study results. When cell proliferation is increased, on average there will be less time available for DNA repair prior to DNA replication. DNA replication converts repairable DNA adducts into permanent alterations in the genetic code. This includes the induction of point mutations, chromosomal aberrations, deletions, insertions, and gene amplification. All of these events represent important mechanisms for the initiation and progression of carcinogenesis. Cell proliferation is also required for clonal expansion of initiated populations of cells, further enhancing the probability of additional genetic change and progression of malignancy (Fig. 7.2). Enhancement of critical effects brought about by increased cell proliferation that is secondary to high dose toxicity will not be present at low doses (Swenberg et al., 1992). It is therefore important that such high-dose effects be taken into account when extrapolating from high to low exposure scenarios.

The effect of saturation of detoxication on the molecular dose of formaldehyde was described earlier in this chapter. As was pointed out, the slope of the molecular dose increases between 2 and 6 ppm formaldehyde. It is important to note, however, that the molecular dose is linear between 6 and 15 ppm formaldehyde, yet this is the proportion of the concentration–exposure curve that exhibits the marked nonlinearity for the induction of squamous cell carcinomas of the nasal cavity (Fig. 7.10). A 2.5-fold increase in exposure and in molecular dose results in a 50-fold increase in nasal cancer. Recent data has shown that sustained increases in cell proliferation of the nasal epithelium exhibit the same change in slope as nasal cancer (Monticello et al., 1993). This provides strong scientific evidence that cell proliferation is the driving factor for the increase in nasal cancer that occurs in rats exposed to 10 and 15 ppm formaldehyde. Since this enhancing factor is not present under conditions of occupational or environmental exposure to formaldehyde, risk estimates for humans should take this important factor into account.

A similar example has been shown for 2-acetylaminofluorene (2-AAF). The molecular dose of 2-AAF DNA adducts has been determined in liver and urinary bladder and found to be linear following dietary exposures to 5–150 ppm (Beland et al., 1988). Liver tumors increased linearly with dose from 30 to 150 ppm 2-AAF, but bladder tumors exhibited a strongly sublinear dose–response that increased in slope at exposures of 60 ppm and greater (Fig. 7.11). This is the same portion of the dose–response curve that epithelial hyperplasia, an indicator of cell proliferation, increases (Cohen and Ellwein, 1990).

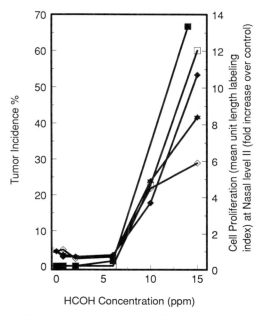

Figure 7.10 A sharp sublinear dose–response is evident for nasal cancer in rats exposed to formaldehyde vapor in (■) the original bioassay and in (□) the second pathogenesis bioassay. The increase in slope occurs at exposure concentrations above 6 ppm formaldehyde, even though the molecular dose is linear between 6 and 15 ppm (Fig. 7.7). The increase in slope for nasal cancer is nearly identical to the increase in slope for sustained cell proliferation at (★) 6, (♦) 12, and (◇) 18 months of exposure. (Drawn from data in Monticello et al., 1993.)

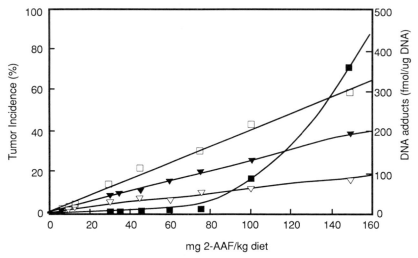

Figure 7.11 Dose–response relationship for DNA adducts in the (▽) liver and (□) bladder and neoplasms of the (▼) liver and (■) bladder for mice exposed to 2-acetylaminofluorine. (Modified from Beland et al., 1988.)

5   SPECIES-SPECIFIC MECHANISMS OF CARCINOGENESIS THAT AFFECT QUALITATIVE EXTRAPOLATION

Several examples of species-specific carcinogens are known for which qualitative differences between the animal model and humans prescribe that it would be inappropriate to extrapolate from animal to human risk. One of the best studied examples is the male-rat-specific α_{2u}-globulin nephropathy and carcinogenesis that is induced by such chemicals as d-limonene (U.S. EPA, 1991). α_{2u}-Globulin is the major urinary protein of male rats. It selectively binds certain chemicals or their metabolites, is freely filtered by the glomerulus, and resorbed by the proximal tubule of the nephron, where it accumulates in cellular lysosomes. The lysosomal digestion of the chemical–protein complex is decreased compared to α_{2u}-globulin alone, leading to a marked accumulation of the complex and subsequent cytotoxicity. In response to cell death, cell replication is specifically enhanced, which in turn promotes spontaneously initiated cells of the renal tubules to neoplasms (Dietrich and Swenberg, 1991; Swenberg et al., 1992). Since humans do not synthesize α_{2u}-globulin and related human proteins do not bind the same metabolites, humans are not at risk from chemicals that cause cancer by this mechanism.

Another well-characterized example of species-specific susceptibility for carcinogenesis is thyroid-stimulating-hormone-mediated induction of thyroid tumors (Hill et al., 1989). The normal means of maintaining hormonal homeostasis involves a feedback system between the hypothalamus-pituitary gland and the thyroid. When circulating thyroxin levels become low, the pituitary releases thyroid-stimulating hormone (TSH), which in turn stimulates hypertrophy and hyperplasia of the thyroid follicular cells in order to synthesize more thyroxin. Exposure to some chemicals interferes with the production or metabolism of thyroxin, leading to sustained increases in TSH release and subsequent cell proliferation. This increase in cell proliferation eventually leads to the induction of thyroid neoplasia. The most obvious species difference between rodents and primates is the lack of thyroid binding globulin (TBG) in the rodent and some other species (Dohler et al., 1979). The half-life of the thyroxine is 12 hr in the rat versus 5–9 days in humans, and serum TSH is 25 times higher in the rodent as compared to man. These differences make rodents much more susceptible to chemicals that produce thyroid neoplasia secondary to simple hypothyroidism. Such agents can be considered species-specific in that simple hypothyroidism in humans is not a known etiologic factor for human thyroid cancer.

Species specificity has also been demonstrated for several nongenotoxic bladder carcinogens, including saccharin, uracil, and melamine (Cohen and Ellwein, 1991). In these cases, urinary calculi or crystalluria lead to increased cell proliferation and hyperplasia in the urinary bladder of rats. The specific factors involved in the formation of calculi/crystalluria must be taken into account when extrapolating risk to humans.

6   POTENTIAL OF MOLECULAR EPIDEMIOLOGY TO PROVIDE HUMAN DATA ON CARCINOGEN EXPOSURE AND RISK

Molecular epidemiology is a new field of investigation that utilizes a variety of techniques to identify the presence of early markers of neoplastic disease, susceptibility factors that predispose individuals to the induction of cancer, as well as to detect and quantify human exposure to chemical agents (Hulka, 1991). Determining the molecular dose present in target or surrogate tissues or proteins of individuals with occupational or environmental exposure should provide a much better understanding of interindividual differences in susceptibility to cancer induction. When exposures are known and similar, differences in carcinogen biotransformation, DNA repair, and genetics may be detectable. When exposures are not known, molecular epidemiology may be able to quantify exposure using DNA or protein adducts. Mutational spectra for individual chemicals may have characteristic "hotspots" that permit identification of causal chemical exposures. These studies may also demonstrate that widespread exposure to agents inducing identical adducts or mutations occurs through diet and so forth and that some occupational exposures contribute little additional risk. While it is difficult to predict the impact of molecular epidemiology, it is highly likely that it will greatly improve our understanding of human risk to chemical carcinogens.

REFERENCES

Ames, B. N. and L. S. Gold (1990). *Proc. Natl. Acad. Sci., USA*, **87**, 7772–7776.

Ashby, J. and R. W. Tennant (1991). *Mutat. Res.*, **257**, 209–227.

Beland, F. A., N. F. Fullerton, T. Kinouchi, and M. C. Poirier (1988). In *Methods for Detecting DNA Damaging Agents in Humans: Applications in Cancer Epidemiology and Prevention,* H. Bartsch, K. Hemminki, and I. K. O'Neill, Eds., IARC Scientific Publications No. 89, pp. 175–180.

Belinsky, S. A., C. M. White, J. A. Boucheron, F. C. Richardson, J. A. Swenberg, and M. W. Anderson (1986). *Can. Res.*, **46**, 1280–1284.

Belinsky, S. A., C. M. White, T. R. Devereux, J. A. Swenberg, and M. W. Anderson (1987). *Can. Res.*, **47**, 1143–1148.

Belinsky, S. A., J. F. Foley, C. M. White, M. W. Anderson, and R. R. Maronpot (1990). *Can. Res.*, **50**, 3772–3780.

Belinsky, S. A., T. R. Devereux, J. F. Foley, R. R. Maronpot, and M. W. Anderson (1992). *Can. Res.*, **52**, 3164–3173.

Berenblum, I. and P. Shubik (1947). *Br. J. Cancer*, **1**, 384–391.

Boucheron, J. A., F. C. Richardson, P. H. Morgan, and J. A. Swenberg (1987). *Can. Res.*, **47**, 1577–1581.

Bovari, T. (1914). *Zur Frage der Entstehung Maligner Tumoren*, Vol. 1, Gustave Fischer Verlag, Jena.

Casanova, M. and H.d.'A. Heck (1987). *Toxicol. Appl. Pharmacol.*, **89**, 105–121.

Cohen, S. M., and L. B. Ellwein (1990). *Science*, **249**, 1007–1011.

Cohen, S. M. and L. B. Ellwein (1991), *Can. Res.*, **51**, 6493–6505.

Dietrich, D. R. and J. A. Swenberg (1991). *Can. Res.*, **51**, 3512–3521.

Dohler, K. D., C. C. Wong, and A. von zur Muhlen (1979). *Pharmacol. Ther.*, **5**, 305–318.

Fennell, T. R., S. C. J. Sumner, and V. E. Walker (1992). *Can. Epidemiol., Biomarkers Prev.*, **1**, 213–219.

Harris, C. C. (1991). *Can. Res.*, **51s**, 5023s–5044s.

Haseman, J. K. and S. K. Seilkop (1992). *Fundam. Appl. Toxicol.*, **19**, 207–213.

Heck, H.d'A, M. Casanova, W. H. Steinhagen, J. I. Everitt, E. T. Morgan, and J. A. Popp (1989). In *Nasal Carcinogenesis in Rodents: Relevance to Human Health Risk*, V. J. Feron, and M. C. Bosland, Eds., Pudac Wageningen, The Netherlands, pp. 159–164.

Heck, H.d.'A, M. Casanova, and T. B. Starr (1990). *CRC Crit. Rev. Toxicol.*, **20**, 397–426.

Hill, R. N., L. S. Erdreich, O. E. Paynter, P. A. Roberts, S. L. Rosenthal, and C. F. Wilkinson (1989). *Fundam. Appl. Toxicol.*, **12**, 629–697.

Hoel, D. G., R. A. Merrill, and F. P. Perera, Eds. (1985). In *Banbury Report*, Vol. 19, Cold Spring Harbor Laboratory, Cold Spring Harbor, New York.

Huff, J. and J. Haseman (1991). *Environ. Health Perspect.*, **96**, 23–31.

Hulka, B. S. (1991). *Can. Epidemiol., Biomarkers, Prev.*, **1**, 13–19.

IARC (1987). *IARC Monographs on the Evaluation of Carcinogenic Risks to Humans, Suppl. 7*, Lyon, France.

IARC (1991). *Mechanisms of Carcinogenesis in Risk Identification*, 91/002, Lyon, France.

Kastan, M. B., O. Onyekwere, D. Sidransky, B. Vogelstein, and R. W. Craig (1991). *Can. Res.*, **51**, 6304–6311.

Kensler, T. W., P. A. Egner, N. E. Davidson, B. D. Roebuck, B. D., A. Pikul, and J. D. Groopman (1986). *Can. Res.*, **46**, 3924–3931.

Loeb, L. A. (1989). *Can. Res.*, **49**, 5489–5496.

Loeb, L. A. (1991). *Can. Res.*, **51**, 3075–3079.

Monticello, T. M., E. A. Gross, and K. T. Morgan (1993). *Environ. Health Perspect.*, **101 supplement 5**, pp. 121s–124s.

Nakamura, J. and J. A. Swenberg (1992). In *Proceedings of the Eighty-Third Annual Meeting of the American Association for Cancer Research*, **33**, 886.

Nowell, P. C. (1976). *Science*, **194**, 23–28.

Pegg, A. E. and G. Hui (1978). *Biochem. J.*, **173**, 739–748.

Perera, F. P. and I. B. Weinstein (1982). *J. Chronic Dis.*, **35**, 581–600.

Poirier, M. C., J. M. Hunt, B. True, B. A. Laishes, J. F. Young, and F. A. Beland (1984). *Carcinogenesis*, **5**, 1591–1596.

Purchase, I. F. H., J. Stafford, and G. M. Paddle (1987). *Food Chem. Toxicol.*, **25**, 187–202.

Ruddon, R. W., Ed. (1987). *Cancer Biology*, 2nd ed., Oxford University Press, New York.

Shelby, M. D. (1988). *Mutat. Res.*, **204**, 3–15.

Shelby, M. D. and E. Zeiger (1990). *Mutat. Res.*, **234**, 257–261.

Singer, B. and J. M. Essigmann (1991). *Carcinogenesis*, **12**, 949–955.

Skipper, P. L. and S. R. Tannenbaum (1990). *Carcinogenesis*, **11**, 507–518.

Swenberg, J. A., N. Fedtke, T. R. Fennell, and V. E. Walker (1990). In *Progress in Predictive Toxicology*, D. B. Clayson, I. C. Mucro, P. Shubik, and J. A. Swenberg, Eds., Elsevier Science Publishers, Biomedical Division, Amsterdam, pp. 161–184.

Swenberg, J. A., D. R. Dietrich, R. M. McClain, and S. M. Cohen (1992). In *Mechanisms of Carcinogenesis in Risk Identification*, H. Vainio, P. N. Magee, D. B. McGregor, and A. J. McMichael, Eds., IARC, Lyon, 467–490.

U.S. Environmental Protection Agency (1991). *Alpha₂ᵤ-Globulin: Association with Chemically Induced Renal Toxicity and Neoplasia in the Male Rat*, EPA/625/3-91/019F.

U.S. National Research Council (1989). *Drinking Water and Health*, Vol. 9, National Academy Press, Washington D.C.

Vogelstein, B., E. R. Fearon, S. R. Hamilton, S. E. Kern, A. C. Preisinger, M. Leppert, Y. Nakamura, R. White, A. M. Smits, and J. L. Bos (1988). *N. Engl. J. Med.*, **319**, 525–532.

Walker, V. E., T. R. Fennell, P. B. Upton, J. P. MacNeela, and J. A. Swenberg (1993). *Environ. Health Perspect.*, 99, 11–17.

Walker, V. E., T. R. Fennell, P. B. Upton, T. R. Skopek, V. Prevost, D. E. G. Shuker, and J. A. Swenberg (1992a). *Can. Res.*, **52**, 4328–4334.

Walker, V. E., J. P. MacNeela, J. A. Swenberg, M. J. Turner, Jr. and T. R. Fennell (1992b). *Can. Res.*, **52**, 4320–4327.

Weinstein, I. B. (1991). *Can. Res.*, **51**, 5080s–5085s.

CHAPTER EIGHT

Cancer Risk Assessment

Colin N. Park, Ph.D., and Neil C. Hawkins, Ph.D.

1 INTRODUCTION

Risk assessment, in its broadest sense, is defined as the process of estimating the probability of an adverse impact on human health or the environment resulting from some activity (1). This broad definition covers everything from the potential for global warming, habitat destruction and loss of species diversity, the estimation of the probability of rail accidents, and the potential for human effects resulting from chronic chemical exposures. Each of these dimensions of risk assessment are important in themselves, and the relative priority attached to the broad areas is a key question that the U.S. Environmental Protection Agency (EPA) is currently dealing with in an attempt to optimally allocate public resources (2).

This chapter will be restricted to the question of estimating carcinogenic risk to humans resulting from exposures to specific compounds. In particular, the focus will be on the use of carcinogenic risk assessment as currently practiced by regulatory agencies. Developing trends and future directions will also be discussed. The *interpretation* of risk assessment results will also be discussed but not in detail. Risk communication and public perception of risk are important evolving topics that have been discussed by Slovic, Covello, and the Chemical Manufacturers Association (3–5).

In the regulatory context Arthur Hayes, past commissioner of the Food and Drug Administration (FDA), defined risk assessment as the determination of "safe" levels of exposure where safe is defined as: "A reasonable certainty of no significant risk, based upon adequate scientific data."

Patty's Industrial Hygiene and Toxicology, Third Edition, Volume 3, Part B, Edited by Lewis J. Cralley, Lester V. Cralley, and James S. Bus.
ISBN 0-471-53065-4 © 1995 John Wiley & Sons, Inc.

This definition neatly summarizes a number of contentious issues. The first issue is that there is never absolute certainty of a negative result. The absence of risk can never be absolutely proven. The best that can be done is to show beyond some reasonable doubt that no adverse outcome is expected. Similarly, the definition refers not to the absolute lack of risk but the existence of no *significant* risk. If some high level of exposure results in a particular risk, then lowering the exposure will decrease the risk, but it is not possible to define a specific exposure for which there is a guarantee of absolutely no risk for any person(s), unless the exposure approaches zero. However, some levels of exposure will result in risks that are zero for all practical purposes. Thus, the regulatory application of risk assessment methods attempts to define a practical equivalent of zero risk. The third point to be made in the above definition of safety is that the conclusions are based upon *adequate* scientific data. Never will it be possible to have done every possible experiment and to have collected every piece of data. There will always be some degree of incompleteness. However, decisions are made, and must be made, based on data sets that are adequate for the decision process.

There are two general approaches to health risk assessment. The first approach involves the use of safety factors (or uncertainty factors) applied to the no observed adverse effect level (NOAEL) in animals to derive a safe level for humans. This is the approach used by the Threshold Limit Values (TLV) Committee of the American Conference of Governmental Industrial Hygienists (ACGIH) to set occupational exposure limits and by the EPA to set regulatory limits for noncarcinogens. The other approach, which will be the focus of this chapter, attempts to model probability of tumor formation as a function of dose. This conceptual approach has been the subject of some controversy. Because of the large uncertainty in extrapolating empirical dose–response models orders of magnitude below the observed dose range, any empirical methodology will result in considerable uncertainty at low doses. In the face of this considerable uncertainty, regulatory agencies have adopted, by policy, very conservative default assumptions to ensure that risk is not being underestimated. This conservative default approach has also been controversial.

The current regulatory application of risk assessment relies on a set of default assumptions that, in theory, should be used only in the absence of adequate alternative scientific assumptions. In practice, however, the default methodology is used for almost all regulatory applications of risk assessment (6). The major issue in replacing default assumptions with alternative approaches is that the exact mechanisms for cancer etiology, including the shape of the dose–response functions in animals and humans, are rarely known. Thus, the question becomes that of defining *how much* scientific certainty is necessary before default assumptions are replaced. This, to a large extent, is the central question in the controversy surrounding most risk assessments.

Proposed and finalized risk assessments on specific compounds have, in some cases, utilized additional information beyond the default methodology, and it is expected that future data generation will be used in the evaluation of other compounds. Examples to date include the incorporation of pharmacokinetic information on methylene chloride (7) and perchloroethylene (8) and mechanistic information on formaldehyde and unleaded gasoline. There is currently an extensive review on the carcinogenic mechanisms of dioxin(s) and the possible incorporation of these data in the risk assessment of this family of compounds. There is also considerable interest in incorporating more mechanistic considerations into the *classification* step in risk assessment. Many believe that a more comprehensive classification procedure would aid in the clarification of the relevance of animal data to the potential for human carcinogenicity (9).

2 THE RISK ASSESSMENT PARADIGM

Discussion of the concept of quantitative dose–response modeling of animal data for estimating the probability of tumor development in humans goes back at least as far as 1961 papers by Mantel and Bryan (10); however, it was not until 1979 when the Interagency Regulatory Liaison Group (IRLG) issued its guidelines on cancer risk assessment (11) that the concept took hold. That 1979 methodology largely replaced the traditional regulatory approach of using safety factors to determine safe or allowable levels of exposure for presumed carcinogens.

Since the formal quantitative risk estimation methods were proposed by the IRLG in 1979, the methodology has met with resistance from a number of different quarters. Those who were comfortable with the traditional safety factor approach, for example, classical toxicologists, argued against the models on the grounds that the data were too simplistic for the relatively sophisticated mathematical methods being used, and the results then implied far more precision than was inherent in the raw data. They also argued strongly that the use of only the tumor frequencies takes all scientific judgment out of the process of safety assessment. The proponents of quantitative modeling have argued that the methodology provides more information than the uncertainty factor margin of safety approach, and that this information is useful in the regulatory process. They have also noted that the automatic use of a 100-fold safety factor, applied to the no observed effect level (NOEL), uses no more scientific judgment than the use of computer-generated "risk" results.

Today, the original IRLG methodology, with minor modifications, is being used by EPA and other agencies. Guidelines from EPA (12) and the Office of Science and Technology Policy (OSTP) (13) lay out the general principles to be used in quantitative risk assessment and endorse the use of mathematical models to estimate *upper bounds* on risk for regulatory use.

In 1983 the National Academy of Sciences (NAS) issued a report describing the risk assessment process (14). This report, commonly referred to as the "red book," describes the elements of a risk assessment, the interrelationship of the components, and the specific decision points in each of the components.

The risk assessment paradigm described in the NAS study can be briefly described as follows. The risk assessment model is a process involving a series of four steps beginning with (1) hazard identification and (2) dose–response assessment, followed by (3) exposure assessment. These first three components are integrated into (4) a risk characterization that is followed by a risk management decision. The results are then communicated to relevant audiences. In one sense the communications step is not part of the risk assessment process; in another sense it is the most important part. It is this part that ultimately either convinces the public that the system is working and that its health is being protected or leads to distrust of the message and the involved parties.

The NAS report heavily emphasizes the need for separation of risk assessment and risk management activities. This separation is currently under examination, however (15,16). The separation of risk assessment and risk management is appropriate, in one sense, by not allowing a desired management outcome to influence the methodology used in the scientific component. From another point of view, however, the complete separation of the scientific and management aspects of the risk assessment process has led to a problem. The problem is that, as will be discussed, there is considerable uncertainty in risk assessment methodology that is generally resolved by making policy decisions designed not to underestimate the risk. The degree of uncertainty and the degree of possible overestimation in the scientific phase of the assessment are often not communicated to, nor understood by, the risk manager. The result is that risk managers may believe that the estimates they have received are both precise and unbiased. If actions are based on this assumption, the result may be the misallocation of resources, time and money to trivial problems, whereas these same resources could be allocated to potentially more important societal problems.

Ruckelshaus, and more recently Reilly (2), as administrators of EPA, have discussed this resource allocation problem in some detail. Ruckelshaus points out that in his first tenure as director of the EPA, after its inception in 1970, recourse allocation was not a major issue. The big environmental problems were fairly obvious and the solutions were more clear-cut. Since then, resources spent on these big issues have yielded results, and programs are underway to continue the progress (17,18). When Ruckelshaus returned to the EPA in 1983, one of his major efforts was to establish methods of prioritizing by risk so that the EPA was focused on the most important problems. This effort resulted in the publication in 1987 of the ''Unfinished Business'' report (19) in which senior scientists prioritized by risk.

Risk assessment was a tool that Ruckelshaus believed would help determine these priorities by *quantifying* risk. While it is true that the tool has helped, it has also led to the potential to introduce bias into the decision-making process when uncertainties in the estimates are not explicitly considered. Finkel (20) has extensively discussed how uncertainty can influence decision making and how uncertainty should be considered in the process. He notes that risk managers should explicitly consider uncertainties and the *distributions* of the likely risk estimates, and that these uncertainties and distributions should be communicated to the public along with the effect that uncertainties have on specific decisions. Sielken (21) similarly advocates not just point estimates of risk but also the complete distribution of risk estimates so as to encourage more informed decision making.

3 HAZARD IDENTIFICATION

As described in the NAS paradigm of risk assessment, the first step in the process is to identify which compounds have the potential to cause cancer in humans. This is a qualitative decision reflecting the *presence* of a hazard. The existence of a hazard does not imply the existence of a risk, however, since risk implies exposure at a level sufficient for the hazard to be expressed. In theory, hazard identification should be a simple step; in practice it has not been so simple. The primary database for cancer hazard identification has been the National Toxicology Program (NTP), formerly the National Cancer Institute (NCI) testing program, augmented by industry and academic studies. The underlying concept for the original experimental protocol for the NCI bioassay was that the bioassay was designed to (qualitatively) identify the few compounds that caused cancer in animals and then apply stringent controls to minimize risk to humans from these compounds. The bioassay was then designed as a screening study; maximum doses with demonstrated toxicity and gavage dosing of sensitive animals. The intent was to provide a maximally sensitive screen to identify those compounds that could be a problem.

Unfortunately, the results did not turn out as planned. Since the early 1970s, over 400 compounds have been tested with almost 60 percent having some positive evidence of carcinogenicity (22). What went wrong? Some have argued (23) that the extremely high rate of positives is due to judicious selection of the most hazardous compounds to test. Most others (24), however, believe that the bioassay has not accomplished what was intended. It functions as a highly sensitive screen; it has excellent sensitivity, but unfortunately, it has poor selectivity, and therefore identifies compounds that probably do not pose a hazard to humans at expected exposure levels. It has been suggested that a primary confounder in the interpretation of these studies has been the use of maximum tolerated doses (MTD) that exceed the animals' natural detoxification mechanisms and leads to cancer through secondary mechanisms or pathways. The high doses used in the studies are designed to maximize the potential response in the animals and are operationally defined on the basis of resulting in demonstrated toxicity to the test animals. This dose selection criteria, therefore, maximizes statistical power for identifying potential carcinogens, but it also may be identifying compounds which do not have carcinogenic potential at lower doses, that is, false positives may be identified.

The bioassay was designed believing that direct acting primary carcinogens, following the somatic mutation theory of carcinogenesis, accounted for the majority of carcinogenic compounds, and that these compounds could be easily identified. It now appears that many compounds act entirely or primarily, through secondary mechanisms (25–27).

As a result of the large number of compounds testing positive, the focus has shifted to the question of dose–response. If a compound causes cancer in animals at high doses, what is the risk to humans at low doses? Unfortunately, as will be seen, the answer is not definitive.

There is an additional step, however, in going from the qualitative acceptance of the bioassay to the quantitative question of dose–response. That step deals with the *relevance* of the high-dose animal bioassay to low-dose human risk.

Different agencies attempt to deal with this question in different ways; EPA and the International Agency for Research on Cancer (IARC) have formal classification schemes, whereas FDA appears to use more of a case-by-case decision mechanism. Other agencies such as the Consumer Product Safety Commission (CPSC) seems to generally use the EPA and IARC criteria without explicitly adopting them.

The EPA classification scheme (12) is as follows:

Group A: Human Carcinogens

This group is used only when there is sufficient evidence from epidemiologic studies to support a causal association between exposure to the agents and cancer.

Group B: Probable Human Carcinogens

This group includes agents for which the weight of evidence of human carcinogenicity based on human studies is "limited" (B1) and also includes agents for which the weight of evidence of carcinogenicity based on animal studies is "sufficient" (B2).

Group C: Possible Human Carcinogens

This group is used for agents with limited evidence of carcinogenicity in animals and in the absence of human evidence for carcinogenicity. It includes a wide variety of evidence, including studies with marginal statistical significance, benign tumors only, and increased tumor incidence only in tissues with high spontaneous tumor rates.

Group D: Not Classifiable as to Human Carcinogenicity

This group is generally used for agents with inadequate human and animal evidence of carcinogenicity or for which no data are available.

Group E: Evidence of Noncarcinogenicity for Humans

This group is used for agents that show no evidence for carcinogenicity in at least two adequate animal tests.

Group A, "known human carcinogens," is a small and generally accepted list. It includes about 26 compounds and processes for which there is adequate epidemiological data to demonstrate an observed relationship between (qualitative) exposure and increased cancer incidence. Examples include benzene, bis-chloromethyl ether, asbestos (by inhalation), chromium VI, and vinyl chloride.

Group B1, "probable human carcinogens" is also a small list and is somewhat more controversial by nature of the definitions. B1 includes compounds for which there is limited evidence of carcinogenicity in humans. Usually these compounds have epidemiology studies in which the results are ambiguous. The data generally are not sufficiently precise to be labeled as positive, either because of weak statistical results or the possibility of extraneous confounders. Typically, the data do show some indication of a positive finding, however, which could be related to the particular chemical; thus, the results are ambiguous. Examples include acrylonitrile and ethylene oxide, as well as cadmium.

Groups B2, "probable human carcinogens," and C, "possible human carcinogens," are by far the largest groupings, containing over 200 compounds in each. The dividing line between these two categories and category D has been a source of discussion and controversy for some time. Note that the criteria generally, although not completely, refer to the *number* of positive animal studies that have been reported. The issue then revolves around the relevance of positive animal results, particularly when results are at least partially contradictory. What is missing in these simplistic classification schemes is a more complete analysis of the "interface" compounds. Recently, the U.S. EPA Scientific Advisory Board issued the opinion that a particular compound, tetrachloroethylene, was "on the continuum" between B2 and C (28). They recommended that EPA not classify tetrachloroethylene as a B2 carcinogen but instead devise a new classification scheme that more completely deals with the relevance of animal results.

Group E does not exist because of the stringency of the criteria and the difficulty in proving a negative.

IARC uses a similar classification, but there are a few exceptions. One of the exceptions is that IARC refers to category 2-B (approximately equivalent to B2 in the EPA scheme) as possible human carcinogens rather than probable human carcinogens. This may appear to be little more than a semantic difference, but the difference in wording much more accurately reflects the supporting animal database. It is not entirely clear exactly what is meant by these two terms, probable and possible, for the classification schemes, but some research directed at that particular question indicates that in common everyday usage, probable and possible correspond roughly to a 70 or 30 percent chance of occurrence, respectively (29). Thus, the public interpretation of class B compounds in the EPA system is that there is substantially more than a 50/50 chance that these compounds are human carcinogens. For most of the same compounds, the public perception is that IARC views these as having less than a 50/50 chance of posing a human carcinogenic risk.

In June of 1991, IARC took a step toward incorporating a more complete toxicological review in their classification process. The IARC system now allows upgrading or downgrading of a classification based on mechanistic data regarding the relevance of animal studies to human hazard.

In keeping with the concept of more completely addressing the relevance of animal results, a classification scheme has been proposed (9) in which there is a more clear recognition and discussion of the relevance of animal results. The essence of the Ashby et al. (9) proposal is that a distinct gradation is made among the compounds for which there is sufficient evidence of carcinogenicity in animals. These compounds are subdivided into probable human carcinogens, possible human carcinogens, probably not human carcinogens, and compounds considered not to be human carcinogens, depending on the strength of the data for or against the relevancy of the animal data. As can be seen, this classification system clearly recognizes that positive bioassays in animals are not all indicative of the same potential for human hazard. In some cases the results are likely to predict human hazard or have an uncertain potential for human hazard or, in some cases, are judged (based on *adequate* scientific data) to be not relevant for humans. This differentiation is important. Without this differentiation, very different compounds are treated in regulation with the same degree of stringency and certainty, with the result that priorities and resources may be seriously misallocated.

Historically, ACGIH has categorized carcinogens into one of two groups: A1— Confirmed Human Carcinogens and A2—Suspected Human Carcinogens. Recently (30), they have created a third category for experimental carcinogens with negative evidence of human carcinogenicity, based on the following criterion:

"Evidence of increased incidence of cancer under experimental conditions which are not considered to reflect circumstances normally found in the work environment. Consideration is given to dose, method of administration and site of cancers" (30).

This classification group reflects a central concept of the proposed classification of Ashby et al. (9); namely that experimental conditions are evaluated with respect to relevance to human exposures. This evaluation is likely to be judgmental but is also likely to more appropriately separate the real carcinogenic hazards from those much less likely to pose any risk.

The European Economic Community (EC) also classifies chemicals for the purposes of labeling and warnings. Their system depends much more heavily on a judgmental evaluation of the total toxicological database. In particular, genotoxicity data are used heavily along with a review of the human database and ancillary mechanistic information to determine if the compound under consideration is likely to be a risk to humans *at expected exposure levels*. The result is that a number of compounds, such as methylene chloride, are not classified as potential human carcinogens by the EC, although they are by EPA.

Also of concern in some areas is the *automatic* acceptance of IARC/EPA classification in regulations or pseudo regulations. For example, in California a state referendum requires that if members of the public are exposed to compounds that cause cancer, they must be warned (31). The wording in the referendum is vague as to what compounds might cause cancer. In the implementation of this referendum, EPA's *entire* list of compounds (categories A, B1, B2, and C) are used, along with the required wording in the warning that the compound is ''known to the State of California to cause cancer.'' The criteria for inclusion of a compound in EPA category C and the strength of the required warning label under Proposition 65 are clearly not consistent. Thus, there is the potential for misuse of lists, and the importance of the most relevant descriptors and communication material can be seen.

If a compound has been classified as a known, probable, or possible carcinogen it passes to the next step in the risk assessment process. This is a weakness in the current application of risk assessment. Not all of these chemicals have the same potential for causing cancer in humans; yet this is ignored. The compound is subject to regulation in air, ground, and water as a potential human carcinogen, and regulatory limits are calculated in terms of ''risk'' rather than by defining ''safe'' levels. In theory, group C carcinogens are reviewed on a case-by-case basis relative to the use of quantitative risk assessment. In practice, however, most EPA program offices regulate class C compounds as potential human carcinogens. The Office of Pesticides is the major exception in that they appear to use the case-by-case approach.

4   DOSE–RESPONSE ASSESSMENT

This next step in the risk assessment process involves the extrapolation from results at high doses in animals to low doses in humans. In some instances dose–response assessment has been used directly with human data, but these instances are rare due to the general lack of good human exposure data for most retrospective epidemiology studies. This extrapolation step is necessary because of the high doses used in animal bioassays as compared to human exposure levels. The extrapolation presents a logical problem, however; namely that data are being extrapolated many orders of magnitude using empirical models that do not reflect the underlying carcinogenic mechanisms.

A basic principle in fitting of empirical models is not to extrapolate outside the range of the observable data. This empirical extrapolation results in large uncertainties, both from a statistical point of view and from a biological point of view.

The mechanism by which chemicals cause cancer is not completely known and may involve a variety of mechanisms occurring at various stages in the carcinogenic process. A chemical may act at a single stage or at more than one stage. Currently, for most if not all chemicals, data are not available to determine the exact mechanism by which they cause cancer. As a result, significant controversy exists regarding the shape of the dose–response curve in animals and humans. One particular issue is the question of the existence of thresholds for carcinogens. The term threshold is not used here in an absolute sense but instead describes a point on the dose–response continuum at which the response approaches zero in a distinctly nonlinear (concave up) manner. The term does not necessarily mean an *absolute* threshold dose below which no response at all will occur since, for threshold phenomena, there is presumably a distribution of thresholds across individuals. The carcinogenic process is considered by some to be nonthreshold, since it is believed to be initiated by a mutation in a single cell, and a single molecule of a chemical could theoretically produce this mutation. Some chemicals, however, appear to cause cancer through different mechanisms that are likely to have practical thresholds (32). For the majority of cases in which it is not possible to determine the mechanism by which a chemical causes cancer, the EPA recommends as a *policy* decision that these chemicals be treated as if they act by a nonthreshold mechanism. It is stressed that this is the *default* assumption that can be overcome with adequate data.

The concept of threshold doses for certain chemicals that produce tumors is being debated actively as data are being generated for many compounds in support of such mechanisms. Threshold exceeding doses may induce nonphysiological changes in the test animal or there may be levels that overwhelm natural detoxification capacity. These levels are likely to vary between individuals within a given population, and thus an absolute threshold is difficult to define, although a *practical* threshold, a level at which no response is expected in any *finite* population, is likely to exist.

There has been considerable discussion on the possible existence of thresholds for carcinogens in the toxicological literature (33,34), but no comprehensive mechanistic approaches appear to have been proposed. It is generally believed that there are many mechanisms responsible for the induction of cancer and, reflecting this piecemeal approach, the discussion of thresholds has followed many different paths.

Several chemicals of *endogenous* origin have been shown to induce tumors when tested to conventional lifetime rodent bioassays. Some of these are genotoxic, such as formaldehyde, others are nongenotoxic, such as estrogen. Either individual threshold levels for neoplasia produced by these chemicals must exist or the concentrations of these compounds are present in a narrow range of the dose–response curve that then contributes to the general background rate of cancer in the population. In the case of formaldehyde, the critical cellular concentration above which cytotoxicity and tumor formation occur is that concentration at which formaldehyde can no longer be detoxified or genotoxic events can no longer be repaired.

If a threshold mechanism and an approximate threshold level can be identified for a particular carcinogen, uncertainty factors (safety factors) should be used to set regulatory criteria. The application of safety factors in regulation is a methodology that has been used for many years and is still being used for noncarcinogens. Various approaches have been developed for determining appropriate uncertainty factors to derive safe or allowable levels of exposure. All of these have in common identification of the no observed adverse effect level (NOAEL) or lowest observed adverse effect level (LOAEL) or an equivalent such as a benchmark dose (35), from the most appropriate animal test, and dividing that dose by a series of factors to establish safe levels for human exposure. The choice of factors is inherently dependent on professional judgment. Several considerations affect the final judgment. These considerations include appropriateness of the animal model and exposure route, exposure duration relative to lifespan, the quality of the study, and the availability of related studies including human epidemiology. Although there is no set range of safety factors, they are often multiples of 10 ranging from 1 to 10,000 with 100 being typically used as an uncertainty factor in extrapolating an appropriate, well-defined animal NOAEL to the comparable human exposure scenario. Where appropriate data exists, however, more specific safety factors should be and are being used. For example, when animal and human target tissue doses are known through pharmacokinetic evaluation, the exact ratio of the relevant target doses can be used, rather than order-of-magnitude conversion factors.

An interesting modification of the uncertainty factor approach has been proposed (36) that breaks the components of the uncertainty down into specific individual factors. The methodology clearly decomposes the overall factor into components that reflect uncertainty, components reflecting scientific extrapolation, and societal (risk management) components reflecting prudent conservatism.

4.1 Mathematical Models Used in Dose–Response Assessment

Mathematical models that have been used to describe the relationship between the animal dose, time, and the response (percent animals with tumors) area based on either mechanistic or tolerance distribution assumptions:

Mechanistic Models	Tolerance Distribution Models
One hit	Probit
Multistage	Logit
Linearized multistage	Weibull
Multihit	

These models are typically formulated in terms of both dose and time. For regulatory purposes, however, *lifetime* risk is used and time disappears from the models. The use of assumptions of these models have been discussed in detail elsewhere (37–39) and will be reviewed only briefly here.

The selection of an appropriate model, based on the empirical data generally available, is an indeterministic task. The statistical approach to selecting the best model would be to design the bioassay to optimally discriminate among these models, then identify the most useful candidate model(s) based on statistical goodness-of-fit criteria. Unfortunately, as Crump (40) points out, it usually is not possible to design such experiments, and therefore the goodness-of-fit criteria does not provide the answer. For almost all of the data sets, most or all of the models fit equally well in the observed range due to the limited number of doses and animals (two to three doses with 50–100 animals per dose). The models will tend to disagree the most in the low-dose region, but there is no information (data) in this region upon which to discriminate among the models (i.e., no excess tumors are observed). The models diverge several orders of magnitude (41) at exposures representative of typical environmental level, but with no practical or theoretical guidance on model selection.

4.2 Low-Dose Linearity

It has been recognized for a long time that different mathematical models, which all fit the observed data reasonably well, give extrapolated safe doses at *de minimis* risk levels that differ by several orders of magnitude (41).

These differences depend on the models being fit, as well as the spontaneous tumor incidence and whether it is assumed that the mechanism of action for the particular chemical is independent of the background rate of additive onto background. Differences between extrapolated mathematical models are largest when the background tumor incidence is assumed to arise from a mechanism that is *independent* of that of the carcinogenic substance (13). If, on the other hand for a specific chemical, complete additivity onto existing background processes is assumed, by definition there is no threshold dose. Since, in this case, there is already a spontaneous incidence, the incremental incidence will show an approximately linear dose response at very low doses, since even highly nonlinear functions can be approximated by a straight line over a small region when the first derivative exists and is positive (42–44). However, if observed background risk is low, then the dose-response relationship might not approach low-dose linearity until extremely low doses are reached and the slope may be very small (44). This concept is demonstrated graphically in Figures 8.1(a) and 8.1(b). In each case it is assumed that there is a nonlinear dose–response relationship, but the applied dose of chemical is additive onto an existing background "dose" that causes the background tumor response. In Fig. 8.1(a) the spontaneous incidence is high, whereas in Figure 8.1(b), the background rate is low. In either case it can be seen that the assumption of additive dose results in an approximately (locally) linear response, although the slopes are quite different. It must be noted that additivity to background and the ensuing low dose linearity is a strong biological assumption that can rarely if ever be proven or disproven. Also, the *slope* of the linear component, if it exists, cannot be estimated using statistical approaches. All that can be done is to estimate an upper bound with no estimate of the uncertainty in the bound. Nevertheless, the possibility of background additivity has given theoretical support to extrapolation procedures that are characterized by low-dose linearity, although many toxicologists appear to believe that linear nonthreshold behavior is highly unlikely and is a conservative assumption.

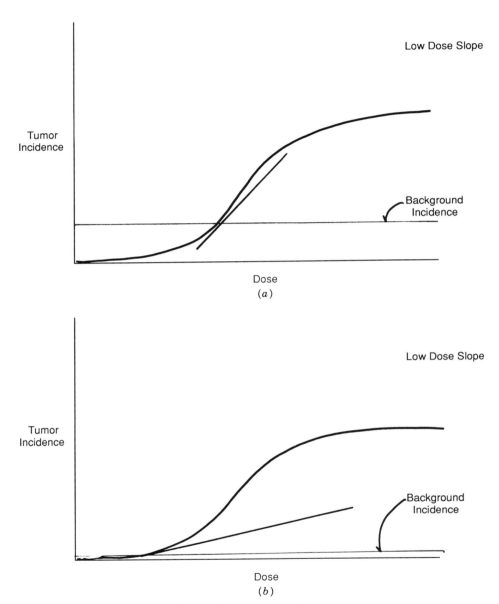

Figure 8.1 Low dose linearity.

Krewski and Van Ryzin (44) showed mathematically, and by example, that the models tend to have the following relationship to each other at low doses:

Model	Estimate of Potential Risk
Linearized multistage	
One hit	Highest
Multistage	
Weibull	
Multihit	
Logit	
Probit	Lowest

4.3 Tolerance Distribution Models

The tolerance distribution models assume that each member of a population will develop a tumor if exposure to the carcinogen exceeds a critical level (threshold). This threshold level varies from individual to individual and can be modeled by various tolerance distributions.

Tolerance distribution models have been found to adequately model many types of biological dose–response data (e.g., acute lethality), but it is an overly simplistic expectation to represent the entire carcinogenic process by one tolerance distribution. A tolerance distribution model may give a good description of the observed data, but from a mechanistic point of view there is no reason to expect extrapolation to low dose to be universally valid. The probit model extrapolation has, however, fit well in some instances (45).

4.3.1 Log Probit Model

The log probit model assumes that the individual tolerances follow a lognormal distribution. This is likely true for individual steps in the complex chain of events that lead to carcinogenesis. For example, it is reasonable to assume that in many cases the distribution of a population of kinetic rate constants for detoxification, metabolism, and elimination, in addition to the distribution of immuno-suppression surveillance capacity and DNA repair capacity, can be adequately approximated by normal or lognormal distributions. But, the total process is likely much more complex than what can be modeled by any single distribution.

A modification of the log probit model was proposed by Mantel and Bryan in 1961 (10). This modification dealt with the uncertainty in the observed dose–response by suggesting the use of a slope of one in the log probit model when extrapolating to low doses. It was felt that a slope of one was sufficiently conservative based on historical experience and judgment.

The Mantel–Bryan procedure was used by government agencies briefly in the mid-1970s (46). Originally, the procedure was regarded as conservative in the sense of protecting the public health because it employed a presumably conservative slope and because it extrapolated from an upper confidence limit on excess risk. However, the Mantel–Bryan procedure has been shown to lack conservatism near zero dose (47,48), since the normal distribution falls off to zero faster than any fixed function (e.g., linear, quadratic, power, etc.) at very low doses. This and legal procedural problems in the application of the Mantel–Bryan methodology lead to its demise as a regulatory tool in the mid-1970s.

4.3.2 Logit and Weibull Models

Other tolerance distributions that have been used to model carcinogenicity dose-response data include the logit and Weibull models (49,50). The log probit, logit, Weibull, and gamma distributions all have potentially similar shapes between tumor frequencies of 2–98 percent; hence it is not surprising that these models often give essentially identical fits to the observed data, but again, the models differ widely at low doses. These models are currently not used by regulatory agencies in North America for this application.

4.4 Time-to-Tumor Models

During the course of a bioassay there are sometimes observations made on the time that individual tumors are observed. This information can be used in a number of ways in the risk assessment process. One utilization of the time-to-tumor information involves recasting the response metric in terms of decreased expected lifetime rather than the percentage of the population affected (risk). Sielken (51) and Albert (52) have advocated the use of time, rather than dose, as the metric of interest in the regulatory use of cancer risk assessment models. This reformulation of the problem utilizes the tumor incidence data as well as the time-to-tumor information but requires policy assumptions as to the shape of dose response curves analogous to the linearity assumption made in the linearized multistage (LMS) modeling procedure. The methodology has not been utilized by regulatory agencies in North America.

4.5 Mechanistic Models

Mechanistic models are so called because, in theory, they represent underlying biological mechanisms of cancer. In practice, most of the early applications of mechanistic models are highly simplistic representations of biology and do not represent fundamental mechanisms.

The traditional mechanistic models most often used for dose–response modeling and extrapolation in carcinogenesis risk assessment are based on two fundamental assumptions: (1) there is no threshold dose for the carcinogenic effect and (2) carcinogenic effects of chemicals are directly proportional to dose at low dose levels, that is, the dose–response is linear. The justification for these assumptions results from the theory that a tumor originates from a single cell that has been damaged (13).

One class of mechanistic models derives from assumptions about age-specific tumorigenicity rate (the hazard function), which is defined as the proportion of people in any specific age group (e.g., 46–50 years) developing cancer. If the age-specific tumor rate (r) is a function of dose and age, and if the dose and age components can be separated, then the cumulative lifetime risk for a given dose is found by integrating over age. The resulting class of models is $P = 1 - \exp[-g(\text{dose})]$, where P is the probability of tumor and g (dose) is a mathematical function of dose.

4.5.1 One-Hit Model

The simplest of the traditional mechanistic models is the one-hit model. It is based on the biological theory that a single "hit" of a carcinogen at a cellular target, for example DNA, can initiate a series of irreversible events that eventually lead to tumor formation. It is assumed (53) that the age-specific hazard rate, the relationship between dose and hazard at a given age, is linear in dose

$$\text{hazard} = \alpha + (\beta \cdot \text{dose})$$

where α represents a background incidence and β is a chemical intensity parameter. That is, the probability of cancer development in a normal cell is proportional to concentration. Then the lifetime risk is

$$P(d) = 1 - \exp - (\alpha + \beta \cdot d)$$

which at low doses (small values of d) may be approximated by the linear model

$$p = \alpha + \beta \cdot \text{dose}$$

The one-hit model, and variations on it utilizing upper statistical limits (54), represent a highly conservative approach to the extrapolation problem (55). For example, a linear extrapolation of the Chemical Industry Institute of Toxicology formaldehyde study predicted that an average lifetime dose of less than 0.66×10^{-3} ppm was necessary to keep lifetime risk to less than 10^{-6} (56). Such an estimate appears contradictory as an estimate of the risk to humans when viewed in light of about 100 years of experience with human exposures to formaldehyde that generally are less than 0.1 ppm but have often been in the 0.1- to 5-ppm range (57) with no definable increased carcinogenic risk.

4.5.2 Multi-Hit Model

The multihit model is discussed in some detail in the Food Safety Council Report (38). One derivation of this model follows from the assumption that K hits or molecular interactions are necessary to induce the formation of a tumor, and the distribution of these molecular events over time follows a Poisson process. In practice, the model appears to fit some data sets reasonably well and to give low-dose predictions that are similar to the other models. There are cases, however, in which the predicted values are inconsistent with the predictions of other models by many orders of magnitude. For instance, the virtually safe dose as predicted by the multihit model appears to be too high for nitrilotriacetic acid and far too low for vinyl chloride (38).

4.5.3 Multistage Model

The mechanistic model most frequently used in the regulatory process as the basis for low dose extrapolation is a variation of the multistage model of Armitage and Doll (58). A historical perspective on the evolution of the multistage model has been provided by Whittemore and Keller (53). The multistage model is based on the concept that a tumor develops from a single cell in an organ as a result of a number of biological events (e.g., mutations) that occur in order. For continuous exposure to a constant target tissue dose level, d, of a carcinogen, if $\alpha_i + q_i d$ represents the instantaneous hazard for the ith event or stage, where a total of k events are required for a tumor to develop, then, for an organ comprised of a large number, N, of independently acting cells, the probability of a tumor by time t is given by

$$g\ (\text{dose}) = (\alpha_1 + q_1\ \text{dose})\ (\alpha_2 + q_2\ \text{dose}) \cdots (\alpha_n + q_n\ \text{dose})$$

and then

$$P(d) = 1 - \exp\left\{ - (\alpha_1 + q_1\ \text{dose})\ (\alpha_2 + q_2\ \text{dose}) \cdots (\alpha_n + q_n\ \text{dose})\right\}$$

where $\alpha_i, q_i, \geq 0$ are parameters that vary from chemical to chemical. This model has been used to describe incidence of human cancer resulting from smoking (59) and may be useful for describing other observed data sets, but is unlikely to be useful in mechanistically describing the dose–response function at doses below the observed range. In practice, the model is fit with the exponent expressed as a general polynomial,

$$q_0 + q_1 \cdot d + q_2 \cdot d^2 + q_3 \cdot d^3 + \cdots$$

rather than in the more restrictive factored form shown above (42,47,60). This polynomial hazard function will fit almost any observed data set as long as the dose–response is not markedly concave downward at low responses. Therefore, the model results in a curve-fitting exercise and has limited applicability to the estimation of real risk at low doses. The limitations arise first because the model cannot reflect changes in kinetics, metabolism, and mechanisms at low doses and second because low-dose estimates are highly sensitive to a change of even a few observed tumors at the lowest experimental dose.

In practice, the dose is usually the administered dose, in which case it is assumed to be proportional to target tissue dose; however, occasionally a measure of target tissue dose is obtained from pharmacokinetic modeling (61,62) and is used in the model.

4.5.4 Linearized Multistage Model

As will be seen, the multistage model is very sensitive to minor changes in the tumor incidence data. To account for this statistical variability, a bounding procedure was derived under EPA contract (63). The linearized multistage model, developed by Crump and utilized by many of the regulatory agencies, fits the general multistage model, as defined above, then replaces the linear term (q_1) of the polynomial function by its upper 95 percent confidence limit to account for statistical variability in the observed tumor frequencies. The dose–response predicted by this model is approximately linear at low doses since the higher-order terms approach zero faster than the linear term and become insignificant at exposures of regulatory interest. This results in estimates of potential risk that are almost identical to those of the one-hit model. Even for extremely nonlinear data, for example, nitrilotriacetic acid (38), the linear term dominates low dose extrapolation and estimated doses corresponding to low dose risk levels differ only by a factor of less than 6 from estimates from a one-hit model extrapolation. Thus, for almost all applications, there is no appreciable difference between the one-hit model and the linearized multistage model.

The linearized multistage model has been used routinely by the EPA in setting safe levels of environmental carcinogens (12,64). Implementation of the methodology usually is easily accomplished with a computer program such as GLOBAL82 (65), which employs maximum-likelihood estimation of the parameters and generates confidence limits based on the asymptotic distribution of the likelihood ratio. Ordinarily the value of k is chosen as the total number of dose groups less one, to ensure enough degrees of freedom for a unique fit, although this is not strictly required (66). The procedure is called the linearized multistage (LMS) model because of the fact that only upper confidence limits on excess risk above background, and not point estimates of excess risk, are utilized by most regulatory agencies, and because these upper confidence limits are linear at low doses, regardless of how nonlinear the best fitting maximum-likelihood model might be. That is, even when the best fitting model does not have a linear term in the exponent, the upper confidence limit at low risk levels (e.g., 10^{-5}) will reflect a linear coefficient (67), which will dominate the dose–response at low doses.

In the linearized model, the upper bound for q_1 is designated by Q^*, and is referred to as *potency* (oral ingestion, units of mg/kg day) or *unit risk* (inhalation, units of $\mu g/m^3$). Although the LMS model utilizes a computer algorithm to calculate the parameters, Q^* can be reasonably approximated by calculating the slope of the line between the lowest increased tumor incidence and the control incidence, then doubling this slope to obtain an upper bound.

In application, excess low-dose risk is calculated from the equation

$$\text{risk} = Q^* \cdot \text{dose}$$

Low-dose linearity is believed to be a generally conservative assumption since a linear upper bound on the low-dose portion of the dose–response will be a bound for any concave up function. This has prompted the development of various other linear low-dose extrapolation procedures. For example, the procedure of Gaylor and Kodell (54) involves simply extrapolating linearly downward to zero from an upper confidence limit on excess risk at a low-dose point (either the lowest experimental dose level or the dose corresponding to a predicted 1 percent response level, whichever is larger). The upper confidence limit is found using either the multistage model or some other reasonable dose–response model, and the results are essentially identical to LMS model results. The FDA's Center for Food Safety and Applied Nutrition employs the Gaylor–Kodell procedure (68) and the FDA's Center for Veterinary Medicine has recommended the procedure for extrapolating risks from animal drugs (69). Recently, Krewski (70) has proposed a "model-free" approach for linear, low dose extrapolation based on the assumption that the true dose response relationship at low doses is either linear or sublinear. The Krewski approach yields results similar to the LMS model and to the Kodell methodology.

The predictions from the linear models are essentially independent of the shape of the observed dose–response curve, give very little weight to NOELs, and may not be predictive of potential risk at low levels. When good epidemiology data are available for comparison, it has in some cases been found that linear models are not compatible with the human experience (45,71,72). Carlborg (50) concluded that the one-hit model did not give an adequate fit to most of the 31 data sets he analyzed.

With appropriate species conversion, the one-hit model does, however, estimate an upper limit on the potential risk and may be useful in situations where an upper bound is of interest. For example, if the potential risk calculated by the one-hit model is not unacceptable, then there would be less need to consider other models. While it is likely that the linear models substantially overestimate low-dose risk, the information necessary to fit other models is generally not available, except for a few well-studied compounds.

4.6 Maximum-Likelihood Estimates (MLEs) vs. Upper Confidence Limits (Q*)

Although a number of empirical dose–response models have been proposed for use in fitting animal dose–response data, the linearized multistage model has been the most widely used by the regulatory agencies. It has been suggested, however, that the maximum-likelihood estimates, that is, the "best" estimates of the multistage model, should be used either along with, or in preference to, the linearized upper 95 percent bounds. To some extent, best estimates are reported along with upper bounds, but they do not appear to be used to any extent by the agencies. In general, MLEs will be less than the upper bound by two- to threefold in some cases and up to 1000-fold in extreme cases, depending on the degree of curvature in the observed dose–response.

MLEs *do* more accurately reflect the dose–response curve in the vicinity of the observed data and can validly be used to model the data down to the 10^{-1}, or 10^{-2} response range. At lower doses, however, the MLEs become progressively more sensitive to small changes in the tumor frequencies. For instance, a change in the tumor incidence of one tumor at the low dose can make a substantial difference, in some cases, of up to 100-fold in the estimation of the virtually safe dose (the dose corresponding to a risk of 10^{-6}. An example is shown in Table 8.1.

Linearized multistage results for data sets A and B in Table 8.1 are similar, but MLE differ by almost 100-fold (at 10^{-6}), with data set B showing less risk. This extreme sensitivity of the MLE is a result of extrapolating an empirical model orders of magnitude below the range of the observed data. The upper bound confidence limit, on the other hand, has the advantage and disadvantage of being relatively insensitive to the shape of the dose–response curve, even to the extent of calculating results from negative studies. The upper bound is not unduly influenced by minor

Table 8.1 Hypothetical Tumor Counts in Two Bioassays

		Tumor Bearing Animals
	Dose (ppm)	Number of Animals Observed
Set A	30	10/50
	10	3/50
	3	0/50
	0	0/50
Set B	30	10/50
	10	1/50
	3	0/50
	0	0/50

perturbation in the tumor frequencies, but, as a result, it is also insensitive to the *shape* of the observed dose–response and thus does not discriminate between widely different dose–response curves. For example, the presence of an apparent threshold in the observed data has little or no effect on the LMS risk estimates, since low-dose linearity is assumed in that model. This phenomenon can be seen by comparing various risk estimates from the formaldehyde and methylene chloride data sets. The relevant tumorigenicity data for each of these compounds is shown in Table 8.2.

As can be seen, inhalation exposures to formaldehyde of 14.3 ppm for 6 hr per day for a lifetime resulted in tumors of the nasal turbinate in approximately 50 percent of the animals, but the dose–response is well defined, falling off sharply. For the case of methylene chloride, inhalation exposure to 2000 and 4000 ppm caused tumors of the lung and liver in mice, but the low-dose effects have not been well defined. The methylene chloride bioassay is typical of the NCI/NTP experimental protocol, an MTD is used together with a single dose at one-half the MTD, in case the MTD is incorrectly estimated. This protocol does not define the low-dose effects. For both data sets, LMS (linear) estimates were calculated as well as multistage (best or MLE) estimates. The LMS results indicate formaldehyde to be 100 times more potent than methylene chloride based on the fact that tumors occurred at a dose approximately

Table 8.2 Observed Tumors in Two Bioassays

		Tumor Bearing Animals
Compound	Dose (ppm)	Number of Animals Observed
Formaldehyde	14.3	103/199
	5.6	2/200
	2.0	0/200
	0	0/200
Methylene	4000	46/47
Chloride	2000	37/48
	0	5/50

100-fold lower. In contradiction, however, the multistage, or best estimates indicate that methylene chloride is 80 times more potent than formaldehyde (at a risk of 10^{-6}). This latter conclusion results from the fact that the observed dose response for formaldehyde is highly nonlinear, whereas the observed data for methylene chloride can only be fit to a linear (or supralinear) curve. Thus, comparative results using MLEs of empirical models can be misleading.

The upper 95 percent confidence interval, or any linear model, behaves very much like an uncertainty factor approach. An observed tumor frequency of 10 percent would be an effect level for most tumor sites in a modern protocol, whereas a 1 percent incidence would be undetectable in any standard assay. Under the LMS model, the dose corresponding to a predicted 10^{-6} risk level is 10,000-fold less than the predicted 1 percent level and 100,000-fold less than the predicted 10 percent level. Therefore, the predicted 10^{-6} dose in a linear model corresponds roughly to an uncertainty factor of 10,000 compared to the NOAEL and approximately 100,000-fold compared to a LOAEL. Krewski (71), starting from the same premise, has correlated potency (Q^*) with the maximum dose tested and found a high correlation. He shows that, to a first approximation, $Q^* = MTD/384,000$.

The surface area animal-to-human extrapolation factor, discussed later, adds approximately another order-of-magnitude safety factor as compared to the traditional body weight conversion used for other toxicological endpoints. It can be seen that the magnitude of the above uncertainty factors is far larger than those used for other toxicological endpoints, and this has led to much of the controversy in the application of the LMS model. Thus, MLEs are more appropriate in the observable range, while below those levels, MLEs are highly unstable and can be misleading. On the other hand, results from the LMS model are very similar to a safety factor approach and should only be interpreted as such.

4.7 Interspecies Conversion

Once upper bounds on the potency have been determined from the animal data, the results must be extrapolated to humans. In theory, relative potencies for animals and humans could be estimated for each compound, and a unique interspecies conversion factor could be used for each application. In practice, however, relative potencies are almost never known for chronic health effects such as cancer, thus standard interspecies conversion factors are used. Historically, interspecies scaling for non-cancer endpoints has been done on the basis of body weight (73). The Food and Drug Administration still assumes equal sensitivity between animal and human on an equal mg/kg/day dose metric, but EPA assumes equal sensitivity across species is equal on a mg/surface area/day basis across species. The scientific rationale for the surface area assumption is that it has been found empirically that basal metabolic rates are roughly proportional to the surface area of different species. To a first approximation, steady-state body burdens of chemicals with a constant rate of input to the body will be inversely proportional to metabolic rates, that is, the higher the rates of metabolism, the shorter the half-life in body compartments. Therefore, species with proportionally larger surface areas will eliminate compounds more quickly, and will therefore be at less risk. There is some empirical evidence for this idea, at least for the parent compound, but in cases where metabolites are likely the ultimate carcinogen, the relationship is not expected to hold (72,74).

To a first approximation, surface area (SA) α (body weight)$^{2/3}$ (BW), therefore species sensitivity is assumed to be approximately proportional to body weight to the two-thirds power. This species extrapolation factor is usually calculated by scaling the animal potency from the LMS model as follows:

$$Q^*_{human} = Q^*_{animal} \times \frac{SA_{human}}{SA_{animal}}$$

If Q^* has already been calculated in dose units of (mg/kg day)$^{-1}$, as is usually the case, then the surface area correction reduces to

$$Q^*_{human} = Q^*_{animal} \times (BW_{human}/BW_{animal})^{1/3}$$

The result is that humans are assumed to be 6 times more sensitive than a rat and 13 times more than a mouse on a mg/kg day basis. In effect, the surface area correction amounts to an additional safety factor of approximately one order of magnitude for most compounds compared to a body weight conversion.

4.8 Distributions of Risk

The process, as it is currently used, has a number of decision points for which there is no clear scientific guidance as to the appropriate generic choice (14). The regulatory approach to choosing among these decision points has been to enunciate a set of "science policy" principles or guidelines (12,13) that dictate, as a default, use of conservative choice at each step. The result is a methodology that is protective of public health but that overstates the true value of the risk by an undetermined amount; hence the description of the estimates as upper bounds, with the qualifier that "the true value of the risk is unknown and may be as low as zero" (12).

A refinement of this would be to use decision analysis methods to estimate the *probability distribution* of risk estimates that could be derived under different reasonable assumptions (21, 75–77). By considering all possible *combinations* of assumptions and their individual likelihoods, an overall probability distribution can be developed to represent the complete range of risk estimates within the bounds of the allowable assumptions. Then, a specific percentile of this distribution could be used for regulatory purposes. For example, the upper 99th percentile could be used as a conservative estimate of risk, and the median (50th percentile) could be used to represent a more central estimate. The difference between these two values gives an indication of the sensitivity of the estimates to the specific assumptions being made. The use of this methodology has been promoted as a method for showing the complete distribution of risk estimates that could be derived as the various assumptions are given different relative likelihoods of being correct. The specific assumptions that are allowed to vary can be controlled, for instance, to stay within the class of low-dose linear models or not.

The major quantitative factors that could be allowed to vary include:

- Different species scaling factors for the particular compound under consideration. This may be surface area, body weight, or the inverse of surface area (72).
- In the absence of information on the most appropriate study upon which to base the LMS estimates, potency estimates from all comparable studies could be used, including upper bounds on the negative studies. This methodology has been incorporated to some extent in assessments done by the Department of Health Services in California (78).
- Basing potencies on different tumorigenic endpoints; malignant tumors only, malignant plus benign tumors, and tumors of different origins.
- Different dose–response models, linear or not.

In estimating exposure, there is generally much less scientific uncertainty in the assumptions. Instead the uncertainty has more to do with the appropriate *percentile* of the particular population distribution to use. People vary in terms of height, weight, pulmonary volume, water consumption, and food consumption as well as particular food fractions consumed, such as meat, vegetables, and so forth. They also vary as to how long they work on one job and how long they live in one place. The issue then becomes that of picking the appropriate percentile of these distributions to use for regulatory purposes. Should the median be used or the 90th percentile or the 99th percentile? This problem can again be solved by using distributions instead of point estimates for the individual parameters, then combing the distributions using Monte Carlo computer techniques, resulting in an overall distribution of possible exposures across the population or subpopulation of interest. This information can be carried forward through the risk assessment process and the decision maker can decide on the appropriate percentile of the possible distribution to use for each application.

4.9 Limitations of the Models

The models described are generally simplifications of a complex system and apply only to the chronic animal toxicity studies that are but one input to the risk assessment process. The models have little biological relevance, have been shown to provide poor extrapolation estimates in some cases, and generally are not validated either in animals or with respect to the human experience. The models can be useful, however, as one tool in the regulatory process if used appropriately.

A major problem with the use of formalized modeling approaches is the inability of models to incorporate much of the qualitative biological information that must be used to arrive at logical decisions.

The importance of qualitative data is evidenced in the process by which both governmental agencies and industry determine how to handle a potential carcinogen. It is initially a two-step procedure in which a qualitative decision is made as to whether the compound is an animal and/or a potential human carcinogen. If the answer is yes, then the risks to humans are estimated, and a decision is made about how to handle the compound. If a complete risk estimation model existed, there would be no need for the first step. The model would be used to compute the risks directly, and noncarcinogens would be assigned very small or zero risks. Unfortunately, our current state of knowledge does not permit us to take this approach. Some of the qualitative data that would be included in a judgmental evaluation of the potential for risk are discussed below.

4.9.1 Tumor Type

The B6C3F1 mouse, for example, is highly sensitive to hepatic tumors, and the majority of male Fischer 344 rats develop testicular tumors independent of any chemical exposure. Such chemicals may be much less likely to cause human cancer than a potential carcinogen that produces a spectrum of histogenically different tumors including those tumors with low spontaneous background rates. The mechanisms of action for some of these promoters of spontaneous tumors is entirely different from those of complete carcinogens, and certain steps in the assessment of risk should reflect the different mechanisms.

4.9.2 Genotoxic and Nongenotoxic Mechanisms

Probably the most important qualitative difference between animal carcinogens is the distinction between carcinogens that operate primarily through direct genotoxic mechanisms and those that produce tumors through mechanisms other than direct interaction with DNA (i.e., epigenetic, nongenotoxic, or nongenetic mechanisms). The subject of tumorigenic mechanisms has received a great deal of discussion (27,32). Briefly stated, some chemicals that are carcinogenic in animals also interact directly with DNA as indicated by results of short-term in vitro tests or by direct measurement of in vivo DNA alkylation and repair rates. Other chemicals that are tumorigenic in animals show virtually no activity in the genotoxicity tests and show no propensity to bind or interact with DNA. Research on some of these chemicals (e.g., saccharin, chloroform, trichloroethylene, and perchloroethylene) show that little to no direct interaction with DNA occurs. On the other hand, in vivo tests indicate a dose-dependent acceleration in the rate of DNA synthesis at doses that correlate well with tumorigenicity and demonstrate an apparent tumorigenic mechanism and threshold (27,79).

In some of these cases, the likely mechanism of action is cytotoxicity, resulting in cellular regeneration accompanied by an increased rate of DNA synthesis. DNA is constantly undergoing a low background rate of damage and subsequent repair. If this background rate is increased (e.g., when cellular damage necessitates an increase in the rate of DNA synthesis), the increased demands on the repair surveillance systems may lead to an increased probability of faulty DNA repair or to the possibility of replication before repair is completed. This phenomenon can be demonstrated by the production of skin tumors following burns and the repeated freezing of skin with dry ice (80,81) and liver tumors following partial hepatectomy.

Cytotoxicity and direct cellular damage leading to the tumorigenesis is only one nongenotoxic mechanism for carcinogenesis. Weisburger and Williams (32) point out that other nongenotoxic mechanisms exist. Examples include solid-state carcinogens (polymers, asbestos), hormonal imbalance carcinogens (estradiol, DES), immunosuppressors (azothioprine), and promoters (phorbol esters, saccharin).

4.9.3 Use of Mechanistic Information

The subject of tumorigenic mechanisms is extremely complex. For example, when classifying chemicals with respect to carcinogenic mechanisms, it must be kept in mind that genotoxicity is a continuous spectrum rather than a dichotomous classification. Some chemicals clearly have a direct genotoxic component while others are at or near the nongenotoxic end of the scale. The importance of the distinction between genotoxic and nongenotoxic mechanisms is that according to current theory, all that is needed for directly genotoxic chemicals to initiate a tumor is a single molecular event; thus, thresholds may not exist, although the existence of protective mechanisms, such as DNA repair systems that are saturable, would imply the existence of a practical threshold.

On the other hand, for nongenotoxic mechanisms, more than a single molecular interaction is necessary to produce a tumor. The consensus of scientific opinion is that for some of these mechanisms, as with other toxicological phenomena, either an absolute threshold exists or the dose–response curve is so flat as to be indistinguishable from zero slope at low doses (i.e., a practical threshold exists). In some cases the existence of an observable precursor related to tumorigenicity (e.g., cellular toxicity, necrosis, and hyperplasia), or a change in the rate of DNA synthesis may provide a marker variable that can be used to predict the threshold below which the nongenotoxic mechanism ceases to pose a risk.

4.10 Interpretation of Risk Estimates

The risk assessment methodology, as applied, is a useful regulatory tool, but it has definite limitations as to how the estimates can be interpreted. Upper bounds on risk, resulting from the default assumptions, have often been misunderstood or misapplied by decision makers and risk managers. These estimates have been used as though they were actual estimates of risk, rather than estimates of the *upper bound* on the risk. This misperception continues (82), even though the explanation of potencies that often accompanies the estimated value clearly states that the numbers provided are upper bounds and that "the true risk is unknown and may be as low as zero." (12). In spite of increasing emphasis on the use of weight-of-evidence information, these caveats often go unheeded, and the upper bounds are used as though they represent an approximation of the risk as it really exists. For instance, upper bound estimates of population burdens have been reported in the media as though they were estimates of actual or predicted "body counts."

The estimates do not estimate actual risk because of the upper bound nature of the statistical methodology, as described earlier, and because of the biological default positions built into the process. The published risk assessment guidelines (12,13) clearly state a preference for risk estimates based on the most sensitive study, species, sex and strain, as is often done in other safety assessments. But, this bias, together with the inherent statistical biases, results in a conservative process that does not estimate risk. On the other hand, the methodology provides a comparative framework within which to evaluate different risk scenarios within the risk management process. As has been seen, there is no viable generic alternative to the existing LMS procedure. Future improvements are likely to be in the incorporation of compound specific biological information, rather than in the development of alternative models.

4.11 Pharmacokinetic Modeling

A major assumption in linear risk assessment extrapolations is that the internal biologically effective dose of a compound is proportional to administered dose over the entire dose range. This is a simplifying assumption that is likely not true in most cases. Simplistic pharmacokinetic models have been used in the past to predict the concentration of parent compound and metabolites in the blood and at reactive sites, where possible. Cornfield (83), Gehring and Blau (84), and Anderson, Hoel, and Kaplan (85) have extended this concept to include rates for macromolecular events (e.g., DNA damage and repair) involved in the carcinogenic process.

Recent extensions of pharmacokinetic models (61) to include physiologically realistic descriptions of species compartments and measured physiological parameters in the models have greatly increased the usefulness of pharmacokinetic models for risk assessment purposes. These models (physiologically based pharmacokinetic models, or PB-PK models) allow generic assumptions in the risk assessment process to be replaced by compound-specific and species-specific experimental information. These models can then improve the reliability of actual risk estimates by providing quantitative information on the relationship between administered dose and internal (delivered) dose. This information is critical in extrapolating from high doses to low doses and in extrapolating across routes of exposures as well as across species. Apart from explaining perceived discrepancies between the results of bioassays utilizing different routes of exposures or different species, PB-PK models provide considerable insight into the *actual* risk for specific exposure scenarios.

The use of internal dose estimates in place of administered dose does not necessarily decrease low dose risk estimates. The degree to which pharmacokinetically derived internal doses change predicted low dose risks depends on the specific internal dose versus administered dose relationship. In some cases the relationship between the two dose functions will be linear, in which case the risk estimates will not change. In other cases, such as exposure to very high levels of vinyl chloride (45), a supralinear internal dose function could result from the proximate carcinogen being formed in a saturable metabolic pathway. This phenomenon does not happen at low doses, in fact, the opposite may be true (86). In the case of methylene chloride, a sublinear internal dose function demonstrates that the low-dose hazard from the compound is not as great as would have been predicted from the default assumptions. EPA has utilized part of this PB-PK information in the derivation of their unit risk for methylene chloride.

In the future as more is understood about the mechanisms of carcinogenesis, formalized quantitative approaches incorporating pharmacokinetic data will likely become more useful in risk assessment.

4.12   Other Types of Models

Dissatisfaction with the linearized multistage procedure, particularly in the case of perceived nongenotoxic chemical carcinogens, has prompted the search for more flexible alternatives. Much attention has focused on stochastic models that include only two genetic events but allow for tissue growth and for proliferation of cells that have experienced the first of these events (87–89). The models consider three types of cells—normal cells, initiated cells, and malignant cells—and use birth and death processes to model two-stage clonal expansion of the cell pool. These models are variously called two-stage clonal expansion models, stochastic birth–death–mutation models, and initiation–promotion–progression models. Initiation refers to the occurrence of the first genetic event, which predisposes a cell to malignancy, and progression to the occurrence of the second genetic event, which transforms the initiated cell to a malignant one. Promotion generally refers to expansion of the pool of initiated cells. Without the promotional component, this model reduces to the ordinary two-stage Armitage–Doll model.

These models have the flexibility to explain different phenomena such as genetic predisposition to cancer, increased risk of some childhood cancer rates, initiation-promotion, and antagonism or synergism. As with any model, however, there are biological assumptions that must be validated. In particular, the relationship between cellular proliferation and the oncogenic process must be, and is currently being, explored further. Cellular proliferation, which expands the population of initiated cells and promotes the phenotypic expression of cancer, has a powerful effect on the model.

The validity and use of these models in regulatory applications will depend on the ability to measure or estimate the mutation and proliferation parameters, which is in turn dependent on the incorporation of the necessary mechanistic studies in testing protocols.

5   EXPOSURE ASSESSMENT

The risk assessment components of hazard identification and dose–response assessment have been described. These are primarily carried out by toxicologists and risk assessors in government agencies or corporate settings. This section is devoted to exposure assessment, the third leg of risk assessment and the part of risk assessment most familiar to industrial hygienists. Exposure assessment is often thought to be the component of risk assessment with the least inherent uncertainty because of the ability to monitor exposures in some cases and the availability of relatively well validated exposure models (90). This is only partially true, however, since most exposure assessments are not conducted in ways that take full advantage of the range of tools and available information. For that reason, there is usually a great deal of room for improving exposure distribution estimates. In this section the performance of exposure assessments for use in assessing cancer risk is discussed, and some distinctions are drawn with respect to the current practice of most industrial hygienists.

5.1 Issues in Environmental Exposure Assessments

While occupational exposure assessments for use in quantitative cancer risk assessments are less common, environmental exposure assessments for such purposes have become very widely performed. These usually occur under various U.S. EPA and state programs including Superfund, RCRA Corrective Action, incinerator evaluations, air toxics evaluation, multimedia risk assessments, and drinking water evaluations (82,91,92). A virtual culture for performing these types of assessments has evolved, and the EPA has codified its recommended practices in environmental exposure assessment in many different volumes (93,94). These assessments also focus on the long-term average exposure, but several key issues have emerged in performing such assessments.

Environmental exposure assessments are almost always performed using *modeling* techniques rather than measurement. There is a recognized need to do more measurement of environmental exposures and to validate models, but the difficulties in making meaningful environmental exposure measurements have made this area lag occupational exposure assessment. The modeling of environmental exposure generally follows several steps:

1. *Emission rates* of substances are estimated using either mass balance models or selected measurements of concentrations in stacks, pipes or bagged valves.
2. These emission rates serve as inputs into *transport and fate models* that predict concentrations in various microenvironments (personal breathing zones, drinking water sources, etc.). Representative of these models are air dispersion models, water dispersion models, groundwater models, multimedia models, and other transport models. Often these models do not include incorporation of chemical degradation processes that can significantly reduce long-term concentrations of substances in the various environmental media. All of these models require extensive inputs for various parameters including meteorology, hydrology, chemical characteristics, and geography. When data are not available for these, it is common to make conservative assumptions that tend to multiply throughout the calculations and often lead to gross overestimates of environmental concentrations (95).
3. Once microenvironmental concentrations are estimated, *human activity and behavior* must be modeled in order to estimate exposures. This would include parameters related to physiology (skin surface area, body mass, etc.), human intake rates for inhalation, various food and beverage consumption rates including water, and length of time at given locations in terms of residence duration and within-day and between-day mobility (96).

5.2 Regulatory Environmental Exposure Assessments

Many of the environmental exposure assessment programs involve the evaluation of what has become known as the *maximum exposed individual,* or MEI (97). The MEI has been used as a surrogate for evaluating greatest potential individual risk in an exposed population, but in practice it has become an almost meaningless exposure measure due to conservative assumptions (98). As an example, in the air toxics MEI the commonly used model assumes that a person resides at the fenceline (or at nearest residence) for every hour of a 70-year lifetime, indoor and outdoor air concentrations are the same, the person breathes a rather large amount of air per day, and the plant emits at a given, unchanging level over the entire 70-year period. An alternative analysis (98) suggested that this procedure overestimates the exposure of the MEI by an order of magnitude. One should always keep in mind that this is the upper bound estimate of the exposure of the most exposed person in the whole population—not a measure of the individual risk faced by an average person around a site.

A suggested solution to this exposure assessment problem is the use of *statistical distributions of exposures* around a given facility (98,99). With distributional information about population exposures, risk managers can use summary measures (such as the mean exposure) as well as select upper percentiles of the exposure distribution to protect individual risks to an appropriate level. To date, such population assessments have been rarely performed, but with growing requirements such as the hot spots rule in California (100,101), the computer technology and software is growing more sophisticated, and estimation of population distributions is becoming more common.

Beyond the logging of exposure estimates by census tracts (exposure variability), there is sometimes a need to evaluate uncertainties in estimates of exposures. Such uncertainties arise due to sampling error, variability in concentrations, uncertainty in applied models and their parameters, and other sources. When necessary, such uncertainties can be dealt with using computer simulation methods including Monte Carlo and Latin Hypercube methods. The use of simulation involves the specification of probability distributions describing all the parameters in a given exposure equation. For example, with a particular Superfund exposure assessment, ingestion of dirt, one would have the following distributions to consider: concentration distribution in soils, degradation rates in soil and transport rates from soil to groundwater and soil to air, ingestion rates per age category, body masses per age category, uptake rates from the digestive system, and averaging time information describing how long people live around the site in question. For each of these parameters an input distribution would be specified based on existing data (e.g., see the EPA *Exposure Factors Handbook*) (96), and the simulation would involve random selection from each distribution over some specified number of trials. Out of the simulation one determines the statistical distribution of exposure, which accounts for variability in the various input parameters. A recent technical report described the application of Monte Carlo simulation to most of the common EPA exposure models (Versar/CMA Report) (102). The same methodology could be used to more accurately characterize occupational exposures.

Intermediate to the full characterization of distributions of exposure, one can select more scientifically reasonable assumptions than are seen in the typical MEI scenario. The uncertainty can be managed by selecting appropriate percentiles of the particular parameter distributions rather than relying on unreasonable upper bounds for each exposure assumption. People vary in terms of height, weight, pulmonary volume, water consumption, and food consumption as well as particular food fractions consumed, such as meat, vegetables, and so forth. They also vary as to how long they work on one job and how long they live in one place. The issue then becomes that of picking the appropriate percentile of these distributions to use for regulatory purposes. Should the median be used or the 90th percentile or the 99th percentile?

In a manner analogous to the estimate of potency, a "more central" exposure estimate would also be prepared as part of a sensitivity analysis. In this way a more likely estimate would be available to provide the decision maker with two estimates in the exposure range. An upper bound of exposure would be presented, which would be unlikely to be exceeded; as well as a more central estimate.

Some assumptions used to characterize the MEI include (102):

- Seventy years of continuous exposure at one location, the fenceline in the case of air permit applications.
- 24 hr/day exposure
- 2 liters/day of water consumption
- 20 m^3/day of air inhaled per 24 hr
- 100 percent of the inhaled compound is bioavailable and is absorbed
- All crops are treated with a specific pesticide and all consumed food has residues at the maximum allowable level.

More central estimates could be based on modifications to all or some of the above assumptions. Examples include:

- Exposure occurs at the nearest residence or potential point of contact, rather than at the fenceline of a specific facility.
- Length of exposure is dependent on the type of source, e.g., 20 years for an agricultural product, 35 years for an incinerator.
- Lower ventilation rates, for instance 16 m^3 per day, are more appropriate for lifetime averages unless constant, moderate to heavy physical exertion is involved.
- Water consumption of much less than 2 L per day would be appropriate for compounds that would volatilize in cooking or tea and coffee preparation.
- Use analytically determined food residues, rather than the tolerance level.
- Other assumptions to be modified would depend on the exposure being estimated. For example, in estimating values for ingestion of soil by children, bioavailability, human activity patterns, and so forth, an estimate closer to the median of the distribution curve, rather than an extreme, would be selected.

5.3 Special Considerations in Occupational Exposure Assessment

For most industrial hygienists, the assessment of risk involves the comparison of exposure measurements with guidelines based on a threshold concept (e.g., the threshold limit values, or TLVs) (30). Because of the threshold concept, most monitored exposures below the threshold are considered safe and reported as such. When one is conducting exposure assessments for use in risk assessment, the assessed exposure is often fed into an algorithm that predicts an upper bound on cancer risk; hence the risk result is usually a very small number such as "one in a million," but it will not be a safe level so much as a level for comparison with a risk standard. The second major difference in the common assessment performed by hygienists is the collection of monitoring over an 8-hr or 15-min period with comparison to a TLV of the same averaging time. For cancer risk assessment the usual exposure metric is *lifetime average exposure* rather than the typically used short-term exposure measures.

5.4 Why Lifetime Average Exposure?

Many hygienists question why they need to switch the focus of their exposure assessments from daily assessments to longer-term assessments. Many chemicals that can cause cancer in humans and animals seem to act in a chronic fashion—that is, the long-term interaction of the substance with the physiology dictates the effect rather than short-term peaks of exposure (103). Because of this and due to practical considerations such as logistics, most animal bioassays are conducted in a way that mimics a chronic lifetime of exposure at a specified level. Further, epidemiology studies that seek to define dose–response relationships for cancer often use cumulative exposure as the metric of dose (104,105) largely because the detailed pattern of exposure is unknown. When these animal or epidemiological data are used to specify a unit risk of cancer (risk per unit exposure), the subsequent use of the unit risk assumes that the estimated human exposure is the lifetime average exposure.

While there are some problems with the use of lifetime average exposure (e.g., it assumes virtually no dose rate dependence), it is the appropriate metric when using unit risk estimates based on chronic studies. Information about peaks of exposure are most vital for determining control strategies for reducing overall exposures during identifiable tasks (106); however, for cancer risk assessment, the current science is not able to make much use of this more complete information. The exception is a physiologically based pharmacokinetic model that is time dependent and can handle short term exposure information (107).

For the hygienist the initial task in quantitative risk assessment is to estimate the average occupational exposure (i.e., the average exposure to a compound during the time period spent on the job during a working lifetime). This can be a difficult task and will be discussed more fully in the following sections. One additional correction is needed beyond the estimation of average occupational exposure: it must be converted to lifetime average exposure. A commonly used correction has the following form:

$$\text{Lifetime average exposure} = \text{average occupational exposure} \times \frac{40 \text{ years}}{70 \text{ years}}$$

$$\times \frac{250 \text{ days}}{365 \text{ days}} \times \frac{10 \text{ m}^3}{20 \text{ m}^3}$$

The lifetime average exposure reflects the average exposure experienced during the time at work with an assumed exposure away from work of zero. (This may not be true in all cases, but if one is estimating the incremental cancer risk due to exposures that occurred during work periods, then it is appropriate to make this assumption. However, if one's goal is to estimate total risk from exposure to a given chemical, then off the job contact should also be averaged into the lifetime number). The 40/70 correction reflects that people typically work at most 40 years in a lifetime of average duration 70 years. The 250/365 correction reflects an assumption that a typical worker works 250 days per year in a year that he is on the job. And last, the 10/20 correction reflects that, of the 20 m^3 that the typical adult breathes in a day, half is actually breathed at the workplace where one is using more energy. (An alternative correction for this last one is 8/24, which is a simple reflection of the hours worked in a typical day of work.) Collectively, these corrections reduce the average occupational exposure by a factor of 5, which in turn would reduce predicted risk by a factor of 5 assuming a linear risk model. All of these corrections could be implemented under other assumptions, particularly where more specific information is available such as actual lifetime durations of exposure. However, these are customary values used by the U.S. EPA and other government agencies as defaults when no more detailed information is available.

5.5 The Homogeneous Exposure Group Concept

Another idea worth mentioning in this section is the concept of the homogeneous exposure group (HEG). When one is working with large populations of workers (or even small ones), it is indeed rare for groups of workers to experience identical exposure profiles or distributions. In occupational exposure assessment, the HEG concept has evolved to help permit more accurate exposure and risk assessments. Homogeneous exposure groups are groupings of workers that have the same or very similar profiles of exposures; these subdivisions of workers will be assigned exposure estimates or distributions that are assumed applicable to all in the group. It is possible to have HEGs of only one person. Homogeneous exposure groups are determined using either data analysis (e.g., analysis of variance methods) or using subjective methods following exposure task analysis. Homogeneous exposure groups are described in more detail elsewhere and are equivalent to stratified random sampling in statistics (106). They are mentioned here because it may be important to split populations into groups of similar exposure so that risk for each group can be accurately predicted. Lumping of mixed exposure profiles would lead to risk estimates that may be conservative for some, but for others may be underestimates of actual risk.

5.6 Methods for Assessing Occupational Exposure

There are essentially three ways to assess occupational exposures: monitoring, mathematical modeling, and professional judgment. Professional judgment is probably the most commonly used technique among industrial hygienists, but it is also the most highly uncertain. Typically it is used to group workers into HEGs, and then each group is assigned a degree of exposure usually based on fractions of the TLV. This technique is useful for screening risk assessments. The validity of the subjectively assigned exposure increases with additional exposure monitoring results. This technique is also commonly used for assessing exposures of workers in retrospective epidemiology studies (108,109).

Mathematical modeling of exposures is also a commonly used technique, particularly when the exposures have already occurred or when the exposures have yet to happen such as in a planned facility (110,111). Modeling encompasses almost any mathematical representation of a process, and would include indoor air models, time–task activity models, and dermal absorption modeling. There have been many reviews of available models (112). The key is to use these models to address the issue of average exposure during the worker's working lifetime, preferably among homogeneous exposure groups. Also, it is important to account for uncertainties in models and their assumptions by sensitivity analysis or computer simulation.

Direct monitoring of exposures is the most direct and accurate method for assessing the concentration of a contaminant at the interface of a worker's physiology over some interval of time. The three main categories of worker exposure monitoring are: inhalation/air sampling, dermal exposure monitoring, and biological monitoring (106).

Inhalation/air monitoring is clearly the most commonly used technique of industrial hygienists. Personal air samples are usually taken in workers' breathing zones to monitor either for task exposures or for time-weighted average (TWA) exposures over longer time periods such as 8 hr. Also, one might perform area air monitoring in conjunction with time activity models to estimate average worker exposures (111).

Dermal exposure monitoring is the least sophisticated technique, although efforts have been made to improve it, especially for assessing dermal exposures during pesticide applications (110,113). Techniques include patch monitoring, skin washes, wipe testing, and other methods.

Biological monitoring has promise for assessing long-term exposures to some chemicals, but it should only be used in those cases where a well-established pharmacokinetic relationship has been determined in humans and for which known background levels in the population have been determined. Biological monitoring seems particularly well suited to assessment of long-term metal exposures, such as lead. Guidance on use of these tools can be found in the threshold limit value and biological exposure indices (TLV–BEI) documentation (30,114,115).

5.7 Statistics of the Average Exposure

Most often, exposure assessors are asked to estimate lifetime average inhalation exposure. Concentrations of most contaminants in air seem to follow a lognormal distribution, regardless of averaging time and sampling method. Lognormality is often observed in other data such as dermal and biological monitoring, but the following section will focus primarily on the inhalation case.

For purposes of estimating lifetime average concentration, one needs to focus on estimating the arithmetic mean of the lognormal distribution. There has been some confusion on this point in the industrial hygiene literature as some have suggested that the geometric mean be used. But for purposes of estimating the cumulative exposure over a lifetime of work, the arithmetic mean is the appropriate parameter (103,106,116). Biologically, total exposure (dose) is assumed to be the relevant parameter; this can only be estimated using arithmetic totals (i.e., based on the arithmetic mean).

5.8 Lognormal Statistics

In a lognormal distribution which is skewed to the right, there are several parameters of importance: the mode (most common value), the geometric mean (which is also the median, or 50th percentile), the arithmetic mean (center of mass of the distribution), and the geometric standard deviation. These are shown graphically in Figure 8.2. While most data probably are not lognormal in the truest sense of goodness of fit, lognormality closely approximates most industrial hygiene data sets and has an excellent property: logarithms of values of a lognormal population follow the normal distribution. This property makes working with the lognormal distribution very advantageous and facile (117,118). Lognormality of industrial hygiene data has been extensively discussed elsewhere (103,106). These two references give more detail on the use of lognormal statistics with industrial hygiene data.

Several key assumptions need to be at least acknowledged when performing industrial hygiene data analysis:

Data should be randomly sampled if possible or should be representatively sampled with no conscious investigator biases (e.g., worst-case samples).

Data should only be combined for statistical analyses if the averaging periods were similar (15-min samples should not be mixed with 8-hr samples).

Nondetectable samples should not be discarded; they should be included in the

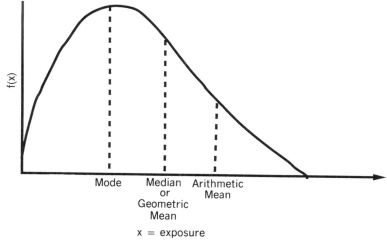

Figure 8.2 Low dose linearity.

analysis at some fraction of the Limit of Detection (LOD), typically selected at $\frac{1}{2}$ LOD or randomly included in the Not Detected (ND) interval using, for example, simulations from a lognormal distribution.

The most commonly used technique for estimating long term exposures is to estimate annual average exposures from samples within a year long period and then combine these annual averages across years. It is almost always true that the exposure assessor will necessarily have to extrapolate to time periods with no sampling data (either retrospectively or prospectively) (109). In these cases, the quality of the estimates will depend on the competency of the exposure assessor to make such judgments. Many exposure assessors will supplement time periods of no data with estimates from mathematical models for exposure in order to arrive at reasonable estimates of lifetime average worker exposures (111).

The need to estimate the *lifetime* average exposure is probably the major difference that the hygienist needs to be aware of in making estimates for purposes of human cancer risk assessment.

6 RISK CHARACTERIZATION

Risk characterization is the fourth and final step in the risk assessment paradigm. It is the most important step because from it comes application of the risk assessment results to decision making and public perception; however, it traditionally has been given the least amount of attention in this process (16).

Risk characterization is the culmination of information from the first three steps of risk assessment: hazard identification, dose–response assessment, and exposure assessment. In a risk characterization, it is the risk assessor's task to document clearly and concisely all that is known (and how little may be known) about the potential for health effects in an exposed population. This risk characterization forms the basis of a communication tool. Will it be used to brief risk managers who are charged with making a decision [e.g., permissible exposure limits (PEL) level? new control technology on an existing stack? clean up levels for an abandoned Superfund site?]. It often will be used to communicate risk information to the general public around a facility or to plant workers. Because of the communication aspect to risk characterization and because it is where the scientific experts usually remove themselves from the process, it is vital that the risk characterization give a clear summary of what the expert (or group of experts) determined and inferred during the months or years of data evaluation and what simplifying assumptions were made (119,120). All too often, risk characterizations boil down to simple summary statements such as: given the estimated maximum exposure levels in the population, the upper bound cancer risk would be one in one hundred thousand (20 cancer deaths expected due to exposure in the population). As described in detail in previous sections, this kind of summary is wholly inadequate for scientific communications with risk managers and the public. The following section describes ways for improving the content of risk characterization. More detailed ideas can be found in other references (16).

6.1   Ideas for More Useful Risk Characterization Reports

Executive Summary is the first section of the report. It must not be so concise as to conceal the wealth of information contained in the report. It must give a summary of the conclusions but must also describe the uncertainties in the assessment and the data gaps. Discussion of uncertainty is often missing in summaries of risk assessments.

Purpose of the Risk Assessment serves as an introduction to the risk characterization report. The use of the chemicals should be described to lend some perspective on why the chemical is important in commerce or use. The population exposed should be briefly described and the potential exposure pathways mentioned. Finally, the types of health effects of concern should be briefly described.

Hazard Identification Summary begins a more detailed section of the report, which forms the main body of its content. At least the following should be included:

Review of the evidence of noncancer health effects.

Review of the evidence of carcinogenic effects in humans, animals and in vitro test results.

Discussion of the mechanisms of cancer in the positive studies, differentiating what is known from what is conjecture, but including both.

Determination of the potential for human carcinogenicity at expected exposure levels based on all the results (in some cases, classification determination).

Summary statements of toxicological and epidemiological experts supporting or refuting the classification (121–123); statements as to the need for quantitative dose–response assessment at current levels of human exposure.

Dose Response Assessment Summary should include the following:

Discussion of noncancer health effect threshold determinations method applied, health endpoint, and safety/uncertainty factors with rationale (36,124).

Discussion of threshold determination if there is evidence that a confirmed animal carcinogen is believed not to cause cancer in humans. The health endpoint used, possible safety factors to be applied, and mechanisms of action should be described in detail.

Discussion of the determination of cancer potency for presumed human carcinogens. Description of the data selected for dose–response (epidemiological or animal) as well as the alternative sets not used and the dose–response model fitted to the data. Distributions of potencies should be described when appropriate.

If physiologically based pharmacokinetic modeling is available and used, its adjustments to the curve-fitting procedure should be described in detail along with the uncertainty in the models and their application. Alternative biologically based models should be described when appropriate.

For any combination of the above dose–response curve fitting methods, a *statement of uncertainty in dose response* is required. In this section the strengths and weaknesses of the applied methods need to be explained. A presentation of the probability distribution of potencies or virtually safe doses should be estimated. This could involve an analysis across data sets using various fitting methods or could involve an information analysis that weights the values estimated out of different procedures and with varying data sets as to their plausibility as a human model. The key is to estimate a distribution or range of potencies or virtually safe doses (VSDs), which reflects uncertainty in the dose–response assessments (125).

The *Exposure Assessment Summary* should also focus on the presentation of a probability distribution of exposures for the population as a whole and for important subpopulations (such as highly exposed individuals or sensitive subpopulations). As an introduction to the section, the exact exposure metric used should be defined as well as the exposed population. The methods to assess exposure should be fully described including strengths and weaknesses (e.g., modeling versus monitoring). Consideration of population exposure versus individual exposure (equity issues) should be accommodated for later presentation to risk managers.

The ultimate *Risk Characterization* is the combination of exposure information and dose–response information to estimate health risks in an exposed population (simple summary has risk = exposure × potency or risk = fraction exceeding a safe level). In keeping with the need to express uncertainty openly and credibly, the risk characterization should be presented in probability distribution form, combining the distribution on potency with the distribution on exposure in a joint distribution on risk (125–127). This should be done for the population as a whole and for important subpopulations. If a threshold has been estimated, the appropriate probabilistic interpretation would be an estimation of the fraction of the population exceeding the threshold (and also for various subpopulations).

An important section often not included in a risk characterization is the specification of *Research Options* available to reduce uncertainty in risk estimates. It is important to state these recommendations while the work is still fresh in the minds of the experts conducting the risk assessment. For each research option, the estimated costs, potential impacts, and time horizon should be given for consideration by the risk manager.

Finally, an appendix should be included containing *Visual Summaries* of the risk assessment. For each of the components of risk assessment some visual accounting can be given either in tabular form or through figures. For example, in the dose–response, curves for the various curve fitting procedures could be given as well as a decision tree if information analysis has been used. For exposure, maps with isopleths and distributions of exposure should be given. For risk characterization, distributions of risk should be presented in several forms so that the most effective visual presentation will be available whether it be for risk managers or the general public.

7 SUMMARY

This chapter has attempted to describe the quantitative risk assessment process as currently practiced in U.S. regulatory agencies. Strengths and weaknesses of the system have been discussed together with changes that are possible and changes that are likely to occur. The present system is a useful tool as part of the regulatory process, but it is too simplistic and does not include much of the available data. Current practices are changing, however, on a compound-by-compound basis as more complete data become available. There is considerable controversy surrounding the use of default assumptions that are to be used unless there are adequate scientific data to demonstrate an alternative position. The EPA and OSTP guidelines discuss the defaults and reference the fact that these defaults can be overcome with sufficient data. A criticism of this system has been that the hurdles to overcome the default positions are too high; it is too difficult to overcome the default methodology.

The issues of how much data are necessary to replace default assumptions, and the reliance on upper bounds on risk rather than more central estimates, particularly for exposure estimates, are probably the two biggest controversial issues in current risk assessment methodology.

REFERENCES

1. D. J. Paustenbach, *The Risk Assessment of Environmental Hazards*, Wiley, New York, 1989.
2. "Future Risk: Research Strategies for the 1990s," U.S. Environmental Protection Agency Science Advisory Board, SAB-EC-88-040.
3. P. Slovic, B. Fishhoff, and S. Lichtenstein, "Facts and Fears: Understanding Perceived Risk," in R. Schwing and W. A. Albers, Jr., Eds., *Societal Risk Assessment: How Safe is Safe Enough?*, Plenum Press, New York, 1980.
4. V. T. Covello, D. von Winterfeldt, and P. Slovic, "Communicating Scientific Information about Health and Environmental Risks: Problems and Opportunities from a Social and Behavioral Perspective," in V. Covello, A. Moghissi, and V. R. R. Uppulori, Eds., *Uncertainties in Risk Assessment and Risk Management*, Plenum Press, New York, 1987.
5. V. T. Covello, P. M. Sandman, and P. Slovic, "Risk Communication, Risk Statistics, and Risk Comparisons: A Manual for Plant Managers," Chemical Manufacturers Association, Washington, D.C., 1988.
6. "Current Regulatory Issues in Risk Assessment and Risk Management," in Regulatory Program of the United States Government, April 1, 1990–March 31, 1991, Executive Office of the President, Washington, D.C.
7. "Update to the Health Assessment Document and Addendum for Dichloromethane (Methylene Chloride): Pharmacokinetics, Mechanism of Action, and Epidemiology," U.S. Environmental Protection Agency, EPA/600/8-87/030A, July 1987.
8. "Addendum to the Health Assessment Document for Tetrachloroethylene (Perchloroethylene)," U.S. Environmental Protection Agency, EPA/600/8-82/005FA, April 1986.
9. J. Ashby, N. G. Doerrer, F. G. Flamm, J. E. Harris, D. H. Hughes, F. R. Johannsen, S. C. Lewis, N. D. Krivanek, J. F. McCarthy, R. J. Moolenaar, G. K. Raabe, R. C. Reynolds, J. M. Smith, J. T. Stevens, M. J. Teta, and J. D. Wilson, "A Scheme for Classifying Carcinogens," *Regul. Toxicol. Pharmacol.*, **3**, 224–238 (1983).
10. N. Mantel and W. Bryan, "Safety Testing of Carcinogenic Agents," *J. National Can. Inst.*, **27**, 455–470 (1961).

11. "Scientific Bases for Identification of Potential Carcinogens and Estimation of Risks," *Fed. Reg.*, **44**, 39858 (1979).

12. U.S. Environmental Protection Agency, "Guidelines for Carcinogen Risk Assessment," *Fed. Reg.*, **51**, 33993–34003 (1986).

13. Office of Science and Technology Policy, "Chemical Carcinogens: Notice of Review of the Science and its Associated Principles," *Fed. Reg.*, **49**, 21594–21661 (1984).

14. National Research Council, Committee on the Institutional Means for Assessment of Risks to Public Health, *Risk Assessment in the Federal Government: Managing the Process*, National Academy Press, Washington, D.C., 1983.

15. Presentation at the American Industrial Health Council Annual Meeting, Washington, D.C., 1990.

16. Presentation of "Risk Assessments of Carcinogens," available from the American Industrial Health Council, Suite 300, 1330 Connecticut Avenue NW, Washington, D.C. 20036, July 1989.

17. W. D. Ruckelshaus, "Science, Risk, and Public Policy," *Science*, **221**, 1026–1028 (1984).

18. "Risk, Science, and Democracy," *Issues Sci. Tech.*, NAS, **1**(3), 19–38 (1985).

19. U.S. Environmental Protection Agency, Office of Policy Analysis and Office of Policy, Planning, and Evaluation, "Unfinished Business: A Comparative Assessment of Environmental Problems," U.S. EPA, Washington, D.C., February 1987.

20. A. M. Finkel, "Confronting Uncertainty in Risk Management: A Guide for Decision Makers," Resources for the Future, Washington, D.C., 1990.

21. R. L. Sielken, "Evaluation of Chloroform Risk to Humans," presented at the Toxicology Forum, 1993 Annual Winter Meeting, February 15–17, 1993, Washington, D.C.

22. J. K. Haseman, D. D. Crawford, J. E. Huff, G. A. Boorman, and E. E. McConnell, "Results from 86 Two-Year Carcinogenicity Studies Conducted by the National Toxicology Program," *J. Toxicol. Environ. Hlth.*, **14**, 621–639 (1984).

23. D. P. Rall (letter), "Carcinogens and Human Health: Part 2," *Science*, **251** (January 4, 1991), pp. 10–11.

24. B. N. Ames and L. Swirsky Gold, "Too Many Rodent Carcinogens: Mitogenesis Increases Mutagenesis," *Science*, **249**, 970–971 (1990).

25. G. M. Williams and J. H. Weisberger, "Chemical Carcinogens," in Casarett and Doull's *Toxicology: The Basic Science of Poisons*, C. D. Klaassen, M. O. Amdur, and J. Doull, Eds., 3rd ed., Macmillan, New York, 1986, pp. 99–173.

26. B. E. Butterworth and T. Slaga, Eds., *Nongenotoxic Mechanisms in Carcinogenesis* (Banbury Report 25), Cold Spring Harbor Laboratory, Cold Spring Harbor, NY, 1987.

27. W. T. Stott, R. H. Reitz, A. M. Schumann, and P. G. Watanabe, "Genetic and Nongenetic Events in Neoplasia," *Food Cosmetics Toxicol*, **19**, 567–576 (1981).

28. Letter from N. Nelson, R. A. Griesemer, and J. Doull to L. M. Thomas, Director of EPA, March 9, 1988, SAB-EHC-88-011.

29. J. D. Graham, "Communicating about Chemical Hazards," *J. Policy Analysis Manag.*, **8**, 307–313 (1989).

30. *Threshold Limit Values for Chemical Substances and Physical Agents and Biological Exposure Indices*, American Conference of Governmental Industrial Hygienists, pp. 41–42, 1993–1994.

31. *California Code of Regulations* (CCR), Division 2, Chapter 3, Safe Drinking Water and Toxic Enforcement Act of 1986, Articles 6 and 7, April 12, 1991.

32. J. H. Weisburger and G. M. Williams, "Chemical Carcinogens," in Casarett and Doull's *Toxicology, The Basic Science of Poisons*, 2nd ed., Macmillan, New York, 1980, pp. 84–138.

33. M. A. Pereira, "Mouse Liver Tumor Data: Assessment of Carcinogenic Activity," *Toxicol. Ind. Hlth.*, **1**(4), 311–333 (1985).

34. R. D. Bruce, "Low Dose Extrapolation and Risk Assessment," *Chemical Times and Trends*, October, 20–23 (1980).

35. C. A. Kimmel, "Quantitative Approaches to Human Risk Assessment for Noncancer Health Effects," *NeuroToxicology*, **11**, 189–198 (1990).

36. S. C. Lewis, J. R. Lynch, A. I. Nikiforov, "A New Approach to Deriving Community Exposure Guidelines from No-Observed-Adverse-Effect-Levels," *Regul. Toxicol. Pharmacol.*, **11**, 314 (1990).

37. C. Brown, "Statistical Aspects of Extrapolation of Dichotomous Dose Response Data," *J. Natl. Canc. Ins.*, **60**, 101–108 (1978).

38. Food Safety Council, "Quantitative Risk Assessment," *Food Cosmetics Toxicol.*, **18**, 711–734 (1980).

39. D. A. Krewski and C. C. Brown, "Carcinogenic Risk Assessment: A Guide to the Literature," *Biometrics*, **37**, 353–366 (1981).

40. K. S. Crump, "Designs for Discriminating Between Binary Dose-Response Models with Applications to Animal Carcinogenicity Experiments," *Comm. Stat., Theory, Meth.*, **11**, 375–393 (1982).

41. FDA Advisory Committee on Protocols for Safety Evaluation, Panel on Carcinogenesis, "Report on Cancer Testing in the Safety Evaluation of Food Additives and Pesticides," *Toxicol. App. Pharmacol.*, **20**, 419–438 (1971).

42. K. S. Crump, D. G. Hoel, C. H. Langley, and R. Peto, "Fundamental Carcinogenic Processes and Their Implications for Low Dose Risk Assessment," *Can. Res.*, **36**, 2973–2979 (1976).

43. D. G. Hoel, "Incorporation of Background in Dose-Response Models," *Fed. Proceed.*, **39**, 73–75 (1980).

44. D. Krewski and J. Van Ryzin, "Dose Response Models for Quantal Response Toxicity Data," in *Current Topics in Probability and Statistics*, M. Csorgo, D. Dawson, J. N. K. Rao, and E. Saleh, Eds., North Holland, New York, 1991.

45. P. J. Gehring, P. G. Watanabe, and C. N. Park, "Risk of Angiosarcoma in Workers Exposed to Vinyl Chloride and Predicted for Studies in Rats," *Toxicol. Appl. Pharmacol.*, **49**, 15–21 (1979).

46. F. R. Johannsen, Risk Assessment of Carcinogenic and Noncarcinogenic Chemicals, *Toxicology*, **20**, 341–367 (1990).

47. H. O. Hartley "Estimation of 'Safe Doses' in Carcinogenic Experiments," *Biometrics*, **33**, 1–30 (1977).

48. H. A. Guess, K. S. Crump, and R. Peto, "Uncertainty Estimates for Low Dose Extrapolations of Animal Carcinogenicity Data," *Can. Res.*, **37**, 3475–3483 (1977).

49. F. C. Lu, "Toxicological Evaluations of Carcinogens and Noncarcinogens: Pros and Cons of Different Approaches," *Regul. Toxicol. Pharmacol.*, **3**, 121–132 (1983).

50. F. W. Carlborg, "Dose-Response Functions in Carcinogenesis and the Weibull Model," *Food Cosmetics Toxicol.*, **19**, 255–263 (1981).

51. R. J. Sielken, Jr., "The Use of the Hartley-Sielken Model in Low Dose Extrapolation," in *Toxicological Risk Assessment*, D. Clayson, D. Krewski, and I. Munro, Eds., CRC Press, Boca Raton, 1985.

52. R. E. Albert, and B. Altshuler, "Considerations Relating to the Formulation of Limits for Unavoidable Population Exposures to Environmental Carcinogens," in *Radionuclide Carcinogenesis*, J. E. Ballou, R. H. Busch, D. D. Mahlum, and C. L. Sanders, Eds., AEC Symposium Series, CONF-72050, NTIS, Springfield, VA, 1973, pp. 233–253.

53. A. Whittemore and J. B. Keller, "Quantitative Theories of Carcinogenesis," *SIAM Rev.*, **20**, 1–30 (1978).

54. D. W. Gaylor, and R. L. Kodell, "Linear Interpolation Algorithm for Low Dose Risk Assessment of Toxic Substances," *J. Environ. Pathol. Toxicol.*, **4**, 305–312 (1980).

55. D. G. Hoel, "Carcinogenic Risk: Comment on Regulation of Carcinogens," by E. Crouch and R. Wilson, *Risk Anal.*, **1**, 63–64 (1981).

56. J. E. Gibson, "Risk Assessment Using a Combination of Testing and Research Results," in *Proceedings of the Third Annual Chemical Industries Institute of Toxicology Conference, 'Formaldehyde Toxicity,'* J. E. Gibson, Ed., Hemisphere, New York, 1982.

57. J. A. Todhunter, "Review of Data Available to the Administrator Concerning Formaldehyde and Di (2-Ethylhexyl) Phthalate (DEHP)," U.S. Environmental Protection Agency Memorandum, Office of Pesticides and Toxic Substances, Washington, D.C., February 10, 1982.

58. P. Armitage, and R. Doll, "Stochastic Models for Carcinogenesis," in *Proceedings of the Fourth Berkeley Symposium on Mathematical Statistics and Probability*, Vol. 4, Berkeley, University of California Press, 1961, pp. 19–38.

59. R. Doll, "The Age Distribution of Cancer: Implications for Models of Carcinogenesis," *J. Roy. Statist. Soc.*, **A134**, 133–166 (1971).

60. MULTWEIB Risk Assessment Software for Cancer Endpoints, Clement International Corporation, Ruston, LA, 1982.

61. M. E. Andersen, H. J. Clewel III, M. L. Gargas, F. A. Smith, and R. H. Reitz, "Physiologically Based Pharmacokinetics and Risk Assessment Process for Methylene Chloride," *Toxicol. Appl. Pharmacol.*, **86**, 341 (1986).

62. T. B. Starr, "Quantitative Cancer Risk Estimation for Formaldehyde," *Risk Anal.*, **10**, 85–91 (1990).

63. K. S. Crump, "An Improved Procedure for Low-Dose Carcinogenic Risk Assessment from Animal Data," *J. Environ. Pathol. Toxicol.*, **5**, 675 (1981).

64. E. L. Anderson and the Carcinogen Assessment Group of the U.S. Environmental Protection Agency, "Quantitative Approaches in Use to Assess Cancer Risk," *Risk Anal.*, **3**, 277–295 (1983).

65. R. B. Howe and K. S. Crump, "GLOBAL82: A Computer Program to Extrapolate Quantal Animal Toxicity Data to Low Doses," prepared for the Office of Carcinogenic Standards, OSHA, U.S. Department of Labor, Contract 41USC252C3, 1982.

66. GLOBAL86 Risk Assessment Software for Cancer Endpoints, Clement International Corporation, Ruston, LA, 1986.

67. K. S. Crump and R. B. Howe, "A Review of Methods for Calculating Statistical

Confidence Limits in Low-Dose Extrapolation," in *Toxicological Risk Assessment: Biological and Statistical Criteria*, Vol. 1, D. B. Clayson, D. Krewski, and I. Munro, Eds., CRC Press, Boca Raton, 1985, pp. 187–203.

68. W. G. Flamm and J. S. Winbush, "Role of Mathematical Models in Assessment of Risk and in Attempts to Define Management Strategy," *Fundament. Appl. Toxicol.*, **4**, S395.1 (1984).

69. Food and Drug Administration, "Sponsored Compounds in Food Producing Animals; Proposed Rule and Notice," *Fed. Reg.*, **50**, 45530–45556 (1985).

70. D. A. Krewski, D. W. Gaylor, M. Bickis, and M. Szyszkowicz, "A Model Free Approach to Low-Dose Extrapolation," *Environm. Hlth. Perspect.*, **90**, 279–285 (1991).

71. D. A. Krewski, Cancer Dose-Response Extrapolation, Workshop #3, Sponsored by the Cancer Dose-Response Working Group of the International Life Sciences Institute, Washington, D.C., March 23–24, 1992.

72. P. J. Gehring, R. H. Reitz, and C. N. Park, "Carcinogenic Risk Estimation for Chloroform: An Alternative to EPA's Procedures," *Food Cosmetics Toxicol.*, **16**, 511–514 (1978).

73. M. L. Dourson and J. F. Stara, "Regulatory History and Experimental Support of Uncertainty (Safety) Factors," *Regul. Toxicol. Pharmacol.*, **3**, 224–238 (1983).

74. C. T. Travis and R. K. White, "Interspecies Scaling of Toxicity Data," *Risk Anal.*, **8**, 119–126 (1988).

75. D. E. Burmaster and K. Von Stackelberg, "Using Monte Carlo Simulations in Public Health Risk Assessments: Estimating and Presenting Full Distributions of Risk," *J. Exposure Anal. Environ. Epidemiol.*, **1**(4), 491–512 (1991).

76. N. C. Hawkins, "Conservatism in Maximally Exposed Individual (MEI) Predictive Exposure Assessments: A First-Cut Analysis," *Regul. Toxicol. Pharmacol.*, **13**, 107–117 (1991).

77. S. R. Hayes, "Addressing Indoor/Outdoor Differences and Population Activity/Mobility in Air Toxics Risk Decisions," Invited Paper for Presentation at 84th Annual Meeting, Air & Waste Management Association, Vancouver, B.C., Canada, Paper 91-170.9, pp. 107–117, June 1991.

78. J. P. Christopher, "Applied Action Levels in Air and Water for Trichloroethylene," presented at the 28th Annual Meeting of the Society of Toxicology, Atlanta, GA, March 1, 1989.

79. A. M. Schumann, J. F. Quast, and P. G. Watanabe, "The Pharmacokinetics and Macromolecular Interactions of Perchloroethylene in Mice and Rats as Related to Oncogenicity," *Toxicol. Appl. Pharmacol.*, **55**, 207–219 (1980).

80. I. Berenblum, "Tumour Formation Following Freezing with Carbon Dioxide Snow," *Brit. J. Experim. Pathol.*, **10**, 179 (1929).

81. G. J. Laroye, "How Efficient Is Immunological Surveillance Against Cancer and Why Does it Fail?" *Lancet*, **1**, 1097 (1974).

82. "Cancer Risk from Outdoor Exposure to Air Toxics," External Review Draft, U.S. Environmental Protection Agency, Office of Air Quality Planning and Standards, Research Triangle Park, NC 27711, September 1989.

83. J. Cornfield, "Carcinogenic Risk Assessment," *Science*, **198**, 693–699 (1977).

84. P. J. Gehring and G. E. Blau, "Mechanisms of Carcinogenesis: Dose Response," *J. Environm. Pathol. Toxicol.*, **1**, 163–179 (1977).

85. M. W. Anderson, D. G. Hoel, and N. L. Kaplan, "A General Scheme for the Incorporation of Pharmacokinetics in Low-Dose Risk Estimation for Chemical Carcinogenesis: Example—Vinyl Chloride," *Toxicol. Appl. Pharmacol.*, **55**, 154–161 (1980).

86. R. H. Reitz, A. L. Mendrala, C. N. Park, M. E. Andersen, and F. P. Guengerich, "Incorporation of In Vitro Enzyme Data into the Physiologically-Based Pharmacokinetic (PB-PK) Model for Methylene Chloride: Implications for Risk Assessment," *Toxicol. Lett.*, **43**, 97–116 (1988).

87. S. H. Moolgavkar and D. J. Venzon, "Two-Event Models for Carcinogenesis: Incidence Curves for Childhood and Adult Tumors," *Math. Biosci.*, **47**, 55–77 (1979).

88. S. H. Moolgavkar and A. G. Knudson, Jr., "Mutation and Cancer: A Model for Human Carcinogenesis," *J. Natl. Can. Instit.*, **66**, 1037–1052 (1981).

89. R. E. Greenfield, L. B. Ellwein, and S. M. Cohen, "A General Theory of Carcinogenesis," *Carcinogenesis*, **5**, 437–445 (1984).

90. Michigan State University, Center for Environmental Toxicology, "Reducing Uncertainty in Risk Assessment: Proceedings of a Conference and Workshop," Michigan State University, East Lansing, MI, 1989.

91. U.S. EPA, "National Emission Standards for Hazardous Air Pollutants; Benzene Emissions from Maleic Anhydride Plants, Ethylbenzene/Styrene Plants, Benzene Storage Vessels, Benzene Equipment Leaks, and Coke By-Product Recovery Plants, Final Rule," *Fed. Reg.*, **54**(177), 38044–38139 (1989).

92. U.S. EPA, "Proposed Rule: Corrective Action for Solid Waste Management Units (SWMUS) at Hazardous Waste Facilities," *Fed. Reg.*, **55**(145), 30798–30884 (1990).

93. U.S. EPA, "Proposed Guidelines for Exposure Related Measurements," *Fed. Reg.*, **53**, 48830 (1988).

94. U.S. EPA, "Methods for Assessing Exposure to Chemical Substances," Exposure Assessment Branch, Office of Health and Environmental Assessment, 9 Volumes, EPA 560/5-85-001 to EPA 560/5-85-009, 1985.

95. T. McKone and K. Bogen, "Predicting Uncertainty in Risk Assessment," *Environm. Sci. Tech.*, **25**(11), 112–118 (1991).

96. U.S. EPA, *Exposure Factors Handbook,* Exposure Assessment Group, Office of Health and Environmental Assessment, Washington, D.C., EPA/600/8-89/043, 1989.

97. B. D. Goldstein, "The Maximally Exposed Individual: An Inappropriate Basis for Public Health Decision Making," *Environ. Forum*, November/December 1989, p. 13.

98. N. C. Hawkins, "Conservatism in Maximally Exposed Individual (MEI) Predictive Exposure Assessments: A First-Cut Analysis," *Regul. Toxicol. Pharmacol.*, **14**, 107–117 (1991).

99. P. S. Price, J. Sample, and R. Strieter, "PSEM: A Model of Long-Term Exposures to Emissions From Point Sources," AWMA Vancouver Meeting, Paper 91-172.3, June 1991.

100. State of California Air Toxics "Hot Spots" Information and Assessment, Health and Safety Code, Part 6, pp. 183–191.

101. California Air Pollution Control Officers Association (CAPCOA), Air Toxics "Hot Spots" Program, "Risk Assessment Guidelines," January 1, 1991.

102. VERSAR Risk Focus, "Analysis of the Impact of Exposure Assumptions on Risk Assessment of Chemicals in the Environment," Phase Reports I, II, and III, Available from the Chemical Manufacturers Association, Exposure Assessment Task Group, 2501 M Street, Washington, D.C. 20036, 1991.

103. S. M. Rappaport, "Evaluation of Long-Term Exposures to Toxic Substances in Air," *Annal. Occupat. Hyg.*, **35**, 61–121 (1991).

104. H. Checkoway, J. M. Dement, D. P. Fowler, R. L. Harris, Jr., S. H. Lamm, and T. J. Smith, "Industrial Hygiene Involvement in Occupational Epidemiology," *Am. Ind. Hyg. Assoc. J.*, **48**(6), 515–523 (1987).

105. T. J. Smith, S. K. Hammond, M. Hallock, and S. R. Woskie, "Exposure Assessment for Epidemiology: Characteristics of Exposure," *Appl. Occupat. Environ. Hyg.*, **6**(6), 441–447 (1991).

106. N. C. Hawkins, S. K. Norwood, and J. C. Rock, (editors), "A Strategy for Occupational Exposure Assessment," American Industrial Hygiene Association, Exposure Assessment Strategies Committee, Akron, OH, 1991.

107. National Academy of Sciences, "Pharmacokinetics in Risk Assessment," in *Drinking Water and Health*, Richard D. Thomas, Ed., Vol. 8, National Academy Press, Washington, D.C., 1987.

108. N. C. Hawkins and J. S. Evans, "Subjective Estimation of Toluene Exposures: A Calibration Study of Industrial Hygienists," *Appl. Ind. Hyg.*, **4**(3), 61–68 (1989).

109. M. P. Dosemeci, A. Stewart, and A. Blair, "Three Proposals for Retrospective Semi-quantitative Exposure Assessments and Their Comparison with the Other Assessment Methods," *Appl. Occupat. Environ. Hyg.*, **5**(1), 52–59 (1990).

110. M. A. Jayjock and N. C. Hawkins, A Proposal for Improving the Role of Exposure Models in Risk Assessment. *Am. Ind. Hyg. Assoc. J.*, **54**(12), 733–741, 1993.

111. N. C. Hawkins, M. A. Jayjock, and J. R. Lynch, "A Rationale and Framework for Establishing the Quality of Human Exposure Assessments," *Am. Ind. Hyg. Assoc. J.*, **53**(1), 34–41 (1992).

112. M. A. Jayjock, "Assessment of Inhalation Exposure Potential from Vapors in the Workplace," *Am. Ind. Hyg. Assoc. J.*, **49**(8), 380–383 (1988).

113. B. A. Shurdut, "Dermal Exposure Assessments for Pesticide Applications," International Conference of the International Society for Exposure Analysis, Atlanta, GA, November 1991.

114. V. Fiserova-Bergerova, "Development of Biological Exposure Indices (BEIs) and Their Implementation," *Appl. Ind. Hyg.*, **2**, 87–92 (1987).

115. National Research Council (NRC), Report from the Committee on Biological Markers, 1987.

116. R. O. Gilbert, *Statistical Methods for Environmental Pollution Monitoring*, Van Nostrand Rheinhold, New York, 1987.

117. J. Aitchison and J. A. C. Brown, *The Lognormal Distribution—with Special Reference to Its Use in Economics*, Cambridge University Press, Cambridge, UK, 1957.

118. J. S. Evans and N. C. Hawkins, "The Distribution of Student's *t*-Statistic for Small Samples from Lognormal Exposure Distributions," *Am. Ind. Hyg. Assoc. J.*, **49**(10), 512–515 (1988).

119. N. C. Hawkins, "Health Risk Assessment: The Right Tool for the Job?" *Proceedings of the International Specialty Conference on How Clean is Clean?* Cleanup Criteria for Contaminated Soil and Groundwater, Air and Waste Management Association, 1991, pp. 130–142.

120. National Research Council, Committee on Risk Perception and Communication, *Improving Risk Communication*, National Academy Press, Washington, D.C., 1989.

121. J. D. Graham, N. C. Hawkins, and M. J. Roberts, "Expert Scientific Judgment in

Quantitative Cancer Risk Assessment," *Proceedings of the Banbury Conference on New Directions in Qualitative and Quantitative Aspects of Cancer Risk Assessment*, October 1987, pp. 231–244.

122. N. C. Hawkins and J. D. Graham, "Expert Scientific Judgment and Cancer Risk Assessment: A Pilot Study of Pharmacokinetic Data," *Risk Anal.*, **8**(4), 615–625 (1988).

123. S. K. Wolff, N. C. Hawkins, S. M. Kennedy, and J. D. Graham, "Selecting Experimental Data for Use in Quantitative Risk Assessment: An Expert Judgment Approach," *Toxicol. Ind. Hlth.*, **6**(2), 275–291 (1990).

124. U.S. EPA, "Interim Methods for Development of Inhalation Reference Doses," Office of Health and Environmental Assessment, Washington, D.C., 1987.

125. J. S. Evans, N. C. Hawkins, and J. D. Graham, "The Value of Monitoring for Radon in the Home: A Decision Analysis," *J. Air Poll. Cont. Assoc.*, **38**, 1380–1385 (1988).

126. A. M. Finkel, "Confronting Uncertainty in Risk Management: A Guide for Decision Makers," Center for Risk Management, Resources for the Future, 1616 P Street, NW, Washington, D.C. 20036, 1990.

127. A. M. Finkel and J. S. Evans, "Evaluating the Benefits of Uncertainty Reduction in Environmental Health Risk Management," *J. Air Poll. Cont. Assoc.*, **37**, 1164 (1987).

CHAPTER NINE

Biological Rhythms, Shift Work, and Occupational Health

R. D. Novak, Ph.D., M.P.H., M.P.A., and M. H. Smolensky, Ph.D.

1 INTRODUCTION

Humans are classified as a diurnally active species. This means that a majority of people prefer a routine of daytime (diurnal) work alternating with nighttime sleep. In spite of a strong preference for diurnal work schedules, a significant percentage of persons today in the United States and other industrialized nations are engaged in shift work requiring nighttime work (1–4). Even in developing countries, shift work is gaining popularity (5–12). Although shift work schedules are common in many industries, not all individuals are satisfactory candidates for these. It has been estimated that of all persons involved in shift work, approximately 10 percent enjoy shift work and 60 percent tolerate it well (13). Therefore, only 20–30 percent of all shift workers have difficulty with shift work and may be the population most likely at risk for various maladies and difficulties associated with working shifts. In the past the complaints of shift work intolerance were either viewed by management as organized labor's attempt to justify concessions or accepted as a manifestation of shift work. Over the last 40 years, however, researchers have become increasingly interested in the potential effects of shift work on the health and safety of workers. Only recently have the biological and psychosocial bases of the problems of shift workers been well understood.

Today, it is clear that human functions undergo periodic or cyclic alterations during each 24 hr in support of the predictable variation in human activity and rest

Patty's Industrial Hygiene and Toxicology, Third Edition, Volume 3, Part B, Edited by Lewis J. Cralley, Lester V. Cralley, and James S. Bus.
ISBN 0-471-53065-4 © 1995 John Wiley & Sons, Inc.

during this span. These 24-hr bioperiodicities often referred to as circadian (or about 24-hr) rhythms represent an important aspect of our genetic inheritance. The timing of the circadian peaks and troughs of our biological functions is largely determined by the daily scheduling of sleep and activity. That is, changes in the sleep–wake cycle impact the timing of biological functions. For example, persons who work only during the day and sleep during the night exhibit a different timing, with regard to clock hour, of their biological rhythms than do permanent night workers who are active during the nighttime and sleep during the daytime. When workers rotate between day and night shifts, their circadian rhythms become disorganized leading to a set of symptoms that are familiar to most rotating shift workers. At one time health professionals were concerned that employment on rotating shift work schedules resulted in increased morbidity and mortality. The results of some epidemiological studies suggest that this risk may exist for some individuals.

Taylor (14), a celebrated shift work researcher who conducted extensive epidemiologic investigations of morbidity and mortality patterns in day workers and shift workers, stated that ''man is a diurnal animal and society is designed for the day worker.'' The human preference for diurnal activity and nocturnal rest relates to both psychosocial and biological factors. Most people desire a work schedule that enables sufficient meaningful leisure time for family and friends and for attending cultural, sporting, or other events of interest. With regard to the latter factor, most if not all biological functions are circadian rhythmic in support of the alternating work–rest routine, typically including activity during the day and sleep at night. When working night shifts, the total adjustment of all circadian rhythms would be requisite. In many persons only partial adjustment is possible. During this time many unpleasant symptoms, similar to jet lag, are often experienced. The symptoms of jet lag and shift work, both involving a change in the clock hours of sleep and activity, commonly include fatigue, reduced mental and physical efficiency, and impaired performance as well as sleepiness and hunger at atypical times. In summary, the physiological and emotional conflicts caused by shift work include the disorganization of one's biological rhythms and the disruption of socialization patterns. The result is that people who may otherwise find their work agreeable become less productive, more fatigued, less happy, and perhaps more vulnerable to the adverse effects of exposure to chemical and physical agents in the workplace.

Shift work researchers attempt to minimize potential health hazards or other difficulties of shift work by a variety of methods. One approach involves reducing the adverse physiological consequences of rotating between day and night shifts (15–23). Others involve the use of mathematical models to adjust the limits for workplace exposures to hazardous materials so that persons engaged in the usual rotating 8-hr as well as the so-called unusual (e.g., 10- to 12-hr) work schedules will have the same degree of protection as persons working routine, 8-hr straight day schedules (24–33). Additional approaches involve methods that minimize the effect of shift work on employee morale, job productivity, accident occurrence and family life (21, 34–36). Recent findings from multidisciplinary investigations have resulted in a set of recommendations for minimizing certain difficulties of shift work. These are discussed by Rutenfranz (37,38), Reinberg (17), and Rosa et al. (39).

In attempting to more fully comprehend and alleviate the difficulties experienced by shift workers, a greater number of multidisciplinary research efforts are required. One of the biological disciplines providing useful solutions to some of the difficulties of shift work is chronobiology—that branch of science that investigates biological rhythms. The science of chronobiology is relatively new and unfamiliar to most industrial hygienists, toxicologists, and other occupational health professionals. For the most part, when these professionals were receiving their formal education, human biology was based solely on the theory of homeostasis. This concept was first introduced by Claude Bernard in the nineteenth century. According to homeostatic theory, a set of regulatory systems maintains the cells of the body in relative constancy. Although homeostatic theory was extremely useful for understanding the mechanisms of human physiological function during the first part of this century, new findings from chronobiologic research during the past three to four decades provide a different and broadened insight into the functioning of biological processes. We now know that the physiologic and metabolic activities of all living organisms are predictably rhythmic over time rather than static. Modern biologists now know that the functions of human beings and lower animals are strongly structured not only in space, as in anatomy, but also in time. It is important that occupational health professionals recognize the existence of biological rhythms since rhythm changes affect how human beings respond to various types of chemical challenges, medicines, and shift work schedules. In recent years there has been rapid growth in our knowledge and understanding of human biological rhythms. This knowledge is now being applied to clinical medicine (36,40–49) and to the occupational health aspects of shift work (15–17,19–21,23,37,50), which is the focus of this chapter.

Shift work is of such importance that it has been the subject of several international conferences. The objective of this chapter is to discuss the topic of shift work and chronobiology in relationship to the physical and emotional health of workers. The nature of shift work schedules, their popularity and the subject of human biological rhythms and their history (51) are reviewed. The significance of biological rhythms with regard to the differential biological tolerance of employees for various shift work schedules as well as the effects of shift work on the physiology, health, and well-being of employees is discussed in a manner that is pertinent to the occupational community. Although the authors have attempted to cite as many of the pertinent investigations as possible, the vastness of the literature on this topic precludes the referencing of all those who have made contributions to our knowledge base. For this we apologize.

2 HISTORY OF SHIFT WORK

Shift work may be alluded to as the oldest profession. Before recorded history warriors, guards, watchmen, and prostitutes worked unusual hours. Night operations in early warfare are well documented in the readings of Sun Tzu and the well-known sacking of Troy.

Even Hippocrates noticed a 24-hr variance in the symptoms of his patients. The father of occupational medicine, Bernardo Ramazzini noted in the sixteenth century that bakers and miners often worked odd hours and acted "more like bats than men" (52).

The early, primarily agrarian, societies were dictated by natural cycles of light and darkness. While the advent of the industrial revolution in the eighteenth century changed the way work was performed, natural cycles of light and darkness still determined the timing of work. Most labor was done in one 11-hr shift during the winter, and as daylight expanded the length of work was increased to a 14-hr shift during the summer. As these changes occurred gradually, the internal rhythms of workers could easily adjust.

In 1883 Edison developed the incandescent lamp and banished the dictation of work scheduling by natural light. The incandescent lamp allowed for a constant source of light and increased the capability of industry to run around the clock. Management was quick to notice that the capability of constant operation would allow for an increased return on investment in capital intensive industries. As industries embraced 24-hr operation, an increased need for social services (police and fire) developed, thus increasing the prevalence of shift workers. Most early shift workers were recruited from agriculture and were accustomed to working long hours. The usual schedule at that time dictated that workers remain in the plant 12 hr per day, 7 days a week. Typical shift schedules ran from 0600–1800 and 1800–0600 with rotations occurring each 2-week period. In order to maintain a smooth transition between shifts one employee had to work for 24 consecutive hours while the other had the entire day off.

Many immigrants who were recruited for these positions received poor wages, and eventually poverty and brutal scheduling led to the development of modern unionism. In 1892 a strike by steelworkers ended in a fight between Pinkerton guards and workers, leaving 13 workers dead. The strike gained national attention and brought the problem of shift work before the federal government.

In light of this attention and the importation of new continuous processing methods to the steel industry, the shift schedule was changed to an 8-hr system that resulted in increased productivity (53). The Ford Motor Company also adopted the 8-hr system but many other industries were reluctant to change and continued with the 12-hr shift schedule.

The number of shift workers increased again during World War I. This increase fostered the first study of shift worker problems. After the war, the unions began a crusade to reduce the 84-hr workweek. After 4 years of turmoil, the ferrous metal industries changed to the 8-hr rotating shift schedule that reduced the number of hours worked by each employee. To compensate for lost salary, wages were increased to make up for the deficit in hours (53).

The Great Depression brought about further changes. Since unemployment was high, the workweek was reduced to 40 hr by the Walsh–Healy Act of 1932 in an attempt to increase the number of employed individuals. The advent of World War II increased the number of individuals performing shift work. The increased prosperity of the U.S. economy fueled by a requirement for manufactured goods again raised the number of shift-working employees. In the mid-1950s pharmaceutical industries began to develop 12-hr shift systems to increase productivity. The number of persons working 12 hr continues to grow in the service sectors, while increased utilization of automated procedures has reduced the number of shift workers in manufacturing.

The increases in service sector jobs may continue for the next decade and suggest that even more individuals will be performing shift work. Also, the advent of the global economy may encourage increases in shift work in the manufacturing sector. For these reasons occupational health persons should acquaint themselves with the principles of chronobiology and problems of shift work presented in the balance of this chapter.

3 WORK SCHEDULING

Work schedules are the time patterns that employees conform to while working. They represent an attempt to match industrial production to the availability, needs, and wishes of workers. Shift work is only one type of work schedule. When night work duty is required, specific physiological and psychosocial adjustments are necessary.

3.1 Traditional 40-Hr Workweek

The most common work schedule is the standard 5-day, 40-hr week. Typically, morning arrival to work is at 8 A.M. and evening departure is at 5 P.M. Among all nonfarm workers, more than 80 percent work 5 days a week. By far the largest percentage of employees (53.5%) works exactly 40 hr weekly (see Fig. 9.1). Over 62 percent work between 35 and 40 hr. For example, the mean and median duration

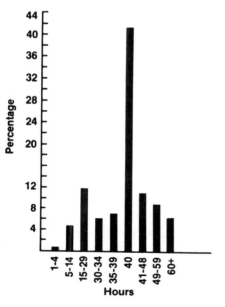

Figure 9.1 Distribution of weekly work hours. From *Employment and Training Report to the President.* U.S. Government Printing Office, Washington, D.C., 1979.

of the workweek for all nonfarm wage earners was 48.4 and 38.4 hr, respectively, in 1985. The mean value was less than 40 hr because of the inclusion of part-time employment in the estimate. Yet there is considerable variation in the length of the workweek depending on the demographics of the employees. Employees of corporations worked on average 38 hr per week, while self-employed individuals worked 43 hr per week in 1985. Approximately 60 percent of all full-time workers still report working a 5-day 40-hr schedule. Therefore, the average number of hours worked per day remains approximately 8 hr, which is virtually unchanged since 1973. However, for part-time workers, work shift compression has increased the average number of hours worked per day from 4.2 to 5.2 hr since 1973. It is expected that compressed work schedules will begin to lengthen the average workday for full-time employees over the next decade.

The need to operate certain manufacturing processes 24 hr/day has made it necessary that the operation be attended around the clock. Many continuous process operations, such as those found in the petroleum, chemical, metal refining, pharmaceutical, and health care industries cannot be shut down without causing serious problems and thus require around-the-clock staffing. Consequently, the inability to halt certain chemical and physical processes coupled with economic factors, such as the high cost of equipment and increased local and global competition, have forced these capital-intensive industries to make shift work a necessary part of manufacturing. This competition has led to the development of various shift schedules.

3.2 Shift Work Classification Systems

Shift work generally refers to employment that requires the presence of employees in the workplace between 4 P.M. and 8 A.M. A great variety of shift schedules are utilized in industry. The Knauth–Rutenfranz (54) classification system has been modified (Table 9.1) for the purpose of discussing the different types of shift work patterns.

Category I refers to permanent shift systems. These are rather common in the United States. With these systems the start and end times of work are always the same for a given employee. This is in contrast to the situation for rotating schedules in which a crew systematically varies its work and off times at specific—for example weekly—intervals. Permanent systems are those in which a given crew consistently works the same shift, whether it be the morning, evening, or night shift. Permanent systems also include the nonvarying split-shift pattern of the Merchant Marine in which the work "day" is divided into two portions or watches with a significant duration of time between each.

Category II pertains to rotating shift work systems. These may be continuous or discontinuous in form. Discontinuous shift work systems are defined as work patterns that require the presence of employees Monday through Friday or Saturday with at least one or two days off-time, usually over the weekend. These are quite common in industrial and commercial plants where continuous operation is not a necessity. Continuous-shift systems, on the other hand, refer to those schedules that include weekend work. These predominate in the continuous-process industries and in the public service sectors, such as fire, ambulance, and police departments. Rotating

Table 9.1 Categories of Shift Work Systems[a]

I. Permanent shift systems
 A. Permanent morning shift
 B. Permanent afternoon shift
 C. Permanent night shift
 D. Split shifts with each consistently timed for a
  given worker
II. Rotating shift systems
 A. Systems without night shifts
  1. Discontinuous
   a. Nonoverlapping (e.g., crew 1, 6 A.M. to
    2 P.M.; crew 2, 2 P.M. to 10 P.M.)
   b. Overlapping (e.g., crew 1, 6 A.M. to 2
    P.M.; crew 2, 1:30 P.M. to 9:30 P.M.)
  2. Continuous
   a. Nonoverlapping (e.g., crew 1, 6 A.M. to
    2 P.M.; crew 2, 2 P.M. to 10 P.M.)
   b. Overlapping (e.g., crew 1, 6 A.M. to 2
    P.M.; crew 2, 1:30 P.M. to 9:30 P.M.)
 B. Systems with night shift
  1. Discontinuous
   a. Two-team (e.g., each crew works 12-hr
    shifts)
   b. Three-team (e.g., each crew works 8-hr
    shifts)
  2. Continuous (regular)
   a. Two-team (e.g., each crew works 12-hr
    shifts)
   b. Three-team (e.g., each crew works 8-hr
    shifts)
  3. Continuous (irregular)
   a. Varying numbers of teams
   b. Varying numbers of cycle lengths

[a]Modified from Knauth and Rutenfranz (54).

shift systems do not necessarily include nighttime hours. This is the case both for the so-called two-team double-day pattern that typically consists of morning and afternoon shifts as well as the three-shift (morning, afternoon, and evening) systems. The evening shift of the latter commonly involves part-time workers. However, many industries require rotating shift work with regularly scheduled nighttime duty. When this is the case, different crews, usually two (each working 12 hr) or three (each working 8 hr), rotate their hours of work at set intervals. In certain industries the speed of crew rotation between shifts is slow, every 7 days or more; in others it may be much shorter, every 2 or 3 days; this is termed rapid rotation. In many cases the order of rotation between shifts is in a forward direction of morning to afternoon (or evening) to nighttime work schedules; however, for some persons the order of rotation is in a backward direction from nighttime to afternoon (or evening) to morning work schedules. In most large industries a variety of shift schedules are likely to be in use simultaneously to meet the needs of different operations.

3.3 Traditional Work Schedules

One popular schedule in use is the three-team rotating shift system, which includes nighttime hours. The use of three crews, each working 8 hr, enables uninterrupted production throughout the 24-hr period, so that a process never need be shut down due to a shortage of workers. Frequently, with this and other forms of shift work, additional overtime hours are necessary to accommodate routine requirements, meet seasonal loads, or to replace absent workers (57–60).

3.4 Employee Dissatisfaction With 8-Hr Shift Work

Wilson and Rose (61) as well as many other researchers (62,63) have reported that one of the major reasons for the move toward the 12-hr and other unusual work shifts was concern about the sociological problems caused by standard 8-hr rotating shift work. In general, the study by Wilson and Rose revealed that the complaints of workers centered around three major problem areas:

1. The necessity to periodically work unusual hours with few weekends off
2. The imposition of a work schedule that requires performance of activities at a time that is contrary to the optimum functioning of bodily processes with respect to the organization of circadian rhythms
3. The decay of family and social life because shift workers, although averaging approximately 40-hr weekly, find their leisure time occurring during hours when friends are at work, children are at school, or most other people are sleeping

Moreover, in addition to these problems, shift workers commonly express concerns about difficulty with sleep, fatigue, digestion, anxiety, and feelings of social isolation (37,54–56).

Rotating 8-hr shift schedules involving nighttime work are likely to disrupt both the biological and social organization of employees. From a sociological perspective, both the standard rotating and, to a lesser extent, unusual work schedules have been shown to affect family life and socialization. Rotating shift work schedules require frequent reorganization of social patterns. Often when shift workers are at home, their friends and family may be at work, school, or asleep. Also, rotating shift workers may have only one full weekend free per month. As a result, shift workers and their family members are frequently disappointed with the quantity and quality of time spent with their families.

Shift work is likely to affect the well-being and health of employees as well. The effects include trouble sleeping, excessive fatigue, and irritability (14,17,37,50, 51,55,57–63). Other problems that have been reported include constipation, gastritis, duodenal and peptic ulcers, high absenteeism, reduced productivity, and others (15–17,37,60). Because of these and other difficulties, numerous kinds of modified work schedules have been developed and implemented by companies seeking alternatives to the standard rotating 8-hr schedule. Modified or unusual work schedules include all those that do not involve an 8-hr/day, 5-day workweek. Generally, these schedules involve a shortened workweek and lengthened workdays.

3.5 Unusual Work Shifts

An unusual work shift is different from the standard 8-hr one that generally begins at 8 A.M. and ends around 5 P.M. The unusual work shift may be longer or shorter in duration with regard to the hours worked per day. They are often shorter (compressed) with regard to the number of days worked per week. Over the past 15 years, "other-than-standard" work schedules—which have been variously termed or classified, for example, as unusual, modified, altered, abnormal, exceptional, abbreviated, nonnormal, novel, extraordinary, odd, nontraditional, special, compressed, and even weird—have received increased attention both by labor and management of many industries. A permanent committee of the American Industrial Hygiene Association (AIHA) was convened in 1981 to review and promote research on the occupational health aspects of these work schedules. This committee decided that the term *unusual work schedule* seemed to best describe those schedules that consist of workdays that are either longer or shorter than 8 hr. Thus, this term is used to describe any schedule that is markedly different in length (either longer or shorter) than the standard 8 hr/24 hr, generally 5-day workweek, pattern.

A compressed workweek schedule is one type of unusual work shift that has been used in many nonmanufacturing settings. It refers to full-time employment accomplished in less than 5 days/week. Many compressed schedules are used but the most common are: (1) workweeks of four separate 10-hr shifts; (2) workweeks of three separate 12-hr shifts; (3) workweeks of four separate 9-hr shifts plus one 4-hr span (usually on Fridays); and (4) the 5/4–9 plan of alternating 5-day (or night) with 4-day (or night) workweeks with each shift being 9 hr in duration (1).

In 1980, 2.2 percent of all full-time nonfarm wage and salaried employees, or 1.7 million people, were on unusual (compressed) workweeks. Of this number, two-thirds were working 4-day weeks. Compressed workweeks were used more in some industries than in others. Initially, their heaviest use was in certain city service departments and in small manufacturing firms (Table 9.2).

Unusual work shifts do not always involve compressed workweeks. They may be in the form of a great variety of fairly complicated schedules, which have been implemented in manufacturing facilities over the past two decades.

Unusual (including compressed) work shifts began to gain widespread popularity first in Canada and the United States in the early 1970s. Later they gained the attention of the petrochemical industry in Europe. The use of the compressed workweek grew rapidly from 1970 when very few persons worked these types of schedules to 1.3 million workers, or 2.2 percent of the labor force, in 1975. Until 1981 the usage rate has remained steady (Fig. 9.2). In 1981, as a result of a desire to conserve energy, more companies began utilizing compressed workweeks. The statistics on

Table 9.2 Use of the Compressed Workweek in the United States According to Industry and Occupation as of 1980[a]

	Employees on <5-day Workweek	
	Number (1000s)	Percent
Industry		
Mining	17	2.0
Construction	148	3.4
Manufacturing	422	2.2
Transportation, public utilities	143	2.7
Wholesale, retail trade	267	2.4
Finance, insurance, real estate	76	1.8
Professional services	382	2.8
Other services	148	3.6
Federal public administration, except postal service	34	2.1
State public administration	31	3.6
Local public administration	180	10.8
Occupation		
Professional and technical personnel	303	2.5
Managers and administrators	83	1.1
Sales workers	63	1.9
Clerical workers	225	1.7
Craft workers	244	2.4
Operatives	290	3.3
Laborers	73	2.5
Service workers	439	6.7

[a]All figures refer to nonfarm wage and salary workers who usually work full time.
Source: U.S. Bureau of Labor Statistics.

these types of work schedules can be deceiving. Although it is true that the total number of employees (mostly office personnel) working unusual shifts has lessened or leveled off, the number of employees in the chemical industry being placed on unusual shifts is apparently still growing (1,24,26,64).

Many unusual work schedules make use of a rapid shift rotation between two 10- or 12-hr shifts to minimize the number of nights employees must be present on the job. In these schedules, persons transfer between day and night work such that the number of consecutive nights worked is limited to two or three before changing to day work or off-time (17,61,62). One type, for example, involves three daytime 12-hr shifts followed by four days off, then three 12-hr night shifts followed by three days off. This rotation takes 13 weeks to complete and provides for five weekends off (64). While this schedule has admirable features, not all employees find the rapid rotation between night and day work to be physically or emotionally agreeable. Some persons simply find it troublesome and too taxing. Continuing research is being directed at evaluating the psychological and physiological effects of the rapid rotating

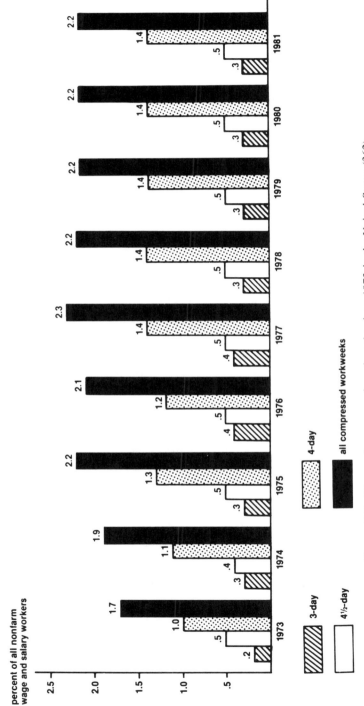

Figure 9.2 Usage of compressed workweeks since 1973 in the United States (362).

shifts, especially with regard to the potential effects of these schedules on circadian rhythms (17,62–64) and ways to predetermine which persons will be most compatible for this system of work.

In summary, unusual shifts have been devised and are now being used as a method to decrease some of the undesirable aspects of shift work. These schedules consist of shift durations that are generally longer than 8 hr and workweeks shorter than or equal to 4 days. Many of these schedules involve rotation between night and day shifts on alternating weeks. The two most common types of unusual schedules used in larger manufacturing facilities are the 10-hr, and 4-day workweek and the 12-hr, 3- or 4-day workweek (1,61).

3.6 Advantages and Disadvantages of Unusual Schedules

The industrial interest in unusual work shifts and the compressed workweek is understandable. Many firms find that their 24-hr continuous processes demonstrate better consistency of operation since two rather than three different crews of employees are involved. Occasionally, fewer persons may be needed to complete the same set of tasks. Younger workers usually desire the longer work hours/shift due to the higher annual income resulting from the scheduled overtime hours, more full days off from work, and less time spent per week commuting.

As with any work schedule, however, there are some drawbacks to the 12-hr shift (Table 9.3). It has been reported that older employees who have become accustomed to 8-hour work durations and who are more susceptible to fatigue are often less than enthusiastic about schedules involving longer shift durations (62,63). Some companies have found that younger workers should be available and trained to fill the jobs of older ones so that they can be transferred as needed to ''normal'' shift work schedules. Indeed, some firms, for example, Eli Lilly and Company, have recognized that an aging workforce may not prefer the 12-hr work schedule and have kept a fraction of their production lines on 8-hr shifts to provide these employees an opportunity to work a schedule that best suits their needs. Other research has suggested that long-term fatigue is greater during 12-hr shifts when compared to 8-hr shifts (65).

Eventually, federal and/or state legislation concerned with protecting worker health may make the 12-hr shift less attractive since stricter occupational health standards might be recommended to protect these persons (61). For example, the American Conference of Governmental Industrial Hygienists (ACGIH) has considered adjusting the threshold limit value (TLV) for the occupational exposure to noise from the present 85 dBA for 8 hr of work to something less than this for those on the 12-hr shift (66). Several groups, including the Occupational Safety and Health Administration (OSHA), have suggested that the time-weighted average (TWA) exposure to airborne toxicants should be reduced for persons working unusual work schedules (27–32,66–67). The 12-hr shift possesses specific problems since physical and emotional fatigue is likely when persons work longer than a 12 hr/shift for several consecutive days.

The Wharton Business School at the University of Pennsylvania conducted a comprehensive evaluation of the 12-hr shift, the most popular of the extraordinary

Table 9.3 Main Factors Affecting Usage of Compressed Workweeks and Unusual Work Shifts

Constraints	Incentives
Stringent work technology requirements may disrupt production operations; interface and coverage problems	Can smooth out production operations in some cases; output and productivity can increase
Potential fatigue of workers could cause productivity losses and increase risk of injury through accidents	Utility costs are reduced if buildings are shut down during off days
Supervision must often be stretched over a longer day; management more difficult	Unpleasant aspects of shift work can be partially alleviated with the compressed or unusual shift schedules
Labor law requires premium pay for hours worked in excess of 8 per day for workers on government contract or under some collective agreements	Morale, absenteeism, and recruiting often improved; income is usually greater each year because of scheduled overtime, shift differential, and unscheduled overtime
Many employees, especially women, older workers, and parents with young children, do not like them	
Family life and weekday social and civic activities often are disrupted	Commuting trips reduced, saving time and money for workers
Sometimes no improvement in employee–employer relationship is observed; does not always improve quality of work life	Leisure time redistributed into longer blocks, which most workers like
High failure rate among one-time user companies and the need to maintain a fairly automated process discourages many firms	Rapid rotation schedules can eliminate a lifetime of night shift work

work shifts (61). This evaluation involved the polling of 50 plant sites that had or were currently using this shift schedule. The investigation examined the costs of implementation, impact on productivity, maintenance of safety standards, and compliance to the Fair Labor Standards and Walsh–Healey Acts. Overall, the investigators concluded that in many industries, such as the highly automated and clerical ones, there were distinct advantages to adopting the 12-hr schedule. For others, such as heavy manufacturing and research, the 12-hr shift system was less advantageous and/or inappropriate (1,61).

3.7 Flexible and Staggered Work Hours

A schedule of flexible work hours, or *flexitime* as it is termed, is one in which employees choose their starting and quitting times within limits set by management. In general, flexitime schedules differ with regard to: (a) daily versus periodic (e.g., weekly or monthly) differences in the starting and quitting times; (b) variable versus constant length workday (whether or not credit and debit hours are allowed); and (c) core time—the hours of the day when all employees are required to be present (1). In 1980 it was estimated that about 11 percent of all organizations and 9 percent of all workers, or 7–9 million people, were using flexitime in the United States (1). If professionals, managers, salespeople, and self-employed persons, who have long set their own hours without calling it flexitime, were to be included, the usage rate might be as great as 81 percent. It has been used by all major businesses and industries, but with a somewhat heavier concentration in financial and insurance companies and government agencies than in manufacturing facilities. There has been a roughly equal popularity of the three major flexitime models with perhaps a slightly greater use of gliding time, which involves flexibility not only in the times of reporting to and leaving work but also in the number of hours worked daily. At present, flexitime is more common in Europe than in the United States. For example, in Germany and Switzerland more than one-third of the workforce has flexible hours (2,68).

The number of employees or percentage of the workforce engaged in shift work in the United States is not trivial (1–3). In general, rotating shift work schedules are especially prevalent in industries requiring uninterrupted services, manufacturing operations that cannot be routinely halted, and those in which the monetary investment is so very great that the noncontinuous operation of equipment is economically prohibitive. Shift work is very common, for example, in the automotive, electronics, health, petrochemical, and pharmaceutical industries.

Dependence on shift work schedules in the United States has increased steadily since World War II (4) and has recently increased as the number of service sector jobs outpace manufacturing. Frequently, the bulk of shift work occurs in industries located in metropolitan centers. Table 9.2 gives the distribution of shift workers in the various types of manufacturing and service industries. This table, which utilizes data obtained in a survey conducted in the mid-1980s reveals that shift work varies from between 2.5 and 60 percent, depending on the type of industry or service. In 1986, 4.9 million persons were working evening shifts, while an additional 2.0 million were working night shifts (358). Overall, during 1986 it was estimated that 15.9 percent of all employees were shift workers (359).

4 THE RELATIONSHIP BETWEEN CHRONOBIOLOGY AND SHIFT WORK

Understanding chronobiology is fundamental in understanding the nature of the biological adjustments to shift work. Biological adjustments to shift work involve an alteration in the sleep–wake pattern with respect to clock hour changes in the timing of the peaks and troughs of circadian and possibly other rhythms. The topic of chronobiology is reviewed below in preparation for the discussion of the chronobiologic aspects of shift work.

4.1 Introduction to Chronobiology

The science of chronobiology explores mechanisms of biological time structure and the temporal changes in physiology, psychology, and behavior that occur because of biological time structure. Although rhythm variance is currently a comparatively young investigative science, ancient civilizations were well aware of rhythmic processes, for example, in the seasonal breeding of wild and domestic animals and the planting and harvesting of agricultural or wild plant foods. Early writers such as the poets, were fascinated with rhythmic events, particularly as they pertained to the leaf and petal movements of plants. Nonetheless, western society was slow in understanding rhythms, especially in higher animals and humans, until the middle of this century.

Rhythms of many frequencies have been demonstrated at all levels of animal and plant life (18,48,69–79). As a result of many studies, rhythmicity has been firmly established as a fundamental property of all living things such that today chronobiology represents a new and powerful field of scientific endeavor.

The range of frequencies that has been found in living systems extends from cycles of less than one second to cycles of one year or more (71–72). Many, but not all, biological rhythmicities clearly correspond to environmental frequencies, for example, the natural 24-hr light–dark and 365.25-day annual cycles. Strong evidence exists that many rhythms are adaptive and serve to adjust organisms to predictable and regularly occurring environmental changes. The day–night and lunar cycles, as well as the seasons of our planet, were in existence before life began; thus, from the beginning, life was subjected to these periodic events. It is safe to say that the rhythmicity of life has been imprinted throughout the eons on living organisms by continuous exposure to this periodic environment during evolution. Throughout evolutionary history, in order to survive, organisms had to be biologically prepared for rather drastic differences in the environment occurring over 24 hr, and the solar year. The numerous examples of annual cycles in reproductive behavior in animals and hibernation represent examples of a rhythm of 1-year duration in the level of activity and metabolism in certain species. Biological rhythmicity implies predictable temporal changes in bodily functions and processes and constitutes a significant adaptation to expected cyclic alterations in the geophysical environment in which one lives.

Chronobiology should not be confused with the fad of biorhythms that was popular in the 1960s and 1970s (80–84). Biorhythm forecasting proposes to predict mental, physical, and emotional prowess based on cycles rooted in the individuals birth date. The predicted cycles create a potential suggestive power, similar to a placebo effect, for employees who have confidence in its purported predictive ability. Because of this placebo effect, accident rates may have been reduced in some workers; however, there is no scientific basis for the existence of the biorhythm phenomena. The biorhythm concept is without scientific merit and has no place in the management of either industrial safety or shift work schedules.

4.2 Naturally Occurring Biological Rhythms

As living beings are organized in three-dimensional space anatomically, they are also rhythmically organized in time as well, and biological rhythms demonstrate predictable variation over specific time domains. Biological processes exhibit several different types of bioperiodicities. These are categorized according to their duration, the amount of time required for a single repetition.

In general, rhythms are divided into three broad classes (Fig. 9.3): (a) Ultradian rhythms exhibit frequencies of less than 1 cycle per 20 hr. Examples of ultradian

CHRONOBIOLOGY

ILLUSTRATIVE SPECTRUM OF BIOLOGICAL RHYTHMS

DOMAIN	HIGH FREQUENCY $T<0.5h$	MEDIAL FREQUENCY $0.5h<T<6d$	LOW FREQUENCY $T>6d$
MAJOR RHYTHMIC COMPONENTS	$T\sim0.1s$ $T\sim1s$ et cetera	ULTRADIAN ($0.5<T<20h$) CIRCADIAN ($20<T<28h$) INFRADIAN ($28<T<6d$)	CIRCASEPTAN ($T\sim7d$) CIRCAMENSUAL ($T\sim30d$) CIRCANNUAL ($T\sim1yr$)
EXAMPLES Rhythms in	Electroencephalogram Electrocardiogram Respiration	Rest-Activity Sleep-wakefulness Responses to drugs Blood constituents Urinary variables Metabolic processes, generally	Menstruation 17-Ketosteroid excretion with spectral components in all regions indicated above and in other domains

Domains and regions [named according to frequency (f) criteria] delineated according to reciprocal f, i.e., period (T) of function approximating rhythm. s - second, h - hour, d - day.

Several variables examined thus far exhibit statistically significant components in several spectral domains.

Figure 9.3   Illustrative spectrum of biological rhythms.

rhythms include the basal activity levels of neurons, which typically exhibit one cycle per 0.1 sec or so, the pulsatile secretion of hormones, for example, from the adrenal cortex, and the repetition of sleep stages such as rapid eye movement (REM) during nightly rest at about 90-min intervals, to mention just a few. (b) Infradian rhythms categorize those bioperiodicities that are greater in duration than 28 hr. Examples of infradian cycles include the circaseptan (about 7-day), circamensual (about monthly, also termed circatrigintan or about 30-day), and circannual (about one-year) rhythms. (c) In between are rhythms that have bioperiodicities in the range of 20–28 hr. Rhythms having these durations are termed circadian (meaning about one day, or 24 hr in length) after Halberg (70,71).

Circadian rhythms in human beings and animals have received more attention by researchers than any of the others, and thus a great deal is known regarding the 24-hr temporal organization of biological processes. Circadian rhythms exist at all levels of biological organization from single-cell to complete organ systems. All life forms, whether plant or animal are made up of a multitude of interrelated circadian physiological, metabolic, and behavioral rhythms that act in harmony in the healthy organism. Thus, chronobiologists commonly refer to these 24-hr periodicities collectively as a circadian system in much the same way in which biologists speak of the nervous, circulatory, or digestive systems. Rhythms of many other frequencies are intermingled and may even modulate or be modulated by the circadian frequency.

The biological consequences of rotating shift work schedules and jet lag (often referred to as desynchronosis) on industrial performance and accidents among others have been studied by chronobiologists with particular emphasis on the role of the human circadian system. In rodents, the circadian system has been examined for the occurrence of susceptibility–resistance rhythms in the effects of hazardous chemical, physical, and microbial substances as a means of modeling predictable changes over the 24 hr in human vulnerability. The findings of these studies are critical to those involved in medicine, toxicology, and occupational health. The results and significance of these studies will be reviewed in subsequent sections of this chapter.

4.3 Synchronizers of Biological Rhythms

A discussion of the concept of synchronization is essential to appreciating the temporal integration of the circadian system and for understanding the chronobiologic aspects of shift work. In short, a synchronizer is a cue or signal from one's environment that serves to lock circadian and other bioperiodicities into fixed frequencies. For example, many animals in nature are synchronized to the natural light–dark cycle, while those in the laboratory are usually synchronized to an artificial 24-hr light–dark cycle (85). The synchronizer (70,86), sometimes is called a Zeitgeber (87), an entraining agent (88), and clue or cue; all are used synonymously. It is important to recognize that synchronizers do not cause and are not the source of rhythmicity. They are capable only of influencing certain characteristics, most prominently the timing of the peak and trough of a given bioperiodicity.

4.4 Significance of Synchronization

For organisms (plants or animals) in the synchronized state, rhythmic variables normally have a fixed time relationship. One would expect to find that body temperature for both humans and rodents is highest during the active or awake stage of the sleep–rest cycle. This fact suggests that temperature and activity rhythms are internally synchronized, their timing or staging under normal conditions being nearly the same. However, the temporal staging of other variables in the normal synchronized state could be very different, being as much as 12 hr out of phase. This is the case, for example, of the phase relationship between the human circadian rhythms of serum cortisol concentration and the number of white cells in the blood. The circadian peak of the former occurs around the commencement of the daily activity span, while that of the latter occurs later around bedtime. Staging of the various circadian functions could be as much as 4, 6, or 12 hr out of phase with one another. This fact can be attributed to the temporal organization of biological processes acquired through adaptation to predictable variations in the geophysical environment that occurred during evolution. This temporal organization represents an important aspect of the body's efficiency in conserving and optimizing its resources and capabilities.

When organisms are synchronized and the synchronizer schedule is known, it is possible to determine and then predict with relative accuracy the staging of a great many circadian rhythms. For example, Figure 9.4 presents certain aspects of the circadian system of rodents synchronized to light from 6 A.M. to 6 P.M. alternating with 12 hr of darkness (72). Figure 9.5 presents selected aspects of the circadian system of diurnally active human beings (72). In both these figures the circadian peak (termed acrophase) of each rhythm is represented by a filled circle; the extensions to the right and left of the symbol indicate graphically the 95 percent confidence

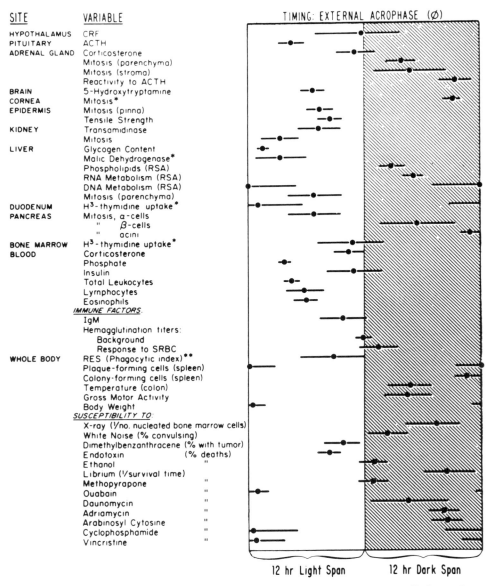

Figure 9.4 The circadian temporal structure of the mouse is shown by a so-called acrophase map—a display by graphic means of the circadian acrophase or crest (designated by ϕ for a phase reference corresponding to the middle of light—animal rest—span) of different biological functions. The acrophase, black dot, and the 95 percent confidence limits, lateral extensions, depict the timing with respect to the 24-hr domain of highest expected values for each designated function. By knowing the light–dark synchronizer schedule, one can determine with confidence the biological time structure at a given clock hour of sampling or experimentation. From Halberg and Nelson (72).

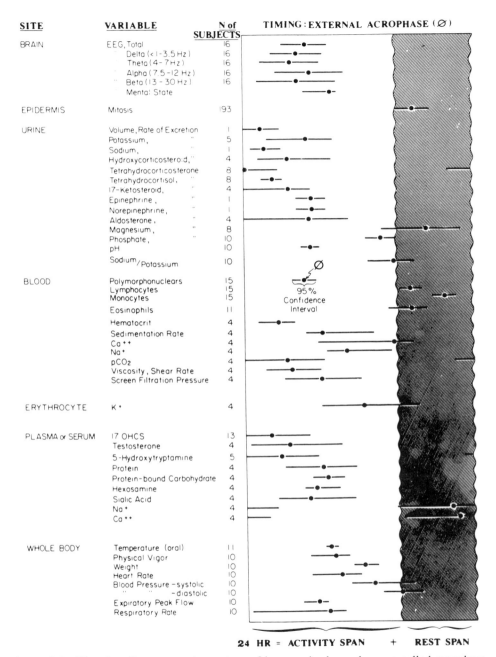

Figure 9.5 The circadian temporal structure of humans is shown by a so-called acrophase map—a display by graphic means of the circadian acrophase or crest (designated by ϕ for a phase reference corresponding to the middle of the sleep span) of different biological functions. The acrophase, filled circle, and the 95 percent confidence limits, lateral extensions, depict the timing with respect to the 24-hr domain of highest expected values for each designated function. By knowing the rest–activity synchronizer schedule, one can determine with confidence the biological time structure at a given clock hour of sampling or experimentation. From Halberg and Nelson (72).

range of each acrophase. These so-called acrophase charts provide a temporal mapping of the circadian crests of various functions. Just as an anatomical atlas is useful for relating the functional integration of biological structure in space so it is that acrophase charts, such as those in Figures 9.4 and 9.5, are useful for relating and knowing the functional integration of circadian (or other period) processes in time.

The two acrophase charts reveal there is a definite temporal staging for each variable with regard to the light–dark schedule in rodents and the activity–rest routine of human beings. The information in these charts defines the fixed temporal relationships between biological variables. As previously discussed, all of the circadian rhythms attain their peak (acrophase) or trough values at the same identical times. Acrophase charts such as these have been very helpful in revealing many rhythms in humans and are useful in biomedical studies, including shift work research. Similar types of acrophase charts have been constructed for circamensual (89,90), and circannual (41,90) systems in humans. If one has knowledge of the circadian synchronizer schedule, it is possible to predict accurately for a given clock hour, when a medical test or research procedure is conducted, the temporal staging of a great many rhythmic variables of the circadian system. This type of information can be critical for accurate interpretation of biological data obtained from a single clinical test on patients or single time point biological samplings. Time-qualified reference values for blood and urine components have been developed that consider circadian rhythms and synchronizer schedules in relation to the clock hour of blood or urine sampling.

The phasing of circadian and other rhythms in human beings is organized temporally to anticipate the different metabolic requirements of activity and rest when adhering to a fixed life routine (91). This routine is disturbed when either rotating shift work or rapid geographic displacement across time zones occurs and the synchronized schedule is changed. When inversion of the synchronizer schedule takes place, the acrophase of each circadian function eventually shifts in an amount equal to that of the change in the synchronizer schedule (92). Generally this adjustment requires several to many days. Some persons seem to be able to adjust rather quickly, while others adjust slowly (93). The individual biological variability makes the estimation of the time needed for resynchronization difficult. Those individuals who have difficulty are biologically or emotionally intolerant and may be incapable of performing rotating shift work (94–97). Consequently, it is important that occupational health professionals understand the scientific basis for shift work intolerance so that fellow employees and managers comprehend the difficulties inherent in adhering to rotating shift work schedules requiring nighttime duty.

During the span of the chronobiologic adjustment to a new work–rest routine, the rate at which each circadian process phase shifts may differ (92,98–101) for each individual. For a few days, alterations in the usual phase relationships between a large number of circadian rhythms occur. This transient state of altered phase relationships within a circadian system is termed *internal desynchronization* (92,98, 100,101). The transient state of internal desynchronization between circadian functions results in temporal disorganization that can result in decreased biological efficiency (34,92,99,100–102). This period of transition of less than optimal biologic efficiency is believed to be responsible for a specific type of fatigue, different from that due to sleep loss. It is commonly experienced by travelers and shift workers during the first few days following a sudden change in their sleep–activity routine and is often referred to as jet-lag.

5 FREE-RUNNING RHYTHMS

Free-running rhythms are defined as rhythmic functions, which in the absence of synchronizers exhibit durations other than expected. In the case of circadian rhythms, the period may be either shorter or longer than exactly 24 hr. Scientists once believed that biological rhythms represented nothing more than direct responses to periodicities in the geophysical environment, such as in the light–dark cycle, temperature and relative humidity, or to nutrient uptake. However, as early as 1729, DeMarian (103), the mathematician and astronomer, found that the leaves of plants would continue their periodic movements when kept in a darkened cave protected from both sunlight and open air. It was suggested by the scientists of that time that some endogenous mechanism within the plant must be responsible for this behavior.

Interestingly, we now know that when a rodent or other animal is removed from the influence of the synchronizer, that is, the light–dark cycle, which can be accomplished by blinding or subjecting the animal to continuous light or darkness, circadian rhythms continue to persist (104–108). The circadian rhythms of body temperature, serum corticosterone, cell proliferation, as well as feeding, rest, and activity are all examples of this phenomenon. However, it is rare that the frequencies of these rhythms will be exactly 24 hr. In humans, circadian rhythms are generally slightly longer than 24 hr. Human beings dwelling in experimental settings without time cues experience elongated sleep–wake patterns of almost 25 hr (108). Women who remained in such settings for several months also exhibited elongated menstrual cycles as well as longer circadian periodicities (109,110). Under constant conditions, each bioperiodicity exhibits a cycle duration that differs from the others. Because the duration of these cycles is not exactly 24 hr, the clock time of the peaks and troughs of individual circadian functions differs from one day to the next. In an environment free of synchronizers, clock time no longer can be used to interpret the staging of specific circadian functions. These circadian, non-24 hr, bioperiodicities are examples of free-running rhythms that have been observed in both plant and animal life (73,106–108,110–115). These findings constitute strong evidence for the endogenous nature of biological rhythms and their genetic origin.

The frequency of a free-running rhythm can be influenced by several factors. In animals and perhaps human beings these include light intensity and magnetic fields (111). These factors have been shown to be capable, at least in certain organisms, of delaying or advancing the stage of free-running rhythms. The period length of free-running rhythms is remarkably resistant to chemical perturbation. For example, if the free-running period has a fixed duration of say 25 hr 4 min, it is difficult to change. However, there is a class of drugs used in psychiatry, such as lithium, imipramine, and clorgyline (116,117), that can alter the periodicity. It is perhaps important to note that these agents are also known to influence cell membrane function.

In the early 1960s a number of investigations focused on the free-running rhythms of humans. Many of these early studies were conducted in caves. Frequently, the subject was a speleologist who would perform some type of exploration or endurance record in the depths of caves while chronobiologists would monitor and interpret certain rhythmic behavior (109,110,113,118). Mills (119) has provided a review of many of these cave studies. Recently, chronobiologists have used specially constructed isolation chambers where humans can be studied in comparative comfort yet remain separated from all social and recognizable environmental cues (108,120–122).

When a human is isolated from all time cues, certain rhythmic variables, even though they are free running, remain entirely internally synchronized; that is, they retain the same phase relationship demonstrated under the normal synchronized state. On the other hand, some bioperiodicities such as the sleep–activity cycle and body temperature rhythm may become internally desynchronized. In isolation experiments performed under constant conditions, this may occur spontaneously anytime during the period of isolation. This suggests that the staging of certain circadian functions are more strongly integrated than others and are subject to individual variability. Such a phenomenon almost certainly implicates the existence of more than one control center for biological rhythms (76,108,123,124).

For some persons in a free-running state, the sleep–wake rhythm may be out of phase from the body temperature rhythm. Simply stated, the time relationship of these two rhythmic variables is continuously changing in relation to one another. Wever, an active investigator of this phenomenon, believes that certain susceptible persons appear to exhibit free-running rhythms that contribute to the biological intolerance of some individuals to rotating shift work schedules (108).

Yet, it is interesting to note that free-running rhythms are relatively independent of ambient temperature. One can increase the ambient temperature within a wide range, but the period of the free-running rhythm will change very little, if at all. However, the 24-hr average or the rhythm's amplitude may be altered. This suggests that ambient temperature probably does not play a significant role in synchronizing humans. Ambient temperature cycles may serve as a dominant synchronizer for a number of cold-blooded animals or even sometimes for warm-blooded animals in the absence of the usual dominant synchronizer, e.g., day–night cycle (125).

The interplay between free-running rhythms and synchronization has developed to assist the organism in adapting to a changing environment. However, evolutionary evidence suggests that this mechanism has not evolved sufficiently to cope with rapid change. In humans, performing rotating shift work or rapid travel between time zones creates drastic changes in the synchronizer schedule that usually results in transient desynchronization of various rhythms.

Occupational health personnel need to be aware of the chronobiological aspects of this adjustment and the consequences of this change. Chronobiology offers useful tools for documenting internal desynchronization and suggesting alternatives to minimize individual worker difficulties. The need to consider the chronobiologic aspects of adjustment to different activity–rest routines that occur with shift work has gained increasing attention in industry. Chronobiologists have become more involved in assisting companies in the development and identification of appropriate shift schedules. Methods have been devised to identify those persons who are likely to be intolerant of shift work and procedures to minimize the consequences of rotating time patterns of work (17,54,93).

6 MECHANISMS OF BIOLOGICAL RHYTHMS

Many experiments have been performed on mammals in an attempt to identify the origin and mechanisms controlling rhythmicity. Some studies employed the classic endocrinological approaches of adrenalectomy, hypophysectomy, cerebral ablation, and pinealectomy in an effort to solve this problem. A great deal of attention has been given to the possible role of the suprachiasmatic nucleus, located in the brain, as a rhythm generator (76,123,124). Currently, the results of these studies indicate that no single regulator can account for the control of all rhythmic variables (91). Halberg has suggested that hormonal secretions from the adrenal cortex deserve serious consideration as one principal mechanism that possibly controls human biological adaptation to a daily activity–rest cycle. It has been shown that the adrenal as well as the pineal gland will continue to secrete and respond for several cycles, even when isolated in vitro preparations. Halberg (46) reviewed the role of the adrenal gland and other possible regulators in mammals. Although there have been numerous attempts to locate a single central regulator of circadian phenomenon at the unicellular level, thus far these have been unsuccessful. Several researchers believe that temporal rhythm control lies in the communication between cell membranes. Some (123,126–128) have postulated that the circadian oscillation may be generated and synchronization effected by the temporal variation within cell membranes and that regulation of all systems starts with the cycles of the organelles within the cells since mitochondria and the endoplasmic reticulum have both been shown to exhibit rhythms in many processes (129). Edmunds (130) reviewed the many diverse mechanisms that have been proposed as regulators. However, it is still unclear whether bioperiodicities are driven by a single master oscillator or by a population of noncircadian biochemical oscillators (76). Even though the majority of chronobiologists accept the endogenous nature of rhythms, one school postulates that circadian rhythms are generated by an interaction of several external forces, such as light, temperature, electromagnetic variation, and possibly yet unknown and subtle geophysical forces (131). In this chapter we cannot adequately comment on this interesting concept, but we do wish to call attention to it for the interested reader. Clearly, evidence exists for geophysical forces affecting life.

As soon as the mechanisms underlying biological rhythms are identified and well understood, it will be possible to better comprehend and hopefully alleviate the health-related problems of rotating shift workers. It may even be possible to hasten the chronobiologic adjustments of rotating shift workers and travelers to changes in their sleep–activity routine through the use of special medications called chronobiotics (132,133). Some reports have suggested the use of fast-acting, short-endurance benzodiazepines to alter the sleep–wake cycle. However, problems with some of these agents have negated their functionality. Nonetheless, investigators are continuing their search for a useful chronobiotic.

6.1 Sleep–Wake Schedules and Circadian Rhythm Adjustments

As discussed earlier in this chapter, persons who are employed on rotating shift schedules that include nighttime work must periodically change their sleep–activity pattern in order to accommodate their job. Although the number of hours worked per week may remain unchanged, the clock time of sleep and activity is transformed. Similarly, travelers and airline crews who are rapidly transferred across several time zones face the same situation—a resulting alteration in the sleep–activity pattern relative to clock time at home. A major alteration in the clock hours of the sleep–activity pattern represents a change in the synchronizer schedule. Such a change, in turn, requires compensatory adjustment of the circadian system, should the synchronizer schedule remain altered for some critical duration, usually considered to be 2 or 3 days for a typical person. It is important to note that not all persons can tolerate rotating between nights and days, thereby giving rise to intolerance to shift work.

Under normal circumstances the biological rhythms of most persons do not immediately adjust following an alteration in the sleep–activity synchronizer schedule. The jet-lag syndrome experienced by vacationers after being rapidly transported across several time zones and the dull lackluster feeling of lassitude and fatigue experienced by shift workers for a few days after the start of night duty represent in large part a reduction in biological efficiency resulting from a disturbed circadian system. This is not surprising since as a minimum at least several days are usually required for circadian and other bioperiodicities to undergo complete acrophase adjustment to a new synchronizer schedule. The exact duration of time needed, however, varies with the biological function, individual, extent of alteration in the synchronizer schedule, and whether the initial alteration occurs by a delay in the timing of sleep and activity (as would be the case if rotating in sequence and direction from an afternoon to a nighttime shift or for a traveler going from Europe to the United States) or by an advance in the timing of the sleep–activity schedule (as would be the case in rotating in sequence and direction from a morning (e.g., working 6 A.M. to 2 P.M.) to a nighttime shift (e.g., working 11 P.M. to 6 A.M.) or for a traveler going from the United States to Europe. In the sections that follow, the chronobiologic factors of shift work are considered in detail.

6.2 Phase Shifting of Biological Rhythms

The primary synchronizer of circadian rhythms in plants, animals, and insects is the environmental light–dark cycle. When the synchronizer is changed with regard to clock time, the circadian system eventually compensates in that a multitude of interrelated rhythmic variables undergoes acrophase shifts equal in amount to the alteration of the synchronizer. In controlled investigations of acrophase shifts in nonhuman organisms, plants, insects, or rodents are initially standardized to an environment in which the hours of lights on and darkness are regulated. For example, light in the case of rodent studies has traditionally been from 6 A.M. to 6 P.M. alternating with 12 hr of darkness. Thereafter, either the light or the dark span may be lengthened or shortened by several hours to study the acrophase adjustments of the circadian system to the alteration of the synchronizer.

6.3   Phase Shifting of Biological Rhythms: Rodent Models

Much data on phase shifting comes from studies on rodents. If the light–dark synchronizer schedule is altered in an animal colony, the acrophase of each rhythmic function ultimately adjusts to the new schedule by the same amount. This is exemplified by the variable body temperature continuously monitored in groups of rodents (92) (Fig. 9.6). When, after 12 days of exposure to a standardized 12 hr light and 12 hr dark regimen, a delay or advance in the synchronizer light–dark schedule is initiated, such as by either prolonging or shortening the light schedule by 6 hr, respectively, the clock hours of the synchronizer become displaced. In the first case, activity and rest are now initiated each day 6 hr later than before; in the second case they are now initiated 6 hr earlier than before. Under usual conditions, when animals

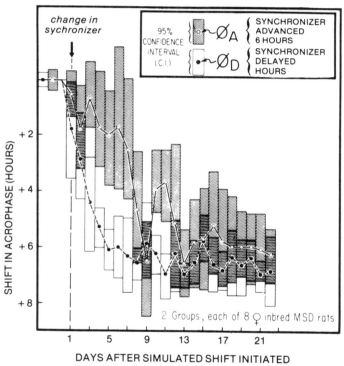

Figure 9.6   Rate of acrophase adjustment ($\Delta\phi$) by the circadian rhythm in body temperature subsequent to a synchronizer alteration in the form of (1) an advance (ϕ_A) by 6 hr, an earlier timing of the light and dark cycle by 6 hr causing the rest and activity cycle to occur earlier by 6 hr as compared to the reference synchronizer schedule and (2) a delay (ϕ_D), a later timing of the light and dark cycle by 6 hr, causing the activity and rest cycle to be delayed by 6 hr as compared to the reference synchronizer schedule. Note that the acrophase shift of 6 hr was completed and "locked" in much quicker, within 5 days, following a delay as compared to an advance, which is quite slow, taking 9–13 days. Adapted from Halberg et al. (114).

are housed under light from 6 A.M. to 6 P.M. alternating with darkness, the acrophase of the circadian rhythm in body temperature occurs around midnight. As shown in Figure 9.6, when the synchronizer is changed by 6 hr, the acrophase of the body temperature rhythm changes with regard to clock time by this same amount; however, this adjustment requires several days. It took only 3–5 days when the synchronizer was delayed; it required two to three times as long as when the synchronizer was advanced (92). The speed at which circadian rhythms adjusts varies with the direction of the change in the synchronizer schedule, that is, advancing (comparable to beginning rest or going to sleep earlier) versus delaying (comparable to beginning rest or going to sleep later) the clock times of rest and activity. This difference in the rate of adjustment by the circadian system is termed *polarity*. It represents an asymmetry in the rate of circadian processes to adjust their staging as a function of the direction of the alteration in the synchronizer schedule. The polarity of circadian rhythms can exhibit differences between functions in the same individual, among individuals and, in addition, among species (92,100,101).

6.4 Phase Shifting of Biological Rhythms: Human Beings

Thus far, all of the data indicate that human beings react to an alteration in their sleep–activity pattern much like the rodent (92). Although there are individual differences, human beings tend to adjust more rapidly to a delay, that is, a later onset of sleep and activity, than to an advance, that is, an earlier onset of sleep and activity, in their schedules. This point is illustrated by the data of Table 9.4. Klein and Wegmann (101,134), through a large number of investigations dealing with jet lag, have quantified the average rates of which biological functions of human beings resynchronize following a delay or an advance in the sleep–activity schedule. Westbound flight constitutes a delay in the synchronizer schedule in that retiring to sleep and also awakening occur later relative to the local clock time before travel. Conversely, eastbound flight constitutes an advance in the synchronizer schedule in that

Table 9.4 Rates of Acrophase Shift (min/day) in Selected Circadian Functions Following Either Westbound or Eastbound Flights Corresponding to Delays and Advances of Synchronizer Schedule[a]

Study Variable	Westbound (Delay)	Eastbound (Advance)
Urinary adrenalin	90	60
Urinary noradrenalin	180	120
Urinary 17-OHCS	47	32
Psychomotor performance	52	38
Reaction time	150	74
Heart rate	90	60
Body temperature	60	39

[a]From Klein and Wegman (101,134).

retiring to bed for sleep and also awakening occur earlier relative to the local clock time before travel.

The findings presented in Table 9.4 substantiate in human beings two phenomena of biological rhythms. First, the rate of adjustment of circadian processes, such as those listed, vary such that it is faster when the sleep–wake is delayed than when it is advanced. Second, individual circadian functions can differ in their rate of adjustment following a change in the hours of sleep and activity. For example, after either a westbound or eastbound flight, the circadian rhythms of psychomotor performance, body temperature, and adrenocortical hormone secretion adjust rather slowly while those of urinary noradrenalin and reaction time adjust at relatively rapid rates. Because interdependent rhythmic variables shift at differential rates, a transient state of internal desynchronization, characterized by atypical and changing phase relationships between circadian processes from one day to the next until adjustment is complete, results. During the span when circadian rhythms are in the process of adjustment, the transient state of internal desynchronization, along with sleep disruption, leads to increased feelings of mental and physical fatigue, the elevated likelihood of disrupted performance, and perhaps susceptibility to illness (135).

The endogenous polarity in the adjustment of the circadian system favors the utilization in traditional 8-hr rotating shift work systems involving periodic nighttime duty of work sequences that represent a delay rather than an advance of the synchronizer schedule (136,137). Thus, traditional 8-hr rotating shift work schedules that incorporate a later timing of duty hours, that is, a forward rotation, from one shift to the next tend to be better biologically tolerated than those that incorporate an earlier timing, that is, a backward rotation. It must be emphasized that because of the possibility of individual differences between persons with regard to the polarity of their circadian systems there will be exceptions to any generalization. Moreover, certain workers may prefer, for nonbiological reasons, to rotate in a backward rather than a forward direction.

Knauth and Rutenfranz (136) recommend forward rotation especially for the discontinuous three-shift systems having a 5-day work and 2-day off pattern. With typical shift changes of 6 A.M., 2 P.M., and 10 P.M., there is a significant difference in the total amount of time off between successive shifts when rotation is in a forward sequence in comparison to when it is in a backward sequence. A forward rotation sequence allows 72 hr off-time between morning and afternoon shifts, 72 hr off between afternoon and night shifts, and 48 hr between night and morning shifts (without a post-night shift day sleep). A backward rotation, in contrast, allows only 56 hr between night and afternoon shifts (without a post-night shift day sleep), 56 hr off between afternoon and morning shifts, and 80 hr off between morning and night shifts. With the backward rotation pattern, the immediate transfer from an afternoon to a morning shift or from a night to an afternoon shift results in between-shift intervals that are too short. Since forward rotation results in only one short between-shift interval (48 hr between the night and morning shifts), while the backward rotation results in two (56 hr between the night and afternoon shifts as well as between the afternoon and morning shifts), backward rotation is not recommended.

In summary, results of chronobiologic studies on travelers and shift workers have been useful in gaining insight as to how to better design rotating shift work schedules.

By optimizing the clock hours when shifts begin and end, the interval of time between shift rotations and the direction of shift sequences, some of the undesirable aspects of shift work have been moderated. Nonetheless, chronobiologists continue their research to devise ways to minimize and alleviate continuing shift work problems as well as develop selective criteria specific for predicting long-term tolerance to rotating shift work. In this regard the results of field studies on shift workers, as described in the following sections, are most exciting.

6.5   Tolerance to Shift Work

Even at a young age, not everyone is equally tolerant of the rigors of shift work and dropout is particularly common during the first 6–12 months. It is not unusual for biological intolerance to develop suddenly after 10, 25, or even 30 years of shift work, around the age of 50 years or after. An employee may suddenly find it difficult to sleep during the daytime and may complain of fatigue due to the inability to sleep. Many individuals request transfer to permanent day work.

Just why intolerance to shift work develops is not completely known. Reinberg and his colleagues (94) have suggested that intolerance may be related to a disorganization of the circadian temporal structure in susceptible persons. This problem of intolerance will be considered in detail from a chronobiologic (biological rhythm) point of view in the next section.

7   INTOLERANCE TO SHIFT WORK

Most occupational health professionals are aware that not all persons are capable of tolerating shift work and that there are various differences between those who do shift work and those who find it intolerable. Some persons can adhere to rotating shift schedules that include periodic nighttime duty throughout their working life, 35 or more years, without health problems or complaints of any kind. On the other hand, some young healthy workers may exhibit intolerance to the rigors of shift work even at a very early age.

Both young as well as older workers may develop intolerance to rotating shift work schedules. Older workers in particular, for as yet unknown reasons, after being successfully engaged in rotating shift work for several decades can around the age of 40–50 years suddenly exhibit biological intolerance to shift work. This intolerance is manifested by the onset or exacerbation of the typical shift work-associated medical complaints. Sleep alterations such as shortened durations, frequent awakenings, and insomnia are commonly reported. Persisting fatigue, not always the result of these sleep difficulties, is another. The intolerant shift worker often complains of tiredness immediately after awakening, during weekends away from work, and during vacations. This type of persisting fatigue is different from the physiological fatigue resulting from physical and/or mental effort on the job since the latter nearly always dissipates after an adequate night's rest. Intolerant shift workers may also exhibit changed behavior. This can be manifested as an increase in irritability and excitability with malaise and lowered performance. Digestive complaints such as dyspepsia, epigastric pain, and peptic ulcer are also common difficulties of the intolerant shift worker.

Over the years, methods to evaluate and minimize intolerance to shift work have been addressed by many investigators. Some have identified or postulated biological attributes that might be responsible for intolerance while others have focused on nonbiological factors. Figure 9.7 presents a summary of the various factors either known or thought to affect one's capacity for long-term tolerance to rotating shift work schedules (138). This schematic is intended to introduce the subsequent sections on shift work tolerance as well as place in perspective those variables that have been identified or proposed as affecting shift work intolerance.

For the purpose of discussion, the factors that can affect tolerance have been divided into two categories: endogenous, that is, biological in nature, and exogenous, that is, nonbiological in nature. With regard to the endogenous factors, age, health status, sleep variables, physiologic tolerance to night work, genotype, and personality have been discussed in the literature (37,54). At this time there is no convincing evidence indicating a differential capability between men and women for rotating shift work schedules. However, it has been shown that women with children tend to find shift work very demanding. In particular, married women usually have less available time to catch up on lost sleep, especially when working the night shift, and may have an increased risk of injury due to having major responsibility for organizing the everyday activities of the family (35,54,139,140). The significance of polarity with regard to the rapidity of acrophase adjustment, such as found in studies of travelers, has already been discussed. Chronotype refers to proposed slight differences in the temporal staging of biological rhythms between persons. It has been postulated that some persons actually prefer to be active late into the night and sleep later into the day; they have been called "owls" by scientists. It is postulated as well that others prefer to retire early to bed in favor of arising very early in the morning; they have been called "larks." The question of whether there is a true biological difference in tolerance for rotating shift work between so-called owls, and so-called larks, has been suggested but not resolved with certainty (141–144). Few measurable and quantifiable biologic differences between these persons have been discovered. However, we wish to point out that only about 10–20 percent of the population can be classified into either one of these two (owl or lark) categories. Most persons cannot be classified into either.

8 INTERNAL DESYNCHRONIZATION OF CIRCADIAN RHYTHMS AND INTOLERANCE FOR SHIFT WORK

Internal desynchronization of biological rhythms is defined as an alternation in the circadian period and/or phase relationship between interdependent rhythmic functions. Generally, circadian (and other) rhythms are synchronized such that a fixed phase relationship exists between interdependent bioperiodicities of a given frequency (see Figs. 9.4 and 9.5). Transient internal desynchronization of circadian rhythms, in terms of an alteration of the phase relationship of biological functions, occurs during the chronobiologic adjustment of shift workers after changing to nighttime

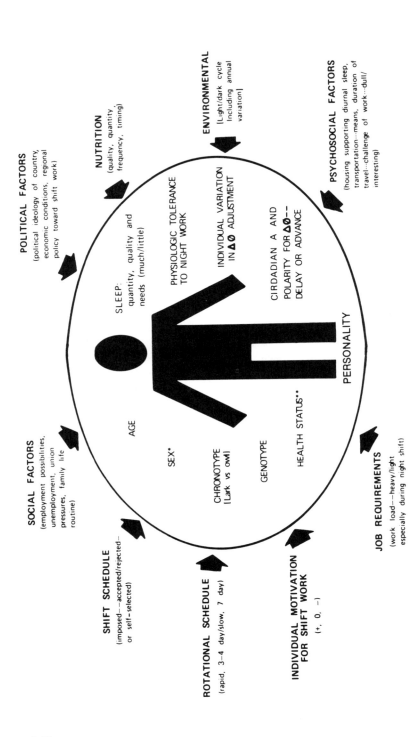

POLITICAL FACTORS
(political ideology of country, economic conditions, regional policy toward shift work)

NUTRITION
(quality, quantity frequency, timing)

ENVIRONMENTAL
(Light/dark cycle including annual variation)

PSYCHOSOCIAL FACTORS
(housing supporting diurnal sleep, transportation—means, duration of travel—challenge of work—dull/interesting)

SOCIAL FACTORS
(employment possibilities, unemployment, union pressures, family life routine)

SHIFT SCHEDULE
(imposed—accepted/rejected—or self-selected)

ROTATIONAL SCHEDULE
(rapid, 3–4 day/slow, 7 day)

INDIVIDUAL MOTIVATION FOR SHIFT WORK
(+, 0, –)

JOB REQUIREMENTS
(work load—heavy/light especially during night shift)

SLEEP: quantity, quality and needs (much/little)

PHYSIOLOGIC TOLERANCE TO NIGHT WORK

INDIVIDUAL VARIATION IN Δ∅ ADJUSTMENT

CIRDADIAN A AND POLARITY FOR Δ∅—DELAY OR ADVANCE

PERSONALITY

AGE

SEX*

CHRONOTYPE (Lark vs owl)

GENOTYPE

HEALTH STATUS**

*In certain countries such as France, night and shift work is forbidden by law, except for nurses, airline stewardesses and aircraft personnel. A difference between males and females for tolerance to shift work has not been documented experimentally.

**Personal and/or family history for certain diseases such as peptic ulcer, epilepsy, depressive illness, diabetes, etc.

Figure 9.7 Major factors underlying tolerance to shift work.

duty involving daytime sleep. Why only certain, perhaps prone, employees develop intolerance to a rotating shift work schedule remains to be explained. It has been proposed by Reinberg et al. (97) that some persons may be predisposed genetically and thus may be susceptible to rapid internal desynchronization when subjected to recurring, periodic alteration in their sleep–activity schedule.

Reinberg et al. (97) found in a sample of biologically intolerant shift workers, evidence of a persisting internal desynchronization of the circadian system. The period duration of at least certain rhythmic variables are longer than 24.0 hr. For example, Table 9.5 shows the findings for studies of the oral temperature circadian rhythm in workers, who at the time of being studied reported good (subjects 1–11) or poor (subjects 12–15) biological tolerance for rotating shift work involving the periodic alteration of the sleep and activity pattern. With the exception of subject No. 1, the circadian period of the oral temperature circadian rhythm in tolerant shift work employees was 24.0 hr. Moreover, the amplitude of this rhythm in the tolerant workers was large, at least 0.21°C. In contrast, for the intolerant workers of this sample, the circadian period for the oral temperature rhythm was longer, being

Table 9.5 Characteristics of Circadian Rhythm of Body Temperature in Tolerant and Nontolerant Shiftworkers[a]

Subject No. (Age in Years)		History of Rotating Shift Work	Dominant Circadian Period[b] (hr)	Circadian Amplitude[c] (°C)
		Tolerant		
1	(20)[d]	1 month	25.1	0.21
2	(22)	2.5 years	24.0	0.23
3	(22)	2.5 months	24.0	0.27
4	(25)	3 years	24.0	0.31
5	(25)	4 years	24.0	0.25
6	(27)	1 month	24.0	0.35
7	(27)	3 months	24.0	0.19
8	(31)	1.5 years	24.0	0.21
9	(36)	2.5 years	24.0	0.27
10	(35)	4 years	24.0	0.36
11	(43)	2 years	24.0	0.42
		Nontolerant		
12	(20)	1 month	25.7	0.04
13	(28)	3 months	25.1	0.04
14	(30)	3 months	24.9	0.13
15	(31)	3 years	25.2	0.09

[a] From Reinberg et al. (97).
[b] Estimated by power spectrum.
[c] Estimated by the single cosinor method.
[d] Following publication of these data, subject 1 had to be transferred from rotating shift work to permanent day work because of a developed intolerance to the former.

between 24.9 and 25.7 hr. [In some cases shift work intolerance was associated with circadian periods that are shorter than 24 hr (Reinberg, personal communication).] Too, the circadian amplitudes were very small, between 0.04 and 0.13°C. It is of interest that since these findings were published it was necessary to transfer subject No. 1 from rotating shift work to permanent daywork (Reinberg, personal communication) because of a developed biological intolerance to shift work. Thus, it appears that one important component of shift work tolerance relates to the strength of the circadian rhythm temporal organization, as best gauged at present by the magnitude of the amplitude of the oral temperature circadian rhythm. According to Reinberg and his colleagues (97), shift workers with large-amplitude rhythms appear to be less prone to developing an internal desynchronization of circadian functions and intolerance to shift work.

The phenomenon of internal desynchronization in an intolerant as compared to a tolerant shift worker is clearly apparent in Figure 9.8 (97). The upper left-hand portion of the figure shows the clock hour timing of the circadian acrophase (indicated by closed circles) of the oral temperature rhythm for one subject (No. 3) who exhibited good biological tolerance to shift work. The timing of the acrophase of this rhythm was quite stable from day to day, despite the regular rotation to night work (shown by the asterisked horizontal bars) during the weekdays. The bottom left-hand portion of this figure shows the so-called power spectrum of the oral temperature data for this worker. This method was used by Reinberg et al. (97) to indicate through a partitioning of the variance in the data set of oral temperature the existence of prominent bioperiodicities for each of the various shift workers studied. It can be seen for subject 3 that the rhythm of oral temperature was exactly 24.0 hr in duration. The upper right-hand portion of Figure 9.8 gives an example of a desynchronized circadian rhythm of an intolerant subject (No. 13). It can be seen that the timing of the circadian acrophase (closed circles) of the oral temperature rhythm occurred progressively later from one day to the next. Such a finding would be expected if the circadian period of his oral temperature rhythm was longer than 24.0 hr. In this nontolerant shift worker, the period of this rhythm as shown by the power spectrum analyses (lower right-hand portion of Fig. 9.8) was 25.1 rather than the expected 24.0 hr in duration. Thus, in this biologically intolerant shift worker, the later timing of the circadian acrophase resulted from a prolongation of the period length of at least the oral temperature circadian rhythm.

In this and other presumably susceptible persons, internal desynchronization apparently results from recurring changes in the hours of sleep and activity associated with shift work schedules that include a rotation of work during the nighttime and sleep during the daytime. Although only reports dealing with desynchronization of the oral temperature circadian rhythm have been published, other circadian functions appear to become internally desynchronized in nontolerant shift workers (Reinberg, personal communication). Presumably, alteration in the temporal organization, as manifested as an internal desynchronization of the circadian system, constitutes an important aspect of the biological nontolerance to shift work.

It has been suggested that in predisposed persons, an internal desynchronization is associated with a set of symptoms that are common both to clinical depression and shift work intolerance (94). In connection with this, Michel-Briand et al. (145)

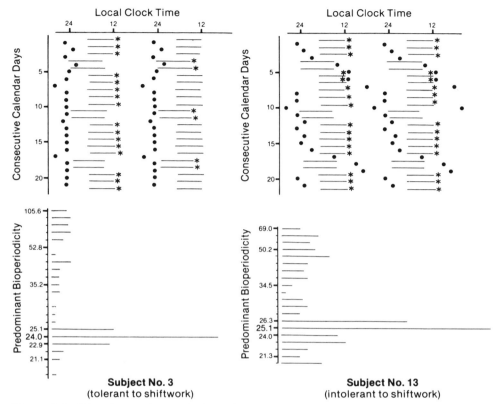

Figure 9.8   The plots at the top indicate the timing of the circadian acrophase of the body temperature rhythm of a person exhibiting either good tolerance (subject 3, left side) or poor tolerance (subject 13, right side) to rotating shift work. The acrophase of the rhythm is consistently timed from one day to the next in subject 3 when sleeping during the day (shown by asterisks) for 5 of 7 days each week. The acrophases for subject 13 exhibit a systemic delay with each consecutive day. This latter phenomenon suggests a period length for this rhythm greater than 24 hr. This is documented statistically by spectral analysis (bottom right). The most prevalent period for the body temperature of subject 13 was shown to be 25.1 rather than 24 hr as shown for subject 3 in the lower graph to left. From Reinberg et al. (97).

reported a greater incidence of depression and affective illness in retired shift workers than in retired day workers in whom cardiovascular and locomotor illnesses and disorders were the more common. In patients suffering from depressive illness, effective treatment with specific medications such as lithium and other antidepressant drugs, which are known to influence the circadian period of certain rhythmic variables, moderates or corrects the internal desynchronization. Often nontolerant shift workers resort to using medications of another type—sleep-inducing aids such as barbiturates, tranquilizers, diazepam derivatives, and antihistamines. These medications, which lack the ability to affect the period of circadian functions, frequently are over used, leading to addiction, although typically without effective resolution of the employee's sleep disorders or persisting fatigue.

It is important to note that persons intolerant to rotating shift work do recover from their untoward symptoms when they are returned to a regular pattern of activity and sleep. Permanent day work with nighttime sleep leads to a synchronization of biological rhythms and usually alleviates or modulates the associated clinical symptoms of intolerance to shift work. However, complete regression of the symptoms may require in certain cases as long as 6 months or so. In this regard, the decision to discharge an employee from shift to permanent daytime work must be done early when the signs and symptoms of biological intolerance first become apparent. This is important in order to increase the likelihood that the biological effects of shift work intolerance will be minimal and that recovery will be rapid.

9 DIFFERENCES BETWEEN WORKERS IN THE RATE OF BIOLOGICAL RHYTHM ADJUSTMENTS WHEN ON THE NIGHT SHIFT

Aschoff (146) and Wever (108) have conducted many studies on volunteers living in isolation under controlled constant conditions that provide no time cues or synchronizers. Their research was significant in that their findings suggested that persons exhibiting a high-amplitude oral temperature circadian rhythm were less disturbed chronobiologically when exposed to conditions lacking synchronizers. Based on these findings, Reinberg reexamined his data from field studies on shift workers to determine if the rate of change in the acrophase (designated as (Δ, ϕ) for a set of circadian rhythms) varied with the magnitude of the respective rhythm amplitudes. Reinberg et al. (93,147) found for the circadian rhythms of oral temperature and urinary 17-hydroxycorticosteroids was negatively related to the respective circadian amplitude. Specifically, the greater the amplitude of the rhythm exhibited by a worker, the slower the rate of change in his acrophase following a shift in the sleep–activity synchronizer schedule. Figures 9.9 and 9.10 show the association between the circadian amplitude and rate of acrophase shift for these two rhythms.

Data from field studies of both tolerant and nontolerant shift work employed in the steel, chemical, oil, and coal industries suggest that individuals who exhibit a large-amplitude oral temperature circadian rhythm possess a better capability for the long-term biological tolerance of rotating shift work that includes nighttime duty (93, 95–97,147,148). As shown in Table 9.6, shift workers with oral temperature circadian rhythms greater than 0.30°C were found to display good tolerance to rotating shift work without medical problems. On the other hand, employees who were found to exhibit a small-amplitude oral temperature circadian rhythm, 0.23°C or less, were those who experienced biological intolerance to rotating shift work schedules (97).

The overall findings of chronobiologic field studies on tolerant and nontolerant shift workers, according to Reinberg and his co-workers (97), suggest the following conclusions. First, workers differ chronobiologically in their ability to do rotating shift work. Second, persons who have demonstrated a good tolerance for rotating

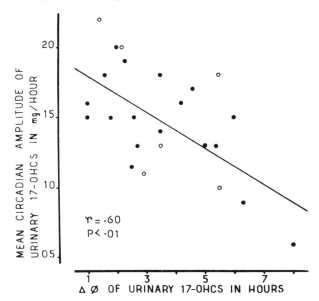

Figure 9.9 Correlation between individual mean amplitude A (in milligrams per hour) and acrophase shift $\Delta\phi$ (in hours) of the urinary 17-hydroxycorticosteroid (17-OHCS) circadian rhythm; $\Delta\phi$ corresponds to the difference of acrophase location on the 24-hr scale between control days and the first night shift. The graph shows that the greater a person's circadian amplitude, the slower the acrophase shift of the 17-OCHS circadian rhythm. Symbols: ●, 18 shift workers; Reichstett study; ○; 6 shift workers, Petit–Couronne study. From Reinberg et al. (93).

shift schedules that require night work, for perhaps as long as 15–30 years, are likely to possess a high-amplitude oral temperature circadian rhythm. This trait indicates that they adjust their circadian system slowly following alteration in their sleep–activity pattern such as when working nights. In short, these shift workers have a strong (high-amplitude) circadian temporal structure that is evidenced by the fact that their circadian system seems to show little ill effect by repeated short-term alterations in the sleep–activity schedule during many years of rotating shift work duty. Third, persons who exhibit a small-amplitude oral temperature circadian rhythm and who rapidly adjust their circadian system following alteration in the sleep–activity pattern appear to have a weaker circadian temporal organization. It is this type of individual who tends to be at risk of becoming shift work intolerant because his or her rhythms so quickly invert when changed from a day to night schedule of work. These rapid "inverters" are likely to be at greater risk for suffering from those medical maladies typical of shift work intolerance and at an early age.

Should these findings be validated in prospective field studies on large groups of shift workers, they have the following implications. Employees exhibiting large-amplitude circadian rhythms, such as in the rhythms of oral temperature and urinary

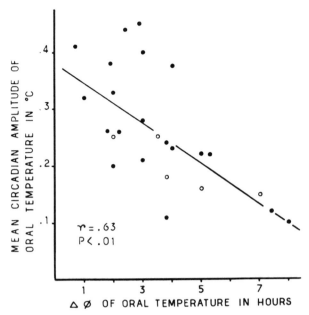

Figure 9.10   Correlation between mean amplitude A (in degrees Celsius) and acrophase shift $\Delta\phi$ (in hours) of the oral temperature circadian rhythm; $\Delta\phi$ corresponds to the difference of the acrophase locations on the 24-hr scale between control days and the first night shift. Symbols: ●, 20 shift workers, Reichstett study; ○, 5 shift workers, Petit–Couronne study. From Reinberg et al. (93).

17-hydroxycorticosteroid concentration, appear to be best suited for rapid, 2- to 3-day duration, rotating shift work schedules. This is because those having such high-amplitude rhythms tend to adjust the acrophases of their circadian system so slowly that by the last ''day'' of the nighttime shift, day 2 or 3, the alteration of the circadian acrophase is minimal. Thus, when rotating back to a pattern of nighttime sleep alternating with daytime work, the chronobiologic organization of such employees only needs to undergo slight readjustment, if any. Individuals with this type of temporal organization, according to Reinberg and his co-workers (93,147), seem to be most appropriate for rapidly rotating shift work systems since their circadian rhythms are not likely to be much affected by a short-term change in their sleep–wakefulness schedule. On the other hand, persons exhibiting low-amplitude circadian rhythms, such as in oral temperature and 17-hydroxycorticosteroid concentration, are likely to undergo comparatively rapid shifts in their circadian system following a change to nighttime work requiring daytime sleep. For these individuals chronobiologic adjustment is likely to be complete by the end of the first or second day of the nighttime shift. Thus, for this type of person, rapid, 2- to 3-day, rotational shift work is less appropriate than are 7-day or longer rotations.

Table 9.6 Circadian Rhythm in Oral Temperature of Subjects with Good, Adequate, or Poor Tolerance to Shift Work: Single Cosinor Summary[a,b]

Group	Mean Age (Years)	Tolerance to Shift Work (Number of Subjects)		Rhythm Detection p Value	Mesor (24-hr Rhythm Adjusted Mean ± SE) (°C)	Circadian Amplitude (95% Confidence Limit)	Acrophase φ in hr and min, Referenced to Midnight (95% Confidence Limit)
I	25.3	Good	(6)	<0.005	36.53 ± 0.08	0.37 (0.29–0.45)	1549 (1438–1700)
II	50.0	Good	(10)	<0.005	36.51 ± 0.07	0.35 (0.30–0.40)	1534 (1438–1627)
III	50.2	Adequate	(6)	<0.005	36.53 ± 0.11	0.30 (0.24–0.36)	1657 (1541–1905)
IV	47.4	Poor	(7)	<0.005	36.52 ± 0.12	0.23 (0.17–0.29)	1711 (1607–1814)

[a]From Reinberg et al. (94).

[b]All subjects were shift workers at the same oil refinery. Groups differed mainly by a good (e.g., group II) versus a poor (group IV) tolerance to shift work as well as by a large (group II) versus a small (group IV) amplitude. Group II (good tolerance) involved 10 senior operators (no history of shift work difficulty; mean age = 25.1 years, range 15–32 years). Group IV involved seven senior operators (who were to be discharged from shift work due to nontolerance; mean age = 47.4 years, range 30–56 years; mean shift work duration = 22.9 years, range 9–29). In any group, amplitude differs from zero with $p < 005$; mesor, no difference between groups; amplitude (one-half of total peak to trough variability), large in groups I and II (good tolerance), small in group IV (poor tolerance to shift work); acrophase φ (crest time), no difference between groups.

10 SHORT-TERM DIFFICULTIES ASSOCIATED WITH SHIFT WORK SLEEP DISRUPTION AND FATIGUE

Shift workers often complain about the inadequacy of the quantity and quality of their sleep. This subject has been studied by many investigators (57,64,149–167). The need for sleep, independent of the type and schedule of shift work varies greatly between individuals and may be associated with age. However, the quantity of sleep required is often less than that obtained by individuals performing night work as evaluated by Rutenfranz et al. (166). These investigators used self-assessment questionnaires to sample a group of 329 shift-working locomotive engineers. Figure 9.11 shows the results of their survey. Inquiries pertaining to the duration of sleep needed are plotted in the form of a frequency distribution. It is clear that a majority of the workers (85%) expressed a preference for between 7 and 9 hr. When the same persons were questioned about the amount of sleep attained when working nights, the distribution of sleep durations was very different. When compared to the sleep-needed curve, the obtained sleep curve was skewed to the left, meaning there was a rather large percentage of the workers who obtained less than the quantity of sleep required.

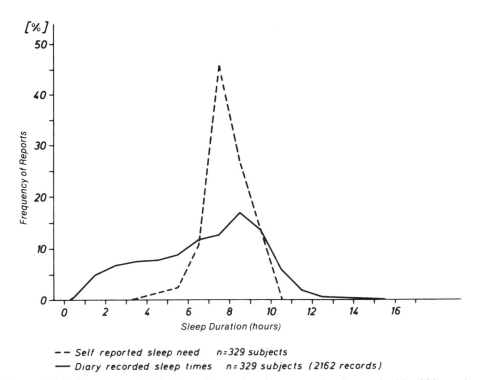

Figure 9.11 Frequency distribution of sleep duration reported to be required by 329 rotating shift workers employed as locomotive engineers compared to the frequency distribution of their sleep durations (based on 2162 sleep records) between two consecutive shifts. Sleep duration data were obtained by self-assessment diaries. From Rutenfranz et al. (166).

Results of various surveys indicate that as many as 90 percent of all workers complain about their sleep when working nights. In contrast, only 5–20 percent of day workers and shift workers not having to work nights complain of sleep problems. However, it has been shown (168) that those shift workers who prefer working nights actually get less sleep than those workers who dislike night work. It is important to point out that there are a number of factors that can contribute to sleep difficulties in shift workers. These include housing conditions, times of going to bed, age, noise, marital status, the presence and age distribution of children in one's household and the regularity of shift rotation. In particular, normal noise levels can make daytime sleep difficult when working nights. Even when marital status, the number of children in the home, age, and experience were controlled, sleep impairment was found to be related to night work in 1505 nurses (169). Figure 9.12 from Knauth and Rutenfranz (56) provides factual information about sleep duration in relation to the type of work schedule based on a thorough study of almost 5000 shift workers engaged

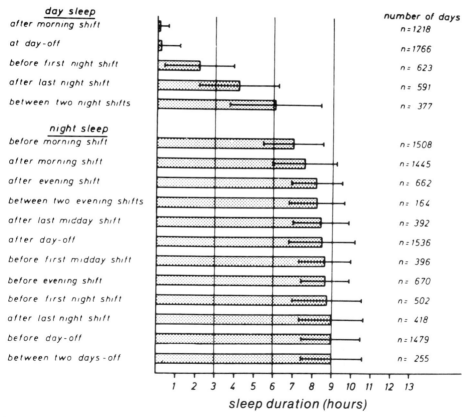

Figure 9.12 Average (shown with the standard deviation) duration of sleep as a function of the shift schedule worked and the timing of rest during either daytime or nighttime. The data summarized in the figure represent the findings from sleep diary records from 1230 persons encompassing a total of 9840 24-hr spans. See text and Knauth and Rutenfranz (100) for details.

in various occupations. For the purpose of classifying the different work schedules, those commencing between 5 and 9:59 A.M. were termed morning shifts; between 10 A.M. and 1:59 P.M., midday shifts; between 2 and 6 P.M., evening shifts; and between 10 P.M. and 3:30 A.M., night shifts. The data in the upper portion of the figure suggest that shift workers either did not choose to sleep during the daytime unless it was absolutely necessary or if they did, to take less than that typical amount of nighttime sleep. In particular, day sleep during a day off or after returning home from working a morning shift was taken only by about 10 percent of the rotating shift workers, apparently as a nap averaging less than an hour. This was not unexpected; shift workers prefer to partake in activities with friends and family who are diurnally active. Only when persons worked the night shift was sleep more commonly taken during the day. In this instance more than two-thirds took sleep during the daytime prior to the first night duty and especially after the last night shift when prior day sleep averaged about 4 hr. Between successive night shifts, sleep had to be scheduled during the daytime; however, the duration was found to average only about 6 hr. The duration of nighttime sleep when working morning, midday, and evening shifts does not vary greatly. Only as a consequence of working the morning shift was the average duration of nightly sleep somewhat shorter than 8 hr. Since human beings tend to be rather inflexible about the time (clock hours) when they go to bed (56), a reduction in sleep duration often occurs when the morning shift has an early commencement.

The inability to obtain sufficient sleep during the daytime when working the night shift results from at least two types of problems. First, achieving restful sleep during the daytime is difficult due to noises emanating from diurnally organized activities (56,166). Workers most often complain that their daytime sleep is disrupted because of noise (Fig. 9.13) made by children, neighbors, or road traffic (56). Second, for

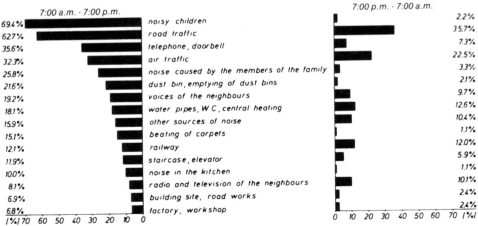

Figure 9.13 Relative frequency of sources of noise resulting in interrupted daytime (7 A.M. to 7 P.M.) or nighttime (7 P.M. to 7 A.M.) sleep. Data from 808 European shift workers who complained about frequent noise-induced sleep disruptions. From Rutenfranz et al. (56).

biological reasons, sleep taken during the daytime, especially around midday (Fig. 9.14) tends to be of shorter duration than nighttime sleep (100,149,151). When taken during the daytime, sleep is forced to occur at a biological time that is inappropriate and incompatible with the staging of circadian rhythms. The adjustment of circadian rhythms to nighttime duty in a majority of shift workers requires at least a few days. Until this adjustment is complete, the ability to sleep during the daytime is poor. The subject of biological rhythm adjustments in relation to shift work has already been addressed further in previous sections of this chapter.

In their review Rutenfranz et al. (56) found that the frequency of complaints about sleep disturbances varied with the type of task performed and the shift schedule worked (Fig. 9.15). In examining the data from a study of 5766 employees, it was found that sleep disturbances were not uncommon. They were a complaint even for as many as 10–20 percent of the day workers. In comparison, as many as between 70 and 80 percent of those employees who rotated between day and night work experienced sleep disturbances. An interesting and important finding is that the incidence of sleep disturbances in former shift workers while engaged in rotating shift work could be as great as 90 percent. Shift workers frequently suffer from sleep deprivation. While much work has been done on total sleep deprivation and continuous work, most shift workers suffer from partial sleep deprivation. Naitoh (170) notes that partial sleep deprivation differs significantly from selective sleep depri-

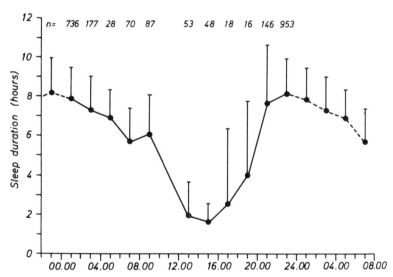

Figure 9.14 Sleep duration ($\overline{X}$ and SD) based on data from 304 shift workers as a function of the time during the day and night when it began. Indicated across the top is the number of reports for sleep onset according to clock hour from which the means and standard deviations were calculated. The revealed rhythmic trend found by Knauth and Rutenfranz (100) is similar to that shown by field studies on train drivers by Foret and Lantin (151) and through laboratory studies by Akerstedt and Gillberg (149). From Knauth and Rutenfranz (100).

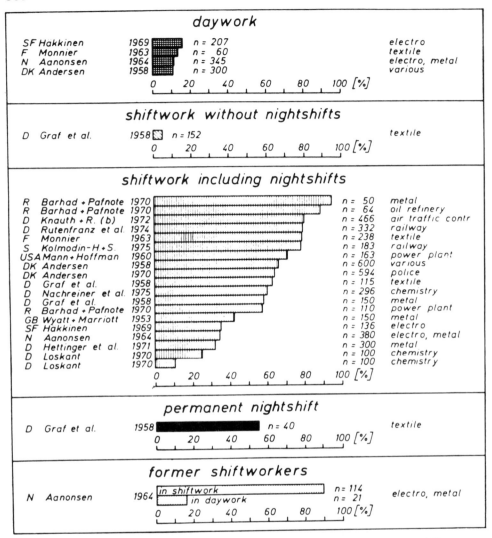

Figure 9.15 Frequency of complaints about sleep disturbances according to shift system, industry, and investigator. From Rutenfranz et al. (56).

vation that occurs only in sleep laboratories and is not found naturally. The effects of partial sleep deprivation as they occur in shift workers are not well documented.

Using ambulatory polysomnography, Akerstedt et al. (171) evaluated the sleep architecture of 14 shift workers performing a rotating 8-hr shift schedule. Results indicated that the shortest sleep times occurred when working the day shift and the longest when working the evening shift. Similarly, Stage 2, REM (rapid eye movement) and SWS (slow wave sleep) periods were shortened when working the day shift. In most cases, total sleep periods were significantly shorter while working day and night shift when compared to the evening shift. Frequency of napping was also increased after day and night work. However, the content of the sleep cycles did not significantly differ between shifts. This finding suggests that sleep deprivation per se may not cause the feeling of sleepiness that routinely occurs during night shift work. Another study by Akerstedt and Kecklund (172) suggested that sleep architecture did not change over a 2-year study period in rotating shift workers.

In summary, it may be concluded that permanent day work and shift work schedules that exclude nighttime duty do not ordinarily cause problems with sleep. However, rotating shift work schedules that incorporate a nighttime rotation or that involve permanent night work are more likely to cause sleep problems in some persons. The major concerns regarding sleep disturbance and reduction occur when night work is performed (during the usual span of sleep). Usually workers perceive reduced physical well-being, a potential increased risk of illness (56), and increased fatigue with loss of vigor (Fig. 9.16). The loss of vigor coupled with insufficient sleep has been shown to contribute to a reduction in worker performance and productivity as well as the potential of increased risk of an accident in the workplace (34, 100, 101).

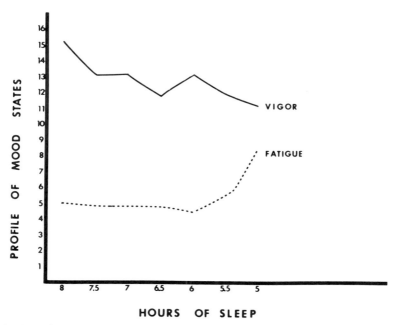

HOURS OF SLEEP

Figure 9.16 Effect of sleep reduction on self-assessed feelings of vigor and fatigue. The figure indicates a definite elevation in the self-perceived level of fatigue when sleep duration is reduced to less than 6 hr. The curve of vigor exhibits a trend of decline as sleep becomes reduced in length from 8 hr. From Johnson (57).

11 CIRCADIAN RHYTHMS, SHIFT WORK, AND HUMAN PERFORMANCE

Before a discussion of performance rhythms and shift work can begin, performance must be defined. In laboratory studies performance is often measured using standardized neuropsychological tests. Parameters such as reaction time, logical reasoning, tracking ability, vigilance, and mood are commonly taken using either paper or computerized tests. Experimental data collected at the work site may differ for a variety of reasons including changes in environment, subject motivation, and differences in the types of tests presented.

More recent studies (173) have used portable computers and software at the work site to obtain measures of performance similar to that collected in laboratories, although paper tests and questionnaires are still frequently used. In some studies, particularly those from the field of nursing, performance may be defined as a subjective measure of worker performance made by a supervisor rather than neuropsychological evaluation (174). Two questions need to be addressed when assessing the appropriateness of the methodology used for performance testing. First, what is the relationship between the neuropsychological task selected for performance evaluation and the actual type of work being performed by each individual worker? Second, what is the value of the supervisor's subjective measure of an individual's performance given that other factors may bias this evaluation. Therefore, the way performance has been measured must always be carefully evaluated when determining the relevance of any study to the actual work environment.

Colhoun (175) and Lavie (176) have extensively reviewed the early studies of performance and noted that this research was frequently concerned with variance in memory during different times of the day. It was commonly believed that mental fatigue developed throughout the work period and decreased mental capabilities.

Kleitman (177) discounted much of the early experimental work because of the lack of carefully controlled experimental conditions. His work indicated that an apparent correlation existed between body temperature rhythms and performance on simple noncognitive tasks. It was suggested that a casual relationship existed. However, later studies (178) confirmed the existence of a correlation between simple tasks and body temperature. However, cognitive tasks did not correlate well with rhythmic changes in body temperature (179).

Instead, Colhoun and co-workers (180,181) suggested that performance is related to the rhythm of basal arousal (the opposite of sleepiness). This rhythm was never well defined but is believed to be related to the temperature rhythm in an "inverted U relationship," such that performance increases with arousal to a point and then declines with further increases in arousal.

Until the early 1970s much work focused on a single performance rhythm. Later work suggested that a variety of circadian performance rhythms exist and field and laboratory investigation of these differences began. Laboratory investigations of circadian differences in human mental and physical performance are relatively common (21,34,101,102,134,182–184). During the last four decades only the findings of a small number of field studies have been published on temporal changes in human performance or accidents over the 24 hr. Results of these investigations indicate that there is a problem of impaired efficiency when working the night shift. Folkard (102) has reviewed the findings of six field studies; these are summarized in Figure 9.17.

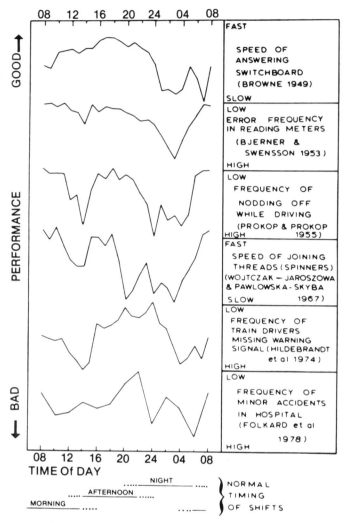

Figure 9.17 Temporal patterns in performance, vigilance, and accidents over 24 hr as indicated from a review of six studies. Time is shown in military fashion: 1600 = 4 P.M., 2000 = 8 P.M., 2400 = midnight. Indicated at the bottom is the timings of the night, afternoon, and morning shifts. In general, performance and vigilance are reduced during the nighttime as well as for a short duration after midday. From Folkard (102).

The findings of each of these studies are presented in this figure from top to bottom, in the order of their date of publication. In this figure, arbitrary scales have been selected to make the amplitudes of the six curves approximately equal. For each curve, the lower the reading the poorer the performance. In apparently the earliest study, Browne (185) examined the speed with which shift-working switchboard operators answered calls at different times of the day or night (top panel of the figure). The data have been corrected by Browne to take into account the number of calls at any given time of the day. Performance speed improved in a fairly linear manner from 8 A.M. to 6 P.M. and dropped sharply after about 10 P.M. such that it was slower during the night than at any other time.

Bjerner et al. (188) studied the 24-hr variation of committing errors by shift-working meter readers (second panel from the top). They found that performance decreased only slightly over most of the normal working day with evidence of a slight "postlunch dip." There was a fairly sharp drop in accuracy after about 10 P.M. with a minimum during the nighttime hours. Prior to the publication of this study the existence of a postlunch dip in performance generally was not recognized.

The third panel from the top presents the frequency with which professional truck drivers reported falling asleep ("nodding off") while driving at different times during the day or night (186). In this data set the postlunch dip is very apparent, with the frequency of nodding off being almost as high around 2–3 P.M. as it is during the night period. The reason for this is uncertain, although it has been suggested that the postlunch dip is more marked in people who are relatively sleep-deprived (187). Again, however, there was clear evidence of an impairment during the night hours.

Wojtczak-Jaroszowa and Pawlowska-Skyba (22) examined the speed with which five spinners joined broken threads (middle panel). This study was a particularly detailed one, with about 5000 measurements being taken. The studied employees had all been on shift work for at least 10 years and were highly proficient at their task. They were studied on the morning (5:30 A.M. to 1:30 P.M.), afternoon (1:30 to 9:30 P.M.), and night (9:30 P.M. to 5:30 A.M.) shifts, although it was not reported how rapidly they rotated. In the figure the data from these three shifts have been combined as a continuous 24-hr curve. As the authors pointed out, performance speed was about 10 percent slower when the employees worked the night as compared to when they worked either the morning or afternoon shifts.

The results of another detailed study are shown in the next to the bottom panel. In this study by Hildebrandt et al. (187) automatic recording devices were fitted into the cabs of 10 train locomotives. Approximately every 20 min a warning light appeared for 2.5 sec followed by an auditory warning signal for an additional 2.5 secs. If neither of these signals was heeded, there was a 30-sec sounding of a loud horn making mandatory that the engineer operate the safety gear to avoid automatic braking of the locomotive. Hildebrandt et al. (187) recorded a total of 2238 occurrences of signal sounding, an indication that neither of the two warning signals had been heeded. Despite the fact that the warning light was more visible at night, more warning signals were missed, that is, the signal sounded more often during the night than during the day. Also apparent was a decrease in vigilance during the period immediately after midday.

The bottom portion of Figure 9.17 shows the findings of Folkard (179) who studied the frequency with which patients incurred minor accidents while hospitalized. In this study the records of a large modern hospital were examined for a 5-year period. A total of 1854 "unusual incidents" occurred during this period, of which 1576 were minor accidents involving individual patients, and for which there was a clear indication of the time of its occurrence. Only 30 percent of these accidents resulted in injury to patients, and in the majority of these cases (80%) only minor scratches or bruises were sustained. The frequency of the incidents tended to decrease throughout the normal waking period and to increase during the night, despite the fact that the patients were asleep during most of the nighttime. The two sharp peaks at 10 P.M. and 8 A.M. apparently are associated with the patients' activity related to the need to urinate before going to sleep at night and on awakening in the morning. Taken together the results shown in Figure 9.17 suggest that there is a problem of impaired efficiency during the nighttime. More recent studies of 8-hr rotating shift workers (189) also suggest that performance declines during the night shift, but performance deterioration also may occur as a function of the number of days spent in a particular shift rotation and a cumulative sleep deficit.

As previously stated, most workers do not suffer from total sleep deprivation. However, some workers who must participate in extended work shifts must perform acceptably for periods of 24 hr or more. The majority of research in this area was performed on medical residents and interns and has shown that on-the-job performance and mood declines (190,191) over the work period. One study has suggested that this is not the case (192) but the methodology of this study has been severely criticized.

In a previous section it was noted that the 12-hr shift has gained in popularity over the last two decades. While there are advantages to this shift system, some recent research has suggested that 12-hr/4-day workweeks are more detrimental to performance and are more fatiguing than an 8-hr/6-day workweek (65) while other studies have suggested that no difference exists (193). Again, the importance of evaluating the relationship between the selected performance test and the actual work being performed must be emphasized.

In summary, a variety of investigations reveal circadian variation in performance. A majority of studies reveal decreased efficiency during the night shift as well as during the afternoon. Knowledge of these predictable differences is fundamental not only to the development of an effective industrial safety program but in addition to maintaining quality control and productivity from one shift to the next. However, comparisons made between shift work schedules must account or control for dissimilarities between workers, tasks, environment and the relevancy of testing before validity can be accepted.

12 SHIFT WORK, ACCIDENTS, AND INJURIES

While most persons consider the terms *accident* and *injury* to be synonymous, in the context of occupational health they must be defined separately. The term accident may imply that an event is unavoidable; more commonly, it is classified as an unexpected error or event with some catastrophic consequence. However, Heinrich (194) proposed that any injury consists of two distinct events, the error or accident that is made and the injury that is caused by the error. Thus, reduced performance may cause the occurrence of an accident as demonstrated by recent environmental disasters at Three Mile Island, Chernobyl, Bhopahl, and the Exxon *Valdez* disaster in Alaska. Indeed, the effects of shift work and sleep deprivation have been implicated as a cause in each of these mishaps (195). Incidents of this type are often caused by reduced performance in ''sentinel workers.'' This would suggest that the chronohygiene of workers charged with prevention of potential environmental disasters should be closely monitored. The multifactorial nature of an injury, however, does not allow for such a cohesive correlation.

Although there is no direct correlation between decreased performance and the occurrence of an injury, a connection remains suspect. Impaired performance may lead to an injury but may not necessarily be its cause. The difficulty in establishing causality may be due in part to the multifactorial nature of the injury process and may be related to other confounding variables in the workplace.

Several investigations have been performed on shift workers and their injuries in the last 70 years. While the results of these studies do not propose that one shift work system is superior in reducing injuries, they do identify various conditions at individual plants where shift work was employed that may have contributed to an increased risk of injury.

In an effort to increase production, studies of productivity and work-related injury were performed at British munitions factories during World War I and published in a government report after the war (196). The results of this study determined that reduced work time (shift length) and increased rest periods improved performance as measured by output. It was also discovered that the best performance (output) was measured during the afternoon and that a greater number of severe injuries occurred during the night shift.

The ensuing Great Depression and World War II eliminated further shift work studies between 1929 and 1946. There was a renewed interest in shift work related performance after the end of the war (185–188). Studies of injury as related to shift work were not conducted until later. Recent research suggests that the number of injuries is greatest during the day shift (140,197) especially during or just prior to the lunch hour (198,199). It has been suggested that the higher frequency of injury during the day shift is related to increased worker density and noise level during the day or lapsed attention on the job just prior to the lunch break. This increase in injury may also be associated with the postlunch performance dip previously described.

In a study of a chemical manufacturing plant utilizing a backward rotation 8-hr shift schedule, Novak et al. (140) determined that the greatest number of injuries occurred in shift workers during the day shift, but that injury rates were increased during the first four work periods of the day and night shift. It was suggested that the increased injury frequency early in the shift rotation was caused by worker desynchronization caused by the phase advance shift schedule. However, in a study of paint factory workers (200) performing 8-hr rotating shift work, it was discovered that the greatest frequency of injury and the most severe injuries occurred during the night shift. Similar results were obtained by Wagner (201) in miners. Circannual rhythms in work-related injuries have also been discovered (202).

Laundry and Lees (197) investigated the injury rates at a manufacturing plant that had changed from an 8-hr rotation to a 12-hr rotation system. Overall injury severity rates were decreased after transition to the 12-hr system; however, the nonemployment-related injury rate increased. Further, when classified by gender and corrected for age, male injury rates declined after transfer to the 12-hr shift work system while the rate remained the same for females. In another study female shift workers at a chemical manufacturing plant were found to have an injury rate almost three times greater than shift-working males in similar job categories (140).

While these research findings appear to be contradictory, it may be related to the differences between work sites where the various studies were performed. Research has suggested that at different work sites, even the same job title may carry different risks of injury. The risk for injury is influenced by the type of tasks performed on each shift by each worker, the level of safety training and enforcement, and a variety of other social, physical, and managerial elements (203). Thus, finding worker populations with different shift scheduling but equitable risk for injury is extremely difficult. Therefore generalization of the research findings from one work site to another is not recommended. Rather, evaluation of each individual work site for an accurate appraisal of the effect of shift work on injury causality is required.

13 PSYCHOSOCIAL PROBLEMS OF SHIFT WORKERS

Workers who perform either permanent or rotating night work experience a lifestyle that differs markedly from their day-time-working friends and also from their diurnally active family members. The majority of businesses, recreational facilities, cultural events, and various other types of activities are scheduled mainly to meet the needs of diurnally active persons. Even though a certain number of businesses nowadays have evening hours, or round-the-clock hours, the choices available for rotating shift workers are less than those of permanent day workers. Although generalization is difficult, a variety of studies have found that work schedules involving nighttime duty results in at least some degree of difficulty (141,204–211).

Some of the drawbacks of shift work have been exemplified by the findings of a large study of British steelmill shift workers (211). Complaints centered around not having weekends free or not having enough time for pursuing a full social life, watching sports, attending social functions, planning social engagements, and following a regular TV series (211). Nachreiner and Rutenfranz (141) found vast disparities between shift workers and day workers with regard to the type and number of complaints about not having sufficient time for pursuing social and spare time desires. Figure 9.18 indicates that from four to six times as many shift workers as compared to day workers in the chemical industry complained of not having enough time for social activities, including friends and family. Although spouses of shift workers have rarely been involved as respondents in studies, it is suggested that they often assess the attitudes of shift workers as more negative than the workers themselves (212). However, another study of British shift workers by Wedderburn (211) suggested that relative to day workers they seem better off with regard to pay, freedom during the daytime, time for one's self, free time, and variety of working hours.

Shift work has significant impact not only for the employees themselves but on members of their families as well (37,17,55,206,207). Time budget studies clearly point out that wives of shift workers have a more difficult time trying to keep the family together. Among other things, they find it necessary to change the timings of meals to fit the husband's shift schedule, which may change weekly. Moreover, in

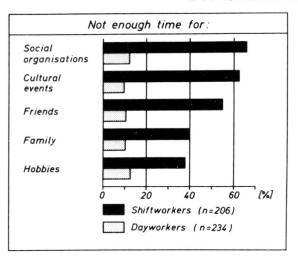

Figure 9.18 Frequency distribution of reports of insufficient time for designated social activities in a group of 206 rotating shift workers and in a group of 234 permanent day workers. From Nachreiner and Rutenfranz (360).

families in which one spouse is a rotating shift worker and the other a day worker, great strain is placed upon the marriage. Communication between husband and wife, as well as children, becomes restricted or curtailed. The shift-working parent may find it difficult or impossible to attend important school and/or other functions. In certain situations children become resentful toward the shift-working parent. However, some studies have shown that while more time may be available for child activities, social life can be restricted. Spouses also can become disenchanted; they may feel alone and neglected and marital discord may result. Thus, many researchers suggest that rotating shift work can increase the likelihood of divorce and contribute to the development of personality disorders in children. Moreover, when it is the wife who is a rotating shift worker, off-the-job demands and responsibilities, such as tending to the children and home, further add to the stresses of the rotating shift work pattern. These responsibilities usually take priority over making up sleep deficits due to working nights. This contributes to the high level of fatigue experienced by many women shift workers who have families (206). Shift work may also have a powerful impact on the social activities and fatigue levels of the increasing number of single parent families.

Shift work, particularly rotating shift work, is also commonly viewed as a stressful activity (213). However, the stress caused by shift work may induce psychosocial and physical difficulties only if other stressors are present (214). Others suggest that individual, organizational, and psychosocial variables can influence the degree of psychosocial well-being in shift workers (215,216). Each shift rotation system possesses its own particular set of advantages and disadvantages that effect workers (217). This may signify that each working situation is different and individual evaluation is required. Occupational health personnel must be cognizant of the potential psychosocial problems that may occur in individual workers and worker populations and work to resolve any difficulties that exist.

14 SHIFT WORK AS A POTENTIATOR OF ILLNESS

The relationship between shift work and ill health remains controversial, although some studies indicate that illness is more common in shift workers when compared to day workers. The findings of early epidemiologic investigations on the morbidity and mortality of such employees have been somewhat inconclusive (134,218–221). Some have reported increased morbidity or mortality in shift workers, while others have reported the opposite or no difference. A failure to select appropriate control groups has weakened the conclusions of many studies. Some epidemiologists believe that shift workers represent a type of survival population, since employees who are incapable of tolerating shift work, for one reason or another, eventually withdraw to day work schedules or other employment. As a result, the dropouts from rotating shift work schedules become part of the daytime controls, which are utilized in typical case comparison epidemiologic investigation.

A few investigators, believing that shift work dropouts who have been switched to daywork might represent a special group, have separately compared the morbidity patterns of these to other day workers who have never attempted shift work, as well as to rotating shift workers (165,222).

Some studies have indicated that former rotating shift workers who have transferred to permanent day work in general reveal a greater fragility to illness as compared either to other day workers without previous history of shift work or to rotating shift workers who exhibit good tolerance to this type of work schedule (165,222). It is now thought that dropouts from shift work exhibit a greater susceptibility to disease than do most day and tolerant shift workers. It has been proposed that much of this may be related to on-the-job stress (214). Some researchers believe that for certain types of persons shift work poses a special risk in that it may promote the development of certain diseases. Thus, studies that compare separately the health profiles of dropouts from shift work to those of tolerant shift workers and permanent day workers are preferable (165,222). Earlier findings were perhaps compromised because of methodological drawbacks. These results cloud our understanding of whether shift work contributes to or promotes the development of illness.

Today, it is generally accepted that shift work may be a risk factor in work exacerbating or potentiating certain ailments or disorders (64). Changes in the timing of work, meals, and sleep affect appetite during night work and may be related to the observation of an elevated incidence of dyspepsia and certain other gastrointestinal disturbances or ailments of shift workers (64,165,218,223,224). Figure 9.19 summarizes the frequency of complaints of altered appetite in relationship to shift system and industry. Disturbance of eating habits seems to be relatively infrequent in day workers and shift workers without nighttime duty. In comparison, it occurs in about 35–75 percent of the workers who rotate with a night shift, depending on the industry and study, and in about 40 percent of those engaged in permanent night work.

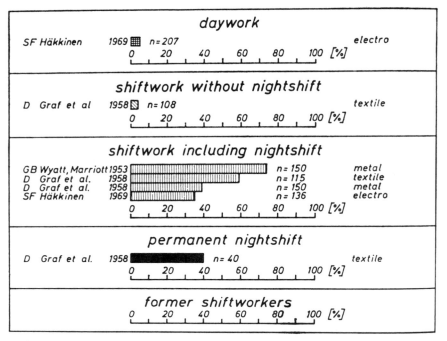

Figure 9.19 Frequency of complaints about disturbed eating habits related to shift system, industry, and investigator. From Rutenfranz et al. (56).

Altered meal timing and disruption of biological rhythms caused by shift work may be the cause of an increased incidence of dyspepsia. Figure 9.20 illustrates the frequency of gastrointestinal disturbances in relation to shift system and industry (37, 64). Overall, gastrointestinal complaints were found in 10–25 percent of the day workers and about 17 percent of the shift employees not having to do night work. Such complaints were found in from 5 to 35 percent of the employees engaged in rotating shift work involving a night rotation, depending on the industry, and in about 50 percent of those permanently working nights.

Figure 9.21 summarizes the findings of studies pertaining to the incidence of ulcers according to shift system and industry. Gastric ulcers were reported among 0.3–7.0 percent of the day workers, about 5 percent of the shift workers not having a night shift, and approximately 2.5–15 percent of those having to rotate to a night shift. Gastric ulcers were reported to occur in from 10–30 percent of former shift workers who had changed to day hours. The findings reveal a rather wide overlap in the incidence of this health problem. It has been suggested that differences in findings between studies on comparable groups might represent methodological biases of the various investigations (37). Based on the large differences in incidence of this disorder between the various groups studied, Rutenfranz et al. (37) question whether rotating shift work involving nighttime duty represents a true risk factor. Nonethe-

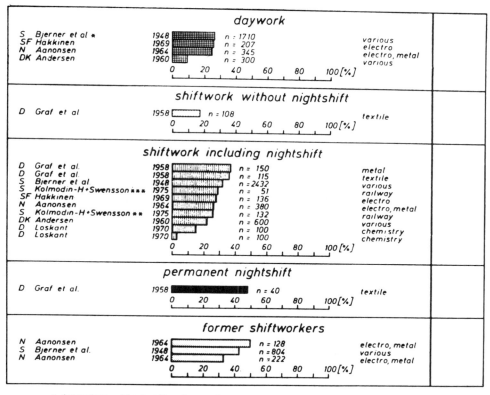

Figure 9.20 Frequency of complaints about gastrointestinal disturbances related to shift system, industry, and investigator. From Rutenfranz et al. (56).

less, it is the opinion of a majority of researchers that ulcers are a risk of certain susceptible shift workers.

Tarquini et al. (225) have suggested that changes in the gastro/acidopepsin secretion system during shift work may cause increased gastrointestinal complaints and may lead to the pathogenesis of peptic ulcers. Surprisingly, gastrointestinal complaints were a problem for 30–50 percent of those persons who had switched from rotating shift work to permanent day work, thus implicating worker physiology and not work scheduling as a cause. Nonetheless, it is striking that former shift workers exhibited such a high incidence of gastrointestinal disturbances.

While the majority of epidemiological studies focused on sleep deprivation and the increased incidence of gastrointestinal complaints, Akerstedt et al. (226) investigated associations between shift work and coronary heart disease (CHD). The results of these studies indicate a dose-responsive relationship between years per-

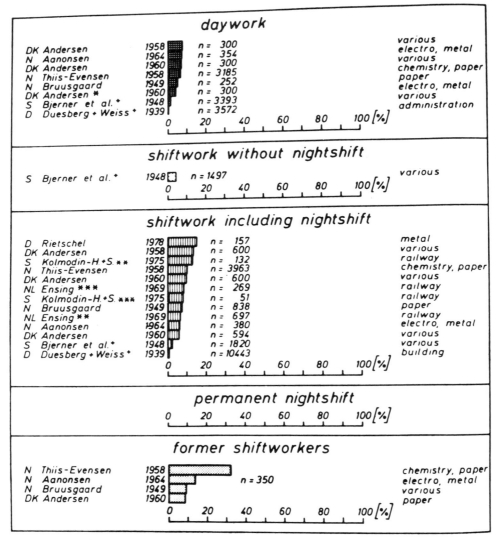

Figure 9.21 Frequency of gastric and duodenal ulcers related to shift system, industry, and investigator. From Rutenfranz et al. (56).

forming shift work and an increased risk of heart disease. Further, cross-sectional studies revealed shift-working cohorts smoked more frequently and had higher serum triglyceride levels when compared to day-only workers, when obesity, smoking, and alcohol consumption were controlled. The diet and activity level of the shift-working cohort were implicated as potential causal factors.

The results of gender-related disease is less clear. Some studies (227–229) have implicated unusual work schedules as a risk factor in premature abortions or low birth weight (230) while others have not (231,232). Other factors, including an increased risk of pregnancy-induced hypertension were found in female shift workers working nights in environments with noise levels of 80 dB or more (233).

In summary, it appears that shift workers, in comparison to permanent day employees who have never worked nights, are at elevated risk of sleep irregularities as well as gastrointestinal difficulties, the latter manifested as dyspepsia, appetite disturbances, and/or duodenal or peptic ulcers. It is of interest that as a group, former rotating shift workers, even after being transferred to permanent daytime employment with nighttime rest, exhibit the highest incidence of these health-related complaints. While the studies of other diseases are less conclusive, current research indicates that long-term shift work may increase the risk of some types of disease. Further epidemiologic research in this area is necessary.

15 ENCOURAGING ADJUSTMENTS TO SHIFT CHANGES

Interest in medications or procedures to facilitate the chronobiologic adjustments to change in the sleep–activity routine have been long sought (99,132,133). Ehret (234–239) and his co-workers, based on results from a series of investigations on rodents involving the careful manipulation of several known synchronizers for these animals, including food and drugs, were able to accelerate phase shifts of selected circadian functions. Based on their findings on animal investigations, Ehret and his co-workers proposed a diet plan to hasten the chronobiologic adjustments of travelers and shift workers (240).

15.1 Diet and Meal Timing

The Argonne anti-jet-lag diet has been derived from research mainly on rodents. The objective and scientific substantiation of the diet plan in human beings, as of this writing, is inadequate. Outside of several subjective evaluations or volunteered testimonials, little properly controlled human research on this diet plan has been conducted. The work of Graeber and his co-workers (98,241), represents apparently the only published study on human beings that purportedly has evaluated the diet. In their research several synchronizers—including the use of the Argonne anti-jet-lag diet, methyl-xanthines (caffeine), and a sleep-inducing medication—were used to enhance the speed of adjustment of circadian functions of military personnel undergoing rapid deployment across many time zones. In this study, complicated by several logistic problems, it is uncertain whether the diet plan alone was the sole reason for a better tolerance of the experimental group of soldiers to their rapid deployment. In this study each member of the experimental group received a 100-mg dose of dimenhydrinate, a hypnotic, to induce sleep at an appropriate time in addition to adhering to the anti-jet-lag diet. Moreover, more recent chronobiologic studies suggest that with the exception of but a few circadian variables, meal timing is only a weak synchronizer of 24-hr rhythms in human beings (242).

In rodents the situation is somewhat different in that meal timing has been shown to be capable of strongly affecting certain circadian functions; on the other hand certain others may be only affected to a slight degree or not at all (242,243). The subject of meal timing and its role as a synchronizer in rodents and human beings has been reviewed by Reinberg and his colleagues (242,244) with the conclusion that it is of relatively minor significance as a synchronizer in human beings.

15.2 Scheduling and Napping

Shift scheduling has been proposed to assist in adaptation of workers to changing schedules. As previously discussed, chronobiologic research suggests that forward rotation schedules in an 8-hr workday system and short rotation periods in 12-hr systems would provide the best acclimatization for shift workers. However, both social forces and worker habit may dictate a preference for either backward rotation schedules or long rotations. Thus habit or individual preference may be of greater importance than the selection of a chronobiologically appropriate shift rotation system. Individual differences and preferences in each work situation should be evaluated by management and the workforce prior to the initiation of changes in the shift schedule.

Sleep or prophylactic napping during work has been suggested by some (39) as a method to reduce progressive sleep deprivation in shift workers. Japanese industries have provided facilities for sleep during the night shift for some time. It is possible that napping during the night shift may provide some benefit (245) but the benefit may be short and insignificant (246).

Harma and co-workers (247,248) have investigated the benefit of physical training on shift-working nurses. Nurses who exercised had statistically significant improvements in heart rate and maximum oxygen consumption, fatigue level, and sleep length. It remains unclear if these changes, while statistically significant, are clinically relevant, especially since an earlier study (249) did not show any benefit from physical conditioning.

15.3 Light Therapy or Phototherapy

A variety of recent studies have suggested that circadian rhythms can be manipulated by exposure to bright light (250–254). Although it was once believed that light was not a strong zietgeber (synchronizer) for humans, the discovery by Lewy (250) that bright light (>2500 lux) could suppress melatonin secretion led to the application of this therapy for seasonal affective disorder, or SAD, a recurrent form of depression that usually strikes each winter. While it has been suggested that approximately 25 percent of the population may suffer from this disease, it has also been suggested that more than half of this number obtain relief from exposure to bright light for 1–3 hr each day (251). Work by Czeisler et al. (253) and Eastman and Miescke (252) suggest that rhythms of night workers may be relatively quickly entrained with great efficiency by artificially produced light. This is accomplished by exposure to bright light at some time during the phase response curve (PRC). Pulses of bright light can have markedly different effects depending on the time of administration. For example, a pulse of light introduced during the middle of the sleep span will have a drastic effect on circadian rhythms while another provided during the middle of the active span will have little effect. Light pulses initiated at the onset of subjective night for most workers will tend to delay rhythms while light at the end of night will advance them. Eastman (356) describes two techniques of rhythm modification as either "squashing" (application of light at midsleep to affect both a phase delay and phase advance simultaneously) or "nudging" (application of light at a later period before sleep each day). It has been suggested that nudging is preferable to squashing. However, while these therapies may be useful for some workers, the timing of the light exposure is critical. Many of the previous studies required workers to wear welding goggles during certain periods to prevent unwanted exposure to bright light from other sources that attenuate phase shifts. Unfortunately, exposure to external bright light may easily occur in real-world situations and interfere with the rhythm entrainment process. It has also been suggested that external 24-hr zeitgebers (natural sleep–wake and/or other zeitgebers) may also hinder temperature rhythm entrainment because for some individuals, these time clues may be in conflict with the phototherapy regimen (357). More field experimentation in the workplace is necessary to validate the utility of this therapy.

Overall, any method of intervention will require considerable work on the part of management and the individual worker, thus indicating that an accurate cost–benefit analysis be performed prior to the initiation of any change. Changes in shift scheduling may be beneficial, but extensive evaluation of the popularity of the current system and the potential effectiveness of the proposed schedule must be performed prior to initiation to obtain compliance. Management and employees must be in complete agreement prior to the commencement of any schedule changes.

In summary, the countermeasure involving the anti-jet-lag diet has not been proven to be effective. Similarly, the expectation of either improved physical fitness, light therapy, or napping on augmented adaptation to shift work remains unclear. More research in these areas is necessary.

15.4 Selection Criteria for Shift Workers

Currently, there is no reliable method for predicting the long-term biological capability of employees for shift work. Although the individual differences in the manner in which persons adjust to shift work have been linked to personality characteristics (141–144), the significance of these has not been appropriately validated through controlled field studies. Through analyses of a series of field studies on both tolerant and intolerant shift workers, Reinberg et al. (17) have detected chronobiologic differences between the two types of shift workers. Reinberg and his colleagues believe measurement of these differences have relevance for improving the selection of employees who are likely to have a high chronobiologic tolerance for specific rotating shift work schedules. However, a recent study by Knauth and Harma (255) did not identify any correlation between acrophase shifting and the amplitude of oral temperature rhythms suggested by Reinberg as measures of tolerance to rotating shift work. Therefore, at this time, previous shift-working experience continues to be the best indicator of one's tolerance to shift work.

16 MEDICAL IMPLICATIONS OF BIOLOGICAL RHYTHMS

A great deal of research on rodents and human beings indicates that biological rhythms are so significant they must be taken into account in clinical medicine (41–43,46,69,256–260,330–333). For example, in the diagnosis of allergy using intradermal injections of antigens, the timing of tests, that is, morning, afternoon, or night, appears to be almost as important as what is tested! On the average, the cutaneous sensitivity of patients to several common allergens, such as house dust or mixed pollen, evidences a 3.5-fold difference depending on whether the timing of the diagnostic test is done in the morning or evening. Some patients show as great as a 7- or even 11-fold variation (261–263) (Fig. 9.22). Moreover, in certain ailments, the occurrence or exacerbation of symptoms is predictable over time as biological rhythms (264–266). This is the case for allergic, cardiovascular, neurologic and inflammatory diseases, among others. In addition, the pharmacokinetics and pharmacodynamics of a large number of medications vary significantly according to their timing with respect to the circadian system (40,41,46,71,267,268). For some medications, such as those used to treat arthritis, synthetic corticosteroids or a variety of nonsteroid anti-inflammatory preparations (269–272), the timing of the medication is critical. The side effects of synthetic corticosteroids in day-active persons may be so great should these medications be taken before bedtime that the patient may have to give up their use. Yet when taken in the morning there is little or only minor side effect, depending on drug form and dose (269). Similarly, many arthritic patients do not tolerate nonsteroid anti-inflammatory medications if taken in the morning but do so with good relief when taken before bedtime (270–272). We wish to add that biological rhythms in the pharmacokinetics and the effects of most drugs are not dependent on day–night differences in posture nor meal timing and composition.

17 BIOLOGICAL RHYTHMS AND HUMAN ILLNESS

The significance of biological rhythms with regard to the occurrence or intensification of symptoms of human disease was recognized even in biblical times when it was noted that the illness we now call asthma tended to worsen or not occur at all until nightfall. Today, a division of chronobiology, chronopathology, is concerned with understanding the relationships between alterations of rhythmic processes and the development of human illness as well as the predictable, rhythmic, exacerbation of symptoms.

Because of the high incidence of cardiac problems that occur primarily in the

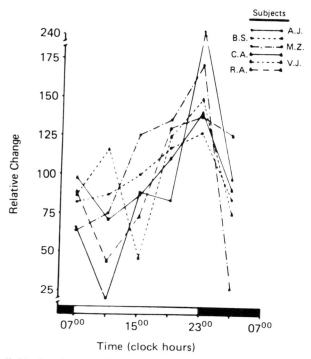

Figure 9.22 Individual and mean chronograms (time plots) for the circadian susceptibility rhythm of the skin studied by intradermal injection of house dust extract at different times of the day and night in six house-dust-allergic adult patients. The temporal changes are shown as percentage deviations from each person's 24-hr average response. There exists large variation in cutaneous reactivity with respect to clock hour of testing as well as between persons. Clock hour time is given in military fashion: 1500 = 3 P.M., 2300 = 11 P.M., and 0700 = 7 A.M. The patients were synchronized with sleep (shown as shading along horizontal axis) from 11 P.M. to 7 A.M. From Reinberg and colleagues (262, 263).

evening, these ailments were some of the first to be studied. The onset of symptoms (as opposed to the admission time to the hospital emergency room) of myocardial and cerebral infarct, cerebral hemorrhage, and angina pectoris is strongly circadian rhythmic. For example, in a group of more than 1200 diurnally active patients, myocardial infarct (Fig. 9.23) was found to be most frequent between 8 A.M. and 10 A.M. (264). There was a secondary peak 12 hr later. However, for cerebral infarction (Fig. 9.24), the phasing was different; the occurrence was very much more frequent during sleep with a peak at 3 A.M. (273). Also, it was found that spontaneous intracerebral hemorrhage (264) (Fig. 9.25) was primarily an evening event with a peak around 7 P.M. The clinical signs of Prinzmetal's variant angina as evidenced by ST-segment elevations of the electrocardiogram (Fig. 9.26) were more common during sleep, between 2 and 4 A.M. and quite uncommon around midday and during the afternoon (274).

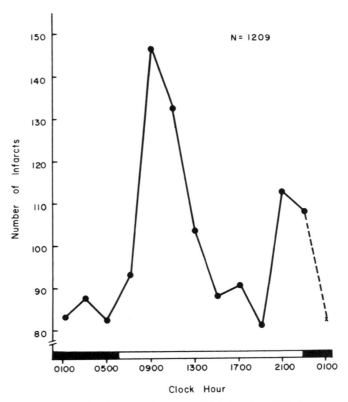

Figure 9.23 The temporal distribution of myocardial infarction (MI) is nonrandom; MIs are more common in occurrence around 0900 and 2100 (9 P.M.) than 0500 and also between 1500 (3 P.M.) and 1900 (7 P.M.). The shaded portion of the horizontal axis shows the presumed sleep span for the sample of patients. From Smolensky (264).

When biological rhythms are found in the occurrence or exacerbation of human illnesses, frequently the question arises as to whether they result from cycles in ambient conditions, rest–activity, and/or from rhythms in related biological processes. For example, asthmatic patients exhibit a high-amplitude circadian rhythm of airway patency and dyspnea. The breathing of asthmatic persons tends to be considerably easier during midday than at night or upon awakening. With regard to peak expiratory flow, the rhythm represents a 25 percent variation in airway function over the 24 hr (275). The acrophase chart of Figure 9.27 also confirms that the dyspnea of asthma in day-active patients is a nocturnal event with the symptoms being worse between 11 P.M. and 5 A.M. This acrophase chart (266) reveals that the staging of several circadian rhythms, which are known to influence airway caliber, contribute to the nocturnal predilection of asthma. In particular, the dyspnea of asthma worsens during the nighttime when the urine and blood levels of anti-inflam-

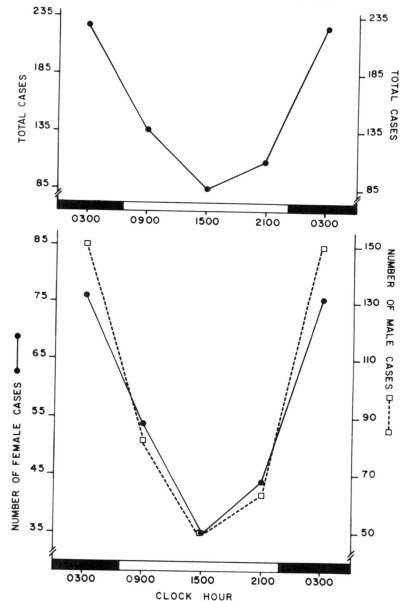

Figure 9.24 Circadian rhythm in the occurrence of cerebral infarction (mobidity). In presumably diurnally active males and females infarction was considerably more common at night around 0300 (3 A.M.) than during the afternoon at 1500 (3 P.M.). Shaded portion at bottom shows the presumed usual sleep span. Data from Marshall (273).

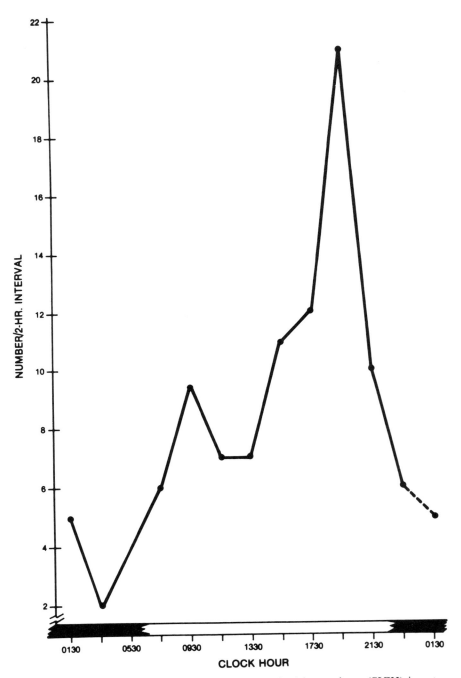

Figure 9.25 The occurrence of spontaneous intracerebral hemorrhage (SICH) is not evenly distributed over the 24 hr. In a sample of 100 cases for which the time of the event was known, the susceptibility was considerably greater between 1730 (5:30 P.M.) and 2130 (9:30 P.M.) than between 2330 (11:30 P.M.) and 1330 (1:30 P.M.). Shaded portion at bottom depicts the presumed sleep span. From Smolensky (264).

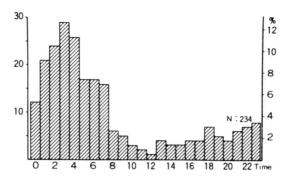

Figure 9.26 Circadian variation in the occurrence of 234 episodes of ST-segment elevation, a characteristic sign of Prinzmetal's variant angina, as detectable from the continuous recording of the electrical activity of the heart muscle in a sample of diurnally active patients suffering from this form of heart disease. The number of episodes of ST-segment anomaly per hour is greatest during the sleep span with the peak around 0300 (3 A.M.). Few episodes of ST-segment anomaly are detected throughout the daytime span of activity. From Kuroiwa (274).

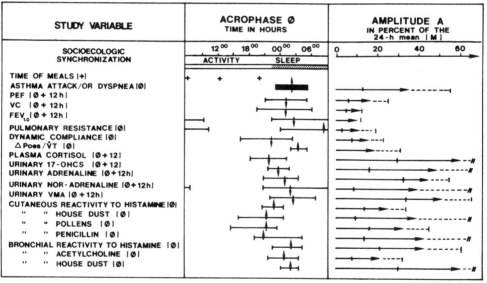

Figure 9.27 Temporal occurrence in the peak or trough of circadian functions affecting the 24-hr pattern of airway function in diurnally active asthmatic patients. The arrows in the middle column pointing upward indicate the timing of the circadian peaks (termed acrophases), whereas those arrows pointing downward indicate the timing of the circadian troughs. A horizontal line extending to the right and left of the acrophase or trough marker represents the 95 percent confidence limit of each. The amplitude, a measure of the 24-hr rhythmic variability, shown in the column to the right (the length of the arrow is proportional to the circadian rhythmic variation), is expressed as a percentage of the 24-hr time-series mean, termed the mesor. The timing of the peak and trough values of these and other biological rhythms is believed to contribute to the heightened susceptibility of patients to asthma nocturnally. Reproduced from Smolensky et al. (265).

matory hormones—adrenal corticosteroids (serum cortisol and urinary 17-hydroxy-corticosteroids) and catecholamines (noradrenalin, adrenaline, and vanillylmandelic acid, VMA)—are at their circadian minimum. The nocturnal worsening of the dyspnea of asthma corresponds in time also to poorest airway function [as determined by the 1-sec forced expiratory volume ($FEV_{1.0}$), vital capacity (VC), and peak expiratory flow (PEF)]. It corresponds as well to greatest susceptibility of the airways and skin to antigens, such as house dust, and to the nonspecific chemical irritants of histamine and acetylcholine. Acrophase charts such as the one for asthma help explain why these patients become worse at night. This type of information also indicates that temporal variations in one's physiology may be equally or even more important than those of the ambient environment in explaining the cyclic nature of certain human illnesses and also in defining their etiology. Evidence for the importance of biological rhythms in human illness was obtained in an exemplary study by Halberg and Howard (276) on a man exhibiting prominent circadian variation in the occurrence of his overt epileptic seizures (Fig. 9.28). During the years when he was active between 6 A.M. and 9 P.M., seizures were most commonly experienced after awakening, especially between 7:30 A.M. and 1:30 P.M. After altering his sleep–activity routine to accommodate nighttime work, the clock hours during which his symptoms were most frequent had changed to between 12 and 7 P.M. However, with reference to the sleep–activity schedule of this individual, the occurrence of the majority of his seizures was comparable under both sleep–activity routines; the occurrence was always most frequent during the several hours immediately following the termination of sleep.

Information pertaining to circadian and other rhythms in symptoms of disease is important for three reasons. First, occupational as well as clinical practitioners can improve their treatment of certain ailments through an understanding of chronopathology. Second, appropriate steps may be taken to minimize or prevent the occurrence of disease or exacerbation of existing conditions. Third, such information can be utilized by physicians to determine whether a disease might be occupationally induced. For example, a frequent difficulty for some workers employed in certain industries is occupationally induced recurrent nocturnal asthma (277–280). Patients suffering from this disorder quite often exhibit marked circadian patterns in the status of their airway function. Figure 9.29 presents one example of circadian changes of airway patency in an individual who became sensitized to budgerigar serum due to exposures arising from breeding parakeets. It is readily apparent that in this day-active person a brief exposure to the offending antigen around 10 A.M. resulted initially in only a small effect on airway patency as indicated by the pulmonary function $FEV_{1.0}$ measurements. However, around midnight, well after the single brief morning exposure, the effect was fully manifested. Moreover, it was only at night, for several subsequent 24-hr spans, that the symptoms and effects of the asthmatic condition were worsened even though no subsequent exposure to the antigen occurred.

Occupationally induced recurrent nocturnal asthma is not uncommon and is often observed in employees of several different industries, including those who work with wood, plastics, resins, grains, and pharmaceuticals (277). The prior example illustrates how susceptible workers who are exposed during daytime work need not

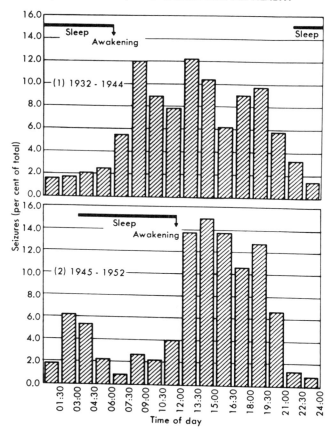

Figure 9.28 Circadian pattern in the initiation of epileptic seizures in one patient studied longitudinally over several years while residing in an institution. The temporal distribution with regard to the clock hour of the seizures changed following transfer to a different work–rest pattern when the sleep schedule was delayed from one of 9 P.M. to 6 A.M. to one of 3 A.M. to 11 A.M. Clock time along horizontal axis is given in military fashion, for example, 1500 = 3 P.M.; 2100 = 9 P.M. From Halberg and Howard (276).

develop significant symptoms until much later in the evening, only to reoccur again on several consecutive nights even without further exposure to the offending antigen. The recurrent nocturnal exacerbation of the asthma is believed to represent, at least in part, the circadian organization of processes that affect airway patency. This example serves to illustrate that epidemiologic studies of occupationally induced diseases should take into account rhythmic variation in the symptoms or occurrences of disease. Quite often, patients and physicians do not associate the etiology of certain diseases with employment when complaints occur many hours after leaving the workplace. This appears to be especially true for occupationally induced asthma for which the interval between daytime exposure at work and the manifestation of the nighttime illness at home can be quite lengthy.

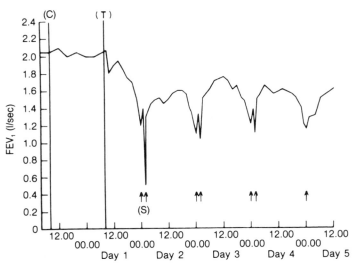

Figure 9.29 Temporal variation in airway patency of a presumably diurnally active asthmatic patient following a single pulmonary challenge on the test day (numbered 1) with an antigen to which hypersensitivity had developed (budgerigar serum) through occupational exposure. Although the FEV$_{1.0}$, a measure of airway patency, was close to normal during the day, it was extremely low during the night, giving rise to severe episodes of dyspnea over several consecutive nights. From Taylor et al. (278).

It is important to remember that chronopathologies can show frequencies different from the circadian one. For example, the occurrence of asthma and epilepsy in women can be circamensual (281) (Fig. 9.30) as well as circadian rhythmic, and the occurrence of asthma and cardiovascular disorders is known to be circannual as well as circadian rhythmic (264).

18 BIOLOGICAL RHYTHMS AND MEDICATIONS

Chronopharmacology is the study of the rhythmic aspects of the absorption, distribution, metabolism, excretion and action of therapeutic chemical agents in animals, including humans. Research conducted primarily during the last three decades reveals that the behavior or fate of medications and other xenobiotics can vary rather dramatically as a function of the time of their administration with respect to the scale of 24 hr and, for some, the time of year (40,267,268,282–285). The emergence of new findings from chronopharmacologic research has resulted in a different perspective about the metabolism and response of not only therapeutic medications but toxic chemical substances as well (135,286,287). These findings have, in turn, given rise to a set of new concepts and terms, such as chronokinetics, chronesthesy, and chronergy (40,285) for describing the chronopharmacologic aspects of medications or chronotoxicologic properties of other chemical agents. In the subsequent sections the

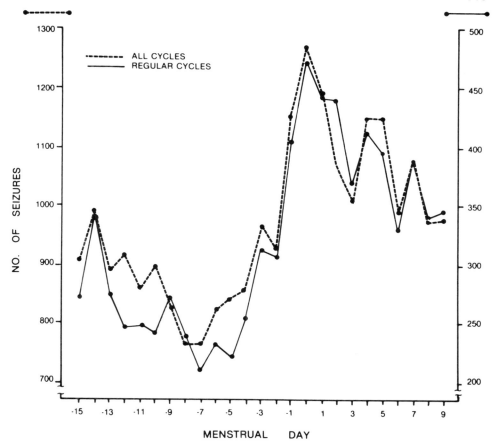

Figure 9.30 In this sample of 50 epileptic but otherwise normal menstruating women, the incidence of overt epileptic seizures varied with the phase of the menstrual cycle. The incidence just before (day −1) and on the first day (0) of menstruation was nearly twice that 7 days before. Similar patterns were exhibited independent of the regularity or irregularity in the menstrual cycle duration. From Laidlaw (281).

concepts of chronopharmacology are reviewed in preparation for discussing the topic of chronotoxicology and shift work.

18.1 Chronokinetics of Medications

Many medications exhibit important differences in the time course of their concentration in the blood or urine, its pharmacokinetics, depending on when treatment is given during the day or night. For example, the time to achieve the peak blood concentration, magnitude of peak plasma concentration, half-life of elimination, and the area under the time–concentration curve for a given drug can vary predictably in the same person depending on when in the course of a day (and perhaps year) it is administered due to the influence of bioperiodicities. Rhythms in pharmacokinetic phenomena due to the effect of naturally occurring bioperiodicities on chemical substances is termed chronokinetics. When chronokinetic findings were first detected and reported, most clinical pharmacologists were skeptical. Most believed the findings were the result of improper research methods, effects of posture, or the influence of meal timings and composition. However, as more and more findings have been published, it has become clear that for a majority of the medications the circadian differences represent the result of endogenous rhythmic processes.

A wide range of drugs is sensitive to the timing, with regard to rhythms, of administration (40,282–285). For example, as shown in Figure 9.31, the chro-

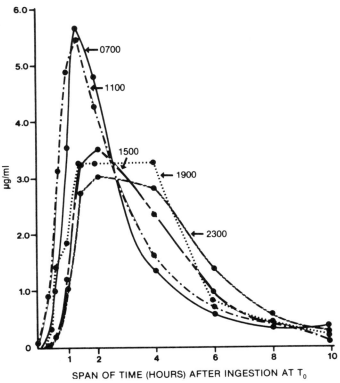

Figure 9.31 In random order and at weekly intervals, nine healthy subjects 19 to 29 years of age, synchronized with activity from 7 P.M. to 0000, received a single oral dose (100 mg) of indomethacin at fixed hours: 0700 (7 A.M.), 1100 (11 A.M.), 1500 (3 P.M.), 1900 (7 P.M.), and 2300 (11 P.M.). Venous blood (sampled at 0, 0.33, 0.67, 1.0, 1.5, 2.0, 4.0, 6.0, 8.0, and 10.0 hr postingestion) was obtained for plasma drug determinations. Ingestion at 1900 and/or 2300 led to the smallest peak height and longest time to attain the peak concentration, whereas ingestion at 0700 and/or 1100 led to the largest peak height, shortest time to attain the peak height, greatest area under the time–concentration curve, and fastest disappearance rate from the blood. A circadian rhythm of both peak height and time to peak was statistically validated. From Clench et al. (288).

nokinetics of indomethacin, a nonsteroid anti-inflammatory drug used to treat arthritis and other tissue inflammatory disorders, can be seen to differ according to its administration time when studied under carefully controlled conditions (288). When given to diurnally active patients in the morning, either at 7 or 11 A.M., the peak plasma concentration is achieved much more rapidly than it is when administered at 3, 7, or 11 P.M. Moreover, the peak concentration varies also as a function of its timing. It is greater for the 7 and 11 A.M. dosings than it is for the others. These findings show that indomethacin when taken in the morning quickly reaches very high plasma concentrations. It is pertinent that patients' complaints of indomethacin intolerance are much more frequent when the drug is taken in the morning, when the peak height is greatest, than when taken midday or in the evenings, when the peak height is more moderate.

The disposition and efficacy of propanolol, a widely used medication for treating certain cardiovascular ailments, also have been shown to be sensitive to human circadian rhythms (289). Figure 9.32 shows that the time to reach peak plasma concentration, the concentration at the peak, the elimination rate, and the area under the plasma–time concentration curve all vary significantly according to when propanolol is given. Predictable circadian differences in the pharmacokinetics of many commonly used medications, such as aspirin, ethanol, theophylline, and antihistamines, as well as analgesics and carcinostatics, among others, are now known (41,

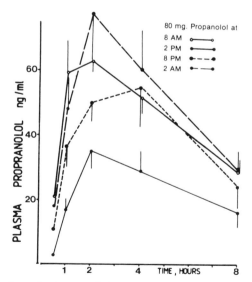

Figure 9.32 Mean and standard error of plasma propranolol concentrations after acute single administrations of 80 mg at selected clock hours. The same six diurnally active healthy persons were studied at each time point. Note the large difference in the pharmacokinetics of propranolol as a function of the treatment time. Statistically significant differences were detected in the peak height concentration, time to peak, and the area under the curve (AUC) pharmacokinetic indices. From Markiewicz et al. (289).

285). The urinary kinetics of medications also exhibit important differences dependent on treatment time. This is the case for aspirin (290) in human beings. The shape of the time–concentration curve for the excretion of salicylates, the major constituent of aspirin, differs, for example, for treatment at 7 A.M. vs. 7 P.M. (Fig. 9.33). The time to reach the peak and the peak concentration of salicylates in the urine are administration time dependent as is their elimination from the body. The shortest time to achieve highest urinary concentration, the greatest peak height, and the fastest elimination occur when 1 g of pure aspirin is taken at 7 or 11 P.M. in comparison to when taken at 11 or 7 A.M. The results of studies on mequitazine, an antihistamine, as well as beta-methyl-digoxin (291), a medicine used to treat heart disease, indicate

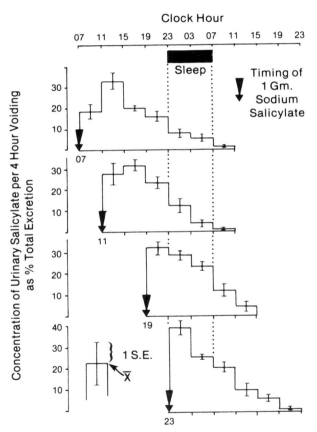

Figure 9.33 Circadian differences in the urinary chronopharmacokinetics of aspirin (sodium salicylate). Six subjects in good health synchronized to sleep from 11 P.M. to 7 A.M., meal timing of composition controlled. For a single 1-g oral administration of aspirin the time to achieve the peak urinary concentration is fastest for a 1900 (7 P.M.) ingestion; the greatest peak height concentration coincides with ingestion at 2300 (11 P.M.). Slowest elimination of salicylate follows ingestion at 0700 (7 A.M.); fastest elimination follows ingestion at 1900 (7 P.M.). From Reinberg et al. (290).

that the shape of the time–blood or time–urinary concentration curve may be so radically different for the same group of persons treated at different times, during the morning or around midday in comparison to during the evening or midnight, that the data must be dealt with in a different manner. For example, some drugs when studied under rigorously controlled conditions may exhibit patterns that can be best described by a one-compartment model if given in the morning but by a two-compartment model if administered in the afternoon or evening. This phenomenon has been shown to exist for the time–urinary excretion curve of mequitazine (Fig. 9.34). In a carefully controlled experiment, including the randomization of subjects for the sequence of treatment times, a group of persons were given a single dose at 7 A.M. on one occasion and again at 7 P.M. a week later (285). The shape of the time–urinary excretion curve following the 7 A.M. treatment suggested a two-compartment model while the shape of the curve following the 7 P.M. one suggested a simple one-compartment model. Although there are many possible explanations for this difference in behavior, it is clear that circadian differences in one or more of the following processes—absorption, metabolism, distribution, and elimination—are involved.

18.2 Chronesthesy

It frequently has been assumed that in the same person the effect produced by a given dose of a drug will be approximately the same independent of the time when it is administered. Studies by many investigators have shown that the quantitative and qualitative effects of a given medication, however, can differ greatly according to the timing of treatment. The differential effect of a medication or toxicity of a chemical contaminant as a function of its timing is termed chronesthesy. Chronesthesy is defined as circadian and other rhythmic susceptibilities to chemical agents arising from bioperiodicities primarily related to receptor sites, enzyme activity, and metabolic processes.

Lidocaine-induced analgesia is one response for which a circadian chronesthesy has been demonstrated (Fig. 9.35). The data show that a dose of lidocaine when injected either to the forearms or the apical region of the gum above a tooth with decay at midday is twice as effective as it is when given around 8 A.M. or 4 P.M. (26). Other medications that exhibit circadian chronesthesys are antihistamines, analgesics, synthetic hormones, hypnotics, and stimulants (40,271,284,285).

Another type of chemically induced chronesthesy produced by a xenobiotic is illustrated by the response of the airways to histamine aerosols when briefly inhaled by asthmatic, emphysemic, or bronchitic persons at various times during the day or night. In these types of patients the airways are known to exhibit a hyperreactivity to a variety of chemical agents, including histamine, causing an increase in the resistance of airflow in the lungs. The results of studies by DeVries and his colleagues (293) (Fig. 9.36) indicate that there is large variation in the threshold concentration required to provoke a 15 percent decrease of airway patency using this nonspecific irritant chemical. The airways are least susceptible, that is, they exhibit the highest threshold around noontime and are most susceptible around midnight, that is, they exhibit the lowest threshold. The difference in the threshold concentration between

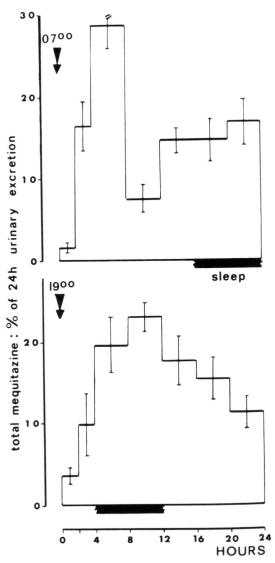

Figure 9.34 Circadian time dependence of pharmacokinetic models. A single 5-mg dose of mequitazine (an antihistaminic agent) was given orally to six healthy subjects at 0700 (7 A.M.), and 1900 (7 P.M.), 1 week apart. Participants were synchronized with diurnal activity from approximately 7 A.M. to about 11 P.M., alternating with nocturnal rest. Determinations of mequitazine were made from urine voidings collected at first every 2 hr (twice) and thereafter at 4-hr intervals during 24 hr. The results are graphed as the percentage of mequitazine excreted per urinary collection interval with reference to the entire 24-hr collection span. The kinetics were characterized by two peaks (at two-compartment model) when the antihistamine was ingested at 0700 and by 1 peak (a one-compartment model) only when given at 1900. Values for mequitazine and the occurrence of sleep are shown with respect to time posttreatment at 0700 (top) or 1900 (bottom) in hours. From Reinberg (285).

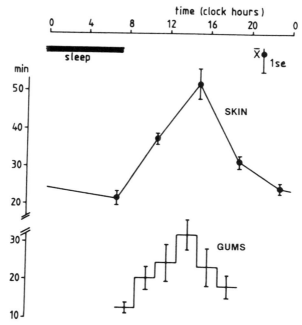

Figure 9.35 Circadian changes in the duration (min) of local anesthesia produced by lidocaine. Top curve: In six apparently healthy adults synchronized with diurnal activity from 7 A.M. to midnight and nocturnal rest, 0.1 mL of a 2 percent lidocaine solution was injected intradermally every 4 hr during 24 hr, at specified clock hours. The flexor surface of both forearms was used exclusively. The duration of anesthesia was determined by measuring the time in minutes from injection to the recovery of cutaneous sensitivity. The mean duration was only about 20 min at 0700 (7 A.M.): it was 52 min at 1500 (3 P.M.) and about 25 min at 2300 (11 P.M.). Differences between the longest and shortest durations are statistically significant ($p < 0.0005$). Bottom curve: for rigorous standardization, the study was restricted to selected patients suffering from decay (dentin caries) in a living, single-rooted upper front tooth. Lidocaine (2 ml 2 percent solution) was injected in the para-apical region. A stopwatch was started at the end of the injection. Thereafter, the tooth was drilled to remove decay, but the cavity was left unfilled until the return of sensitivity as determined by a set of tests. A group of 35 subjects (apparently healthy apart from their tooth decay) was investigated. Each patient was treated only once. The timing of treatment was randomized between 7 A.M. and 7 P.M. There were six subgroups of five to seven patients studied at 2-hr intervals. All subjects had diurnal activity and nocturnal rest. The duration of local anesthesia was about 12 min for the test interval from 7 A.M. to 9 A.M., about 32 min for that of 10 A.M. to 2 P.M. ($p < 0.005$), and about 19 min for that of 3 to 7 P.M. ($p < 0.025$). For both of these experiments illustrating the chronesthesy of lidocaine, differences between the longest and shortest effect are statistically significant. From Reinberg and Reinberg (292).

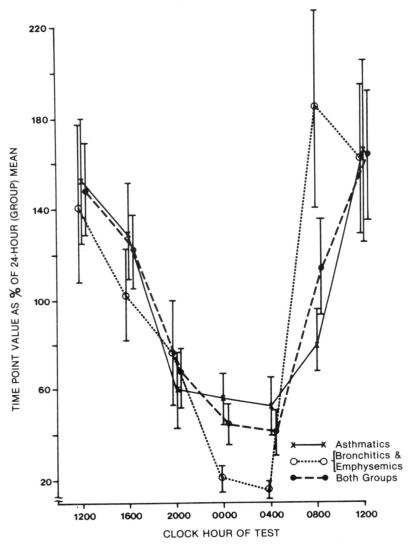

Figure 9.36 The circadian difference in hyperreactivity to inhalation of histamine aerosols by the airways of asthmatic, bronchitic, and emphysemic patients is great. At midnight (0000) or 4 A.M. (0400), times of heightened responsiveness, the concentration required to induce a 15 percent decrease in airway patency relative to the time-specific control baselines is from 100 to 160 percent less than that required 12 hr later, at noon or 1600 (4 P.M.), when the airways are least reactive. Note that the data of each curve are plotted as deviations from the given group's 24-hr mean reactivity, which for the purpose of graphing has been set equal to 100 percent. From Smolensky et al. (265).

the times of greatest and lowest reactivity can be as great as 100 percent or more (293,294). These findings suggest that persons with small or large airway disease who are also involved in shift work and are exposed to irritant chemicals may experience more airway reactivity and provocation during one shift (the evening or night shift) in comparison to another. In other words a given concentration of a chemical irritant that may be rather well tolerated during the day shift may not be tolerated at all when encountered during the night shift.

18.3 Chronergy

Chronergy is defined as rhythmic variations in the effect of medications or other chemical agents, whether they be desired or undesired ones. The chronergy of a chemical substance is dependent on both its chronokinetics as well as its chronesthesys. Homeostatically designed, single time point, pharmacology and toxicology research generally has found that the effects of a chemical agent are related directly to its concentration in the serum or tissue. However, chronobiologic studies reveal that this need not be the case; in fact, it is rare that the chronesthesy of a medication coincides with its chronokinetics. This is exemplified in Figures 9.37 and 9.38, which review the results of a chronopharmacologic study of ethanol (295,296).

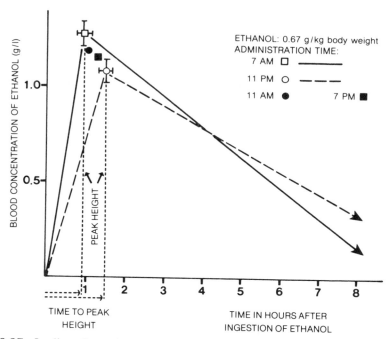

Figure 9.37 In diurnally active persons, the greatest peak height, shortest span to reach peak height, and fastest elimination for ethanol occur when this substance is given at 7 A.M. In comparison, ingestion at night or 7 or 11 P.M. by the same persons is associated with lowest peak height, longest time to reach this peak, and slowest elimination. Redrawn from Reinberg et al. (296).

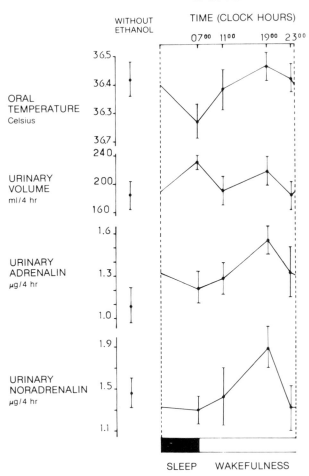

Figure 9.38 The 24-hr rhythm-adjusted mean ($M \pm 1$ SE) of physiological variables after the ingestion of ethanol (0.67 g/kg body weight) at different test times [0700 (7 A.M.), 1100 (11 A.M.), 1900 (7 P.M.), and 2300 (11 P.M.)]. Control data were collected over 24 hr without ethanol ingestion. Timed ethanol ingestions were performed at least 1 week apart, in random order. Subjects were six healthy young adult males synchronized with activity from 0700 to 0000 alternating with nocturnal rest. Subjects had fasted about 7 hr before and 7 hr after each ethanol ingestion. Measurements and integrated urine samples were gathered at 4-hr intervals. Relative to the respective control values, ethanol ingested at 0700 induced a decrease of the temperature 24-hr mean—M, but no change in the catecholamine M values; ethanol ingested at 1900 induced a rise of the catecholamine M values but no change of the temperature M. The temperature decrease due to ethanol at 0700 is presumably related to the fact that the ethanol-induced peripheral vasodilation is not counteracted at this time by a change in catecholamines. On the contrary, ethanol ingestion at 1900 is followed by higher levels of catecholamines that persumably resulted in peripheral vasoconstriction compensating for the ethanol-induced vasodilation with no change of body temperature. From Reinberg et al. (296).

In this study it was shown that ethanol (0.67 g/kg body weight) ingested by diurnally active healthy persons demonstrated significant chronopharmacokinetics differences. Ethanol, when taken at 7 A.M., resulted in the greatest serum peak height, shortest span of time to reach this peak, and fastest disappearance rate (296). In comparison, ethanol ingestion during the evening, at 7 or 11 P.M., resulted in the lowest peak ethanolemia, longest span of time to reach this peak, and slowest disappearance rate. It is of interest that the chronesthesy of ethanol defined as the level of ebriety self-estimated 60 min following its consumption also differed as a function of the time of intake (285,295,296). In short, ebriety was greatest following an 11 P.M. ingestion and least following the 7 or 11 A.M. ingestions. This study shows that although ethanol was always consumed in the same dosage both the pharmacokinetics and effect were different depending on the time of its ingestion. The effects on the brain were less at 7 A.M., when the serum peak height concentration was greatest and the disappearance rate from the blood fastest, than they were at 7 or 11 P.M., when the peak height was lowest and the disappearance rate was slowest (285,296).

Chronergic effects of ethanol on the oral temperature circadian rhythm in humans have been examined by Reinberg et al. (295,296). The findings are related here as a further example of how kinetic data may not always be accurately predictive of systemic effects. In these studies ethanol in the same dosage (0.67 g/kg body weight) was given at 7 A.M., 11 A.M., 7 P.M., or 11 P.M., with each study time being done in a random order and on different days. A placebo liquid was also given to obtain baseline data on several biological rhythms, including that of oral temperature. Figure 9.38 summarizes the circadian-stage-dependent effect of ethanol on oral temperature (285,296). Shown are the 24-hr average oral temperature values calculated using data collected following each of the separately timed ethanol ingestions. The results indicate that only when ethanol was ingested at 7 A.M. was a statistically significant posttreatment hypothermia induced, relative to control conditions. When ethanol was ingested at the other test times, no hypothermia occurred. The hypothermia that followed the ingestion at 7 A.M. could have been related to a drug-induced peripheral vasodilation not being counteracted at this time by an appropriate catecholamine (adrenalin and noradrenalin) secretion as suggested by the data of Figure 9.38. On the other hand, ethanol ingestion at the other test times was followed by an elevation of catecholamine secretion presumably causing peripheral vasoconstriction and compensation for the ethanol-induced vasodilation so that there was no net change in the body temperature.

Although the examples put forward to describe chronergies do not have direct occupational relevance, the concept that blood levels need not always be predictive of the same level of biological effect because of chronobiologic factors is very important. A potentially harmful concentration of a toxicant may or may not lead to symptoms of toxicity, depending on when during the 24 hr with respect to the circadian organization of biological functioning it occurs. Liver enzymes at one circadian time may exhibit greater efficiency in detoxification than at others. Similarly, target issues may be less susceptible to toxicants at one circadian time than another. This is discussed in greater detail in the subsequent sections of this chapter.

In summary, an awareness of the potential impact of chronokinetics, chronesthesy, and chronergy on the human response to medications is useful in understanding and improving the treatment of human illnesses, including ones experienced by shift workers. More pertinent to the focus of this chapter, these phenomena have relevance to predicting temporal differences in the pharmacokinetics and adverse effects of exposure to contaminants in the workplace. The implications to the field of industrial hygiene, toxicology, and occupational medicine are abundant, especially with respect to evaluating the potential adverse effects of shift work on human health. Persons may exhibit different degrees of risk from exposure to chemical or physical agents depending on the shift worked.

19 EFFECT OF BIOLOGICAL RHYTHMS ON TOXIC RESPONSES

Chronobiologic studies on rodents conducted during the late 1950s and 1960s revealed that the effects of many chemical agents differ in the intensity of their acute toxicity depending on the time of exposure with respect to circadian and perhaps circannual rhythms. F. Halberg (70) initially referred to this as the "hours of diminished resistance" (Fig. 9.39); he pioneered the concept and research of susceptibility-resistance rhythms in rodents as well as human beings.

HOURS OF DIMINISHED RESISTANCE

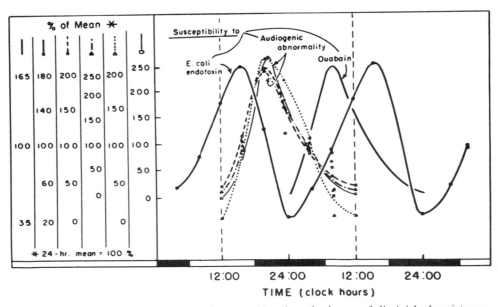

Figure 9.39 Chronotoxicology was first considered as the hours of diminished resistance. Experimental results reported by Halberg demonstrated the existence of circadian rhythms in the susceptibility of mice exposed to various potentially noxious agents, for example, a fixed "dose" of *E. coli* endotoxin, ouabain, or white noise. The endpoint of response in the studies on mice herein summarized was the number of deaths per treatment group per time of testing. From Halberg (70).

19.1 Toxicology Evaluation and Susceptibility-Resistance Rhythms in Rodents

Following many years of investigation, it became clear that the acute and chronic toxicity of a chemical is influenced by the time of administration (70,135,297–307). Numerous studies have shown that the most common and fundamental test to evaluate toxicity, the LD_{50}, can be dramatically affected by rhythmic processes. The inconsistencies between laboratory reports for LD_{50} values often result from differences in the timing of studies and/or variations in the synchronization of the test animals. In certain instances the influence of rhythm changes may vary the LD_{50} by one or two orders of magnitude (297). In addition, more subtle indicators of toxicity such as the concentrations of serum liver enzymes as well as tests of behavioral toxicity also may vary depending on the timing of dosing (135,287,295,297). Other types of toxicity tests can be affected by biological rhythms as well. They include tests for teratogenicity, target organ toxicity, mutagenicity, and carcinogenicity (297,307). Research suggests that the timing for evaluation of toxicity is of critical importance due to the influence of circadian and other rhythms.

Chronobiology suggests that toxicologists should account for temporal variation in their studies of chemical agents since the results of toxicologic studies have tremendous impact on a firm's ability to register, manufacture, and market chemicals. It is surprising that industrial and research laboratories as well as regulatory agencies have not been sensitive to the methods and findings of chronobiology.

Representative examples of circadian susceptibility-resistance rhythms in rodents are contained in Figures 9.40–9.42. As shown in these figures, the acute toxicity of many agents such as nicotine, strychnine, amphetamine, and ethanol as well as urethane, paraquot, malathion, and mercury chloride is characterized by large-amplitude circadian patterns. The mortality induced by exposure to x-ray irradiation, bacterial endotoxin, and 100 percent oxygen also are strongly circadian rhythmic.

Rhythms have also been shown to influence tests used to determine the teratogenic potential of chemicals (308,309). Sauerbier (309) reported that in pregnant mice injected once with cyclophosphamide (20 mg/kg) or Th-R (N-mustard, 2 mg/kg) at one of four different circadian stages (7 A.M., 1 P.M., 7 P.M., or 1 A.M.) on the twelfth day of gestation at precisely the same number of hours from copulation at each time point, there was a strong circadian-stage dependence. The highest incidence of malformations due to maternal treatment with cyclophosphamide was found to be associated with the dark-to-light transition (7 A.M.), whereas the lowest occurred at 1 A.M. In contrast, the teratogenic action of Th-R was strongest at the onset of darkness (7 P.M.) and the lowest at 7 A.M. Interestingly, the mean embryo toxicity of both compounds was reported to be subjected to seasonal modifications, being highest during the spring and summer and lowest during the winter. The authors conclude in the evaluation of teratogenic potentials of drugs that rhythms cannot be neglected. Also, Wegner and Neu (345) determined that circadian changes in foliate concentrations in mouse embryo and plasma may be related to prevention of valproic acid teratogenesis. Their research suggests that appropriately timed foliate supplementation may help to prevent teratogenic damage.

Chaudhry and Halberg (310) and Halberg (311) reported that the induction of submandibular gland sarcomas in hamsters using dimethylbenzanthracene was shown to be remarkably circadian-stage dependent (Fig. 9.43). Iversen and Kauffman

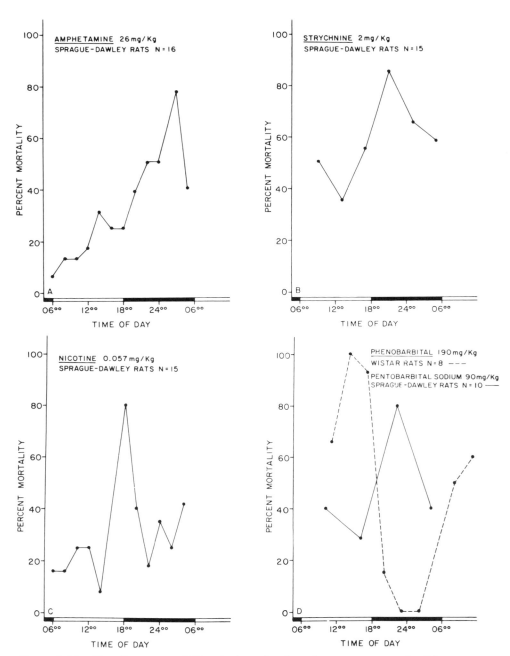

Figure 9.40 Circadian susceptibility rhythm of rodents in response to each of four agents, using mortality as the endpoint. [For details, see Scheving et al. (364) for amphetamine data, Tsai et al. (304) for strychnine and nicotine data, and Müller (365) for phenobarbital sodium data.] Note in the case of the long-acting barbiturate (phenobarbital) that at one circadian stage 100 percent of the rats survived, whereas at another they all died. In all studies animals were synchronized with 12 hr of light alternating with 12 hr of darkness, the timing of the latter indicated by shaded portion of horizontal axis.

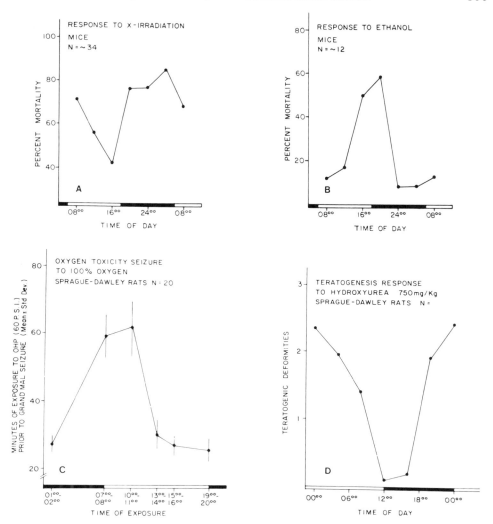

Figure 9.41 Circadian susceptibility rhythm of rodents in response to each of four agents: (A) mortality from x-irradiation [Haus et al. (324)]; (B) mortality from ethanol [Haus (363)]; (C) seizures due to oxygen toxicity [Hof et al. (361)]; and (D) teratogenesis from hydroxyurea [Clayton et al. (366)]. The response to each agent using the indicated endpoints is markedly circadian rhythmic. Shaded portion of horizontal axis indicates the timing of darkness and activity for the nocturnally active rodents.

(312), with only two time-point sampling and using the chemical carcinogens methylnitrosourea (MNU) and β-propiolactone (BPL) on hairless mice, concluded that the skin was more sensitive to these two quick-acting carcinogens at 1200 than at 0000. The differences were statistically significant. The influence of circadian rhythms on the acute toxicity of various medications used to treat human cancers is sum-

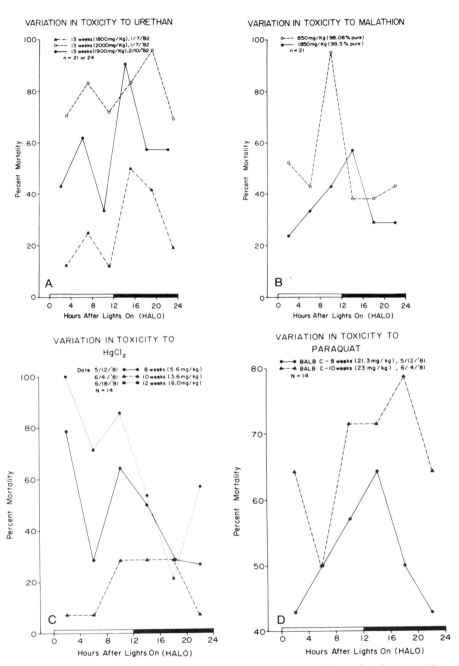

Figure 9.42 Circadian susceptibility rhythm in response (mortality) of rodents to (*A*) urethane, three dosages (Tsai et al., unpublished); (*B*) malathion, two dosages (Tsai et al., unpublished); (*C*) mercuric chloride, three dosages [Tsai et al. (305)]; and (*D*) paraquat, two dosages [Tsai et al. (305)]. The synchronizer schedule was 12 hr of light alternating with 12 hr of darkness (shown by shading on the horizontal axis). For (*A*) to (*D*), the data are plotted with reference to the hours after lights on (HALO) in the animal colonies.

400

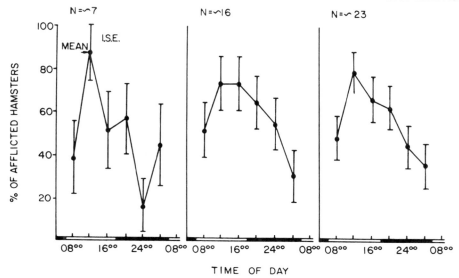

Figure 9.43 Circadian variation in tumor development from a carcinogen (dimethylbenzanthracene) injected into the salivary glands of hamsters. Note that the highest incidence of tumors was recorded in animals injected during the light period rather than the dark period; the latter is indicated by shading along the horizontal axis. (Note clock hour given in military designation with 1600 = 4 P.M.; 2400 = midnight, and 0800 = 8 A.M.). From Scheving (297).

marized at the bottom of the acrophase map shown in Figure 9.4. Clearly, circadian changes within the cells of the target organ and/or in the metabolism of the chemical agent give rise to the observed circadian differences. For example, the duration of hexobarbital sodium-induced sleep in rodents (Fig. 9.44) is markedly influenced by circadian rhythms. Nair (313) found that sleep was longest in rodents treated at 2 P.M., a time corresponding to the middle of the animal's rest span and was shortest in rats treated at 10 P.M. The time of longest induced sleep duration coincided with the trough of the circadian rhythm of hepatic hexobarbital oxidase, the enzyme responsible for the metabolism of hexobarbital sodium. It has been shown in a number of studies that many subcellular organelles, which are responsible for the manufacture or elaboration of enzymes involved in metabolism, are circadian rhythmic (306). An extensive review of the subject of chronopharmacology has been prepared by Belanger (346).

19.2 Circadian Rhythms in Liver and Brain Enzymes

Critical to understanding and evaluating the potential harmful effects of human exposures to chemical and physical agents when engaged in rotating shift work is a recognition of circadian differences in the levels of metabolizing enzymes. A review

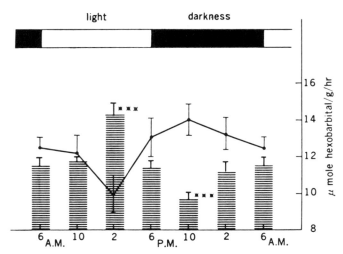

Figure 9.44 Circadian variation in the duration of hexobarbital-induced sleep (vertical bars) compared to the circadian changes in hepatic oxidase activity (solid line). The duration of induced sleep at each clock hour of study represents the average of 6–10 animals. Hexobarbital sodium (150 mg/kg) was given intraperitoneally. The enzyme activity is expressed as micromoles of hexobarbital metabolized per gram of tissue per hour; each datum represents the mean of 6–8 animals. The shorter vertical lines represent the standard errors; the triple asterisks (***) = $p < 0.001$. The timing of highest and lowest hexobarbital oxidase activity at 10 and 2 P.M. coincides with shortest and longest hexobarbital-induced sleep. Redrawn from Nair (313).

of the findings from rodent studies conducted by North et al. (314) is summarized by the acrophase chart found in Figure 9.45. Circadian differences in the activity level of enzymes of both the liver and brain are well known. The acrophase chart denotes statistically significant ($p \leq 0.05$) circadian rhythmicity if the 95 percent confidence limits, shown by extension to the right and left of the acrophase symbol (a triangle), are shown. It can be seen that circadian rhythms are depicted for 4 of the 6 listed brain enzymes and 11 of the 13 listed liver enzymes. It should be mentioned that subsequent investigations have revealed statistically significant circadian differences in the brain enzymes of pyruvate kinase and malate dehyrogenase as well as the liver enzymes of fatty acid synthetase and aldolase fructose-1-phosphate. Almost all the liver enzymes reveal greatest activity during either the beginning, middle, or end of the nocturnal hours of activity. The brain enzymes have a tendency to peak somewhat later than those of the liver. It is important to point out that these rhythms of enzyme activity in both the liver and brain are susceptible to inversion. The acrophases of these circadian rhythms undergo phase shifts of 12 hr but at variable rates with some of the enzyme rhythms taking as long as 2 weeks after the light–dark synchronizer in the animal colony is altered by 12 hr (315).

   The existence of circadian changes in enzymes such as these very likely contribute to the temporal differences in the susceptibility-resistance rhythms of rodents, and

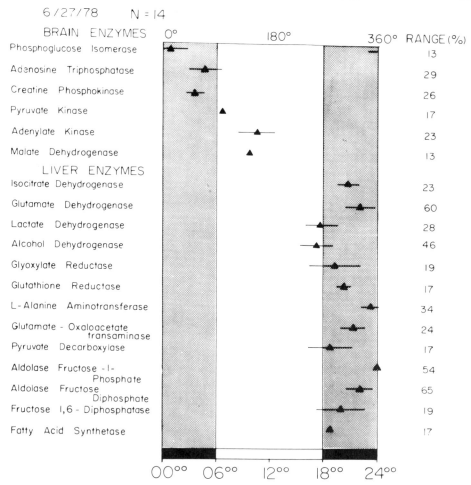

Figure 9.45 Acrophase mapping of selected enzymes of the brain and liver in CD2F, male mice that were standardized to alternating 12-hr durations of light and darkness (from midnight to 6 A.M.). There were two replicate studies done within 30 days with the findings exhibiting remarkable reproducibility. From North et al. (314).

presumably, to chemical agents to which persons are exposed during rotating 8- or 12-hr shifts. One early toxicology study on rodents that quantitatively evaluated the changes in acute histopathology and clinical chemistry due to circadian rhythms was conducted by Craft (135). His research was designed to evaluate the circadian variation in the toxic response of the male albino rat to carbon tetrachloride (CCl_4). He measured changes in the susceptibility to CCl_4 following a shift in the lighting regimen to which the experimental animals had been previously adapted.

Craft (135) found that the susceptibility of the albino rat to a given dose of CCl_4, as measured by histopathological evaluation of liver tissue and by changes in certain enzymes—serum glutamic oxaloacetic transaminase (SGOT) and serum lactic dehydrogenase (LDH)—varied significantly depending on the time of day the dose was received. Different subgroups of animals all living under conditions standardized for periodicity analysis were injected (intraperitoneal) at 4-hr intervals throughout a 24-hr period with 1.0 mL CCl_4/kg body weight. This dose produced only small effects at one time of day but resulted in extensive hepatic necrosis, marked serum enzyme elevations, and even death when administered at another time of day. Histopathological review confirmed the liver damage suggested by the increased levels of the liver enzymes. The reported differences were much greater than that which could be attributed to the increased blood flow and ventilation rate, which is associated with increased physical activity during the night in these nocturnally active animals.

In Craft's study, the mean serum LDH, SGOT, calcium and potassium levels of normal, untreated albino rats were shown to vary in a circadian manner. However, the phase and magnitude of the rhythms in the untreated animals were different from those of animals treated with CCl_4. The isoenzyme patterns of LDH revealed that the liver is the organ principally affected by a dose level of 1.0 mL/kg and that there is no apparent variation with time of day in the response of other tissues. Interestingly, the serum enzyme results revealed an increased hepatic susceptibility on the day the animal's synchronizer, lighting, had been changed. Then, for a period of 6 days after the shift, there was a progressively diminished susceptibility response. Within 10 days after the lighting shift, the rhythm was reversed from its preshift phase relationship with reference to the clock hour. These findings suggest that shift work transitions in themselves may alter hepatic function resulting in the increased risk to hepatotoxicity in those exposed to chemical hazards. These findings await confirmation both in prospective animal and human investigations.

Scheving et al. (316) have demonstrated that the response in activity of certain enzymes in the liver and brain of mice to certain peptides known to have significant roles in metabolism is circadian-stage dependent. For example, a single intraperitoneal injection of insulin increased by 23 percent the response in activity of the pivotal enzyme, hepatic pyruvate kinase (PK) when administered toward the end of the dark span and determined 4 hr later. No statistically significant effect on enzyme activity was seen if the identical dose of insulin was injected toward the end of the animals' rest span and determined 4, 8, or 12 hr later. When glucagon was administered toward the end of the light span, there was a 17 percent increase in the enzyme activity 12 hr after treatment, but when administered toward the end of the dark span no statistically significant response was seen 4, 8, or 12 hr later. When epidermal growth factor (EGF) was administered toward the end of the dark span, there was a statistically significant increase of 23 percent in PK activity when determined 12 hr later; no such response was seen when administered at the end of the light span. Similar findings were found for many other enzymes. In fact some enzymes showed a positive response at one circadian stage, a negative one at another, or no response at still another. One such example was recorded in the response of hepatic malic enzyme to glucagon; when administered at one time, the response was a statistically significant increase, whereas at another time it was a statistically significant decrease.

Remarkably, over the last 10 years, the number of studies evaluating the adverse effects of toxic substances at different times of the day have been few. However, recent studies (347–350) have determined that the toxicity of some organic solvents are associated with peak levels of serum glutamic pyruvate transaminase (GPT) activity that occurs during the active phase of rodents. Motohashi et al. (349) determined that the toxicity of trichloroethylene (TRI), a common degreasing agent, was greatest at the beginning of the resting phase (0900) in rodents, which is significantly associated with GPT activity. However, it was also determined that for free-running animals kept in total darkness, the peak toxicity of TRI occurred at 2100 hr and was also correlated to the free-running peak in GPT activity. The authors suggest that this finding implies that individuals who are desynchronized may be at greater risk of toxic effects of TRI exposure at times when other synchronized individuals would be significantly protected. These findings are an example of the numerous difficulties that could be encountered in attempting to develop exposure criteria for shift-working individuals.

In comparing day and evening exposure to ozone, Van Bree et al. (351) determined that active-phase (night) exposure produced significant increases in protein, albumin, and inflammatory cell production in rats but not in guinea pigs. It was determined that this increased inflammation during the night was related to increased respiratory function during the nocturnal period of the rat, resulting in inhalation of a larger dose of ozone. No effect was found in the guinea pig because of its random activity schedule. Also, Harbuchi et al. (350) determined that performance decrements in behavior in the rat occur when exposure to pre-LD_{50} levels of toluene during the rest (light) phase were compared to controls and performance during the dark or active phase. Also it was determined that concentrations of toluene were higher after exposure during the light phase. Although the mechanism of this effect remains unknown, it has been suggested that the rate of liver metabolism of toluene is a potential mediator.

In recent years investigations have begun to review the effects of pesticides and other toxic substances on brain hormone levels that are related to the sleep–wake cycle. Results from Attia et al. (352) suggest that the pesticide carbaryl stimulates a pineal production of melatonin in a dose-related response that was circadian rhythmic. However, while pineal levels of melatonin rose and fell according to an altered 24-hr rhythm, serum levels of melatonin remained unaffected. Along these same lines Pohjanvirta and others (353,354) have determined that TCDD (2,3,7,8 tetrachorodibenzo-p dioxin) depresses serum melatonin in rats. However, the mechanism of the effect has not been determined as no morphologic differences were noted in the pineal glands of treated rats when compared to untreated controls. Similarly, Matsumura et al. (355) determined that inorganic selenium compounds directly administered to the third ventricle of the rat brain inhibited synthesis of prostaglandin D_2 and consequently inhibited their sleep entirely.

In summary, based on results of chronobiologic studies, circadian difference in the effects of toxic and tumor-inducing chemicals are to be expected. Differences in toxic effect over the 24 hr appear to be dependent on circadian variations in pharmacokinetic phenomena and in enzyme activity in the liver and target issues. Circadian rhythms in enzyme activity help explain temporal patterns in the toxic effects of chemicals and seemingly are important in helping to explain 24-hr patterns of metabolism in general.

19.3 Significance of Circadian Chronotoxicity for Medicine and Occupational Health

Schering and others have carried out a series of studies that document the fact that the susceptibility to several anticancer agents which are cell-cycle specific is strongly circadian rhythmic (Fig. 9.46). One of our interests has been how can the administration of these treatments be optimized in experimental cancer chemotherapy (48, 307,317–323). Figure 9.47 presents one such example. In this particular case, cyclophosphamide (100 mg/kg) and adriamycin (5 mg/kg) were found to be synergistic in treating mice that had been inoculated with 1×10^5 L1210 leukemia cells 4 days prior to treatment. With only one course of treatment, there was a dramatic circadian variation in response as monitored by mean survival time and cure rate. The variation in cure rate (mice alive and apparently free of disease 75 days posttumor inoculation) as a function of treatment timing ranged from 8 to 68 percent in male animals standardized to 12 hr of light alternating with 12 hr of darkness. Similarly, in female mice standardized to 8 hr of light alternating with 16 hours of darkness, the cure rate ranged from 0 to 56 percent, depending on when the drugs were injected during the 24-hr span. No cures were obtained with either drug alone. The maximum cure rate was recorded when the two drugs were administered in the early part of the dark portion of the light–dark cycle (whether the animals were synchronized to 12 hr of light and 12 hr of darkness or 8 hr of light and 16 hr of darkness); maximum mortality occurred following treatment in the light span. The data also documented that maximal therapeutic advantage was obtained when the two drugs were separated by 2- or 3-hr intervals and that this effect of drug sequencing was strongly circadian-stage dependent (307,322,323). The data from experimental models demonstrate clearly that one should not ignore circadian variation in tolerance when dealing with chemotherapy or sequencing studies. The evidence is compelling that the temporal organization in general carries with it significant implications, not only for cancer chemotherapy and radiotherapy, but equally important for basic research on normal as well as abnormal growth (307,297,319,324). Perhaps we can come to better grips with the mechanism of such diseases as cancer if we consider in the first place the rhythmic nature of cell proliferation. We have just reached the stage where most scientists accept without question the existence of and the potential importance of such rhythms, even though some may still ignore them in experimental design. Based on the facts, we must abandon the erroneous concept that somehow sampling at the same time of day takes care of the rhythm problem (307).

With regard to shift work it can be expected that exposure to chemical agents may result in different degrees of toxicity or irritation, depending on the timing of the exposure relative to the circadian system and the dose. Based on the fundamental research of the late H. von Mayersbach (306), it is now well understood that the functional and structural organization of the hepatocyte, a primary site of chemical biotransformation, differs tremendously throughout the 24 hr. Microscopic studies

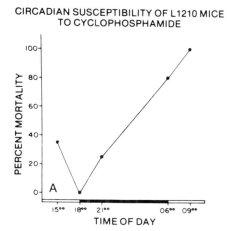

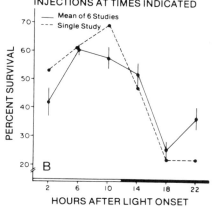

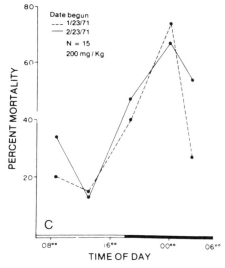

Figure 9.46 Circadian variation in the toxicity (measured by mortality) from three different carcinostatic agents: (*A*) cyclophosphamide [Haus et al. (319) and Cardoso et al. (321)], (*B*) adriamycin [Lévi et al. (367)], and (*C*) arabinosylcytosine [Kühl et al. (368) and Scheving et al. (320)] among several that have been studied [Lévi et al. (367)]. The timing of darkness is indicated by shadings along each horizontal axis. The peak susceptibility of each of the agents varies, occurring between the end of the dark and the beginning of the light span for cyclophosphamide, 6–10 hr after light onset for adriamycin and during the middle of the dark span for ara-C.

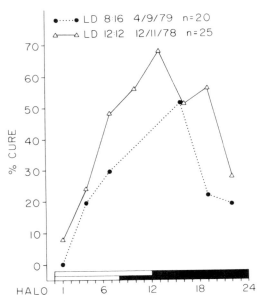

Figure 9.47 Variation in cure rate of "leukemic" mice depending on the timing of combined treatment with cyclophosphamide (CTX) and adriamycin (ADR). Study 1 (solid line) on male mice synchronized to 12 hr of light alternating with 12 hr of darkness; study 2 (dotted line) on female mice synchronized to 8 hr of light alternating with 16 hr of darkness. HALO = hours after lights on. Reproduced from Scheving and colleagues (307, 322).

have shown that the ultrastructure of the liver cell during the middle to later portion of the activity span is mainly organized to support glycogen deposition, while 12 hr later the ultrastructure appears remarkably different in support of enzyme synthesis. If it is acceptable to assume the existence of a similar circadian rhythm in hepatocyte function in human beings, then for day workers during the early to middle hours of daytime work it should be expected that the hepatocyte is involved primarily in protein and enzyme synthesis, while during the late evening it is primarily involved in glycogen deposition. The findings of von Mayersbach (306) on circadian rhythms in liver ultrastructure and function are consistent with the results of studies showing circadian variation in the potency of hepatotoxin. Von Mayerbach's findings support the supposition that rhythms are a very important component of biological adaptability representing a readied capacity during activity to react to challenges and changes within our geophysical environment. Chronobiologically, organisms differ in their tolerance and reactivity to various types of challenges; we are not the same, biochemically and physiologically, in the morning in comparison to the evening or night.

19.4 Methods for Conducting Chronotoxicology Evaluations in Rodents

Conducting chronobiological studies on animals in the past has required sampling round the clock, thus creating logistical problems and an enormous demand on the stamina of the personnel involved (325). This demanding effort more than anything else has impeded research along this line. Earlier we discussed the fact that the circadian system of the experimental animal could be synchronized to the light–dark cycle, and we believe an explanation of how to exploit this in carrying out toxicologic research and evaluations in industry warrants some comment in this chapter.

The ability to synchronize a rhythm to the light–dark cycle permits an investigator to carry out 24-hr rhythmic studies during the span of an 8-hr working day. We have spent a great deal of time exploring the reliability of the various techniques used to accomplish this. One that has been successful is to subject half of the animals to be studied to the conventional light–dark cycle (assume, for the purpose of explanation, that the light is on from 0600 to 1800 and that darkness is from 1800 to 0600) while the other half is subjected to the opposite schedule (with light on from 1800 to 0600 and darkness from 0600 to 1800). Thus, if one wanted to sample at six circadian stages, animals need only to be taken from both environments at three different clock hours. Figure 9.48(a) shows a model illustrating the laborious conventional approach of "staying up" round the clock to sample at six different circadian stages (297), and Figure 9.48(b) represents the inverted light–dark technique. The same type of

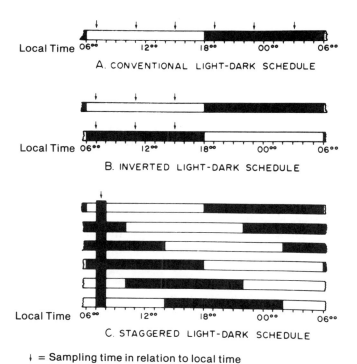

A. CONVENTIONAL LIGHT-DARK SCHEDULE

B. INVERTED LIGHT-DARK SCHEDULE

C. STAGGERED LIGHT-DARK SCHEDULE

↓ = Sampling time in relation to local time

Figure 9.48 Models of three different ways of sampling animals for studies of circadian rhythms. Arrows indicate the times of designated sampling for each design. Note that sampling done by a conventional design (*A*, top of figure) requires round-the-clock work. The inverted (*B*, middle of figure) or staggered (*C*, bottom of figure) designs enable the completion of toxicity studies within the regular 8-hr workday (see text for details). From Scheving (297).

sampling as illustrated in Figure 9.48(a) can be obtained from the two different environmental light–dark schedules shown in Figure 9.48(b) by conducting studies on the animals at approximately the beginning of light or darkness, in the middle of light or darkness, and toward the end of light or darkness.

It must be kept in mind that to standardize animals in this manner does take time, and one should not attempt short cuts. In animals subjected to an inverted light–dark cycle, some rhythmic variables may shift completely in 1 week; others may take as long as 3 or 4 weeks. While the phase shifting is going on, the circadian system is in a state of transient internal desynchronization. Therefore, to avoid the effects of such desynchronization, one should standardize animals to the respective light–dark synchronization schedules for at least 3 weeks prior to each experiment. Data should be collected only after one is convinced that the rhythmic variables to be studied are likely to have phase shifted; this is done in a preliminary study using this time span. Essentially, if the data are not already available in the literature, one has to determine just how long it takes to invert the variable under question. Also it is very important that the animals, such as the rodent, are not subjected to any light perturbation during the dark span; this could, under the right circumstances, cause a phase shift or at least a disturbance in the rhythm that could be seen on subsequent days. If it is necessary to view or to attend to the animals during their dark span, a dim red light should be used; 0.5 lux at the level of the mouse eye is acceptable.

Figure 9.48(c) represents still another way of sampling all six circadian stages, by subjecting six separate groups of animals to six different light–dark schedules; such a technique is commonly referred to as a "staggered" light–dark schedule. With such a technique, one must allow sufficient time for all rhythms to phase shift within the various groups, on all different light–dark schedules. The phase shifting of various rhythms may be quick or slow, but the complexity must be recognized; and the tendency is for it to be greater in the staggered light–dark scheme. In addition, how fast a rhythm phase shifts will depend on whether the process involves a phase advance or phase delay. Certainly, if enough time is allowed, the staggered light technique can be effective in synchronizing the circadian system. Nonetheless, the simple inverted light–dark cycle as illustrated in Figure 9.48(b) when carrying out circadian rhythm studies is preferred by many researchers.

Some investigators have argued that keeping animals in continuous light each 24 hr is one way to avoid the "nuisance" of biological rhythms. There are many reasons why animals should not be kept in continuous light, and these have previously been discussed in detail (105,325). It is clearly documented that continuous light has a deleterious effect. Even a short exposure to such an environment may significantly alter a variable in comparison to one measured in animals synchronized to a light–dark schedule or even in animals housed in continuous darkness. For example, animals drink and eat less in continuous light (325). The not uncommon practice of keeping rodents in continuous light should be abandoned. It is our suspicion that some of the controversy that arose in the first place about the so-called lack of rhythmicity in certain studies resulted from this practice. Editors have not always required that the conditions under which the animals were standardized be documented. Such should be a sine qua non for the reporting of data in the future.

The implementation of the different models shown in Figure 9.48 by keeping subgroups of animals in simple light-tight boxes such as seen in Figure 9.49. These

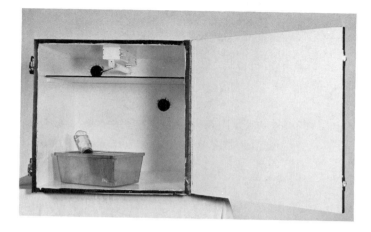

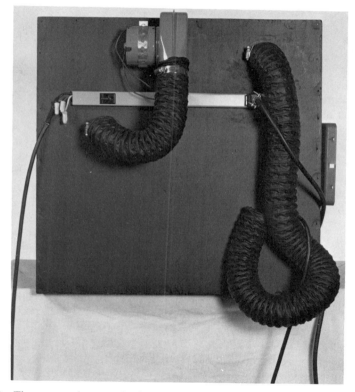

Figure 9.49 The upper photograph shows the interior of the box with the air inlet below and the outlet above; a glass partition separates the two to prevent accumulation of heat from the light. Each box can accommodate five cages such as the one shown. The lower photograph is of the same box showing the back and how the equipment is attached to it. An automatic timer is attached to the side of each box to control the lighting schedule.

are sound attenuated boxes having their own ventilation system as well as a mechanism for programming the light–dark cycle. They are relatively cheap to construct. Others use separate rooms; some prefer the boxes because there is less total disturbance. For example, animals used at the end of the sampling span need not be disturbed in any way by the taking of animals from the room at the beginning of the study; this eliminates a lot of experimental "noise" (297–306,308–319).

19.5 Susceptibility-Resistance Rhythms of Human Beings

Susceptibility-resistance rhythms in human beings, although less studied because of ethical reasons, have been shown to exist. Much of our understanding is based primarily on studies of patients allergic to substances such as dust and pollen or of responses to chemical irritants (262,263,326,327). Circadian rhythms have been identified for the susceptibility-resistance of the airways of asthmatic patients sensitive to histamine (293,294) (Fig. 9.36) and acetylcholine (328) as well as house dust, in house-dust-sensitive asthmatic patients (329). Inhalation of an allergen by asthmatic patients whose airways are specifically sensitized to it typically induces an allergic response of the airways characterized by an elevation in the resistance to airflow. The degree to which airway resistance is increased due to a standardized house dust provocation was found to be dependent on when inhalation occurred as shown in Figure 9.50. It is clear that a given concentration of house dust extract when inhaled in aerosol form around 3 P.M. was well tolerated since there was only an 8 percent transient increase in airway resistance. In contrast, the same persons when exposed in the identical manner to the exact same concentration of the antigen at 11 P.M. had a much more severe reaction; the increase in airway resistance was nearly 23 percent. At this time the patients exhibited chest rales and dyspnea that persisted for several hours. In other words exposure to house dust in house-dust-sensitive asthmatic patients was well tolerated as long as it was encountered during the middle of the daytime activity span. When exposure to the same concentration occurred late at night, it was poorly tolerated.

The presence of rhythms is sometimes so obvious to the affected persons that scientific research involving quantitative data gathering is not necessary. For example, some house-dust-sensitive asthmatic patients often recognize a circadian pattern in their tolerance to moderate concentrations of dust. Some persons set up schedules for dusting and vacuuming their homes during the middle of the day because they have found it difficult or troublesome to breath as a consequence of conducting these activities at night. Occasionally, workers in certain jobs within the cotton, pharmaceutical, fiberglass, woodworking, grain, and foundry industries recognize that there are periods of the day during which they are better able to tolerate chemical or dusty environments and therefore try to schedule "dirty" tasks for these times of the day.

19.6 Chronotoxicity and Shift Work

Human susceptibility-resistance rhythms have been described and exemplified in the previous section. The concepts of chronokinetics, chronesthesy and chronergy were

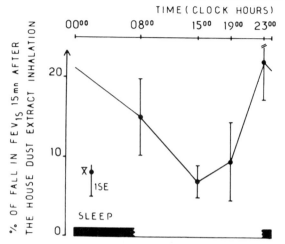

CHANGES IN THE BRONCHIAL PATENCY (FEV$_{15}$) AS FUNCTION
OF THE TIMING OF A BRONCHIAL CHALLENGE TO HOUSE DUST EXTRACT

Figure 9.50 Circadian changes in the susceptibility of the airways to an aerosol of house dust in diurnally active house-dust-sensitive asthmatic patients. The pre- to posttest difference in the FEV$_{1.0}$ at 1500 (3 P.M.) was relatively minor compared to that at 2300 (11 P.M.) when the airways exhibited marked hyperreactivity. From Gervais et al. (329).

introduced and the method in which biological rhythms affect the disposition of medications was discussed. In shift work industries involving the manufacture or processing of chemicals, the concepts and principles of chronopharmacology and chronotoxicology require attention for very practical reasons.

19.7 Chronotoxicity and Rapid Rotation Shift Schedules

For employees engaged in rapid rotational shifts of 2–3 days duration, it is conceivable that the acrophases of most circadian processes apparently change only slightly in timing, if at all, when working during the nighttime and resting during the daytime for the rather brief span of a few days. When this is the case, work during the nighttime, in the absence of significant acrophase shifts, represents exposure to workplace contaminants taking place at biologically different times in comparison to when working the day or afternoon shifts. The data from rodent investigations (40–42) clearly indicate the potential for different levels of resistance-susceptibility as a function of the biological timing of hazardous exposures. One would hope that the already promulgated threshold limiting values (TLV) or permissible exposure limits (PEL) are sufficiently conservative to be protective, although studies of the chronobiology of workplace toxicants in human beings have yet to be conducted and apparently appropriate time- or shift-qualified data are not available from medical records. From a theoretical, experimental, and occupational health basis, it is necessary to take into account circadian susceptibility-resistance rhythms, especially in the case when concentrations of harmful substances exceed safety limits, for example, because of industrial accidents or the mishandling of materials. It can be expected that toxicologists and pharmacologists will continue to examine on a larger scale the significance of circadian differences in the kinetics and pharmacodynamics of toxic substances found in industries utilizing rotating shift work schedules of 8 or 12 hr, for example.

19.8 Chronotoxicity and Unusual Shift Schedules

Unusual work shifts pose at least two special kinds of problems. The first problem relates to the necessity of adjusting TLVs for unusual work schedules to ensure employees will not be at greater risk than those on regular shifts. The second problem involves the circumstance that the promulgation of all workplace standards has been guided by the biological concept of homeostasis. This concept infers that all bodily functions and processes are maintained in relative constancy with variations occurring at random or resulting from environmental causes. This concept infers also that the biological response to toxic chemicals in individuals does not vary over time—over 24 hr, menstrual cycle, and year. Homeostasis does not take into account the existence of biological rhythmicity that, as has been discussed, can be quite significant, especially to occupational health concerns. It is only in recent years that investigations have begun to model in rodents workplace exposures, taking into consideration the role circadian rhythms play in the metabolism and effects of workplace and other toxicants (135,138,286,287). In connection with this, one of the objectives of chronobiologists working with the Unusual Work Shifts Subcommittee of the American Industrial Hygiene Association is to evaluate the importance of biological rhythms on the toxicity of workplace chemicals in persons employed on unusual shifts.

With regard to unusual as with most shift schedules, emphasis is placed upon the peak body burden of the chemicals to which the worker is exposed by the end of the last day of, for example, 10 to 12-hr in duration shifts. Biological monitorings, when conducted, are done typically on shift workers when they are on daytime duty, when the occupational health department is well staffed to carry out such samplings. Inferences are then drawn, assuming homeostatic principles, about the safety of the workplace environment for employees when they work the night shift. When from week-to-week employees alternate between day and night work, chronobiologic factors can be significant. In the case of unusual work schedules, with work periods of 10 to 12 hr or more in length, exposures of workers to chemicals and other potentially hazardous substances not only occurs for longer durations than they do during the standard 8-hr shift, but they occur as well during different circadian stages. The concern of chronobiologists is that circadian susceptibility-resistance phenomena are important considerations for workers on unusual shift schedules. Moreover, from a chronobiologic perspective, when chemical exposures are involved, biological rhythms in their disposition have to be considered. This entails the chronokinetics of chemicals, that is, circadian or other rhythmic influences on the time to reach peak concentration, peak height, clearance, and area under the time–concentration curve. As is the case for special chemical agents such as medications, these chronokinetic parameters can be vastly different, depending on the biological time of exposure (40,41,284,285). Although at this time sufficient data do not exist either from animal or human beings to assess the significance of circadian chronokinetic phenomena with respect to the need for adjusting TLVs for unusual work shifts, current research efforts are directed at evaluating for which types of exposures circadian chronokinetic changes are critical.

The biological effects of xenobiotic exposures in the case of both usual and unusual work schedules are dependent not only on their chronokinetics but, in addition, on their chronesthesy. Exposures to substances either during differently timed 8-hr shifts or 10- to 12-hr (or longer) unusual work shifts may result in different biological effects due to circadian differences in the chemical susceptibility of target issues. As in the case of medications, circadian chronesthesys may represent among others 24-hr variations in the number of receptor sites and/or the capacity of various biochemical pathways to metabolize, transform, or incorporate chemical agents. Taking into consideration the findings from chronopharmacologic investigations of medications as well as knowledge about the chronotoxicology of chemicals, circadian rhythms in liver and brain enzymes, and known circadian resistance-susceptibility rhythms of human beings, the possibility that the biological response to or effect of the peak body burden of chemical agents when working, for example, the daytime in comparison to the nighttime shift being different, especially for workers on rapid rotation shift work schedules, cannot be ignored.

Currently, except for medications, circadian chronesthesys of chemical substances in human beings have yet to be demonstrated. This is due primarily to the fact so few investigations of this type have been attempted. Nonetheless, they are known for substances used in the diagnosis of human diseases, for example, the acetylcholine, histamine, and house-dust airway challenge tests, which serve to evaluate the hyperactivity of airways in patients exhibiting obstructive airways disease. In rodents, additional data are available, for example, for the effect of CCl_4 on liver enzyme activity and toxicity (24–26,135,287,313). Moreover, other findings from studies on rodents reveal circadian differences in the effects of a large variety of toxic substances as exemplified in Figures 9.40–9.42. Since circadian variation occurs at all levels of biological organization, circadian chronesthesys are possible at each. Keeping in mind that clinically meaningful chronesthesys are common with medications, it behooves occupational health researchers and physicians to evaluate findings from laboratory models and medical records for temporal patterns of susceptibility and toxicity to chemicals found in the workplace to determine the need for altering exposure limits for unusual, as well as standard 8-hr, rotational shift work schedules.

20 METHODS FOR CONDUCTING CHRONOBIOLOGIC STUDIES

20.1 Studies of Shift Workers

To a large extent, our knowledge about the chronobiologic adjustments to 8-hr and other rotating shift work schedules comes from investigations on volunteers studied under simulated shift work conditions. While the findings from these are helpful, their usefulness for deriving general principles is limited. Frequently, the volunteers of such investigations have been studied under atypical conditions. Although they are subjected to simulated synchronizer shifts, they may not be required to perform tasks similar to those found in an industrial setting. Furthermore, the social and experimental milieu of a simulated study can be very different from that of shift work. For these reasons actual field studies of shift workers are necessary.

Field studies of the chronobiologic adjustments of rotating shift workers have been conducted on selected groups of volunteering employees (17,62,63,76). For investigation of the biological rhythms of workers, several measurements separated by more or less equal intervals over the activity span are necessary. Recent developments have made available a variety of portable, light-weight instruments for conducting self-measurements on a variety of health-related functions. Moreover, numerous questionnaires useful for studying aspects of shift work have been developed, field tested, and validated.

The utilization of medical instrumentation to self-monitor biological functions repeatedly over time for the purpose of detecting and quantitatively describing biological rhythms is termed autorhythmometry. Autorhythmometry (334) has been used to study and screen blood pressure in children and adults, evaluate airway patency in asthmatic persons, quantitatively assay the irritant effects of air pollutants, and evaluate the biological rhythm-dependent effects of medications (92,266,275,329, 335,336). Autorhythmometry has also been used to intensively study various chronobiologic aspects of shift work in employees of certain industries (17,337). Also it has been used for the purpose of teaching certain principles of health education as discussed by Glasgow et al. (336).

20.2   Criteria for Conducting Chronobiologically Oriented Field Studies on Shift Workers

When conducting chronobiologic field studies on shift workers, three major types of difficulties must be overcome. These usually can be placed in one of the following categories: social, methodological, or analytical.

20.3   Social Obstacles to Shift Work Research

The first problem to solve when conducting field studies on shift workers involves social obstacles. For many reasons few companies or unions wish to pursue research on shift workers. Therefore, in the design of research protocols, both the aims and benefits that can be expected from the investigation must be stated clearly and publicly to management, union, and participating employees. The priorities relative to the aims and methods of the study should be noted. If the study could pose any risks, these too must be defined. Usually, the development of a workable protocol requires several meetings between the interested parties if they are to achieve close cooperation and trust. At the conclusion of such field studies, it is the responsibility of the investigators to report pertinent and unbiased information to management, organized labor, and select others with the objective of improving the health and well-being of workers while maximizing safety and productivity.

20.4 Methodological Difficulties of Shift Work Research

The methodological problems of chronobiologic field studies are mainly those related to data gathering. The simultaneous investigation of a set of behavior physiologic and metabolic variables usually requires the participation of a large number of trained investigators and the use of rather large and cumbersome medical instrumentation. Meeting these experimental requirements is difficult. Not only is the appropriate equipment expensive, but it generally is available only in well-equipped but all too often remote laboratories. One reasonable solution to the problems of data gathering, that is, cost, manpower, and instrumentation, is to select a set of easy-to-carry, moderately priced, yet medically and socially acceptable instruments and to educate participants in their proper use so that they can accurately self-measure certain important biological functions. In principle, such an approach is workable, but at the present time in the United States and certain other countries it is abundant with practical problems. To succeed, dedicated investigators are required to devote a great deal of time and effort to assure that each participant is properly trained as well as motivated so that the research procedures are carried out correctly.

In doing field studies, according to Reinberg and his colleagues (337) who have conducted several successful large-scale investigations on shift workers in European industries, selection of biological measurements and instrumentation should be based on the following criteria: (1) measurements should be biologically meaningful, (2) instruments should be easy to use, calibrate, and transport as well as reasonably priced; and (3) measurements should not require too much time to conduct. Reinberg, one of the most respected researchers in the field of shift work chronobiology, in a series of studies with his colleagues on more than 100 shift workers obtained data by autorhythmometry on the following biological and psychological variables: mood and activity level, time estimation, heart rate, oral temperature, card sorting (a measure of eye–hand skill), peak expiratory flow (a measure of airway patency), and grip strength of the left and right hands (using a dynamometer) (17,337). In addition, urine samples were collected so that electrolytes as well as adrenocorticoid and catecholamine hormones could be determined. Self-measurements and self-assessments as well as the collection of urine samples were done at 4-hr intervals, or less, only during the waking hours, for several consecutive days, in some cases, or on selected days of the week, in others. Some of the long-term studies involved the sampling of rhythms only on the first and last days of each shift rotation. Information pertaining to the onset and end times as well as the qualitative aspects of sleep also were obtained using specifically designed and field-tested questionnaires.

In Reinberg's studies (17) selected participants also were invited to record what, how much, and when meals and snacks were consumed in order to learn of possible differences in the intake of carbohydrates, proteins, lipids, and calories according to shift. Such data provide basic important information on the eating habits of shift workers. This information is relevant to exploring the etiology of the elevated incidence of digestive complaints and also peptic and duodenal ulcers found in rotating shift workers. In general, it is the experience of Reinberg and his co-workers that persons who volunteer to participate in shift work field studies are dependable and accurate in carrying out their self-assessments.

In summary, autorhythmometric methods can be useful under careful supervision for investigating the chronobiologic adjustments of shift workers to alteration in their activity–sleep routine. Self-assessments also can be evaluated for alterations in function and these are likely to give insight into the type and degree of biological adaptations that occur in shift workers. Autorhymometry also can be used to conduct biological monitoring as a means to evaluate the health status of employees at special risk to workplace contaminants.

21   STATISTICAL METHODS FOR BIOLOGICAL RHYTHM DETECTION AND DESCRIPTION

As chronobiology developed, it became increasingly evident to some investigators that the classic statistical procedures, such as analysis of variance (ANOVA) and Student *t*-test, were not adequate because they did not always provide the desired quantitative numerical endpoints. Thus a number of mathematical models for evaluating and quantifying time series were developed.

21.1   Periodogram

Among them was the adaptation of the periodogram introduced by Schuster (338) for the detection and analysis of geophysical time series. A periodogram provides quantitative estimates for both the period and amplitude of a given periodicity. It can be an appropriate technique for the detection of bioperiodicity when there exists a sufficiently long series of data with a consistent rhythm overlain by noise (random events that distort the detection of periodic functions); but its indiscriminant use in the study of biological time-series data may lead to the description of spurious as well as real periods, and real periods sometimes may not be revealed. For a more detailed treatment of the computational aspects of this method and its usefulness and limitations for describing biological time-series data, see Koehler and others (339).

21.2   Spectral Analyses

Generalized harmonic or power spectral analyses also have been used for the study of periodic phenomena. Spectrum analysis is based on the assumption that a time series represents the sum of infinite sinusoidal functions, so-called harmonics with different periods. The variance spectrum or so-called power spectrum, which may be obtained by electronic computer methods, shows the prominence of different periodicities in terms of the variance accountable by each. This method was made practical for application to limited, noisy time series found in biological material through the development by Tukey (340) of methods for determining the statistical reliability of spectral estimates. The variance spectrum method was adopted for use on biological time series by Halberg and Panofsky (341) and Halberg et al. (342). This method detects different frequencies in a noisy time series and evaluates their statistical significance; it is particularly useful if a biological variable exhibits changes in the form of several frequencies, some of which may be ill-defined. Nonetheless, the method of spectral analysis has its limitations. It does not give information on phasing (the occurrence of peak values) of the rhythms detected, nor does it provide a numerical endpoint for the amplitude. Another shortcoming, especially for conducting chronobiologic studies on human beings, is that the technique requires sampling at regular intervals over a relatively long time span.

There are other ways of analyzing data, one among them being the multiple complex demodulation technique (343) and cluster analysis (344), which we will not discuss at this time, but that might be considered as methods of choice in some instances, although even these analytical tools have limitations.

21.3 Cosinor Analysis

Since the data from field studies are collected only during the hours of activity, so as not to disturb the sleep span, the time series consists of measurements derived at unequal intervals. Special statistical methods must be used to analyze these since most conventional so-called time-series analyses require that the collected data be spaced at equal intervals. One very useful method is the so-called cosinor. This technique frequently is employed by chronobiologists; it has been the method chosen to analyze some data included herein, and it is widely encountered in the literature. One should be familiar with the method if the data analyzed by it are to be fully appreciated. The cosinor method was described originally by Halberg et al. (86) and modified by Halberg and others (334). A brief explanation of what the cosinor is and what it does follows.

Cosinor analyses consist of the least-squares approximations of one or more cosine curves to time-series data in order to objectively detect and quantify rhythms. The quantitative endpoints used to describe rhythms include the period (the duration of time needed to complete one cycle of a rhythmic variation), acrophase (the crest time relative to a defined reference such as some aspect of the synchronizer schedule, e.g., the middle of the rest span), amplitude (a measure of the prominence of the rhythm and equivalent numerically to one-half the extent of rhythmic change between the peak and trough), and mesor (the rhythm-determined average, e.g., in the case of a single cosine approximation, the midway value between the peak and trough). The cosinor indicates the existence of statistically significant rhythmicity if by F test the amplitude is found to be significantly different from zero. Cosinor analyses have been used to evaluate the chronobiologic adjustments of various biological rhythmic parameters in humans due to shift work as well as jet travel across several time zones. They have been used as well to substantiate circadian patterns of industrial performance and accidents as well as the response to chemical substances. A more recently developed analysis enables one to determine the exact period length for a given rhythm when the data of the time series are of unequal interval. This represents an important step forward in that it allows one to determine, for example, if an employee's intolerance to shift work is related to an internal desynchronization of circadian processes.

22 SUMMARY

The facts clearly demonstrate that practically every physiological or behavioral variable amenable to measurement is characterized by being rhythmic, at least in the healthy organism. The spectrum of rhythms is broad. The one frequency that has received the greatest attention in the past few years is that which approximates the frequency of the rotation of the earth, or the circadian. It is this frequency, primarily in the mammalian organism, upon which we have focused. Biological rhythms in most physiological variables are not apparent to us in the same sense that the respiratory or menstrual rhythm is; they only become overt when they are properly measured at frequent intervals over the 24-hr time scale. Because of their somewhat "invisible" nature, there is a tendency on the part of some investigators to slight or to ignore them. In spite of all that is known they simply have not been accorded the attention they deserve; this undoubtedly is in large part due to the fact that the science of chronobiology is relatively young. Perhaps this explains why many occupational health professionals are not aware of practical developments in this emerging science. On the other hand, many health professionals are confused about the significance of biological rhythms as reflected by their thoughts about them.

Some rationalize that biological rhythms represent no more than minor fluctuations around a daily mean and consequently do not warrant the additional amount of work and expense required to properly explore for their significance. Those who believe this simply are ignoring the scientific facts available to them. Admittedly, oral or rectal temperature may change only one or two degrees centigrade over the 24-hr time scale; there are, however, few physiologists who would minimize the effect that this small-amplitude rhythm has on metabolic or physiological activity. On the other hand, the corticosterone concentration in the plasma of human beings may vary three- or fourfold; the same applies to many enzymes in the liver and plasma of rodents and probably human beings as well. When 5-hydroxytryptamine levels in the whole brain of the rodent are measured at frequent intervals over a 24-hr period, the difference between the lowest and highest recorded mean may represent only an 18 percent change, but the same substance in the pineal may vary as much as 900 percent. When similarly compared, the cell proliferation in a number of mitotically labile tissues, such as the cornea, may represent over a 1200 percent increase. Generally, fluctuations in humans are equally as great as the daily change in rodents; the evidence to support the above statements is extensively documented. Because the biological system is continually changing, the organism is a different biochemical entity at different stages of its circadian system. If an organism such as the rodent is subjected to a potentially noxious agent at one time, it may succumb, whereas at another phase of the circadian system it may evidence very little harm. Circadian differences in the response of human beings, although less studied, are well documented as well. This circadian differential in response has led to the concept of the "hours of changing resistance." The overwhelming evidence that supports this concept necessitates total rejection of the idea that circadian time structure represents no more than minor and, hence, unimportant fluctuations around a daily mean. Furthermore, the facts render completely untenable the concept of homeostatic balance as is presently taught in many freshman medical school physiology courses as well as industrial toxicology and hygiene curricula.

Other health professionals have expressed the opinion that circadian rhythms represent nothing more than day–night changes or that they are some kind of direct response to the ingestion of a meal, and that if one simply samples once during the day and once during the night in some vague way one will have taken care of this nuisance variable. We wish to point out that sampling only twice per 24 hr does not account for the possible influence of circadian temporal structure. One may happen to sample at times that approximate the 24-hr mean level, the mesor, even though the times of sampling are separated by 12 hr. Also, sampling by clock time only, without regard to synchronizer schedule, especially when shift work is concerned, is meaningless. This attitude toward rhythmicity simply cannot be defended, yet it still is held by most investigators! Although facts demonstrate that bioperiodicities are responsive to changes in the environment, they represent endogenous and inherited attributes. Teleologically, their properties provide the mechanisms that enable the organism to adjust not only to predictable geophysical changes over time but, in particular, to shift work.

A large number of investigators (especially toxicologists) conduct research or carry out biological samplings each day at the same time, ostensibly to avoid or minimize dealing with circadian rhythms. In view of what has been learned about their properties, such as their endogenous nature and in rodents their ability to phase shift with alteration in the environmental light–dark synchronizer schedule or in humans with changes in the sleep and activity pattern and their tendency to freerun in continuous light or darkness, it is this widespread practice that is most difficult for the chronobiologist to accept as the means for avoiding the effects of rhythms. All that this practice of conducting procedures at the "same time of day" does is to assure an investigator that he or she is sampling at one particular clock hour, perhaps indicative of a single circadian stage of one or more rhythms—in the trough, on the incline, at the peak, or on the decline. However, the investigator can only be sure of this if he or she has first gone to the trouble of mapping the rhythm(s) under controlled conditions, that is for rodents, the establishment of a carefully controlled synchronizing cycle such as that of light and darkness on a fixed alternating schedule; even then the circadian system phase may systematically shift with the changes in season so that sampling at 0900 June 1 may not be the same in terms of biological time as sampling at 0900 January 1. This practice of sampling once daily always at the same time of day can lead the investigator into pitfalls.

There are many excellent toxicologists who have at least an intuitive appreciation of the importance of rhythms, for example, in industrial toxicology, and would explore their effects in depth but find that proper control of the animal quarters simply cannot be obtained. Many animal facilities are geared more to housekeeping activities and the work schedules of animal caretakers than to careful control of the light–dark synchrony. We believe this is a major reason for staying away from this type of investigation; it certainly is the most frequent excuse given.

No longer is it necessary for one to have to "stay up round the clock" to get a handle on circadian variation, and we have pointed out how this can be done in Section 9.5. If the determination of an LD_{50} is important, it should be reliable. We are at a stage where an effort must be made to come to grips with biological oscillation, and we offer the following suggestions as to how this might be done. The leadership in the biological scientific community, including industrial toxicologists and occupational health physicians, must critically examine the body of knowledge upon which chronobiology presently is founded. We assume that if this were done, it would lead one to enthusiastically recognize the potential that this science has to contribute to the better design of shift work schedules and protection of shift workers from hazardous exposures, keeping in mind the existence of circadian differences in resistance-susceptibility.

With regard to improving shift work conditions in industry, Knauth and Rutenfranz (54), reviewing a large number of investigations, including chronobiologic ones, suggest a number of recommendations. For the good of the worker:

1. A shift system should have few successive night shifts to minimize jet-lag-like shift work symptoms and sleep problems with resulting fatigue.

2. The morning shift should not commence too early, relative to clock hour, keeping in mind the sleep schedules of workers and the local traffic patterns, for example.

3. The shift change times should allow individuals some degree of choice or flexibility, for example, by being allowed to participate in the planning of shift timing changes.

4. The length of the shift should depend on the mental and physical demands of the job and when possible the duration of the night shift should be shorter than the morning and afternoon shifts.

5. Short intervals of time (e.g., 7–12 hr) between two successive shifts, which is often common to irregular shift systems, should be avoided since they result in too much fatigue.

6. Continuous shift systems should include some free weekends with a minimum of two successive full days off to enable socialization with friends who also are free at this time.

7. In the case of continuous shift systems, a forward rotation rather than a backward rotation is preferred.

8. The duration of the shift cycle (the pattern of shifts and days off until the sequence repeats on the same day of the week) should be reasonable in length and not too long in order to favor planning for leisure activity.

9. Shift rotas should be regular to facilitate the planning of activities away from work.

These suggestions are offered with the understanding that complying with all of them is difficult since many factors must be considered simultaneously. Nonetheless, these recommendations are offered as a means for improving the conditions of many shift work schedules now in operation in industrial plants worldwide. In recent years rapid progress has been realized in the chronobiology of industrial hygiene and occupational health; yet, much more remains to be done. Chronobiologists, as do other health professionals, look forward to continued collaboration with management, labor, and scientists of various disciplines to alleviate the many still troublesome problems associated with rotating shift work schedules.

REFERENCES

1. S. D. Nollen, *Ind. Eng.*, 58–63 (1981).
2. S. D. Nollen, Work Schedules, in *Handbook of Industrial Engineering*, G. Salvendy, Ed., Wiley-Interscience, New York, 1982.
3. M. Colligan, Methodological and Practical Issues Related to Shiftwork Research, in *Biological Rhythms, Sleep and Shift Work*, L. C. Johnson, D. I. Tepas, W. P. Colquhoun, and M. Colligan, Eds., SP Scientific and Medical Books, New York, 1981, pp. 197–204.
4. W. B. Webb, Work/Rest Schedules: Economic, Health and Social Implications, in *Biological Rhythms, Sleep and Shift Work*, L. C. Johnson, D. I. Tepas, W. P. Colquhoun, and M. Colligan, Eds., SP Medical and Scientific Publications, New York, 1981, pp. 1–10.
5. R. Mahathevan, *J. Human Ergol.*, **11**(Suppl), 139–145 (1982).
6. A. Manuba, *J. Human Ergol.*, **11**(Suppl), 147–153 (1982).
7. A. Khaleque and A. Rahman, *J. Human Ergol.*, **11**(Suppl), 155–164 (1982).
8. F. M. Fischer, *J. Human Ergol.*, **11**(Suppl), 177–193 (1982).
9. S. E. G. Perera, *J. Human Ergol.*, **11**(Suppl), 201–208 (1982).
10. C. N. Ong and B. T. Hoong, *J. Human Ergol.*, **11**(Suppl), 209–216 (1982).
11. M. Wongphanich, H. Saito, K. Kogi, and Y. Temmyo, *J. Human Ergol.*, **11**(Suppl), 165–175 (1982).
12. C. N. Ong and K. Kogi, *Occ. Med.*, **5**, 417, (1990).
13. S. Folkard, D. S. Minors, and J. M. Waterhouse, *Chronobiologia*, **12**, 31–54 (1985).
14. P. J. Taylor, The Problems of Shift Work, Proceedings of an International Symposium on Night and Shiftwork, Oslo, Sweden, 1969.
15. A. N. Nicholson, Ed., *Sleep, Wakefulness and Circadian Rhythm, AGARD* Lecture Series **105**, 1979.
16. L. E. Scheving and F. Halberg, Eds., *Chronobiology: Principles and Applications to Shifts in Schedules*, NATO Advanced Study Institutes Series D: Behavioral and Social Sciences, No. 3, Sijthoff and Noordhoff, The Netherlands, 1980.
17. A. Reinberg, Ed., Chronobiological Field Studies of Oil Refinery Shift Workers, *Chronobiologia*, **6**(Suppl), 1979.
18. L. C. Johnson, D. I. Tepas, W. P. Colquhoun, and M. Colligan, "Preface," in *Biological Rhythms, Sleep and Shift Work*, L. C. Johnson, D. I. Tepas, W. P. Colquhoun, and M. Colligan, Eds., SP Scientific and Medical Books, New York, 1981.
19. K. Kogi, T. Miura, and H. Saito, Eds., *Shiftwork: Its Practice and Improvement*, Center for Academic Publications, Tokyo, 1982.
20. A. Reinberg, N. Vieux, and P. Andlauer, Eds., *Night and Shift Work: Biological and Social Aspects*, Pergamon Press, Oxford, 1981.
21. W. P. Colquhoun, Ed., *Biological Rhythms and Human Performance*, Academic Press, London, 1971.
22. J. Wojtczak-Jaroszowa and K. Pawlowska-Skyba, *Medycyna Pracy*, **18**, 1 (1967).
23. T. H. Monk, *Occ. Med.*, **5**, 183 (1990).
24. D. J. Paustenbach, "Occupational Exposure Limits, Pharmacokinetics and Unusual Work Schedule," in *Patty's Industrial Hygiene and Toxicology, IIIA, Industrial Hygiene Aspects*, 2nd ed., Wiley-Interscience, New York, 1985.

25. D. J. Paustenbach, G. S. Born, G. P. Carlson, and J. E. Christian, "A Comparative Study of the Pharmacokinetic and Toxic Effects of Exposing Rats to Carbon Tetrachloride Vapor During a Standard or Novel Workweek," American Industrial Hygiene Conference, Portland, Oregon, 1981.

26. D. J. Paustenbach, "The Effect of the Twelve-Hour Workshift on the Toxicology, Disposition and Pharmacokinetics of Carbon Tetrachloride in the Rat," Doctoral Dissertation, Purdue University, West Lafayette, IN, 1982.

27. R. S. Brief and R. A. Scala, *Am. Ind. Hyg. Assoc. J.*, **36**, 467–471 (1975).

28. J. W. Mason and H. Dershin, *J. Occup. Med.*, **18**, 603–607 (1976).

29. J. W. Mason and J. Hughes, *Scand. J. Work Env. Health*, 1984.

30. J. L. S. Hickey, *Amer. Ind. Hyg. Assoc. J.*, **41**, 261–263 (1980).

31. J. L. S. Hickey and C. Reist, *Am. Ind. Hyg. Assoc. J.*, **38**, 613–621 (1977).

32. J. L. S. Hickey and P. C. Reist, *Am. Ind. Hyg. Assoc. J.*, **40**, 727–734 (1979).

33. I. Eide, *Ann. Occ. Hyg.*, **34**, 13–17 (1990).

34. S. Folkard, "Shiftwork and Performance," in *Biological Rhythms, Sleep and Shift Work*, L. C. Johnson, D. I. Tepas, W. P. Colquhoun, and M. Colligan, Eds., SP Medical and Scientific Publications, New York, 1981, 283–306.

35. B. Kolmodin-Hedman, *J. Human Ergol.*, **11**(Suppl), 447–456 (1982).

36. F. Brown and R. C. Graeber, Eds., *Rhythmic Aspects of Behavior*, LEA, New Jersey, 1982.

37. J. Rutenfranz, *J. Human Ergol.*, **11**(Suppl), 67–86 (1982).

38. J. Rutenfranz, M. Haider, and M. Koller, Occupational Health Measures for Nightworkers and Shiftworkers, in *Hours of Work: Temporal Factors in Work-Scheduling*, S. Folkard and T. H. Monk, Eds., Wiley, Chichester, 1985, 199–210.

39. R. Rosa, M. Bonnet, R. Bootzin, C. Eastman, T. Monk, P. Denn, D. Tepas, and J. Walsh, *Occ. Med.*, **5**, 391–415 (1990).

40. A. Reinberg and M. H. Smolensky, *J. Clin. Pharmacokinetics*, **7**, 401–420, 1982.

41. A. Reinberg and M. H. Smolensky, *Biological Rhythms and Medicine*, Springer-Verlag, New York, 1983.

42. J. P. McGovern, M. H. Smolensky, and A. Reinberg, Eds., *Chronobiology in Allergy and Immunology*, Thomas, Springfield, IL, 1977.

43. M. H. Smolensky, H. Reinberg, and J. P. McGovern, Eds., *Recent Advances in the Chronobiology of Allergy and Immunology*, Pergamon Press, Oxford, 1980, p. 37.

44. L. E. Scheving, *Endeavour*, **35**, 66–72 (1976).

45. L. E. Scheving, *Trends Pharmacological Sci.*, 303–307 (1980).

46. F. Halberg, *Amer. J. Anat.*, **168**, 543–594 (1983).

47. F. Halberg, E. Haus, S. S. Cardoso, L. E. Scheving, J. F. W. Kl, R. Shiotsuka, G. Rosene, J. P. Pauly, W. Runge, J. F. Spalding, J. K. Lee, and R. A. Good, *Experientia*, **29**, 909–1044 (1973).

48. E. Haus, F. Halberg, L. E. Scheving, and H. Simpson, *Int. J. Chronobiol.*, **6**, 67–107 (1979).

49. W. J. M. Hrushesky, *Am. J. Ant.*, **168**, 519–542 (1983).

50. L. C. Johnson, D. I. Tepas, W. P. Colquhoun, and M. Colligan, Eds., *Biological Rhythms, Sleep and Shift Work*, SP Medical and Scientific Publications, New York, 1982.

51. J. Rutenfranz and W. P. Colquhoun, The Scientific Committee on Shiftwork: A Short Account of Its History, Aims and Achievements, in *Night and Shift Work: Biological and Social Aspects*, A. Reinberg, N. Vieux, and P. Andlauer, Eds., Pergamon Press, Oxford, 1981, pp. 11–12.

52. B. Ramazzini, *De Morbus Artificum Diatriba*, W. C. Wright, Eds., University of Chicago, Chicago, 1940.

53. R. Coleman, *Wide Awake at 3 A.M. by Choice or by Chance*, Wiley, New York, 1986.

54. P. Knauth and J. Rutenfranz, *J. Human Ergol.*, **11**(Suppl), 337–367 (1982).

55. J. Rutenfranz, W. P. Colquhoun, P. Knauth, and J. N. Ghata, *Scand. J. Work, Environ., Health*, **3**, 165–192 (1977).

56. J. Rutenfranz, R. Knauth, and D. Angerbach, Shift Work Research Issues, in *Biological Rhythms, Sleep and Shift Work*, L. C. Johnson, D. I. Tepas, W. P. Colquhoun, and M. J. Colligan, Eds., SP Medical and Scientific Books, New York, 1981, 165–196.

57. L. C. Johnson, On Varying Work/Sleep Schedules: Issues and Perspectives as Seen by a Sleep Researcher, in *Biological Rhythms, Sleep and Shift Work*, L. C. Johnson, D. I. Tepas, W. P. Colquhoun, and M. J. Colligan, Eds., SP Medical and Scientific Books, New York, 1981, pp. 335–346.

58. P. G. Rentos and R. D. Shepard, Eds., *Shift Work and Health—A Symposium*, U.S. HEW PHS, National Institute of Occupational Safety and Health, Washington, D.C., 1976.

59. J. Wojtczak-Jaroszowa, *Physiological and Psychological Aspects of Night and Shift Work*, National Institute of Occupational Safety and Health, Cincinnati, 1977.

60. D. L. Tasto, M. J. Colligan, E. W. Skjei, and S. J. Polly, *Health Consequences of Shift Work*, U.S. DHEW PHS, National Institute of Safety and Health, Washington, D.C., 1978.

61. J. T. Wilson and K. M. Rose, *The Twelve-Hour Shift in the Petroleum and Chemical Industries of the United States and Canada: A Study of Current Experience*, Wharton Business School, University of Pennsylvania, Philadelphia, Pennsylvania, 1978.

62. G. D. Botzum and R. L. Lucas, Slide Shift Evaluation—Practical Look at Rapid Rotation Theory, in *Proceedings of the Human Factors Society*, 1981, 207–211.

63. T. A. Yoder and G. D. Botzum, The Long-Day Short-Week in Shift Work—A Human Factors Study, in *Proceedings of the Human Factors Society*, Human Factors Society, Santa Monica, Vol. 1, 1971.

64. E. D. Weitzmann and D. F. Kripke, Experimental 12-Hour Shifts of the Sleep-Wake Cycle in Man: Effects on Sleep and Physiologic Rhythms, in *Biological Rhythms, Sleep and Shift Work*, L. C. Johnson, D. I. Tepas, W. P. Colquhoun, and M. J. Colligan, Eds., SP Medical and Scientific Books, New York, 1982, pp. 93–110.

65. R. R. Rosa, D. D. Wheeler, J. S. Warm, and M. J. Colligan, *Behav. Res. Methods, Instrum. & Comput.*, **17**, 6–15 (1985).

66. S. A. Roach, *Am. Ind. Hyg. Assoc. J.*, **39**, 345–364 (1978).

67. Occupational Safety and Health Administration, Compliance Officers: Field Manual, Department of Labor, Washington, D.C., 1979.

68. H. Allenspach, *Flexible Working Hours*, International Labor Office, WHO, Geneva, Switzerland, 1975.

69. L. E. Scheving and J. E. Pauly, Chronopharmacology: Its Implication for Clinical Medicine, in *Annual Reports in Medicinal Chemistry*, F. H. Clarke, Ed., Academic Press, New York, Vol. 11, 1976, pp. 251–260.

70. F. Halberg, *Cold Spring Harbor Symp. Quant. Biol.*, **25**, 289–310 (1960).

71. F. Halberg, *Ann. Rev. Physiol.*, **31**, 675–725 (1969).

72. F. Halberg and W. Nelson, Chronobiologic Optimization of Aging, in *Aging and Biological Rhythms*, H. V. Samis, Jr., and S. Capobianco, Eds., Plenum Press, New York, 1978, pp. 5–56.

73. F. Halberg, M. Engeli, C. Hamburger, and D. Hillman, *Acta Endocr.*, **103**(Suppl), 5–54 (1965).

74. L. E. Scheving, F. Halberg, and J. E. Pauly, Eds., *Chronobiology*, Igaku Shoin Ltd., Tokyo, 1974.

75. A. Reinberg and J. Ghata, *Biological Rhythms*, Walker, New York, 1964.

76. M. C. Moore-Ede, F. M. Sulzman, and C. A. Fuller, *The Clocks That Time Us*, Harvard University Press, Cambridge, MA, 1982.

77. I. Assenmacher and D. S. Farner, Eds., *Environmental Endocrinology*, Springer-Verlag, Berlin, 1978.

78. G. Luce, *Biological Rhythms in Psychiatry and Medicine*, U.S. DHEW Publ. # (ADM) 78–247, 1970.

79. E. T. Pengelley, Ed., *Circannual Clocks*, Academic Press, New York, 1974.

80. G. Thommen, *Is This Your Day?* Universal Publishing and Distributing Corp., New York, 1969.

81. K. E. Klein and H. M. Wegmann, Appendix: Circadian Rhythms of Human Performance and Resistance: Operational Aspects, in *Sleep, Wakefulness and Circadian Rhythm*, A. N. Nicholson, Eds., AGARD Lecture Series No. 105, 1979, pp. 2:10–2:17.

82. G. Schholzer, G. Schilling, and H. Mler, *Schwerz Z. Sportmed.*, **1**, 7 (1972).

83. H. R. Willis, Biorhythm and Its Relationship to Human Error, Proc. 16th Ann. Meeting, Human Factor Society, Beverly Hills, Calif., 1973, pp. 274–282.

84. H. R. Willis, The Effect of Biorhythm Cycles. Implication for Industry, American Industrial Hygiene Association Conference, Miami Beach, FL., 1974; cited in T. M. Khalil and Ch.N. Kurveg, *Ergonomics*, **20**, 397 (1977).

85. F. Halberg, *Zeit Vitamin-Hormone—und Fermentforschung*, **10**, 225–296 (1959).

86. F. Halberg, Y. L. Tong, and E. A. Johnson, Circadian System Phase, An Aspect of Temporal Morphology: Procedures and Illustrative Examples, in *The Cellular Aspects of Biorhythms*, H. von Mayersbach, Ed., Springer-Verlag, Berlin, 1967, pp. 20–48.

87. J. Aschoff, *Naturwissenschaften*, **41**, 49–56 (1954).

88. C. S. Pittendrigh, *Cold Spring Harbor Symp. Quant. Biol.*, **25**, 159–184 (1960).

89. A. Reinberg and M. H. Smolensky, Secondary Rhythms Related to Hormonal Changes in the Menstrual Cycle: General Considerations, in *Biorhythms and Human Reproduction*, M. Ferin, F. Halberg, R. M. Richart, and R. L. Vande Wiele, Eds., Wiley, New York, 1974, pp. 241–258.

90. F. Halberg, M. Lagoguey, and A. Reinberg, *Inter. J. Chronobiol.*, **8**, 225–268 (1983).

91. L. E. Scheving, T. H. Tsai, and L. A. Scheving, *Amer. J. Anat.*, **168**, 433–465 (1983).

92. F. Halberg, A. Reinberg, and A. Reinberg, *Waking and Sleeping*, **1**, 259–279 (1977).

93. A. Reinberg, N. Vieux, J. Ghata, A.-J. Chaumont, and A. Laporte, Consideration of the Circadian Amplitude in Relationship to the Ability to Phase Shift Circadian Rhythms of Shift Workers, in *Chronobiological Field Studies of Oil Refinery Shift Workers*, A. Reinberg, Ed., *Chronobiologia*, **6**(Suppl), 57–63 (1979).

94. A. Reinberg, N. Vieux, P. Andlauer, and M. Smolensky, *Adv. Biol. Psychiatr.*, **11,** 35–47 (1983).

95. A. Reinberg, N. Vieux, J. Ghata, A. J. Chaumont, and A. Laporte, *Ergonomics,* **21,** 763–766 (1978).

96. A. Reinberg, N. Vieux, P. Andlauer, P. Guillet, A. Laporte, and A. Nicolai, Oral Temperature Circadian Rhythm Amplitude, Aging and Tolerance to Shift-Work (Study 3), in *Chronobiological Field Studies of Oil Refinery Shift Workers*, A. Reinberg, Eds., *Chronobiologia,* **6**(Suppl), 67–85 (1979).

97. A. Reinberg, P. Andlauer, P. Teinturier, J. DePrins, W. Malbecq, and J. Dupont, *C.R. Acad. Sc. (Paris),* **296,** 267–269 (1983).

98. C. R. Graeber, Alternations in Performance Following Rapid Transmeridian Flight, in *Rhythmic Aspects of Behavior*, F. M. Brown and R. C. Graeber, Eds., LEA, New Jersey, 1982, pp. 173–212.

99. G. A. Christie and M. Moore-Robinson, *Clin. Trials J.*, **7,** 45 (1970).

100. P. Knauth and J. Rutenfranz, Duration of Sleep Related to the Type of Shift Work, in *Night and Shift Work: Biological and Social Aspects*, A. Reinberg, N. Vieux, and P. Andlauer, Eds., Pergamon Press, Oxford, 1981, pp. 161–167.

101. K. E. Klein and H. M. Wegmann, The Effects of Transmeridian and Transequatorial Air Travel on Psychological Well-Being and Performance, in *Chronobiology: Principles and Applications to Shifts in Schedules*, NATO Advanced Study Institutes Series D: Behavioural and Social Sciences, No. 3, Sijthoff and Noordhoff, The Netherlands, 1980, pp. 339–352.

102. S. Folkard, Circadian Rhythms and Human Memory, in *Rhythmic Aspects of Behavior*, F. Brown and R. C. Graeber, Eds., LEA Publishers, New Jersey, 1982, pp. 241–272.

103. J. DeMarian, Observations Botaniques, *Hist. Acad. Roy. Sci. (Paris)*, 35–36 (1729).

104. L. E. Scheving, J. E. Pauly, H. von Mayersbach, and J. D. Dunn, *Acta Anat.*, **88,** 411–423 (1974).

105. L. E. Scheving, G. Sohal, C. Enna, and J. E. Pauly, *Anat. Rec.*, **175,** 1–6 (1973).

106. E. Haus, D. Lakatua, and F. Halberg, *Exp. Med. Surg.*, **25,** 7–45 (1967).

107. J. Ghata, F. Halberg, A. Reinberg, and M. Siffre, *Ann. Endocr. (Paris)*, **30,** 245–260 (1969).

108. R. Wever, *The Circadian System of Man. Results of Experiments under Temporal Isolation*, Springer-Verlag, New York, 1979.

109. A. Reinberg, Eclairment et cycle menstruel de la femme, in *La Photorulation chez les Oiseaux et les Mammifes*, J. Benoit and I. Assenmacher, Eds., Coll. Internat. C.N.R.S. Publ. #172, Paris, 1970, pp. 529–546.

110. A. Reinberg, F. Halberg, J. Ghata, and M. Siffre, *Compt. Rend. Acad. Sci.*, **262,** 782–785 (1966).

111. J. Aschoff, *Cold Spring Harbor Sym. Quant. Biol.* **25,** 11–28 (1960).

112. A. P. de Candolle, *Physiologie Vale*, Bhet Jeune, Paris, 1832.

113. M. Siffre, *Hors du Temps*, Tulliurd, Paris, 1963.

114. F. Halberg, A. Reinberg, E. Haus, J. Ghata, and M. Siffre, *Bull. Nat. Speleol. Soc.*, **32,** 89–115 (1970).

115. M. Siffre, A. Reinberg, F. Halberg, J. Ghata, G. Perdriel, and R. Slind, *Presse Med.*, **74,** 915–919 (1966).

116. W. Englemann, *Z. Naturf.*, **28c**, 733–736 (1973).

117. A. Wirz-Justice, M. S. Kalka, D. Naber and T. A. Wehr, Life Sci., **27**, 314–347 (1980).

118. F. Halberg, M. Siffre, M. Engeli, D. Hillman, and A. Reinberg, *Compt. Rend. Acad. Sci.*, **260**, 1259–1262 (1965).

119. J. N. Mills, *Trans. Brit. Cave Res. Assoc.*, **2**, 95 (1975).

120. C. A. Czeisler, Human Circadian Physiology: Internal Organization of Temperature, Sleep-Wake and Neuroendocrine Rhythms Monitored in an Environment Free from TimeCues, Ph.D. Dissertation, Stanford, 1978.

121. C. A. Czeisler, G. S. Richardson, J. C. Zimmerman, M. C. Moore-Ede, and E. D. Weitzman, *Photochem. Photobiol.*, **34**, 239–247 (1981).

122. L. E. Scheving and W. S. Kals, *Time and You*, Doubleday, New York, in press.

123. H.-G. Schweiger and M. Schweiger, *Int. Rev. Cytol.*, **51**, 315–342 (1977).

124. W. J. Reitveld and G. A. Gross, The Role of the Suprachiasmatic Nucleus: Afferents in the Central Regulation of Circadian Rhythms, in *Biological Rhythms in Structure and Function*, H. von Mayersbach, L. E. Scheving, and J. E. Pauly, Eds., A. R. Liss, New York, 1981, pp. 205–211.

125. K. Hoffmann, *Z. Vergl. Physiol.*, **37**, 253 (1955).

126. B. M. Sweeney, *Int. J. Chronobiol.*, **2**, 25–33 (1974).

127. D. Njus, F. M. Sulzman, and J. W. Hastings, *Nature*, **248**, 116–120 (1974).

128. R. D. Burgoyne, *Fed. Exp. Biol. Soc. Lett.*, **94**, 17–19 (1978).

129. H. von Mayersbach, *Arzneim.-orsch.*, **28**, 1824–1836 (1978).

130. L. N. Edmunds, Jr., *Am. J. Anat.*, **168**, 389–431 (1983).

131. F. A. Brown, The Exogenous Nature of Rhythms, in *Chronobiology: Principles and Applications to Shifts in Schedules*, L. E. Scheving and F. Halberg, Eds., NATO Advanced Studies Institutes Series D: Behavioural and Social Sciences, No. 3, Sijthoff and Noordhoff, The Netherlands, 1980, pp. 127–135.

132. H. M. Simpson, N. Bellamy, J. Bohlen, and F. Halberg, *Int. J. Chronobiol.*, **1**, 287–311 (1973).

133. H. W. Simpson, Chronobiotics: Selected Agents of Potential Values in Jet Lag and Other Dyschronisms, in *Chronobiology: Principles and Applications to Shifts in Schedules*, L. E. Scheving and F. Halberg, Eds., NATO Advanced Study Institutes Series D: Behavioural and Social Sciences, No. 3, Siithoff and Noordhoff, The Netherlands, 1980, pp. 433–446.

134. K. E. Klein and H. M. Wegmann, in *Sleep, Wakefulness and Circadian Rhythm*, A. N. Nicholson, Ed., AGARD Lecture Series No. 105, NATO, 1979, pp. 2:1–2:9.

135. B. J. Craft, The Effects of Circadian Rhythms on the Toxicological Response of Rats to Xenobiotics, Ph.D. Dissertation, University of Michigan, Ann Arbor, 1970.

136. P. Knauth and J. Rutenfranz, The Effects of Noise on the Sleep of Nightworkers, in: *Studies of Shiftwork*, W. P. Colquhoun and J. Rutenfranz, Eds., Taylor and Francis Ltd., London, 1980, pp. 111–120.

137. C. A. Czeisler, M. Moore-Ede, and R. M. Coleman, *Science*, **217**, 460–463 (1982).

138. M. H. Smolensky, The Conceptual Implications of Chronotoxicology and Chronopathology for Occupational Health and Shift Work, in *Chronobiology; Principles and Applications to Shifts in Schedules*, L. E. Scheving and F. Halberg, Eds., NATO Ad-

vanced Study Institutes Series D: Behavioural and Social Sciences, No. 3, Sijthoff and Noordhoff, The Netherlands, 1980, pp. 325–337.

139. D. Brown, *J. Human Ergol.*, **11**(Suppl), 475–482 (1982).

140. R. D. Novak, M. H. Smolensky, E. J. Fairchild, and R. Reves, *Chronobiol. Internat.*, **7**, 155–164 (1990).

141. F. Nachreiner and J. Rutenfranz, Sozialpsychologische, arbeitspsychologischeund medizinische Erhebungen in der chemischen Industrie, in *Schichtarbeit bei Kontinuierlicher Produktion*, F. Nachreiner et al., Eds., Wirtschaftsverlag Nordwest GmbH, Withelmshaven, 1975, pp. 83–117.

142. M. J. K. Blake, Treatment and Time of Day, in *Biological Rhythms and Human Performance*, W. P. Colquhoun, Ed., Academic Press, London, 1971, pp. 109–148.

143. W. P. Colquhoun and S. Folkard, *Ergonomics*, **21**, 811–817 (1978).

144. O. Oestberg, *Ergonomics*, **16**, 203–209 (1973).

145. C. Michel-Briand, J. L. Chopard, A. Guiot, M. Paulmeier, and G. Struder, The Pathological Consequences of Shift-Work in Retired Workers, in *Night and Shiftwork Studies: Biological and Social Aspects*, A. Reinberg, N. Vieux, and P. Andlauer, Eds., Pergamon Press, Oxford, 1981, pp. 399–407.

146. J. Aschoff, *Ergonomics*, **39**, 739–754 (1978).

147. A. Reinberg, N. Vieux, P. Andlauer, P. Guillet, and A. Nicolai, Tolerance to Shift-Work, Amplitude of Circadian Rhythms and Aging, in *Night and Shift-Work Studies: Biological and Social Aspects*, A. Reinberg, N. Vieux, and P. Andlauer, Eds., Pergamon Press, Oxford, 1981, pp. 341–354.

148. R. Leonard, Amplitude of the Temperature Circadian Rhythm and Tolerance to Shift Work, in *Night and Shiftwork Studies: Biological and Social Aspects*, A. Reinberg, N. Vieux, and P. Andlauer, Eds., Pergamon Press, Oxford, 1981, pp. 323–329.

149. T. Akerstedt and M. Gillberg, The Circadian Pattern of Unrestricted Sleep and Its Relation to Body Temperature, Hormones and Alertness, in *Biological Rhythms, Sleep and Shift Work*, L. C. Johnson, D. I. Tepas, W. P. Colquhoun, and M. Colligan, Eds., SP Medical and Scientific Books, New York, 1981, pp. 481–498.

150. J. Foret and O. Benoit, in *Chronobiological Field Studies of Oil Refinery Shift Workers*, A. Reinberg, Ed., *Chronobiologia*, **6**(Suppl), 45–53 (1979).

151. J. Foret and G. Lantin, The Sleep of Train Drivers: An Example of the Effects of Irregular Work Schedules on Sleep, in *Aspects of Human Efficiency: Diurnal Rhythms and Loss of Sleep*, W. P. Colquhoun, Ed., English University Press, London, 1972, pp. 273–282.

152. H. Fukuda, S. Endo, T. Yamamoto, Y. Saito, and K. Nishihara, *J. Human Ergol.*, **11**(Suppl), 245–257 (1982).

153. N. Kleitman, *Sleep and Wakefulness*, University of Chicago Press, Chicago, 1963.

154. K. Kogi, *J. Human Ergol.*, **11**(Suppl), 217–231 (1982).

155. D. F. Kripke, B. Cook, and O. F. Lewis, *Psychophysiology*, **7**, 377–384 (1971).

156. F. Lille, *Travail Humain*, **30**, 85 (1967).

157. D. Minors and J. M. Waterhouse, Anchor Sleep as a Synchronizer of Rhythms on Abnormal Routines, in *Biological Rhythms, Sleep and Shift Work*, L. C. Johnson, D. I. Tepas, W. P. Colquhoun, and M. Colligan, Eds., SP Medical and Scientific Books, New York, 1981, pp. 399–414.

158. D. I. Tepas, *J. Human Ergol.*, **11**(Suppl), 325–336 (1982).

159. D. I. Tepas, J. Walsh, and D. Armstrong, Comprehensive Study of the Sleep of Shift Workers, in *Biological Rhythms, Sleep and Shift Work*, L. C. Johnson, D. I. Tepas, W. P. Colquhoun, and M. Colligan, Eds., SP Medical and Scientific Books, New York, 1981, pp. 347–356.

160. D. I. Tepas, *J. Human Ergol.*, **11**(Suppl), 1–12 (1982).

161. S. Torii, N. Okudaria, H. Fukuda, H. Kanamoto, Y. Yamashiro, M. Akiya, K. Nomoto, N. Katayama, M. Hasegawa, M. Sato, M. Hatano, and H. Hemoto, *J. Human Ergol.*, **11**(Suppl), 233–244 (1982).

162. W. J. Price and D. C. Holley, *J. Human Ergol.*, **11**(Suppl), 291–301 (1982).

163. J. Walsh, D. I. Tepas, and P. Moss, The EEG Sleep of Night and Rotating Shift Workers, in *Biological Rhythms, Sleep and Shift Work*, L. C. Johnson, D. I. Tepas, W. P. Colquhoun, and M. Colligan, Eds., SP Medical and Scientific Books, New York, 1981, pp. 371–382.

164. W. B. Webb and H. W. Agnew, Jr., *Aviation, Space Environ. Med.*, **49**, 384–389 (1978).

165. G. S. Tune, *Brit. J. Ind. Med.*, **26**, 54–58 (1969).

166. J. Rutenfranz, P. Knauth, G. Hildebrandt, and W. Rohmert, *Int. Arch. Arbeitsmed.*, **32**, 243–259 (1974).

167. M. Freese and C. Harwich, *J. Occ. Med.*, **26**, 561–566 (1984).

168. D. I. Tepas and A. B. Carvalhis, *Occ. Med.*, **5**, 199–208 (1990).

169. M. Estryn-Behar, M. Kaminiski, E. Peigne, N. Bonnet, E. Vaichere, C. Gozlan, S. Azoulay, and M. Giorgi, *Br. J. Ind. Med.*, **47**, 20–28 (1990).

170. P. Naitoh, T. Kelley, and C. Englund, *Occ. Med.*, **5**, 209 (1990).

171. T. Åkerstedt, G. Kecklund, and A. Knutsson, *Scand. J. Work Environ. Hlth.*, **17**, 330–336 (1991).

172. T. Åkerstedt and G. Keklund, *Sleep*, **14**, 507–510 (1991).

173. R. R. Rosa, M. J. Colligan, and P. Lewis, *Work Stress*, **3**, 21–32 (1990).

174. R. R. Alward and T. H. Monk, *Int. J. Nurs. Stud.*, **27**, 297–302 (1990).

175. W. P. Colquhoun, Biological Rhythms and Performance, in *Biological Rhythms, Sleep and Performance*, W. B. Webb, Ed., Wiley, Chichester, 1982, pp. 59–86.

176. P. Lavie, *Chronobiologia*, **7**, 247 (1980).

177. N. Kleitman, *Sleep and Wakefulness*, University of Chicago, Chicago, 1963.

178. M. J. F. Blake, *Psychonom. Sci.*, **9**, 349 (1967).

179. S. Folkard, T. H. Monk, and M. C. Lobban, *Ergonomics*, **21**, 785–799 (1978).

180. W. P. Colquhoun, *Depart. Employment Gazette*, **86**, 682 (1978).

181. G. R. J. Hockey and W. P. Colquhoun, Diurnal Variation in Human Performance: a Review, in *Aspects of Human Efficiency: Diurnal Rhythm and Loss of Sleep*, W. P. Colquhoun, Ed., English Universities Press, London, 1971.

182. R. T. Wilkinson, The Relationship between Body Temperature and Performance Across Circadian Phase Shifts, in *Rhythmic Aspects of Behavior*, F. Brown and R. C. Graeber, Eds., LEA Publications, New Jersey, 1982, pp. 213–240.

183. W. P. Colquhoun, Circadian Variations in Mental Efficiency, in *Biological Rhythms and Human Performance*, W. P. Colquhoun, Ed., Academic Press, London, 1971, pp. 39–107.

184. M. I. Hma, J. Ilmarinen, and I. Yletyinen, *J. Human Ergol.*, **11**(Suppl), 33–46 (1982).

185. R. C. Browne, *Occupat. Psych.*, **21**, 121 (1949).

186. O. Prokop and L. Prokop, *D. Zeitschrift Gesamte Gerichtlichen Medizin*, **44**, 343 (1955).

187. G. Hildebrandt, W. Rohmert, and J. Rutenfranz, *Int. J. Chronobiol.*, **2**, 175–180 (1974).

188. B. Bjerner, A. Holm, and A. Swensson, *Br. J. Ind. Med.*, **12**, 103–110 (1955).

189. A. J. Tilley, R. T. Wilkinson, P. S. G. Warren, B. Watson, and M. Drud, *Hum. Factors*, December, 629–641 (1982).

190. R. C. Friedman, J. T. Bigger, and D. S. Kornfeld, *New Eng. J. Med.*, **285**, 201–203 (1971).

191. R. Rubin, P. Orris, S. L. Lau, D. O. Hryhorczuk, S. Furner, and R. Letz, *J. Occ. Med.*, **33**, 13–18 (1991).

192. T. F. Deaconson, D. P. O'Hair, M. F. Levy, M. B. F. Lee, A. L. Schueneman, and R. Condon, *JAMA*, **260**, 1721–1727 (1988).

193. W. L. Fields and C. Loverage, *Nurs. Econ.*, **6**, 189–191 (1990).

194. H. W. Heinrich, *Industrial Accident Prevention*, McGraw-Hill, New York, 1950.

195. M. H. Smolensky and A. Reinberg, *Chron. Intnl.*, **2**, 61 (1985).

196. M. Greenwood and H. M. Woods, The Incidence of Industrial Accidents upon Individuals with Special Reference to Multiple Accidents, Industrial Fatigue Research Report No. 4, H.R.M. Printing Service, London, 1919.

197. B. R. Laundry and R. E. M. Lees, *J. Occ. Med.*, **33**, 903–906 (1991).

198. J. Wotczak-Jarasozwa and D. Jarosz, *J. Saf. Res.*, **18**, 33–41 (1988).

199. S. H. Lee and K. S. Cho, *J. Hum. Ergol.*, **11**(Suppl.), 87 (1984).

200. L. Levin, J. Oler, and J. R. Whiteside, *Accident Anal. Preven.*, **17**, 67 (1985).

201. J. Wagner, Time of Day Variations in the Severity of Injuries Suffered by Mine Workers, in *Proceedings of the Human Factors Society*, 32nd Annual Meeting, 1988, pp. 608–611.

202. R. D. Novak and M. H. Smolensky, *Prog. Clin. Biol. Res.*, **341B**, 355–362 (1991).

203. J. Pimble and S. O'Toole, *Ergonomics*, **5**, 967–979 (1982).

204. B. Barhard and M. Pafnotte, *Trav. Hum.*, **33**, 1–19 (1970).

205. D. L. Bosworth and P. J. Dawkins, Private and Social Costs and Benefits of Shift and Night Work, in *Night and Shift Work: Biological and Social Aspects*, A. Reinberg, N. Vieux, and P. Andlauer, Eds., Pergamon Press, Oxford, 1981, pp. 207–214.

206. C. Gadbois, Women on Night Shift: Interdependence of Sleep and Off-the-Job Activities, in *Night and Shift Work: Biological and Social Aspects*, A. Reinberg, N. Vieux, and P. Andlauer, Eds., Pergamon Press, Oxford, 1981, pp. 223–228.

207. M. Kundi, M. Koller, R. Cervinka, and M. Hiader, Job Satisfaction in Shift Workers and Its Relation to Family Situation and Health, in *Night and Shift Work: Biological and Social Aspects*, A. Reinberg, N. Vieux, and P. Andlauer, Eds., Pergamon Press, Oxford, 1981, pp. 237–245.

208. M. Maurice, Shiftwork: Economic Advantages and Social Costs, International Labor Office, WHO, Geneva, 1975.

209. P. E. Mott, F. C. Mann, Q. McLoughlin, and D. P. Warwick, *Shift Work: The Social, Psychological and Physical Consequences*, University of Michigan Press, Ann Arbor, 1965.

210. A. A. I. Wedderburn, *Occup. Psychol.*, **41**, 85–107 (1967).

211. A. A. I. Wedderburn, How Important Are the Social Effects of Shiftwork? in *Biological Rhythms, Sleep and Shift Work*, L. C. Johnson, D. I. Tepas, W. P. Colquhoun, and M. Colligan, Eds., SP Medical and Scientific Books, New York, 1981, pp. 257–269.

212. H. Thierry and B. Jansen, *J. Human Ergol.*, **11**(Suppl.), 483–498 (1982).

213. M. J. Colligan and D. Tepas, *Am. Ind. Hyg. Assoc. J.*, **47**, 686–695 (1986).

214. M. Freese and N. Semmer, *Ergonomics*, **29**, 99–114 (1984).

215. P. Bohle and A. J. Tilley, *Ergonomics*, **32**, 1089–1099 (1989).

216. D. Milne and F. Watkins, *Int. J. Nurs. Stud.*, **23**, 139–146 (1986).

217. M. Colligan and R. R. Rosa, *Occ. Med.*, **5**, 315–322 (1990).

218. T. Äkerstedt and L. Torsvall, *Ergonomics*, **21**, 849–856 (1978).

219. P. J. Taylor, *Br. J. Ind. Med.*, **24**, 93–102 (1967).

220. P. J. Taylor and S. J. Pocock, *Br. J. Ind. Med.*, **29**, 201–207 (1972).

221. J. M. Harrington, *Shift Work and Health: Critical Review of the Literature*, Her Majesty's Stationery Office, London, 1978.

222. M. Koller, M. Kundi, and R. Cervinka, *Ergonomics*, **21**, 835–847 (1978).

223. D. Angersbach, P. Knauth, H. Loskant, M. J. Karvonen, K. Undeutsch, and J. Rutenfranz, *Int. Arch. Occup. Environ. Hlth.*, **45**, 127–140 (1980).

224. R. Doll and F. A. Jones, Occupational Factors in the Aetiology of Gastric and Duodenal Ulcers, Medical Research Council, Special Report Series No. 276, His Majesty's Stationery Office, London, 1951.

225. B. Tarquini, M. Cecchettin, and A. Cariddi, *Int. Arch. Occup. Environ. Hlth.*, **58**, 99–103 (1986).

226. A. Knuttson, *Scand. J. Soc. Med.*, **1**, 5–36 (1989).

227. T. Uehata and N. Saskawata, *J. Hum. Ergol.*, **11**, 465–474 (1982).

228. G. Axelson, C. Lutz, and R. Rylander, *Br. J. Ind. Med.*, **41**, 305–312 (1984).

229. A. D. McDonald, J. C. McDonald, and B. Armstrong, *Br. J. Ind. Med.*, **45**, 148–157 (1988).

230. B. G. Armstrong, A. D. Nolin, and A. D. McDonald, *Br. J. Ind. Med.*, **45**, 56–62 (1989).

231. G. Axelson, R. Rylander, and I. Molin, *Br. J. Ind. Med.*, **46**, 393–398 (1989).

232. N. Mamelle, B. Laumon, and P. Lazar, *Am. J. Epidemiol.*, **119**, 309–322 (1984).

233. T. Nurmeinen, *Scand. J. Work Environ. Hlth.*, **15**, 395–403 (1989).

234. C. F. Ehret, K. R. Groh, and J. C. Meinert, Circadian Dyschronism and Chronotypic Ecophilia as Factors in Aging and Longevity, in *Aging and Biological Rhythms*, H. V. Samis and S. Capobianco, Eds., Plenum Press, New York, 1978, pp. 185–214.

235. C. F. Ehret, V. R. Potter, and K. W. Dobra, *Science*, **188**, 1212–1215 (1975).

236. C. F. Ehret and V. R. Potter, *Int. J. Chronobiol.*, **2**, 321–325 (1974).

237. A. L. Cahill and C. F. Ehret, *J. Neurochemistry*, **37**, 1109–1115 (1981).

238. A. L. Cahill and C. F. Ehret, *Am. J. Physiol.*, **243**, R218 (1982).

239. N. D. Horseman and C. F. Ehret, *Am. J. Physiol.*, **243**, R373 (1982).

240. C. F. Ehret and L. W. Scanlon, *Overcoming Jet Lag*, Berkley Books, New York, 1983.

241. R. C. Graeber, H. C. Sing, and B. N. Cuthbert, The Impact of Transmeridian Flight on Deploying Soldiers, in *Biological Rhythms, Sleep and Shift Work*, L. C. Johnson, D. I. Tepas, W. P. Colquhoun, and M. J. Colligan, Eds., SP Medical and Scientific Books, New York, 1981, pp. 513–537.

242. A. Reinberg, Chronobiology and Nutrition, in *Biological Rhythms and Medicine*, A. Reinberg and M. H. Smolensky, Springer-Verlag, New York, 1983, pp. 266–300.

243. J. F. K. Kuhl, E. Haus, F. Halberg, L. E. Scheving, J. E. Pauly, S. Cardoso, and G. Rosene, *Chronobiologia*, **1**, 316–317 (1974).

244. A. Reinberg, C. Migraine, M. Apfelbaum, L. Brigant, J. Ghata, N. Vieux, A. Laporte, and A. Nicolai, in *Chronobiological Field Studies of Oil Refinery Shift Workers*, A. Reinberg, Ed., *Chronobiologia*, **6**(Suppl), 89–102 (1979).

245. M. Gilberg, *Biol. Psychol.*, **19**, 45–54 (1984).

246. A. S. Rogers, M. B. Spencer, B. M. Stone, and A. N. Nicholson, *Ergonomics*, **32**, 1193–1205 (1989).

247. M. I. Harma, J. Ilmarinen, P. Knauth, J. Rutenfranz, and O. Hanninen, *Ergonomics*, **31**, 39–50 (1988).

248. M. I. Harma, J. Ilmarinen, P. Knauth, J. Rutenfranz, and O. Hanninen, *Ergonomics*, **31**, 51–63 (1988).

249. R. Moog and G. Hildebrandt, Effects of Physical Training on Adaptation to Night Work in *Contemporary Advances in Shiftwork Research: Theoretical and Practical Aspects in the Late Eighties*, A. Oginski, J. Pokorski, and J. Rutenfranz, Eds., Krackow Medical Academy, Krackow, 1987, pp. 81–85.

250. A. J. Lewy, T. A. Wehr, F. K. L. Goodwin, D. K. Newsome, and S. P. Markey, *Science*, **210**, 1267–1269 (1980).

251. R. J. Cole, R. T. Loving, and D. F. Kripke, *Occup. Med.*, **5**, 301–314 (1990).

252. C. I. Eastman and K. J. Miescke, *Am. J. Physiol.*, **259** (6 Pt 2), R1189–1197 (1990).

253. C. A. Czeisler, M. P. Johnson, J. F. Duffy, E. M. Brown, J. M. Ronda, and R. E. Kronauer, *N. Engl. J. Med.*, **322**, 1253–1259 (1990).

254. D. Dawson and S. S. Campbell, *Sleep*, **14**, 511–516 (1991).

255. P. Knauth and M. Harma, *Chronobiol. Intl.*, **9**, 46–54 (1992).

256. D. F. Kripke, D. J. Mullaney, M. Atkinson, and S. Wolf, *Biol. Psychiat.*, **13**, 335–351 (1978).

257. T. A. Wehr, A. Wirz-Justice, F. K. Goodwin, W. Duncan, and J. C. Gillin, *Science*, **206**, 710 (1979).

258. T. A. Wehr, A. Wirz-Justice, and F. K. Goodwin, Advanced Circadian Rhythms and a Sleep-Sensitive Switch Mechanism in Depression, in *Circadian Rhythms in Psychiatry*, F. K. Goodwin and T. A. Wehr, Eds., Boxwood Press, Los Angeles, 1982.

259. E. Haus, D. J. Lakatua, J. Swoyer, and L. Sackett-Lundeen, *Am. J. Anat.*, **168**, 469–517 (1983).

260. E. Haus, D. J. Lakatua, L. Sackett-Lundeen, and J. Swoyer, Chronobiology in Laboratory Medicine, in *Clinical Aspects of Chronobiology*, W. Reitveld, (Ed.), Madition in Buaen, 1984.

261. R. E. Lee, M. H. Smolensky, C. S. Leach, and J. P. McGovern, *Ann. Allergy*, **38**, 231–236 (1977).

262. A. Reinberg, E. Sidi, and J. Ghata, *J. Allergy*, **36**, 273–283 (1965).

263. A. Reinberg, *Perspect. Biol. Med.*, **11**, 111–128 (1968).

264. M. H. Smolensky, Aspects of Human Chronopathology, in *Biological Rhythms in Medicine*, A. Reinberg and M. H. Smolensky, Eds., Springer-Verlag, New York, 1983, pp. 131–209.

265. M. H. Smolensky, A. Reinberg, and J. Queng, *Ann. Allergy*, **47**, 234–252 (1981).

266. M. H. Smolensky, A. Reinberg, R. J. Prevost, J. P. McGovern, and P. Gervais, The Application of Chronobiological Findings and Methods to the Epidemiological Investigations of the Health Effects of Air Pollutants on Sentinel Patients, in *Recent Advances in Chronobiology of Allergy and Immunology*, M. H. Smolensky, A. Reinberg, and J. P. McGovern, Eds., Pergamon Press, New York, 1980, pp. 211–236.

267. A. Reinberg and F. Halberg, *Ann. Rev. Pharmacol.*, **11**, 455–492 (1971).

268. R. Takahashi, F. Halberg, and C. A. Walker, Eds., *Toward Chronopharmacology*, Pergamon Press, New York, 1982.

269. M. H. Smolensky and A. Reinberg, *Nurs. Clin. N. Am.*, **11**, 609–620 (1976).

270. I. C. Kowanko, R. Pownall, M. S. Knapp, A. J. Swannell, and P. G. C. Mahoney, *Br. J. Clin. Pharmacol.*, **11**, 477–484 (1981).

271. F. Levi, C. LeLouarn, and A. Reinberg, *Ann. Rev. Chronopharmacol.*, **1**, 345–348, (1984).

272. V. Rejholec, V. Vitulova, J. Vachtenheim, N. Pickvance, and R. Pownall, *Ann. Rev. Chronopharmacol.*, 357–360 (1984).

273. J. Marshall, *Stroke*, **8**, 230–231 (1977).

274. A. Kuroiwa, *Jpn. Circ. J.*, **42**, 459–476 (1978).

275. A. Reinberg, P. Guillet, P. Gervais, J. Ghata, D. Vignaud, and C. Abulker, *Chronobiologia*, **4**, 295–312 (1977).

276. F. Halberg and R. B. Howard, *Postgrad. Med.*, **24**, 349–358 (1958).

277. J. Pepys and R. J. Davies, Occupational Asthma, in *Allergy, Principles and Practice*, E. Middleton, C. E. Reed, and E. F. Ellis, Eds., Mosby, New York, 1978, pp. 812–842.

278. A. N. Taylor, R. J. Davies, D. J. Hendrick, and J. Pepys, *Clin. Allergy*, **9**, 213–219 (1979).

279. B. Gandevia and J. Milne, *Br. J. Ind. Med.*, **27**, 235–244 (1970).

280. A. Siracusa, F. Curradi, and G. Abbritti, *Clin. Allergy*, **8**, 195–201 (1978).

281. J. Laidlaw, *Lancet*, **2**, 1235–1237 (1956).

282. B. Lemmer, Chronopharmacokinetics, in *Topics in Pharmaceutical Sciences*, D. O. Breimer and E. Speiser, Eds., Elsevier/North-Holland Biomedicine Press, 1981, pp. 49–68.

283. B. Lemmer, *Chronopharmakologie. Tagesrhythmen und Arzneimittel-Wirkung*, Wissenschaft Verlagsgeselschft. MbH, Stuttgart, 1983.

284. A. Reinberg, M. H. Smolensky, and G. Labrecque, Eds., *Annual Review of Chronopharmacology*, Pergamon Press, New York, 1984.

285. A. Reinberg, Clinical Chronopharmacology: An Experimental Basis for Chronotherapy, in *Biological Rhythms and Medicine*, A. Reinberg and M. H. Smolensky, Springer-Verlag, New York, 1983, pp. 211–263.

286. J. V. Bruckner, R. Luthra, G. M. Kyle, S. Muralidhara, R. Ramanathan, and D. Acosta, *Ann. Rev. Chronopharm.*, **1**, 373–376 (1984).

287. J. G. Lavigne, P. M. Belanger, F. Dore, and G. Labrecque, *Toxicology*, **26**, 267–273 (1983).

288. J. Clench, A. Reinberg, Z. Dziewanowska, J. Ghata, and M. H. Smolensky, *Eur. J. Clin. Pharmacol.*, **20**, 359–369 (1981).

289. A. Markiewicz, K. Semenowicz, J. Korczynska, and H. Boldys, Temporal Variations

in the Response of Ventilatory and Circulatory Functions to Propranolol in Healthy Man, in *Recent Advances in the Chronobiology of Allergy and Immunology*, M. H. Smolensky, A. Reinberg, and J. P. McGovern, Eds., Pergamon Press, New York, 1980, pp. 185–193.

290. A. Reinberg, J. Clench, J. Ghata, F. Halberg, C. Abulker, J. Depont, and Z. Zagula-Mally, *C.R. Acad. Sci.*, **280**, 1697–1700 (1975).

291. L. Carosella, P. DiNardo, R. Bernabei, A. Cocchi, and P. Carbonin, Chronopharma-cokinetics of Digitalis: Circadian Variations of Beta-Methyl-Digoxin Serum Levels after Oral Administration, in *Chronopharmacology*, A. Reinberg and F. Halberg, Eds., Pergamon Press, New York, 1980, pp. 125–134.

292. A. Reinberg and M. Reinberg, *Naunyn Schmiedebergs Arch. Pharmacol.*, **297**, 149–159 (1977).

293. K. DeVries, J. T. Goei, H. Booy-Noord, and N. G. M. Orie, *Int. Arch. Allergy*, **20**, 91–101 (1962).

294. G. J. Tammeling, K. DeVries, and E. W. Kruyt, Circadian Pattern of Bronchial Reac-tivity to Histamine in Healthy Subjects and in Patients with Obstructive Lung Disease, in *Chronobiology in Allergy and Immunology*, J. P. McGovern, M. H. Smolensky, and A. Reinberg, Eds., C. Thomas, Springfield, IL, 1977, pp. 139–149.

295. A. Reinberg, J. Clench, N. Aymard, M. Gaillot, R. Bourdon, P. Gervais, C. Abulker, and J. Dupont, *C.R. Acad. Sci.*, **278**, 1503–1505 (1974).

296. A. Reinberg, J. Clench, N. Aymard, M. Gaillot, R. Bourdon, P. Gervais, C. Abulker, and J. Dupont, *J. Physiol. (Paris)*, **70**, 435–456 (1975).

297. L. E. Scheving, Circadian Rhythms in Cell Proliferation: Their Importance when In-vestigating the Basic Mechanism of Normal Versus Abnormal Growth, in *Biological Rhythms in Structure and Function*, H. von Mayersbach, L. E. Scheving, and J. Pauly, Eds., A. R. Liss, New York, 1981, pp. 39–79.

298. L. E. Scheving, H. von Mayersbach, and J. E. Pauly, *Eur. J. Toxicol.*, **7**, 203–227 (1974).

299. L. E. Scheving, D. F. Vedral, and J. E. Pauly, *Nature*, **219**, 612–622 (1968).

300. J. Fisch, A. Yonovitz, and M. H. Smolensky, *Ann. Rev. Pharmacol.*, **1**, 385–388 (1984).

301. H. Bafitis, M. H. Smolensky, B. Hsi, S. Mahoney, and H. Kresse, *J. Pharmacol. Toxicol.*, **11**, 251–258 (1978).

302. N. K. Synder, M. H. Smolensky, and B. P. Hsi, *Chronobiologia*, **8**, 33–44 (1980).

303. G. M. Kyle, M. H. Smolensky, and J. P. McGovern, Circadian Variation in the Susceptibility of Rodents to the Toxicity Effects of Theophylline, in *Chronopharma-cology*, A. Reinberg and F. Halberg, Eds., Pergamon Press, Oxford, 1979, pp. 239–244.

304. T. H. Tsai, L. E. Scheving, and J. E. Pauly, *Jpn. J. Physiol.*, **20**, 12–29 (1970).

305. T. H. Tsai, L. E. Scheving, and J. E. Pauly, Circadian Variation in Host Susceptibility to Mercuric Chloride and Paraquat in Balb/Cann Female Mice, in *Toward Chrono-pharmacology*, R. Takahashi, F. Halberg, and C. A. Walker, Eds., Pergamon Press, New York, 1982, pp. 249–255.

306. H. von Mayersbach, An Overview of the Chronobiology of Cellular Morphology, in A. Reinberg and M. H. Smolensky, Springer-Verlag, New York, 1983, pp. 47–78.

307. L. E. Scheving, Chronotoxicology in General and Experimental Chronotherapeutics of

Cancer, in *Chronobiology: Principles and Applications to Shifts in Schedules*, L. E. Scheving and F. Halberg, Eds., NATO Adv. Study Inst. Series D: Behavioral and Social Sciences, No. 3, Sijthoff and Noordhoff, The Netherlands, 1980, pp. 455–480.

308. D. L. Clayton, A. W. McMullen, and C. C. Barnett, *Chronobiologia*, **2**, 210–217 (1975).

309. I. Sauerbier, Circadian System and Teratogenicity, in *Progress in Clinical and Biomedical Research*, H. von Mayersbach, L. E. Scheving, and J. E. Pauly, Eds., A. R. Liss, New York, 1980, pp. 143–149.

310. A. P. Chaudhry and F. Halberg, *J. Dent. Res.*, **39**, 704 (1960).

311. F. Halberg, *Mkurse. Arztl. Forbild.*, **14**, 67 (1964).

312. O. H. Iversen and S. L. Kauffman, *Int. J. Chronobiol.*, **8**, 95–104 (1982).

313. V. Nair, Circadian Rhythm in Drug Action: A Pharmacologic, Biochemical and Electromicroscopic Study, in *Chronobiology*, L. E. Scheving, F. Halberg, and J. E. Pauly, Eds., Igaku Shoin, Tokyo, 1974, pp. 182–186.

314. C. North, R. J. Feuers, L. E. Scheving, J. E. Pauly, T. H. Tsai, and D. A. Casciano, *Am. J. Anat.*, **162**, 183–199 (1981).

315. R. J. Feuers, L. A. Scheving, R. R. Delongchamp, T. H. Tsai, D. A. Casciano, J. E. Pauly, and L. E. Scheving, *Chronobiologia*, **10**, 125–126 (1983).

316. L. A. Scheving, R. J. Feuers, L. E. Scheving, R. R. DeLongchamp, T. H. Tsai, D. A. Casciano, and J. E. Pauly, *Chronobiologia*, **10**, 155–156 (1983).

317. E. Haus, F. Halberg, L. E. Scheving, S. Cardoso, A. Kl, R. Sothern, R. Shiotsuka, D. S. Hwang, and J. E. Pauly, *Science*, **177**, 80–82 (1972).

318. F. Li, W. Hrushesky, E. Haus, F. Halberg, L. E. Scheving, and B. J. Kennedy, Experimental Chronooncology, in *Chronobiology: Principles and Applications to Shifts in Schedules*, L. E. Scheving and F. Halberg, Eds., NATO Adv. Study Inst. Series D: Behavioral and Social Sciences, No. 3, Sijthoff and Noordhoff, The Netherlands, 1980, pp. 481–512.

319. E. Haus, G. Fernandes, J. Kl, E. J. Yunis, J. K. Lee, and F. Halberg, *Chronobiologia*, **3**, 270–277 (1974).

320. L. E. Scheving, S. S. Cardoso, J. E. Pauly, F. Halberg, and E. Haus, Variations in Susceptibility of Mice to the Carcinostatic Agent Arabinosyl Cytosine, in *Chronobiology*, L. E. Scheving, F. Halberg, and J. E. Pauly, Eds., Igaku Shoin Ltd., Tokyo, 1974, pp. 213–217.

321. S. S. Cardoso, T. Avery, J. M. Venditti, and A. Goldin, *Europ. J. Can.*, **14**, 949–954 (1978).

322. L. E. Scheving, J. E. Pauly, T. H. Tsai, and L. A. Scheving, Chronobiology of Cellular Proliferation: Implications for Cancer Chemotherapy, in *Biological Rhythms and Medicine*, A. Reinberg and M. H. Smolensky, Springer-Verlag, New York, 1983, pp. 79–130.

323. L. E. Scheving, E. R. Burns, J. E. Pauly, and F. Halberg, *Can. Res.*, **40**, 1511 (1980).

324. G. Haus, F. Halberg, M. K. Loken, and Y. S. Kim, Circadian Rhythmometry of Mammalian Radiosensitivity, in *Space Radiation Biology and Related Topics*, C. A. Tobias and P. Todd, Eds., Academic Press, New York, 1974, pp. 435–474.

325. L. E. Scheving and J. E. Pauly, Several Problems Associated with the Conduct of Chronobiological Research, in *Die Zeit und das Leben*, J. H. Scharf and H. von Mayersbach, Eds., *Nova Acta Leopoldina*, **46**, 237–258 (1977).

326. J. P. McGovern, M. H. Smolensky, and A. Reinberg, Circadian and Circamensual

Rhythmicity in Cutaneous Reactivity to Histamine and Allergenic Extracts, in *Chronobiology in Allergy and Immunology*, J. P. McGovern, M. H. Smolensky, and A. Reinberg, Eds., C. Thomas, Springfield, IL, 1977, pp. 76–116.

327. S. Flannigan, Cutaneous Reactivity to Contact Irritants, Master's Thesis, The University of Texas School of Public Health, Houston, 1981.

328. P. Gervais, A. Reinberg, C. Gervais, M. H. Smolensky, and O. DeFrance, *J. Allergy Clin. Immunol.*, **59**, 207–213 (1977).

329. E. Haus and F. Halberg, Endocrine Rhythms, in *Chronobiology: Principles and Applications to Shifts in Schedules*, L. E. Scheving and F. Halberg, Eds., *NATO Adv. Study Inst. Series D: Behavioural and Social Sciences*, No. 3, Sijthoff and Noordhoff, The Netherlands, 1980, pp. 137–188.

330. F. Ungar and F. Halberg, *Science*, **137**, 1058 (1962).

331. A. Reinberg, W. Dupont, Y. Touitou, M. Lagoguey, P. Bourgeois, C. Touitou, G. Murianx, D. Przyrowsky, S. Guillemant, J. Guillemant, L. Briere, and B. Zean, *Chronobiologia*, **8**, 11–31 (1981).

332. A. Reinberg, S. Guillemant, N. J. Ghata, J. Guillemant, Y. Touitou, W. Dupont, M. Lagoguey, P. Bourgeois, L. Briere, G. Fraboulet, and P. Guillet, *Chronobiologia*, **7**, 513–523 (1980).

333. F. Halberg, E. A. Johnson, W. Nelson, W. Runge, and R. Sothern, *Physiol. Teacher*, **1**, 1–11 (1972).

334. P. Gervais, A. Reinberg, C. Fraboulet, C. Abulker, O. Vignaud, and M. E. R. Delcourt, Circadian Changes in Peak Expiratory Flow of Subjects Suffering from Allergic Asthma Documented Both in Areas of High and Low Air Pollution, in *Chronopharmacology*, A. Reinberg and F. Halberg, Eds., Pergamon Press, Oxford, 1979, pp. 203–212.

335. D. R. Glasgow, L. E. Scheving, J. E. Pauly, and J. A. Bruce, *J. Arkansas Med. Soc.*, **79**, 81–91 (1982).

336. A. Reinberg and M. H. Smolensky, Investigative Methodology for Chronobiology, in *Biological Rhythms and Medicine*, A. Reinberg and M. H. Smolensky, Eds., Springer-Verlag, New York, 1983, pp. 23–46.

337. A. Reinberg, N. Vieux, A.-J. Chaumont, A. Laporte, M. Smolensky, A. Nicolai, C. Abulker, and J. Dupont, Aims and Conditions of Shift Work Studies, in *Chronobiological Field Studies of Oil Refinery Shift Workers*, A. Reinberg, Ed., *Chronobiologia*, **6**(Suppl. 1), 7–23 (1979).

338. A. Schuster, *Trans. Cambridge Phil. Soc.*, **18**, 107–135 (1900).

339. F. Koehler, F. K. Okano, L. R. Elbeback, R. Halberg, and J. J. Bittner, *Exp. Med. Surg.*, **14**, 5–30 (1956).

340. J. W. Tukey, The Sampling Theory of Power Spectrum Estimates, in *Proceedings of Symposium on Applications of Autocorrelation Analysis to Physical Problems*, Woods Hole, Massachusetts, Office of Naval Research, Washington, D.C., 1949, pp. 46–67.

341. F. Halberg and H. Panofsky, *Exp. Med. Surg.*, **19**, 284–309 (1961).

342. F. Halberg, H. Panofsky, and H. Mantis, *Ann. N.Y. Acad. Sci.*, **117**, 254–270 (1964).

343. J. DePrins and G. Cornelissen, Methods Workshop, in *Chronobiology: Principles and Applications to Shifts in Schedules*, L. E. Scheving and F. Halberg, Eds., *NATO Advanced Study Institutes Series D: Behavioural and Social Sciences*, No. 3, Sijthoff and Noordhoff, The Netherlands, 1980, pp. 249–260.

344. J. D. Veldhuis and M. L. Johnson, *Amer. J. Physiol.*, **250**, E486–493 (1986).

345. F. Wegner and H. Nau, *Reprod. Toxicol.*, **5**(6), 465–471 (1991).

346. P. M. Bélanger, Chronopharmacology in Drug Research and Therapy, *Adv. Drug Res.*, **24**, 1–80 (1993).

347. M. Desgagne and P. M. Bélanger, *Annu. Rev. Chronopharmacol.*, **3**, 103–106 (1986).

348. M. Desgagne, M. Boutet, and P. M. Bélanger, *Annu. Rev. Chronopharmacol.*, **5**, 235–238 (1988).

349. Y. Motohashi, T. Kawakami, Y. Miyazaki, T. Takano, and W. Ekataskin, *Toxicol. Appl. Pharmacol.*, **104**, 139–148 (1990).

350. I. Harabuchi, R. Kishi, T. Ikeda, H. Kiyosawa, and H. Miyake, *Brit. J. Indus. Med.*, **50**, 280–286 (1993).

351. L. VanBree, M. Marra, and P. J. A. Rombout, *Toxicol. Appl. Pharmacol.*, **116**, 209–216 (1992).

352. A. M. Attia, R. J. Reiter, K. O. Nokana, M. H. Mostafa, S. A. Soliman, and A. H. El-Sebae, *Toxicology*, **65**, 305–314 (1990).

353. R. Pohjanvirta, J. Tuomisto, J. Linden, and J. Laitinen, *Pharmacol. Toxicol.*, **65**, 239–240 (1989).

354. J. Linden, R. Pohjanvirta, T. Rahko, and J. Tuomisto, *Pharmacol. Toxicol.*, **69**, 427–432 (1991).

355. H. Matsumura, R. Takahata, and O. Hayaishi, *Proc. Natl. Acad. Sci.*, **88**, 9046–9050 (1991).

356. C. I. Eastman, *Persp. Biol. Med.*, **34**, 181–195 (1991).

357. L. C. Gallo and C. I. Eastman, *Physiol. Behav.*, **53**, 119–126 (1993).

358. E. F. Mellor, *Monthly Labor Rev.*, **109**, 14–21 (1986).

359. S. J. Smith, *Monthly Labor Rev.*, **109**, 7–13 (1986).

360. F. Nachreiner and J. Rutenfranz, Sozialpsychologische, arbeitspsychologischeund medizinische Erhebungen in der chemischen Industrie, in *Schichtarbeit bei kontinuierlicher Produktion*, F. Nachreiner et al., Eds., Wirtschaftsverlag Nordwest GmbH, Withelmshaven, 1975, pp. 83–117.

361. D. G. Hof, J. D. Dexter, and C. E. Mengel, *Aerospace Med.*, **42**, 1293–1296 (1971).

362. *Employment Report.*, United States Bureau of Labor Statistics, United States Government Printing Office, 1982.

363. E. Haus, Biological Aspects of a Chronopathology, Ph.D. Dissertation, University of Minnesota College of Medicine, 1970.

364. L. E. Scheving, D. F. Vedral, J. E. Pauly, *Nature*, **219**, 612–622 (1968).

365. O. Müller, Circadian Rhythmicity and Response to Barbiturates, in *Chronobiology*, L. E. Scheving, F. Halberg, and J. E. Pauley Eds., Igaku Shoin, Tokyo, 1974, pp. 187–190.

366. D. L. Clayton, A. W. McMullen, and C. C. Barnett, *Chronobiologia*, **2**, 210–217 (1975).

367. F. Lévi, W. Hrushesky, E. Haus, F. Halberg, L. E. Scheving, and B. J. Kennedy, Experimental Chrono-oncology, in *Chronobiology: Principles and Applications to Shifts and Schedules*, L. E. Scheving and F. Halberg, Eds., NATO Advanced Study Institutes Series D: Behavioral and Social Sciences, No. 3, Sijthoff and Noordhoff, The Netherlands, 1980, pp. 481–512.

368. J. F. K. Kühl, E. Haus, F. Halberg, L. E. Scheving, J. E. Pauly, S. Cardoso, and G. Rosene, *Chronobiologia*, **1**, 316–317 (1974).

CHAPTER TEN

Applied Ergonomics

Katharyn A. Grant, Ph.D., Vernon Putz-Anderson, Ph.D., and Alexander Cohen, Ph.D.

1 INTRODUCTION

During the past 10 years, ergonomics has grown and matured as a research discipline and become a subject of major interest in the field of occupational health and safety. This trend is reflected in the ever-increasing amount of ergonomics-related information now available in books and journal reports. Publication of the *Ergonomics Abstracts* has grown from four issues per year to six issues. At least four new scientific publications* have been inaugurated since 1986. Membership in ergonomics professional societies has more than doubled in the last decade (1). Both labor and management groups, recognizing the role of ergonomics in improved job design have begun to include ergonomic information in their safety and health programs (2). The Occupational Safety and Health Administration (OSHA) has issued guidelines for controlling ergonomic hazards in the meatpacking industry and has taken initial steps toward developing an ergonomics standard having more general application (3). Perhaps the most compelling argument supporting the coming-of-age of ergonomics is its use by the advertising community to market everything from automobiles to garden tools.

The demand for ergonomic information is a response, at least in part, to a growing recognition by employers, labor groups, and employees of the importance of accommodating the *human factor* in job design. Until recently job design for safety and

---

*International Journal of Industrial Ergonomics, International Journal of Human Factors in Manufacturing, International Journal of Occupational Safety and Ergonomics, and *Ergonomics in Design.*

Patty's Industrial Hygiene and Toxicology, Third Edition, Volume 3, Part B, Edited by Lewis J. Cralley, Lester V. Cralley, and James S. Bus.
ISBN 0-471-53065-4   © 1995 John Wiley & Sons, Inc.

health was limited to reducing the risk of exposure to chemicals, airborne particles, and acute traumatic events or accidents. Although work accidents continue to be responsible for the vast majority of occupational injuries and illness (nearly 95% in 1992), the incidence of reported occupational illness has increased by more than 350 percent from 126,100 to 457,400 cases since 1981 (4,5). In 1992 the majority of the reported occupational illnesses (63%) were attributed to "repeated trauma from job-related activities." The health conditions resulting from repeated trauma are recorded in Category 7f of the OSHA Form 200. Figure 10.1 shows the increase in "disorders associated with repeated trauma" from 1981 through 1992. Disorders in this category included, among others, diagnoses of carpal tunnel syndrome, tenosynovitis, and vibration white-finger. Cases of noise-induced hearing loss are also recorded in Category 7f of OSHA Form 200. In the United States these disorders, excluding hearing loss conditions, are commonly referred to as repetitive motion disorders or cumulative trauma disorders (CTDs). Concern for the prevention of CTDs has been the impetus for the attention now being given by occupational health professionals to ergonomics.

Ergonomics is an applied science concerned with the design of tools, workplaces, and tasks to match human physiological, anatomical, and psychological characteristics and capabilities. Ergonomics as a discipline draws from the fields of physiology, anatomy, psychology, and engineering for theory and methodology (6). While the military and aerospace industries have long been interested in ergonomic principles and applications, the benefits to other industries are just being realized. There is increasing evidence that the benefits of well-designed jobs, equipment, and workplaces include improved efficiency, safety, health, and increased satisfaction for employees. Industries faced with staggering workers' compensation costs and rising disability insurance premiums for musculoskeletal disorders have been motivated to seek out and implement ergonomic solutions. Many have found ergonomic job rede-

ILLNESS DUE TO REPEATED TRAUMA

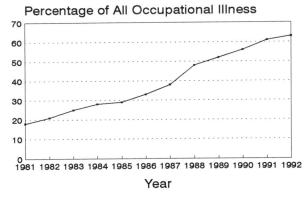

Figure 10.1 Increase in cumulative trauma disorders (CTDs). Based on entries to OHSA Form 200, Column 7f, 1981–1992.

sign effective for reducing the losses associated with musculoskeletal injuries (7–9). Further, as pressure increases for companies to enhance quality and remain competitive in a global marketplace, there is hope that ergonomic improvements will ultimately translate into greater productivity and profitability.

1.1 Content/Scope of Chapter

The treatment of ergonomics in this chapter is intended to illustrate the application of ergonomic knowledge to industrial hazard prevention and control, with primary attention given to the risk factors for work-related musculoskeletal disorders and means for their alleviation. This chapter differs in important ways from the discussions by Tichauer in Chapter 40 of the *General Principles* volume and by Cohen and Dukes-Dobos in the 1985 2nd ed. (10,11). The Tichauer version provides an in-depth discussion of a subarea (biomechanics) and emphasizes basic mechanisms underlying human movement processes and measures of the forces involved. The Cohen and Dukes-Dobos chapter addressed a broader range of task, equipment, workstation, and environmental issues, highlighting design guidelines as derived from the existing literature. This chapter includes presentations of case studies and other field research involving the application of ergonomic concepts to the study of physical job demands of consequence to musculoskeletal disorders. Different investigations were chosen to highlight various methods for identifying problems, ascertaining causal or risk factors, and developing control measures. This information is considered beneficial to a reader seeking guidance in dealing with contemporary ergonomic issues in the workplace.

2 APPLICATION OF ERGONOMICS TO OCCUPATIONAL SAFETY AND HEALTH PROBLEMS IN INDUSTRY

Abatement of occupational or environmental hazards is classically defined in terms of the following sequence of events: (a) recognition of health problems created by the industrial environment, (b) evaluation of the long- and short-range effects of working conditions on health, and (c) development, dissemination, and implementation of control technology (12). In terms of resolving health problems associated with poorly designed workplaces and tools (i.e., *ergonomic hazards*), progress is most clearly evident in the identification stage of this process. That is, there is a growing recognition that the improper design of tools, equipment, workstations, and job tasks can cause errors, tax or limit human capabilities, and trigger pathogenic processes resulting in a musculoskeletal impairment. Although significant steps have been made in the evaluation and control of ergonomic hazards, many unknowns remain. For example, there is little quantitative dose–response information to indicate safe levels of exposure to various biomechanical factors that can stress the musculoskeletal system. The ergonomic approach is based largely on the assumption that work activities involving less force, repetition, vibration, weight, and constrained postures are less likely to have adverse health effects (13).

In Section 3 examples will be presented to demonstrate that ergonomic interventions can reduce musculoskeletal injuries and enhance productivity in industry. However, some preliminary discussion of a basic methodology for identifying, evaluating, and controlling ergonomic hazards is appropriate. The following sections introduce the industrial hygienist or safety practitioner to concepts and methods of ergonomic hazard identification, evaluation, and control in the workplace. In the discussions that follow, the reader is assumed to be in a *retrofit* situation, that is, the techniques are to be applied to an existing workplace. The authors have made an effort to focus discussion on methods of practical value in a workplace or field setting, rather than the more esoteric methods sometimes used in the ergonomic research laboratory.

2.1 Identification of Ergonomic Hazards

The results of poor job design can be described in terms of health effects and operational effects (14). Cumulative trauma disorders and back injuries are but a few of the adverse health effects associated with poor job design. Accidents and traumatic injuries are frequently the result of design-induced errors (15). Operational effects impact quality and cost. Poor design leads to increased error; this in turn can lead to increased scrap, lower quality, and decreased production. Similarly, absenteeism and turnover often result from tasks that are difficult, unpleasant, or dangerous. As a consequence, there can be increased training and administrative costs (16).

Operational and health data can sometimes provide clues in the search for ergonomic deficiencies in the workplace. If enough information exists, these sources can generally provide some indication of the location and magnitude of suspected problems. However, the identification of ergonomic hazards usually depends on the level of training and awareness of workers, managers, and supervisors, and their willingness to undertake activities to locate or isolate ergonomic hazard sources.

Surveillance activities are a primary source of information for identifying jobs in need of further study regarding ergonomic problems. Surveillance methods generally fall into one of two categories: passive surveillance techniques and active surveillance techniques. Discussion of these two concepts follow.

2.1.1 Passive Surveillance Techniques

Passive surveillance data can be defined as any source of information used for surveillance that is normally collected by in-plant personnel for other purposes (17). Common sources of passive surveillance data include accident reports, dispensary or clinic records, OSHA 200 logs, and worker compensation records. Production records can also be useful to identify areas with high scrap or rework rates. Reviewing historical records can sometimes expose problem areas that remain unresolved.

The clear advantage of passive surveillance data is its availability and low cost. Unfortunately, many factors limit the usefulness of passive surveillance data. Record-keeping practices can vary significantly. Some plants record only those injuries that are clearly work-related; others record all visits to the medical department. Local management and state laws can also impact record-keeping and reporting practices. Some injuries may not be recognized as being work-related; CTDs frequently go unreported in plant records. OSHA 200 logs tend to be much better at identifying acute injuries than chronic disorders resulting from day-to-day exposure to poor working conditions. Therefore, passive surveillance data should be reviewed carefully and interpreted with caution. In general, a review of historical records should be considered a starting point for subsequent analysis and investigation (6).

2.1.2 Active Surveillance Techniques

Active surveillance is specifically designed and carried out to monitor and identify high prevalences of disease. An effective method of active surveillance involves use of questionnaires, interviews, and physical examinations to elicit information from workers about potential problems in the workplace.

Questionnaires are usually easy to administer and provide a quick method for identifying worker's perceptions of hazards and sources of discomfort (18). One particularly common and easy-to-use format is the Body Part Discomfort Survey depicted in Figure 10.2. The worker is given a picture of the body and asked to rate the level of comfort/discomfort experienced in different parts (19).

The chief advantage of questionnaires and interviews is that they are often successful at eliciting information about job-related complaints and symptoms that would otherwise go unnoticed. If large numbers of workers in a specific job or department report job-related discomfort, an investigation of tool, workstation layout, or job design may be indicated. Written questionnaires are relatively inexpensive to administer, workers can complete them at their convenience, and responses can be kept anonymous. Questionnaires do, however, have limitations. For example, symptom surveys are usually sensitive to CTDs but are poor at discriminating specific disorders or indicating the cause of the complaint. Tolerance of pain varies considerably from person to person and may affect responses (6). In the early stages of disease, workers may not recognize symptoms or be willing to report discomfort. Finally, factors such as the length of the questionnaire, the wording of the instructions, and the time and method of administration may all affect questionnaire results as well.

Physical examinations have two potential applications in the workplace. Simple, low-cost tests are often suitable for providing disease prevalence information and identifying high-risk worker populations. More sensitive and specific screening tests administered by medical personnel may be useful for identifying individuals in the early stages of disease who can then be referred for treatment. Although screening procedures can result in earlier identification and treatment of problems, the consequences of inaccurate test results should be examined carefully before physical examination procedures are incorporated into a surveillance program (20). The National Institute for Safety and Health (NIOSH) has recently proposed a surveillance case definition of work-related carpal tunnel syndrome (21). The NIOSH case definition is based on occupational history, reported symptoms, and physical findings and has proven suitable for epidemiologic surveillance.

2.2 Evaluation of Ergonomic Hazards

The purpose of an ergonomic hazard evaluation is to determine the link between adverse health effects and working conditions. The ergonomic evaluation process

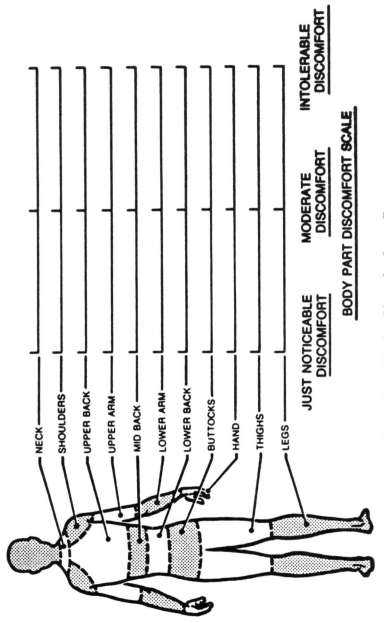

Figure 10.2   Body Part Discomfort Survey Format.

can be divided into two parts. The first stage involves an evaluation of job demands to identify the physical, mental, and perceptual requirements of the task. In the second stage, job demands are compared to known human capacities. If task requirements do, in fact, exceed the capabilities of a substantial fraction of the workforce, control measures may be indicated.

2.2.1 Evaluation of Job Demands

The physical, mental, and perceptual demands of most industrial jobs are a function of the work environment. The work environment can be described in terms of three basic components. These are:

1. The tools, machines, parts, and materials used to perform the job.
2. The workstation and the physical environment.
3. The task, including its content and the organizational environment in which it is performed.

These components can be arranged in a hierarchy, depicted in Figure 10.3. The hierarchy illustrates the increasing complexity of human interactions with the basic components of the workplace. The interactions between the human and his/her tools are most easily observed and best understood. A generic definition of tools may include hand tools, powered tools, machines, computer terminals and keyboards, instruments, and their component parts. Traditionally, ergonomic evaluations begin with an investigation of the tools and equipment used in the workplace. That improperly designed tools can increase the risk of accidents and traumatic injuries is readily appreciated. However, that improper tool design can have more insidious adverse consequences is only now being recognized. Specifically, tools that require awkward postures and repeated forceful exertions or transmit vibration to the hand have been implicated in the development of CTDs (6).

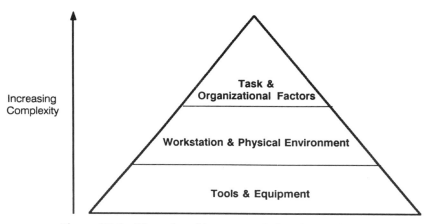

Figure 10.3 Hierarchy of human–workplace interactions.

The workers' interactions with the workstation and the physical environment are usually somewhat more complex. The workstation can include tables and benches, stools and chairs, controls and displays, storage bins, and so forth. The physical environment includes lighting, noise levels, air quality, temperature, ventilation, and so forth. Both workstation and environmental factors can have significant effects on comfort and functional ability as well as health. Ergonomic deficiencies in either of these two conditions may not be as obvious as tool design deficiencies, and special measurements (e.g., sound, illumination) may be required to identify problematic aspects. Further, correcting these problems may require greater capital expense (e.g., major facility renovations) than changes in tool design (22).

Finally, task and organizational factors are increasingly recognized as important to the health, safety, productivity, and satisfaction of workers. Job content (i.e., simple, routine vs. complex, varied duties), work scheduling, work pacing, management style and climate, worker autonomy, feedback, worker support, opportunity for advancement, training, and so forth are variables that can contribute to a positive work environment or, alternatively, produce stress. These factors are often the most difficult part of the work environment to evaluate. Although work rate is usually easy to measure, other problems emanating from job/organization factors are usually less evident from a physical inspection of the workplace. Furthermore, these problems are often far more difficult to correct. Indeed, solutions can require significant organizational/management changes that in and of themselves can be disruptive or cause other difficulties (22). Employers are understandably reluctant to implement these kinds of interventions.

Although some studies may be limited to an investigation of tool and workstation factors, a thorough ergonomic hazard evaluation should examine the interaction of the human with all three components of the work environment. Some hazards result from interactions between tool, workstation, and job design characteristics. To accurately characterize the severity of the hazard, an investigation of all three components is necessary. For example, poor workplace design, involving poor chair design or visual display problems, may have only modest consequences for workers with moderate production demands or for professionals able to exercise control over the job regimen. The same design flaws may have far more important implications for workers with more stringent performance demands or little control over their job situation.

There are no generic procedures for conducting an ergonomic evaluation of the workplace; the specifics of an investigation are dependent on a number of constraints, and procedures must be tailored to the individual workplace. However, the protocol for conducting an ergonomic evaluation usually follows one of two formats (6). A classical approach involves adaptations of traditional work measurement methods for the purpose of documenting and measuring exposures to ergonomic stressors. These techniques are referred to as task analysis methods. A second approach involves use of an ergonomic checklist. A brief description of each approach is provided below.

2.2.1.1 Task Analysis. Task analysis refers to a broad spectrum of methods used to analyze observable and covert human behavior for the purpose of identifying the performance demands of jobs and job tasks (23). Once task elements and job demands are determined, the analyst can decide whether these demands fall within the capabilities of workers.

Many of the techniques used by ergonomists to analyze tasks and work activities are based on industrial engineering principles of time and motion study. By far, the most common type of analysis used in ergonomic hazard evaluations is a timed activity analysis (24). The objective of timed activity analysis is to identify what the worker is doing and how it is being done in time. Timed activity analysis can be used to describe tasks at varying levels of detail, to characterize very broad, irregular activities, or very repetitive, short-cycle tasks. Examples of different procedures for performing timed activity analyses are discussed in Chapanis (25), Barnes (26), Armstrong et al. (27), Drury (28), and Rohmert and Laundau (29).

2.2.1.2 Checklists.

Ergonomic checklists are often used as an alternative or supplement to task analysis methods. Investigators with limited formal training can often use checklists to identify common hazard sources in a fairly short period of time. Checklists also provide reminders to investigators during data collection activities. In wide-ranging studies, they also ensure that systematic and standardized procedures are followed by different investigators.

Examples of items that might be found on an ergonomic checklist are found in Table 10.1 (30).

Although most checklists are problem oriented, some risk factors may be overlooked if they are not specifically described by the checklist. For example, ergonomic hazards in an office environment are likely to be different than those presented in a manufacturing facility. Therefore, existing checklists should be customized and evaluated in a walk-through survey to ensure that the questions are appropriate to the work site of interest (6).

2.2.2 Evaluation of Human Capacities

Often the most difficult task in an ergonomic hazard evaluation is to determine if job demands exceed acceptable limits of human capacity. In many cases prior research has provided data to assist in that judgment. For example, anthropometric tables can help investigators determine if workstations provide sufficient clearance to accommodate users in the working population. Strength data for different populations and muscle groups have been published in a number of sources (24,31). If the force requirements of a task are known, it may be possible to compare these requirements against existing strength data to estimate the percentage of the population for which the job may be difficult. Recently, computerized two- and three-dimensional biomechanical models have been developed to predict the percentage of males and females capable of exerting static forces in certain postures (32). The recently revised (1991) NIOSH lifting equation is based on studies that indicate that a number of variables, including job factors and personal factors, influence the amount of weight a person can lift without back injury. Based on analyses of biomechanical stresses on the lower back, data on aerobic capacities and lifting strength capabilities of the working population, and psychophysical studies of acceptable exertion levels, formulas for calculating recommended weight limits for lifting tasks are defined (33).

Table 10.1 Example Checklist Questions

Risk Factors	Yes	No
1. Physical Stress:		
Can the job be done without hand/wrist contact with sharp edges?		
Does the tool operate without vibration?		
Are worker's hands exposed to temperatures >21°C (70°F)?		
Can the job be done without using gloves?		
2. Force:		
Does the job require more than 4.5 kg (10 lb) of force exertion with the hands?		
Can the job be done without using finger pinch grips?		
3. Posture:		
Can the job be done without flexion or extension of the wrist?		
Can the tool be used without flexion or extension of the wrist?		
Can the job be done without deviating the wrist from side to side?		
Can the tool be used without deviating the wrist from side to side?		
Can the worker sit while performing the job?		
Can the job be done without "clothes wringing" motion?		
4. Workstation Hardware:		
Can the orientation of the work surface be adjusted?		
Can the height of the work surface be adjusted?		
Can the location of the tool be adjusted?		
5. Repetitiveness:		
Is the cycle time longer than 30 seconds?		
6. Tool Design:		
Are the thumb and finger slightly overlapped in a closed grip?		
Is the span of the tool handle between 5 and 7 cm (2–2¾ inches)?		
Is the handle of the tool made from material other than metal?		
Is the weight of the tool below 4 kg (9 lb)?		
Is the tool suspended?		

In other cases existing data may be insufficient to indicate the magnitude of hazards. More specialized techniques may be needed to determine the impact of job demands on workers. Three indicators of task difficulty (i.e., stress) are task performance, physiological response, and the worker's subjective assessment of the work load (34).

The use of multiple indicators of job demand is dictated by a number of factors. As job demands increase, the adaptive response of different individuals may be influenced by various considerations (Fig. 10.4). Elements such as motivation, skill, and individual tolerance for working conditions play a major role in determining the impact of job demands. Response also depends on the individual characteristics,

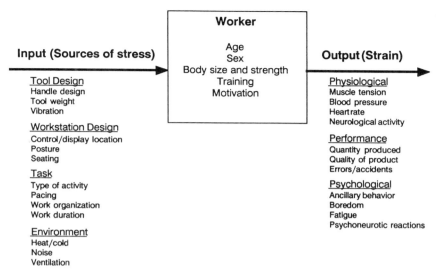

Figure 10.4 Factors affecting worker response to job demands.

needs, and abilities of the working person, which may change with time. This relationship between task demands, individual capacities, and worker responses is represented by the block diagram in Figure 10.4. Because a worker's response to job demands is usually multidimensional, it may be possible to make trade-offs between the response variables to maintain one at a desired level. For example, as tasks become more difficult, it may be possible for some workers to maintain satisfactory performance, but with greater physiological cost. Where motivation is low, and the consequences of failure are minor, workers may allow productivity to lapse rather than suffer increased physical strain. Finally, where job demands overwhelm the worker's capacity to compensate, strains may be manifest along several dimensions (e.g., increased fatigue accompanied by increased error and reduced productivity, etc.). Descriptions of some common performance, physiological, and subjective measures are provided below.

2.2.2.1 Performance Measures. Performance measures quantify the productivity and quality of output by the worker. Common performance measures are listed in Table 10.2 (34). Job demands that exceed workers' capacities may be reflected in decrements in performance measures (24).

Generally, the best performance measures are those that are objective, quantitative, unobtrusive, and easy to collect without specialized instrumentation (34). Time and errors are usually easy to measure; however, these measures can be difficult to interpret. Unless time is critical to a task, the relationship between time and job demand may be very weak. Errors may be indicative of performance quality, but in well-trained personnel or at moderate stress levels, significant errors may be so rare as to render them insensitive for evaluation purposes. Finally, if tasks are primarily perceptual and cognitive, performance measures may be inherently subjective.

Table 10.2 Common Performance
Measures

---

1. Time
    Reaction time
    Activity duration time
2. Accuracy
    Observation errors
    Response errors
3. Frequency of Occurrence
    Number of responses per unit or
        interval
    Number of errors per unit or interval
4. Amount Achieved or Accomplished
    Percent of activities accomplished
    Degree of success
5. Consumption or Quantity Used
    Units consumed to accomplish activity
    Units consumed per unit time

One method for evaluating worker performance in a variety of work situations is described by Rosa et al. (35). A computerized battery of standardized tests was designed to evaluate a range of behavioral functions, including cognitive abilities, perceptual-motor functions, motor skills, and sensory acuity. The system was designed to allow workers to self-administer the tests and scales during short work breaks. Decrements in performance over the course of a work shift may indicate decreased alertness and increased fatigue due to workplace conditions. The test battery has been used successfully to track variations in alertness and fatigue associated with different work schedules and work durations.

2.2.2.2 Physiological Measures. Physiological measures are frequently used in laboratory studies to evaluate an individual's response to controlled working conditions. Many of the guidelines and design recommendations found in ergonomic textbooks are based on studies of human response to defined stressors. Physiological measurement systems available for use at the work site are limited to those that are noninvasive, do not interfere significantly with job performance, are not expensive, and can be operated by one or two investigators. Physiological parameters that can be readily monitored during job performance are listed in Table 10.3 (24,34). Like performance measures, physiological measures tend to be unidimensional and can be influenced by factors not necessarily related to the job.

2.2.2.3 Subjective Assessment Measures. The worker's subjective assessment of his/her work load has traditionally played an important role in evaluating the impact of job demands in industry. If the worker feels pressured or stressed, then this is what they report, regardless of what performance and physiologic measures show.

Psychophysical ratings of perceived exertion or comfort are probably among the

Table 10.3 Physiological Indicators of Job Demand

System	Measure
Cardiovascular system	Heart rate
	Blood pressure
	Peripheral blood flow
Respiratory system	Respiration rate
	Oxygen consumption
Nervous system	Brain activity (EEG)
	Muscle activity (EMG)
	Pupil size
	Tremor
	Voice changes
	Blink rate
Metabolic system	Catecholamine excretion
	Galvanic skin response
	Body temperature

most widely used subjective measures. Using these techniques, subjects are asked to estimate their effort or exertion and assign them a number. An advantage of perceived exertion ratings is that they integrate a large amount of information into a single measure of physical strain. Cues from the peripheral muscles and joints, cardiovascular and respiratory functions, and the central nervous system may all affect perceived exertion. Perceived exertion scales have been found particularly valuable in studies of short-term static work for which valid physiological measures are difficult to obtain (36).

There are some inherent deficiencies in the use of subjective measurements. A main problem is the lack of fundamental units for measuring perceived exertion (37). The worker may also be unaware of the extent to which he/she is stressed, he/she may confuse mental and physical effort, and his/her estimates may change over time (38). Nonetheless, psychophysical scales have been used successfully in a number of ergonomic investigations of work tasks, and high correlations have been demonstrated between subjective ratings and physiological variables (37,38).

2.3 Control of Ergonomic Hazards

Once job demands are identified as overly stressful, the final step in the process of ergonomic hazard abatement is to develop and implement an intervention plan. Resolving ergonomic hazards is frequently difficult for a number of reasons. In some cases several factors combine to create a hazard. Overlapping problems can include high production demands, faulty work methods, awkward workstation layouts, and ill-fitting tools (6). Thus, improvements addressing one factor may not eliminate the overall risk. Further, problematic work methods and management practices can be more difficult to correct than problems relating to workstation or tool design (22). Also, interventions effective in one situation may be inappropriate in other settings.

Most control plans involve compromise and trade-offs to arrive at the most appropriate solution.

Approaches for controlling ergonomic hazards may include the following (39):

1. Engineering design changes to tools, equipment, or other aspects of the workplace, often called engineering controls.
2. Changes in work practices or organizational and management policies, sometimes called administrative controls.
3. Use of personal protective equipment.

The following paragraphs explore each of these concepts in greater detail.

2.3.1 Engineering Controls

The best ergonomic solutions are those in which safe, healthful working conditions are a natural result of the design of the job, workstation, and tools and are independent of specific worker capabilities or work techniques (6). The objective of engineering controls is to make the job fit the person, that is, to improve the design of tools, equipment, workstations, and work methods to reduce job demands and alleviate sources of ergonomic stress.

Some common engineering strategies for controlling ergonomic hazards are listed in Table 10.4. Specific recommendations for designing tools, equipment, and workstations are described by Tichauer in Chapter 40, of Patty's Volume I, 4th rev. ed. (10). The reader is also referred to information contained in *Ergonomic Design for People at Work* (Volumes I and II) and the *Human Factors Design Handbook* (24,40,41).

Table 10.4 Examples of Engineering Design Interventions

---

1. Use mechanical aids and assist devices to reduce repetition and manual force requirements.
2. Decrease the weight of tools, containers, and parts.
3. Redesign handles to optimize shape, size, orientation, and friction for the range of users and the application.
4. Balance hand-held tools and containers.
5. Use torque control devices to minimize transmitted forces.
6. Enlarge corners and edges.
7. Provide pads and cushions to avoid high contact forces.
8. Provide adjustability in the workplace to accommodate the range of users in the population.
9. Provide jigs and fixtures to support tools and parts.
10. Select tools to minimize vibration transmission.
11. Select work processes to minimize surface and edge finishing.

2.3.2 Administrative Controls

Administrative controls can be defined as policies or work practices used to prevent or control exposure to ergonomic stressors that can result in work-related injury or disease. Examples of administrative controls include the following (39):

1. Providing frequent rest breaks to offset undue fatigue in jobs requiring heavy labor or high performance/production rates.
2. Limiting overtime work and periodically rotating workers to less stressful jobs.
3. Varying work tasks or broadening job responsibilities to offset boredom and sustain worker motivation.
4. Training workers to use work methods that improve posture and reduce stress and strain on the extremities.

The effectiveness of administrative controls has been examined by a number of researchers. One investigation of keyboard operators found that operators who were provided short but frequent rest breaks were more productive than operators receiving only the traditional midmorning, midafternoon, and lunch breaks (42). An electromyographic study of five jobs where job rotation had been introduced concluded that job rotation may be more useful for reducing stress associated with heavy dynamic tasks than for reducing static muscular load in "light" work situations (43). The effectiveness of training programs has been difficult to evaluate. Some authors have attributed significant reductions in low back disability and lost time injuries to worker training programs (44,45). Other studies indicate that well-planned training programs can have small but significant effects on lifting behavior (46,47).

In many cases the effectiveness of administrative controls is dependent on management commitment. Regular monitoring and reinforcement is necessary to ensure that control policies and procedures are not circumvented in the name of convenience, schedule, or production. An advantage of administrative controls is that they can usually be implemented quickly and easily without capital expense. However, because administrative controls fail to eliminate the source of the hazard, they should be considered temporary solutions for controlling exposure until engineering controls can be implemented.

2.3.3 Personal Protective Equipment

In the standards adopted under the Occupational Safety and Health Act, personal protective equipment is considered a "last resort" for hazard control, that is, it is appropriate only in situations where effective engineering or administrative controls cannot be implemented (12). Personal protective equipment seldom provides complete protection from exposure to a significant hazard; rather it seeks to reduce the exposure to an acceptable level (48).

Devices commonly used in the workplace to protect workers against harmful physical agents include hearing protectors, vibration-attenuating gloves, protective eye wear, safety shoes, and special thermal protective clothing. Braces, wrist splints, back belts, and similar devices purported to reduce biomechanical stress on the musculoskeletal system have questionable value. Indeed, there is little research evidence to demonstrate that these devices limit the risk of injury. In some instances these devices restrict motion and interfere with task performance, thereby increasing the physical stress of the activity. Further, there is some evidence that braces and lifting belts give workers a "false sense of security" and may actually encourage overexertion (49).

Generally, the role of the ergonomist in the selection of personal protective equipment is not in evaluating whether the device will be effective. In short, the ergonomist's concern is that personal protective equipment does not compound the job demands or health risks imposed by the work situation (39,48). The ergonomist must ensure that if personal protective equipment is to be used, it is provided in a variety of sizes to fit the user group and does not contribute to extreme postures and forces on the job.

3 CASE EXAMPLES

Although a general methodology for identifying, evaluating, and controlling ergonomic hazards in industry has been described, the process of applying ergonomic principles to the design or redesign of tools, workstations, jobs, and work environments requires adaptability and creativity. Each situation presents its own unique set of circumstances, and factors such as technology, cost, management commitment, and employee cooperation must be considered in devising a procedure to evaluate and resolve ergonomic problems.

In this light the following case studies are offered as examples where systematic procedures were applied to identify, evaluate, and control ergonomic hazards in industry. The cases described are largely the result of recent investigative activities sponsored by NIOSH to gain insight into ergonomic aspects of workplace hazards and their control. The scope of these activities varies widely. In some cases the evaluation was limited to the most basic aspects of the work environment, that is, the tools and equipment used by workers in their everyday activities. In other cases a more extensive investigation of the workstation and the organizational demands of the job was required to identify and resolve hazards. Table 10.5 is designed to

Table 10.5 Case Studies

Issue	Case Example
Tool/equipment design	Carpet laying
	Jewelry making
Workstation design	Grocery checking
	VDT use
Methods/organization design	Meatpacking
	Grocery warehousing

provide guidance to the reader interested in selecting case examples that focus on specific hazards in the workplace.

3.1 Case 1: Investigation of Tool Design in Carpet Laying Tasks

3.1.1 Background

Workers whose jobs involve work in a kneeling position may inflict chronic trauma to their knee joints. For example, carpet and floor layers report substantially more knee morbidity than either the general U.S. white male population or a blue-collar working population of comparable age, sex, and race (50). Carpet layers also submit a disproportionately large fraction of workers' compensation claims for knee joint inflammation. While comprising only 0.06 percent of the total workforce, carpet installers account for approximately 6.2 percent of such claims, a rate that is 108 times that expected in the general working population (51).

Apart from kneeling, a suspected added risk factor for knee trauma among carpet layers is the use of a device called a "knee kicker." Knee kickers have been used for more than 80 years for stretching carpet during wall-to-wall installation tasks. Workers using this tool generate force by striking the suprapatellar area of the knee against the instrument. Since the late 1950s, the mass production of carpet for wall-to-wall installation has increased dramatically. Therefore, use of the knee kicker has become widespread among carpet layers who install carpets in homes and offices.

Identification of the high incidence of chronic knee injuries in carpet and floor layers resulted in an investigation to define the nature and magnitude of knee morbidity in carpet layers, and to identify causative work-related factors that could be eliminated and controlled.

3.1.2 Study Design

The plan to evaluate the link between work activities and knee morbidity in carpet layers took a two-pronged approach. First, questionnaires and medical examinations were administered to union tradespeople to determine the prevalence of knee problems in three populations. The sampled groups included carpet layers who kneel frequently and use a knee kicker, tile and floor layers who kneel frequently but do not use a knee kicker, and bricklayers who seldom kneel and do not use knee kickers. The questionnaire inquired about symptoms of knee disease, personal attributes, and work practices. The examination included a physical inspection of the knee, range-of-motion tests, and an x-ray of the knee joint.

Second, an ergonomic analysis of the carpet installation task was performed to identify and quantify potential sources of biomechanical stress to the knee. Nine carpet layers were studied while performing normal job activities at an apartment building work site. Typical work tasks and associated work postures are shown in Figure 10.5. To identify trauma-producing aspects of the task, investigators performed a task analysis to determine:

1. The number of activities carpet layers perform
2. The percentage of time required by each activity

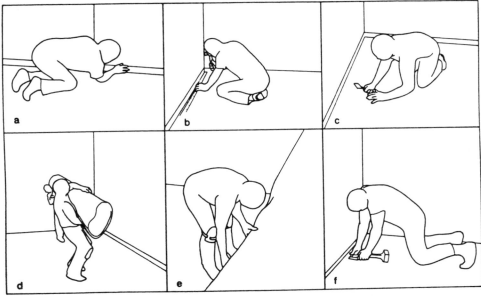

Figure 10.5   Worker postures in carpet-laying tasks.

3. The percentage of time carpet layers kneel
4. The percentage of time carpet layers use the knee kicker during carpet-stretching tasks.

Investigators recorded additional measurements for analysis in the laboratory. Knee-kicking activities were recorded on film. The film was analyzed using a stop-action projector, a digitizer, and a computer system to calculate angular changes in body segments associated with knee-kicker use. Also, miniature piezoresistive accelerometers were attached to the kicking leg of each worker to measure impact acceleration during the kicking task. Measurements were made during light kicks (approximately half force) and normal to heavy kicks. Finally, load cells were mounted in the middle of a knee-kicker shaft to measure peak impact forces during knee-kicking activities (52).

3.1.3   Data Collected

Based on medical questionnaire results, investigators found carpet layers reported knee symptoms and disorders three to five times more frequently than workers who seldom kneel and never use a knee kicker. Specifically, carpet layers reported an increased prevalence of bursitis, fluid buildup under the knee cap, and skin infections involving the knee (51).

Observation of carpet layers at the work site revealed that during the carpet stretching task, workers spent about 75 percent of their time on hands and knees using the knee kicker. During these activities, workers executed up to 140 strong knee kicks per hour. Investigators noted that few workers used knee pads or other forms of knee protection during this activity.

Film analysis revealed that workers frequently assumed postures requiring near-maximum knee flexion. Knee-flexion angles at impact during the knee-kicking task averaged 58°. Previous research indicates that knee flexion beyond 60° produces shearing forces that can cause ligament damage (53). Load cell measurements indicated that at impact, forces averaging 3019 N (about four times body weight) were exerted on the knee. The investigators noted that periodic impact loading causes more rapid damage to cartilage than static loading, and that the visco-elastic properties of knee cartilage are not effective for attenuating impulsive loads. It was hypothesized that high, compressive load-induced squeezing of the patella against the femur during knee kicking could result in development of bone osteophytes, destruction of cartilage, and narrowing of joint spaces in the knees of carpet layers (52).

3.1.4 Conclusions

This evaluation concluded that both kneeling and the use of the knee kicker contribute significantly to the high frequency of carpet layers' knee disorders (51). The primary recommendation to emerge from this study was to replace the knee-kicking tool with another device appropriate for carpet-stretching tasks. Specifically, a hand/arm-operated device known as a power stretcher was recommended as an alternative carpet-stretching tool. Use of this device requires no knee action; therefore, there is no impact trauma to the knee. Improved knee kickers that require less impact force to use provide another alternative. Other recommendations included the following:

1. Carpet layers should wear protective knee pads while working on hard floor surfaces. Knee pads reduce pressure on the knee by distributing the weight of the body over a larger surface area. They also provide protection against penetrating wounds resulting from kneeling on sharp objects.
2. Crew members should take turns performing the carpet-stretching task to limit exposure to excessive knee loading.
3. If the knee kicker must be used, impact forces should be kept perpendicular to the impact surface of the knee kicker. In all cases the minimum amount of force necessary to perform the task should be used (54).

3.2 Case 2: Investigation of Tool Design in a Jewelry Making Facility

3.2.1 Background

Investigations of health hazards presented to jewelry workers have focused largely on exposures to harmful dusts and toxic chemicals. Silicosis and silicotuberculosis, due to exposure to mixed dusts, have been described among jewelry foundry workers (55). Workers' exposures to metal fumes and a number of potentially toxic acids and solvents have also been described (56). However, the potential biomechanical hazards presented by jewelry making tasks are not well described.

In 1990 a jewelry manufacturing facility was investigated to evaluate possible hazards for upper extremity CTDs among their employees (57). The facility employed approximately 100 workers in the manufacture of custom gold jewelry for wholesale and retail distribution. A preliminary investigation of the plant's OSHA 200 logs identified seven recorded cases of carpal tunnel syndrome among the plant's employees in a 3-year period.

3.2.2 Study Design

Because little was known about the prevalence or source of upper extremity musculoskeletal complaints, medical and ergonomic investigations of the work site were launched simultaneously. To determine the prevalence of upper extremity musculoskeletal complaints, a questionnaire was administered to all current employees of the plant.

The ergonomic evaluation focused on tool and workstation design features that could potentially contribute to upper extremity complaints. Work activities in 10 production departments were videotaped. Task descriptions were subsequently developed by two ergonomists who observed the videotape and confirmed through discussions with the plant employees (Table 10.6).

3.2.3 Data Collected

Because all jewelry was custom-crafted, very few automated processes were used in the facility. Production quotas were not strictly set or enforced, although employees were expected to be busy at their workstations unless at lunch or on break. Annual turnover in the facility was very low (<5%). Discussions with employees found that most were satisfied with their work and took pride in their abilities and craftsmanship.

Management acknowledged that tool and workstation design had changed little in the previous 10 years. Workstations in most departments were set up using long benches or tables that accommodated a number of employees. The investigators identified a number of biomechanical hazards directly attributable to hand tool use. Tweezers, small hammers, files, awls, wire cutters, and soldering torches were widely used in the casting, soldering, bright cutting, and repair departments. In many instances these instruments were used frequently for prolonged periods of time. In some cases high force exertions were required to remove metal fragments or engrave designs into metal surfaces. These activities were especially problematic in the bright cutting and wriggling departments. Analysis of questionnaire results revealed at least two-thirds of employees in these areas experienced hand–wrist CTD symptoms (Table 10.7). In other instances tool design combined with factors such as the tool location and the workbench height resulted in awkward wrist, elbow, and shoulder postures (Fig. 10.6). For example, the design of the soldering torch and holding fixture in the soldering department required employees to work with arms and elbows in a raised position for long periods of time. Hand–arm vibration was identified as an additional hazard presented by tools in the grinding and wriggling departments.

Table 10.6 Jewelry Manufacturing Tasks

Department	Number of Workers	Job Description
Casting	8	Workers rotate between two tasks: (1) workers create wax casts for new jewelry designs (seated); (2) workers create plaster molds for jewelry pieces from wax casts (stand).
Stamping	1	Worker stamps decorative leaves from sheets of gold foil using foot-pedal-operated press (seated).
Soldering	30	Workers apply designs to rings, earrings, and pendants with soldering torches, tweezers, files, and bench-mounted spot weld machines (seated).
Grinding	8	Workers buff and polish ring surfaces with mechanically driven grinding wheels (seated).
Plating	2	Workers immerse jewelry in electroplating solutions (stand).
Waxing	2	Workers mount jewelry pieces atop wooden handles with wax (seated).
Wriggling	9	Workers chip away rough edges, engrave designs into gold using a linear actuator mounted on top of the work bench (seated).
Bright cutting	9	Workers use small hand tools to remove fragments and etch designs into jewelry surfaces (seated).
Repair	9	Workers inspect jewelry and perform necessary repairs using hand-held tools (seated).
Packing and shipping	12	Workers perform final inspection, place jewelry in boxes, and wrap boxes for shipping (stand).

Table 10.7 Work-Related CTDs by Job Title

Job Title	N	Neck	Shoulder	Elbow	Hand/Wrist
Cast	8	0 (0%)	0 (0%)	0 (0%)	1 (13%)
Solder	31	1 (4%)	0 (0%)	1 (4%)	4 (13%)
Grind/Polish	8	1 (13%)	1 (13%)	2 (25%)	3 (38%)
Plate	2	0 (0%)	0 (0%)	0 (0%)	0 (0%)
Wax	2	1 (50%)	1 (50%)	1 (50%)	1 (50%)
Wriggle	9	2 (23%)	3 (34%)	3 (34%)	7 (78%)
Bright Cut	9	5 (56%)	7 (78%)	4 (45%)	6 (67%)
Repair/Design	9	1 (12%)	0 (0%)	0 (0%)	0 (0%)
Pack/Ship	12	1 (9%)	0 (0%)	0 (0%)	2 (17%)
Supervisor	4	0 (0%)	0 (0%)	0 (0%)	0 (0%)
Total	94	12 (13%)	12 (13%)	11 (12%)	24 (26%)

Figure 10.6   Examples of awkward postures in jewelry making tasks.

3.2.4   Conclusions

On the basis of medical and ergonomic evaluations, it was concluded that numerous biomechanical hazards existed at this plant. Specific engineering, administrative, and medical management interventions were recommended to reduce the severity of these hazards. These included the following:

1. Redesign the soldering torch and provide an adjustable fixture at the soldering station to reduce wrist deviation and elbow abduction during soldering tasks. Specifically, the fixture should allow the user to freely adjust the angle and position of the workpiece. The torch handle should be angled to allow the user to direct the flame at the workpiece without raising the arms. Additional light fixtures and magnifying glasses should be provided to improve visual conditions at the workstation.

2. Redesign grinding and wriggling tools to allow workers to perform tasks with less force and the wrists in a more neutral posture. Two specific design modifications were recommended. First, a fixture with a handle and a clamping mechanism to grip the workpiece (jewelry item) was needed. This fixture would allow workers to hold the workpiece with a power grip rather than a pinch grip. Placing the jewelry piece in a fixture would also reduce vibration transmission from the grinding tool to the hand. Second, grinding and wriggling machines should be mounted on adjustable pedestals. The pedestal would allow workers to adjust the height of the grinding spindle and wriggling tip. Finally, a magnifying glass was recommended to improve viewing conditions at the workstation.

3. Replace the bright cutting tool with a power tool to reduce the manual force requirements of the task. A power tool would be easier to manipulate and would likely improve precision in cutting and etching tasks. The tool should also have a padded handle extending the length of the palm to reduce pressure on the palm of the hand and prevent hand–arm vibration transmission. A fixture mounted to the top of the work bench was recommended to hold the workpiece during cutting tasks. This fixture would free the nondominant hand.

4. Train new employees in proper craftsmanship, tool use, and maintenance. For example, new employees should be trained to keep cutting tools sharp (to reduce force requirements) and power tools balanced and lubricated (to minimize vibration). Providing new employees with frequent rest breaks was recommended to prevent fatigue and overexertion.

5. Rotate employees to jobs that require use of different muscle or tendon groups. For example, employees who use small hand tools should be rotated to inspection tasks that have fewer manual requirements.

6. Provide more frequent rest breaks to employees who engage in manual tasks for long periods of time (e.g., employees in the bright cut, wriggle, and grinding departments).

7. Perform regular employee health evaluations to identify individuals with CTD symptoms (e.g., pain, swelling, tingling or numbness) for treatment.

3.3 Case 3: Investigation of Workstation Design for Grocery Checking Tasks

3.3.1 Background

One of the most revolutionary changes in the domestic supermarket has been the introduction of the laser scanner workstation. Since laser scanning technology was introduced in the late 1970s, more than 95 percent of all grocery stores have implemented laser scanners in their checkstand designs (58).

The retail food industry employs more than 25 million workers. Of these, approximately one million are employed as cashiers. Grocery checkers can handle more than 500 items per hour and fill as many as 80 bags per hour (59). In general, there are three major factors that may place grocery checkers at increased risk for a number of neurologic and musculoskeletal disorders. These include the checkstand configuration, work practices used by grocery cashiers, and work rate (60).

Checkstand design can greatly affect the cashier's posture during work. Although most checkstands are composed of six basic elements (feeder belt, scanner, produce scale, cash register, bagging platform, and take-away belt), workstation dimensions and the location of elements can vary significantly. Therefore, the layout of the checkstand can have a substantial effect on the magnitude of reaches, lifts, and bends required by grocery checkout tasks.

In August 1988 an investigation at a large grocery store was undertaken to determine the source of employee-reported symptoms of shoulder, neck, and upper back pain following the introduction of a new express checkstand (61). The new checkstand was smaller than a conventional checkstand and was designed specifically for processing express orders (Fig. 10.7). From management's standpoint the new checkstand was advantageous since two registers could be installed in the space formerly occupied by one.

3.3.2 Investigative Plan

Data collection in this evaluation included the use of employee interviews, body part discomfort surveys, and an inspection of the physical features of the checkstand to identify problems associated with the workstation design.

First, a baseline questionnaire/survey was administered to all cashiers at the store at the beginning of the study. The baseline questionnaire/survey was designed to collect demographic and work history information and to elicit musculoskeletal complaints ascribed to (1) work and nonwork activities and (2) the regular checkstand and the express checkstand. Symptoms were clustered into three groups: (1) upper

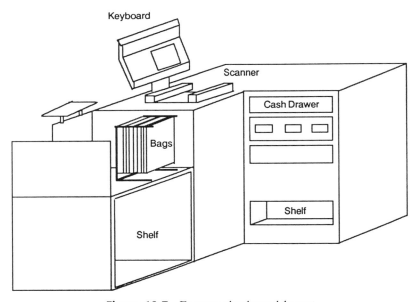

Figure 10.7   Express checkstand layout.

back, neck, and shoulder; (2) lower back, buttocks, and leg; and (3) arm, forearm, and wrist.

Second, an ergonomic assessment of the physical design features of the checkstand was initiated. Measurements were made of checkstand dimensions, reach distances, and lifting heights. Cashier work activities and movement patterns were recorded on videotape for detailed analysis. The videotapes were used to determine average cycle time and document the frequency of sweeping movements and awkward postures.

3.3.3 Data Collected

The baseline survey administered to the cashiers revealed that symptoms were localized to the neck, upper back, and shoulder, with fewer complaints in the lower back, buttocks, and legs. Less than half of employees complained of discomfort in the arm, forearm, and wrist. Several checkers felt their discomfort was related to the introduction of the express checkstand and that certain design characteristics were directly responsible. Specifically, employees reported that (a) the far corner of the checkstand was difficult to reach, (b) the keyboard was awkward to operate, (c) the configuration of the checkstand required frequent twisting movements, (d) the checkstand design encouraged use of one hand only during checking, instead of distributing the load over both hands, and (e) a shelf built into the checkstand introduced a lift into the bagging process.

Investigators observing cashiers working at the express checkstand also identified several design characteristics that introduced stressful work postures. These were:

1. Excessive reaches were required to collect groceries from the far corner of the checkstand. As a result, frequent trunk flexion and shoulder abduction were observed.
2. The height and distance of the keyboard from the scanner resulted in trunk twisting, shoulder flexion, and prolonged awkward postures.
3. Based on extrapolation from the videotape, it was estimated that the checking task required approximately 1442 hand movements per hour. Because of poor workstation design, approximately half of these movements involved excessive wrist deviation or shoulder flexion.

3.3.4 Conclusions

In this study the findings of the ergonomic evaluation corroborated the information gathered from the employees. Three ergonomic interventions were suggested and implemented to reduce stressors associated with the express checkstand. First, a physical barrier was placed at the far corner of the checkstand to limit the reach distance to retrieve groceries for bagging. Second, a base that allowed adjustment of the keyboard height was added to the checkstand. Finally, cashiers were shown an instructional video that provided information about proper work methods.

Four months after these modifications were completed, a follow-up questionnaire/survey was conducted. The survey was similar to that administered prior to the investigation and elicited information about musculoskeletal complaints. Employees

Table 10.8 Follow-up Survey Results

Variable	Before	After
Any symptom (% employees)	100	79
Neck/upper back/shoulder symptoms (%)	90	63[a]
Arm/forearm/wrist symptoms (%)	42	47
Lower back/buttock/leg symptoms (%)	79	63[a]
Employees requiring medication (%)	78	26[a]

[a] Indicates percentages are significantly different.

were also asked to evaluate the effectiveness of the interventions in reducing discomfort and improving checkstand function. The results of the follow-up survey were compared to the baseline results (Table 10.8).

The follow-up survey found significant improvements in neck, shoulder, and upper back discomfort and lower back, buttock, and leg discomfort after intervention. Overall, 74 percent of employees indicated the adjustable keyboard was an effective modification for reducing musculoskeletal discomfort. The barrier was also judged to be an improvement by 53 percent of employees. Seventy-one percent of employees found the training tape to be informative, although only 21 percent felt it reduced their discomfort.

Despite improvements, a substantial number of employees continued to complain of musculoskeletal symptoms in the follow-up survey. Therefore, additional interventions were recommended for further study. These included installing a conveyor belt in the checkstand to eliminate reaches to the far end of the checkstand, placing the cash drawer and the keyboard next to each other to eliminate trunk twisting, and rotating cashiers to different checkstands or tasks at more frequent intervals.

3.4   Case 4: Investigation of VDT Workstations in an Office Environment

3.4.1   Background

The rapidly accelerating trend toward the application of video display terminal (VDT) technology in the American workplace has, more than any other factor, generated an awareness of the physical design of the office. The evolution of computer hardware and the development and implementation of software has made it possible to perform an ever-increasing number of tasks electronically. Memos can be typed and sent; databases can be compiled; complex financial transactions can be executed; and timetable schedules can be made with just a few keystrokes. The widespread implementation of such technology has resulted in a workforce that spends more and more time at a VDT workstation.

The long-term implications of prolonged VDT use are not yet clear. However, VDT work has been blamed for a variety of health problems, ranging from mild headaches to perinatal mortality (62). Among the most common complaints are visual fatigue, stress, and musculoskeletal discomfort (63,64). Principles of physiology and anthropometry suggest there may be a basis for a relationship between ergonomic stress factors in VDT work and musculoskeletal complaints. However, a consensus report released by the World Health Organization states that psychosocial factors are at least as important as the physical ergonomics of workstations and the working environment in determining the users' health and well-being (62). In any case adverse effects are believed to be the result of inadequate planning, inappropriate work demands, and poor design of the workplace and environment.

3.4.2 Study Design

The relationship between biomechanical factors and musculoskeletal discomfort in VDT work was examined in a study of 905 VDT operators in Wisconsin (63). Specifically, the investigation focused on operator posture and the interface between VDT operators and their workstations.

To gather data for this investigation, surveys were distributed to experienced VDT users in two state agencies. The group consisted of 539 data entry operators and 259 workers who performed a number of different tasks. The data entry workers entered handwritten, alphanumeric information from tax forms or traffic citations into a computer using a VDT with a detached "Qwerty" keyboard. Eighty-seven percent of the sample was female.

The survey contained questions about age, height, weight, use of corrective lenses, hours of VDT use per week, and length of time in current job. Using the method of Corlett and Bishop (19), participants were also asked to record the frequency of musculoskeletal discomfort at 18 body locations (Fig. 10.8). The second stage of data collection involved direct measurement of worker postures and workstation parameters. A subsample of 40 data entry workers was selected randomly from the original study group, and a detailed ergonomic evaluation was performed at each workstation. Information was collected on the parameters listed and shown in Figure 10.9.

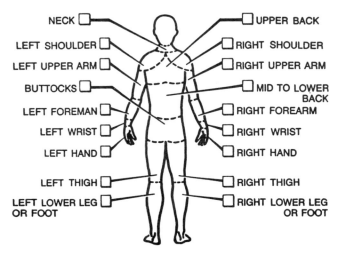

Figure 10.8 Body part discomfort scale for VDT study (62).

```
Chair type - tilt/swivel or non-tilt/swivel.

Seated posture - erect, stooping or reclining.

Popliteal height - seat pan height (cm).

Pan compression - compression of chair seat padding.

resulting from pressure application at the center.

Seat back height relative to 7th cervical vertebra.

Trunk angle (A).

Upper arm angle (B).

Forearm angle (C).

Shoulder abduction (D).

Hand ulnar deviation (E).

Hand flexion (F).

Relative document distance (G).

Gaze angle to document and to VDT (H).

Head tilt to document and to VDT (I-J).
```

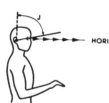

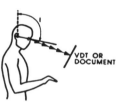

Figure 10.9   VDT workstation variables (62).

3.4.3   Data Collected

The ergonomic evaluation revealed remarkable similarities in the workstation design. In all cases the keyboard was positioned immediately in front of the worker, and the document was placed either on the left side or between the keyboard and display. Wrist rests were not provided and there was little opportunity to adjust the keyboard or VDT screen. Work tables and chairs were not readily adjustable.

After survey data was collected from all participants, the 18 original musculoskeletal discomfort ratings were collapsed into five discomfort scales representing five body regions (trunk, left arm, right arm, legs, and buttocks). Scale scores were computed by summing the discomfort scores for the body sites within each region. The scale scores were used as dependent variables in five multiple regression models to predict discomfort in each body region. Demographic information (i.e., age, sex, height, etc.), VDT exposure information (job tenure, hours of VDT use per week), and the workstation measures were used as independent variables in the models.

Results of the regression analyses revealed that none of the demographic or VDT exposure variables provided any meaningful predictions of discomfort. In contrast, ergonomic factors were significantly related to all discomfort measures except buttocks discomfort. Specifically, the following results were obtained:

1. Leg discomfort was predicted by chair type, pan compression, and popliteal height. Tilt-swivel chairs resulted in less frequent discomfort than chairs that did not tilt or swivel. Discomfort was more pronounced when operators sat on low chairs with soft cushions.

2. Trunk discomfort was predicted by seated posture and seat back height. Reduced discomfort was associated with an erect sitting posture and increased back rest height.

3. The regression analysis indicated arm discomfort became less frequent as elbow height increased relative to keyboard height. Left arm discomfort was less frequently reported as the reach distance to the document decreased. As reach distance to the keyboard increased, the frequency of right arm discomfort increased. Increased wrist flexion was also associated with increased arm discomfort.

3.4.4 Conclusions

Investigators concluded the results provided compelling evidence of the influence of workstation design on musculoskeletal discomfort during VDT use. Workstation factors accounted for significant amounts of variance in arm, leg, and trunk discomfort. By contrast, none of the demographic variables significantly predicted musculoskeletal discomfort. If musculoskeletal discomfort is a precursor to more serious disability, the significant associations between posture and workstation measures and musculoskeletal discomfort indicate reductions in morbidity can be achieved by improvement in workplace design. The investigators noted that special control measures may be necessary to relieve static loads on shoulder and neck muscles. Providing movement opportunities for the neck and shoulders is difficult through workstation design; therefore, job rotation, physical exercise, or rest breaks were recommended to minimize constrained postures. These recommendations are supported by a recently completed study of data entry operators that examined the performance effects of periodic 3-min minibreaks and frequent 30-sec pauses in the work schedule (42). The results found frequent rest pauses to have a positive effect on keying performance. On average, the adjusted keystroke rate was approximately 7 percent higher among typists who received the frequent rest breaks compared to those who followed the more traditional midmorning, lunch, and midafternoon break routine. These results should dispel concerns that frequent rest breaks would reduce productivity and would not be cost-effective.

3.5 Case 5: Investigation of Repetitive Motion Disorders in a Meatpacking Plant

3.5.1 Background

Working conditions in meat processing plants first drew widespread media attention in the late 1980s. Sometimes described as ''the most dangerous industry in America,'' meat processing industries consistently report the highest injury and illness rate of any industry in the United States (4). Red meat processing plants have also been the target of multimillion dollar fines by OSHA for high rates of CTDs, record-keeping violations, and failure to implement meaningful ergonomic programs (65).

In March 1988 an investigation of upper extremity CTDs was undertaken at a meatpacking plant in a midwestern city (66). A multidisciplinary team consisting of a physician, epidemiologist, industrial engineer, and industrial hygienist was assembled to carry out this investigation.

3.5.2 Study Design

The objectives of the evaluation were as follows (66):

1. To determine the prevalence and incidence of CTDs among employees working in selected departments and jobs
2. To identify jobs with known ergonomic risk factors for developing CTDs
3. To develop recommendations to eliminate CTD hazards

To accomplish these objectives, two study plans were developed. The medical/epidemiological study plan called for the determination of upper extremity CTD incidence and prevalence rates for the entire plant and for individual departments. The intent was to target jobs with high incidence and prevalence rates for follow-on study during the ergonomic evaluation. The incidence rate of CTDs in the previous year was calculated from entries to the plant's OSHA 200 logs. Standardized questionnaires and physical examinations were administered to plant employees to determine the prevalence rate of CTDs at the time of the investigation. A work-related upper extremity CTD was defined using the following criteria:

1. Symptoms of pain, numbness tingling, aching, stiffness, or burning reported in the neck, shoulder, elbow, hand, or wrist.
2. Symptoms did not result from accidents or sudden trauma to the joint.
3. Symptoms began after employment at the plant.
4. Symptoms occurred within the past year.
5. Symptoms lasted more than one week or occurred more than three times in the previous year.

The ergonomic study plan called for a walk-through survey of the entire plant, followed by a more detailed study of departments where jobs appeared to present a higher risk. Videotape was used to record the movement patterns of workers performing normal job activities. A strategy for assessing exposure to ergonomic risk factors was devised. Attention was focused on two previously identified risk factors for upper extremity CTDs: repetition rate and manual force requirements. Task repetitiveness was assessed by counting the number of cuts or hand movements occurring within a specific time frame and extrapolating to an 8-hr workday. Low repetitive jobs were defined as jobs requiring less than 10,000 movements or cuts per work shift or having cycle times more than 30 sec long. High repetitive jobs were defined as jobs requiring more than 20,000 movements or cuts per work shift.

Similarly, manual effort was visually estimated as low, medium or high. Using these two estimates, a measure of overall CTD risk exposure was derived.

3.5.3 Data Collected

Incidence rates for upper extremity CTDs were calculated from OSHA 200 log entries for the one-year period preceding the start of the investigation. The investigators found a high incidence rate of upper extremity CTDs and carpal tunnel syndrome among workers at the plant. The plantwide incidence rate was calculated as 41.7 CTDs per 100 full-time workers per year. This compares with an incidence rate of 6.7 cases per 100 full-time workers per year reported by the meatpacking industry as a whole, and an incidence rate of 0.1 per 100 full-time workers in general industry in 1987.

Departments with particularly high incidence rates included the Hog Kill/Pork By, Ham Bone, Pork Cut, Beef Fab and Pork Trim departments. These departments also had the highest prevalence of upper extremity CTDs determined from the questionnaires and physical examinations. Hands and wrists were the most frequently affected areas, followed by elbows, shoulders, and neck. Tendinitis and strain were the most common diagnoses, although carpal tunnel syndrome represented 17 percent of all recorded CTDs.

A total of 185 jobs representing 14 departments were videotaped during the ergonomic evaluation. A review of work activities revealed the vast majority of jobs required highly repetitive movements and high force exertions. Of the 185 jobs examined, 16 (9%) were classified as low repetitive; 23 (12%) were classified as medium repetitive; and 146 (79%) were classified as high repetitive. Similarly, 21 jobs (11%) were classified as low force, 122 (66%) were classified as medium force, and 43 (23%) were classified as high force. In addition, the tools (e.g., powered trimmers and saws) used in 21 (11%) jobs exposed workers to some form of hand–arm vibration. Forty percent of jobs in the Hog Kill/Pork By, Ham Bone, Pork Cut, Beef Fab and Pork Trim departments were identified as both highly repetitive and requiring forceful hand–wrist movements. Only two jobs involving four workers had low force *and* low repetition requirements. Although definitive data relating particular ergonomic risk factors and upper extremity CTDs could not be established in this study, the results of the ergonomic investigation agree with others, which suggest repetitive motion and force exertion contribute to a high incidence of upper extremity CTDs (67).

3.5.4 Conclusions

Based on health outcome data and the results of the ergonomic evaluation, NIOSH investigators concluded that an upper extremity CTD hazard existed at this plant. Therefore recommendations for specific engineering, administrative, and medical actions were provided to abate the CTD hazards. Recommendations included the following (66):

1. Changes in the design of knives and tool handles were recommended to reduce awkward hand/wrist postures and grip force exertion during cutting tasks. Specifically, changing the orientation of the handle with respect to the blade

was recommended in several instances to maintain the wrist in a neutral position during cutting tasks. Handles with textured surfaces were also recommended to reduce the force required to grip the knife. The investigators emphasized that handles should be made available in a variety of sizes for both left- and right-handed workers.

2. Use of automated equipment was recommended to reduce worker involvement in tasks requiring repetitive motion. For example, investigators recommended using mechanical devices to separate large pieces of meat or machines equipped with sensors to detect and remove bones from meat products.

3. Adjustable workstations, fixtures, and rotating tables were recommended to allow workers to access and manipulate the meat with less force and fewer awkward postures during cutting and trimming tasks.

4. Power tools were recommended to reduce manual force requirements of different cutting and deboning tasks. An example of a power tool used for various cutting and trimming tasks is the Whizzard knife. Frequent maintenance of power tools was recommended to minimize vibration output by the tool.

5. Management was advised to provide safety and ergonomics training to new employees. Suggested training topics included proper cutting and knife sharpening techniques to reduce manual force requirements and CTD symptom recognition. Management was also encouraged to start new employees in less strenuous cutting jobs, to allow time for muscle strengthening and conditioning.

6. Job rotation was recommended to limit worker exposure to manual stresses that could not be eliminated. The investigators cautioned, however, that different jobs may actually impose similar physical demands. For example, although packing jobs require less force than cutting jobs, both require repetitive reaching and grasping. Additional rest pauses were also recommended to allow recovery of fatigued muscle–tendon groups.

7. Initiation of a comprehensive ergonomics program was recommended. The ergonomics program would be administered by a committee consisting of management representatives, union and nonunion employees, supervisors, and the plant medical staff. The committee would be responsible for the following program components:

   a. Training workers in ergonomic and medical principles of CTD prevention

   b. Evaluating jobs systematically for risk factors related to CTD development

   c. Evaluating the program's effectiveness by surveying workers and reviewing trends in the CTD incidence rate

3.6 Case 6: Investigation of Physiological and Biomechanical Demands in a Grocery Warehousing Task

3.6.1 Background

The physical stress–strain a job produces is related to the rate of work and the ratio of work and rest cycles (18). Production standards in industry are frequently established using work measurement techniques. Work measurement is the application of methods developed to estimate the time needed by suitably qualified and motivated workers to perform a specified task at a specified level of performance (68). The objective of work measurement is to increase productivity and lower unit cost, thus allowing more goods and/or services to be produced for more consumers (69).

To achieve higher productivity, many grocery companies with warehouse operations require workers to meet production or time standards. In some cases failure to meet these standards results in disciplinary action or termination of employment (70). Time standards are based on the concept of "normal time" or the time required for the "standard operator" to perform an operation when working at a pace that is neither too fast nor too slow, without delay for personal reasons or unavoidable circumstances. These standards are established using time and motion studies, work sampling, or predetermined motion–time systems. Predetermined motion–time systems are a collection of time standards assigned to fundamental motions and groups of motions (69). Allowed times are determined by listing the basic motions required for a task, finding the time values associated with the basic motions, adding the time values to obtain the standard time, and adjusting for allowances needed by the operation.

Most predetermined motion–time systems were developed in the manufacturing sector for setting time standards for highly repetitive bench or light assembly tasks. In contrast, grocery operations, particularly in warehouses, are nonrepetitive and physically demanding. For physical work involving whole body exertion, several studies have demonstrated that work standards based on traditional techniques are not always consistent with the capacities of workers (71, 72). Specifically, many predetermined time standards provide no allowance for fatigue, since standards are supposedly based on a work rate that can be sustained for an 8-hr working day by the average healthy employee. However, it is recognized that fatigue is influenced by many conditions and that physiological and psychological factors may limit the ability to continue work.

The physical stresses posed by grocery warehouse operations were investigated to determine the effects of daily work rates on cardiovascular response and metabolic energy expenditure, and to identify stressful work postures and motions associated with order selection tasks (73). The study was initiated in response to evidence that overexertion and back pain among workers in grocery warehouse operations was a significant problem. The wholesale grocery industry is one of the 15 industries with the highest prevalence of work-related back pain. As an occupation, stock handlers have an annual estimated prevalence of back pain caused by work activities of 17.8 percent (74). Workers' compensation data from 1990 to 1992 also reveals that manual materials handling and push–pull activities cause more than 20 percent of all injuries reported by warehouse workers (75).

3.6.2 Study Design

An assessment of physiological and biomechanical stresses associated with grocery warehousing tasks was performed in a large grocery warehouse employing 145 order selectors (73). Order selectors load cases of grocery items from warehouse shelves onto pallets according to a set of written instructions known as a picking order. The picking order identifies the items and quantities to be picked, the order of picking the items, and their locations in the warehouse. The order selector drives a pallet jack to the location of items in the warehouse, lifts items from shelves, carries them to the pallet jack, lowers or lifts the items onto the pallet, and places a label on the item. The order selector then proceeds to the next item on the list. After the entire order is picked, the order selector wraps or tapes the stacked cases together and places the loaded pallet in the loading dock area for transport out of the warehouse.

Prior to the evaluation, a standard incentive program was installed by the company to establish a "fair amount of time" for order selecting activities. Achieving 100 percent of the standard was defined as a "day's work." Performance was averaged over a week. Employees were disciplined for performance below 95 percent of standard. Workers who exceeded the standard received additional pay or time off.

To evaluate the potential risk associated with manual lifting tasks, investigators collected the following information:

1. Weight of objects lifted
2. Posture of the worker in reference to the loads lifted
3. Dynamics of the lifting motion that affect spinal forces
4. Frequency and duration of manual lifting activities, that is, the temporal pattern of manual lifting, including work–rest cycles
5. Energy demands posed by work activities

To evaluate the biomechanical demands of lifting tasks, information on load weights and body postures were systematically recorded for five representative lifting tasks judged by workers and investigators as having a high potential for injury. Joint postures were recorded using electrogoniometers, tape measures, videotape, and still photographs. A triaxial lumbar motion monitor (LMM) was also worn by a small group of selectors to record the motion of the trunk about the L5/S1 intervertebral joint during their job activities. Measurements were input to the Michigan 3D Static Strength Prediction Program (SSPP) and the NIOSH lifting equation (32, 33). The Michigan 3D SSPP model estimates the compressive forces placed on the L5/S1 disc during lifting activities and provides information on muscle strength requirements for designated lifts. The NIOSH lifting equation provides a recommended weight limit (RWL) based on the characteristics of a specific lifting task. The ratio of the RWL to the actual load lifted [known as the lifting index (LI)] can be used as an indicator of the risk for developing lifting-related back pain as a result of specific lifting activities (33).

Energy demands were assessed using various physiological measures and an energy expenditure model (76). Three participants were fitted with a portable heart rate monitor and an Oxylog oxygen consumption meter to continuously record heart rate and oxygen consumption during routine order selecting activities. During the study, workers were told to use their normal work method for selecting orders and to work at a pace close to the standard pace. Workers were videotaped while performing order picking tasks, and the time required to complete orders was determined using a stopwatch. Investigators were also provided with a computer printout showing the number of items in each order, the weight and volume of each item, and its location in the warehouse. Wet- and dry-bulb temperature and air velocity were measured in different locations of the warehouse at different times during the shift.

3.6.3 Data Collected

The results of the study are summarized in Tables 10.9 and 10.10. Based on computer data, investigators determined that selectors performed an average of 4.1 lifts per minute during their normal work activities. Measurements indicated that 53 percent of these lifts required extreme trunk flexion, and 40 percent required reaches above shoulder height. The investigators noted that many of the storage racks were double-tiered and that retrieving grocery items from these racks frequently required stooping, kneeling, or crawling (to access items on the lower rack) or climbing over boxes and reaching overhead (to access items on the upper rack). Furthermore, order selectors were often required to stack items up to 90 inches high on their pallets, which also resulted in additional climbing and reaching.

Evaluation of five representative lifting tasks using the NIOSH lifting equation (33) indicated that all of the loads exceeded the RWL by significant margins. Furthermore, evaluation of the same tasks using the Michigan 3D SSPP indicated that four of the five tasks generated compressive forces on the L5/S1 disc in excess of 770 lb. Compressive force values of 770 lb and greater have been associated with increasing rates of low back pain and lost work time (77). The model also indicated that only a fraction of the U.S. male workforce would have the strength capacity at both the hip and torso necessary to perform the five lifting tasks in the attendant postures.

Mean metabolic rates, as measured by oxygen consumption and predicted using Garg's metabolic model, exceeded 5.0 kcal/min, recommended in the literature as an upper limit for young healthy male workers during an 8-hr workday (78–81). Average heart rate levels for two of the three workers examined exceeded 110 beats per minute, also suggested as the maximum acceptable for the majority of healthy workers (80, 82). Based on energy expenditure and heart rate criteria, the investigators concluded that energy demands of the job would pose a significant risk of overexertion injury from excessive physical fatigue for the majority of healthy workers who would perform this type of job.

Table 10.9 Grocery Warehouse Study Results—Biomechanical Analyses (73)

Task	Load (lb)	NIOSH RWL (lb)	NIOSH LI	Disc Compression Force (lb)	% Capable Hip	% Capable Torso
1	30	7.2	4.2	930	63	94
2	38	6.8	5.6	830	70	95
3	42	5.2	8.1	896	16	76
4	38	5.2	7.3	662	52	59
5	58	7.2	8.0	801	55	86

Table 10.10 Grocery Warehouse Study Results—Physiological Data (73)

Participant	Variable	Value
1	Cases/order	167
	Weight/order (lb)	2198
	Allowed time/order (min)	34.9
	Working metabolic rate (kcal/min)	5.4
	Working heart rate (beats/min)	122
	Predicted metabolic rate (kcal/min)	6.0
2	Cases/order	138
	Weight/order (lb)	4220
	Allowed time/order (min)	36.7
	Working metabolic rate (kcal/min)	5.9
	Working heart rate (beats/min)	104
	Predicted metabolic rate (kcal/min)	5.0
3	Cases/order	101
	Weight/order (lb)	3862
	Allowed time/order (min)	25.8
	Working metabolic rate (kcal/min)	8.0
	Working heart rate (beats/min)	131
	Predicted metabolic rate (kcal/min)	7.6

3.6.4 Conclusions

Based on quantitative biomechanical and physiological data collected during this study, the investigators concluded that order selectors in the warehouse examined are at substantial risk for low back injury. Furthermore, the investigators concluded that the performance standards encouraged and contributed to excessive levels of exertion. To avoid excessive fatigue, injury, and illness, the investigators recommended several changes in the layout of the warehouse and the organization of the work. These included the following (73):

1. Use of objective measures of physical effort (such as heart rate and oxygen uptake) to determine reasonable performance standards that do not increase the workers' risk of injury.

2. Arranging grocery items on racks such that heavy items are stored at or above knee height, and only light, nonbreakable items are stored in top racks.

3. Restricting pallet heights to 60 inches or less.

4. Use of administrative controls, including rotating order selectors to less physically demanding jobs, following heavy or difficult orders with less difficult orders, and limiting overtime work.

4 SUMMARY

It has been shown in this chapter that no industry is immune to the effects of poor ergonomic workplace design. The application of ergonomic principles to the design of tools and equipment, workstations, and work methods can significantly impact the health, safety, and productivity of workers in virtually all work settings. The key to success in making ergonomic interventions is securing the commitment of management and employees to a continuing process of identifying, evaluating, and controlling ergonomic hazards. No ergonomic investigation is complete without periodic follow-up or review. Periodic surveys should be conducted to identify previously unnoticed risk factors or deficiencies in work practice or engineering controls. Finally, factoring ergonomic considerations into the early design stage of a tool or piece of equipment, a workstation layout, a job task, or a work process remains the optimal approach to preventing biomechanical and other stress-related injuries to which this chapter has alluded.

ACKNOWLEDGMENT

Major Phoebe C. Fisher, U.S. Air Force, assisted in first efforts at developing the content of this chapter during her fellowship appointment to the National Institute for Occupational Safety and Health. Her fellowship was sponsored by the Air Force Institute of Technology for the period 1990–1991.

REFERENCES

1. L. Strother, Ed., *Human Factors and Ergonomics Society Directory and Yearbook*, Human Factors and Ergonomics Society, Santa Monica, CA, 1994.
2. *Ergonomics*, UAW/GM Assembly Division, Oklahoma City.
3. U.S. Department of Labor, Occupational Safety and Health Administration, *Ergonomic Safety and Health Management; Proposed Rule*, 29 CFR 1910, August 3, 1992.
4. U.S. Department of Labor, Bureau of Labor Statistics, *Workplace Injuries and Illnesses in 1992*, USDL-93-553, December, 1993.
5. *Accident Facts*, National Safety Council, Chicago, 1990.
6. V. Putz-Anderson, Ed., *Cumulative Trauma Disorders: a Manual for Musculoskeletal Diseases of the Upper Limb*, Taylor & Francis, Philadelphia, 1988.
7. D. T. Ridyard, A Successful Applied Ergonomics Program for Preventing Occupational Back Injuries, in *Advances in Industrial Ergonomics and Safety II*, B. Das, Ed., Taylor & Francis, Philadelphia, 1990, pp. 125–132.
8. F. McKenzie, J. Storment, P. Van Hook, and T. Armstrong, A Program for Control of Repetitive Trauma Disorders Associated with Hand Tool Operations in a Telecommunications Manufacturing Facility, *Am. Ind. Hyg. Assoc. J.*, **46**, 674–678 (1985).
9. G. Lutz and T. Hansford, Cumulative Trauma Disorder Controls: The Ergonomics Program at Ethicon, Inc., *J. Hand Surg.*, **12A**(2Pt2), 863–866 (1987).
10. E. R. Tichauer, Ergonomics, in *Patty's Industrial Hygiene and Toxicology*, 4th rev. ed., Vol 1B, *General Principles*, G. D. Clayton and F. E. Clayton, Eds., Wiley, New York, 1991, Chapter 40.
11. A. Cohen and F. N. Dukes-Dobos, Applied Ergonomics, in *Patty's Industrial Hygiene and Toxicology*, 2nd rev. ed., Vol 3B, *Biological Responses*, L. Cralley and L. Cralley, Eds., Wiley, New York, 1985, Chapter 8.

12. *The Industrial Environment—Its Evaluation and Control*, National Institute for Occupational Safety and Health, U.S. Government Printing Office, Washington, D.C., 1973.

13. V. Putz-Anderson, T. Pizatella, and S. Tanaka, *A Proposed National Strategy for the Prevention of Musculoskeletal Injuries*, National Institute for Occupational Safety and Health and Associations of Schools of Public Health, Washington, DC, 1986, pp. 17–34.

14. B. S. Joseph, Ergonomic Considerations and Job Design in Upper Extremity Disorders, in *Occupational Medicine: State of the Art Reviews*, M. Kasdan, Ed., Hanley & Belfuls, Philadelphia, 1989, pp. 547–557.

15. M. S. Sanders and E. J. McCormick, *Human Factors in Engineering and Design*, McGraw-Hill, New York, 1987.

16. D. C. Alexander and B. M. Pulat, *Industrial Ergonomics: A Practitioner's Guide*, Institute of Industrial Engineers, Norcross, GA, 1985.

17. L. J. Fine, Active Surveillance for Cumulative Trauma Disorders, in *Proceedings—Occupational Cumulative Trauma Disorders of the Upper Extremities*, P. Boman, Ed., Center for Occupational Health and Safety Engineering, University of Michigan, Ann Arbor, MI, 1983.

18. D. J. Habes and V. Putz-Anderson, The NIOSH Program for Evaluating Biomechanical Hazards in the Workplace, *J. Safety Res.*, **16**, 49–60 (1985).

19. N. Corlett and R. Bishop, The Ergonomics of Spot Welders, *Appl. Ergon.*, **9**, 23 (1978).

20. J. N. Katz, M. G. Larson, A. H. Fossel, and M. H. Liang, Validation of a Surveillance Case Definition of Carpal Tunnel Syndrome, *Am. J. Public Hlth.*, **81**, 189–193 (1991).

21. Centers for Disease Control, "Occupational Disease Surveillance—Carpal Tunnel Syndrome," *MMWR*, **38**, 485–489 (1989).

22. T. B. Snyder, J. Himmelstein, G. Pransky, and J. D. Beavers, "Business Analysis in Occupational Health and Safety Consultations," *J. Occupat. Med.*, **33**, 1040–1045 (1991).

23. C. G. Drury, B. Paramore, H. P. Van Cott, S. M. Grey, and E. N. Corlett, Task Analysis, in *Handbook of Human Factors*, G. Salvendy, Ed., Wiley, New York, 1987, pp. 370–401.

24. Eastman Kodak Company, *Ergonomic Design for People at Work*, Vol. 2, Van Nostrand Reinhold Company, New York, 1986.

25. A. Chapanis, *Research Techniques in Human Engineering*, Johns Hopkins University Press, Baltimore, 1962.

26. R. M. Barnes, *Motion and Time Study, Design and Measurement of Work*, 7th ed., Wiley, New York, 1983.

27. T. J. Armstrong, J. A. Foulke, B. S. Joseph, and S. A. Goldstein, Investigation of Cumulative Trauma Disorders in a Poultry Processing Plant, *Am. Ind. Hyg. Assoc. J.*, **43**, 103–116 (1982).

28. C. G. Drury, Task Analysis Methods in Industry, *Appl. Ergon.*, **14**, 19–28 (1983).

29. W. Rohmert and K. Landau, *A New Technique for Job Analysis*, Taylor and Francis, London, 1983.

30. Y. Lifshitz and T. Armstrong, A Design Checklist for Control and Prediction of Cumulative Trauma Disorders in Hand Intensive Manual Jobs, *Proceedings of the 30th Annual Meeting of Human Factors Society*, 1986, pp. 837–841.

31. V. Mathiowetz, N. Kashman, G. Volland, K. Weber, M. Dowe, and S. Rogers, Grip

and Pinch Strength: Normative Data for Adults, *Archives Phys. Med. Rehab.*, **66**, 69–72 (1985).

32. D. B. Chaffin and G. B. J. Andersson, *Occupational Biomechanics*, 2nd ed., Wiley, New York, 1991.

33. T. R. Waters, V. Putz-Anderson, A. Garg, and L. J. Fine, Revised NIOSH Equation for the Design and Evaluation of Manual Lifting Tasks, *Ergonomics*, **36**, 749–776 (1993).

34. D. Meister, *Behavioral Analysis and Measurement Methods*, Wiley, New York, 1985.

35. R. R. Rosa, D. D. Wheeler, J. S. Warm, and M. J. Colligan, Extended Workdays: Effects on Performance and Ratings of Fatigue and Alertness, *Behav. Res. Methods, Instru. Comput.*, **17**, 6–15 (1985).

36. G. Borg, Psychophysical Scaling with Applications in Physical Work and the Perception of Exertion, *Scandinavian J. Work Environ. Hlth.*, **16**, 55–58 (1990).

37. F. Gamberale, Perceived Exertion, Heart Rate, Oxygen Uptake and Blood Lactate in Different Work Operations, *Ergonomics*, **15**, 545–554 (1972).

38. G. Borg and D. Ottoson, Eds., *The Perception of Exertion in Physical Work*, Mcmillan Press, London, 1986.

39. *Ergonomics Program Management Guidelines for Meatpacking Plants*, OSHA 3123, U.S. Department of Labor, Occupational Safety and Health Administration, Washington, D.C., 1990.

40. Eastman Kodak Company, *Ergonomic Design for People at Work*, Vol. 1, Van Nostrand Reinhold, New York, 1983.

41. W. E. Woodson, *Human Factors Design Handbook*, McGraw-Hill, New York, 1981.

42. S. L. Sauter and N. G. Swanson, Having Your Cake and Eating It Too: Increased Rest Breaks Yield Increased Productivity in Repetitive VDT Work, 35th Annual Meeting of Human Factors Society, San Francisco, CA, 1991.

43. B. Jonsson, The Static Load Component in Muscle Work, *Europ. J. Appl. Physiol.*, **57**, 305–310 (1988).

44. J. R. Glover, Prevention of Back Pain, in *The Lumbar Spine and Back Pain*, M. Jayson, Ed., Grune and Stratton, New York, 1976.

45. M. Bergquist-Ullman and U. Larsson, Acute Low Back Pain in Industry: A Controlled Prospective Study with Special Reference to Therapy and Confounding Factors, *Acta Orthoped. Scandin.*, **170**, 1–117 (1977).

46. D. B. Chaffin, L. S. Gallay, C. B. Woolley, and S. R. Kucimeba, An Evaluation of the Effect of a Training Program on Worker Lifting Postures, *Internat. J. Ind. Ergon.*, **1**, 127–136 (1986).

47. S. Varynen and U. Kononen, Short and Long-Term Effects of a Training Programme on Work Postures in Rehabilitees: A Pilot Study of Loggers Suffering from Back Troubles, *Internat. J. Ind. Ergon.*, **7**, 103–110 (1991).

48. J. B. Moran and R. M. Ronk, Personal Protective Equipment, in *Handbook of Human Factors*, G. Salvendy, Ed., Wiley, New York, 1987, pp. 876–894.

49. A. Amendola, An Investigation of the Effects of External Supports on Manual Lifting, Ph.D. Dissertation, Texas A&M University, College Station, TX 1989.

50. M. Thun, S. Tanaka, A. B. Smith, W. E. Halperin, S. T. Lee, M. E. Luggen, and E. V. Hess, Morbidity from Repetitive Knee Trauma in Carpet and Floor Layers, *Brit. J. Ind. Med.*, **44**, 611–620 (1987).

51. National Institute for Occupational Safety and Health, *NIOSH Alert: Preventing Knee*

Disorders in Carpet Layers, DHHS (NIOSH) Publication No. 90-104, Cincinnati, OH 1990.

52. A. Bhattacharya, M. Mueller, and V. Putz-Anderson, Traumatogenic Factors Affecting the Knees of Carpet Installers, *Appl. Ergon.*, **16**, 243–250 (1985).

53. B. G. Ariel, Biomechanical Analysis of the Knee Joint During Deep Knee Bends With Heavy Load, in *Biomechanics IV*, R. C. Nelson and C. A. Moorehouse, Eds., University Park Press, Baltimore, MD, 1974.

54. A. Bhattacharya, D. Habes, and V. Putz-Anderson, "Suggested Preliminary Guidelines for the Development of Work Practices Procedures for Carpet Layers," American Industrial Hygiene Conference, Las Vegas, NV, April 1985.

55. L. Parmeggiani, Technical Ed., *Encyclopedia of Occupational Health and Safety*, International Labor Office, Geneva, Switzerland, 1983.

56. G. M. Martin, Hidden Jeopardy for Jewelry Workers, *Job Safety Hlth.*, **6**, 22–31 (1978).

57. T. Hales, K. Grant, W. Daniels, and D. Habes, F. L. Thorpe & Co., Inc., HETA 90-273-2130, National Institute for Occupational Safety and Health, Cincinnati, OH, 1991.

58. *Checkstand Usage in the United States*, Research Requirements Sub-Committee, FMI Ergonomics Task Force, 1990.

59. S. Barnhart and L. Rosenstock, Carpal Tunnel Syndrome in Grocery Checkers: a Cluster of Work-Related Illnesses, *Western J. Med.*, **147**, 37–40 (1987).

60. C. C. Rodrigues, A Technical Analysis of the Supermarket Laser Scanning Workstation, in *Advances in Industrial Ergonomics and Safety I*, A. Mital, Ed., Taylor & Francis, Philadelphia, 1989.

61. D. L. Orgel, M. J. Milliron, and L. J. Frederick, Kroger Company, HETA 88-345-2031, National Institute for Occupational Safety and Health, Cincinnati, OH, 1990.

62. C. J. MacKay, Work with Visual Display Terminals: Psychosocial Aspects and Health, *J. Occupat. Med.*, **31**, 957–968 (1989).

63. S. L. Sauter, L. M. Schleifer, and S. J. Knutson, Work Posture, Workstation Design and Musculoskeletal Discomfort in a VDT Data Entry Task, *Human Factors*, **33**, 151–168 (1991).

64. L. M. Schleifer and S. L. Sauter, Controlling Glare Problems in the VDT Work Environment, *Library Hi Tech*, **3**, 21–25 (1985).

65. K. Thompson, Action Update: Association, Union and Government Safety Activities, *Meat Poultry*, **35**, 24 (1989).

66. T. Hales, D. Habes, L. Fine, R. Hornung, and J. Boiano, John Morrell & Co., HETA 88-180-1958, National Institute for Occupational Safety and Health, Cincinnati, OH, 1989.

67. B. A. Silverstein, L. J. Fine, and T. J. Armstrong, Hand Wrist Cumulative Trauma Disorders in Industry, *Brit. J. Ind. Med.*, **43**, 779–784 (1986).

68. A. M. Genaidy, A. Mital, and M. Obeidat, The Validity of Predetermined Motion Time Systems for Setting Production Standards for Industrial Tasks, *Internat. J. Ind. Ergon.*, **3**, 249–263 (1989).

69. B. W. Niebel, *Motion and Time Study*, Richard D. Irwin, Homewood, IL, 1988.

70. A. Garg, G. Hagglund, and K. Mericle, Physical Fatigue and Stresses in Warehouse Operations, NIOSH Contract Report No. 210-81-6008, National Institute for Occupational Safety and Health, Cincinnati, OH, 1981.

71. A. Garg and U. Saxena, Maximum Frequency Acceptable to Female Workers for One-Handed Lifts in the Horizontal Plane, *Ergonomics*, **25**, 839–853 (1982).

72. A. Mital, S. S. Asfour, and F. Aghazadeh, Limitations of MTM in Accurate Determination of Work Standards for Physically Demanding Jobs, in *Trends in Ergonomics/ Human Factors IV*, S. S. Asfour, Ed., Elsevier Science Publishers, Amsterdam, 1987.

73. V. Putz-Anderson, T. Waters, S. Baron, and K. Hanley, Big Bear Grocery Warehouse, HETA 91-405-2340, National Institute for Occupational Safety and Health, Cincinnati, OH, 1993.

74. H. Guo, S. Tanaka, L. L. Cameron, P. J. Seligman, V. J. Behrens, J. Ger, D. K. Wild, and V. Putz-Anderson, Back Pain Among Workers in the United States: National Estimates and Workers at High Risk, Presentation at the Epidemiologic Intelligence Service Conference, Centers for Disease Control and Prevention, Atlanta, GA, April 1993.

75. *Voluntary Ergonomics Program Management Guidelines*, National American Wholesale Grocers' Association and International Food Service Distributors Association, 1992.

76. A. Garg, D. B. Chaffin, and G. D. Herrin, Prediction of Metabolic Rates for Manual Materials Handling Jobs, *Am. Ind. Hyg. Assoc. J.*, **39**, 661–674 (1978).

77. G. D. Herrin, M. Jaraiedi, and C. K. Anderson, Prediction of Overexertion Injuries Using Biomechanical and Psychophysical Models, *Am. Ind. Hyg. Assoc. J.*, **47**, 322–330 (1986).

78. G. Lehmann, Physiological Measurements as a Basis of Work Organization in Industry, *Ergonomics*, **1**, 328–344 (1958).

79. B. Bink, The Physical Working Capacity in Relation to Working Time and Age, *Ergonomics*, **5**, 25–28 (1962).

80. S. H. Snook and C. H. Irvine, Psychophysical Studies of Physiological Fatigue Criteria, *Human Factors*, **11**, 291 (1969).

81. P. Astrand and K. Rodahl, *Textbook of Work Physiology: Physiological Bases of Exercise*, McGraw-Hill, New York, 1986.

82. L. Brouha, *Physiology in Industry*, Pergamon Press, Oxford, 1967.

CHAPTER ELEVEN

Abnormal Pressure

Richard D. Heimbach, M.D., Ph.D., and Alfred A. Bove, M.D., Ph.D.

1 INTRODUCTION

Persons are exposed to decreased barometric pressure relative to sea level, at altitude in the atmosphere and in space flight, and to increased barometric pressure in undersea work, exploration, and recreation. An emerging use of pressure greater than sea level is the use of high-dose, short-term oxygen inhalation therapy (hyperbaric oxygen) in treatment of a limited group of medical disorders. Once the realm of flight surgeons, diving medical officers, divers, and tunnel and bridge building engineers, altered barometric pressure today affects the lives of workers, passengers in aircraft, recreational scuba divers and aviators. Excursions of sea level dwellers to mountainous regions for business or recreation is another important exposure.

Medical uses of hyperbaric oxygen have introduced a new dimension to the need for full engineering safety efforts related to exposures of patients and medical staff to altered barometric pressure as well as associated gas purity standards and fire and structural safety concerns.

The central theme of this chapter is represented by the interrelated physiological effects of the continuum of altered barometric pressure—higher or lower than sea level.

1.1 Physics of Abnormal Environmental Pressure

The weight of atmospheric air and water at any point on or above the surface of the earth or beneath the surface of the water may be expressed in various units of

Patty's Industrial Hygiene and Toxicology, Third Edition, Volume 3, Part B, Edited by Lewis J. Cralley, Lester V. Cralley, and James S. Bus.
ISBN 0-471-53065-4 © 1995 John Wiley & Sons, Inc.

Table 11.1 Pressure Equivalents to Altitude and to Depth in Seawater[a]

	Feet	ATM	mm Hg	psi
Altitude	16000	0.542	412	8.0
above	12000	0.636	483	9.3
sea level	8000	0.742	564	10.9
	4000	0.863	656	12.7
Sea level	0	1	760	14.7
	33	2	1520	29.4
Depth	66	3	2280	44.1
in	99	4	3040	58.8
seawater	132	5	3800	73.5
	165	6	4560	88.2

[a]Sea level is considered zero depth and zero altitude. ATM, pressure in atmospheres absolute, psi, pressure in pounds/square inch.

pressure. The major difference between pressure changes at altitude and at depth is the curvilinear change of pressure at altitude compared to the linear pressure changes at depth. On ascent to altitude, we pass through the blanket of air that extends from sea level to an altitude of about 430 miles, where the number of molecules is so small that essentially no collisions occur and pressure is absent. At sea level the atmosphere weighs 14.7 pounds per square inch (psi) or 760 mm Hg (760 torr). Pressure at any altitude is the weight of air above. However, due to the compressibility of gas, the pressure change per foot of altitude change decreases as altitude increases. For example, with ascent from sea level to 3048 m (10,000 ft), a pressure reduction of 237 mm Hg from 760 to 523 mm Hg is seen, whereas the same 3048-m altitude change from 9146 m (30,000 ft) to 12,192 m (40,000 ft) gives a pressure reduction of only 85 mm Hg from 226 to 141 mm Hg. At depth pressure change is linear because each foot of seawater weighs 0.445 psi, thus 14.7 ÷ 0.445 = 33 ft of seawater or about 10 m, which is equivalent to a full atmosphere of pressure. The barometric pressure (P_B) at depth can be referred to in terms of gauge pressure (pressure in excess of sea level pressure) or more commonly, as absolute pressure (the total of sea level atmospheric pressure plus the water pressure). The most commonly used pressure units are millimeters of mercury (mm Hg), feet of seawater (FSW), meters of seawater (MSW), or atmospheres absolute (ATA). Examples are shown in Table 11.1.

2 DECREASED PRESSURE

2.1 Altitude Physiology

Air is a mixture of gases, and the percentage of gases in this mixture remains relatively constant with increasing altitude. Oxygen makes up about 21 percent and nitrogen about 79 percent of this mixture. Traces of carbon dioxide, water vapor, and other inert gases are so small that they are included with the nitrogen percentage for calculation. Dalton's law states that in a mixture of gases, the total pressure is the sum of partial pressures of gases in the mixture. For example, in air at sea level, partial pressure of oxygen (PO_2) equals 0.21×760, or 160 mm Hg and partial pressure of nitrogen (PN_2) equals 0.79×760, or 600 mm Hg. The key point is that at altitude, gas percentages remain constant but partial pressures of gases change. For example, at 18,000 ft where barometric pressure is 380 mm Hg or one-half an atmosphere, $PO_2 = 0.21 \times 380$ mm Hg or 80 mm Hg and $PN_2 = 300$ mm Hg. At normal body temperature the vapor pressure of water is 47 mm Hg; thus by the time inspired dry air reaches the trachea, it has equilibrated at a PH_2O of 47 mm Hg, which then remains constant in the alveolar air. Venous blood returns carbon dioxide from the body's normal metabolism to the pulmonary circulation at PCO_2 of 40 mm Hg. Admixture of carbon dioxide diffused across the alveolocapillary membrane into the alveolus results in an aveolar PCO_2 of about 40 mm Hg in the normal person. Thus to calculate the approximate alveolar partial pressure of oxygen (PAO_2) at sea level or at any altitude, the simplified alveolar gas equation may be used if we assume the respiratory quotient (RQ) to be 1:

$$RQ = \frac{CO_2 \text{ output (mL/min)}}{O_2 \text{ consumption (mL/min)}}$$

The simplified alveolar gas equation is

$$PAO_2 = FIO_2(PB - PAH_2O) - PACO_2$$

where $\quad PAO_2$ = alveolar PO_2
$\quad\quad\quad FIO_2$ = fraction of inspired O_2
$\quad\quad\quad PB$ = total barometric pressure
$\quad\quad PAH_2O$ = alveolar PH_2O
$\quad\quad PACO_2$ = alveolar PCO_2

For example, at sea level breathing air:

$$PAO_2 = 0.21(760 - 47) - 40$$

$$PAO_2 = 109 \text{ mm Hg}$$

and breathing air at 18,000 ft,

$$PAO_2 = 0.21(380 - 47) - 40$$
$$PAO_2 = 30 \text{ mm Hg}$$

The same equation can be used to determine the FIO_2 needed to remain above hypoxic levels at altitude.

For example, breathing 100 percent O_2 at 18,000 ft

$$PAO_2 = 1.0(380 - 47) - 40$$

$$PAO_2 = 293 \text{ mm Hg}$$

Nitrogen is inert and maintains equilibrium between alveolar, blood, and tissue PN_2. At sea level, for example,

$$\text{Alveolar } PN_2 = 760 - (PAO_2 + PACO_2 + PAH_2O)$$

$$PN_2 = 760 - (109 + 40 + 47)$$

$$PN_2 = 654 \text{ mm Hg}$$

2.2 Oxygen Transport—Lung

The transfer of gases at the alveolocapillary membrane occurs by diffusion and can be represented by Figure 11.1, a schematic model of an alveolus and capillary. The ambient PO_2 of 160 mm Hg falls to an alveolar PO_2 of 103 mm Hg after dilution with water vapor and CO_2 returned by venous blood. With passage through pulmonary capillaries, arterialized blood leaves the lungs with a PAO_2 of about 100 mm Hg.

Oxygen diffusion occurs as the alveolar PO_2 is exposed to the respiratory membrane of some 0.36–2.5 mm thickness over its 70 m^2 of surface area. By means of the almost solid network of capillaries that fill the respiratory membrane, the 140 mL of blood contained in the pulmonary capillaries at any one time is exposed to alveolar oxygen. Red blood cells "squeeze" through the narrow 8 μm-diameter capillaries providing close proximity to alveolar gases for rapid diffusion from the

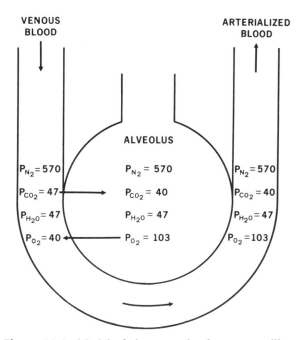

Figure 11.1 Model of alveous and pulmonary capillary.

higher 104 mm Hg alveolar PO₂ to the lower O₂ concentration of about 44 mm Hg in returning venous blood.

2.3 Oxygen Transport from Lungs to Tissue

The unique properties of hemoglobin and the way it combines with and releases oxygen constitute a key element in altitude and diving physiology. The presence of hemoglobin allows blood to transport 30–100 times as much oxygen as would be possible by physical solution of dissolved oxygen alone.

Normally, 97 percent of oxygen is carried by hemoglobin and 3 percent is dissolved in plasma and cells. The so-called oxygen buffer function of hemoglobin is based on the shape of the hemoglobin–oxygen dissociation curve (Fig. 11.2). Thus with hypoxia at altitude, PAO₂ can drop to 60 mm Hg and hemoglobin saturation is maintained at about 90 percent. No matter how high the PAO₂ goes with oxygen breathing at depth, only 100 percent hemoglobin saturation is possible and additional oxygen is carried only by its inefficient physical solution in plasma.

Hemoglobin concentration is expressed as grams-percent or grams of hemoglobin per 100 mL of blood (1). An average, healthy adult male has about 15 g of hemoglobin per 100 mL of blood or 15 grams-percent (g %). Each gram of hemoglobin can combine with 1.34 mL of oxygen. Thus a person with 15 g % of hemoglobin can carry $1.34 \times 15 = 20.1$ mL of oxygen per 100 mL of blood or 20.1 volumes-percent (v %).

Hemoglobin uptake and release of oxygen is described by the hemoglobin–oxygen dissociation curve (Fig. 11.2) whose features are of great significance in altitude physiology (2).

The flat portion of the curve between 60 and 110 mm Hg allows a significant drop in alveolar PO₂ before there is a physiologically important drop in blood oxygen

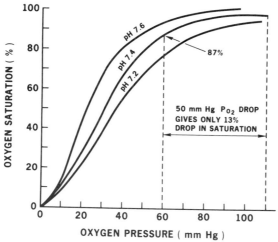

Figure 11.2 Hemoglobin–oxygen dissociation curve.

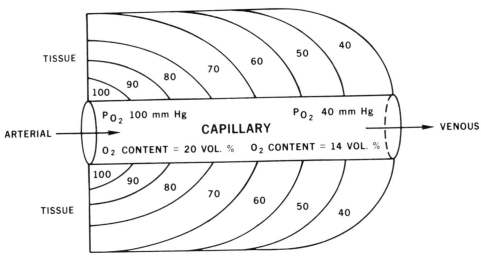

Figure 11.3 Capillary–tissue oxygen model.

saturation. It is this characteristic of the curve that allows ascent to 3048 m (10,000 ft) altitude breathing air, where alveolar PO_2 is 60 mm Hg, without symptoms of hypoxia. Further ascent above 3048 m (10,000 ft) results in a rapid drop in arterial oxygen saturation as small drops in oxygen pressure result in large decreases in arterial saturation along the steep portion of the curve.

The curve also shows the profound effect of acidosis or alkalosis as the shape of the curve is altered by changes in pH. Thus, for example, hyperventilation, a normal compensatory mechanism to hypoxia, results in a lowering of PCO_2 and a resultant rise in pH. The shift of the curve to the left illustrates the body's attempt to increase efficiency of oxygen uptake at the lung level and thus transport increased amounts of oxygen to tissues. On the other hand, the elevated PCO_2 and acidosis seen in exercise at altitude aggravates hypoxia by shifting the curve to the right.

Figure 11.3 is a simplification of the Krough–Erlang tissue oxygen model (3). The cylinder of tissue supplied by a given capillary is seen as a spectrum of oxygen tensions regulated by the arterial oxygen partial pressure supplied, distance from the capillary, and location along the capillary as oxygen is extracted during flow of blood toward the venous end.

3 PHYSIOLOGIC EFFECTS OF EXPOSURE TO ALTITUDE

Several important physiologic changes occur with exposure to altitude and are outlined below. Some of these changes are adaptive while others underlie pathologic consequences (4–7).

1. Increase in pulmonary ventilation mediated by the effects of hypoxia on the carotid bodies.

2. Alkalosis and a decrease in arterial PCO_2 as a result of hyperventilation.
3. Decrease in resting PO_2 and arterial oxygen saturation.
4. Increase in cardiac rate in response to exercise at altitude but decrease in maximal cardiac rate.
5. Decrease in stroke volume and cardiac output.
6. Decrease in left ventricular dimensions.
7. Movement of fluid from the vascular bed to the intracellular space evidenced by a decrease in plasma volume and probably the result of cellular hypoxia.
8. Acute increase in sympathetic activity gradually subsiding in 1–2 weeks usually followed by a decrease in sympathetic activity to levels below normal.
9. Increase in red cell mass secondary to hypoxic stimulation of renal erythropoietic production.

4 HYPOXIA

Without supplemental oxygen, the physiological effects of the reduced atmospheric PO_2 at altitude are described as *hypoxic hypoxia*. Other factors that lead to tissue hypoxia can coexist and be additive. Thus a patient whose oxygen transport capacity is impaired by anemia or carbon monoxide bound hemoglobin (*hypemic hypoxia*) or whose tissue oxygen utilization is impaired by cyanide or ethanol (*histotoxic hypoxia*) will be more susceptible to hypoxia at altitude. A special potentiating factor in aviation is *stagnant hypoxia* in an aviator exposed to high head-to-foot acceleration forces as in recovering from an aircraft dive. The increased weight of the column of blood reduces cerebral perfusion with resultant cerebral hypoxia, and redistributes blood in the lung, which causes arterial hypoxemia. Although these factors can be important contributors to hypoxia in aviation, this discussion centers on hypoxia caused by reduction in PO_2 at decreased barometric pressure.

As previously discussed, the shape of the hemoglobin–oxygen dissociation curve is beneficial up to 3048 m (10,000 ft). Except for a decrement in night vision at about 1220–1829 m (4000–6000 ft) and above, there are no significant effects of hypoxia in healthy people below 3048 m (10,000 ft). In general, pressurized aircraft cabins are maintained well below that altitude (see below) and no supplemental oxygen is required. However, patients with reduced cardiac reserve may have difficulty tolerating even this small drop in arterial oxygen saturation and will require supplemental oxygen during flight.

4.1 Manifestations

Table 11.2 summarizes the major manifestations of altitude hypoxia with corresponding PAO_2 breathing ambient air. The time required for the onset of symptoms listed in Table 11.2 varies depending on age, fitness, acclimatization, and individual susceptibility. For example, at the altitude of 3658 m (12,000 ft) the symptoms shown may not appear in a subject at rest for several hours. At higher altitudes the progress

Table 11.2 Responses to Hypoxic Hypoxia

Altitude	Symptoms	Alveolar PO_2 (mm Hg)
3,048 m (10,000 ft)	Impaired judgment and ability to perform calculations; increased heart rate and respiratory rate	60
3,658 m (12,000 ft)	Shortness of breath, impaired ability to perform complex tasks, headache, nausea, decreased visual acuity	52
4,573 m (15,000 ft)	Decrease in auditory acuity, constriction of visual fields, impaired judgment, irritability; exercise can lead to unconsciousness	46
5,486 m (18,000 ft)	Threshold for loss of consciousness in resting unacclimatized individuals after several hours exposure	40
6,706 m (22,000 ft)	Almost all individuals unconscious after sufficient exposure time	30
12,802 m (42,000 ft)	Inability to perform useful function in 15 sec or less	—

of symptoms to the point of inability to perform useful functions [time of useful consciousness (TUC)] is quite rapid and within a narrower range of individual variation. For example, the TUC at 9146 m (30,000 ft) is about 90 sec and at 13,106 m (43,000 ft) about 15 sec or less.

4.2 Prevention

The use of pressurized cabins to stay within the physiologically comfortable zone below 3048 m (10,000 ft) is the main preventive factor in modern aviation. When aviators must be exposed to higher cabin altitudes, the percentage of supplemental oxygen required to maintain an alveolar PO_2 no less than that at 3048 m (10,000 ft) breathing air can be derived by varying the FIO_2 in the simplified alveolar gas equation. For example, 100 percent oxygen is required at 12,192 m (40,000 ft) where the barometric pressure is 141 mm Hg and $PACO_2$ is somewhat lower because of hyperventilation:

$$PAO_2 = 1.0(141 - 47) - 35$$

$$= 59 \text{ mm Hg}$$

This is equivalent to breathing air at 3048 m (10,000 ft) and is the maximum altitude at which 100 percent oxygen breathing is adequate to prevent hypoxia. At higher altitudes positive pressure breathing and, finally, pressure suits are required (8).

4.3 Aircraft Cabin Pressurization

The most commonly employed method to avoid the risks of exposure of aircrews and passengers to the low barometric pressure and partial pressures of oxygen at altitude is to pressurize the aircraft cabin. There are two basic factors in cabin pressurization; (a) air compressors powered by the aircraft engines force outside air continuously into the cabin whose structural integrity (b) is able to safely withstand differential pressures of 8 psi or more with compensatory relief valves to continuously dump air overboard (9). A commonly used example in aviation is a differential pressure (cabin pressure above ambient pressure) of 8.19 psi, so that at flight level (FL) 12,000 m (40,000 ft; 2.72 psi) the addition of the 8.19 psi pressurization gives a total cabin pressure of 10.91 psia (pounds per square inch absolute) equivalent to 2400 m (8000 ft). From previous discussions, it is clear that 2400 m (8000 ft) is within the physiologically safe zone where no supplemental oxygen is required to maintain sufficient arterial oxygen saturation.

4.4 Limitations of the Pressurized Cabin

As the atmosphere density decreases at great altitude, a limitation is finally reached where it is no longer feasible to compress the ambient gas molecules efficiently to pressurize the aircraft cabin. This practical limitation is reached at altitudes of about 21,000–24,000 m (70,000–80,000 ft) where only very high aircraft velocity with sufficient ram pressure could achieve adequate aircraft pressurization (9). Even so, adiabatic compression of gas so rare results in very high cabin temperatures so that at FL above 24,000 m (80,000 ft) the sealed cabin with self-contained life support systems becomes necessary. Thus spacecraft carry onboard gas supplies to pressurize the craft to habitable cabin altitudes and gas composition.

4.5 Cabin Decompression

The remote but ever-present risk in aircraft flying at high altitude with cabins pressurized as just described is that of accidental decompression. This could occur in case of loss of aircraft structural integrity as in failure of a door or window. The severity of cabin decompression is dependent on several factors, including volume of the cabin, area of the opening into the pressurized compartment, and cabin pressure–ambient pressure differential. Hazards include physical harm to occupants by being sucked out of the opening or being struck by loose objects flying about the cabin because of the rush of pressurized air out of the cabin, hypoxia on exposure to ambient pressure, low temperature, pulmonary barotrauma with possible cerebral air embolism, and decompression sickness.

Excellent quality control in construction of modern aircraft has made this event quite rare, but provision of emergency oxygen supplies for crew and passengers wearing restrain belts or harnesses during flight, routine securing of loose objects, and training of flight crews in emergency procedures are necessary backup provisions.

4.6 Acute Adaptation to Altitude Exposure

On rapid exposure to altitude, whether in an unpressurized aircraft or balloon or because of accidental decompression of a pressurized aircraft, decompression in an altitude chamber, or rapid ascent of a mountain, a sequence of physiological mechanisms attempts to compensate for the acute effects of the exposure.

With decrease in arterial PO_2 there is hyperventilation caused by hypoxic stimulation of the chemoreceptors, the carotid and aortic bodies. Hyperventilation decreases alveolar and arterial PCO_2 producing tissue alkalosis, and the favorable shift to the left of the hemoglobin–oxygen dissociation curve. There is also unfavorable cerebral vascular vasoconstriction as arterial PCO_2 drops since carbon dioxide is a potent cerebral vasodilator. This effect is overridden by the hypoxia-induced vasodilation that occurs when venous PO_2 drops to about 50 mm Hg (7).

Also, with decreased PAO_2 the carotid and aortic bodies mediate tachycardia, vasoconstriction in the extremities, systemic hypertension and increased pulmonary vascular resistance. The net effect is hyperventilation, pulmonary hypertension, increased heart rate, and beneficial redistribution of blood from extremities to the brain and heart.

5 MOUNTAIN SICKNESS

A number of disorders including retinal hemorrhage, high-altitude edema of the face and extremities, chronic mountain polycythemia, thromboembolism, and pulmonary hypertension have been seen in association with high altitude exposure. Additionally a number of well-defined syndromes can be initiated by such exposure and these are described below.

5.1 Acute Mountain Sickness

Soon after rapid ascent to 3660 m (17,000 ft) symptoms will develop in nearly all people while few will develop symptoms with exposure below 2400 m (8000 ft). The most common presenting complaint is bilateral headache usually worse in the mornings and intensified by exercise. Additionally cognitive function is usually impaired and memory lapses are common. Gastrointestinal and sleep disturbances are sometimes experienced.

Physical signs are variable and diagnostically unreliable. Cyanosis of the lips and extremities can be present along with pallor and periorbital edema.

The symptoms of acute mountain sickness are probably secondary to cerebral hypoxia and cerebral edema. Factors that increase hypoxia tend to increase the severity of symptoms while those that decrease hypoxia relieve symptoms.

Symptoms of acute mountain sickness are relieved by bedrest, light diet, caffeine-containing substances, aspirin, and acetaminophen. Without treatment, symptoms tend to abate after 2–4 days at altitude. Prevention of acute mountain sickness is best accomplished by stage or gradual ascents. Acetazolamide (10) or dexamethasone (11) have also been successfully used to prevent symptoms.

5.2 Chronic Mountain Sickness

This rare form of high-altitude illness occurs after continued exposure to altitudes usually higher than 3050 m (10,000 ft) for months to years. It is characterized by progressive weakness, fatigue, dyspnea, and impairment of mental function. Patients demonstrate low arterial oxygen saturations, pulmonary hypertension, and severe polycythemia. The pathogenesis of this disorder is unclear but may be related to an increasing insensitivity to hypoxia and carbon dioxide in susceptible people.

Treatment consists of descent to lower altitudes and complete recovery can be expected. However, return to high altitude will usually result in recurrence of the problem.

5.3 High-Altitude Pulmonary Edema

This disorder is the most serious of the mountain sickness syndromes. It is seen within 24–72 hr in unacclimatized persons who rapidly ascend to altitudes greater than 2440 m (8000 ft) and then engage in heavy physical activity (12). With rapid ascents to 3660–4270 (12,000–14,000 ft) the incidence is 0.5 percent in adults and 8 percent in children younger than 16 years (13).

Initial symptoms include dyspnea, fatigue, and cough with anorexia, with nausea and vomiting seen particularly in children. Symptoms become progressively more severe, particularly at night although orthopnea is only seen in 10 percent of cases. Finally symptoms can become so severe that patients are unable to stand. Obtundation progressing to coma can occur and once consciousness is lost death will occur in 6–12 hr unless oxygen is administered or the patient is moved to lower altitude. Death during the night with few premonitory symptoms sometimes occurs.

High-altitude pulmonary edema is a disorder of the pulmonary circulation that results in "leakiness" of pulmonary capillaries without left ventricular failure. Increased pulmonary artery pressure, the result of acute hypoxia, compounds hypoxia damage to capillaries. When this is coupled with a lessening of surfactant activity leading to decreased intraalveolar osmotic pressure, fluid rapidly collects in the alveoli resulting in the symptom complex.

Early recognition is the key to effective treatment since prompt descent while the patient is still able to walk with assistance is essential. Oxygen should be administered as soon as it is available by a tight fitting mask at a high flow rate (6–8 L/min). Early results of the use of nifedipine in the treatment of high-altitude pulmonary edema are promising (14).

5.4 Hyperbaric Oxygen

The use of hyperbaric oxygen in the treatment of acute high-altitude illness has been known for many years to hold theoretical promise. However, since hyperbaric chambers were not available at the high-altitude locations where the problems occurred, this promise could not be tested. Recently, a practical, portable hyperbaric chamber, called a Gamow bag, has been developed. This fabric device will accommodate one patient, and early clinical trials in patients with acute mountain sickness and high-altitude pulmonary edema have demonstrated rapid clinical improvement (15). In the future the Gamow bag may be used instead of oxygen alone for the initial treatment of high-altitude illnesses.

6 EFFECTS OF HIGH ALTITUDE ON PREEXISTING DISORDERS

6.1 Pulmonary Disorders

Primary pulmonary hypertension is severely adversely affected by ascent to altitudes as low as 1520 m (5000 ft). Exercise tolerance is diminished, dyspnea increased, and syncope is more frequent. An increase in pulmonary vascular resistance and pulmonary arterial pressure secondary to hypoxia is probably the cause. Descent improves symptoms.

Patients with chronic obstructive pulmonary disease, but in whom blood gas abnormalities are minor, usually tolerate high altitude fairly well. In fact some patients with severe obstructive disease experience decreased dyspnea at altitude because of the lower air density. However, those patients with significantly decreased arterial PAO_2 and elevated PCO_2 at sea level tend to be more symptomatic at altitudes greater than 1520 m (5000 ft). In these patients supplemental oxygen may be indicated during aircraft flights or ascents to high-altitude areas.

6.2 Cardiac Disorders

Patients with mitral stenosis and pulmonary hypertension as well as patients with cyanotic congenital heart disease tend to have an increase in their symptoms at high altitude. This is due to an increase in their right-to-left shunt and reduction in cardiac output as a response to exercise and/or an increase in pulmonary arterial pressure.

The symptoms of congestive heart failure will be exacerbated at altitude. Temporary relief can be provided by supplemental oxygen and restriction of activity.

Patients with symptomatic coronary artery disease may experience an increase in symptoms acutely with exposure to high altitude. This is secondary to increased sympathetic stimulation from hypoxia with a concomitant increase in cardiac rate and output. Limitation of activity at altitude is advisable to control symptoms and prevent unstable angina until acclimatization is achieved.

6.3 Systemic Hypertension

While prolonged residence at high altitude tends to decrease both systolic and diastolic blood pressure (16), acute exposures result in an increase in systolic and diastolic pressures (17). This rise appears to be secondary to hypoxic stimulation of carotid body chemoreceptors producing an increase in sympathetic activity.

Although hypertensive crises related to high-altitude exposure have not been reported, patients with angina or congestive failure may experience an increase in symptoms. Because of this, hypertensive patients may need to increase their medication during such exposures.

6.4 Sickle Cell Disease

Exposure to high altitude is contraindicated in patients with sickle cell disease because such exposure will often precipitate a sickle cell crisis. However, exposure of people heterozygous for the sickle cell gene, that is, people with sickle cell trait, are also at greater risk for sickle cell crisis with altitude exposures exceeding 2440 m (8000 ft). While the true incidence of altitude-induced crisis is not known, the incidence of crisis increases with increasing altitude. If exposure to high altitude is mandatory, supplemental oxygen may prevent crisis.

7 ALTITUDE ACCLIMATIZATION

The acute physiological adaptations to altitude exposure presented above are ineffective in preventing serious symptoms of hypoxia upon acute exposure to altitude. Therefore, the main protection lies in ensuring adequate oxygen equipment to provide supplemental oxygen. Yet people born in mountainous regions of the world live and work for a lifetime at altitudes that would produce serious symptoms in the sea-level dweller exposed to such altitudes. In addition near-sea-level dwellers can acclimatize to altitude although this acclimatization is never as complete or efficient as it is for those who reside at altitude.

The process of acclimatization involves several physiologic adjustments directed at minimizing the decrease in intracellular oxygen tension (18). Initially and promptly on exposure to high altitude pulmonary ventilation increases and remains increased for the duration of the stay at altitude. More capillaries are recruited at the tissue level and there is an increase in myoglobin. Additionally intracellular enzymes are increased and mitochondria enlarge. Increased pulmonary ventilation decreases alveolar carbon dioxide thus producing respiratory alkalosis. This, in turn, shifts the hemoglobin–oxygen dissociation curve to the left enhancing alveolar oxygen uptake by hemoglobin.

From a practical standpoint acclimatization can be achieved by two methods. The staging method involves transitional stays at intermediate altitudes before progressing to higher altitudes. Various staging schedules have been recommended, but in general staged ascents of 1200–1830 m (4000–6000 ft) with 2- to 4-day stops at each altitude will effectively acclimatize most people up to an altitude of 4270 m (14,000 ft). Graded ascent involves a gradual daily increase in altitude over several days to weeks. No matter which approach is used, it is advisable that above 4270 m (14,000 ft) a daily ascent of no more than 300 m (1000 ft) be done with a day of rest between every 1–2 days of ascent.

The time required for near complete acclimatization is variable. While a few days to a week of gradual ascent will usually prevent the symptoms of acute mountain illness, it may be necessary for athletes to train for 1–2 months at altitude to achieve peak performance.

8 INCREASED PRESSURE

Humans are exposed to increased pressure when submerged underwater and when subjected to increased pressures in closed pressure vessels. Underwater exposure is experienced by divers who are subject to increasing pressure directly proportional to depth (Table 11.1). Pressure vessels include hyperbaric chambers used for treatment of medical and diving related diseases, pressurized tunnels and caissons used for underwater construction, and underwater habitats (19).

Divers entering the aquatic environment are also subject to limited propulsion and risk of heat loss. As depth and pressure increase, the diver breathes gas of increased density using breathing equipment that provides oxygen and removes carbon dioxide from the breathing gas.

8.1 Direct Pressure Effects

As pressure increases, the volume of gas in exposed spaces diminishes according to Boyle's law (pressure × volume = constant). Volume in the lungs, middle ear, paranasal sinuses, gut, and so forth all are reduced as pressure increases. Displacement of tissues into the diminishing volume of these spaces causes a phenomenon called *squeeze*, which damages tissues and may cause dysfunction of the organ involved. Ear squeeze (middle-ear barotrauma) occurs when the eustachian tube is blocked, and the middle ear space cannot equilibrate with the increasing ambient pressure (20). The tympanic membrane is displaced inward and ultimately ruptures. The middle ear may fill with blood from engorged mucous membranes. Infection and reduced hearing loss are complications. Prevention is achieved by assuring that the eustachian tube is patent and the middle ear can be equilibrated at surface pressure before descending. A gentle Valsalva maneuver is commonly used to accomplish this goal. Similar events can occur in a paranasal sinus with a blocked orifice (sinus squeeze), in a small residual air pocket left between a tooth filling and the base of the tooth (tooth squeeze), in the air space within a diving mask (mask squeeze), or between the skin and a fold in a dry diving suit (suit squeeze). All will produce tissue injury due to displacement of tissues into the diminishing air space. Tooth fillings may also be loosened if air has diffused into a space beneath a filling while under pressure, and expands upon ascent. Although lung squeeze is theoretically possible by breath-hold diving to a depth that reduces the lung volume below the residual volume (21), in practice this has not been observed. Squeeze is a common, usually minor, consequence of diving. Middle-ear barotrauma is the most common diving-related disorder encountered in divers.

8.2 Pulmonary Barotrauma

Injury to the lungs is a potentially lethal complication of poor diving practices. When a diver subject to increasing pressure breathes from an air supply at depth, the breathing gas is pressurized to the same pressure as the ambient, so that pressure gradients from breathing supply to the airways are not altered as the diver descends. A diver breathing compressed air at 33 ft (2 ATA) will experience expansion of the lung volume on ascent due to Boyle's law. If the diver breathes normally and properly vents the increasing gas volume, no lung injury will occur; however, if the diver ascends while breathholding, at the surface (1 ATA), the lung gas will attempt to expand to twice the volume at 33 ft. When the lung volume expands to the chest capacity, further ascent will cause intrapulmonary pressure to increase to levels that damage lung tissue and pulmonary barotrauma will result. Overpressure 95–110 cm H_2O will initiate damage to the lung (22,23). Pressures of 300–400 mm Hg are possible in the lung subjected to gas expansion against a closed glottis. Lung damage occurs immediately upon surfacing, and air is injected into the pulmonary veins. Embolization of the heart, brain, and other organs follows. The most commonly involved organ is the brain where strokelike symptoms are found within minutes of surfacing with a closed glottis. Divers in swimming pools breathing compressed air at depths greater than 4 ft are susceptible. Prevention is accomplished by training. Treatment is accomplished by recompression in a hyperbaric chamber (see below).

8.3 Inert Gas Kinetics

Pressure determines the amount of gas dissolved in tissues and fluids of the body. A person subjected to increased ambient pressure will take up gases into the tissues based on an exponential kinetic model. Dissolved gas content of tissues depends on the partial pressure of the gas, and the solubility of the gas in the specific tissue (see Table 11.3). Of the gases involved in diving, oxygen is not a problem with decompression sickness due to its low partial pressure at the cellular level (24): CO_2 does not cause concern for decompression sickness because of its high solubility and low partial pressure. Inert gas (nitrogen, helium), however, is a limiting factor in diving because of its ability to supersaturate quickly as the diver ascends with increased dissolved gas in the tissues. Excess supersaturation will allow dissolved gases to change phase to the gaseous form, and bubbles will form in blood and tissues (25). Expansion of gases in blood and tissues upon ascent results in damage and dysfunction of the tissues, and venous gas embolism to the lungs. Decompression

Table 11.3 Characteristics of Several Inert Gases[a]

GAS	Molecular Weight	Lipid Solubility[b]	Water Solubility[b]	Narcotic Potential[c]
Helium	4	0.015	0.009	0.23
Neon	20	0.019	0.009	0.28
Hydrogen	2	0.036	0.018	0.55
Nitrogen	28	0.067	0.013	1.00
Argon	40	0.140	0.026	2.32

[a]Proportional solubility of the various gases in lipid is related to their narcotic potential relative to nitrogen. Helium is the least and argon the most narcotic gas in the list. Adapted from Bennett (56).
[b]Expressed as gas volume/solute volume at 1 bar.
[c]Values relative to Nitrogen.

sickness results from damage to various organs and tissues as a result of bubble production and growth (26,27).

Gases dissolve in tissues, fats, and water according to Henry's law:

$$V = KP$$

where V, the volume of dissolved inert gas is determined by K, the solubility coefficient, and P, the partial pressure of the gas. Solubility coefficients for water and lipid are provided in Table 11.3. Of interest is the volume of inert gas (nitrogen when breathing air) dissolved in various tissues. Although Henry's law determines the amount of gas in the tissue, there is a finite time required for the equilibrium state to be achieved (Fig. 11.4). Gases entering tissues are thought to follow a first-order difference relationship where rate of gas entry into tissue is proportional to the pressure difference between blood and tissue. Factors that affect the rate of entry include blood flow to the tissue, and the rate of diffusion of gases into the tissue (29). In the body, blood flow is the major factor controlling rate of gas exchange in tissues. The concentration of gas in tissues follows the kinetic equation:

$$\frac{dC}{dt} = k \,\Delta C$$

Solution of this equation results in the exponential relationship

$$C = C_d(1 - e^{-kt}) + C_s(e^{-kt})$$

where C is the concentration of gas in the tissue at time t, C_d is the equilibrium concentration at the depth of the dive, and C_s is the concentration of gas in the tissues on the surface at one atmosphere. This equation describes an asymptotic curve [Fig. 11.4(a)] in which the tissue gas concentration approaches the maximum concentration for the given pressure after about 5 half-times (time to reach one-half of maximum tissue pressure) have elapsed. There is a family of curves for different values of k. These curves represent different tissue compartments, which have different gas exchange characteristics (30). Similar kinetics control the washout of inert gas from tissues when ambient pressure is reduced [Fig. 11.4(b)], however, washout and uptake are not necessarily symmetrical.

Decompression tables are designed to prevent excess supersaturation of tissues and bubble formation. Decompression stages are selected to avoid excess supersaturation within specific tissues (Fig. 11.5). Decompression schedules have been designed for different inert gases, continuous or staged decompression, saturation exposure, caisson or tunnel exposure, and types of diving such as commercial or recreational.

Operations involving exposure to increased partial pressure of inert gas must always consider the decompression time as well as the exposure time. For certain exposures, decompression time far exceeds the exposure time and renders the exposure impractical. For long or deep exposures, the technique of saturation diving

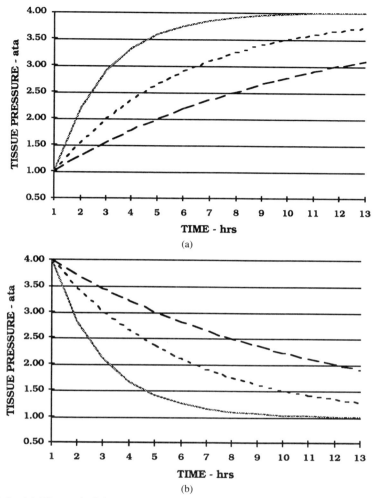

Figure 11.4 (a) Theoretical inert gas uptake curves for several tissue compartments when the tissues undergo an ambient pressure change from 1 to 4 atmospheres. Upper curve represents a tissue compartment that reaches saturation at 11 hr, while other curves represent slower filling compartments. (b) Washout curves for tissue compartments with the same kinetic properties as shown in (a). The curves depict the tissue inert gas content after excursion from 4 to 1 atmosphere ambient pressure. Both figures assume a single excursion with no stops.

is used. In this type of exposure, workers are maintained under pressure for days to weeks until the work is completed. During this time, tissue partial pressures become equilibrated with the ambient pressure, and inert gas reaches its saturation concentration. Once tissues are saturated, no further gas uptake occurs, and a single decompression is performed upon completion of the work (31). Pressurized habitats have been constructed specifically to house workers who must live under increased

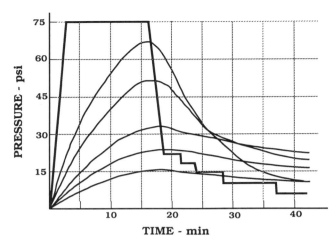

Figure 11.5 Combined uptake and washout curves of 5 tissue compartments resulting from an experimental pressure exposure and decompression. The heavy line outlines the actual dive profile. Most tissue compartments do not reach saturation at the end of the dive. Stages of decompression are designed to prevent excess supersaturation of any tissue compartment during the decompression. Adapted from Boycott et al. (35).

pressure for prolonged periods (19). Decompression procedures are well defined for air, mixtures of nitrogen and oxygen; helium and oxygen, nitrogen, helium and oxygen (trimix); hydrogen and oxygen; and the noble gases argon and neon, which are only used in experimental exposures.

8.4   Decompression Sickness

A second important aspect of increased environmental pressure in diving is decompression sickness (DCS). In the nineteenth century, Paul Bert described the pathophysiology of decompression sickness (26). Hill and others in the late nineteenth century, concluded from autopsies on divers and caisson workers that decompression sickness was caused by inert gas bubbles in the blood and tissues (32–34). Based on bubble volume and location, they were able to explain the variety and severity of the symptoms. Paralysis resulted from bubbles in the spinal cord, unconsciousness often resulted from bubbles in the brain, and dyspnea was associated with bubbles in the pulmonary circulation. The causes of the minor forms of decompression sickness are less clear. Muscle and joint pain may be due to bubbles in ligaments, fascia, periosteum, or nerve sheaths. Bubbles also have biochemical effects (34), which are independent of their mechanical effects.

There is a spectrum of injury in this illness (Table 11.4) that can mimic a variety of other disorders (35,36). The disease develops on ascent from depth when a diver does not follow established procedures for returning to the surface, which are designed to prevent effervescence of dissolved inert gas (nitrogen). The formation of free gas in the body has several consequences (35,38,39). First, free gas entering

Table 11.4 Classification of Decompression Sickness

Type I	Type II
Pain only decompression sickness	Serious decompression sickness
Limb or joint pain	Central nervous system disorder
Itch	Pulmonary (chokes)
Skin rash	Systemic (hypovolemic shock)
Fatigue	Inner ear/vestibular

the vascular system from the peripheral tissues will transit the veins and cause pulmonary vascular obstruction (40,41). When the free gas volume is large and significant pulmonary obstruction occurs, a classical syndrome (chokes) is described (36,40). This syndrome is manifest by chest pain, dyspnea, and cough. Decompression sickness is often associated with free gas in the blood and with tissue injury from expanding gas. Bubbles and tissue injury result in activation of acute inflammation (41). The inflammatory response alters vascular permeability, causing fluid to leak into the interstitial tissues of the systemic and pulmonary vascular beds (42,43). In severe cases pulmonary edema can occur and hypovolemia with significant plasma loss and hemoconcentration will result (44,45). With severe DCS, endothelial damage to blood vessels occurs (45), and significant focal regions of tissue ischemia are evident. A frequent target organ is the spinal cord in humans (46,47). A common manifestation of DCS in divers is evidence of spinal cord dysfunction usually at levels below the diaphragm (47). Symptoms (Table 11.5) include paresthesias, muscle weakness, paralysis of the lower extremities, bowel or bladder incontinence, urinary retention, and sexual impotence. In cases of sudden ascent from the deep depths associated with commercial diving exposures (blowup), a severe decompression sickness syndrome can occur with both cerebral and spinal neurological symptoms, unconsciousness, hypovolemic shock, pulmonary edema, and a high mortality rate (48).

A less serious type of DCS is manifest by minor pains in the extremities and joints (37,40). Symptoms of local joint pain are often confused with pain from

Table 11.5 Frequency of Decompression Sickness Symptoms in 100 Cases

Symptom	Percentage of All DCS Symptoms
Skin itch	4
Headache	11
Fatigue/malaise	13
Bone/joint pain	54
Spinal/back pain	11
Spinal/neurological	22
Respiratory	21

Source: Adapted from U.S. Navy Diving Manual (52).

injuries, and the diagnosis of decompression sickness may be missed. In some populations a high incidence of aseptic bone necrosis is found in divers who have experienced decompression sickness of the joints in the distant past, or who have experienced deep, prolonged dives in commercial operations (49), and in caisson workers (50). A rare but important symptom of DCS is sudden acute neurological hearing loss or vestibular dysfunction. Decompression sickness of this type usually occurs from deep, prolonged commercial diving exposures and, if untreated, will result in permanent deafness (51).

8.5 Treatment of Decompression Sickness

Treatment of DCS requires extensive knowledge of the use of hyperbaric chambers and hyperbaric oxygen for therapy (37,52), and is usually provided by specialists in diving and hyperbaric medicine. Divers with joint pains or neurological abnormalities following a diving exposure require consultation with a diving medicine expert and may need recompression treatment in a hyperbaric chamber. Symptoms of DCS may appear up to 24 hr following exposure (Table 11.6). Mistreatments have resulted in permanent brain and spinal cord injury or death because of misdiagnosis. Since there are only a few centers in the United States which can provide expert consultation in this area, it is wise to learn of the treatment facilities available in specific regions of the United States and other countries. Persons suspected of having a diving-related accident involving neurological abnormalities should not be retained in a hospital without hyperbaric treatment facilities since prolonging the delay between onset of symptoms and recompression treatment increases the risk of permanent neurological injury (53).

Bubbles from inadequate decompression form in blood and tissues and activate the inflammatory process (41,44,45). Adjunctive therapy with drugs has been developed to provide therapy for the secondary (i.e., nonobstructive) effects of bubbles. Individuals with neurological DCS or arterial gas embolism may benefit from fluid therapy to prevent hemoconcentration, use of intravenous steroids to inhibit cerebral or spinal cord edema, and antiplatelet agents to prevent platelet aggregation by bubble surfaces in the blood (41). In severe DCS, disseminated intravascular coagulation may develop (54) and anticoagulation may be useful (41,43). Individuals with permanent neurological injury from decompression sickness or air embolism require

Table 11.6 Time of Onset of Decompression Sickness Following Diving

Cumulative Percentage	Time of Onset
50	30 min
85	1 hr
95	3 hr
99	6 hr
100	12–24 hr

physical therapy to regain musculoskeletal function. Unlike traumatic cord injury or stroke, prompt treatment of neurological injury from decompression sickness or air embolism often results in excellent recovery of function.

It is sometimes difficult to distinguish between severe decompression sickness and cerebral air embolism. These two conditions may coexist when a diver who has been underwater for a prolonged period of time ascends rapidly and develops pulmonary barotrauma (55). Because treatment of both illnesses involves recompression in a hyperbaric chamber, hyperbaric oxygen, and adjunctive drug therapy, it is less important to make a precise diagnosis and more important to institute therapy quickly. Decompression sickness usually develops sometime after diving. Symptoms may develop within minutes of ascent or may be late, appearing 12–24 hr after a dive has been completed. On the other hand, pulmonary barotrauma with arterial gas embolism occurs immediately upon ascending and may produce initial unconsciousness. Subtle symptoms of pulmonary barotrauma may go undetected in the initial few hours after ascent; thus timing of symptom onset should not be the only criteria for differentiating this disorder from decompression sickness.

Recompression therapy involves one of several available protocols for application of pressure and oxygen. Commonly used protocols are U.S. Navy Table 6 for decompression sickness, and Table 6-A for arterial gas embolism (52,53). These tables describe the course of pressure and oxygen to be provided to the afflicted diver.

9 INERT GAS NARCOSIS

Inert gas narcosis (nitrogen narcosis) results from breathing air at a depth greater than 100 ft of seawater (56,57). The symptoms are similar to alcohol inebriation and can range from loss of fine motor control and high-order mathematical skills to bizarre and inappropriate behavior, improper response to emotional stress, hostility, loss of coarse and fine motor control, and unconsciousness. Symptoms increase with depth from 100 ft. At 200–250 ft, severe symptoms occur; at depths of 300–400 ft, unconsciousness occurs due to the general anesthetic effect of nitrogen at this pressure. In commercial and military diving, helium replaces nitrogen as the inert gas at depths below 150 ft, and narcosis does not occur (58). The recommended safe depth limit using scuba gear is 130 ft. Some individuals, however, are more susceptible to nitrogen narcosis and may manifest severe symptoms at 130 ft. Fatigue, heavy work, and cold water can augment the narcotic effect of nitrogen. Symptoms disappear immediately upon surfacing and often there is amnesia for the events that occurred below (56). The greatest danger from nitrogen narcosis is to the scuba diver who insists on exceeding recommended depth limits. As depth increases, judgment and motor skills diminish. More than one diver was last seen by his partner swimming downward beyond 200 ft with no evidence of distress, oblivious to the lethal consequences of his action. Narcotic effects of nitrogen are prevented by using helium–oxygen mixtures for breathing or by adhering to safe depth limits. Oxyhelium mixtures are not readily available to sport divers but are commonly used in commercial diving. Persons found to be exceptionally susceptible to nitrogen narcosis must limit diving to shallower depths where no symptoms are evident.

Despite the potential for unique injuries related to the diving and hyperbaric environment, the vast majority of dives are done safely. The key to safe diving is training. A thorough understanding of the physiology and physics of diving, and intimate knowledge of the diving equipment, will minimize the risk of a diving-related injury.

REFERENCES

1. F. M. G. Holmstrom, in *Aerospace Medicine*, 2nd ed., H. W. Randel (Ed.), Williams and Wilkins, Baltimore (1971), p. 60.

2. F. J. W. Roughton, in *Handbook of Respiratory Physiology*, W. M. Boothby, Ed., USAF School of Aerospace Medicine, Randolph Air Force Base, Texas, 1954, p. 51.

3. A. Krough, *J. Physiol.*, **52,** 409 (1919).

4. R. F. Grover, J. V. Weil, and J. T. Reeves, *Exerc. Sport Sci. Rev.*, **14,** 269 (1986).

5. J. K. Alexander, L. H. H. Hartley, M. Modelski et al., *J. Appl. Physiol.*, **23,** 849 (1967).

6. R. E. Fowles and H. N. Hultgren, *Am. J. Cardiol.*, **52,** 862 (1983).

7. P. J. Sheffield and R. D. Heimbach, in *Fundamentals of Aerospace Medicine*, R. L. DeHart, Ed., Lea and Febiger, Philadelphia, 1985, p. 95.

8. J. Ernsting, in *A Textbook of Aviation Physiology*, J. A. Gillies, Ed., Pergamon Press, London, 1965, p. 374.

9. R. D. Heimbach and P. J. Sheffield, in *Fundamentals of Aerospace Medicine*, R. L. DeHart, Ed., Lea and Febiger, Philadelphia, 1985, p. 111.

10. W. Evans, S. M. Robinson, D. H. Horstman, et al., *Aviat. Space Environ. Med.*, **47,** 512 (1976).

11. J. B. Jobe, B. Shukitt-Hale, L. E. Banderet et al., *Aviat. Space Environ. Med.*, **62,** 727 (1991).

12. N. D. Menon, *N. Engl. J. Med.*, **273,** 66 (1965).

13. H. N. Hultgren and E. A. Marticorena, *Chest*, **74,** 372 (1978).

14. O. Oely, M. Maggiorini, M. Ritter et al., *Lancet*, **2,** 1241 (1989).

15. J. F. Kasic, M. Yaron, R. A. Nicholas et al., *Ann Emerg. Med.*, **20,** 1109, 1991.

16. E. Marticorena, L. Ruiz, J. Severino et al., *Am. J. Cardiol.*, **23,** 364 (1969).

17. S. R. Kamat and B. C. Banerzi, *Am. Rev. Respir. Dis.*, **106,** 404 (1972).

18. J. R. Sutton, J. T. Reeves, R. D. Wagner et al., *J. Appl. Physiol.*, **64,** 1309 (1988).

19. J. W. Miller and I. G. Koblick, *Living and Working in the Sea*, Van Nostrand Reinhold, New York, 1984, p. 227.

20. H. H. Vail, *Arch. Otolaryngol.*, **10,** 113 (1929).

21. K. E. Schaefer, R. D. Allison, J. H. Dougherty et al., *Science*, **162,** 1020 (1968).

22. K. E. Schaefer, W. P. Nulty, C. Carey et al., *J. Appl. Physiol.*, **13,** 15 (1958).

23. M. C. Malhotra and C. A. M. Wright, *Proc. R. Soc. Med.*, **B154,** 418 (1960).

24. B. R. Dulling and R. M. Berne, *Circ. Res.*, **27,** 669 (1970).

25. E. N. Harvey, *Bull. New York Acad. Med.*, **21,** 505 (1945).

26. P. Bert, Barometric Pressure Undersea Medical Society, Bethesda, 1978.

27. S. Erdman, *JAMA*, **49,** 1665 (1907).

28. P. B. Bennett, in *Inert Gas Narcosis*, P. B. Bennett and D. H. Elliott, Eds., Williams and Wilkins, Baltimore, 1975, p. 205.

29. J. Piper and M. Meyer, *Adv. Exp. Med. Biol.*, **169**, 457 (1984).

30. S. Kety, *Pharmac. Rev.*, **3**, 1 (1951).

31. R. Larson and W. Mazzone, in *Underwater Physiology*, Vol. III, C. J. Lambertsen, Ed. Williams and Wilkins, Baltimore, 1966, p. 241.

32. L. Hill, Arnold, London, 1912.

33. E. Levy, Dept. of the Interior, Bureau of Mines, Tech. Paper 285. U.S. Government Printing Office, Washington, D.C., 1922.

34. J. M. Hallenbeck and J. C. Andersen, in *The Physiology and Medicine of Diving*, 3rd ed., P. B. Bennett and D. H. Elliott, Eds., Bailliere Tindall, London, 1982, p. 435.

35. A. E. Boycott, G. C. C. Damant, and J. B. Haldane, *J. Hyg.*, *Cambridge*, **8**, 342 (1908).

36. D. H. Elliott, J. M. Hallenbeck, and A. A. Bove, *Lancet*, **2**, 1193 (1974).

37. R. D. Workman, *Aerosp. Med.*, **39**, 1076 (1968).

38. T. S. Neuman, R. G. Spragg, P. D. Wagner, and K. M. Moser, *Respir. Physiol.*, **41**, 143, (1980).

39. A. A. Bove, J. M. Hallenbeck, and D. H. Elliott, *Undersea Biomed. Res.*, **1**, 207 (1974).

40. A. Erde and C. Edmonds, *J. Occup. Med.*, **17**, 324 (1975).

41. A. A. Bove, *Undersea Biomed. Res.*, **9**, 91 (1982).

42. A. A. Bove, J. M. Hallenbeck, and D. H. Elliott, *Aerosp. Med.*, **45**, 49 (1974).

43. R. B. Philp, *Undersea Biomed. Res.*, **1**, 117 (1974).

44. A. T. K. Cockett and R. M. Nakamura, *Lancet*, **1**, 1102 (1964).

45. L. L. Levin, G. J. Stewart, P. R. Lynch, and A. A. Bove, *J. Appl. Physiol.*, **50**, 944 (1981).

46. J. M. Hallenbeck, D. H. Elliott, and A. A. Bove, *Neurology*, **25**, 308 (1975).

47. W. A. Nix and H. C. Hope, *Deutsch Med. Wochenschr.*, **105**, 302 (1980).

48. J. N. Norman, C. M. Childs, C. Jones, J. A. R. Smith, J. Ross, G. Riddle, A. MacIntosh, N. I. P. McKie, J. I. MacCauley, and X. Fructus, *Undersea Biomed. Res.*, **6**, 209 (1979).

49. D. H. Elliott, *Proc. Roy. Soc. Med.*, **64**, 1278 (1971).

50. R. I. McCallum and D. N. Walder, *J. Bone J. Surg.* **48B**, 207 (1966).

51. J. C. Farmer, W. G. Thomas, D. G. Youngblood, and P. B. Bennett, *Laryngoscope*, **86**, 1315 (1976).

52. U.S. Navy Diving Manual NAVSEA 00994-LP001-9010, U.S. Government Printing Office, Washington, D.C., 1985, p. 815.

53. J. C. Davis, in *Diving Medicine*, A. A. Bove and J. C. Davis, Eds., W. B. Saunders, Philadelphia, 1990, p. 249.

54. R. B. Philp, P. Schacham, and C. W. Gowdey, *Aerosp. Med.*, **42**, 494 (1971).

55. T. S. Neuman and A. A. Bove, *Undersea Biomed. Res.*, **17**, 429, 1990.

56. P. B. Bennett, in *Diving Medicine*, A. A. Bove and J. C. Davis, Eds., W. B. Saunders, Philadelphia, 1990, p. 69.

57. A. R. Behnke, R. M. Thompson, and E. P. Motley, *Am. J. Physiol.*, **112**, 554 (1935).

58. C. B. Momsen, U.S. National Experimental Diving Report 2-42, AD728758 U.S. Government Printing Office, Washington, D.C., 1942.

CHAPTER TWELVE

Biological Agents

James P. Hughes, M.D.

1 INTRODUCTION

1.1 Occupational vs. Environmental Risk

Hazardous exposure to biological agents in the course of one's work is recognized as a significant risk of producing adverse health effects among some occupational groups. Workers in human health care, veterinary medicine, biomedical and biotechnical laboratories, agriculture, animal handling, meat processing, outdoor construction, resource conservation, commercial fishing, and public safety may be cited as being at special risk under some circumstances. Some aspects of biological hazards are discussed in other sections of this work: *Patty's Industrial Hygiene and Toxicology*, Vol I, Chapters 13 (Health Care), 14 (Laboratories), 17 (Indoor Air), 20 (Agriculture), and 32 (Environmental Control).

This chapter undertakes to list and categorize specific biological agents of occupational and environmental concern, updating and expanding the subject from previous editions, while identifying occupations at particular risk.

Differentiating traditional occupational risk from environmental risk is sometimes unsatisfactory. An outbreak of Legionnaires' disease among clerical workers in an office building, malaria in business travelers to an endemic area, or cholera among crew and passengers of a commercial aircraft from contaminated food or water taken aboard come to mind. In each instance individuals exposed in the course of their work would certainly regard their disease as occupational in origin, while others exposed as casual visitors to a contaminated office site, or as recreational travelers, are likely to be regarded as having been at "environmental risk." Reported cases of

Patty's Industrial Hygiene and Toxicology, Third Edition, Volume 3, Part B, Edited by Lewis J. Cralley, Lester V. Cralley, and James S. Bus.
ISBN 0-471-53065-4 © 1995 John Wiley & Sons, Inc.

leishmaniasis among American troops in the Middle East during the Gulf War of 1992 further blur the distinction between occupational and environmental risk.

Health disorders that may be caused by hazardous exposure to biological agents include infections, contact dermatoses, hypersensitivity reactions such as asthmatic attacks, trauma resulting from animal bites or scratches, and a few malignant tumors.

Infectious diseases arising out of occupational exposure deserve special attention. Of historical significance is pulmonary tuberculosis in stonecutters, Weil disease (leptospirosis) in miners and sewer workers, undulant fever (brucellosis) in husbandmen, glanders in muleteers, and anthrax in wool sorters, to name a few.

The development of effective antibiotic therapy for these predominantly bacterial infections, together with improved preventive measures, produced a satisfactory decline in frequency of infections of occupational origin over many decades. However, growing complacency over this period has been shattered by the recent emergence of newly identified risks, such as the transmission of viral hepatitis and acquired immunodeficiency syndrome in the health care setting and Legionnaires' pneumonia among the employed occupants of contaminated buildings.

A 1992 authoritative report by an Institute of Medicine Committee included this warning about infectious diseases in the United States: "Pathogenic microbes can be resilient, dangerous foes. Although it is impossible to predict their individual emergence in time and place, we can be confident that new microbial diseases will emerge" (1).

This prediction has been borne out in a number of instances, including the 1993 emergence in the southwestern United States of a highly lethal disease termed hantavirus pulmonary syndrome. Human exposure to the aerolized excreta of a small rodent, the deer mouse (*Peromyscus maniculatus*), the recognized reservoir of the causative viral agent, has led to reported cases of the disease. While person-to-person transmission has not been demonstrated, the potential for infection of agricultural workers, pest control personnel, and others regularly exposed to rodents is evident. (See Section 2.2.1.6.)

Additionally, the clinical prognosis of some other infections of occupational significance, notably tuberculosis in health care providers and laboratory personnel, has markedly worsened due to an increase in strains of causative organisms that have become resistant to available antibiotics.

The public health infrastructure for the surveillance of infectious diseases in the United States, which had slowly eroded during the period of growing complacency, is being revived and expanded as a result of increased recognition of the potential for rapid spread of lethal infections in today's world of international travel and exposure, and of the warnings expressed in the Institute of Medicine Committee report on emerging infections. Relevance to the prevention and control of infections due to exposures in the occupational and environmental settings should not be overlooked.

1.2 Modes of Exposure

1.2.1 Respiratory

The respiratory tract constitutes the main portal of entry for most of the biological agents, either as viable organisms or as products of their multiplication on various fibers. Larger particles are filtered through the nose or deposited on the mucous membranes of the bronchi. Particles of 5 μm or less mean diameter will reach alveoli, whose total surface, approximately 70 m^2, provides a huge area of exposure to allergenic material. Dependent on the size of particulate matter, therefore, the site of reaction in the respiratory tract may be primarily bronchial, with symptoms suggestive of asthma, or pulmonary, with various types of response within the alveoli.

1.2.2 Cutaneous

Continuous exposure of the skin to some of the agents just cited is likely to occur under usual working conditions. Dusts and particulates of various kinds settle out and stick to sweaty skin, where they may set up inflammatory or allergic processes. The immediate agent may not be readily identifiable. Viable pathogenic organisms may be inoculated through minor wounds or through preexisting abrasions or other breaks in the integument. Inoculation by means of stings and bites, as with the many vector-borne diseases, or in envenomization may also occur.

1.2.3 Gastrointestinal

The gastrointestinal portal of entry is probably the least likely under occupational conditions. Improper cleaning of the hands or contamination of food or water by infectious agents may permit entry by way of the gastrointestinal tract.

1.2.4 Parenteral

Occupational exposure to infectious agents, such as the virus causing hepatitis B and that causing acquired immunodeficiency syndrome, or AIDS, are of primary importance to health care providers and a few other worker categories. A few additional viruses, as well as certain other blood-borne infectious agents, may also be transmitted by the parenteral route in the health care setting, as described in a subsequent section.

1.3 Sources of Biological Agents

1.3.1 Soil

Soil is the immediate source of infections with the spore-forming mode *Histoplasma capsulatum*, the causative agent of histoplasmosis, a disease common in the western United States. Droppings from birds, fowl, or bats produce organic rich soil in which molds proliferate. Spores lie dormant for long periods until disturbed by humans or by winds. The spore-contaminated dust may then be inhaled by outdoor workers. Soil is also the source of *Coccidioides immitis*, the causative agent of Valley Fever, common as well in western United States. In this case the soil is contaminated by excreta or carcasses of infected burrowing animals. Spores in inhaled dust result in pulmonary infections in humans. Soil contaminated with spores of *Clostridium tetani*, the tetanus bacillus, on entering a skin wound, may produce this deadly disease of occupational significance. Likewise, infections caused by tetanus and the clostridial organisms originate in soil that has already been contaminated by the manure of sheep, horses, and cattle. In the case of the two fungi mentioned, distribution of the infectious particles into the atmosphere occurs in part as a result of human activity, but winds may also be a sufficient cause of disturbance of contaminated soil. Infections by the spore formers and by other fungi most commonly occur in connection with puncture wounds or injuries that devitalize tissue.

1.3.2 Water

Mycobacterial infections such as that caused by *Mycobacterium marinum* and possibly other mycobacteria arise from human contact with contaminated water. In the instance of *M. marinum* it appears that the organism may survive and propagate in interstices of concrete and stone in swimming pools and be inoculated into minor skin abrasions producing ulceration. The source of the contamination of water is unknown, but it has been observed that organisms are present in the mouths of fish, and injuries sustained in an aquarium may also produce typical skin lesions. Certain parasites, such as the schistosomes, also infect humans as a result of wading or swimming in snail-infested water.

1.3.3 Air

The agents of disease of the general type known as farmer's lung are airborne. Particles containing antigenic substances of fungi are the causative agent. Although contaminated soil constitutes the reservoir of *C. immitis* and *H. capsulatum*, their distribution is airborne. A similar statement applies to infections such as brucellosis and Q fever, which may be acquired from exposure to diseased animal carcasses in preparation for human consumption, although inoculation through the skin undoubtedly represents an alternate pathway.

1.3.4 Insect Bites

All blood-sucking insects possess the potential of transmitting infectious agents. Ticks, lice, fleas, and mosquitoes are the commonest vectors. The variety of diseases transmitted in this manner is extensive. The more familiar diseases are malaria, Rocky Mountain spotted fever, typhus, yellow fever, and dengue. Many other diseases of limited geographic distribution are encountered elsewhere in the world.

1.4 Types of Agents

Two types of agents considered are:

1. *Living organisms,* including viruses, bacteria, other microbial agents, fungi, protozoa, helminths, ectoparasites, and venom-producing species.
2. *Nonliving substances* of biological origin, including products of fungal or bacterial metabolism, and perhaps degradation products of material fibers altered by fungal or bacterial activity: textile and cordage fibers; bagasse; moldy hay, grain or feed; wood dusts; other dusts such as tobacco, coffee, and subtilin; and certain gums and resins.

2 LIVING AGENTS OF DISEASE

Exposure to certain living agents may result in infection in susceptible individuals. Some definitions may be helpful in interpreting the discussion and tables that follow.

2.1 Definitions

Infection. The entry and development of multiplication of an infectious agent in the body of humans or animals. Infection is not synonymous with infectious disease; the result may be inapparent or manifest. The presence of living infectious agents on exterior surfaces of the body, or upon articles of apparel or soiled articles, is not infection but represents contamination of such surfaces and articles.

Infectious agent. An organism (virus, rickettsia, bacteria, fungus, protozoa, or helminth) that is capable of producing infection or infectious disease.

Infectious disease. A clinically manifest disease of humans or animals resulting from an infection.

Infestation. For persons or animals, the lodgment, development, and reproduction of arthropods on the surface of the body or in the clothing. Infested articles or premises are those that harbor or give shelter to animal forms, especially arthropods and rodents.

Nosocomial infection. An infection occurring in a patient in a hospital or other health care facility and in whom it was not present or incubating at the time of admission or the residual of an infection acquired during a previous admission. Includes infections acquired in the hospital but appearing after discharge and also such infections among the staff of the facility.

Reservoir (of infectious agents). Any person, animal, arthropod, plant, soil, or substance (or combination of these) in which an infectious agent normally lives and multiplies, on which it depends primarily for survival, and where it reproduces itself in such manner that it can be transmitted to a susceptible host.

Source of infection. The person, animal, object or substance from which an infectious agent passes to a host. Source of infection should be clearly distinguished from source of contamination.

Transmission of infectious agents. Any mechanism by which an infectious agent is spread from a source or reservoir to a person. These mechanisms are:

> A. *Direct transmission*. Direct and essentially immediate transfer of infectious agents to a receptive portal of entry through which human or animal infection may take place. This may be by direct contact as by touching, biting, kissing, or sexual intercourse or by the direct projection (droplet spread) of droplet spray onto the conjunctiva or onto the mucous membranes of the eye, nose, or mouth during sneezing, coughing, spitting, singing, or talking.

B. Indirect transmission.

1. Vehicle-borne: Contaminated inanimate materials or objects (fomites) such as toys, handkerchiefs, soiled clothes, bedding, cooking or eating utensils, surgical instruments or dressings (indirect contact); water, food, milk, biological products including blood, serum, plasma, tissues or organs; or any substance serving as an intermediate means by which an infectious agent is transported and introduced into a susceptible host through a suitable portal of entry. The agent may or may not have multiplied or developed in or on the vehicle before being transmitted.

2. Vector-borne. (a) Mechanical: Includes simple mechanical carriage by a crawling or flying insect through soiling of its feet or proboscis or by passage of organisms through its gastrointestinal tract. This does not require multiplication or development of the organism. (b) Biological: Propagation (multiplication), cyclic development, or a combination of these is required before the arthropod can transmit the infective form of the agent to humans. An incubation period (extrinsic) is required following infection before the arthropod becomes **infective**. The infectious agent may be passed vertically to succeeding generations of the insect or by its passage from one stage of life-cycle to another, as nymph to adult. Transmission may be by injection of salivary gland fluid during biting or by regurgitation or deposition on the skin of feces or other material capable of penetrating through the bite wound or through an area of trauma from scratching or rubbing. This transmission is by an infected nonvertebrate host and not simple mechanical carriage by a vector as a vehicle. However, an arthropod in either role is termed a **vector**. (c) Airborne: The dissemination of microbial aerosols to a suitable portal of entry, usually the respiratory tract. Microbial aerosols are suspensions of particles in the air consisting partially or wholly of microorganisms. They may remain suspended in the air for long periods of time, some retaining and others losing infectivity or virulence. Particles in the 1- to 5-μm range are easily drawn into the alveoli of the lungs and may be retained there. Not considered as airborne are droplets and other large particles that promptly settle out. (i) Droplet nuclei: Usually the small residues that result from evaporation of fluid from droplets emitted by an infected host. They also may be created purposely by a variety of atomizing devices, or accidentally as in microbiology laboratories or in abattoirs, rendering plants, or autopsy rooms. They usually remain suspended in the air for long periods of time. (ii) Dust: The small particles of widely varying size that may arise from soil (as, e.g., fungus spores separated from dry soil by wind or mechanical agitation), clothes, bedding, or contaminated floors.

Virulence. The degree of pathogenicity of an infectious agent, indicated by case fatality rates and/or its ability to invade and damage tissues of the host.

Zoonosis. An infection or infectious disease transmissible under natural conditions from vertebrate animals to man. [Definitions adapted from Benenson (2).]

Persistence of an infectious agent in nature depends on the maintenance of a chain of transmission. The ability to cause severe disease is neither required nor necessarily advantageous to the infectious agent (3).

Infectious diseases related to occupational activity tend to be sporadic and seldom affect a high proportion of workers. However, three types of causative agent or disease may appear in special situations and affect considerable numbers of exposed workers. The first class is composed of those organisms capable of developing vegetative states that permit long survival and transportation over considerable distances: notably, the spore formers and the fungi. The second class consists of diseases transmissible from animals to humans and, except for occasional cases among veterinarians and animal husbandry workers, appears especially in meat-processing plants. The third type consists of vector-borne diseases: workers are not much more exposed to these diseases than is the general population, except as their activities may require extensive work in the field or travel and exposure in parts of the world where endemic diseases of this type are prevalent.

A fourth type, one of increasing importance in the health care setting, includes diseases transmitted from person to person. These are the human-borne infections.

The living agents of disease are here discussed by microbiological category.

2.2 Viruses

Viruses are the simplest, smallest, and most pervasive of microorganisms. Some are obligate and exclusively infectious in humans. These include the causative agents of the common diseases of childhood, and these often cause asymptomatic infections. Some are more virulent, such as the human immunodeficiency virus (HIV), which causes AIDS. But a large group of important and dangerous viruses are not specific for humans, such as the agents of yellow fever and rabies, which cause primary infections in other hosts.

The classification of viruses has evolved during this century and has been the source of much debate if not confusion. Of the thousands of viruses isolated from humans, animals, plants, insects, and bacteria, the several hundred known to infect humans have been assigned by the International Committee on Taxonomy (ICTV) to families, genera, and species, as well as strains, types, and subtypes, in a rather complex hierarchical system. Viral taxonomy has been described as "a jungle of acronyms, sigla, neologisms, Greco-Latin hybrids and morphologic descriptions" (4).

In this chapter, selected viruses will be mentioned only by species, using the English vernacular name listed by ICTV, such as "rabies virus." The complicated hierarchical system of families, subfamilies, and genera are not considered here. Rather, viruses will be discussed in categories of occupational importance, such as: (1) those of primary concern in the health care setting due to risk of transmission from human to human, (2) those transmitted by the bite of an arthropod vector (the arboviruses), (3) others transmitted from animals to humans (zoonotic agents), and certain miscellaneous agents. These categories are discussed below, and selected viral agents are listed in Tables 12.1 and 12.3–12.5 (5).

2.2.1 Viruses in the Health Care Setting

This section focuses primarily on those viral infections that are of importance due to the risk of transmission to the health care provider as a disease arising out of and in the course of employment. Less emphasis is placed on nosocomial infections in hospitalized patients, especially in those individuals of immunocompromised status due to preexisting infections such as AIDS or to treatment with antineoplastic agents or following therapy with bone marrow suppressants.

Certain human-borne viruses posing special risk to health care providers are listed in Table 12.1. The two most important of these are hepatitis B virus (HBV) and human immunodeficiency virus (HIV), both transmitted primarily by percutaneous exposure to infected blood. The Occupational Safety and Health Administration (OSHA) has estimated that the population at risk of occupational transmission of these blood-borne pathogens may exceed 5 million persons in the United States (6). The types of establishment in which exposure may occur are listed in Table 12.2.

2.2.1.1 Hepatitis B.
The acute and chronic consequences of HBV infection are major problems in the United States. The reported incidence of acute hepatitis B increased by 37 percent from 1979 to 1989, and an estimated 200,000–300,000 new infections occurred annually during the period 1980–1991. The estimated 1.0–1.25 million persons with chronic HBV infection in the United States are potentially infectious to others. In addition, many chronically infected persons are at risk of long-term sequelae, such as chronic liver disease and primary hepatocellular carcinoma; each year approximately 4000–5000 of these persons die from chronic liver disease (7).

Hepatitis B virus infection is the major infectious blood-borne occupational hazard of health care workers. The Hepatitis Branch of the Centers for Disease Control (CDC) estimates that there are approximately 8700 infections in health care workers with occupational exposure to blood and other potentially infectious materials in the United States each year. These infections cause over 2100 cases of clinical acute hepatitis, 400–440 hospitalizations, and approximately 200 deaths each year in health care workers. Death may result from both acute and chronic hepatitis. Infected health workers can spread the infection to family members or, rarely, to their patients.

Health care workers at high risk of occupational exposure include medical technologists, operating room staff, phlebotomists and intravenous therapy nurses, surgeons, pathologists, oncology and dialysis unit staff, emergency room staff, nursing personnel, staff physicians, dental professionals, laboratory and blood bank technicians, emergency medical technicians, and morticians.

Most infected health care workers are unaware that they have been exposed to or infected with HBV. Approximately 1 percent of hospitalized patients are HBV carriers; most HBV carrier patients seen in the health care setting are not symptomatic, are unaware that they are carriers, and their medical charts do not contain this information. Health care workers may take extraordinary precautions when dealing with a known carrier but are often unaware that they may treat five carriers for each one they recognize. This is a key point in understanding the rationale for the concept of *universal precautions* (described in Section 2.3.1.1) and for widespread use of the

Table 12.1 Viral Agents of Significance in Health Care Settings (5)

Virus (v.)	Disease	Transmission	Vaccine available?
Hepatitis B v. (HBV)	Viral hepatitis B	Percutaneous and permucosal exposure to body fluids of infected persons; injuries from needles and sharp instruments	Yes
Human immunodeficiency v. (HIV)	Acquired immunodeficiency syndrome (AIDS)	Percutaneous exposure to blood or tissues, especially by needlestick or scalpel injuries; virus occasionally found in saliva, tears, urine, and bronchial secretions, but transmission after contact with these secretions has not been reported in health care settings	No
Delta agent (HDV)	Delta hepatitis	Same as HBV	No
Hepatitis C v. (HCV)	Viral hepatitis C: Posttransfusion (PT-NANB)	Percutaneous exposure to contaminated blood and plasma derivatives	No
Hepatitis A v. (HAV)	Viral hepatitis A	Person-to-person via fecal-oral route; (rarely transmitted to animal handlers from infected chimpanzees)	No (immune globulin recommended; candidate vaccines under consideration)
Rubella v.	Rubella (German measles)	Direct contact; airborne droplet; infection of nonimmune woman during first trimester of pregancy results in high risk of fetal anomalies	Yes
Measles v.	Measles	Direct contact; airborne droplet	Yes
Mumps v.	Mumps	Direct contact; airborne droplet	Yes
Influenza v. 1, 2, and 3	Influenza	Direct contact; airborne droplet, plus dried mucus	Yes
Herpes simplex v. (HSV type 1)	Herpes simplex: vesicular lesions of lips, oropharynx	Saliva of carriers	No
Varicella–zoster v.	Chicken pox–herpes zoster	Direct contact; airborne droplet	No
Human cytomegalo-v.	Mononucleosis syndrome	Intimate exposure by mucosal contact with infectious tissues, secretions or excretions	No

Table 12.2 U.S. Establishments and Populations at Risk of Occupational Transmission of Blood-Borne Pathogens (6)

Type of Establishment	Population at Risk (thousands)
Hospitals	2386
Physicians' offices	640
Nursing homes	285
Law enforcement	342
Dentists' offices	316
Fire & rescue	252
Home health	212
Health units in industry	179
Personnel services	163
Correctional facilities	120
Research labs	89
Medical & dental labs	63
Funeral services	57
Government clinics	56
Linen services	50
Residential care	49
Schools	41
Blood/plasma/tissue centers, hemodialysis, waste removal, hospices, drug rehabilitation, medical equipment repair, life saving	74
Total	5374

hepatitis B vaccine in workers with exposure to blood. Although the risk of encountering HBV carriers may vary in the hospital setting, being highest in inner-city referral hospitals dealing with high-risk groups such as drug abusers and homosexual men, risk will be present in any work setting where human blood is encountered.

Percutaneous exposure to blood through needlesticks and cuts with other sharp instruments are visible and efficient modes of transmission, but reported injuries do not account for the majority of infections in health care workers. Some workers doing traumatic procedures get cuts, needlesticks, or large blood exposures so frequently that they do not bother to report them; other workers become infected when the blood of an unsuspected HBV carrier gets into a small preexisting skin lesion or is rubbed into the eye.

Transmission of HBV infection from exposure to contaminated environment surfaces has been documented to be a mode of HBV spread in certain settings, particularly hemodialysis units. The virus can survive for at least one week dried at room temperature on environmental surfaces. HBV-contaminated blood from the surface of dialysis machines and carried on the hands of medical personnel has been postulated as one mechanism of transmission in dialysis units (6).

Prevention of HBV infection by observance of *universal precautions* and other preventive measures is described below, following the discussion of HIV infections. Fortunately, an effective vaccine for HBV is available and should be widely used.

Hepatitis B virus is thought to be far less resistant to sterilization and disinfection procedures than microbial endospores or mycobacteria used as reference criteria. Any sterilization or disinfection procedure or sterilization agent or high-level disinfectant will kill the virus if used as directed. Diluted solutions of sodium hypochlorite (household bleach) are particularly effective, although they may be corrosive or damaging to certain materials. Certain low-level "germicides" such as quaternary ammonium compounds are not considered to be effective against the virus. Unfortunately, soaking medical and dental instruments in these solutions is a common and potentially dangerous procedure since health workers may handle the sharp instruments soaked in these solutions with a false sense of security (6).

Hepatitis B Virus Vaccine In 1982, a safe, immunogenic and effective hepatitis B vaccine derived from human plasma was licensed in the United States and was recommended for use in health care workers with blood or needle exposure in the workplace. A second vaccine, produced in yeast by recombinant technology, was first licensed in 1987. Since the introduction of these vaccines, OSHA estimates a minimum of 2,568,974 persons in the United States have been vaccinated, 2,029,189 of whom are health care workers. Hepatitis B vaccination is the most important part of any HBV control program because gloving and other protective devices cannot completely prevent puncture injuries from needles and other sharp instruments.

Persons planning hepatitis B vaccine programs may consider the need for prevaccination and postvaccination testing for antibody. Prescreening may be cost-effective, depending on the likelihood of prior HBV infection. An algorithm to help assist with this determination has been published by the U.S. Public Health Service (U.S. PHS). Discussions on the issues surrounding the option of postvaccination testing have also been published. At this time postvaccination testing is not considered necessary unless poor response to vaccine is anticipated (such as for those who have received vaccine in the buttock, persons $\geq$ 50 years of age, and persons known to have HIV infection); subsequent patient management depends on knowing the immune status (such as with dialysis patients and staff) or there may be a need to know whether the person ever responded to vaccine for management of postexposure prophylaxis (6).

Hepatitis B Virus: Postexposure Prophylaxis Percutaneous and mucous membrane exposure to blood occur and will continue to occur in the health care setting. Hepatitis B virus infection is the major infectious risk that occurs from these exposures, and needlesticks from HBsAg positive individuals will infect 7–30 percent of susceptible health care workers. Preexposure vaccination is the most effective method for preventing such infection. However, it can be expected that some individuals who initially decline vaccination will experience an exposure incident. Fortunately, effective postexposure prophylaxis exists for HBV exposures if appropriate protocols are followed. The February 9, 1990 recommendations of the U.S. PHS specify that if the source individual is known to test positive for hepatitis B surface antigen

(HBsAg), then the exposed individual should be given hepatitis B immunoglobulin (HBIG) and the hepatitis B vaccine series be initiated. Hepatitis B vaccine is recommended for any previously unvaccinated health care worker who has a needlestick or other percutaneous accident with a sharp instrument or permucosal (ocular or mucous membrane) exposure to blood (6).

2.2.1.2 Non-A, Non-B Hepatitis. Non-A, non-B hepatitis in the United States is caused by more than one viral agent. Studies have shown that parenterally transmitted (PT) non-A, non-B hepatitis accounts for 20–40 percent of acute viral hepatitis in the United States and has epidemiologic characteristics similar to those of hepatitis B.

The principal mode of transmission in the United States is blood borne; therefore, persons at greatest risk for infection include IV drug users, dialysis patients, and transfusion recipients. Over 90 percent of all posttransfusion hepatitis is due to the non-A, non-B viruses. These hepatitis viruses cause not only acute hepatitis but may also lead to chronic hepatitis; an average of 50 percent of patients who have acute PT non-A, non-B hepatitis infection later develop chronic hepatitis with potential for progression to cirrhosis and for infectivity to others for the duration of life. The amount of virus present in the blood of acutely or chronically infected persons is modest, usually less than 1000 infectious doses per milliliter, although occasionally up to 1000 times higher. Thus, relative infectivity of blood is 100- to 100,000-fold lower than with HBV. Relative infectivity of other body fluids is not known.

Some evidence indicates that non-A, non-B hepatitis also presents an occupational risk to health care workers. At least one episode of transmission of non-A, non-B hepatitis from an acutely infected patient to a nurse by needlestick has been reported. One case-control study has shown an increased risk of non-A, non-B hepatitis for patient care and lab workers.

2.2.1.3 Acquired Immunodeficiency Syndrome (AIDS). The risk of occupationally acquired AIDS in the health care setting is much lower than that of hepatitis B but is still of grave concern because of the universal lethality of the disease.

Of great importance to health care workers is the 1.0–1.5 million persons in the United States who are infected with human immunodeficiency virus (HIV), often unknowingly so, and who require medical care for related or unrelated conditions.

Human immunodeficiency virus is a member of a group of viruses known as human retroviruses. Its genetic material is ribonucleic acid (RNA) rather than deoxyribonucleic acid (DNA), the genetic material found in most living organisms. The virus particle is comprised of a core containing the RNA and viral enzymes surrounded by an envelope consisting of lipids and proteins.

Human Immunodeficiency Virus: Serological Testing Infection with HIV may be identified through testing the blood for the presence of HIV antibodies. Tests were first licensed for use in the United States in 1985 and have been used routinely to screen donated blood, blood components, and blood products and by physicians and clinics to diagnosis HIV infection in patients. The military also uses the antibody tests to screen recruit applicants and active-duty personnel for HIV infection. Although the antibodies do not appear to defend or protect the host against HIV, they serve as markers of viral infection. Most people infected with HIV have detectable antibodies within 6 months of infection, with the majority generating detectable antibodies between 6 and 12 weeks after exposure.

The enzyme-linked immunosorbent assay (ELISA or EIA) technique used to detect HIV antibodies is sensitive, economical, and easy to perform. However, as with all laboratory determinations, this test can produce a false-positive result when HIV antibody is not present. Therefore, current recommendations include repeating the ELISA test if the first test is positive. If the second test is also positive, another test, usually employing the Western blot technique, is used to validate the ELISA results. A positive ELISA test and a positive Western blot result include the presence of HIV antibodies and HIV infection (6).

Human Immunodeficiency Virus: Transmission　　The human immunodeficiency virus has been isolated from human blood, semen, breast milk, vaginal secretions, saliva, tears, urine, cerebrospinal fluid, and amniotic fluid; however, epidemiologic evidence implicates only blood, semen, vaginal secretions, and breast milk in the transmission of the virus. Documented modes of HIV transmission include engaging in sexual intercourse with an HIV-infected person; using needles contaminated with the virus; having parenteral, mucous membrane, or nonintact skin contact with HIV-infected blood, blood components, or blood products; receiving transplants of HIV-infected organs and tissues including bone or transfusions of HIV-infected blood; through semen used for artificial insemination and perinatal transmission (from mother to child around the time of birth).

The human immunodeficiency virus is not transmitted by casual contact. Studies evaluating nearly 500 household contacts of individuals diagnosed with AIDS revealed no cases of HIV infection of household members who had no other risk factors for the virus (including no sexual contact with or exposure to blood from the infected person).

A number of epidemiological studies and surveys have been conducted to determine occupational risks for HIV infection. The CDC have conducted a national prospective study since 1983 of the risk of HIV among health care workers exposed to the blood of an HIV-infected individual as a result of parenteral (needlestick or cut with a sharp object), mucous membrane, or nonintact skin exposure (8).

As of July 31, 1988, a cohort of 1201 health care workers with exposure to HIV-contaminated blood had been followed. Of these, 751 (63%) were nurses, 164 (14%) were physicians or medical students, 134 (11%) were technicians or laboratory workers, 90 (7%) were phlebotomists, 36 (3%) were respiratory therapists, and 26 (2%) were housekeeping or maintenance staff (6).

The 1201 subjects had blood samples drawn and tested for the presence of HIV antibodies. Acute blood specimens collected within 30 days after exposure were obtained and tested from 622 subjects. Exposed health care workers were retested at 6 weeks, 3 months, 6 months, and 12 months after the exposure incident to determine if seroconversion had occurred. Seroconversions were defined as health care workers who were seronegative for HIV antibody within 30 days after occupational exposure and seropositive 90 days or more after the exposure incident.

Of 963 subjects followed for at least 6 months, 860 (89%) had sustained either a needlestick injury or a cut with a sharp instrument. Of these, 4 were seropositive, yielding a seroprevalence rate of 4/860 = 0.47 percent (6).

Health Care Workers with Acquired Immunodeficiency Syndrome Further evidence of occupational transmission is provided by reports of health care workers who have AIDS but have no identifiable risk for infection. As of September 30, 1990, there were at least 69 health care workers with AIDS for whom no risk factors have been identified after thorough investigation. This group was comprised of 13 physicians (1 of whom was a surgeon), 2 dental workers, 8 nurses, 14 aides/attendants, 12 housekeeping or maintenance workers, 7 technicians, 2 therapists, 3 embalmers, 1 paramedic, and 7 others. Of these 69 individuals, 35 reported needlestick and/or mucous membrane exposure to the blood or body fluids of patients during the 10 years preceding their diagnosis of AIDS. However, none of the source patients was known to be HIV-infected at the time of exposure and none of the workers was evaluated at the time of exposure to determine HIV infection status or to document seroconversion. While data on these cases are less complete compared to the case reports mentioned earlier, it is reasonable to assume that at least some of them resulted from occupational exposure.

Approximately 94 percent of the subjects reported sustaining accidental "parenteral inoculations with sharp instruments," ranging from one to many within a 5-year period. Serologic test results revealed that at least 21 percent of the subjects who had not received the HBV vaccine had been infected with HBV; however, only one, a male dentist, of the 1309 subjects surveyed, was seropositive for HIV. Since the study represents a point prevalence survey, the HIV seroconversion rate among dental personnel cannot be estimated from it (6).

2.2.1.4 Human Immunodeficiency Virus Type 2.

A case of AIDS in a person from Africa, caused by another human retrovirus, human immunodeficiency virus type 2 (HIV-2), was diagnosed and reported for the first time in the United States in December 1987. Since then, the CDC have received reports of additional cases of HIV-2 occurring in West Africans and diagnosed in the United States. Human immunodeficiency virus type 2 appears to be similar to HIV-1 in modes of transmission and natural history but has not yet been studied in as much detail. Although HIV-2 is unquestionably pathogenic, there is still much to be learned regarding its epidemiology, pathogenesis, and efficiency of transmission. Although only a few cases of HIV-2 have been reported in the United States, the infection is endemic in West Africa, where it was first linked with AIDS in 1986. There have also been cases of HIV-2 infection reported among West Africans living in Europe. Human immunodeficiency virus type 2 surveillance is being conducted in the United States to monitor the frequency of occurrence using specific tests not yet available commercially. The occupational health significance has not been determined (6).

2.2.1.5 Other Viruses of Occupational Significance.

Certain viruses, in addition to those already described as blood-borne agents in the health care setting, are selected for discussion because of (1) relative importance as agents of occupational exposure or (2) distribution worldwide or in the Americas or (3) high risk potential, that is, CDC/NIH (National Institute of Health) Biosafety Level 3 or 4, from handling the agent in typical laboratory functions or activities (9). These viruses, listed in Tables 12.3 through 12.5, are classified by categories of occupational setting in which they may be encountered:

Table 12.3: Health care, child care, developmentally disabled care and industrial medical facilities

Table 12.4: Certain outdoor occupations, travel in endemic areas, and microbiological/biomedical laboratories
12.4a: Mosquito-borne arboviruses
12.4b: Tick-borne arboviruses
12.4c: Phlebotamine-borne arboviruses
12.4d: Selected arenaviruses, filoviruses, and hantaviruses

Table 12.5: Animal handlers

Smallpox is not discussed because of the success of vaccination programs in achieving worldwide eradication, although an exceedingly remote risk of infection in nature or from laboratory accident may persist. Polio virus is also excluded because of the success of widespread protection by vaccination, especially in developed countries, although a small risk of infection by ''wild strains'' (no vaccine available) may remain.

In addition to diseases caused by viruses identified by standard nomenclature, several disorders in animals have been attributed to infection by prion agents, described as filterable, formaldehyde-resistant, self-replicating agents, possibly true viruses. The following encephalopathies in animals due to infection by prion agents have been recognized: (a) scrapie of sheep and goats, (b) transmissible mink encephalopathy, (c) chronic wasting disease of American mule deer and elk, and, (d) bovine spongiform encephalopathy (BSE), especially in the United Kingdom. While there have been no documented cases of human infection acquired from animals with these disorders, the risk of transmission to humans has been hypothesized (10).

2.2.1.6 Hantavirus Pulmonary Syndrome: Occupational Aspects.

An emergent infection of grave consequence, in terms of wide geographic risk of exposure, of high mortality rate, and of absence of effective treatment was first recognized in the southwestern region of the United States in mid-1993. Now termed hantavirus pulmonary syndrome (HPS) and reported in a growing number of western states, this infection has a serious potential for occupational exposure.

In recognition of the lethal quality of this condition—over 50 percent mortality during the early months of case reporting—and relative unfamiliarity with its clinical features and mode of transmission in North America, a panel of expert consultants convened at the CDC in Atlanta, Georgia, in July, 1993 and developed a set of interim recommendations for the reduction of risk of exposure (11). These provisional recommendations were based in part on the basis of knowledge about related hantaviruses, while information was being developed on the causative virus and its epidemiology.

Table 12.3 Other Viruses of Occupational Significance in Health Care (HC), Child Care (CC), Developmentally Disabled Care (DDC), and Industrial Medical Facilities (IMF) (5)

Virus	Disease	Transmission	Risk to
Coxsackie viruses, various groups and types	Enteroviral vesicular pharyngitis (herpangina)	Direct contact: nasopharyngeal discharges and feces; or aerosol droplet spread	CC, DDC
	Enteroviral vesicular stomatitis with exanthem (hand, foot, and mouth disease)[a]	Direct contact: nasopharyngeal discharges and feces; or aerosol droplet spread	CC, DDC
	Enteroviral lymphonodular pharyngitis	Direct contact: nasopharyngeal discharges and feces; or aerosol droplet spread	CC, DDC
	Coxsackie meningitis[b]	Direct contact: nasopharyngeal discharges and feces; or aerosol droplet spread	HC
	Coxsackie carditis	Direct contact: nasopharyngeal discharges and feces; or aerosol droplet spread	HC
	Epidemic myalgia (epidemic pleurodynia; Bornholm disease)[c]	Fecal-oral; respiratory droplet; handling fomites	HC
	Epidemic hemorrhagic conjunctivitis (acute)[d]	Direct or indirect contact with discharge from infected eyes	HC, DDC
Adenoviruses (type 8, etc.)	Adenoviral keratoconjuctivitis (also termed "shipyard conjunctivitis" in industrial settings)	Direct contact; contaminated instruments or solutions	HC, DDC, IMF
Norwalk agent	Epidemic viral gastroenteropathy	Fecal-oral; airborne from fomites	HC, DDC

[a] Vesicular stomatitis may also be caused by the stomatitis virus, normally of cattle and horses, which in humans usually occurs in dairy workers, animal husbandrymen, and veterinarians; to be differentiated from foot and mouth disease.

[b] Viral meningitis may also be caused by other enteroviruses, by echoviruses, and by arboviruses.

[c] Epidemic myalgia may also be caused by echoviruses (types 1 and 6).

[d] Acute hemorrhagic conjunctivitis, including swimming pool conjunctivitis, may also be caused by other enteroviruses and by adenoviruses.

Table 12.4 Viruses of Occupational Significance in Certain Outdoor Workers,[a] Occupational Travelers, and Microbiological/Biomedical Laboratory Workers (9)

Virus(es) (v./vs.)	Human Disease	Reservoir	Biosafety Level[b]
a. Mosquito-Borne Arboviruses (9)			
Eastern equine encephalitis v.	Encephalitis	Birds, rodents	2
Western equine encephalitis v.	Encephalitis	Unknown	2
St. Louis encephalitis v.	Encephalitis, hepatitis	Unknown	3
California encephalitis v.	Encephalitis	Unknown	2
Murray Valley encephalitis v.	Encephalitis	Unknown	3
Venezuelan equine encephalitis v.	Encephalitis	Unknown	3
Japanese encephalitis v.	Encephalitis	Unknown	3
West Nile v.	Encephalitis, hepatitis	Birds	3
Dengue vs.	Fever, rash	Man, monkey (?)	2
Yellow fever v.	Hemorrhagic fever	Man, monkey, marsupials, mosquito	3
Rift Valley fever v.	Hemorrhagic fever, encephalitis	Vertebrates, mosquito	3
Oropouche v.	Fever, meningitis	Vertebrates, mosquito	3
Chikungunya v.	Hemorrhagic fever, polyarthritis	Kangaroo, wild rodents	3
Mayaro v.	Fever arthalgia	Unknown	3
Rocio v.	Encephalitis	Unknown	3
Spondweni v.	Fever	Unknown	3

Table 12.4 *(Continued)*

Virus(es) (v./vs.)	Human Disease	Reservoir	Biosafety Level[b]
b. *Tick-Borne Arboviruses* (9)			
Colorado tick fever v.	Fever	Small mammals	2
Louping ill v.	Encephalitis	Sheep, deer	3
Powassan v.	Encephalitis	Tick, mammals	3
Tick-borne encephalitis v.			
Far Eastern	Encephalitis	Tick, mammals	4
European subtype	Encephalitis	Goats, sheep	?
Thogoto v.	Meningitis	Tick	3
Crimean-Congo hemorrhagic fever v.	Hemorrhagic fever	Hares, birds, ticks	4
Omsk hemorrhagic fever v.	Hemorrhagic fever	Rodents, muskrats, ticks	4
Kyasanur v.	Hemorrhagic fever, meningoencephalitis	Rodents, shrews, monkeys	4
c. *Phlebotomine-Borne Arborviruses* (9)			
Vesicular stomatitis, Indiana and New Jersey vs.	Fever, encephalitis	Man-sandfly complex	2
d. *Selected Arenaviruses, Filoviruses, and Hantaviruses of Significance in Microbiological Laboratories* (9,12)			
Arenaviruses:			
Lassa v.[c]	Hemorrhagic fever, hypotensive crises, encephalopathy	Wild rodents (West Africa)	Deposited rodent excreta, food; personal contact; hospital and laboratory transmission via blood and secretions	4

Virus	Disease	Reservoir/host	Transmission	Biosafety level
Junin v.[c]	Hemorrhagic fever	Wild rodents, house mouse (Argentina)	Rodent saliva and excreta, including airborne excreta dust	4
Machupo v.[c]	Hemorrhagic fever	Wild rodents (Bolivia)	Rodent saliva and excreta, including airborne excreta dust	4
Lymphocytic choriomeningitis v. (LCMV)[c]	Lymphocytic choriomeningitis (infection may be asymptomatic)[c]	House mouse, nude mouse; mouse and hamster colonies (Europe, Americas)	Oral or respiratory contact with excreta, food or dust; contamination of skin lesions	2,3
Filoviruses: Marburg v.[c]	Hemorrhagic fever	African green monkeys	Person-to-person by contact with blood, secretions, organs, or semen; nosocomial infections frequent	4
Ebola v.[c]	Hemorrhagic fever	Unknown (Africa)	Person-to-person by contact with blood, secretions, organs, or semen; nosocomial infections frequent	4
Ebola-like vs.	Only asymptomatic seroconversion reported as of 1991; pathogenicity and frequency of infection undetermined	Cynomolgus monkeys from Philippines, Indonesia, East Africa	Animal handling, including U.S.A.	4

Table 12.4 (*Continued*)

Virus(es) (v./vs.)	Human Disease	Reservoir	Biosafety Level[b]
d. *Selected Arenaviruses, Filoviruses, and Hantaviruses of Significance in Microbiological Laboratories* (9,12) (Cont.)			
Hantaviruses:			
Hantaan v. (Korea)[c]	Hemorrhagic fever with renal syndrome (HFRS)	Striped field mice (Apodemus spp.)	3–4[d]
Muerto Canyon v. (term proposed) (U.S.A.)[c] See Section 2.2.1.6	Hantavirus pulmonary syndrome (HPS)	Deer mice (Peromyscus maniculatus)	3–4[d]

Infective saliva or excreta inhaled as aerosols produced directly from the animal

Infective saliva or excreta inhaled as aerosols produced directly from the animal. Humans appear to be only an accidental host; field workers exposed to infected rodents and laboratory personnel at risk (see text for discussion)

---

[a] Outdoor workers include game wardens, foresters, etc.

[b] Biosafety levels for laboratories as recommended by CDC/NIH. See discussion in text (Section 4.5) for description of different levels.

[c] These viral agents are selected primarily due to marked virulence and high risk of transmission.

[d] Biosafety Level 3 for microbiology; Biosafety Level 4 for animal work.

524

Table 12.5 Viruses of Occupational Significance in Animal Handlers, Including Veterinarians and Abattoir Workers[a] (5)

Virus(es): (v/vs.)	Human Disease	Reservoir	Transmission
Rabies v.	Rabies	Domestic dogs, cats; wild fox, skunk, raccoon; bats; livestock	Saliva of infected animal via bite or scratch
Cercopithecine herpes virus 1	Meningoencephalitis (Simian B disease)	Old world monkeys (macaques, etc.)	Monkey bite or scratch; mucocutaneous exposure to infected saliva or monkey tissue cultures
Orfvirus	Human orf (contagious pustular dermatitis)	Sheep, goats, reindeer, musk oxen	Direct contact with mucous membranes of infected animals; passive transfer from normal animals contaminated by contact with knives, shears, etc.
Milker's nodule v.	Nodule on fingers or hand	Dairy cattle	Direct contact with udder of infected animal
Bovine papular stomatitis v.	Papular stomatitis	Beef cattle	Direct contact with infected animals
Newcastle disease v.	Conjunctivitis	Fowl	Direct contact with infected fowl
Contagious ecthyma parapox v.	Contagious ecthyma	Domesticated camels	Direct contact with infected animals
Monkeypox v.	Smallpoxlike disease	Nonhuman primates; rodents, including squirrels	Multiple animal contact (West and Central Africa)

[a]Laboratory workers also may be at risk.

525

Hantaviruses are negative-stranded RNA viruses belonging to the family Bunyaviridae. Hantaviral disease first became a U.S. public health concern in the 1950s, when soldiers serving in the Korean conflict developed an illness referred to as Korean hemorrhagic fever, or hemorrhagic fever with renal syndrome (HFRS). The causative agent, Hantaan virus, was not isolated until the late 1970s. Hemorrhagic fever with renal syndrome is characterized by severe renal failure and prominent hemorrhage and is therefore quite distinct from the new disease, hantavirus pulmonary syndrome, with its severe, noncardiac pulmonary edema and shock. Other hantaviruses include Puumala virus, which causes a mild form of HFRS common in Europe; Prospect Hill virus, a North American virus that has not been associated with human disease; and Seoul virus, antibodies to which have been associated with hypertensive end stage renal disease in patients in the eastern United States (12).

Rodents are the primary reservoir hosts of recognized hantaviruses. Each hantavirus appears to have preferential rodent hosts, but other small mammals can be infected as well. Available data strongly suggest that the deer mouse (*Peromyscus maniculatus*) is the primary reservoir of the newly recognized hantavirus, Muerto Canyon virus (MCV), in the southwestern United States. Serologic evidence of infection has also been found in Piñon mice (*P. truei*), brush mice (*P. boylii*), and western chipmunks (*Tamais* spp.). *P. maniculatus* is highly adaptable and is found in different habitats, including human residences in rural and semirural areas, but generally not in urban centers. The broad distribution of *P. maniculatus* suggests that sporadic cases may occur over much of the United States if the virus is distributed wherever the reservoir is found, which is to be determined by serologic surveys of captured specimens.

Hantaviruses do not cause apparent illness in their reservoir hosts. Infected rodents shed virus in saliva, urine, and feces for many weeks, but the duration and period of maximum infectivity are unknown. The demonstrated presence of infectious virus in saliva of infected rodents, and the marked sensitivity of these animals to hantavirus following inoculation, suggests that biting may be an important mode of transmission among rodents.

Human infection may occur when infective saliva or excreta are inhaled as aerosols produced directly from the animal. Persons visiting laboratories where infected rodents were housed have been infected after only a few minutes of exposure to animal holding areas. Transmission may also occur when dried materials contaminated by rodent excreta are disturbed.

Known hantavirus infections of humans occur primarily in adults and are associated with domestic, occupational, or leisure activities that bring humans into contact with infected rodents, usually in a rural setting. Patterns of seasonal occurrence differ, depending on the virus, species of rodent host, and patterns of human behavior.

Occupations of special risk include planting or harvesting field crops, cleaning barns and other outbuildings, handling rodents in the laboratory setting, and pest control work. Unpredictable or incidental contact with rodents or their habitats may be encountered by telephone installers, maintenance workers, plumbers, electricians, and certain construction workers. Workers in these jobs may have to enter various buildings, crawl spaces, or other sites that may be rodent infested (11).

2.3 Other Blood-Borne Pathogens

Several additional infectious diseases are characterized by a phase in which the causative agent may circulate in blood for a prolonged period of time, and, therefore, have the potential of occupational transmission to health care and laboratory workers. With the exception of syphilis and malaria, these diseases are rare in the United States (6).

Syphilis Syphilis is caused by infection with *Treponema pallidum*, a spirochete. Syphilis, a sexually transmitted infectious disease, is increasingly prevalent in the United States; 35,147 cases were reported in civilians in 1987. Although syphilis is primarily transmitted sexually and in utero, a few cases of transmission by needlestick, by tattooing instruments, and by blood transfusion have been reported. A reported transmission has occurred by needlestick exposure to the blood of a patient with secondary syphilis, resulting in a chancre on the hand. Preventive treatment of an exposed health care worker with an antibiotic during the incubation period would be expected to prevent serological test positivity and the potential for permanent reactivity on treponemal testing, as well as preventing the manifestations of infection.

Malaria Malaria is a potentially fatal mosquito-borne parasitic infection of the blood cells characterized by paroxysms of fever, chills, and anemia; 944 cases were reported in the United States in 1987. Malaria is an important health risk to immigrants from numerous malaria-endemic areas of the world and to Americans who travel to such areas. Moreover, transmission by mosquito vector has been documented in some areas of the United States. Malaria is characterized by a prolonged erythrocytic phase, during which the causative agent, one of several species of the *Plasmodium* genus, is present in the blood. In many nations malaria is among the most common transfusion-related infectious diseases. In temperate countries it is only occasionally reported. Malaria has also been transmitted by needlestick injury; in one incident malaria was transmitted to a child who received a unit of blood, and to the recipient's physician, who stuck himself with a needle.

Babesiosis Babesiosis is a tick-borne, parasitic disease similar to malaria that is usually caused by the intrarythrocytic parasite *Babesia microti*. It is endemic in certain islands off the northeastern coast of the United States. Transmission by transfusion of fresh blood from asymptomatic donors has been reported.

Brucellosis Brucellosis is a febrile illness caused by members of the genus *Brucella*. It is typically associated with occupational exposure to livestock or with ingestion of unpasteurized dairy products (129 cases were reported in 1987). It is characterized by fever and weakness, sweats and arthralgia. Transmission by blood transfusion has been reported. In one incident, brucellosis and syphilis were transmitted in the same unit of blood to one recipient.

Leptospirosis Leptospirosis, a prolonged illness characterized by fever, rash, and, occasionally, jaundice, is caused by strains of *Leptospira interrogans*, a spirochete.

The septicemic phase, during which leptospira are present in the bloodstream of patients, usually resolves within 1–2 weeks. It is typically acquired by contact with urine of infected animals, including cattle, swine, dogs, and rats (43 cases were reported in 1987). No cases of nosocomial transmission by blood have been reported.

Arboviral Infections Arboviral (arthropod-borne) infections generally do not lead to high or sustained levels of viremia in humans. Therefore, there is relatively little potential for person-to-person transmission of these infections through blood products or needlestick exposure.

The exception is Colorado tick fever (CTF), caused by a tick-borne virus that infects red blood cells. Within 3–14 days following tick exposures, the patient experiences fever, chills, headache, muscle, and back aches. Several hundred cases are reported annually and transmission by blood transfusion has been documented.

Relapsing Fever Relapsing fever is a rare disease, caused by pathogenic *Borreliae*, transmitted by lice or ticks and characterized by recurring febrile episodes separated by periods of well-being. In the United States, a few cases of tick-borne relapsing fever are reported in localized geographic areas (western United States). Though very rare, occupational transmission as a result of patient care practices has been reported. Infections have been attributed to blood from the nose of a technician and, in another incident, splashing from a placental specimen into the eye of an attendant.

Creutzfeldt–Jakob Disease Creutzfeldt–Jakob disease (CJD), a rare disorder with worldwide distribution, is a degenerative disease of the brain caused by a virus. It is believed to be transmitted by ingestion of or inoculation with infectious material, primarily neural tissue. No cases of nosocomial transmission by blood have been reported, although rare instances of transmission have occurred secondary to homologous dura mater implants, receipt of human growth hormone, and insertion of unsterilized stereotactic electrodes, which had been inserted into the brains of Creutzfeldt–Jakob disease patients and then used on others. There is a report of a case of Creutzfeldt–Jakob disease confirmed by autopsy in a neuropathology technician. She had been employed in the neuropathology facility for 22 years and her duties included rinsing formalin-fixed brains and processing, cutting, and staining sections of brain. Log records indicate that during her tenure two individuals with CJD were autopsied, 16 and 11 years prior to the technician's illness. It is not definitely known how this individual became infected.

Human T-Lymphotropic Virus Type I Human T-lymphotropic virus type I (HTLV-I), the first human retrovirus to be identified, is endemic in southern Japan, the Caribbean, and in some parts of Africa, but it is also found in the United States, mainly in intravenous drug users. The virus can be transmitted by transfusion of cellular components of blood (whole blood, red blood cells, platelets). Human T-lymphotropic virus type I has been associated with a hematologic malignancy known as adult T-cell leukemia/lymphoma and with a degenerative neurologic disease known as tropical spastic paraparesis, or HTLV-I-associated myelopathy. There is some evidence that the neurologic disease may be associated in some cases with blood transfusion. No cases of occupational acquisition of HTLV-I infection have been reported.

Viral Hemorrhagic Fever The term viral hemorrhagic fever refers to a severe, often fatal illness caused by several viruses not indigenous to the United States and rarely introduced by travelers coming from abroad. Although a number of febrile viral infections may produce hemorrhage, only the agents of Lassa, Marburg, Ebola, and Crimean-Congo hemorrhagic fevers are known to have caused significant outbreaks of disease with person-to-person transmission, including nosocomial transmission. Blood and other body fluids of patients with these illnesses are considered infectious (6).

Precautions designed to minimize transmission of the more important blood-borne viral diseases, namely AIDS, hepatitis B, and non-A, non-B hepatitis, would be effective in minimizing occupational transmission of all the above agents in the clinical setting.

2.3.1 Prevention of Transmission of Blood-Borne Diseases

Increasing concern over the dire consequences of contracting hepatitis B or AIDS as a result of occupational exposure led in 1991 to the promulgation by OSHA of an important standard for the prevention of transmission of all blood-borne diseases (6).

This standard mandates that each employer having employees with occupational exposure (to blood-borne infection) shall establish a written Exposure Control Plan of defined scope and content to include:

- Exposure determination
- Control methods
  Universal precautions (see below)
  Engineering controls
  Work practice controls
  Personal protective equipment
- HBV vaccination
- Postexposure evaluation and follow-up
- Infectious waste disposal
- Tags, labels, and bags
- Housekeeping practices
- Laundry practices
- Training and education of employees
- Record keeping

The requirements of the standard are based chiefly on the recommendations of the CDC. In 1985, CDC developed the strategy of "universal blood and body fluid precautions" to address concerns regarding transmission of HIV in health care settings. The concept, now referred to simply as *universal precautions* stresses that ***all patients should be assumed to be infectious for HIV and other blood-borne patho-***

gens. In the hospital and other health-care settings, universal precautions should be followed when workers are exposed to blood, certain other body fluids (amniotic fluid, pericardial fluid, peritoneal fluid, pleural fluid, synovial fluid, cerebrospinal fluid, semen, and vaginal secretions), or any body fluid visibly contaminated with blood. Since HIV and HBV transmission has not been documented from exposure to other body fluids (feces, nasal secretions, sputum, sweat, tears, urine, and vomitus), universal precautions do not apply to these fluids. Universal precautions also do not apply to saliva, except in the dental setting, where saliva is likely to be contaminated with blood (13,14).

In this context "exposure" is defined as contact with blood or other body fluids (to which universal precautions apply) through percutaneous inoculation or contact with an open wound, nonintact skin, or mucous membrane during the performance of normal job duties.

The unpredictable and emergent nature of exposures encountered by emergency and public safety workers may make differentiation between hazardous body fluids and those that are not hazardous very difficult and often impossible. For example, poor lighting may limit the workers' ability to detect visible blood in vomitus or feces. Therefore, **when emergency medical and public safety workers encounter body fluids under uncontrolled, emergency circumstances in which differentiation between fluid types is difficult, if not impossible, they should treat all body fluids as potentially hazardous** (14).

2.3.1.1 Universal Precautions. Universal precautions recommended by the CDC include the following provisions:

- All health care workers should routinely use appropriate barrier precautions to prevent skin and mucous membrane exposure when contact with blood or other body fluids of any patient is anticipated. Gloves should be worn for touching blood and body fluids, mucous membranes, or nonintact skin of all patients, for handling items or surfaces soiled with blood or body fluids, and for performing venipuncture and other vascular access procedures. Gloves should be changed after contact with each patient. Masks and protective eyewear or face shields should be worn during procedures that are likely to generate droplets of blood or other body fluids to prevent exposure of mucous membranes of the mouth, nose, and eye. Gowns or aprons should be worn during procedures that are likely to generate splashes of blood or other body fluids.

- Hands and other skin surfaces should be washed immediately and thoroughly if contaminated with blood or other body fluids. Hands should be washed immediately after gloves are removed.

- All health care workers should take precautions to prevent injuries caused by needles, scalpels, and other sharp instruments or devices during procedures; when cleaning used instruments; during disposal of used needles; and when handling sharp instruments after procedures. To prevent needlestick injuries, needles should not be recapped, purposely bent, or broken by hand, removed from disposable syringes, or otherwise manipulated by hand. After they are used, disposable syringes and needles, scalpel blades, and other sharp items should be placed in puncture-resistant containers for disposal; the puncture-resistant containers should be located as close as practical to the use area. Large-bore reusable needles should be placed in a puncture-resistant container for transport to the reprocessing area.

- Although saliva has not been implicated in HIV transmission, in order to minimize the need for emergency mouth-to-mouth resuscitation, mouthpieces, resuscitation bags, or other ventilation devices should be available for use in areas in which the need for resuscitation is predictable.

- Health care workers who have exudative lesions or weeping dermatitis should refrain from all direct patient care and from handling patient care equipment until the condition resolves.

- Pregnant health care workers are not known to be at greater risk of contracting HIV infection than health care workers who are not pregnant; however, if a health care worker develops HIV infection during pregnancy, the infant is at risk of infection resulting from perinatal transmission. Because of this risk, pregnant health care workers should be especially familiar with and strictly adhere to precautions to minimize the risk of HIV transmission.

Implementation of universal blood and body fluid precautions for all patients eliminates the need for use of the isolation category of **"Blood and Body Fluid Precautions"** previously recommended by the CDC for patients known or suspected to be infected with blood-borne pathogens. Isolation precautions should be used as necessary if associated conditions, such as infectious diarrhea or tuberculosis, are diagnosed or suspected.

In addition to universal precautions, detailed precautions have been developed for the following procedures and/or settings in which prolonged or intensive exposures to blood occur: invasive procedures, dentistry, autopsies or morticians' services, dialysis, and the clinical laboratory (15).

2.4 Nonviral Agents

2.4.1 Bacteria, Chlamydiae, Rickettsiae, and Spirochaetes

Microbial diseases of occupational and environmental significance cut across a broad spectrum of infectious agents. These agents are classified and discussed in detail in other texts: see Mandell, Douglas, and Bennett's *Principles and Practice of Infectious Diseases* (16), and Benenson's *Control of Communicable Diseases in Man* (10). In this chapter human infections by selected agents are presented in Tables 12.6 through 12.8.

Certain bacteria elaborate toxins, some of which are of occupational significance, especially in agriculture. These have a biologically active lipopolysaccharide or endotoxin, an essential molecular component of the outer membrane of Gram-negative bacteria. Endotoxin liberated from bacteria may be in turn a component of organic dusts encountered by farm workers, and an etiologic agent in bronchitis and organic dust toxic syndrome, as discussed later in this chapter. Other bacteria, including *C. tetani,* the tetanus bacillus, also elaborate toxins that enhance the virulence of the infectious agent.

Table 12.6 Bacterial and Mycoplasmal Infections of Occupational Significance (10)

Human Disease	Pathogen	Reservoir	Transmission	Occupations at Risk
Anthrax	*B. anthracis*	Cattle, sheep, goats, horses, pigs; impala, etc.	Skin contact with infected animal tissues; contaminated hair, wool, hides; spores in contaminated soil. Inhalation of spores may cause repiratory anthrax.	Animal and hide handlers, wool sorters, agricultural workers
Bartonellosis (Oroya fever)	*B. bacilliforms* (limited to Peru, Ecuador and Colombia)	Human blood	Sandfly bite; blood transfusion	Foresters in endemic mountain valleys
Brucellosis	*B. abortus* *B. melitensis* *B. suis* *B. canis*	Cattle, swine (U.S.); goats, sheep, esp. in Mediterranean; dogs, coyotes, caribou	Contact with tissues, blood, urine, placentas; airborne infection of humans in laboratories and abattoirs	Animal handlers, laboratory and abattoir workers
Captocytophaga (septicemia)	*C. canimorsus* *C. cynodegmi*	Dogs, cats	Bites, licking or scratching	Animal handlers
Cholera	*V. cholerae*	Humans	Ingestion of water or food contaminated with feces of patients or carriers	Airline crews; disaster relief workers
Ehrlichiosis (fever, etc.)	*E. sennutsu*	Unknown	Tick bite suspected	Outdoor workers
Erysipeloid (localized cutaneous infection)	*E. rhusiopathiae*	Fish, animals	Skin contact	Handlers of fish, animals meat, poultry
Glanders	*P. mallei*	Horses, mules, donkeys (Asia, E. Mediterranean)	Skin contact	Animal handlers; laboratory workers

Disease	Organism	Reservoir	Transmission	At risk
			Skin workers	Building maintenance workers; clerical workers in contaminated buildings; ship crews
Legionellosis (Legionnaires' pneumonia; Pontiac fever)	*L. pneumophila*	Primarily aqueous: hot water systems, air conditioning cooling towers, evaporation condensers; hot and cold water taps and showers	Airborne via aerosol-producing devices	
Listeriosis (meningoencephalitis, septicemia; hand and arm lesions)	*L. monocytogenes*	Forage, water, mud, silage; infected domestic and wild mammals and fowl; humans	Direct skin contact with contaminated soil	Farm workers, animal handlers
Melioidosis (rodent glanders)	*P. pseudomallei*	Saprophytic in certain soils and waters; various animals and birds. Clinical disease is uncommon; occurs in Asia, S. Africa, S. America. Not seen in U.S.	Contact with contaminated soil or water through overt or inapparent skin wounds; inhalation of dust from soil	5–20% of agricultural workers in endemic areas may have antibodies but no history of overt disease
Mycoplasma pneumonia	*M. pneumoniae*	Humans	Droplet inhalation; direct contact with infected person or freshly soiled articles	Health care
Meningococcal meningitis	*N. meningitidis*	Humans	Direct contract, esp. respiratory droplets among newly aggregated adults under crowded living conditions—barracks and institutions	Military recruits, health care workers

Table 12.6 (*Continued*)

Human Disease	Pathogen	Reservoir	Transmission	Occupations at Risk
Plague (see also Yersiniosis)	*Y. pestis*	Wild rodents, esp. ground squirrels; rabbits, hares	Bite of infected flea; handling infected tissues; airborne droplets from human patients or household pets	Foresters; health care and laboratory workers
Rat bite fever (streptobacillosis)	*S. moniliformis*	Infected rats; rarely squirrel, weasel	By secretions of mouth, nose, or conjunctival sac of infected animal, usually by bite; infection also occurs in persons living or working in rat-infested buildings; blood from infected laboratory animal	Pest exterminators; laboratory workers
Tetanus	*C. tetani*	Intestines of horses, other animals or humans; soil or fomites contaminated with animal or human feces	Tetanus spores enter through a puncture wound contaminated with soil, street dust, or animal or human feces; through lacerations, burns or unnoticed trivial wounds	Rescue workers, farm workers, ranchers, foresters, pathologists
Traveler's diarrhea	*E. coli* (also rota viruses)	Humans	Contaminated food or water	Business travelers
Tuberculosis	*M. tuberculosis* *M. bovis* (rare in U.S.)	Primarily humans; diseased cattle	Airborne droplet from persons with pulmonary or laryngeal disease; exposure to infected cattle or by ingestion of unpasteurized milk;	Health care workers (17), correctional institution staff; farmers, animal handlers

animal handlers. HIV-infected persons unusually susceptible, resulting in spread in health care facilities and persons

Disease	Agent	Reservoir	Mode of transmission	Occupations at risk
Tularemia (ulceroglandular fever; pneumonia or systemic disease)	*F. tularensis*	Numerous wild animals, esp. rabbits, hares, muskrats, beavers; some domestic animals; hard ticks	Inoculation of skin, conjunctival sac or oropharyngeal mucosa with blood or tissue while handling infected animals, as in skinning or dressing; by fluids of infected flies, ticks or animals; also bites by wood ticks, dog ticks, Lone Star ticks or by deerfly; handling insufficiently cooked rabbit or hare meat; inhalation of dust from contaminated soil, grain or hay; rarely from bite of animal contaminated by eating infected rabbit; laboratory infections	Foresters, game handlers, farmers, domestic animal handlers, veterinarians, butchers, chefs
Yersiniosis (see also Plague)	*Y. pseudotuberculosis*	Many species of avian and mammalian hosts, particularly rodents and other small animals	A zoonotic disease of wild and domesticated birds and mammals; humans are an incidental host	Animal and bird handlers (rare in the U.S.)
Various	Bacterial toxins (see text)	Various	Various	Various

Table 12.7 Rickettsial Infections of Occupational Significance (10)

Human Disease	Pathogen	Reservoir	Vector/Transmission	Occupations at Risk
Q fever	C. burnetii	Cattle, sheep, goats, cats, rodents	Airborne dust from placental tissues, birth fluids, and excreta of infected animals; also by direct contact with infected animals and contaminated materials: wool, straw, fertilizer, and laundry of infected persons	Animal handlers, including laboratory workers; abattoir workers; veterinarians; health care workers; pathologists
Rocky Mountain spotted fever	R. rickettsia	Ticks, dogs, rodents, etc.	Dog tick, wood tick, Lone star tick bites	Field or forestry workers
Typhus, epidemic	R. prowazekii	Humans	Body louse: infected crushed louse or feces into broken skin or by aerosol to mucous membranes	Close contact with louse-ridden individuals: relief workers, military troops
Typhus, murine	R. burnetti	Rats, mice	Rat flea bite or feces: into broken skin or by aerosol to mucous membranes	Close contact with louse-ridden individuals: relief workers, military troops
Typhus, scrub	R. orientalis	Mites	Larval mites	Field or forestry workers
Boutonneuse fever	R. conorii	Ticks, dogs, rodents, etc.	Tick bites	Field or forestry workers

Table 12.8 Spirochaetal Infections of Occupational Significance (10)

Human Disease	Pathogen	Reservoir	Vector/Transmission	Occupations at Risk
Lyme disease	*B. burgdorferi*	Ixodid ticks, esp. *I. dammini*; wild rodents, deer, etc.	Tick bite	Field and forestry workers
Leptospirosis	*L. interrogans*	Rats, swine, cattle, dogs, raccoons, etc.	Contact of skin, especially if abraded, with water, moist soil, or vegetation contaminated with urine of infected animals; direct contact with urine or feces; occasionally through ingestion of food contaminated with rat urine, or by inhalation of droplet aerosols of contaminated fluids	Workers in sewers, in rice, or sugarcane fields, mines, dairies, abattoirs, animal husbandry; veterinarians, military troops
Relapsing fever, louse borne	*B. recurrentis*	Humans	Body louse: crushing infected louse, contaminating a bite wound or abraded skin	Close contact with louse-ridden individuals: relief workers, military troops
Relapsing fever, tick borne	*B. recurrentis*	Wild rodents, ticks	Argasid tick bite (night feeding)	Field and forestry workers

2.4.2 Fungi

Fungi include yeasts and molds. Dimorphic fungi, which include the agents of histoplasmosis, blastomycosis, sporotrichosis, and coccidioidomycosis, grow in hosts as yeastlike forms but grow at room temperature in vitro as molds. Virtually all fungi reproduce by forming spores. Those pathogenic for humans are often found in soil and may be of occupational significance (Table 12.9). In addition to these, certain fungi (alternaria, aspergillis, cladosporium, and mushroom) may produce asthma in susceptible individuals.

2.4.3 Protozoa

Protozoa are unicellular animals, some of which are known to infect humans. Protozoa such as *Plasmodium* sp., *Entamoeba histolytica*, *Trypanosoma* sp., and *Leishmania* are major worldwide pathogens and are among the leading causes of disease and mortality in areas of Africa, Asia, and Central and South America. *Giardia lamblia* and *Cryptosporidium* are frequent causes of diarrhea in developing areas and established Western countries. *Pneumocystis carinii* has become of increasing importance due to the worldwide spread of HIV, which after a long latent period induces the onset of AIDS. Since pneumocystis pneumonia is of interest exclusively as an opportunistic infection in malnourished, chronically ill and premature infants, as well as in AIDS, it is not of occupational significance and will not be considered here. On the other hand, *Toxoplasma gondii*, *Cryptosporidium* sp., and *Leishmania* sp. are also of importance in AIDS patients, and these have the potential of transmission to health care personnel.

At present levels of international travel in endemic regions, including commercial and military operations, a heightened awareness of diseases due to protozoa is desirable in occupational health practice. Transmission to health care workers, animal handlers, and laboratory personnel may also occur, as indicated in Table 12.10.

2.4.4 Helminths

Helminths, described as *macroparasites* (vs. microparasites: viruses, bacteria, fungi, and protozoa) include several species of nematodes (round worms), trematodes (flukes), and cestodes (tapeworms). Worm infestations are exceedingly common, worldwide in occurrence, and often of multiple species in a given host. Only those few of occupational significance are considered in Table 12.11.

2.4.5 Ectoparasites

An ectoparasite is an organism that lives on or in the skin of its host and derives sustenance from that host. This term can include those organisms that live on the host only long enough to obtain a blood meal as well as those that burrow into the superficial layers of the skin and remain there if left untreated. Some biting or vesicating arthropods that inflict injury on humans do not parasitize the host.

Arthropods are classified as either insects (Hexapoda) or arachnids (Arachnida). Insects are six-legged arthropods and include lice, bedbugs, fleas, and flies. Multiple bites by bedbugs, fleas, and ticks may produce skin irritation or eruptions of varying

Table 12.9 Selected Fungal Infections of Occupational Significance (10)

Human Disease	Pathogen	Reservoir	Transmission	Occupations at Risk
Histoplasmosis (respiratory and/or systemic)	H. capsulatum	Soil around old chicken houses, in caves harboring bats and around starling and blackbird roosts	Inhalation of airborne spores	Workers exposed to old bird, fowl, or bat droppings in recently disturbed contaminated soil
Blastomycosis (granuloma of lungs or skin)	B. dermatitidis	Probably soil	Inhalation of spore-laden dust	Outdoor workers
Coccidioidomycosis (Valley fever)	C. immitis	Soil around Indian middens and rodent burrows in certain climatic regions (Lower Sonoran Life Zone); infects humans, cattle, cats, dogs, horses, sheep, swine, desert rodents, coyotes, etc.	Inhalation of infective arthroconidia from soil; laboratory accidents from cultures	Ranchers, migrant workers, archeologists, laboratory workers
Sporotrichosis (skin nodule)	S. schenckii	Soil, decaying vegetation, wood, moss, hay	Skin pricks by thorns or barbs; handling sphagnum moss or by slivers from wood or lumber; baling hay	Gardeners, nursery people, horticulturists
Cryptococcosis (Torulosis)	C. neoformans	Old pigeon nests and nesting places (building ledges, barns, etc.), pigeon droppings; contaminated soil	Probably by inhalation	Demolition crews: abandoned buildings

Table 12.9 (*Continued*)

Human Disease	Pathogen	Reservoir	Transmission	Occupations at Risk
Dermatophytoses (Tinea, ringworm)	Various species of *Microsporum* and *Trichophyton*	Humans, animals, soil	Direct or indirect contact with skin lesions of infected persons or animals; contaminated floors, shower stalls, benches, etc.	Animal handlers, barbers, gymnasium workers, wrestlers
Chromoblastomycosis (Dermatitis verrucosa)	Various fungal species	Wood, soil, decaying vegetation (rare in U.S.)	Minor penetrating trauma, usually with a sliver of contaminated wood or other debris	Primarily rural barefooted agricultural workers in tropical regions
Mycetoma (foot, hand, shoulders)	Various fungal species	Soil and decaying vegetation (rare in U.S., common in Mexico, etc.)	Contaminated debris in penetrating wounds by thorns or splinters: barefoot or carrying loads on shoulders	Agricultural workers

Table 12.10 Selected Protozoal Infections of Occupational Significance (10)

Human Disease	Pathogen	Reservoir	Vector/Transmission	Occupations at Risk
Malaria (fever, hemolysis)	*P. vivax* *P. malariae* *P. falciparum* *P. ovale*	Humans	Anopheline mosquito bite	Residence or travel in endemic areas: tropics and subtropics; laboratory workers
Amebiasis (enteritis, etc.)	*E. histolytica*	Humans	Ingestion of fecally contaminated water containing amebic cysts; by contaminated raw vegetables; soiled hands of food handlers and laboratory personnel	Laboratory workers
Giardiasis (enteritis)	*G. lamblia*	Humans; possibly beaver and other wild and domestic animals	Person-to-person by hand-to-mouth transfer of cysts from feces, especially in institutions and daycare centers; ingestion of cysts in fecally contaminated water or food	Day-care and institutional workers
Naegleriasis (meningoencephalitis)	*Naegleria fowleri*	Aquatic and soil habitats	Exposure of the nasal passages to contaminated water by diving or swimming in fresh water, especially stagnant ponds, thermal springs, hot tubs, water warmed by effluent of industrial plants, or poorly maintained swimming pools	Lifeguards, maintenance workers

Table 12.10 (*Continued*)

Human Disease	Pathogen	Reservoir	Vector/Transmission	Occupations at Risk
Balantidiasis (enteritis)	*B. coli*	Swine, possibly rats and non-human primates	Ingestion of cysts from fecally contaminated water or food	Animal handlers, laboratory workers
Toxoplasmosis (fever, lymphadenopathy)	*T. gondii*	Mammals, esp. rodents; birds; cats and other felines are the definitive hosts, infected by eating infected species; intermediate hosts are sheep, goats, swine, cattle, chickens	Ingestion of cysts in water or food contaminated by cat feces	Laboratory workers
Cryptosporidiosis (enteritis, cholecystitis)	*Crytosporidium* spp.	Humans, cattle, other domestic animals	Fecal/oral: person-to-person, animal-to-person and waterborne	Veterinarians, animal handlers, travelers, workers in health care, day-care centers, laboratories
Babesiosis (fever, hemolysis)	*B. microti* *B. divergens*	Rodents; cattle	Bite of nymphal ticks carried by voles or deer mice	Foresters
Leishmaniasis (fever, anemia, hepatosplenomegaly)	*Leishmania* spp.	Humans, dogs, rodents	Bite of phlebotamine sandfly; handling infected tissues	Laboratory workers, military personnel (Persian Gulf, etc.)
American trypanosomiasis (fever, lymphodenopathy, hepatosplenomegaly)	*T. cruzi*	Humans; over 150 spp. of domestic and wild animals, esp. dogs, cats, rodents; primates	Bite of blood-sucking reduviid bugs: parasites in bug feces passed while feeding, scratched into the skin or mucous membranes	Rural workers, esp. Central and South America; laboratory workers

Table 12.11 Selected Nematodes, Trematodes and Cestodes of Occupational Significance (10)

Human Disease	Pathogen	Reservoir	Vector/Transmission	Geographic Distribution	Occupations at Risk
a. Nematodes					
Ancylostomiasis (hookworm)	*A. duodenale* *N. americanus*	Humans	Eggs in feces deposited on the ground and hatch; infective larvae penetrate the skin, usually the foot	All climates where sanitary disposal of human feces is not practiced	Miners, laboratory workers
Dirofilarias (pulmonary disease)	*D. immitis*	Dogs, other domestic and wild animals	Mosquito bite	Asia, Australia, U.S.A.	Travelers
Filariasis (lymphangitis)	*W. bancrofti*	Humans	Aedes mosquito bite	Tropics	Travelers, military personnel
Loiasis (eyeworm)	*Loa loa*	Humans	Bite of horsefly or deerfly of genus Chrysops	Central Africa	Travelers
Onchocerciasis (River blindness)	*O. volvulus*	Humans	Bite of blackfly of genus *Simulium*	Guatemala, Central and West Africa	Travelers, construction workers on dams, bridges, hydroelectric projects
Strongyloidiasis (enteritis)	*S. stercoralis*	Humans, dogs, cats, primates	Infected larvae developed in feces or moist soil penetrate the skin	Warm wet regions; in institutions with poor hygiene	Travelers, laboratory personnel, institutional workers
Toxocariasis (visceral larva migrans)	*T. canis* *T. cati*	Dogs Cats	Direct or indirect transmission of infected eggs from soil contaminated with animal droppings	Worldwide	Animal handlers

Table 12.11 (*Continued*)

Human Disease	Pathogen	Reservoir	Vector/Transmission	Geographic Distribution	Occupations at Risk
b. Trematodes Schistosomiasis (Snail fever)	*S. mansoni* *S. haematobium* *S. japonicum*	Humans Humans Humans, water buffalo, etc.	Water containing free swimming larval forms (cercariae) that have developed in snails: penetrate intact skin	Asia, West Africa (none in N. America)	Travelers (water recreation), laboratory workers
Paragonimiasis (Lung fluke disease)	*P. mexicanus* *P. peruvianis* *P. kellicoti* (U.S.A.)	Humans, dog, cat, pig, wild carnivores	Water containing free swimming larval forms (cercariae) that have developed in snails: penetrate intact skin	Asia, Americas	Travelers (water recreation), laboratory workers
Swimmer's itch	*Trichobilharzia* spp., etc.	Birds	Cercariae penetrate intact skin	Widespread; common in lakes of Northern U.S.; also in salt water	Clam diggers
c. Cestode Echinococcosis: Cystic hydatid disease Alveolar hydatid disease	*E. granulosus* *E. multi locularis*	Definitive hosts: dog, wolf, dingo. Intermediate hosts: sheep, goats, pigs, cattle, horses	Hand-to-mouth transfer of tapeworm eggs from dog feces; fecally soiled dog hair, harnesses, and environmental fomites	Common where dogs are used to herd graze animals and consume their carcasses	Herders, dog handlers

severity depending in part on the hypersensitivity status of the individual. The arachnids have eight legs and include mites, ticks, and spiders. Selected ectoparasites are listed in Table 12.12.

Although ectoparasitic infestations are no longer a major public health problem in the United States, on a worldwide basis ectoparasitic disease and diseases transmitted by ectoparasites as vectors present very serious health and economic problems, including those of occupational significance.

Many bacterial, spirochetal, viral, rickettsial, helminthic, and protozoal diseases can be transmitted to humans by arthropod vectors. These are listed in Tables 12.1–12.8.

2.4.5.1 Eradication, Control, and Prevention.

For most ectoparasites, eradication is not a practical approach because of the substantial reservoir of wild animals that allow perpetuation of the species. Elimination of some species from domestic animals is reasonable in certain climates but is a virtual impossibility in other regions of the world.

Methods used to control arthropod vectors include the use of insecticides or environmental manipulations that would limit breeding and spread of the undesirable arthropod species. However, the widespread use of insecticides may induce the development of resistance in the arthropods.

Diseases spread by arthropods may be limited not only by controlling the arthropod vector but also by preventing access of the arthropod to its host. When possible, fine screening should be used on windows and doors to prevent entrance of flying arthropods into the dwelling. Insect repellents that are applied to the skin or clothes are also effective in reducing bites (16).

2.4.6 Venom-Producing Species

On land the principal bearers of venoms consist of snakes, lizards, scorpions, centipedes, millipedes, and spiders. In aquatic environments poisonous snakes are to be found in fresh water, while there are numerous poisonous fish and animals in salt water. In North America the dangerous snakes are rattlesnakes, copperheads, cottonmouth, and coral snakes. Habitats and habits vary with the different species; elsewhere in the world many other species of venomous snakes exist and constitute hazards to employees and crews working in the field (18).

Snake venoms are complex mixtures, chiefly proteins, many of which have enzymatic activity. The arbitrary division of snake venoms into neurotoxins, hemotoxins, cytotoxins, and the like is pharmacologically superficial and clinically misleading.

Only two **lizards,** the Gila monster found in Arizona and Sonora and immediately adjacent areas and the beaded lizard of Mexico are known to be venomous.

Scorpion venoms vary considerably with the genus in their chemistry and mode of action. Most North American scorpions are relatively harmless, their stings usually causing no more than some localized pain and paresthesia with minimal swelling.

Some of the larger **centipedes** can inflict a painful bite, with some localized swelling and erythema. The millipedes do not bite but when handled may discharge

Table 12.12 Selected Ectoparasites of Occupational Significance (16)

Agent	Disease	Source	Transmission	Occupations at Risk
Head louse (*P. capitis*) Body louse (*P. humanis*) Crab louse (P. pubis)	Pediculosis	Humans	Direct contact with infested person; indirect contact with personal belongings, especially clothing and headgear. Under wartime conditions or overcrowding, transmission is enhanced.	Head louse: school and institutional personnel; health care. Body louse: military and relief personnel
Itch mite (*S. scabiei hominis*)	Human scabies (acariasis)	Humans	Direct skin-to-skin contact	Military, relief and health care personnel
Sarcoptic mite (*S. scabiei canis*)	Animal scabies (sarcoptic mange)	Dog	Occasional transmission to humans by direct contact with the coat of infested animal. Mites infesting other animals (cats, rabbits, rats, mice, etc.) and fowl may rarely be transmitted to humans	Veterinarians; animal handlers
Grain mites (*P. ventricosis*)	Grain itch, farmers' itch, bakers' asthma	Grain	Mites feed on the larvae of insects attacking grain; may be transmitted to humans from handling infested grain or straw, causing scabieslike condition. Fragments of mites contaminating flour dust may be sensitizing in some individuals ("bakers' asthma")	Agricultural and grain elevator workers; bakers
House-dust mites (*Dermatophagoides*)	Asthmatic episodes in hypersensitive individuals	Organic debris	Inhalation of house dust from carpets, overstuffed	Household workers

Common chigger, harvest mite (*E. alfreddugesi*)	Chigger bite (skin eruption: severity depends on allergic state of the host)	Grasses, shrubs	Larval form attaches itself to skin of humans or animals passing through infested terrain, obtains a blood meal and drops off	Agricultural workers, foresters, etc.
True flies (*Diptera*): botfly, Tumbu fly, bottle flies (green, black, blue), flesh flies	Myiasis (maggot infestation)	Flies, various species	Eggs deposited on intact skin or in wounds, larvae hatch and burrow into adjacent tissue, mature into adult fly, emerging through furunclelike lesion. Eggs deposited on drying clothing may also produce burrowing larvae during skin contact. Some species produce a creeping eruption. Orificial involvement of ear, nose, or mouth may result in migration of larva into sinuses or brain. Ophthalmomyiasis also occurs. Ingested larvae may produce visceral disorders.	Field workers, especially in the tropics.
Burrowing flea (*Tunga penetrans*)	Tungiasis	Flea	Mature flea invades human skin to enter a gravid state that may resemble myiasis	Field workers, especially in the tropics.
Hard ticks (*D. andersoni*; *D. variabilis*; *I. holocyclus*)	Tick paralysis	Dogs, etc.	Tick bite	Outdoor workers, foresters, animal handlers
Hard ticks (*I. dammini*)	Lyme disease		The most common tick-borne disease in the United States. See Table 12.8.	

a toxic secretion that can cause local skin irritation and, in severe cases, some necrosis. Some non-U.S. species can spray a highly irritating repugnant secretion that may cause severe conjunctival reactions.

With the exception of two small groups, all **spiders** are venomous. Fortunately, the fangs of most species are too short or too fragile to penetrate the human skin. Nevertheless, at least 50 species of spiders in the United States have been implicated in bites on humans. Those species, which produce some local or systemic effects and which must be considered dangerous, include the black widow spider, the brown or violin spider, and the tarantula.

While it may take over 100 **bees** to inflict a lethal dose of venom in most adults, one sting can cause a fatal anaphylactic reaction in a hypersensitive person. There are three to four times more deaths in the United States from bee stings than from snake venom poisoning. In the few fatalities that have resulted from multiple bee stings, death has been attributed to acute cardiovascular collapse. **Wasp** and **hornet** stings are other obvious risks to outdoor workers.

Of minor importance is dermatitis and occasional systemic effects caused by contact with certain **caterpillars** and **moths.** Injury occurs from venomous sharp hairs and spines that penetrate the skin.

Of marine animals, **stingrays** cause some injuries each year along North American coasts. The venom is contained in one or more spines located on the dorsum of the animal's tail. Injuries usually occur when the victim treads on the fish while wading in the ocean surf, bay, or slough. Certain fishes, including **scorpion fishes, zebra fishes, lion fishes,** and **stonefishes** also possess venomous spines that may inflict painful injuries, especially in tropical waters (19).

Echinoderms contain several classes known to be venomous. Certain of the **sea urchins** have venomous organs that have calcareous jaws capable of penetrating the human skin, but injuries from these are rare. Far more common are injuries by sea urchin spines, which can break off in the skin and give rise to local tissue reactions.

Coelenterates, including **jellyfish,** contain a highly developed stinging unit (the nematocyst) that in many species is capable of penetrating the skin. These are abundant on the animal's tentacles, and a single tentacle may fire hundreds of nematocysts into the skin following contact. The most dangerous of these is *Physalia*, the Portuguese man-of-war, and *Chironex*, the box jellyfish (19).

Exposure to all of these may occur in unwary skin divers and fishermen.

3 HEALTH HAZARDS OF BIOTECHNOLOGY

The emerging industry of biotechnology is geared to research leading to products produced through molecular biology techniques. These use recombinant DNA and protein engineering processes that markedly increase the precision and shorten the time period required for genetic experimentation (20,21).

Biotechnology workers may be exposed to some traditional chemical hazards (acetonitrile, acrylamide, dichloromethane, methylene chloride, dimethylsulfate, hydrazine, etc.), to physical agents such as ionizing radiation and cold room atmospheres, and to certain biological agents.

Biotechnology processes may require the use of a microorganism, virus, or mammalian cell to manufacture a specific product. Biological considerations in risk assessment include the pathogenicity of the production organism, allergenicity and immunogenicity of surface proteins, and carcinogenicity for mammalian cells.

Hazardous viruses such as HIV, HBV, and HDV (Hepatitis Delta virus) may be selected for study of biological properties leading to the possible development of vaccines. The isolation of specific genes from such agents may be required. Work with the vaccinia virus may result in risk to research personnel who may have skin disorders or an immunosuppressed state. The National Institutes of Health *Guidelines for Research Recombinant DNA Molecules* is an important resource for information about control of biohazards (22).

Ducatman (20) has pointed out that there are no peer review data concerning population outcomes in biotechnology. The science is too new, the populations at individual sites too small, the movement of trained personnel from employer to employer too rapid, and the enthusiasm of scientist-entrepreneurs for nonproduct research too small to have encouraged any systematic looks at health outcomes among biotechnologists. Thoughtfully designed epidemiologic studies are needed.

While infectious diseases have not proved to be an especially important epidemiological problem for recombinant biologists, the introduction of more dangerous organisms to research and production may result in added risk. Even research with noninfectious genome may result in misleading seroconversion in exposed workers. False seroconversion in a worker handling noninfectious pieces of genome material derived from HIV is a serious risk.

The safe disposal of material contaminated with radiolabeled organisms is of special concern. Chemical decontamination or heat sterilization, including incineration, may result in the dispersion of radioactive particulates. A combination of filtration and steam sterilization has been employed (20).

Planning for the disposal of other infectious wastes from biotechnology operations should include the following aspects (20):

- Designation of infectious waste
- Segregation of infectious waste
- Packaging to prevent leaks and spills
- Storage prior to treatment
- Transportation throughout the facility
- Approved treatment methods
- Compliance with disposal regulations
- Contingency plan for emergencies

4 PREVENTION AND CONTROL OF OCCUPATIONAL INFECTIONS

The variety of infectious agents of occupational significance is so great, the conditions of exposure so varied, and the susceptibility of the individual potential host so unpredictable that only a few general preventive measures are mentioned. Diseases for which immunization is available are few in number, chiefly viral infections, but this obviously is of high priority for prevention. Other prophylactic measures, such as the use of immune globulin (IG preparations) and of antimalarial drugs, is of nearly equal importance. Control of reservoir rodent species and prevention of bites by arthropod vectors are obviously desirable. Strict standards of hygiene of premises, such as health care facilities, are imperative, as well as the use of effective sterilization procedures, and the proper handling and disposal of infectious waste. The application of universal precautions for the prevention of blood-borne diseases in the health care setting has been described.

Due to the special risk of transmission of communicable disease in microbiological and biomedical laboratories, preventive techniques recommended by the U.S. Public Health Service are presented here (9).

4.1 Principles of Biosafety

The term *containment* is used in describing safe methods for managing infectious agents in the laboratory environment where they are being handled or maintained. Primary containment, the protection of personnel and the immediate laboratory environment from exposure to infectious agents, is provided by good microbiological technique and the use of appropriate safety equipment. The use of vaccines may provide an increased level of personal protection. Secondary containment, the protection of the environment external to the laboratory from exposure to infectious materials, is provided by a combination of facility design and operational practices. The purpose of containment is to reduce exposure of laboratory workers and other persons to, and to prevent escape into, the outside environment of potentially hazardous agents. The three elements of containment include laboratory practice and technique, safety equipment, and facility design.

4.2 Laboratory Practice and Technique

The most important element of containment is strict adherence to standard microbiological practices and techniques. Persons working with infectious agents or infected materials must be aware of potential hazards and must be trained and proficient in the practices and techniques required for safely handling such material. The director or person in charge of the laboratory is responsible for providing or arranging for appropriate training of personnel.

When standard laboratory practices are not sufficient to control the hazard associated with a particular agent or laboratory procedure, additional measures may be needed. The laboratory director is responsible for selecting additional safety practices, which must be in keeping with the hazard associated with the agent or procedure.

Each laboratory should develop or adopt a biosafety or operations manual that identifies the hazards that will or may be encountered and that specifies practices and procedures designed to minimize or eliminate risks. Personnel should be advised of special hazards and should be required to read and to follow the required practices and procedures. A scientist trained and knowledgeable in appropriate laboratory techniques, safety procedures, and hazards associated with handling infectious agents must direct laboratory activities.

Laboratory personnel, safety practices, and techniques must be supplemented by appropriate facility design and engineering features, safety equipment, and management practices.

4.3 Safety Equipment (Primary Barriers)

Safety equipment includes biological safety cabinets and a variety of enclosed containers. The biological safety cabinet is the principal device used to provide containment of infectious aerosols generated by many microbiological procedures. Three types of biological safety cabinets (Class I, II, III) are used in microbiological laboratories. Open-fronted Class I and Class II biological safety cabinets are partial containment cabinets that offer significant levels of protection to laboratory personnel and to the environment when used with good microbiological techniques. The gastight Class III biological safety cabinet provides the highest attainable level of protection to personnel and the environment.

An example of an enclosed container is the safety centrifuge cup, which is designed to prevent aerosols from being released during centrifugation.

Safety equipment also includes items for personal protection such as gloves, coats, gowns, shoe covers, boots, respirators, face shields, and safety glasses. These personal protective devices are often used in combination with biological safety cabinets and other devices that contain the agents, animals, or materials with which one is working. In some situations in which it is impractical to work in biological safety cabinets, personal protective devices may be the primary barrier between personnel and the infectious materials. Examples of such activities include certain animal studies, animal necropsy, production activities, and activities relating to maintenance, service, or support of the laboratory facility.

4.4 Facility Design (Secondary Barriers)

The design of the facility is important in providing a barrier to protect persons working in the facility but outside the laboratory and those in the community from infectious agents that may be accidentally released from the laboratory. Laboratory management is responsible for providing facilities commensurate with the laboratory's function. Three facility designs are described below, in ascending order by level of containment.

4.4.1 The Basic Laboratory

This laboratory provides general space in which work is done with viable agents that are not associated with disease in healthy adults. Basic laboratories include those facilities described below as Biosafety Levels 1 and 2 facilities.

This laboratory is also appropriate for work with infectious agents or potentially infectious materials when the hazard levels are low and laboratory personnel can be adequately protected by standard laboratory practice. While work is commonly conducted on the open bench, certain operations are confined to biological safety cabinets. Conventional laboratory designs are adequate. Areas known to be sources of general containment, such as animal rooms and waste staging areas, should not be adjacent to patient care activities. Public areas and general offices to which non-laboratory staff require frequent access should be separated from spaces that primarily support laboratory functions.

4.4.2 The Containment Laboratory

This laboratory has special engineering features that make it possible for laboratory workers to handle hazardous materials without endangering themselves, the community, or the environment. The containment laboratory is described as a Biosafety Level 3 facility. The unique features that distinguish the containment laboratory from the basic laboratory are the provisions for access control and a specialized ventilation system. The containment laboratory may be an entire building or a single module or complex of modules within a building. In all cases the laboratory is separated by a controlled access zone from areas open to the public.

4.4.3 The Maximum Containment Laboratory

This laboratory has special engineering and containment features that allow activities involving infectious agents that are extremely hazardous to the laboratory worker or that may cause serious epidemic disease to be conducted safely. The maximum containment laboratory is described as a Biosafety Level 4 facility. Although the maximum containment laboratory is generally a separate building, it can be constructed as an isolated area within a building. The distinguishing characteristic is the provision of secondary barriers to prevent hazardous materials from escaping into the environment. Such barriers include sealed openings into the laboratory, airlocks or liquid disinfectant barriers, a clothing-change and shower room contiguous to the laboratory, a double door autoclave, a biowaste treatment system, a separate ventilation system, and a treatment system to decontaminate exhaust air.

4.5 Biosafety Levels

Four biosafety levels are described that consist of combinations of laboratory practices and techniques, safety equipment, and laboratory facilities appropriate for the operations performed and the hazard posed by the infectious agents and for the laboratory function or activity.

1. *Biosafety Level 1* practices, safety equipment, and facilities are appropriate for undergraduate and secondary educational training and teaching laboratories and for other facilities in which work is done with defined and characterized strains of viable microorganisms not known to cause disease in healthy adult humans. *Bacillus subtilis*, *Naegleria gruberi*, and infectious canine hepatitis virus are representative of those microorganisms meeting these criteria. Many agents not ordinarily associated with disease processes in humans are, however, opportunistic pathogens and may cause infection in the young, the aged, and in immunodeficient or immunosup-

pressed individuals. Vaccine strains that have undergone multiple **in vivo** passages should not be considered avirulent simply because they are vaccine strains.

2. *Biosafety Level 2* practices, equipment, and facilities are applicable to clinical, diagnostic, teaching, and other facilities in which work is done with the broad spectrum of indigenous moderate-risk agents present in the community and associated with human disease of varying severity. With good microbiological techniques, these agents can be used safely in activities conducted on the open bench, provided the potential for producing aerosols is low. Hepatitis B virus, the salmonellae, and *Toxoplasma* spp. are representative of microorganisms assigned to this containment level. Primary hazards to personnel working with these agents may include accidental autoinoculation, ingestion, and skin or mucous membrane exposure to infectious materials. Procedures with high aerosol potential that may increase the risk of exposure of personnel must be conducted in primary containment equipment or devices.

3. *Biosafety Level 3* practices, safety equipment, and facilities are applicable to clinical, diagnostic, teaching, research, or production facilities in which work is done with indigenous or exotic agents where the potential for infection by aerosols is real and the disease may have serious or lethal consequences. Autoinoculation and ingestion also represent primary hazards to personnel working with these agents. Examples of such agents for which Biosafety Level 3 safeguards are generally recommended include *Mycobacterium tuberculosis*, St. Louis encephalitis virus, and *Coxiella burnetti*.

4. *Biosafety Level 4* practices, safety equipment, and facilities are applicable to work with dangerous and exotic agents that pose a high individual risk of life-threatening disease, including epidemic potential. All manipulations of potentially infectious diagnostic materials, isolates, and naturally or experimentally infected animals pose a high risk of exposure and infection to laboratory personnel. Lassa fever virus is representative of the microorganisms assigned to level 4.

Four biosafety levels are also described for activities involving infectious disease activities with experimental mammals. These four combinations of practices, safety equipment, and facilities are designated **Animal Biosafety Levels 1, 2, 3, and 4** and provide increasing levels of protection to personnel and the environment (9).

4.6 Responsibility for Safety in Biomedical Laboratories

According to Public Health Service guidelines, the laboratory director is directly and primarily responsible for the safe operation of the laboratory. Knowledge and judgment are critical in assessing risks and appropriately applying these recommendations. The recommended biosafety level for a specific infectious agent represents those conditions under which the agent can ordinarily be safely handled. Special characteristics of the agent, the training and experience of personnel, and the nature or function of the laboratory may further influence the application of these recommendations.

Work with known agents should be conducted at the biosafety level recommended in Section V of the CDC/NIH document (9), unless specific information is available to suggest that virulence, pathogenicity, antibiotic resistance patterns, and other factors are significantly altered to require more stringent or allow less stringent practices to be used.

4.7 Clinical Laboratories

Clinical laboratories, and especially those in health care facilities, receive clinical specimens with requests for a variety of diagnostic and clinical support services. Typically, clinical laboratories receive specimens without pertinent information such as patient history or clinical findings that may be suggestive of an infectious etiology. Furthermore, such specimens are often submitted with a broad request for microbiological examination for multiple agents (e.g., sputum samples submitted for "routine," acid-fast, and fungal cultures).

It is the responsibility of the laboratory director to establish standard procedures in the laboratory that realistically address the issue of the infective hazard of clinical specimens.

4.8 Importation and Interstate Shipment of Biomedical Materials

The importation of infectious agents and vectors of human diseases is subject to the requirements of the Public Health Service Foreign Quarantine regulations. Companion regulations of the Public Health Service and the Department of Transportation specify packaging, labeling, and shipping requirements for etiologic agents and diagnostic specimens shipped in interstate commerce.

The U.S. Department of Agriculture regulates the importation and interstate shipment of animal pathogens and prohibits the importation, possession, or use of certain exotic animal disease agents that pose a serious disease threat to domestic livestock and poultry (9).

5. NONLIVING SUBSTANCES OF BIOLOGICAL ORIGIN

Nonliving biological substances of occupational interest include certain natural fibers, grains, plant and animal products, wood dusts, and enzymes. A few natural gums and resins may be included. These various substances produce deleterious effects, chiefly through the mechanism of hypersensitivity reaction, although possible pharmacologic effects from exposure to some components of organic dusts have been postulated.

5.1 Pulmonary Reactions to Organic Dusts

Adverse pulmonary reaction may result in a bronchoconstrictive response in 5 percent or more of workers exposed to certain wood dusts, such as Western red cedar, as well as to grain dust (23). Exposure to other organic dusts, such as that of moldy hay, produces in some individuals a form of hypersensitivity pneumonitis.

The products of bacterial and fungal metabolism are complex, consisting of a variety of split proteins, lipoproteins, and mucopolysaccharides, affected by temperature, humidity, and possibly by the substrate on which the organism grows. Although proteins probably produce the greater number of immunologic effects, some terminal sugars may also be antigenic.

5.1.1 Occupational Asthma Due to Organic Dusts

Inhalation of specific allergens is a well-recognized cause of exacerbations of asthma. Allergen inhalation not only causes acute and delayed bronchoconstriction, but it can also lead to an increase in nonspecific bronchial hyperresponsiveness. Animal proteins, encountered by laboratory workers, veterinarians, and farmers, are other examples of allergens that have been implicated both in exacerbating asthma and also in causing new-onset asthma. It has been assumed that exposure to animals, plants, and their degradation products causes asthma on an immunologic basis, because dusts derived from whole plants and animals contain numerous proteins large enough to serve as antigens. However, these dusts may also contain low-molecular-weight chemicals, which in some cases are the constituent responsible for causing asthma (23).

The list of organic dusts suspected as causative agents in asthma is large and growing. It includes products of animals, fish, and parasites as well as plants. The plant products include the dusts of grains, beans, tobacco, natural fibers, vegetable gums, and woods (oak, redwood, mahogany, boxwood, cocabolla, zebrawood, etc.). Certain enzymes, such as pancreatic extracts, papain, trypsin, and pectinase, have also been identified as allergens, and occupations at risk are many (23).

Of special interest to industrial hygienists and occupational physicians is byssinosis, a pulmonary disorder recognized in cotton mill workers. Inhalation of the dusts of cotton, hemp, flax, and perhaps sisal can produce two distinct patterns of pulmonary response. The best-characterized response is the chest tightness and/or shortness of breath that occurs on the first day of the workweek. Pulmonary function studies frequently confirm acute decrements in airway function in affected workers, and these symptoms have been ascribed to bronchoconstriction. Chronic long-term exposure to cotton dust has also been described to be associated with a higher prevalence of chronic productive cough and a higher prevalence of respiratory disability. This condition resembles chronic bronchitis. The term biossinosis has been used to describe both the acute and the chronic pulmonary responses to crude cotton dust, a cause of some confusion (23).

Three approaches to prevention of byssinosis have been proposed: replacement of cotton by synthetic fibers; the use of engineering improvements to enclose dusty operations, such as cotton gins and opening, mixing, and carding machines with efficient dust removal from equipment, mill floors, and work surfaces; and treating cotton plants or raw cotton to eliminate the factor or factors responsible for byssinosis. Washing and steam cleaning of cotton have been shown to decrease the acute bronchomotor effects of dust inhalation, but those procedures have not been instituted on a significant scale. If bacterial endotoxin is proven to be responsible for these effects, other measures designed to decrease bacterial contamination might be effective in preventing them (23).

Medical surveillance of exposed workers provides biologic monitoring of the adequacy of dust control and also allows the identification of specific workers who are adversely affected by dust exposure. Surveillance should include an annual questionnaire and at least annual measurement of spirometry before and after a work shift on the first day of a workweek.

5.1.2 Hypersensitivity Pneumonitis of Occupational Origin

Hypersensitivity pneumonitis, or extrinsic allergic alveolitis, is an immunologically induced disease associated with intense and/or repeated exposure to finely dispersed organic dusts. The clinical and pathologic findings are similar regardless of the inhaled organic dust. Clinically, affected patients have episodes of fever, cough, and dyspnea 4–6 hr after exposure to the appropriate organic dust (e.g., hay, bagasse, pigeon droppings). Symptoms are frequently mistaken for those of bacterial or viral pneumonia. With prolonged exposure to smaller quantities of the antigen, an afebrile chronic form of the disease may occur. This chronic form is associated with cough, dyspnea, malaise, weakness, and weight loss (24).

Various thermophilic and mesophilic actinomycetes, including *Micrypolyspora faeni*, and the *Thermoactinomyces* species *T. vulgaris*, *T. sacchari*, and *T. candidus* may cause disease in such diverse situations as home central air conditioning, farming, and contaminated humidification systems. The most important occupational diseases in this category are farmer's lung (discussed under "Agricultural Hygiene" in *Patty's Industrial Hygiene and Toxicology* Volume I, Part A, Chapter 20), and bagassosis, the hypersensitivity pneumonitis seen in sugar cane workers. Actinomycetes are members of the true bacteria order (Eubacteriales), although they have the morphology of fungi and are often mistaken and identified as such. They grow best in decaying organic matter such as hay and bagasse, under optimal conditions of humidity at temperatures between 37 and 60°C. High numbers of spores are present in contaminated material; the presence of up to 1.6×10^9 actinomycete spores in the air after disturbing moldy hay has been reported. Because particle sizes are smaller than 6 μm, it has been estimated that a farmer working in this environment might inhale and retain in the lung 750,000 spores per minute of exposure (24).

Proteins derived from feathers, serum, and excrement of several avian species have also been shown to produce hypersensitivity pneumonitis. Using soluble, undialyzed pigeon droppings extract yielded four major antigens on immunoelectrophoresis. One of these antigens, a glycoprotein, was shown to be likely to be a pigeon IgA molecule derivative. Another of these antigens appeared to be a mucopolysaccharide protein conjugate (24).

Among the numerous cases of hypersensitivity pneumonitis, the majority of recognized etiologic agents are derived from occupational exposure such as farming, sugar cane harvesting, working with cereal grains or wood products, and packing mushrooms. The disease may also result from exposure to contaminated central heating and humidification units or may be related to hobbies, such as pigeon breeding. Offending antigens may be derived from microorganisms (actinomycetes, bacteria, fungi, amoebas), animal or bird or plant products, and some pharmaceutical products.

5.1.3   Other Clinical Conditions

Other clinical conditions recognized as arising out of occupational exposure to organic dusts include chronic bronchitis and organic toxic dust syndrome. The latter, discussed primarily in the European literature, is characterized by a febrile reaction in farmers without evidence of pneumonitis. Attacks may be recurrent at long intervals, are of a few days duration, and are provoked by handling grain, hay, straw, wood chips, or silocapping material, usually described as extremely moldy.

5.2   Wood Dusts

Wood is a complex biologic and chemical material consisting primarily of cellulose, hemicellulose, and lignin. Woods also may contain a variety of organic compounds, including glycosides, quinones, tannins, stilbenes, terpenes, aldehydes, and coumarins.

Wood dust exposure may cause eye and skin irritation, respiratory effects, and hard wood nasal cancer. Irritation of the skin and eyes resulting from contact with wood dust is relatively common and may result from mechanical action, chemical irritation, sensitization, or a combination of these factors (25).

Primary irritant dermatitis caused by wood contact consists of erythema and blistering, which may be accompanied by erosions and secondary infections. Irritant chemicals typically are found in the bark or the sap of the outer part of the tree. Therefore, loggers and persons involved in initial wood processing are most affected. In most reports of contact dermatitis, hardwoods of tropical origin have been implicated, although other woods, including pine, spruce, Western red cedar, elm, and alder, have been cited.

Allergic dermatitis arising from exposure to wood substances is characterized by redness, scaling, and itching, which may progress to vesicular dermatitis after repeated exposures. Allergic dermatitis may appear after several years' contact, but typically ensues after a few days or a few weeks of contact. Chemicals causing sensitization generally are found in the heartwood; therefore, workers involved in secondary wood processing (carpenters, sawyers, furniture makers) are more often affected than persons involved in initial processing. Numerous sensitizing agents in wood have been identified, including lapachol (teak), usnic acid (Western red cedar), quinones (rosewood), and anthothecol (African mahogany).

Respiratory ailments associated with wood dust exposure include irritation, bronchitis, nasal mucociliary stasis, impairment of ventilatory function, and asthma. The latter condition is mentioned above.

Obstructive lung disease, as measured by pulmonary function tests, has been associated with wood dust exposure. Vermont woodworkers with hardwood or pine dust exposures greater than 10 mg-years/m^3 generally had lower pulmonary function, as determined by FEV_1/FVC, than those with exposure indices of 0 to 2 mg-years/m^3. Higher exposures also significantly lowered pulmonary function values.

A hypersensitivity reaction leading to asthma (defined as reversible airway obstruction) has been reported from exposure to a number of wood dusts, including oak, mahogany, and redwood, as well as more exotic woods, such as cocobolo, zebrawood, and abiruana. Sensitization typically begins as eye and nose irritation, followed by nonproductive cough and difficulty in breathing. In a sensitized individual, exposure may produce an immediate onset of symptoms and rapid reversibility, or a delayed onset of 5–8 hr with a more gradual reversibility.

Extensive studies have been done on a clearly defined asthma syndrome produced by exposure to Western red cedar. Plicatid acid has been identified as the etiologic agent. The Western red cedar asthma syndrome includes rhinitis, conjunctivitis, wheezing, cough, and nocturnal attacks of breathlessness characterized by a precipitous decline in FEV_1. There is no apparent relation between skin sensitivity and respiratory changes. No precipitating IgC antibodies are found in the serum of sensitized individuals, and circulating IgE antibodies are present in about one-third of affected individuals.

It has been estimated that approximately 5 percent of exposed workers are affected. The asthmatic reaction is species-specific; subjects who exhibit asthma with one type of wood dust show no reaction when challenged with another type.

The association between nasal cancer and occupations involving exposure to wood dust has been established from case reports and epidemiologic studies. This relationship was first noted in the late 1960s in Great Britain, where the incidence of nasal adenocarcinoma, a rare type of nasal cancer, among woodworkers in the furniture industry was found to be 10–20 times greater than among other woodworkers and 100 times greater than in the general population. In a 19-year follow-up study of 8141 Swedish furniture workers, nasal adenocarcinoma was 62 times higher than expected, whereas sinonasal adenocarcinoma and sinonasal carcinoma were 44 and 7 times higher than expected, respectively.

Although estimates of the relative risk of nasal adenocarinoma vary considerably because of differences in exposure levels, types of wood dust, latency periods, selection of controls, and other confounding factors, the International Agency for Research on Cancer (IARC) has concluded that there is sufficient evidence that nasal adenocarcinomas have been caused by exposure in the furniture-making industry (26). The carcinogenic agent(s) in wood dust have not been identified.

5.3 Other Agents Derived from Plants

Of primary importance to outdoor workers is the risk of acute eczematous contact dermatitis from exposure to poison ivy and poison oak. Other plants known to cause dermatitis include castor beans and oleander, as discussed under "Occupational Dermatoses" in *Patty's Industrial Hygiene and Toxicology* Volume I, Part A, Chapter 10.

Among plants recognized as inducing photochemical reactions in some individuals are celery that has been infected with pink rot fungus, dill, fennel, and wild parsnip. The chemical photoreceptors in these plants are psoraleus or furocoumarins (27).

Finally, the significance of exposure to certain pollens as a factor in recurrent physical discomfort of hypersensitive outdoor workers should not be overlooked. Hay fever, the acute seasonal form of allergic rhinitis, is generally induced by windborne pollens. The spring type is due to tree pollens (e.g., oak, elm, maple, alder, birch, cottonwood); the summer type, to grass pollens (e.g., Bermuda, timothy, sweet vernal, orchard, Johnson) and to weed pollens (e.g., sheep sorrel, English plantain); the fall type, to weed pollens (e.g., ragweed). Occasionally, seasonal hay fever is due primarily to airborne fungus spores (28).

REFERENCES

1. J. Lederberg, R. E. Shope, S. C. Oaks Jr., Eds., *Emerging Infections: Microbial Threats to Health in the United States*, Institute of Medicine, National Academy Press, Washington D.C., 1992.
2. A. S. Benenson, Ed., *Control of Communicable Diseases in Man*, 15th ed., American Public Health Assn., Washington, D.C., 1990, pp. 501–509.
3. F. A. Murphy and J. M. Hughes in *Encyclopedia of Human Biology*, Vol. 7, R. Dulbecco, Ed., Academic Press, Orlando, FL, 1991, pp. 639–644.
4. E. D. Kilbourne, in *Cecil Textbook of Medicine*, 16th ed., J. B. Wyngaarden and L. H. Smith, Eds., W.B. Saunders, Philadelphia, 1982, p. 1620.
5. A. S. Benenson, Ed., *Control of Communicable Diseases in Man*, 15th ed., American Public Health Assn., Washington, D.C., 1990.
6. *Fed. Reg.*, **56:** 64004–64182, 29 CFR Part 1910.1030, *Occupational Exposure to Bloodborne Pathogens; Final Rule* (1991).
7. Centers for Disease Control, *MMWR*, **40**(No. RR-13) (1991).
8. J. M. Hughes in *Occupational HIV Infection*, (STAR Vol. 4), C. E. Becker, Ed., Hanley & Belfus, Inc., Philadelphia, 1989, pp. 17–18.
9. CDC/NIH, *Biosafety in Microbiological and Biomedical Laboratories*, 3rd ed., HHS Publ. No. (CDC) 93-8395, U.S. Government Printing Office, Washington, D.C., 1993.
10. A. S. Benenson, Ed., *Control of Communicable Diseases in Man*, 15th ed., American Public Health Assn., Washington, D.C., 1990, p. 154.
11. Centers for Disease Control, *MMWR*, **42,** Suppl. No. RR11 (1993).
12. J. M. Hughes et al., *Science*, **26,** 850 (1993).
13. Centers for Disease Control, *MMWR*, **37,** 377 (1988).
14. Centers for Disease Control, *MMWR*, **38,** Suppl. No. 6S (1989).
15. Centers for Disease Control, *MMWR*, **36,** Suppl. No. SS (1987).
16. G. L. Mandell, R. G. Douglas, and J. E. Bennett, Eds., *Principles and Practice of Infectious Diseases*, 3rd ed., Churchill Livingstone, New York, 1990, pp. 2162–2170.
17. Centers for Disease Control, *MMWR*, **39,** Suppl. No. RR-17 (1990).
18. J. H. Chapman in *Patty's Industrial Hygiene and Toxicology*, 2nd ed., G. D. Clayton and F. E. Clayton, Eds., Wiley, New York, 1985, pp. 476–477.
19. H. L. Keegan in *Cecil Textbook of Medicine*, 16th ed., J. B. Wyngaarden and L. H. Smith, Eds., W. B. Saunders, Philadelphia, 1982, pp. 2246–2248.
20. A. M. Ducatman and D. F. Liberman in *Hazardous Materials Toxicology*, J. B. Sullivan and G. R. Krieger, Eds., Williams & Wilkins, Baltimore, 1992, pp. 556–562.
21. U.S. Congress, Office of Technology Assessment, *New Developments in Biotechnology*, OTA-BA-350, U.S. Government Printing Office, Washington, D.C., 1988.
22. *Fed. Reg.*, **51,** 16958–16985, *Guidelines for Research Involving Recombinant DNA Molecules* (1986).

23. D. Sheppard in *Textbook of Respiratory Medicine*, J. F. Murray and J. A. Nadel, Eds., W. B. Saunders, Philadelphia, 1988, pp. 1593–1605.

24. M. Lopez and J. E. Salvaggio in *Textbook of Respiratory Medicine*, J. F. Murray and J. W. Nadel, Eds., W. B. Saunders, Philadelphia, 1988, pp. 1606–1616.

25. G. J. Hathaway et al., *Proctor and Hughes' Chemical Hazards of the Workplace*, 3rd ed., Van Nostrand Reinhold, New York, 1991, pp. 585–588.

26. IARC, *Evaluation of the Carcinogenic Risk of Chemicals to Humans*, Vol. 25, International Agency for Research on Cancer, Lyon, 1981, pp. 99–138.

27. D. J. Birmingham in *Patty's Industrial Hygiene and Toxicology*, 4th ed., Vol. I, Part A, G. D. Clayton and F. E. Clayton, Eds., Wiley, New York, 1991, p. 263.

28. R. Berkow, Ed., *The Merck Manual*, 13th ed., Merck & Co., Rahway, NJ, 1977, p. 229.

CHAPTER THIRTEEN

Evaluation of Exposure to Ionizing Radiation

Robert G. Thomas, Ph.D.

1 INTRODUCTION

Industrial hygiene is the professional field that has responsibility for protection of individuals from potential hazards in the workplace. The field is broad as it must address problems associated with the atmosphere, the drinking water, and/or other sources of toxicants that may be of a chemical, physical, or biological nature. The hazards associated with exposure to ionizing radiation may take many forms, and this chapter presents information on the biomedical effects of radiation, particularly as these potential insults deal with the nature of protection that is required to protect the worker and the population. Ionizing radiation is a source of potential hazard that is known the world over, and the anxiety associated with it has received much attention, primarily from uninformed sources.

Radiation has been known as a toxic agent for a relatively short time (less than 100 years). It has received considerable attention in recent decades, however, primarily as a result of military applications during World War II and the 1986 reactor accident at Chernobyl in the Soviet Union. Compared to chemically toxic agents, the properties of ionizing radiation generally enable its detection at much lower quantities and with much greater accuracy. This condition has contributed greatly to the broadly held awareness of radiation's toxic potential, but in many (or perhaps most) cases this is mainly because its presence can be detected at very low levels. It appears that much of the concern over health hazards due to ionizing radiation is

Patty's Industrial Hygiene and Toxicology, Third Edition, Volume 3, Part B, Edited by Lewis J. Cralley, Lester V. Cralley, and James S. Bus.
ISBN 0-471-53065-4 © 1995 John Wiley & Sons, Inc.

the direct result of this. One of the aspects of health care effects of exposure to ionizing radiation that will be emphasized in this chapter will focus more on the health protection issues and less upon the more esoteric aspects of radiation effects at the discrete molecular biological level. Thus, this chapter is aimed at explaining the current state of health protection from ionizing radiation, the biomedical basis for the guidelines used for protection, and the philosophies that are basic to the underlying application of health physics to industrial hygiene practices.

Radiation protection is a field that evolved from the need to ensure the protection of workers and the population from ionizing radiation. Health physics is the term that was initiated to describe the scientific discipline in which persons were educated and trained to administer and enforce the necessary guidelines to bring about radiation health protection. With the current mode for desiring "complete safety" that permeates the U.S. population, it is not surprising that protection from ionizing radiation has acquired an increasing popularity in the past decade, particularly following the Chernobyl accident. It is an important field and the contents of this chapter will hopefully contain information that will be of assistance in helping to produce a better understanding of radiobiology and health physics and to better tie the necessary guidelines with protection needs and common sense.

Radiobiology is the scientific field (sometimes referred to as radiation toxicology or radiation biology) that produces the data upon which most of the radiation protection rules and guidelines are based. One of the major purposes of this chapter is to sufficiently describe radiobiological information so that its use in establishing health protection standards will become clearer to the average reader. The basic problem with the standards as they exist today, particularly in terms of radiation risk, is that they are primarily based on extrapolation from data obtained from exposures to much higher levels of radiation and under different exposure conditions than those encountered in the workplace or the environment. The largest number of cases of radiation disease in humans that is used for this extrapolative process is from the Japanese experience in World War II following the microsecond (acutely delivered) external radiation exposures, whereas the present-day exposure protection guidelines are for the repeated or continuous exposure regime. The two are not compatible, but they are generally accepted as being so because the data are the only ones on humans that have been exploited and studied on an international basis. Extrapolative processes and interpretation of data are so individualistic that the data from all human exposures are often mistreated to satisfy the desired guidelines of the individual and/or his/her agency.

There are three rather distinct possibilities for human radiation exposure: irradiation from sources external to the body; irradiation from sources deposited internally that have entered the body by various routes, such as inhalation, ingestion, and on or through the skin; and, a third, exposure to significant amounts of both of these internal and external sources. This latter is generally from accident situations, such as with the Chernobyl releases, and though the irradiation may be from a combination of insults the effects can only be interpreted, with current knowledge, as a direct summation of the two types of insult (from internal and external radiations). This chapter devotes considerably more emphasis to the internal sources of potential radiation health problems, primarily because accumulation through the internal routes is often less controllable. Internally acquired radionuclides also represent a more classic toxic poisoning, as we have come to know it. Also, measurement of radiation from external sources is much more reliably related to body exposure than is for instance, the reconstituted air concentration of a radionuclide to the calculated long-term dosage that may be received by an internal organ. There are many difficulties associated with the latter estimates—determining such factors as the particle size breathed, the amount deposited in various sections of the respiratory tract, solubility in body fluids, elemental position in the periodic table—and naturally, the differential metabolism of foreign materials by exposed individuals. The latter factor may involve such basic parameters as sex differences, age, and body and organ weight. Because the degree of injury is probably related to the concentration of a radioactive material in an organ, it is easy to see how organ weight could become such an important factor.

1.1 Types of Radiation Exposure

1.1.1 External Radiation

External radiation, in its less exotic forms, is generally considered to be either from an electromagnetic (wavelike) or a neutron source. There are other sources of external ionizing radiation, but these rarely lead to industrial exposure of the worker. With electromagnetic radiation (gamma rays, x-rays), ionization in a medium may be caused by one of three primary processes, namely, the photoelectric effect, Compton scattering, and pair production. These three types of ionization are illustrated in elementary textbooks on nuclear physics.

The photoelectric effect generally prevails when tissues are affected at photon energies less than 100 keV. In this process an electron is ejected from an atom in the medium, and it subsequently becomes the major source of further ionization within the medium. Ionization by electromagnetic radiation is somewhat "inefficient"; therefore pathways may be very long compared to the length of ionization path of the secondary electron released. Hence ionization events may not occur in close proximity to each other, and effects on tissues may be diffuse. As the secondary electron traverses the tissue, it promulgates additional ionizing electrons known as delta rays. These are discussed in paragraphs to follow. In the photoelectric ionization process all the incident energy is imparted to the ejected electron, which explains the effectiveness of this process at the lower energies.

Between the low (about 100 keV) and very high (about 10 MeV) energy regions, the ionizing process known as Compton scattering prevails. Under these conditions the incident photon is partially degraded by interacting with an orbital electron in an atom of the medium being traversed. The energy remaining in the photon carries it further through the medium, to continue interaction with other atoms in various molecules. The ejected electron (delta ray) continues with its imparted energy (the fraction obtained from the incident photon) to ionize other atoms near to the original event. This "clustering" of events in the immediate vicinity of the first ionization is due to the relatively heavy ionizing density of the particulate radiation (ejected electron), described in Section 1.1.2. Because the principal range of energies for

Compton scattering spans most of the common external radiation sources, this ionizing process is most prevalent in radiobiological effects.

Pair production is confined to high-energy photons (>1.2 MeV). This event has little place in radiobiological reactions because of the energies involved and the atomic structure required. In this process the photon energy, in the proximity of a highly charged nucleus, is transformed into mass because of the strong nuclear electromagnetic forces, thus forming one negatively charged electron mass and one that is positively charged. Each of these newly formed particles is capable of ionizing atoms to an extent that is dependent on their energies, as described.

Neutrons, uncharged particles in the simplest sense, damage the traversed medium in a different manner. The neutral nature of neutrons allows them to have large penetrating distances. Because they are neutral, they generally penetrate the outer electron clouds of atoms and in essence, strike the nucleus. This is due to the relative size of the target. If the neutron is of sufficiently high energy (fast), the nucleus is dislodged and actually is driven some distance through the surrounding tissue, causing damage as it goes. The nucleus ultimately picks up enough electrons to once again become its original "self" as it comes to rest. The elements of greatest abundance in soft body tissues are carbon, hydrogen, oxygen, and nitrogen. The abundance of hydrogen nuclei make this element the most vulnerable target for fast neutron interaction, therefore hydrogen will have a much greater absorbed proportion of the incident neutron than will the less abundant larger atoms. Slower (low energy) neutrons may be captured upon encounter with a nucleus. This can result in an unstable condition in which the nucleus will release energy in many forms, thus becoming a primary source of ionizing energy. This generally limits the radiobiological action to the proximity of the initial event.

Protons and other charged particles are also a source of external radiation and their ionization is similar to alpha and beta particles, as described below.

1.1.2 Internal Radiation

Internal radiation sources, as far as industrial exposure is concerned, are generally comprised of alpha or beta particle emitters. Beta particles, electrons by charge and weight, may interact with orbital electrons of atoms within molecules, or with the nucleus. When the beta particle approaches an orbital electron, energy is imparted, the orbital electron may be ejected as a delta ray, and the incident beta particle will continue at a lowered energy and with an altered direction. This procession by the incident electron will continue until the electron dies by capture somewhere in the medium (tissue). The delta rays (secondary electrons) will traverse the surrounding area and operate in a manner similar to that of the original primary incident electron. Because of the charge and the mass, compared to a photon (Section 1.1.1), the ionizing events associated with electrons are clustered and occur within a very small volume of tissue.

Alpha particles, being helium nuclei of $2+$ charge and 4 mass units, are much more heavily ionizing than are electrons (beta particles). Thus their path is straighter because there is less deflection upon energy loss, and their ionizing events are confined to much smaller volumes of tissue. A 5-MeV alpha particle in soft tissue has a linear path length of approximately 40 μm. Since beta particles from the nucleus of radioactive atoms have a spectrum of energies, their path length is variable. However, the range of the emitted beta particle is many times that of an alpha particle of the same incident energy. Emitted alpha particles are released with discrete energy(ies).

1.1.2.1 Radon. Within the past 5 years there has been an increased interest in the radon-in-homes issue throughout the Western world. This was stimulated when some houses in part of Pennsylvania known as the Reading Prong were found to contain what seemed to be unusually high air concentrations of radon and its daughter products. As a result of the new awareness of this old source of radiation, a recent report compiled by a National Research Council committee on Biological Effects of Ionization Radiation (BEIR IV) devoted one-half of its internal emitter deliberations to radon and its daughter products (1). Radon presents a unique type of hazard because the conventional method of calculating its dose to humans is not used; instead, a manner of describing the radon-containing atmosphere derived for uranium mines is applied to exposure conditions, and the method of calculating the dose to the respiratory tract is being exploited with a host of sophisticated new approaches. Radon is not really an industrial exposure problem except in the particular instances mentioned, but because of the scientific challenge it arouses and its sociopolitical popularity, it is appropriate to give it special consideration in this chapter. It will be a special topic for discussion throughout.

1.2 Radiation Measurement

1.2.1 Electromagnetic Radiation

Electromagnetic radiation represents the easiest form to measure where radiobiological exposures are concerned. It is usually uniform within the location of possible worker exposure, and in most cases one could probably argue that the energy characteristics from a relatively stable source, whether spectral or quite narrow, would fall within a consistent range. This may be speculated because of the more common nature of the source of such "leakage" radiation. If the electromagnetic radiation is from a source of gamma emission, such as from ^{60}Co, the energy distribution is consistent, making interpretation of any measurements, even of a gross nature, much easier and probably more accurate.

Ionization in air has been accepted as the primary measuring standard on which all other (secondary) assessments are based. The amount of ionization in a specific volume of air under precise conditions of temperature and pressure serves to determine this primary standard. When electronic equilibrium is achieved (secondary electron influx into a specific volume of air is exactly equal to secondary electron efflux) within a given measuring device, the amount of current generated by the ionization is proportional to the amount of incoming energy of the photons. It is this incident energy that is directly proportional to the defined units used in radiation studies. The roentgen, which is a measure of this ionization under primary standard conditions, is defined as the quantity of X or γ radiation such that the associated corpuscular emission per 0.001293 g of air produces, in air, under standard conditions of temperature and pressure, ions carrying one electrostatic unit of quantity of electricity of either sign. With this primary basis for measuring incident electromagnetic radiation, it is not difficult to conceive that many instruments have been devised to measure ionization in media that may be related to the primary standard.

In recent years crystalline and solid-state detectors have been most successful for measuring external radiation dosage. The former will yield spectral energy data if sophistication is desired but may be used as constantly recording, full energy monitors without this sophistication. Electronic windows may be applied to the detection device to limit detection to a rather specific range of energies. The solid-state variety works extremely well in sharp spectral resolution, but the need for such equipment in routine monitoring for the worker is often questionable. This is not to imply that solid-state detectors are not widely used, particularly in personnel monitoring.

1.2.2 Neutrons

Neutrons are generally separated by energy into thermal, intermediate, and fast, as indicated in Section 1.1.1. The classical scheme of energies associated with these is less than 0.5 eV, 0.5 eV to 10 keV, and 10 keV to 10 MeV, respectively. Thermal neutrons are primarily degraded in matter by capture, and the capture cross section is inversely proportional to the velocity of the neutron. For instance, when hydrogen ($_1H^1$) captures a neutron ($_0n^1$) the result is a $_1H^2$ atom, plus a gamma ray as excess energy. The intermediate neutrons fall into a range in which there are resonant peaks in the capture cross section, leading to a slowing down of the neutron in matter. This process is generally inversely proportional to the energy of the particle. Fast neutrons interact by scattering and may be elastic or inelastic, depending on whether the energies are toward the low or high end of the range, respectively. As mentioned in Section 1.1.1, the most important interaction in tissue is with hydrogen, each particle having essentially the same mass. Relativistic neutrons, another category, are not really of importance to this chapter.

Methods of detection and measurement of neutron fluxes have been based on neutron properties, as noted earlier. Calorimetry, the measurement of temperature rise due to ionization in a medium, is always very accurate, but it is difficult to measure the small changes in temperature achieved. Ionization depends on many factors, and the medium being ionized within the instrument should have very precise and determinately known qualities. The size (dimension) and structure of the ionization chamber cavity are very sensitive parameters and must adhere to criteria that are related to the energy of the incoming neutron. For instance, an instrument designed to work reasonably well with thermal neutrons will not perform for fast neutrons. As with electromagnetic radiation, chemical means of detection and measurement are choices that give extreme accuracy. The problem in either solid (e.g., photographic) or liquid media is that the dose–response curve is not linear, and impurities play an important role in the resulting estimation. It is difficult to obtain a liquid chemical detection system that is pure enough to give no spurious neutron interactions. Perhaps one of the most reliable methods of detecting and measuring neutron radiation is through its secondary gamma rays following an event. Depending on the energy of the neutron, hence the reaction involved, secondary gamma rays will undoubtedly be emitted. Accurate determination of time of irradiation and magnitude of the gamma rays produced in the reaction of neutrons with the absorbing medium allows calculation back to the incident flux.

All these general principles are operable in their own right. However, since some require equipment that is not suitable for field work, bulkiness becomes an important drawback with most. In recent years the solid-state detection systems have become popular and probably will ultimately replace all other means for detection on personnel. These include the thermoluminescent detectors used routinely in many laboratories for personnel monitoring. They have many advantages and are still under development for more precise detection characteristics.

1.2.3 Particles Other than Neutrons

Assessment of the impact from particle radiation is perhaps the least gratifying of all measurements that can be made in the case of accidental exposure to the worker, although recent advances are encouraging. The use of standard filter sampling at constant volume through small-pore filters is simple and relatively inexpensive. The filter may be constantly monitored, the buildup of activity recorded, and when a certain predetermined radiation level is reached, the result is announced by an alarm. This represents a steady, consistent method of detection of leakage of radioactive materials from a given operation. This type of sampling is usable for both alpha and beta emitters. The difficulty inherent in this method of monitoring arises when an inhalation accident occurs.

Until recently little time and effort have been devoted to methods of determining solubility and particle size of the sample that is collected and is presumed to be representative of the atmosphere breathed by the worker. In the case of external radiation one has some reasonable chance of mocking up the exposure conditions, as was attempted in the Lockport exposures (2), and to closely estimate the worker dose. One generally does not have the opportunity to do this when accidental inhalation exposures to radioactive aerosols are involved; this has been the case with most accidental exposures to plutonium in the atomic energy industry (3,4).

The routine use of continuous air monitors that will allow determination of particle size must be a compromise between sophistication and economics. Ideally, five- or seven-stage cascade impactors of the Anderson (5) or the Mercer (6) variety could be used in a constant monitoring mode so that if an accident occurred, the exposure atmosphere could be carefully determined retrospectively with regard to particle size distribution. However, the constant problem of sophisticated samplers becoming clogged with room dust makes routine sampling sometimes impractical. More important with the use of this type of sampler, coupled with the particle size data, is the ability to do solubility measurements on each size fraction. The combination of these two determinations would enable a reasonably accurate estimation of the quantity deposited in the worker's lower respiratory tract, and the extent to which the deposited particles would be soluble in body fluids. The type of therapy to be applied after an accident could be much more judiciously determined if this kind of information were available at an early time postexposure. If the deposited particles were completely insoluble, one would consider lung lavage (7); but if they were soluble and entering the blood, chelation therapy (see Section 4.2.1) would become an option.

Perhaps the most practical air monitors are those that attempt a relatively crude separation of *respirable* and *nonrespirable* sizes (8–12). Solubility studies from such samplers can be made, and the results perhaps are equally satisfying as those obtained using the more complicated particle sizing instruments, given the spectrum of errors inherent in the measurements.

For certain operations a more specialized system has been devised that enables the detection of various daughter products, by types of emission or energies involved, as in the radium series (13). Collectors may be continuously monitored for various energies and types of emissions so that ratios of two different daughters can be determined, or a total sample can be captured and total daughter activity at a given time determined. Many types of sampler for the uranium mining industry have been developed along these lines with various methods of analysis (e.g., Refs. 14–16). In all cases any counts above "background" may be instrumented to be indicative of an accidental release or an abnormal working environment. These systems have a clear application in the uranium mining industry.

A new aspect of airborne dosimetry was recently reemphasized with the increasing interest in the potential health effects of inhaling radon and its daughters. With the radon-in-homes issue at the forefront, it has been necessary to develop some measuring devices that could be used by the ordinary citizen in the ordinary home to ascertain if his/her family was being "overexposed," according to the current Environmental Protection Agency (EPA) guidelines (4 pCi/L). Thus there have been many new instruments developed to measure radon and its daughters, but they mostly rely on the fundamental process of determining the amount of energy released from the radon collected on charcoal or the particulate daughters being collected on paper or other collection media. The basic principles are the same as in the past.

In addition to actual instrumentation that has been developed, the use of alpha-particle tract etching is receiving more attention as a means to establish long-term exposure to atmospheric radon daughters that have deposited upon plastic, glass, or other etchable materials. This technique was suggested many years ago and has many possibilities as an accurate dosage measurement technique (17).

1.3 Dosimetry

It must again be emphasized that there are situations (generally accidental, like Chernobyl) in which both internal and external radiation exist simultaneously. However, there is a time factor present in these two in which the bulk of the external exposure is probably over, from cloud passage, within a relatively short time. The internal exposure from radionuclides may be occurring simultaneously, particularly from inhalation and skin absorption, but much internal radiation is generally of longer duration, even with short-lived iodine isotopes. Because this scenario is a combination of the discrete sources, no attempt will be made in this chapter to illustrate combination measurements of dose or biological effect information. In addition to the scenarios just described, there is always the situation following an accident involving a radioactive cloud release, in which the soil and water are contaminated and the vegetation they support is consequently contaminated; the extent and longevity of this added exposure is quite involved.

1.3.1 Electromagnetic Radiation

As implied, dosimetry with electromagnetic radiation in the human exposure case may be more straightforward than for exposure to airborne radionuclide particles or vapors, primarily because of the ability to mock up exposure conditions. Historically, the determination of radiation dose has proceeded through a series of changes since the discovery of x-rays by Roentgen in 1895. For quite some time the erythema dose was used to determine proper treatment in radiation therapy. In the 1920s it was recognized that ionization in air was the best approach to estimating dosage for therapeutic purposes. The roentgen was defined in 1928, and this ultimately led to widespread usage of the R unit, followed by the rep, the rem, and finally, in current use today, the rad. The rad is a measure of the energy deposited in any medium, and one rad is equivalent to the deposition of 100 ergs per gram of that medium.

Various kinds of instrumentation have been used to define dosimetrically the energy arising from a gamma- or x-ray source. These include air ionization chambers, semiconductors, thermoluminescent devices, and other instruments using heat, light, and chemical changes as means to quantitate the ionizing events being produced. The most practical and common personnel dosimeters in use are the film badge (optical) and the thermoluminescent dosimeter. In any event, although no dosimetry appears simple and straightforward, electromagnetic radiation lends itself to giving the most consistent results where exposure to humans is concerned.

1.3.2 Neutrons

The accurate determination of neutron dosage is fraught with problems that entail many factors. In general, neutron dosimetry has actually represented a measurement of accompanying protons, through the latter's interaction with some medium that is effective in giving a proportionate relationship to the incident neutron flux. Neutron dosimetry is obviously very energy dependent because the secondary "reactant" serving as the primary detector (of protons) is highly dependent on the type and efficiency of the reaction from which it was derived. Thus, neutron detectors, per se, are very energy dependent, and awareness of the source term is mandatory when detection instrumentation is selected.

The International Commission on Radiation Units and Measurements report of 1977 (18) on neutron dosimetry in biology and medicine separates dosimetric methods and instrumentation into several general types: (1) gaseous devices, (2) calorimeters, (3) solid-state devices, (4) activation and fission methods, and (5) ferrous sulfate dosimeters. The details of these devices may be obtained from Ref. 18, and this chapter does not duplicate that fine summary of the problems inherent in accurate determination of dosage from incident neutrons. The types of device and instrumentation described are quite broad, however, and deserve mentioning here by specific name, for those who may wish to pursue a given type of detection. The more specific devices mentioned are (1) ionization chambers, (2) proportional counters, (3) Geiger–

Müller counters, (4) photographic emulsions, (5) thermoluminescent devices, (6) scintillation devices, (7) semiconductors, and (8) nuclear track recorders. Some of the more general categories listed above (e.g., calorimetry) represent an overall methodology and do not require the individual specificity of a given type of instrumentation.

Neutron dosimetry is not as straightforward as one would desire, and it is certain that research in this field will continue for decades.

1.3.3 Particulate Radiation

The most common dose calculation for internal emitters comprised almost solely of alpha and beta radiation utilizes the *average dose* concept. In the classical sense, if the amount of radioactivity in an organ can be estimated by any means, it is assumed, for simplicity, that the radioactivity is distributed uniformly throughout the organ. This assumption is almost always in error with soft beta emitters and alpha particles, because of their short range in tissues (35–40 μm for the average alpha from heavy radioelements). However, when one attempts to estimate dose–effect relationships on a microdistribution or hot-spot premise, errors are equally forthcoming. The situation with radon and daughters once again has been made different. The scientific community dealing with the radon problem has chosen to bring the dosimetry down to the cell nucleus level in order to estimate the amount of energy that will inflict damage on the DNA structures contained therein. This is an active area of radiobiological research at the present, and it is certain to lead to some new areas for future research. The relatively large research program at the Department of Energy (DOE) has been responsible for most of the good research in this area over the past years, and the accounting of the progress made has been through a series of reports emanating from Dr. Susan Rose's productive management practices in that program (19, 20). These are excellent for maintenance with the advances in the field that deal with the highly sophisticated approaches to respiratory tract dosimetry. For purposes of this chapter, this sophistication will not be expounded as it is in the trial stage, both theoretical and in the laboratory, and therefore it will be assumed that average organ or tissue dose is sufficient, with full knowledge that this concept may be completely dropped at sometime in the future. It must always be remembered however, that such detailed microdosimetric quantitation of respiratory tract dose will probably never find its way into practical routine radiation protection practices.

To arrive at an average tissue dose to an organ from the estimated air concentration breathed, many steps are required. These steps are described below.

1.3.3.1 Particle Distribution.
Airborne particles in real life are generally lognormally distributed by mass and can be described by a mass median diameter—the particle size above and below which lies one-half of the total mass of the sample collected. The aerodynamic diameter of a particle size distribution is becoming a more popular descriptor than it was, say 30 years ago, and it incorporates the density of the material and a correction for very small particles (21). The size spread of the lognormally distributed aerosol particles is defined by the geometric standard deviation, symbolized by σ_g. Thus a distribution of particles collected on a cascade impactor in an industrial situation would be described by $\bar{\mu} \pm \sigma_g$, where $\bar{\mu}$ is the median

diameter defined in mass or aerodynamic terms and is commonly expressed in micrometers.

1.3.3.2 Respiratory Tract Deposition.

It has been long known that site of particle deposition and retention in the respiratory tract during and following inhalation is dependent on particle size. Very simply, the larger particles deposit in the upper areas of the tract (nasopharyngeal region, trachea) and are quickly removed by ciliary action. The smaller particles traverse past the tracheal bifurcation and deposit all the way to the alveolar region. The situation with radon is similar to that with all particulate materials, with exception that the particle characteristics are often those of the dust upon which the radon daughters have deposited before, during, or after the inhalation has occurred, or, of some significant fraction (up to 50%) that may remain unattached to any other material before inhalation. Because of the different half-lives of the daughter products, it depends on when the daughter was formed and when it was inhaled as to where it is apt to be deposited in the respiratory tract. The particle size with radon and daughters is such (so small; angstrom to submicron range) that diffusion is the major means of deposition in the respiratory tract and the usual particle characteristics that are of importance (inertial impaction and gravitational force) are not generally very important.

Respiratory tract deposition characteristics were described in detail in 1966 (22), and despite some modifications this review remains the most comprehensive. New Task Groups have been formed by the International Commission on Radiation Protection (ICRP) and the National Council on Radiation Protection (NCRP) to update this source but the fundamental characteristics will remain about the same. The deposition characteristics for particles in the respiratory tract cannot vary over orders of magnitude as the deposited fraction must be between somewhere above 0 and 100 percent. Therefore, it is difficult to make dramatic changes in the average (for the population) quantitative aspects of particle deposition as they relate to dose.

1.3.3.3 Respiratory Tract Retention.

Respiratory tract retention parameters are also elaborated very thoroughly in the 1966 reference (22). Ciliary activity obviously accounts for a large, rapid clearance of deposited material from the upper respiratory passages. This material does not present a major problem from the viewpoint of toxicity unless the material is readily absorbed during passage through the gastrointestinal tract, or unless it is present in very large quantities (high doses). It is the material that lingers in the deeper (alveolar) spaces of the lung that seems to dictate problems such as carcinogenicity. If the deposited material is very soluble, it will enter the bloodstream through the lung or the gastrointestinal tract and will proceed by way of the circulatory system to deposit in the organ dictated (primarily) by its chemistry. If the material is totally insoluble, and nothing really is in biological fluids, three alternatives present themselves: to remain in the alveoli indefinitely (life of individual), gradually to be transferred upward by ciliary action through an initial step of phagocytosis by macrophages, or to be transported to the pulmonary lymph nodes. Numerous mechanisms and kinetics describing these alternatives have been reported in the literature, and individual subjects of interest can be found in the proceedings of some recent symposia (23–26).

1.3.3.4 Uniform Organ Retention. Once the inhaled material has entered and has been deposited in the organ of choice, its kinetics of loss (retention function) can easily be described if a few facts are known. These include rate of loss from the tissue to blood, and the rate of reentry to the same types of tissue versus rate of excretion in urine or feces. This is rather simply described in terms of a general whole-organ concept, but the next section discusses the more complicated situation of highly localized dose, using bone as an example. The problem is that the rate constants described are seldom known accurately, particularly for humans. If first-order kinetics prevail for loss from the organ, and the organ concentration of the radionuclide is relatively well estimated, average dose calculations are quite straight-forward. Many such data are described for animals, but the extrapolation to humans is, at best, satisfactory. An example of a dose calculation is shown below, using specific units for purposes of demonstration.

$$D \text{ (rads)} = C_0 \times k_1 \times \overline{E} \times k_2 \times k_3 \times \int_0^t R_t \tag{1}$$

where C_0 = organ concentration of the radionuclide at $t = 0$ (μCi/g of tissue)
k_1 = conversion factor (2.22×10^6 dpm/μCi)
$\overline{E}$ = average energy (MeV/disintegration)
k_2 = conversion factor (1.62×10^6 MeV/erg)
k_3 = conversion factor (10^{-2} for the definition, 100 ergs/g = 1.0 rad)
R_t = retention function in the organ of deposition

This equation assumes that all the associated particulate energy is absorbed in the tissue involved. The retention function is obviously the key to accuracy in this equation, and it is the most difficult to estimate for the exposed industrial worker. Only estimations of the assumed inhaled atmosphere, the inhalation-related parameters, and the metabolism of the material involved, based on animal work or suitable related publications (27–29) can be made.

Reversal of the calculation above, using the total dose allowable or recommended in standards will permit solution for other parameters of interest such as C_0, and working backward through the processes just described, a recommended maximum air concentration may be determined. This process is not as simple as it seems, but it does allow for computer computation for any radioelement about which some biological parameters are known.

1.3.3.5 Localized Organ Retention. Skeletal deposition may serve as a good example of localized retention with regard to internal dosimetry problems. The dosimetry of radionuclides deposited in bone is very complicated where alpha- or soft beta-emitting radionuclides are involved. The complications are related to the long-discussed hot-spot problem. With radionuclides (as with any metallic element) the ionic chemistry of the isotope dictates the site in which the particle will locate. Durbin has written an excellent review of the effect of ionic radius on site of deposition (30). One of the chemical groups of elements associated with bone is the bivalent alkaline earths: calcium, barium, strontium, and radium. One of the most important structures in bone is the hydroxyapatite crystal into which mineral calcium is laid down. Alkaline earths are termed volume seekers because they are primarily incorporated into the mineral portion of the bone tissue, where they readily become a part of compact bone. The turnover of these elements is slow once they have become incorporated, even though remodeling of bone occurs throughout life. Thus if an area surrounding a haversian canal is not in the process of remodeling at the time one of the radioactive alkaline earth elements enters the plasma, the chances of that isotope entering that particular haversian system are reduced considerably. The unique quality of bone, however, allows a given haversian system to be in the process of remodeling while at the same time the surrounding systems are in a state of rest. In this case, with a relatively high concentration in the plasma of, say, strontium or radium, there is a finite probability that the ionic form of the alkaline earth will be deposited in the remodeling sites. Based on this general description, it is obvious that hot spots would readily form in the bone as a result of the remodeling system's seeming inconsistency. Autoradiograms of bone observed under these circumstances indicate a very nonuniform distribution, the pattern of which depends on the microanatomic nature of the bone at the time of deposition.

Whereas the alkaline earths are termed volume seekers, the transuranic elements such as plutonium and americium are commonly referred to as surface seekers. These descriptive terms are not entirely all-inclusive because the radionuclides are not strictly confined to the described areas but are deposited to some extent throughout all portions of the bone. The surface seekers tend to attach to the membrane of the osteogenic cell or to enter that cell, attaching to proteins, collagen, or other molecular structures, and remaining in the matrix, particularly if that area is not in a remodeling stage. Thus surface seekers tend to affect the most sensitive part of the bone, that is, the epithelial lining or the areas containing osteogenic cells. Osteogenic cells are stem cells, and because they are proliferative they have a high turnover rate. With surface seekers emitting alpha radiation, damage to osteogenic cells is highly probable. Conversely, if an alpha emitter like radium is buried fairly deep in the mineral portion of bone where the path length of the alpha particle is extremely small (high density), the probability of an ionization occurring in an osteogenic cell to create that particular type of associated radiobiological damage is small.

Some of the foregoing material may be found summarized in ICRP Task Group report published in 1980 (31).

1.3.3.6 Localized Cellular Retention.

1.3.3.6 Localized Cellular Retention. The hot-particle problem in bone, as just described, is extremely important. Uneven distribution is what leads to the difference in quality factors between some of the elements like radium and plutonium, but the biological effects are very difficult to interpret on this basis. The bone example has been used here as it is well known and has been investigated for years. The radon situation is partially the same in that it has been investigated for years, but it currently is undergoing such major research into the localized subcellular dose pattern and resulting projected effects that it will be a few years before it can accurately be described.

Many recent publications describe the current thinking about radon as a potential hazard and describe the theoretical subcellular response, interpreted as radiation dose (1, 19, 20, 32–36). In essence, the radon daughters that are deposited and traverse the respiratory tract via the ciliary escalator form a moving force of radiation. The cells beneath the epithelial lining of the tract are those targeted for carcinogenic insult, and it is the discrete dose to the DNA molecules in these cells that many scientists are theoretically and experimentally expanding their time investigating, either in the laboratory or on computers. The concept of dose as energy absorbed per mass of tissue is no longer used but has been replaced (by the theoreticians) through use of the mass absorption coefficient. If one assumes a flux of radiation striking a ''plate'' across the nucleus, then the energy per square unit of surface of the plate may be theoretically assumed (MeV/cm^2). If the absorption coefficient is assumed (g/cm^2) for this plate of biological matter, then all one needs do is divide to obtain a theoretical $MeV/gram$, and then, can once again use the conventional conversions to classic radiation dose. The remaining question is obvious; what do these new doses really mean, particularly when compared to doses obtained from classic calculations?

The story will be interesting when it unfolds with some accuracy, and right now it is a ripe area for model development and establishment of new concepts in dose–effect relationships (19, 20, 32–36). It is an exciting field in which to work but the basic question of whether radon is a significant problem to the world population is still hovering over the thinking scientific community. There is some place for the radon problem within the industrial hygiene community, but it is a small niche compared to other problems in potential toxicity.

2 GENERAL BIOLOGICAL EFFECTS

2.1 Electromagnetic Radiation and Neutrons

Acute and long-term biological studies with external sources of irradiation have been rather extensively defined, and this is one area in which there is also a sizable amount of human data. One of the best sources of information is from the persons exposed to the atom bomb radiation at Hiroshima and Nagasaki. These data are updated periodically by various United States and Japanese research groups. There are recent publications that review findings through the first 35 years (37, 38); the BEIR III and BEIR V reports and the latest UNSCEAR report use data from these sources to describe many of the general biological effects of radiation on human beings (39–41). Additionally, the epidemiological studies on radiologists have supplied considerable firsthand information (42). A few isolated accidental exposures to gamma and x-rays and/or neutrons may also be cited (38). These human data are considered along with supporting or pertinent animal research data in the material to follow.

2.1.1 Acute External Radiation Effects

The acute radiation syndrome has been described by many authors, and Table 13.1 from Upton (43) provides a clear description. He lists doses of 400, 2000, and 20,000 rems and indicates that with the second two doses death has occurred by at least the

Table 13.1 Effect of Radiation on Humans: Major Forms of Acute Radiation Syndrome

Time after Irradiation	Cerebral and Cardiovascular Form (20,000 rems)	Gastrointestinal Form (2000 rems)	Hematopoietic Form (400 rems)
First day	Nausea Vomiting Diarrhea Headache Erythema Disorientation Agitation Ataxia Weakness Somnolence Coma Convulsions Shock Death	Nausea Vomiting Diarrhea	Nausea Vomiting Diarrhea
Second week		Nausea Vomiting Diarrhea Fever Erythema Emaciation Prostration Death	
Third and fourth weeks			Weakness Fatigue Anorexia Nausea Vomiting Diarrhea Fever Hemorrhage Epilation Recovery (?)

Source: Reproduced with permission, from *Ann. Rev. Nucl. Sci.*, **18**. © 1968 by Annual Reviews, Inc.

third and fourth weeks. Certain symptoms (i.e., nausea, vomiting, and diarrhea) are present at all doses immediately following exposure to these higher levels. The gastrointestinal tract is one of the most sensitive organs to electromagnetic irradiation.

2.1.2 Intermediate Dosage Effects

There is a radiation dosage range, quite individually variable and quite large, at which many deleterious symptoms may appear, but at which recovery may occur and the individual may return to relatively normal health. The hematopoietic system is also one of those most sensitive to radiation, and its symptoms often indicate whether death is imminent or recovery will occur. The lymphocytes are perhaps the most sensitive blood cells and are affected within the circulatory system as well as at the precursor level. (Most standard texts dealing with radiation biology discuss in detail the effects of external irradiation on the hematopoietic system.) Other blood cells are affected, but this is primarily the result of the radiation damage to precursors in the bone marrow. The circulating lymphocytes show a decrease immediately following dosage and, within limits, may serve as an indicator of the severity of the radiation exposure. This phenomenon has been considered by many as a possible dosimeter for radiation exposure. Figure 13.1, from Arthur C. Upton, New York University Medical Center (43), shows a typical pattern of hematopoietic response with subsequent recovery at lengthy times postirradiation.

The testes and ovaries are also very radiosensitive and react to very low doses (a few rems). The testes recover to normal eventually, the time for recovery and severity of damage depending on the dose. The ovary is in a different category because its oocytes are present in entirety, never to be replaced after they have been permanently

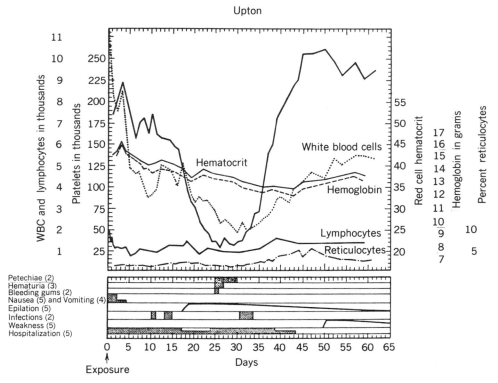

Figure 13.1 Hematological values, symptoms, and clinical signs in five men exposed to whole-body irradiation in a criticality accident. The blood counts are average values; the figures in parentheses denote the number showing the symptoms and signs indicated (43).

damaged. Most other tissue cells, such as those of the central nervous system, heart, and lung, are relatively radioresistant.

2.1.3 Late Effects

Low-level, chronic, or repeated exposure to ionizing radiation is the primary cause of industrial concern. This type of exposure is often difficult to measure accurately because of its quantitative similarity to background radiation; but more important, it is difficult to control. Some types of occupation such as those involving radiographical usage or those associated with reactor sites, inherently are accompanied by a certain low-level radiation. Industry has done a remarkable job in controlling such problems, but reduction of levels from a fraction above normal background to natural background cannot be feasibly accomplished in many cases, particularly when the large variation and range of natural background values is considered.

Late effects of radiation in mammals have generally been categorized as a generalized accelerated aging process or a specific carcinogenic process. *Life shortening* is a relatively nonspecific term and is considered to mean a premature death resulting from a variety of causes, most of which are specifically unidentified in the given individual. Classical animal experiments, such as early work of Lorenz (44), indicate that, in mice, life span may be shortened as a "function of many factors, some of which are unknown." This type of radiation effect is not well understood and merely represents a mammalian system dying before its time. This type of effect on survival would appear to be minimal compared to the more readily diagnosed causes of death, such as carcinogenesis.

The negative life shortening, or lengthening of normal life span, as a result of exposure to subharmful levels of ionizing radiation, has also been shown to be a real radiobiological phenomenon (45). A monograph by Luckey (46) gives many (over 1000) references to this stimulating effect of low levels of radiation, defined as *hormesis*. Spalding et al. have published results of an extensive study showing lengthening of normal life span in mice of various ages exposed to ^{60}Co delivered at many doses and dose rates (47). This stimulation of biological processes by low levels of ionizing radiation merely parallels the effects of many chemical toxicants that are harmful at high doses. The concept has been stymied, however, as there is little to be gained in the media or the classroom by indicating that "small amounts of radiation may be good for you." Perhaps someday, when there is no longer any driving reason for lowering radiation limits further than below background levels, the topic of hormesis will be "rediscovered" and much time and money will be spent pursuing this concept.

The types of cancer observed with low-level radiation have generally been associated with skin or the hematologic system. Radiation dermatitis was common among workers who had chronic irradiation of the skin, primarily of the hands. After latent periods of many years (e.g., 20) it was not uncommon to observe the formation of skin tumors, primarily squamous and basal cell carcinomas (43). Similar lesions have been obtained experimentally with beta particles in rats (48) in which damage to the hair follicles was observed to be implicated in the formation of skin cancer. With modern-day industrial hygiene practices, the observance of radiation-induced skin cancer in workers has essentially disappeared.

Many types of leukemia have been associated with radiation in the human population, mainly in persons who received radiation from nuclear weapons or from radiotherapy techniques (37–43). Many factors seem to be important in the onset of radiation-induced leukemia, including age at exposure, sex, dose, and perhaps dose rate. One of the greatest sources for study of the leukemias is the Hiroshima–Nagasaki survivors, as indicated, although the patients receiving radiation therapy for ankylosing spondylitis (49–52) and from other radiation sources also add significance to the overall interpretation of indication of leukemia from radiation (53). Summaries of many of the related radiobiological findings in the survivors of the bombings in Japan have been compiled (37–41) and cover the major categories of dosimetry, biological effects, future research, and health surveillance. Ishimara and Ishimara stated in the Okada supplement (37) that the intense study of the relationship between radiation dose and incidence of leukemia provides an important link to the effects of external radiation on humans. They quote from other sources (54) that ''the apparent excess incidence (leukemia) of A-bomb survivors is about 1.8 cases per million person-year rads for the period 1950–1970.'' This reference also states that ''those who were in either the youngest or oldest age brackets at the time of exposure were more sensitive to the leukemogenic effects of radiation'' (0–10 and more than 49 years of age). These leukemias were generally of early formation, however, and the risk of chronic granulocytic leukemia among survivors appeared to be greatest at 5–10 years postexposure. Doses calculated to be as low as 50 rads were expressed as being associated with the onset of leukemia, with those exposed at Hiroshima having a higher incidence. This is attributed to a higher relative biological effectiveness (RBE) for neutron than gamma rays and a greater abundance ratio of the former in Hiroshima. These dose estimates were acceptable prior to their report in 1975 (37), but they have since been altered, at least in a lowering of the Hiroshima neutron contribution. Beebe's more recent summary (38) considers the data through 1974 and discusses the relevance of the type of modeling used from high to low dose ranges and compares the cancer incidences found in Hiroshima and Nagasaki with control incidences in the United States. He states that ''even with 550 leukemia cases among the A-bomb survivors in the two cities over the period 1946–1974, it has not been possible to determine what the low-dose risk really is. The question of how to extrapolate from high to low doses in assessing biological effects still remains academically interesting, and uncorroborated with real data.''

The BEIR V Committee reports the relative risk associated with ankylosing spondylitics to be about 0.45 cases of leukemia per 10,000 person-year-Gray (40). Recent treatment of leukemia cases from the Japanese populations by this committee does not alter the results to any great degree as they give average values calculated for leukemia risk of 80–110 excess mortality from leukemia per million person rem and values of 660–730 for risk estimates for mortality due to nonleukemic causes. The value given for leukemia from Okada (37), 1.8 per million person-year-rad would be close to these values from BEIR V (40) depending on the chosen lifetime for substitution.

2.1.4 Dose–Response Relationships

Perhaps one of the most controversial issues in radiobiology has recently centered around the shape of the curve (actually, the function describing the curve) to be used in extrapolating from detectable insult (carcinogenesis, life shortening) at intermediate doses to anticipated consequences of very low doses. These low doses are currently thought by most to be of potential hazard to humans and there is no way in which meaningful biological effects data versus dosage can be collected on a meaningful biological system in this region. The statistical requirements would necessitate a practically infinite population of mammals, and molecular studies have yet to be shown to be relevant to the intact mammalian system. It is possible to use cellular and subcellular systems at these dosage levels to study particular radiobiological effects, but extrapolation of these results to the mammalian organism, with the complications and interactions of a simple circulatory system and hormonal interplay, becomes academic in nature. Thus, there has been great debate regarding a linear versus nonlinear extrapolation, with neither side being supported directly by experimental evidence. One of the treatises on this subject, and perhaps one of the most lucid, was part of a symposium held in 1976 to elucidate the problems (55). In this, Brown discusses the linear quadratic models that have been used to describe various radiobiological responses as a function of dose. Another simplistic description of these proposed processes has been presented in NCRP Report No. 64 (56); Figure 13.2 describing the mathematical formulations is from this reference. Although these references are not recent, they contain almost everything that is used in these same controversial discussions today. The data from some radiation sources (57) indicate a relatively clear threshold dose below which there are no clinically

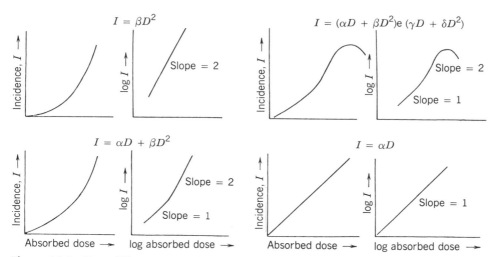

Figure 13.2 Four different types of dose–incidence curves, linear and full logarithmic. The equation for each is also shown (integer values for slopes are approximations since a single overall function is described with contributions from terms having different exponents) (56).

detectable biological effects. As these functions are generally logarithmic in nature (23, 25) and therefore there can be no absolute threshold, the concept of below regulatory concern (58) makes good sense. There are obvious clinical limitations below which (in dose) nothing is detectable, and the radiation protection community really should recognize this fact in establishing its guidelines. The whole area of extrapolation from high to low dose effects is an expanding field, and the computer age has enabled more sophistication than the data may warrant at this time.

The linear and quadratic equations in Figure 13.2 are often interpreted in terms of the single or multihit target phenomenon described years ago by Lea (59). The exponential terms are used to describe the decreased incidence resulting from death at higher doses at a time before the studied endpoint (cancer) has time to develop. Of course, if hormesis (46) is an accepted phenomenon at very low doses, none of these extrapolative processes to zero dose are meaningful. Upton (60) has associated relative dose ranges and related general effects with these curves and indicated that the lower dose term (αD) would be expected to predominate over the quadratic (βD^2) term for mutational or chromosomal effects in mammalian cells at < 50–100 rad at low dose rates. With high linear energy transfer (LET) radiation, the linear term (αD) would predominate and be less dose rate dependent. For carcinogenesis, the relationships differ quantitatively from one type of cancer to another, and no data have been extensive enough to give confidence in any extrapolation to the very low dose region of interest. At this point in time, it is not feasible to reject any of the curvilinear relationships described in Figure 13.2 as being inadequate to fit existing data.

2.1.5 Risk Factors

The acceptance of a given cancer incidence versus dose relationship for Figure 13.2 allows one to proceed one step further in assessing an endpoint for the exposure conditions. From a linear relationship one can obtain a value for the cancer incidence, say fractional incidence, for a given quantity of radiation, in rad. One may have a fraction, say 5 cancers per 10,000 individuals in the population exposed to 10 rad of some radiation. Thus the incidence would be $5/10,000 = 5 \times 10^{-4}$; this may be equilibrated to the 10 rad, as, $5 \times 10^{-4}/10$ rad $= 5 \times 10^{-5}$ cancers per rad. Generally, this type of expression is reported as 50 cancers per million rad-people or 50 cancers per 10^6 person-rad. This is a risk factor, then, to express the carcinogenic effect for this type of cancer following exposure to the prescribed type of radiation. Risk factors have outmoded other means of expressing radiation damage, particularly for use in radiation protection practices.

It was pointed out above that life span shortening is perhaps not a good endpoint for assessing low-level damage because the subjects may well live longer than controls. This should surprise no one. Then, using carcinogenesis allows one to overlook the life shortening (whether negative or positive) aspect and deal in terms of a more independent term. This is also extrapolatable to large populations if one believes in the basic premise of the risk per person-rad. From the example just used, of 5 cancers per 10,000 persons, one could extend this to answer the question of the effect of, say, an additional amount equal to background irradiation, or 100 mrad to the individual per year. Using the risk factor derived above, one could expect in a population of 200 million people, cancers in 1000 persons per year, if the underlying premises were correct. Thus, one in every 200,000 persons exposed to 100 mrad per year above background could expect to develop this hypothetical cancer. One can play interesting games with this concept, and it has the advantage of being applicable to unlimited populations at unlimited smaller and smaller doses. Its practical significance is left to the reader.

The BEIR V report is basically a risk factor report and contains some of the most recent approaches to calculating risk (40). As a summary of some of their findings the information in Table 13.2 is presented.

2.1.6 Dose Rate Effectiveness Factor (DREF)

One of the most controversial factors used in risk assessment in recent years has been the DREF. It is clear that a given external dose of radiation administered acutely (in one single, short, exposure) results in greater cancer incidence and shorter life

Table 13.2 Cancer Excess Mortality by Age at Exposure and Site for 100,000 Persons at Each Age Exposed to 0.1 Sv (10 rem)

			Males			
Age	Total	Leukemia	Nonleukemia	Respiratory	Digestive	Other
5	1276	111	1165	17	361	787
15	1144	109	1035	54	369	612
25	921	36	885	124	389	372
35	566	62	504	243	28	233
45	600	108	492	353	22	117
55	616	166	450	393	15	42
65	481	191	290	272	11	7
75	258	165	93	90	5	—
85	110	96	14	17	—	—
Ave.	770	110	660	190	170	300

			Females			
Age	Total	Leukemia	Nonleukemia	Breast	Respiratory	Digestive	Other
5	1532	75	1457	129	48	655	625
15	1566	72	1494	295	70	653	476
25	1178	29	1149	52	125	679	293
35	557	46	511	43	208	73	187
45	541	73	468	20	277	71	100
55	505	117	388	6	273	64	45
65	386	146	240	—	172	52	16
75	227	127	100	—	72	26	3
85	90	73	17	—	15	4	—
Ave.	810	80	730	70	150	290	220

span than if the same total dose is delivered continuously over the life span. The factor for this from animal experimentation is reported to be between 4 and 6, and this seems to be a suitable value for risk estimation concerning the industrial worker (40). However, the organizations that establish exposure guidelines (ICRP, NCRP) choose to ignore the animal data and to use extrapolative procedures in establishing a factor of 2 for the DREF. Their procedure is to compare the carcinogenicity or life span as a function of dose for these two exposure regimes (acute vs. chronic) at some point and to divide the incidence of one by the other. The problem is that they do this at some arbitrary point at a higher dose than that approaching the industrial situation, and they hence receive a lower value (2 instead of 4–6). Had they used dose points near zero, the ratio would be more realistic. In the radiation protection field, however, there are always several factors introduced into establishment of exposure guidelines that make the outcome very conservative when compared to any actual data from animal experimentation or from human exposures.

2.2 Radionuclide Radiation

As stated in Section 1, the internal exposure to radionuclides is probably the most pertinent to practical toxicity problems in the industrial world. Radionuclides are difficult to control, particularly if their emissions do not allow for detection at reasonable distances from the source. Plutonium is a good example of the latter in that it has alpha-particle emission and only a very weak x-ray component, too weak to detect through the ordinary glove box. Section 1.3.3 covered the difficulties in determining biological dosage following exposure. There has been a wealth of information collected over the past 30 years in animal studies, but extrapolation to humans is not straightforward. We do have the advantage of the excellent epidemiological studies of the radium dial painters (61) and the radium-treated ankylosing spondylitic patients (49, 50). Other studies may also prove useful from the point of view of assessing human exposure to radionuclides [e.g., thorotrast (62, 63) and plutonium (64, 65)]. Workshops on radium, thorium, and the actinides represent an attempt to bring together analytical data from humans and animals and to use these in formulation of usable models (63, 66–68). The problem with studies such as those with the plutonium workers is that since no effects have been observed, there is no way to evaluate dose–effect relationships. Laboratory experimental data are presented in the following sections and where possible, the pertinence to the human situation is discussed. The most pertinent reference on radionuclide behavior, including detailed history and the total picture of dose–effect relationships is contained in Newell Stannard's classic book, *Radioactivity and Health. A. History* (69).

The problems inherent with radiation dose from internally deposited radionuclides (internal emitters) are much more involved than with whole-body electromagnetic radiation. Compartmentalization of the various radioelements is responsible for the complications of dose estimation, and it is the interplay between these sites of deposition in the body that results in the kinetic modeling carried out with internal emitters. The ultimate goal is to arrive at a time-integrated value, in ergs expended per gram of a tissue, and to relate this through the proper constant(s) to the desired radiation dosage unit. The dose arrived at in this manner is an average over the entire tissue or organ and is generally, today, characterized by the rad, equivalent to 100 ergs per gram of exposed medium. For biological samples of soft tissue, where most beta emitters are involved, this is a reasonably good estimate for comparative purposes. This type of radiation distributes rather uniformly because of the relatively long path length in soft tissue, hence average dose is perhaps a reasonable approach to assessing biological damage or effect. With alpha emitters the pathway is very small (Section 1.3.3), and average dose is generally not as suitable for comparison of effects. Except where intricacies of localized dose are of concern, however, the average is still the most acceptable (often the only) means of relating dose to effect.

In biological modeling to arrive at radiation dose and interplay between organs, one generally uses first-order kinetics to define the parameters. This approach allows one to think in terms of the quantity leaving or entering a given depot as being directly related to the amount in that repository at any one time. Through a series of rate constants representing the various compartments of the deposition in the body, half lives (0.693/rate constant) of retention may be calculated, and investigators in this area of radiation biology or biophysics have learned to think and write in these terms. This is an acceptable concept and will be in use in this field for a long time, although many researchers have preferred the power function method of fitting retention kinetics (70–72).

Figure 13.3 gives an example of gross modeling to arrive at the time course of a radionuclide residing in the body. Here the simple model consists of organs A, B, and C, with excretion occurring through two routes to external compartments D and

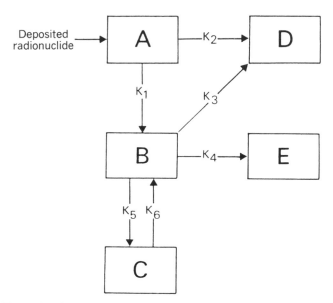

Figure 13.3 Example of first-order kinetics modeling from which the descriptive equations are derived: A = lung, B = blood, C = an internal organ, D = gastrointestinal tract excretion, E = urinary excretion.

E. The development of equations to fit this scheme proceeds according to the following reasoning. Because inhalation appears to be the most practical route of entry to the body, this is a simplified version of a model in which the lung (*A*) is one of the organs involved and *B* is the blood; *C* may represent any other internal organ (kidney, liver, spleen, etc.), and *D* and *E* are gastrointestinal tract and urinary excretion, respectively.

1. The amount of radionuclide entering compartment *D*, gastrointestinal excretion (feces), from the lung is straightforward and is represented by

$$\frac{dD}{dt} = k_2 A \tag{2}$$

2. Compartment *B*, the blood, represents a much more complicated scheme. An expression for the rate of accumulation in the blood may be derived through a series of differential equations as follows:

$$\frac{dB}{dt} = k_1 A = \text{to blood from lung} \tag{3}$$

$$\frac{dB}{dt} = k_6 C - k_5 B = \text{to blood from tissue } C \tag{4}$$

$$\frac{dB}{dt} = -k_3 B - k_4 B = \text{to } D \text{ and } E \text{ from blood} \tag{5}$$

An additional rate constant and equation are applicable for absorption from gastrointestinal tract to blood, but this is assumed to be negligible here for purposes of simplicity.

3. The ultimate goal is to determine the time course of radiation energy released into organ *C*. The process is initiated by combining some of the equations above. The amount in *C* at any time is the integral of the rate of buildup or decline.

$$\frac{dC}{dt} = k_6 B - k_6 C \tag{6}$$

An expression of *B* may be obtained from the integral of

$$\frac{dB}{dt} = k_1 A + k_6 C - k_5 C - k_3 B - k_4 B \tag{7}$$

An expression for *A* at any time may be obtained from a form of equations

$$-\frac{dA}{dt} = k_1 A + k_2 A \tag{8}$$

and

$$A = A_0 e^{-(k_1 + k_2)t} \tag{9}$$

Equation (7) now becomes

$$\frac{dB}{dt} = k_1 A_0 e^{-(k_1 + k_2)t} + C(k_6 - k_5) - B(k_3 + k_4) \tag{10}$$

This may be solved using linear differential equation techniques, giving an expression for B in terms of A_0 and C. The form generally recognized for integration is

$$\frac{dB}{dt} + B(k_3 + k_4) = k_1 A_0 e^{-(k_1 + k_2)t} + (k_6 - k_5) C \tag{11}$$

$$B = \left[\exp - \int_t^0 (k_3 + k_4) \, dt \right] \int_t^0 (k_3 + k_4)$$
$$\cdot \, [k_1 A_0 e^{-(k_1 - k_2)t} + (k_6 - k_5) C] \, dt + \text{constant} \tag{12}$$

4. By proper substitution from Eq. (12), an expression for C can be obtained through integration. To quantitate the scheme, one needs to know the initial quantity of radionuclide deposited in the lung A_0, and the various rate constants k_1 through k_6. These may be obtained using the time course of measured radionuclide in blood and excreta at any time. An analog computer is almost a necessity when the number of compartments is large.

Thus through a series of first-order kinetic manipulations, an expression for dose to an organ may be obtained with which to match biological effect. Although this modeling scheme might appear to belong under Section 1.3.3, it seemed logical to place it in a relationship to the ensuing biological effect.

2.2.1 Effects on Lung

Many radiation effect studies on the respiratory system have been carried out following intratracheal injection (IT) and inhalation (INH). There is always some question of the validity of using IT data inasmuch as this route is a nonpractical means of acquiring a radionuclide; thus the tendency is to utilize information from INH whenever possible. (IT instillation in the lung is generally nonuniform, and the material is in a solution or suspension; an aerosol, used for the INH route, is more uniform and realistic.) Major INH studies using dogs have been carried out in recent years at Battelle Pacific Northwest Laboratories (PNL) (73) and at the Lovelace Foundation (the Inhalation Toxicology Research Institute: ITRI) (74). Doses have been expressed as average energy released to the tissue (rads) or in terms of radioactive unit of material deposited per gram of organ (e.g., nCi/g). Either manner of presentation is acceptable.

In both laboratories the inhalation exposures have been carried out in a manner in which only the nasal region is subject to the aerosol. Such exposures are commonly referred to as "nose-only." With whole-body exposure during inhalation, the animal is covered with particles and obtains extremely large amounts of external contamination. Even when the aerosol is essentially insoluble in body fluids, the small amount that may be absorbed through the gastrointestinal tract because of the animal's licking (cleaning) itself can result in a deposit in internal organs. If the half-time of residence in that organ system is extremely long compared to that in lung, the possibility for appreciable radiation dose to that organ exists. Thus nose-only exposures are desirable. At both laboratories the primary radioactive inhalation exposures now involve alpha emitters, but the ITRI work initially dealt chiefly with beta emitters. At the Los Alamos National Laboratory (LANL) similar studies were carried out using alpha emitters in small animals (75).

Typical dose–effect data from PNL and ITRI for a certain endpoint such as pulmonary neoplasia, indicate that the dose from beta emitters is up to a factor of 20 times that calculated for a comparable effect (carcinogenesis) from alpha emitters (76–78). These carcinogenic doses are in the hundreds and thousands of rads for alpha emitters and in the thousands and tens of thousands of rads for beta emitters. Similar findings (lung carcinogenesis) in the Syrian hamster following inhalation of plutonium particles required at least 2000 rads for induction of lung adenocarcinomas (79). Assuming the concept of linear energy transfer (LET) (or RBE) to be based on firm premises, this difference between types of radiation is to be expected. Some of the types of lesions found do not appear to vary remarkably with the quality of the radiation. Typical of such lesions are:

- Fibrosarcoma
- Squamous cell carcinoma
- Bronchiolar carcinoma
- Adenocarcinoma
- Hemangiocarcinoma
- Bronchioloalveolar carcinoma

Some of these findings described by various authors may be morphologically similar and depend on the individual pathologist's interpretation.

In humans, a limited number of lesions have been attributed to the deposition of radionuclides in the respiratory system. One of the most widely known occupations resulting in such findings is uranium mining. Internal biological lesions in this occupation are primarily the result of breathing radon (^{222}Rn), hence its ensuing decay products (80). The early miners in Schneeburg, Germany, and Joachimsthal, Czechoslovakia, suffered from a disease that was ultimately determined to be lung cancer (81, 82). In the United States the uranium miners in the early part of this century were found to have a significant increase in lung cancer (83). Many publications dealing with the hazards associated with the uranium mining industry carry much of this early history and bring the state of the art up to date (80–84). This includes analytical methods and measurements as well as biomedical effects. The earlier results had shown the lung cancer incidence to correlate with smoking, and a greater incidence of lung cancer has been reported for smokers than for nonsmokers, among the miners (83–87). The lesions are bronchogenic and have been shown to correlate with estimated radiation doses (88). The incidences that have been reported for the increase over that expected from a "normal" nonsmoking population of individuals are variable, and no specific value is significant at this time. It has been estimated that a dose of 360 rads will double the incidence of normal lung cancer, as evidenced in these workers (86). Saccomanno et al. (89) however, have presented convincing evidence that the adenocarcinoma incidence did increase with increasing working level months (WLM) of exposure and with smoking intensity. They also showed (89) an increase in frequency of adenocarcinomas in miners with younger ages at diagnosis and lower ages at the start of mining as an occupation, and state that the type of bronchogenic carcinoma observed with nonsmokers differs from that seen with smoking miners. It would appear that persons who may have added an additional insult, such as poorly controlled mining atmospheres, to their smoking habit, would have created a synergistic background for an increased incidence of pulmonary cancer.

Dr. Saccomanno and associates (85,90) derived a unique method for detection of cancerous lesions in the lung. This method will indicate, in cells taken from sputum samples, the predisposition to an invasive cancerous lesion in the bronchiolar tree. Squamous cell metaplasia of the bronchi is generally accepted as the forerunner to bronchogenic carcinoma. The average time for development of epidermoid carcinoma has been quoted to be approximately 15 years (91). The time of development from early metaplastic changes to marked atypia may be an average of 4 years. Thus cytologic investigation of sputum samples taken during this earlier phase may reflect this atypia and may lead to detection of a carcinomatous condition before it reaches the invasive state. This may allow performance of localized surgery, to eliminate the malignancy that would eventually lead to death.

2.2.2 Effects on Bone

Bone seekers are generally classified as two types, volume or surface, as described earlier. The primary cancerous lesion observed following radionuclide deposition in the skeletal is osteogenic sarcoma. One of the most thorough studies of this lesion versus estimated radiation dose has been carried out at the University of Utah over the past 35 years (92). The beagle dog has been the chief experimental subject in that laboratory, allowing for comparison with the inhalation studies described previously (73,74). In the Utah work the radionuclides (^{90}Sr, ^{241}Am, ^{228}Th, ^{226}Ra, or ^{239}Pu) were injected intravenously (IV) in a soluble form, primarily the citrate. This route has desirable properties because it gives a greater probability of deposition in the organs of "choice" than perhaps would be predicted to occur with material entering the blood from the lung, and the doses thus delivered are administered with great accuracy. The choice of deposition site is determined by the physical-chemical properties of the radionuclide, as is the determination of whether a given material is to be a volume or surface seeker in bone.

Mays et al. have described the finding of bone sarcomas in mice and dogs (93) and have attempted to estimate risk to the skeleton using a linear model. Their study compares the relative risks between [239]Pu and [226]Ra in experimental animals and also presents data from the radium dial painters, with the predicted linear extrapolation to what the risk of bone sarcoma may be in humans from [239]Pu. An estimate of the cumulative risk from [239]Pu using a linear model is about 200 bone sarcomas per 10^6 person-rads. This bone sarcoma risk estimate is based on the results of the German ankylosing spondylitic patients that received [224]Ra as therapy (Fig. 13.4). The work by Spiess and Mays (49) is a thorough summary of bone sarcoma incidence versus radiation dose from many sources and is a recommended reference, not only for the value of the authors' interpretation but for the complete bibliography associated with this field. A supplement to this review, including discussion of Figure 13.4, is contained in another study by Mays and Spiess (94).

More recent coverage has expressed the data from the radium dial painters in three different ways depending on the author and his interpretation. The comparison was made in the BEIR IV report (1) when one chart (reproduced here as Fig. 13.5) was made showing interpretations of the same data from the radium patients, as reflected in the work of Evans (57,95), Mays and Lloyd (96), and Rowland (97). This is a perfect illustration of modern thinking and data analysis in that three quite different interpretations are shown. Evans shows a threshold at about 1000 rad total to the skeleton, and Rowland indicates about the same pattern using a different mathematical approach. In the Rowland case the data will never pass through the 0,0 point as the distribution as described has an exponential term. However, the point at which there are no more tumors for practical reasons is around 1000 rads to the skeleton. Mays on the other hand, touting the linear model, forced the line

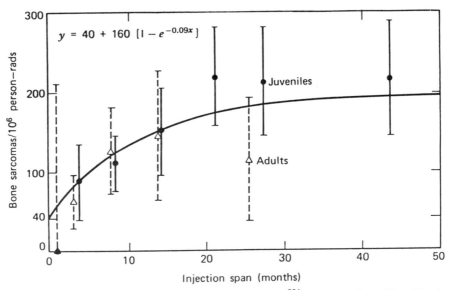

Figure 13.4 Risk of bone sarcomas in humans versus [224]Ra protraction. The risk rises to about 200 bone sarcomas/10^6 person-rad at long protraction. The standard deviation of each point is shown (49).

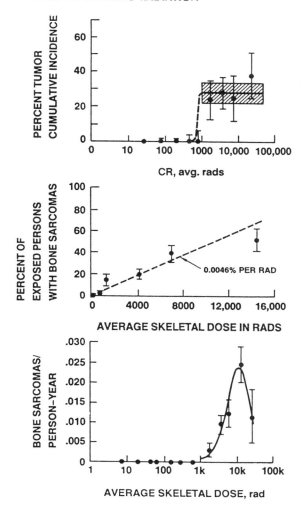

Figure 13.5 Dose–response relationships of (a) Evans et al. (95), (b) Mays and Lloyd, (96), and (c) Rowland et al. (97).

through the 0,0 point using a linear expression. The advantage of the Mays approach is its simplicity of interpretation; the slope is the risk factor for this exposure situation. It is clear that these radium dial painter data produce an ideal case for the use of the Nuclear Regulatory Commission's proposed Below Regulatory Concern concept (58). If the tumor incidence is unmeasurable below 1000 rad, then it is below any concerns for regulatory reasons, as far as looking for guidelines that apply below this practical threshold.

2.2.3 Effects on Soft Tissues

As stated, radionuclides will localize in an organ or tissue to an extent dictated by their physical-chemical properties. Thus by choosing an element, one could feasibly obtain a specific concentration in an organ of choice and confine the resulting biological effect to that area. In fact, if this relationship were as simple as stated, the medical field would be able to localize the proper radionuclide at the site of a tumor, and with the ensuing radiation, destroy the abnormal cells. However, localization is not usually relegated to the tissue of concern alone but also to other, nonaffected tissues. Localization in the tumor-bearing organ subjects normal cells to the radiation as well, and the radiation dose that may cause subsidence of the malignancy will also be destructive to the normally functioning tissue. With this diversion on the behavior of internal emitters as background, it is now feasible to discuss a few examples of specific tissues that are primarily damaged by the entrance of specific radionuclides to the body.

2.2.3.1 The Reticuloendothelial System (RES). The RES broadly consists of the liver, spleen, bone marrow, lymph nodes, and lung. One normally thinks of the RES components of these tissues as containing cells that are available to the circulation (lymph or blood) and have the ability to react to any material that is recognized as foreign to the body. Colloidal or particulate substances are particularly subject to being acted on by cells of the RES. Phagocytosis is the primary means of detoxifying such foreign bodies, with the destructive action occurring within the phagocytic cell. For instance, when colloidal polonium is injected IV into experimental animals, it follows a body distribution pattern that is essentially analogous to the organs of the RES (98). The larger the particulate entity, the more rapidly it is removed from the circulation, and this can take place during the first pass through the circulatory system. The spleen and the bone marrow, for example, are two organs that are highly affected following ^{210}Po administration in this manner (98). The effects on these organs are somewhat comparable, and because they are considered to be at least partly a segment of the hematopoietic system, the resulting damage is included under that discussion.

The lymph nodes are a major sink for the deposition of foreign materials that find their way into the lymphatic channels. Drainage from the lung is one of the more common examples of the defense mechanism exercised by the lymphatics, especially for particles deposited in the alveolar region of the lung. These particles are filtered out in the regional lymph nodes, where they reside for a very long period (23). Although large deposits in lymph nodes have been observed after inhalation of a radioactive aerosol (99), little biological damage of significance to life span, including carcinogenesis, due to the radioactive content of the regional nodes, has been observed. In uranium inhalation studies (100) the lymph nodes of monkeys were seen to concentrate far greater than 50 times the concentration in the lung after repeated exposure to UO_2 aerosol. A similar pattern has been described for many other radionuclide particles (23). The effect is generally one of fibrotic change, with some alteration of the cellularization of the germinal centers. However, the cells involved have shown little tendency for tumor formation, and the lymph nodes under this circumstance are not considered by most to be a critical organ for radiation damage.

2.2.3.2 The Hematopoietic System. It is known that the blood cell forming tissues are subject to radiation when the nuclide is administered in such a form as to deposit in the reticuloendothelial cells associated with this function. There is also a tendency for some elements to localize in the functional cells directly related to blood cell production. Whatever the mode of localization, the general effect is essentially the same. The lymphocytes are the most sensitive of the circulating blood cells, but the precursor cells in the bone marrow and spleen (particularly under conditions of extramedullary hematopoiesis) are also extremely sensitive to radiation. The erythrocyte precursors in the marrow cavity are the key to any subsequent reduction in red cells. Thus what one may readily observe in circulating blood as a result of radionuclide deposition in the hematopoietic system is a rapid decline in the white cell population followed by a much slower decline in circulating erythrocytes (see Fig. 13.1). Reduction in the former, naturally, allows the onset of infection, as the ability to cope with outside disturbances is dramatically curtailed. If the animal (or person) survives this ordeal, the lack of red cell production will result in an anemia and, if the dose is sufficiently high, in the death of the subject.

There are many detailed biological effects inflicted on the hematopoietic system by deposited radionuclides, but the general pattern is as described.

2.2.3.3 The Immune System. The immune system is obviously affected by ionizing radiation, as observed overtly through the effects on the hematologic system. The lymphatic portion of the circulating defense mechanism is one of the most sensitive to whole-body irradiation. Basic questions arise concerning carcinogenesis and the immune system, as to whether a decreased immune function makes way for carcinogenesis or whether the formations of cancerous growth deplete the immune system. It is clear, however, that injury to the immune system from radiation, resulting in reduced capability to respond to foreign body infiltration, leads to illness and death, when the situation attains a serious state. More on the immune system is discussed in Section 6.3.

2.2.3.4 Specific Tissue Localization. The thyroid is an excellent example of a tissue that will localize a particular internal emitter, in this case, iodine. A fine review of this subject with particular regard to the practical aspects of deposition of radioiodine in the thyroid gland has been published by the National Council on Radiation Protection and Measurements (NCRP) (101). The effects of ^{131}I irradiation of the thyroid are aptly described in this report and compared to those observed with x-rays. Because radiation effect on this particular organ has been so thoroughly studied, it is valuable to present two tables from the NCRP reference that generally summarize the findings to date (Tables 13.3 and 13.4). These tables list the risk estimates for various thyroid insults and compare the effects between children and adults.

Table 13.3 Relationship Between Low-Dose Exposure to ^{131}I in Children and Subsequent Hypothyroidism[a]

Number of Subjects	Thyroid Absorbed Dose Range (rads)	Estimated Mean Thyroid Absorbed Dose (rads)	Number of Hypothyroid Subjects	Incidence of Hypothyroidism (%/Year)
146	10–30	18	0	0
146	31–80	52	3	0.15
151	81–1900	233	5	0.23

[a]Preliminary results: Hamilton and Tompkins (102).

Table 13.4 Absolute Individual Risk of Thyroid Abnormalities after Exposure to Ionizing Radiation

Type of Abnormality and Population Surveyed	Mean Absorbed Dose or Dose Range for which Data Were Available (rads)	Absolute Individual Risk $(10^{-6} \text{ rad}^{-1} \text{ y}^{-1})$	Statistical Risk Range[a] $(10^{-6} \text{ rad}^{-1} \text{ y}^{-1})$
Internal Irradiation (^{131}I*)*			
Thyroid nodularity			
Children	9000	0.23	0–0.52
Adults	8755	0.18	0.13–0.23
Thyroid cancer			
Children	9000	0.06	0–0.158
Adults	8755	0.06	0.044–0.075
Hypothyroidism			
"Low-dose" children	10–1900	4.9[b]	3.9–22.9
"High-dose" adults[c]	2500–20,000	4.6[b]	2.8–7.8[d]
External Irradiation			
Thyroid nodularity in children	0–1500	12.4	4–47.4[d]
Thyroid cancer in children	0–1500	4.3	1.6–17.3[d]
Hypothyroidism in adults	1640	10.2	0–24.8

[a]Unless otherwise indicated, the range of risk was determined by assuming that the number of cases, n, out of the population at risk represents the true mean of a Poisson distribution. The range is then estimated by using $\pm 2n$ as the 95 percent confidence level.
[b]Threshold of 20 rads.
[c]See Figure 2 of Ref. 103.
[d]In these cases the risk was determined from the slope of the linear regression line. The range was estimated from the extreme data points, which provide the lowest and highest slopes.

3 STANDARDS AND GUIDELINES

3.1 ICRP Recommendations

In 1977 the ICRP recommended new objectives for radiation protection (104). These entail a new nomenclature and set of terms upon which to base their recommendations. They state that protection of humans must be from detrimental effects against the individual (somatic effects) or against his or her descendants (hereditary effects). They define *stochastic* effects as those for which the probability of occurring is a function of dose, with no threshold; *nonstochastic* effects are those for which the severity is a function of dose and for which a threshold may occur. Carcinogenesis is considered the major somatic effect and is considered stochastic. At the very low doses, heredity changes are also considered stochastic. The aim of the ICRP in radiation protection is to prevent nonstochastic effects by maintaining exposure levels below a threshold dose and to limit stochastic effects by maintaining exposure levels at as low as reasonably achievable (ALARA).

3.1.1 Basic Concepts

The Commission (ICRP) has defined a number of terms, each of which requires understanding before they may be applied. It is not the purpose of this chapter to put forth the detailed parts of the recommendations but rather to describe the overall philosophy.

The ICRP defines the *detriment* concept as a way of identifying and quantifying deleterious effects. It is generally defined as the expected harm that may be incurred in an exposed population, taking into account not only the type of effect but also the severity. The *dose equivalent* is used to predict the degree of the deleterious effect or the probability of attaining it. The dose equivalent involves the quality factor (QF), the absorbed dose, and other modifying factors defined by the ICRP. The QF values recommended are roughly the same as those used by the NCRP (1 for x- and gamma-rays, 20 for alpha particles). The *collective dose equivalent* is really the sum of the dose equivalents in the individuals within a population either in the whole body or the organ system involved. Thus it is the sum of the number of individuals receiving a per capita dose equivalent in the whole body or individual organ system. The *dose equivalent commitment* is the infinite time exposure at the per capita dose equivalent rate in a given tissue or organ system for the given population. The *committed dose equivalent* is the dose equivalent that will be accumulated or integrated over the 50 years of exposure following the intake of a radionuclide into the body. These definitions are not directly applicable to the general purpose of this chapter, but they do serve to indicate the presently accepted guideline boundaries as set forth by the ICRP. A copy of the pertinent radiation protection definitions that were taken directly either from the BEIR V report or ICRP-26 are listed below (40, 104).

Absorbed dose is the mean energy imparted by ionizing radiation to an irradiated medium per unit mass (unit is Gray, Gy = 100 rad).

Background radiation is the amount of radiation to which a member of the population is exposed from natural sources, such as terrestrial radiation due to naturally occurring radionuclides in the soil, cosmic radiation originating in outer space, and naturally occurring radionuclides deposited in the human body.

Collective dose equivalent (S) in a population is defined by the summation of the product of the per capita dose equivalent (H) in the whole body or any specified organ or tissue of the P members of the subgroup of the exposed population.

Detriment to health is an expression of the expectation of the harm endured from an exposure to radiation.

Dose equivalent is a quantity that expresses, for the purposes of radiation protection and control, an assumed equal biological effectiveness of a given absorbed dose on a common scale for all kinds of ionizing radiation. Mathematically the dose equivalent, H, at a point in tissue, is given by the equation $H = DQN$ where D is the absorbed dose, Q is the quality factor, and N is the product of all other modifying factors specified by the ICPP Commission, for example, absorbed dose rate, fractionation.

Effective dose equivalent (EDE) relates dose equivalent to risk; for partial body radiation the effective dose equivalent is the risk-weighted sum of the dose equivalents to the individually irradiated tissues.

Person-Gray is the unit of population exposure obtained by summing individual dose equivalent values for all people in the exposed population. Thus, the number of person-Grays contributed by 1 person exposed to 1 Gy is equal to that contributed by 100,000 people each exposed to 10 μGy.

Relative biological effectiveness (RBE) is the biological potency of one radiation as compared with another to produce the same biological endpoint.

Relative risk is the expression of excess risk relative to the underlying (baseline) risk; if the excess equals the baseline risk the relative risk is 2.

Risk coefficient is the increase in the annual cancer incidence or mortality rate per unit dose: (1) absolute risk coefficient is the observed minus the expected number of cases per person-year at risk for a unit dose; 2) the relative-risk coefficient is the fractional increase in the baseline incidence or mortality rate for a unit dose.

3.1.2 Recommended Limits

The ICRP recommends limits of 50 rem in one year to all tissues (except of the lens of the eye) as the dose equivalent that would prevent nonstochastic effects; it recommends 30 rem for the lens (104). These limits apply regardless of mode of exposure (single organ or summed doses to organs). For stochastic effects, they recommend that the dose equivalent be from whole-body exposure or the sum of various organ systems. For this they use a weighting system that relates the risk attached to individual systems to the total combined risk. In other words, the weighting factors proportion the stochastic risk to that system to the total risk or if it were related to the whole body being irradiated uniformly. Their list of weighting factors is given in Table 13.5.

The weighting factor for the "remainder" tissues is really a recommendation of 0.06 for each of the five remaining tissues or organs that are receiving the highest dose equivalents. This could be, for instance, the spleen, kidney, and pancreas. It is recommended that these weighting factors be used unless the more straightforward approach of limiting whole-body exposure to an annual dosage of 5 rem in any year is feasible. This may be clearly expressed as

$$\Sigma_T W_T H_T \leq H_{WB,L},$$

Table 13.5 Weighting Factors Used for Dosage Calculations, as Recommended by the ICRP

Tissue	w_T
Gonads	0.25
Breast	0.15
Red bone marrow	0.12
Lung	0.12
Thyroid	0.03
Bone surfaces	0.03
Remainder	0.30

Source: Archer et al. (86).

where

H_T = annual dose equivalent in tissue T
W_T = weighting factor for T
$H_{WB,L}$ = recommended annual dose equivalent limit for uniform whole-body radiation, or, 5 rem

These general concepts have been utilized to generate guidelines for limits of intake of radionuclides by the potentially exposed working individual, and the annual limits on intake of radionuclides and derived air concentrations have been calculated for many radioelements and their isotopes with half-lives greater than 10 min (105).

3.1.3 Special Cases

The ICRP also has recommendations (86) for special potential exposure cases such as: (1) occupational women of reproductive capacity, (2) occupational pregnant women, (3) individual members of the public, (4) populations, and (5) accidents and emergencies.

3.1.4 Applications of ICRP Recommendations

The information contained in ICRP-26 may also be used directly to assess any program that is to use radiation, from either an external or internal source (104). The material to follow is paraphrased from this document and this tact is used here to show how explicit ICRP-26 is and how helpful it can be in certain areas.

- Any new installation that will be using products containing radioactive materials or emitting ionizing radiation should be examined at the design stage from the viewpoint of restricting the resulting occupational and general exposure.
- Assessment of exposure of individuals or groups by calculation or measurements is essential for developing procedures for restricting exposure and for intervention. Early on, the focus is on predictive assessments and, where necessary, they are confirmed by monitoring during subsequent operations. Predictive assessments are also needed in the planning of intervention procedures but, in addition, monitoring carried out during normal operations will often provide the basis for any decisions to intervene.
- ICRP-26 points out the importance of distinguishing between distinct types of protection standards: basic limits (dose equivalent limits and secondary limits), derived limits, authorized limits, and reference levels. Dose equivalent limits apply to the dose equivalent or, where appropriate, the committed dose equivalent, in the organs or tissues of the body of an individual or for population exposure, to the average of one of the mean of a group of individuals.
- In practical radiation protection it is often necessary to provide limits associated with environmental conditions. When these are related to the basic limits by a defined model of the situation and are intended to reflect the basic limits, they are called derived limits. Derived limits may be set for quantities such as dose equivalent rate in a workplace, contamination of air, contamination of surfaces,

and contamination of environmental materials. The accuracy of the link between derived limits and basic limits depends on the realism of the model used in the derivation.

- Optimization of the restrictions of exposures and the selection of the optimum level of protection for the given circumstance requires a case-by-case review and the differential cost of going from one strategy to the next are subsequently evaluated.

- It is generally assumed that the detriment to health is proportional to the collective dose equivalent in the exposed group, but two extreme situations are worth considering. In the first, the detriment to health in the exposed group is the predominant component of the total detriment, and the required comparison is the differential change in the cost of protection strategies and the corresponding differential change of the collective dose equivalent. The second extreme situation is where the detriment to health is likely to be very small, for example, where the numbers of individuals are extremely limited. Optimization in this case involves the highest individual dose equivalent as opposed to the collective dose equivalent, and it is necessary to take account of exposure contributions from procedures other than that under review.

- The ICRP Commission's recommended dose equivalent limits for occupational exposure have been in effect for over 20 years. In view of the emphasis that the commission places on risk estimations, it believes it appropriate to assess the levels of risk that are associated with its dose equivalent limits. The commission believes that for the foreseeable future a valid method for judging the acceptability of the level of risk in radiation work is by comparing this risk with that for other occupations recognized as having high standards of safety, which are generally considered to be those in which the average normal mortality due to occupational hazards does not exceed 10^{-4} (0.0001).

This brief inoculation of information on radiation protection seemed appropriate for this chapter and is considered an important aspect of industrial hygiene as related to health physics practices.

3.2 NCRP Recommendations

This section includes the most relevant information presented in the NCRP's most recent report on recommendations for exposure limits (NCRP Report 91) (106). For the person with little experience in radiation protection philosophy, it is recommended that NCRP-39 (107), published in 1971 be used; the use of NCRP-91 is better for a brief rundown on present thinking, but it is not a clear treatise on the subject of basic radiation protection. Although NCRP-91 supercedes NCRP-39, it does not expound on the philosophy of basic radiation protection criteria and merely states the ICRP guidelines as paraphrased under Section 3.1 above. One of the important classic approaches to establishing radiation protection guidelines was through the use of the critical organ concept. In many aspects this old approach is being used today, as partially seen through the use of the weighting factors (Table

13.5), and a brief summary of the concept is given in a section below, followed by a table of the newer NCRP radiation recommendations as taken from NCRP-91.

To clarify the NCRP's position in establishing recommendations on limits of exposure, the following section is quoted verbatim from the Introduction of NCRP-91 (106):

> The NCRP believes that a logical direction in the evolution of the basis of radiation protection standards from the present system based on dose equivalent used by the NCRP and the ICRP is toward the development of an approach based specifically on risk. In such a system, generally acceptable levels of risk form the basis for radiation protection standards according to estimates of risk per unit of absorbed dose (or other convenient measure) for the various types of radiation. Furthermore, although physical measurements would still be used, the limits themselves would be expressed in terms of risk. However, in view of the current degree of uncertainty in risk estimates, for both external and internal exposure, and the fact that the details of the implementation of such a system have yet to be elaborated, the Council believes that additional review and evaluation are required prior to the introduction of a radiation protection system in which the limitation of exposure is based solely on risk. For the present, therefore, the Council adopts, in principle, the effective dose equivalent system used by the ICRP, but has modified and updated this approach in several respects, as discussed in this Report.

3.2.1 Critical Organ Concept

The critical organ concept has been under discussion for many years. As considered in NCRP-39 (107), it might be more intelligent to consider the *critical* organ as the one most highly radiosensitive, and the *dose-limiting* organ as the one that may receive the greatest radiation dose. Thyroid is used as an example of latter. The critical organ concept, as such, is not mentioned in radiation protection circles as an option in terms, but in reality it exists. It is interesting how concepts often remain the same but the term of reference may change, as if the concept had changed. From NCRP-39 (107), again, superseded by NCRP-91 (106), the following critical organ concept was stated:

For the development of protection guides, the concept of identifying the minimum number of organs and tissues that are limiting for dose consideration is the essential simplifying step. For general irradiation of the whole body, such critical organs and tissues are:

1. Gonads (fertility, genetic effects).
2. Blood-forming organs, or specifically red bone marrow (leukemia).
3. Lens of the eye (cataracts).

For external irradiation of restricted parts of the body, an additional critical organ may be:

4. Skin (skin cancer).

For irradiation from internally deposited sources alone or combined with irradiations from external sources, additional limiting organs are determined more by the metabolic pathways of invading nuclides, their concentration in organs, and their effective residence times, than by some inherent sensitivity factors. They include:

1. Gastrointestinal tract
2. Lung
3. Bone
4. Thyroid
5. Kidney, spleen, pancreas, or prostate
6. Muscle tissue or fatty tissues

This is a well-constructed and brief description of the critical organ concept, and it indicates the problems involved in proper organ selection criteria under the many conditions of possible radiation exposure.

3.2.2 Effective Dose Equivalent

Like with the ICRP recommendations the NCRP uses weighting factors and risk coefficients for calculating the recommended limits for exposure. Because these factors have been explained under the ICRP-26 topics above, they will not be elaborated here. It must be remembered that all such values are arbitrary and only represent the best judgment of the committees involved. There are rarely sufficient data to apply to actual derivation of any of these values but they are good within limits; the problem is we do not know what these limits are.

3.2.3 Dose-Limiting Recommendations and Guidance for Special Cases

The topic of dose-limiting recommendations in routine and extreme situations is covered very well in ICRP-91 (106) and values presented in this report are shown in Table 13.6.

4 THERAPEUTIC MEASURES

It is possible to modify the effects of external irradiation by means of certain sulfur-containing compounds, including cysteine and cysteamine, provided the compounds are administered prior to irradiation (108). It is also considered possible to ameliorate radiation effects by bone marrow transfusion, preferably using bone marrow provided by the exposed individual prior to exposure. The practicality of measures such as these is so severely limited, because human exposure or contamination is accidental, that they cannot be considered as viable therapeutic measures.

The decision to apply therapeutic measures to decrease the effects of exposure to radiation usually must be made promptly and under the supervision of a physician and must always be made carefully. This decision must incorporate information on the potential risk due to the radiation and, often, on the potential risk due to therapy.

Table 15.5 Summary of Recommendations

A. Occupational exposures (annual)[c]		
1. Effective dose equivalent limit (stochastic effects)	50 mSv	(5 rem)
2. Dose equivalent limits for tissues and organs (nonstochastic effects)		
a. Lens of eye	150 mSv	(15 rem)
b. All others (e.g., red bone marrow, breast, lung, gonads, skin and extremities)	500 mSv	(50 rem)
3. Guidance: Cumulative exposure	10 mSv × age	(1 rem × age in years)
B. Planned special occupational exposure, effective dose equivalent limit[c]	See Section 15 of report	
C. Guidance for emergency occupational exposure[c]	See Section 16 of report	
D. Public exposures (annual)		
1. Effective dose equivalent limit, continuous or frequent exposure[c]	1 mSv	(0.1 rem)
2. Effective dose equivalent limit, infrequent exposure[c]	5 mSv	(0.5 rem)
3. Remedial action recommended when:		
a. Effective dose equivalent[d]	>5 mSv	(>0.5 rem)
b. Exposure to radon and its decay products	>0.007 Jhm-3	(>2 WLM)
4. Dose equivalent limits for lens of eye, skin and extremities[c]	50 mSv	(5 rem)
E. Education and training exposures (annual)[c]		
1. Effective dose equivalent limit	1 mSv	(0.1 rem)
2. Dose equivalent limit for lens of eye, skin and extremities	50 mSv	(5 rem)
F. Embryo-fetus exposure[c]		
1. Total dose equivalent limit	5 mSv	(0.5 rem)
2. Dose equivalent limit in a month	0.5 mSv	(0.05 rem)
G. Negligible Individual Risk Level (annual)[c]		
Effective dose equivalent per source or practice	0.01 mSv	(0.001 rem)

[a]Excluding medical exposures.
[b]See Table 4.1 (the report) for recommendiation on Q.
[c]Sum of external and internal exposures.
[d]Including background but excluding internal exposures.

The very best measurements of exposure conditions can serve only as good estimates of the radiation dose received, in the case of external radiation, or likely to be received, in the case of internal contamination. Therapeutic risks that must be considered include physical injury due to invasive techniques such as injection or surgery, the toxic effects of certain chemicals, the risk associated with general anesthesia or blood transfusion, and the psychological impact of therapy (particularly heroic measures), on the exposed individual and on the family. A recent joint publication of the Commission of the European Communities and the U.S. Department of Energy will be most helpful in this matter (109).

The basis of therapy is quite different for radiation from external sources as contrasted with internal sources. In the former case therapy is initiated after the total radiation dose has been received. The dose cannot be lessened; therefore the objective of therapy is modification of effect by treatment of radiation sickness, prevention of secondary infection, or supplementation of dwindling hematopoietic elements. When the radiation dose will be protracted because of the internal presence of the radiation emitter, the goal of therapy is reduction of the quantity of the emitter. This may be accomplished by enhancing excretion of the emitter or by other physical means of removal.

4.1 External Radiation Exposure

The use of therapeutic means to combat exposure to external radiation is an area in which there has been little opportunity to gain human experience. Personnel involved in worker protection in this segment of toxicology should be proud of their record for maintaining such a low number of exposure incidents. A summary of the Lockport, New York, accidental exposures (2) will give sufficient insight into the type of care and therapy that appear to be workable for external whole-body or partial-body irradiation.

The Lockport radiation incident provides knowledge of medical treatment following external radiation exposure that is classical insofar as almost everything, therapeutically, appears to have been done properly. Nine persons were exposed to radiation from an unshielded klystron tube at a U.S. Air Force radar site (2). Three of these personnel received doses of x-rays in the range of 1200 to 1500 R over certain areas of the body. Because exposure occurred during a period of 60 to 120 min in the working area, it was extremely difficult to estimate the dose to a given region of any individual.

Table 13.7 lists the basic symptoms shown by the exposed individuals, in descending order of estimated dose received. At the highest exposure level most of the classical signs of acute radiation damage are indicated. Nausea and vomiting were prevalent as well as fatigue and drowsiness. Erythema was a positive indicator of radiation exposure in every case. The first step in therapy was to admit the patients to hospital ward rooms that had been thoroughly cleaned, including the culturing of samples from throughout the room (air, furniture, etc.) for pathogens. This procedure was necessary because of the suspected lowered bacterial resistance of the patients.

All personnel entering the ward were required to wear face masks and to observe other contagion precautions. Visitors were limited to the immediate family. Only

Table 13.7 Signs and Symptoms in Victims of the Lockport Incident

Patient	Tinnitus, Parotid Swelling	Tempero-mandibular Tenderness	Forehead Swelling	Head-ache	Abdominal Pain	Nausea	Vomiting	Anorexia	Chills and Fever	Erythema
1	+	+	0	+	0	+	+	0	+	+
2	0	+	0	+	0	+	+	0	+	+
3	0	0	0	0	+	+	+	0	+	+
4	0	0	0	+	0	+	+	0	0	+
5	0	0	0	+	0	0	0	0	+	+
6	0	0	0	0	0	0	0	0	0	+
7	0	0	0	0	0	+	0	0	+	+

Symptoms

Source: Howland et al. (2).

laboratory studies deemed necessary for making crucial decisions were allowed. Early after admission the decision was made to postpone any bone marrow transplants because the exposures were of the partial body only. Also, no antibiotics or transfusions were administered; it was thought best to hold off these procedures until the advent of a complication of infection or bleeding.

This represents a proper and sensible approach to the medical treatment of individuals following accidental exposure to relatively high levels of external radiation. The methods available to protect the exposed individual if administered prior to exposure, previously mentioned, are hardly applicable to the accidental situation.

4.2 Internal Radiation Exposure

Therapy for exposure from internally deposited radiation emitters consists of reducing the quantity, or body burden, of the deposited radioactive material. Until recently, the use of chelating agents, introduced primarily by IV injection, has been the only productive method of such reduction. Because the effectiveness of chelation therapy is limited to use with internal emitters in a form that is soluble in body fluids, this approach has not been successful for the removal of insoluble materials such as inhaled insoluble particles that are deposited in the lung. In 1972 an accidentally exposed individual underwent bronchopulmonary lavage for removal of inhaled deposited ^{239}Pu (110). The positive results of that treatment, together with results of experiments with animals, indicate that bronchopulmonary lavage is a promising procedure for removing inhaled insoluble radioactive substances. With a choice of effective therapeutic measures, a sometimes difficult problem is that of determining the relative solubility of the internal containment to ensure that the most beneficial treatment is chosen.

4.2.1 Chelation Therapy

Chelation is natural to biological systems. Many metabolically formed compounds, including citric acid and gluconic acid, form chelates, and several amino acids are active chelators. Citric acid chelates calcium and is used by blood banks; the citrate-calcium chelate helps to prevent blood coagulation. Chelation therapy for heavy-metal contamination has been in use for more than 30 years, and a variety of chelating agents have been tried (108, 111). Recent reviews of the subject of chelation therapy express the problems with decorporation and evaluate the more current compounds in use or under investigation (112–115). Whereas citric acid is a very efficient binder of calcium, the well-known chelator ethylenediamine tetraacetic acid (EDTA) has a greater affinity for polyvalent metals such as lead, zinc, tin, yttrium, and plutonium. The effectiveness, thus the proper choice, of a chelator depends on its biological stability and the stability of the materials to be removed. Diethylenetriamine pentacetate (DTPA) forms a more stable chelate with the rare earths than does EDTA, and DTPA has a residual effect that is desirable in terms of lowering the frequency of treatments.

Chelators are generally poorly absorbed from the gastrointestinal tract; thus they are not of much use when administered orally except in cases of ingestion of the contaminant. Although chelator absorption may be quite good following intramuscular (IM) injection, use of the IM route of administration is rare, partly because of the greater likelihood of local irritation associated with this route. Chelator administration by IV injection is most often reported because of its greater efficiency; however, much greater potential for toxicity exists when the IV route is used. The decisions regarding risk–benefit of introducing chelation therapy is not clear-cut; the subject is covered well for plutonium, as an example, in NCRP-65 (113). The greater stability of DTPA and its greater efficiency for removal of heavy metals from the body than EDTA render DTPA more useful at lower dosages, therefore less potentially toxic than EDTA. DTPA has become the chelator of choice for treatment of internal contamination of humans with polyvalent radionuclides. The cation of choice for this chelator may be zinc (Zn-DTPA) as opposed to calcium, as pointed out by Mays et al. (115). The zinc complex appears to be less toxic to the patient by virtue of not depleting necessary cations from the body. Information on the behavior and effects of chelating agents in humans is available because numerous experiments have been carried out using tracer quantities of radioactive rare earths in humans (116, 117).

Chelation therapy has been useful in the removal of plutonium from contaminated individuals, as described in the following brief case studies. In one reported case (118) a worker was sprayed with an acid solution of plutonium chloride and plutonium nitrate. Inhalation and ingestion, as well as skin contamination, occurred. The skin, except for burned areas, was decontaminated with dilute sodium hypochlorite solution. Eleven 1-g, IV, DTPA treatments were administered, beginning 1 hr after the accident and at intervals through 17 days. Burn scabs were removed 2 weeks following the accident and they were found to contain most of the plutonium. The combination of treatment methods used in this case was considered highly effective. Similar effectiveness of prompt DTPA treatment was reported for another plutonium contamination incident involving an acid burn (119). Twenty-seven daily 1-g, IV, DTPA treatments, beginning 1 hr after the accident, resulted in elimination of more than 96 percent of the estimated systemic burden. As in the previous case, much of the contamination was removed with the burn scabs.

The effectiveness of DTPA treatment using a regime similar to those just reported was considered inconclusive in the case of a wound to the thumb from a plutonium-contaminated metal sliver (118). Initially the sliver was removed without tissue excision. Subsequently tissue excisions were performed at the points of entrance and exit of the sliver; excised tissue from the point of sliver entrance contained about 98 percent of the plutonium removed by excision. Wound counting performed over several months indicated movement of the remaining embedded plutonium toward the skin surface. Nodules that formed concurrently with increased wound counts were excised, and these were found to contain essentially all the plutonium estimated to remain in the thumb. Although DTPA was effective in removing plutonium from the blood, the possible influence of DTPA in mobilizing the plutonium from the wound into the blood was noted.

In another puncture wound episode (120), DTPA was administered IV 4 days per week for about 11 weeks followed by a 30-week period of no treatment, then by a 90-week period in which DTPA was administered either IV or by aerosol 2 or 4 days per month. During the early treatment period, oral (tablet) EDTA was substituted for DTPA as a matter of convenience to the employee; the EDTA was found to be very ineffective in enhancing urinary plutonium excretion. Additional cases of puncture wound injury have been reported (119, 121). In each instance of puncture wound, the greatest efficiency of treatment has resulted from prompt DTPA therapy and one or more tissue excisions. In burn cases surface decontamination except in immediate burn areas, and prompt DTPA therapy, are recommended, followed by careful removal of the burn scabs. In all cases wound counting and/or whole-body counting has been used to determine the plutonium burden.

Several instances of human plutonium contamination by inhalation have been reported. In an early study (122) exposed individuals were treated with calcium EDTA, administered IV in 1-g doses twice a day, for the initial treatment, then administered orally. The IV treatment resulted in a 10-fold increase in urinary plutonium excretion; however, the oral doses caused only minor increases and were not considered to be of value. Chelation therapy was deemed successful, although the total urinary excretion amounted to only 10 percent of the body plutonium content. The key to effective chelation therapy is promptness.

4.2.2 Pulmonary Lavage

Pulmonary lavage is a technique that is applied primarily to patients with alveolar proteinosis. Kylstra worked for years in this field and became one of the world's experts (123, 124). In recent years the application of pulmonary lavage to animals has been successful in removing radionuclides after inhalation exposure (7, 125). The combined use of chelating agent and lung lavage has proven very successful as shown in animal studies (126). Some of these studies with dogs utilized inhalation of aerosols from practical uses in the nuclear industry. In addition, one human case of radionuclide exposure has been somewhat successfully treated (110).

The manner of treatment is to place a Carlens catheter (127) into the patient's respiratory tract such that one lung is supplied with oxygen while the other has access to washing fluid. In general, physiological saline is sufficient for the washing process. Approximately a tidal volume of saline is introduced into the open lung, then removed. This procedure may be repeated up to 5–10 times on the same lung during one session. The catheter is withdrawn, and after 24–48 hr the procedure is repeated on the other lung. With this procedure the surfactant that lines the alveolar regions is partially removed, and with it, any cells or other materials that reside there. Thus an insoluble material (radionuclide) in the deep lung, whether free or within macrophages, is removed to some extent. The following experimental data illustrate the effectiveness of the procedure.

At the Inhalation Toxicology Research Institute, Albuquerque, New Mexico, many studies have been performed on the effectiveness of lavage in removing inhaled particles from experimental animals. One aerosol used for this purpose in beagle dogs has been ^{144}Ce in fused aluminosilicate aerosol particles. A study by Boecker (128) showed how effective multiple lavage can be in removing the particles deposited in the respiratory tract. Figure 13.6 gives an example. The reduction in radiation effect on the lung by systematic removal of the radiation source (particles of ^{144}Ce

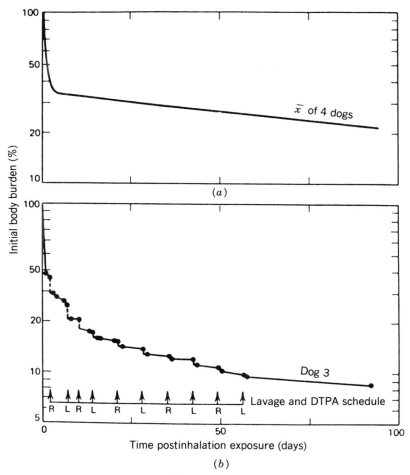

Figure 13.6 Whole-body retention of ^{141}Ce in dogs exposed by inhalation to ^{141}Ce in fused clay. (a) Dogs received no postexposure treatment; dog 3 was lavaged 10 times during the first 56 days after exposure. (b) Treated dogs: DTPA was given intravenously after each lavage: solid points represent whole-body count. The vertical steps represent the ^{141}Ce removed in the lavage fluid with each treatment. Data are uncorrected for physical decay of ^{141}Ce. [From Boecker et al. (128).]

in this case) is obvious. The lavage fluid was physiological saline. DTPA treatment was also used in conjunction with the lavage procedure with some of the dogs, but the overall effectiveness of the added chelator was minimal. A similar inhalation experiment was performed with ^{144}Ce in fused clay aerosol particles in beagle dogs, and those data are expressed in terms of reduction in radiation dose to the lung in Figure 13.7. According to Silbaugh et al. (129) no impairment of health was seen at long times postexposure (14 months) in the lavaged dogs, whereas untreated dogs

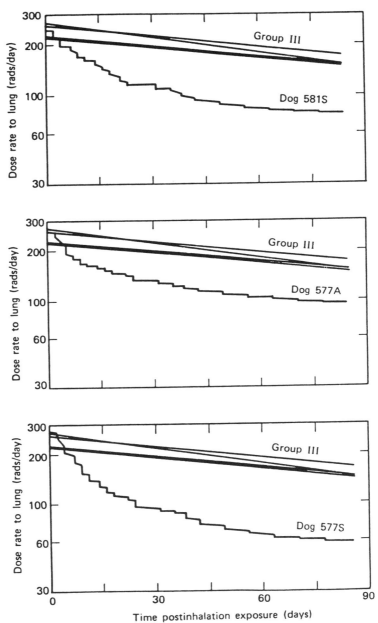

Figure 13.7 Reduction, with time, in the beta dose rate to the lungs of unlavaged group III dogs and lavaged group II dogs 581S, 577A, and 577S. [From Silbaugh et al. (129).]

receiving those initial doses would have been sick or died. An extension of this report to 2600 days (7 years) indicates continued positive results and stresses the lengthening of life span that resulted from use of the combination lavage and chelation therapy (130).

5 BIOASSAY TECHNIQUES

It is difficult to treat irradiated individuals for potential damage if the extent of the projected biological insult cannot be roughly estimated. Bioassay has been one of the descriptive terms applied to this field of making such dosimetric approximations. As will be seen, this is a very difficult task and needs much more experimental advancement before bioassay becomes the exact science that is desirable. It seems to be more effective with internal emitters, as discussed in Section 5.2.

5.1 External Radiation Exposures

There are few techniques for detection of human external radiation exposures, although many have been explored and attempted during the past few decades. Biological indicators of exposure must be sensitive to low-level radiation if the practical ''in-plant'' situation is to be monitored. This criterion eliminates chemical reactions that are quickly restored or require large doses to bring about measurable reaction products. It also essentially eliminates the old practice in therapy of using an erythema reaction to estimate the amount of radiation delivered.

An extremely troublesome factor in biological systems is the tremendous number of different chemical species present in any one finite sphere of interaction (ionization). If one had an ideal chemical oxidation–reduction reaction that would respond to low radiation doses, and if the primary constituents were present in necessary abundance in the volume of tissue of interest (e.g., a cell), the chemical entities would still be overwhelmed by other chemical species. If these ''foreign'' substances were reactive either to the radiation or to the products being formed and measured in the primary event, the accuracy would immediately be challenged. Such is the case with chemically induced bioassay schemes for detecting external exposure to ionizing radiation in vivo.

Perhaps of some potential are schemes that measure actual effects on the biological macrosystems. The genetic materials, chromosomes, offer one possibility as a dosimeter for practical radiation exposure. Although measurement of chromosomal damage through breakage is tedious, this damage nevertheless may prevail for periods of time, lending it practical significance. One drawback is the recombination of chromosomes that have been broken by the radiation, to a normal condition. This can readily occur under the influence of ionizing events or of other toxic reactions that affect chromosomes. However, background quantitation for the probability of recombination is possible (and feasible) and can become part of the overall scheme for dosimetry purposes. Samples of peripheral blood are readily obtainable from workers for use in this technique of quantitating chromosome damage due to ionizing radiation.

Lymphocytes are very sensitive to radiation and prove to be good indicators of damage. In Figure 13.1 it is obvious that many hematological parameters (cell numbers) show a dramatic change as a result of ionizing radiation. Therefore, one may be able to estimate the radiation dose in a rough manner, as was attempted with the Lockport cases (2). One of the more sensitive indicators using blood samples is the increased abundance of bilobed lymphocytes, a technique first explored and implemented by Ingram (131, 132). This rather specific indicator of radiation damage was first applied to workers associated with a cyclotron at the University of Rochester. Bilobed lymphocytes tend to be slightly larger than normal, and they contain two distinct nuclear masses. Each is a typical nucleus of normal lymphocytes, but they occur two to a cell. The main drawback to this type of personnel monitoring is the tedium of microscopic examination of so many cells in a preparation. The low incidence of these altered cells requires that statistical analyses be performed for the detection of low doses. However, the accuracy is such that this indicator provides a suitable bioassay procedure for low doses of either electromagnetic radiation or neutrons (131, 133).

Recently, there has been attention paid to the glycophorin assay techniques developed at the Lawrence Livermore Laboratory (134). These scientists have applied the technique to the accident cases at Goiania, Brazil, and have been quite satisfied with results of estimating dosage to individuals at one year after exposure to the gamma rays from ^{137}Cs (135). They have also been successful in assaying some Japanese atomic bomb survivors (136). Such use of biomarkers for estimating radiation dose at long times after the fact will ultimately prove to be of great use in the event of any future accidental radiation exposures.

5.2 Internal Radiation Exposure

Radionuclides deposited in the body also point up the gap between ease of detection and interpretation of results. A recent review of the subject of bioassay techniques for exposure to radionuclides is divided into four major sections and is an excellent summary of the current status of available techniques (137). The four categories discussed in detail are: (1) cytogenetic procedures; (2) whole-body counting; (3) bioassay procedures; and (4) biological indicators. This reference is excellent for those interested in the general topic of bioassay techniques following exposure to radionuclides.

The radioactivity of a material present in a biological system provides one of the most sensitive means that will ever be known for localizing an element involved in the body's metabolic scheme. Measurement of a radionuclide's effect and its metabolic parameters, as related to the overall in vivo system, is a difficult problem, however, and almost insurmountable under some circumstances. As an example, detection (quantitation) of an alpha particle emitter in a section of tissue can be carried out with great precision. Equally credible measurements can be made on the contaminated excreta resulting from deposition of this material. However, the intervening biological scheme at play between the deposit of the radionuclide and its elimination is generally an unknown entity. It is this factor that leads to frustration where practical bioassay programs for determination of industrial exposures are concerned.

5.2.1 Inhalation Exposures

Many problems are inherent in bioassay of materials deposited in the lung following inhalation. The solubility of the inhaled radionuclide has an immediate, important role in the effectiveness of use of bioassay techniques. A *completely soluble* (in body fluids) radionuclide may enter the blood and either deposit in the tissue of choice or be excreted, primarily in the urine. If the release from the designated site of localization is with constant and known kinetics, appearance of the material in the urine may be an excellent source of information for extrapolation to an estimated body burden. An example of such a material is tritium. When inhaled, this isotope of hydrogen enters the normal body pool and in a short period becomes part of the body's metabolic scheme. Measurement of $^1H_3$ in the urine and application of isotope dilution principles as related to body characteristics (e.g., weight, height), serves as a suitable bioassay technique (138–140). Tritium assay presents some difficulties, which are not unusual with the bioassay of internally deposited radionuclides, in that tritium leaves the body with multiple exponents of excretion that may depend on environmental factors such as season of the year (138). However, tritium represents the ideal case (soluble), and this bioassay method is generally not applicable in inhalation exposure cases.

When the inhaled radionuclide is not in a soluble form, the situation with regard to bioassay becomes nightmarish. If the radionuclide is *completely insoluble* (a misnomer, since it holds true for no material in body fluids), there is no bioassay technique that is quantitative through normal metabolic processes. A radionuclide in this form, which has some sort of detectable radiation (x-ray or gamma ray) outside the body, may be quantitated by chest counting. This is a reasonably accurate method except as the energy of emission approaches 100 keV and lower. At these low energies the rib cage becomes a troublesome factor because it absorbs the photon. The degree of absorption depends on the size of the rib cage and associated tissue, both in thickness and variable density, as well as the photon energy. Accurate calibration for a specific individual is essentially impossible, since mockups rarely match the subject. Thus estimates of the biological parameters are made through height, weight, lean mass–body weight ratios, and so on. Ultrasonic measurements are made at many locations about the chest cavity, but still, determination of the degree of self-absorption for an individual is close, at best. Whole-body counting does remain the only practical scheme, however, for estimating lung burden of an insoluble radionuclide. One problem encountered with this and all chest counting is the inability to distinguish lymph node accumulation of radionuclide from that in lung tissue. Some experimentation has been carried out to evaluate the use of minute detectors that can be lowered into the respiratory tract using a device similar to a bronchioscope, but to date this technique has not found widespread use (141). Pinpoint detection will remain a problem until schemes for precision of measurements have been more thoroughly tested.

The case of exposure to somewhat soluble or slowly solubilized radionuclides is perhaps the most difficult from the viewpoint of bioassay (see discussion of the kinetics of loss from the lung to internal organs in Section 1.3.3). If one knows all the parameters of loss from the lung to blood to urine, a backcalculation readily produces the lung burden. Unfortunately, however, this is rarely the case, particularly in that an aerosol breathed by the same or different persons, at the same location, under seemingly identical conditions, will lead to variable results. Thus any bioassay procedure attempted under conditions of partial solubility will meet with frustration and, probably, erroneous results and erroneous interpretation.

Many references exist on whole-body counting and excretory techniques for determining the radioactivity in man, and these represent a starting place for those interested in pursuing this field (137, 142, 143). Also, directories have been published that contain information on the laboratories that operate whole-body monitors, with information on the technical design and performance of the individual apparatus (137, 144).

5.2.2 Puncture Wounds

One of the most common means of receiving accidental body burdens of radionuclides is through skin absorption or puncture. The situation is similar to inhalation exposure and can be considered in the same light. The wound burden can be assumed to be comparable to the lung burden, and rate of loss from the site is given the same consideration. If there is a detectable radioactive emission, a wound counter can be utilized in the same manner as a chest counter (118). However, the advantage with wounds is that excision or amputation of the contaminated body part can be put to use in relieving the problem; this is less feasible in the case of inhalation exposure. Estimation of the wound radioactivity even can be made through biopsy, a technique that is not as satisfactory where the lung is concerned. Urine analysis may be satisfactory under certain conditions of wound contamination, but external counting, where feasible, still remains the most practical and successful method of assaying for body burdens.

6 RADIATION AND ENERGY SOURCES

Present predictions indicate that fossil forms of energy will allow the United States to satisfy its energy needs for several decades. Also, there appears to be considerable optimism that solar, geothermal, or other sources may increase this period indefinitely, and many research teams are working in these areas. In addition to these energy sources, however, are those from which radiation is an obvious by-product. The sections that follow briefly indicate the possible problems associated with them.

6.1 Radiation Sources

One of the more promising future sources of energy is the practical application of nuclear fusion. If this source is exploited to an economically feasible state, it can be the answer to the future's energy problems. In the fusion process atomic nuclei (light atomic weight) are brought together in such a manner that energy above the amount of the particles' binding energy is released and can be captured and utilized as a practical energy source. Since nuclear fusion would not produce as much atmospheric contamination with radioactive materials as would nuclear fission, the process itself might be acceptable to the factions in the public that are alienated by the term "radiation," regardless of the context in which it is used. However, the fusion process, if and when it is feasibly developed as a practical energy source, may not be without radiation problems. There will be waste disposal problems as with nuclear reactors, and the neutrons that are emitted as part of the energy release could present an industrial radiation problem. The practicability of fusion as a source of energy is a political and socioeconomic topic for discussion, and some relatively recent articles face many of these issues (145–148).

Nuclear fission is the most realistic source of energy at present, and it is currently being used in many countries of the world. An article that appeared in 1976 indicates that France planned an increase from 8 to 70 percent electricity production from nuclear power by 1985, as a result of the embargo by major petroleum producers and increased oil prices (149). Various types of power reactor may be used as sources of energy, some more appealing than others from the standpoint of potential release of airborne radioactivity. Modern design of most proposed nuclear power reactors is such that even an internal malfunction (i.e., within the reactor) would not result in atmospheric contamination. Reutilization of the repaired initial plant, or substitution of a new one, could take place to restore the same site as a source of power. Engineering accomplishments to maintain any radiation problems within the confinement of the reactor complex itself indicate that any exposure to the surrounding area would be of little significance, probably zero. Because this energy source represents a horrifying (Hiroshima-like) scene to some of the American public, it will take considerable time and public education before the nuclear reactor is accepted as a safe major source of energy in this country. Our most realistic view on all of this has come from the incidents surrounding the well-known Three Mile Island nuclear reactor accident. Much has been learned from this mishap. An aspect of concern to proponents of nuclear sources of energy is that 10–20 years may be required from the time that a new power station is begun until it can assume routine operation.

6.2 Mixed Potential Hazards

Potential hazards may be involved in the event that a radiation source combines with an environmental chemical source in the public atmosphere. One must consider the possibility, however slight, that a nuclear power source could give rise to some atmospheric contamination as the result of an accidental release. This release could be in the form of radioactive particles and could be acted on by such factors as local meteorology, temperature, terrain, and altitude.

6.2.1 Radiation and Chemicals

Synergism in inhalation toxicity, particularly concerning a source of radiation and a potential chemical cocarcinogen, has not been studied in depth. A study with experimental animals at the Los Alamos National Laboratory recently indicated a sizable increase in lung cancer incidence when Fe_2O_3 was added to PuO_2 particles deposited in the respiratory tract (150). The latter insult alone (PuO_2) was only slightly more tumorigenic than controls under similar conditions of administration to the animal species used (Syrian golden hamsters) (151, 152). The public is informed almost daily in newspapers and journals of the possible atmospheric contamination arising from common sources. Materials such as asbestos from automobile brakes, oxides of sulfur, oxides of nitrogen, carbon monoxide, dust, and high humidity are popular subjects for news media and other concerned factions. These materials are generally measured and indexed by local (state and city) environmental regulatory organizations, and when certain limits are surpassed, an attempt is made to reduce these to some established ''allowable'' level. Radiation is treated in the same manner, but the measured values are required to be essentially zero (natural background for the area), or the lowest practicable.

6.2.2 External and Internal Radiation

The interesting facet of exposure to mixed radiations is the inability to know, at this time, whether two completely different types of radiation are going to act synergistically, additively, or antagonistically. The cases most likely to be complicated are those in which a whole-body exposure may occur from external radiation while the person is being subjected to solubilized inhaled or ingested radionuclides or to inhaled insoluble particles. Such is the case immediately following a reactor accident and the mishap at Chernobyl in the Soviet Union in 1986 is an example of this mixed radiation exposure.

6.2.2.1 The Chernobyl Accident. The reasons for this accident are important but not for this chapter. It is sufficient to state that high temperatures and pressures were involved to the point that the reactor core was in a state of meltdown and runaway to the point of destruction. The particles released ranged from ''chunks'' to very small, even vaporized, and noble gaseous, particles. The cloud was thus at various heights but fortunately missed the town of Pripyat, the nearest inhabited area. A cloud of this nature will deliver external irradiation in the form of gamma rays and to some degree, beta particles, but probably no neutrons. The debris that falls out of the cloud as it passes, again in various particle sizes, will be made up primarily of beta–gamma emitting fission products; the Chernobyl reactor core contained some inventory of plutonium so there was a limited amount of alpha emitters at ground level.

The current method of estimating the effect of such a disaster is to base the dose–effect relationship upon linear extrapolation to zero effect at zero dose and to use the slope of the line for risk estimation. One calculates the amount of radiation released, the cloud path from meteorological conditions, and so forth, and estimates the external radiation dose to those who may be exposed on the ground. Using the same cloud parameters, it is not too difficult to estimate the inhaled and ingested dose, within very large ranges, and to calculate the potential dose to people from these sources. Combining the doses and using the whole-body equivalent dose to the average individual, the total potential cancers may be calculated. Even the risk factors applied to the final calculation are not known within a wide range, and are generally not known for the given exposure scenario. Thus, given some estimate of the total population affected times the estimated dose to the average individual, a value for total person-dose may be obtained. When multiplied by some arbitrary risk factor as the number of cancers per person-dose, a value for the total number of cancers to be expected in the exposed population is obtained. This is based on a linear extrapolation, and there is no sound reason to apply this type of extrapolation at the low-dose region.

In the case of those estimated to be exposed in Eastern Europe from the cloud passage from Chernobyl the number of cancers expected, *using this approach to calculation*, is probably less than 0.00005 above the normal background incidence or 5 excess cancers per 100,000 exposed persons (153, 154). When it is considered that the values used and the assumptions made are conjectural at best, this number as an insult from the most catastrophic nuclear accident of all time, is unusually small. The genetic effects are considered even more inconsequential.

Books have been compiled on the Chernobyl accident, but the message of this chapter is that, even though the release was extremely high in radionuclide inventory and external dosage, the end result is essentially negligible when compared to the fatalities associated with the mining of coal or the drilling of petroleum in one year.

6.3 Mining as a Factor

Hazards associated with the mining of uranium have been recognized for centuries. Within the past 50 years the origin of lung cancer among these miners has been quite firmly attributed to the radiation associated with the radioactive daughters of radon, the noble radioactive gas from uranium. There is also an external source of radiation associated with the uranium decay products, and this can range from 0.02 to 4 mrem/hr. As Holaday (155) points out, these rates of gamma radiation have been reported for mines in France, Japan, Mexico, Spain, Australia, and the United States. It was determined a few years ago that the respiratory tumorigenicity associated with the miners was closely related to cigarette smoking in conjunction with the deposited radioactive particles of the radon daughters (see Section 2.2.1).

One of the most interesting studies in recent years concerning the mining industry dealt with the immunologic assessment of neoplasia and metaplasia in the lungs of miners (156). The main purposes of the study were to answer the following questions: (1) Can alterations in immunologic function be detected in individuals with preneoplastic pulmonary lesions? (2) If so, will these changes help to identify those individuals who may develop invasive lung cancer? (3) What immunologic test is the most sensitive indicator of abnormalities? (4) Can the relative contributions of carcinogens be determined, such as the relative contributions of radon daughter products and cigarette smoke to lung cancer in uranium miners? The results of this study, performed on blood samples from the mining population of the Western Slope of the Colorado Rockies, were positive in showing defects in the immune system associated with the early development of lung cancer. Rosette inhibition testing was the most sensitive indicator found that would relate to the presence of malignancy in the miners with known neoplasia. Thus, early testing of this nature may be done to detect early changes that are perhaps undetectable by other than simultaneous sputum cytological testing, with the possibility of performing early immunotherapy. There was no evidence in the populations studied (miners and controls) that smoking impaired immune competence. This study, in combination with results of sputum cytological testing (85, 89, 90, 157), shows great promise in associating immune system abnormalities with potentially carcinogenic industrial toxicants. It is included in this chapter because of the obvious relation to the carcinogenic risks associated with irradiation of the pulmonary system by radon daughters.

Mining problems have assumed new importance since the energy crisis has directed industry into evaluating the extensive procurement, refinement, and use of potential sources of underground fossil fuels. Oil shale mining is an excellent example, and certainly there are potential chemical carcinogens associated with the material mined, processed, refined, and utilized for energy. Radon and radon daughters are prevalent in coal mines, and there is every reason to believe that any underground mining, in the Rocky Mountain region particularly, will be a source of radon and its daughters. With newer, stricter regulations and guidelines, the cost of eliminating any such measurable materials can become enormous. In addition to the potentially small radiation problem, the dusty atmospheres associated with mining may contain potential chemical carcinogens that will be inhaled along with the radioactive sources. In fact, it is likely that if workers or the population receive any exposures to potentially toxic levels, the toxicants will be of chemical origin, not the low levels of radiation that are present in the mines.

REFERENCES

1. Health Risks of Radon and Other Internally Deposited Alpha-Emitters. BEIR IV, Committee on the Biological Effects of Ionizing Radiations, National Research Council, National Academy Press, Washington, D.C., 1988.

2. J. W. Howland, M. Ingram, H. Mermagen, and C. L. Hansen, Jr., The Lockport Incident: Accidental Partial Body Exposure of Humans to Large Doses of X-Irradiation, in *Diagnosis and Treatment of Acute Radiation Injury*, Proceedings of a Scientific Meeting Jointly Sponsored by the International Atomic Energy Agency and the World Health Organization, Geneva, October 17–21, 1960, WHO, Geneva, 1961, p. 11.

3. D. M. Ross, A Statistical Summary of United States Atomic Energy Commission Contractors' Internal Exposure Experience 1957–1966, in *Diagnosis and Treatment of Deposited Radionuclides*, H. A. Kornberg and W. D. Norwood, Eds., Excerpta Medica Foundation, Amsterdam (1968). Proceedings of a Symposium, Richland, Wash., May 15–17, 1967, p. 427.

4. H. C. Hodge, J. N. Stannard, and J. B. Hursh, Eds., *Uranium–Plutonium–Transplutonic Elements*, Springer-Verlag, New York, 1973, p. 643.

5. E. C. Anderson, *J. Bacteriol.*, **76**, 471 (1958).

6. T. T. Mercer, M. I. Tillery, and H. Y. Chow, *Am. Ind. Hyg. Assoc. J.*, **29**, 66 (1968).

7. B. A. Muggenburg, S. A. Felicetti, and S. A. Silbaugh, *Hlth. Phys.*, **33**, 213 (1977).

8. K. J. Caplan, L. J. Doemeny, and S. D. Sorenson, *Am. Ind. Hyg. Assoc. J.*, **38**, 162 (1977).

9. K. J. Caplan, L. J. Doemeny, and S. D. Sorenson, *Am. Ind. Hyg. Assoc. J.*, **38**, 83 (1977).

10. M. Lippmann, Aerosol Sampling For Inhalation Hazard Evaluation, in *Assessment of Airborne Particles*, T. T. Mercer, P. E. Morrow, and W. Stober, Eds., Thomas, Springfield, IL, 1972, p. 449.

11. *Amer. Ind. Hyg. Assoc. J.*, **31**, 133 (1970).

12. M. Lippmann, *Am. Ind. Hyg. Assoc. J.*, **31**, 138 (1970).

13. D. A. Holaday, Uranium Mining Hazards, in *Uranium–Plutonium–Transplutonic Elements*, H. C. Hodge, J. N. Stannard, and J. B. Hursh, Eds., Springer-Verlag, New York, 1973, p. 301.

14. R. F. Droullard, Instrumentation for Measuring Uranium Miner Exposure to Radon Daughters, in *Radiation Hazards in Mining: Control, Measurement, and Medical Aspects*, M. Gomez, Ed., Kingsport Press, Kingsport, TN, 1981, p. 332.

15. H. D. Freeman, An Improved Radon Flux Measurement System for Uranium Tailings Pile Measurement, in *Radiation Hazards in Mining: Control, Measurement, and Medical Aspects*, M. Gomez, Ed., Kingsport Press, Kingsport, TN, 1981, p. 339.

16. T. B. Borak, E. Franco, K. J. Schiager, J. A. Johnson, and R. F. Holub, Evaluation of Recent Developments in Radon Progeny Measurements, in *Radiation Hazards in Mining: Control, Measurement, and Medical Aspects*, M. Gomez, Ed., Kingsport Press, Kingsport, TN, 1981, p. 419.

17. K. Becker, *Hlth. Phys.*, **16**, 113 (1969).

18. International Commission on Radiation Units and Measurements, Neutron Dosimetry for Biology and Medicine, ICRU Report 26, 1977.

19. *Radon Research Program, FY-1989*, S. L. Rose, Ed., U.S. Department of Energy, Washington, D.C., DOE/ER-0448P, 1990.

20. *Radon Research Program, FY-1990*, S. L. Rose, Ed., U.S. Department of Energy, Washington, D.C., DOE/ER-0488P, 1991.

21. O. G. Raabe, Generation and Characterization of Aerosols, in *Inhalation Carcinogenesis*, AEC Symposium Series 18, 1970, p. 123.

22. Task Group on Lung Dynamics, *Hlth. Phys.*, **12**, 173 (1966).

23. R. G. Thomas, Uptake Kinetics of Relatively Insoluble Particles by Tracheobronchial Lymph Nodes, in *Radiation and the Lymphatic System*, Proceedings of Symposium, Richland, Wash., ERDA Conference CONF-740930, 1974, p. 67.

24. P. Nettesheim, M. G. Hanna, Jr., and J. W. Deatherage, Jr. Eds., *Morphology of Experimental Respiratory Carcinogenesis*, AEC Symposium Series 21, 1970, p. 417.

25. R. G. Thomas, Tracheobronchial Lymph Node Involvement Following Inhalation of Alpha Emitters, in *Radiobiology of Plutonium*, B. J. Stover and W. S. S. Jee, Eds., J. W. Press, University of Utah, Salt Lake City, 1972, p. 231.

26. W. S. S. Jee, Ed., *The Health Effects of Plutonium and Radium*, J. W. Press, University of Utah, Salt Lake City, 1976, p. 169.

27. U.S. Department of Commerce, National Bureau of Standards, *Maximum Permissible Body Burdens and Maximum Permissible Concentrations of Radionuclides in Air and in Water for Occupational Exposure*, NBS Handbook 69, U.S. Government Printing Office, Washington, D.C., 1959.

28. International Commission on Radiological Protection, Committee II, *Hlth. Phys.*, **3**, 1960.

29. R. G. Cuddihy, Modeling of the Metabolism of Actinide Elements, in *Actinides in Man and Animals*, M. E. Wrenn, Ed., RD Press, Salt Lake City, 1981, p. 617.

30. P. W. Durbin, *Hlth. Phys.*, **8**, 665 (1962).

31. International Commission on Radiological Protection, *Biological Effects of Inhaled Radionuclides*, ICRP Publication 31, Annals of the ICRP, 4/1 (1980).

32. Exposures from the Uranium Series with Emphasis on Radon and Its Daughters, National Council on Radiation Protection and Measurements, NCRP Report No. 77, (1984).

33. Evaluation of Occupational and Environmental Exposure to Radon and Radon Daughters in the United States, National Council on Radiation Protection and Measurements, NCRP Report No. 78 (1984).

34. N. F. Johnson, D. G. Thomassen, A. L. Brooks, and G. J. Newton, *J. Cellular Biochem. Supplement*, **14A**, 67 (1990).

35. S. H. Moolgavkar, F. T. Cross, G. Luebeck, and G. E. Dagle, *Radiat. Res.*, **121**, 28 (1990).

36. N. H. Harley, *Hlth. Phys.*, **55**, 665 (1988).

37. S. Okada, Ed., *J. Radiat. Res.*, *Jpn. Radiat. Res. Soc.*, **16**, Supplement (1975).

38. G. W. Beebe, *Am. J. Epidemiol.*, **114**, 761 (1981).

39. *The Effects of Populations of Exposure to Low Levels of Ionizing Radiation. BEIR III*, Committee on the Biological Effects of Ionizing Radiations, National Research Council, National Academy Press, National Academy of Sciences, Washington, D.C., 1980.

40. *Health Effects of Exposure to Low Levels of Ionizing Radiation. BEIR V*, Committee on the Biological Effects of Ionizing Radiations, National Research Council, National Academy Press, Washington, D.C., 1990.

41. Sources, Effects, and Risks of Ionizing Radiation, United Nations Scientific Committee on the Effects of Atomic Radiation, United Nations Publication, Report ISBN 92-1-142143-8, United Nations, New York 1988.

42. G. M. Matanoski, R. Seltzer, P. E. Sartwell, E. L. Diamond, and E. A. Elliott, *Am. J. Epidemiol.*, **101**, 188 (1975).

43. A. C. Upton, Effects of Radiation on Man, in *Annual Review of Nuclear and Particle Science*, Vol. 18, Annual Reviews, Palo Alto, CA, 1968, p. 495.

44. E. Lorenz, *Am. J. Roetgenol.*, **63**, 176 (1950).

45. H. F. Henry, *J. Am. Med. Assn.*, **176**, 671 (1961).

46. T. D. Luckey, Ed., *Hormesis with Ionizing Radiation*, CRC Press, Boca Raton, FL, 1981.

47. J. F. Spalding, R. G. Thomas, and G. L. Tietjen, Life Span of C57 Mice as Influenced by Radiation Dose, Dose Rate, and Age at Exposure, Los Alamos National Laboratory Report, LA-9528, October, 1982.

48. R. E. Albert, F. J. Burns, and P. Bennett, *J. Nat. Canc. Inst.*, **49**, 1131 (1972).

49. H. Spiess and C. W. Mays, Protraction Effect on Bone-Sarcoma Induction of ^{224}Ra in Children and Adults, in *Radionuclide Carcinogenesis*, Proceedings of the 12th Annual Hanford Biology Symposium, 1973, p. 437. C. L. Sanders, R. H. Busch, J. E. Ballou, and D. D. Mahlum, Eds., AEC Symposium *29*, CONF-720505, National Technical Information Service, Springfield, Virginia (1973).

50. W. M. Court Brown and R. Doll, *Leukemia and Aplastic Anemia in Patients Irradiated for Ankylosing Spondylitis*, Medical Research Council Special Report Series 295, Her Majesty's Stationery Office, London, 1957, p. 21.

51. W. M. Court Brown and R. Doll, *Br. Med. J.*, **II**, 1327 (1965).

52. N. H. Müller and H. G. Ebert, Eds., *Biological Effects of ^{224}Ra. Benefit and Risk of Therapeutic Application*, Martinus Nijhoff Medical Division, The Hague/Boston, 1978.

53. R. H. Mole, *Br. J. Radiol.*, **48**, 157 (1975).

54. S. Jablon and H. Kato, *Radiat. Res.*, **50**, 649 (1972).

55. M. Brown, *Radiat. Res.*, **71**, 34 (1977).

56. National Council on Radiation Protection and Measurements, Influence of Dose and Its Distribution in Time on Dose–Response Relationships for Low–LET Radiations, NCRP Report No. 64 (1980).

57. R. D. Evans, A. T. Keane, R. J. Kolenkow, W. R. Neal, and M. M. Shanahan, Radiogenic Tumors in the Radium and Mesothorium Cases Studied at M.I.T., in *Delayed Effects of Bone-Seeking Radionuclides*, C. W. Mays, W. S. S. Jee, R. D. Lloyd, B. J. Stover, J. H. Dougherty, and G. N. Taylor, Eds., J. W. Press, Salt Lake City, 1969.

58. U.S. Nuclear Regulatory Commission, Below Regulatory Concern. A Guide to the Nuclear Regulatory Commission's Policy on Exemption of Very Low-Level Radioactive Materials, Wastes, and Practices, Report No. NUREG/BR-0157 (1990).

59. D. E. Lea, *Actions of Radiations on Living Cells*, Macmillan, New York, 1947.

60. A. C. Upton, *Radiat. Res,.* **71**, 51 (1977).

61. R. E. Rowland, A. T. Keane, and H. F. Lucas, Jr., A Preliminary Comparison of the Carcinogenicity of ^{226}Ra and ^{228}Ra in Man, in *Radionuclide Carcinogenesis*, AEC Symposium Series 29, CONF-720505, 1973, p. 406.

62. J. D. Abbatt, Human Leukemic Risk Data Derived from Portuguese Thorotrast Experience in *Radionuclide Carcinogenesis*, AEC Symposium Series 29, CONF-720505, 1973, p. 451.

63. D. M. Taylor, C. W. Mays, G. B. Gerber, and R. G. Thomas, Eds., *Risks from Radium and Thorotrast*, British Institute of Radiology, London, Report No. BIR-21 (1989).

64. R. E. Rowland and P. W. Durbin, Survival, Causes of Death, and Estimated Tissue Doses in a Group of Human Beings Injected with Plutonium, in *The Health Effects of Plutonium and Radium*, W. S. S. Jee, Ed., J. W. Press, University of Utah, Salt Lake City, 1976, p. 329.

65. G. L. Voelz, *Hlth. Phys.*, **29**, 551 (1975).

66. Actinides in Man and Animals, Proceedings of the Snowbird Actinide Workshop, October 15–17, 1979, M. E. Wrenn, Scientific Editor, RD Press, Salt Lake City, 1981.

67. Radiobiology of Radium and the Actinides in Man, J. Rundo, P. Failla, and R. A. Schlenker, Eds., *Hlth. Phys.*, **44** (Suppl. 1) (1983).

68. Biological Assessment of Occupational Exposure to Actinides, G. B. Gerber, H. Metivier, and J. Stather, Eds., *Radiat. Protect. Dosimetry*, **26**, 1–4 (1989).

69. J. Newell Stannard, Radioactivity and Health. A. History, R. W. Baalman, Jr., Ed., Department of Energy Publication, Report DOE/RL/01830-T59 (DE 88013791) October 1988.

70. J. H. Marshall, J. Rundo, and G. E. Harrison, *Radiat. Res.*, **39**, 445 (1969).

71. W. P. Norris, T. W. Speckman, and P. F. Gustafson, *Am. J. Roentgenol.*, **73**, 785 (1955).

72. C. E. Miller and A. J. Finkel, *Am. J. Roentgenol.*, **103**, p. 871 (1968).

73. Pacific Northwest Laboratory Annual Report for 1989 to the DOE Office of Energy Research, Part I, Biomedical Science, PNL-4100 PT 1, February 1990.

74. Inhalation Toxicology, Research Institute Annual Report, 1988–89, to the United States Department of Energy, LMF-128, August 1990.

75. R. G. Thomas and D. M. Smith, *Int. J. Canc.*, **24**, 594 (1979).

76. R. O. McClellan, S. A. Benjamin, B. B. Boecker, F. F. Hahn, C. H. Hobbs, R. K. Jones, and D. L. Lundgren, Influence of Variations in Dose and Dose Rates on Biological Effects of Inhaled Beta-Emitting Radionuclides, in *Biological and Environmental Effects of Low-Level Radiation*, Proceedings of the International Atomic Energy Agency Symposium, Chicago, November 3–7, 1975, Vol. 2, 1976, p. 3.

77. C. L. Sanders, G. E. Dagle, W. C. Cannon, D. K. Craig, G. J. Powers, and D. M. Meier, *Radiat. Res.*, **68**, 349 (1976).

78. F. F. Hahn, S. A. Benjamin, B. B. Boecker, T. L. Chiffelle, C. H. Hobbs, R. X. Jones, A. O. McClellan, and H. C. Redman, Induction of Pulmonary Neoplasia in Beagle Dogs by Inhaled ^{144}Ce Fused-Clay Particles, in *Biological and Environmental Effects of Low-Level Radiation*, Proceedings of the International Atomic Energy Agency Symposium, Chicago, November 3–7, 1975, Vol. 2, 1976, p. 201.

79. R. G. Thomas, G. A. Drake, J. E. London, E. C. Anderson, J. R. Prine, and D. M. Smith, *Int. J. Radiat. Biol.*, **46**, 605 (1981).

80. D. A. Holaday, Uranium Mining Hazards, in *Uranium–Plutonium–Transplutonic Elements*, H. C. Hodge, J. N Stannard, and J. B. Hursh, Eds., Springer-Verlag, New York, 1973, p. 296.

81. C. D. Stewart and S. D. Simpson, The Hazards of Inhaling Radon-222 and Its Short-lived Daughters: Consideration of Proposed Maximum Permissible Concentrations in Air, in *Radiological Health and Safety in Mining and Milling of Nuclear Materials*, IAEA Symposium Series, Vol. 1, 1964, p. 333.

82. D. A. Holaday, *Hlth. Phys.*, **16**, 547 (1969).

83. V. E. Archer, J. K. Wagoner, and F. E. Lundin, *Hlth. Phys.*, **25**, 351 (1973).

84. *Radiation Hazards in Mining: Control Measurement, and Medical Aspects, Proceedings*, Conference held at the Colorado School of Mining, Golden, Colorado, October 4–9, 1981, Kingsport Press, Kingsport, TN, 1981.

85. G. Saccomanno, V. E. Archer, R. P. Saunders, O. Auerbach, and M. G. Klein, Early Indices of Cancer Risk Among Uranium Miners with Reference to Modifying Factors, in *Occupational Carcinogenesis*, U. Saffiotti and J. K. Wagoner, Eds., Vol. 271, New York Academy of Sciences, New York, 1976, p. 377.

86. V. E. Archer, J. K. Wagoner, and F. E. Lundin, *J. Occup. Med.*, **15**, 204 (1973).

87. F. E. Lundin, J. W. Lloyd, E. M. Smith, V. E. Archer, and D. A. Holaday, *Hlth. Phys.*, **16**, 571 (1969).

88. V. E. Archer and F. E. Lundin, *Environ. Res.*, **1**, 370 (1967).

89. G. Saccomanno, V. E. Archer, O. Auerbach, M. Kuschner, M. Egger, S. Wood, and R. Mick, Age Factor in Histological Type of Lung Cancer Among Uranium Miners, A Preliminary Report, in Proceedings of International Conference on Radiation Hazards in Mining: Control, Measurement and Medical Aspects, M. Gomez, Ed., Kingsport Press, Inc., Kingsport, TN (1981), p. 675.

90. G. Saccomanno, *Lab. Med.*, **10(9)**, 523 (1979).

91. R. A. Lemen, W. M. Johnson, J. K. Wagoner, V. E. Archer, and G. Saccomanno, Cytologic Observations and Cancer Incidence Following Exposure to BCME, in *Occupational Carcinogenesis*, U. Saffiotti and J. K. Wagoner, Eds., Vol. 271, New York Academy of Sciences, New York, 1976, p. 71.

92. B. J. Stover and C. N. Stover, Jr., The Laboratory for Radiobiology at the University of Utah, in *Radiobiology of Plutonium*, J. W. Press, University of Utah, Salt Lake City, Utah, 1972, p. 29.

93. C. W. Mays, H. Spiess, G. N. Taylor, R. D. Lloyd, W. S. S. Jee, S. S. McFarland, D. H. Taysum, T. W. Brammer, D. Brammer, and T. A. Pollard, Estimated Risk to Human Bone from 23-Pu; in *The Health Effects of Plutonium and Radium*, W. S. S. Jee, Ed., J. W. Press, University of Utah, Salt Lake City, 1978, p. 343.

94. C. W. Mays and H. Spiess, Bone Sarcoma to Man from ^{224}Ra, ^{226}Ra, and ^{239}Pu, in *Proceedings of Symposium on Biological Effects of ^{224}Ra*, W. A. Müller and H. G. Ebert, Eds., Martinus Nijhoff Medical Division, The Hague, Boston, 1978, p. 168.

95. R. D. Evans, A. T. Keane, and M. M. Shanahan, Radiologic Effects in Man of Long-Term Skeletal Alpha-Irradiation, in *Radiobiology of Plutonium*, B. J. Stover and W. S. S. Jee, Eds., J. W. Press, Salt Lake City, 1972, pp. 431–468.

96. C. W. Mays and R. D. Lloyd, Bone Sarcoma Incidence vs Alpha Particle Dose, in *Radiobiology of Plutonium*, B. J. Stover and W. S. S. Jee, Eds., J. W. Press, Salt Lake City, 1972, pp. 409–430.

97. R. E. Rowland, A. F. Stehney, and H. F. Lucas, Jr., *Radiat. Res.*, **76**, 368 (1978).

98. R. G. Thomas and J. N. Stannard, *Radiat. Res. Suppl.*, **5**, 16 (1964).

99. W. J. Bair, J. E. Ballou, J. F. Park, and C. L. Sanders, Plutonium in Soft Tissues with Emphasis on the Respiratory Tract, in *Uranium–Plutonium–Transplutonic Elements*, H. C. Hodge, J. N. Stannard, and J. B. Hursh, Eds., Springer-Verlag, New York, 1973, p. 503.

100. L. J. Leach, E. A. Maynard, H. C. Hodge, J. K. Scott, C. L. Yuile, G. E. Sylvester, and H. B. Wilson, *Hlth. Phys.*, **18**, 599 (1970).

101. National Council on Radiation Protection and Measurements, Protection of the Thyroid Gland in the Event of Releases of Radioiodine, NCRP Report 55, 1977.

102. P. Hamilton and E. A. Tompkins, personal communication (Bureau of Radiological Health, Department of Health, Education and Welfare, Food and Drug Administration, Washington, and Oak Ridge Associated Universities, Oak Ridge, Tenn.), 1975.

103. U.S. Nuclear Regulatory Commission, Reactor Safety Study: An Assessment of Accident Risks in U.S. Commercial Nuclear Power Plants, Appendix VI, Calculations of Reactor Accident Consequences, Report WASH-1400, NUREG-75/014 (U.S. NRC, Washington, D.C.), 1975. [Available from National Technical Information Service, Department of Commerce, Springfield, VA.]

104. International Commission on Radiological Protection, Recommendations of the International Commission on Radiological Protection, ICRP Report 26, Annals of the ICRP 1/3 (1977).

105. International Commission on Radiological Protection, Limits for Intakes of Radionuclides by Workers, ICRP Report 30, Annals of the ICRP 2/3, 4 (1979).

106. Recommendations on Limits for Exposure to Ionizing Radiation, National Council on Radiation Protection and Measurements, NCRP Report No. 91 (1987).

107. Basic Radiation Protection Criteria, National Council on Radiation Protection and Measurements, NCRP Report No. 39, Washington, D.C. (1971).

108. A. Catsch, Radioactive Metal Mobilization in Medicine, Thomas, Springfield, IL, 1964.

109. *Guide Book for the Treatment of Accidental Internal Radionuclide Contamination of Workers*, a joint publication of the Commission of the European Communities and the U.S. Department of Energy, G. B. Gerber and R. G. Thomas, Eds., *Radiation Protection Dosimetry 41/1* (1992) (1991).

110. R. O. McClellan, H. A. Boyd, S. A. Benjamin, R. G. Cuddihy, F. F. Hahn, R. K. Jones, J. L. Mauderly, J. A. Mewhinney, B. A. Muggenburg, and R. C. Pfleger, *Hlth. Phys.*, **23**, 426 (1972).

111. A. Catsch and A. E. Harmuth-Hoene, *Biochem. Pharmacol.*, **24**, 1557 (1975).

112. V. Volf, Treatment of Incorporated Transuranium Elements, International Atomic Energy Agency Report 1984, Vienna, Austria (1978).

113. National Council on Radiation Protection, Management of Persons Accidentally Contaminated with Radionuclides, NCRP Report 65, Washington, D.C., 1980.

114. G. N. Stradling and R. A. Bulman, Recent Research on Decorporation Therapy at National Radiological Protection Board in *Actinides in Man and Animals*, M. E. Wrenn, Ed., RD Press, Salt Lake City, 1981, p. 369.

115. C. W. Mays, G. N. Taylor, R. D. Lloyd, and M. E. Wrenn, Status of Chelation Research: A Review, in Actinides in Man and Animals, M. E. Wrenn, Ed., RD Press, Salt Lake City, 1981, p. 351.

116. A. Soffer, *Chelation Therapy*, Thomas, Springfield, IL, 1964.

117. H. Spencer and B. Rosoff, *Hlth. Phys.*, **11**, 1181 (1965).

118. C. R. Lagerquist, S. E. Hammond, E. A. Putzier, and C. W. Piltingsrud, *Hlth. Phys.*, **11**, 1177 (1965).

119. C. R. Lagerquist, E. A. Putzier, and C. W. Piltingsrud, *Hlth. Phys.*, **13**, 965 (1967).

120. L. Jolly, H. A. McClearen, G. A. Poda, and W. P. Walke, *Hlth. Phys.*, **23**, 333 (1972).

121. F. Swanberg and R. C. Henle, *J. Occup. Med.*, **6**, p. 174 (1964).

122. W. D. Norwood, P. A. Fuqua, R. H. Wilson, and J. W. Healy, Treatment of Plutonium Inhalation: Case Studies, in *Experience in Radiological Protection*, Proceedings of the Second United Nations International Conference on the Peaceful Uses of Atomic Energy, Geneva, September 1–13, 1958, Vol. 23, 1958, p. 434.

123. J. A. Kylstra, D. C. Rausch, K. D. Hall, and A. Spock, *Am. Rev. Respir. Dis.*, **103**, 651 (1971).

124. J. A. Kylstra, W. H. Schoenfisch, J. M. Herron, and G. D. Blenkarn, *J. Appl. Physiol.*, **35**, 136 (1973).

125. K. E. McDonald, J. F. Park, G. E. Dagle, C. L. Sanders, and R. J. Olson, *Hlth. Phys.*, **29**, 804 (1975).

126. B. A. Muggenburg, J. A. Mewhinney, and R. A. Guilmette, Removal of Inhaled Plutonium and Americuim from Dogs Using Lung Lavage and DTPA, in *Actinides in Man and Animals*, M. E. Wrenn, Ed., RD Press, Salt Lake City, 1981, p. 387.

127. E. Carlens, *J. Thorac. Surg.*, **18**, 742 (1949).

128. B. B. Boecker, B. A. Muggenburg, R. O. McClellan, S. P. Clarkson, F. J. Mares, and S. A. Benjamin, *Hlth. Phys.*, **26**, 505 (1974).

129. S. A. Silbaugh, S. A. Felicetti, B. A. Muggenburg, and B. B. Boecker, *Hlth. Phys.*, **29**, 81 (1975).

130. B. A. Muggenburg, R. O. McClellan, B. B. Boecker, J. L. Mauderly, and F. F. Hahn, Long-Term Biologic Effects in Dogs Treated with Lung Lavage after Inhalation of [144]Ce in Fused Aluminosilicate Particles, in *Actinides in Man and Animals*, M. E. Wrenn, Ed., RD Press, Salt Lake City, 1981, p. 395.

131. M. Ingram, Lymphocytes with Bilobed Nuclei as Indicators of Radiation Exposures in the Tolerance Range, in *Legal, Administrative, Health and Safety Aspects of Large-Scale Use of Nuclear Energy*, Proceedings of the International Conference on the Peaceful Uses of Atomic Energy, Geneva, August 8–20, 1955, Vol. 13, 1956, p. 210.

132. M. Ingram, The Occurrence and Significance of Binucleate Lymphocytes in Peripheral

Blood after Small Radiation Exposures, in *Immediate and Low Level Effects of Ionizing Radiations*, A. A. Buzzati-Traverso, Ed., Proceedings of Symposium June 22–26, 1956, UNESCO, IAEA, and CNRN, Venice, 1960, p. 233.

133. R. Lowry Dobson, Binucleated Lymphocytes and Low-Level Radiation Exposure, in *Immediate and Low Level Effects of Ionizing Radiations*, A. A. Buzzati Traverso, Ed., Proceedings of Symposium June 22–26, 1956, UNESCO, IAEA, and CNRN, Venice, 1960, p. 247.

134. R. G. Langlois, W. L. Bigbee, R. H. Jensen, *Hum. Genet.*, **74**, 353 (1986).

135. T. Straume, R. G. Langlois, J. Lucas, R. H. Jensen, and W. L. Bigbee, *Hlth. Phys.*, **60**(1), 71 (1990).

136. R. G. Langlois, W. L. Bigbee, S. Kyoizumi, N. Nakamura, M. A. Bean, M. Akiyama, and R. H. Jensen, *Science*, **236**, 445 (1987).

137. Assessment of Techniques to Determine Previous Radiation Exposures (Emphasizing Exposures Occurring Decades Earlier), G. R. Eisele, Special Edition Coordinator, *Hlth. Phys.*, **60** (Suppl. 1) (1991).

138. H. G. Jones and B. E. Lambert, The Radiation Hazard to Workers Using Tritiated Luminous Compounds, in *Assessment of Radioactivity in Man*, Proceedings of the Symposium by International Atomic Energy Agency, the International Labour Organization, and the World Health Organization, Heidelberg, May 11–16, 1964, Vol. 2, 1964, p. 419.

139. F. E. Butler, Assessment of Tritium in Production Workers, in *Assessment of Radioactivity in Man*, Proceedings of the Symposium by the International Atomic Energy Agency, the International Labour Organization, and the World Health Organization, Heidelberg, May 11–16, 1964, Vol. 2, 1964, p. 431.

140. A. A. Moghissi, M. W. Carter, and E. W. Bretthauer, *Hlth. Phys.*, **23**, 805 (1972).

141. K. L. Swinth, J. F. Park, and P. J. Moldofsky, *Hlth. Phys.*, **22**, 899 (1972).

142. International Atomic Energy Agency, *Assessment of Radioactive Contamination in Man*, Proceedings of a Symposium by the IAEA and the World Health Organization, Stockholm, November 22–26, 1971, IAEA, Vienna (1972).

143. G. R. Meneely and S. M. Linde, Eds., *Radioactivity in Man*, Proceedings of the Symposium on Whole-Body Counting, International Atomic Energy Agency, Vienna, June 12–16, 1961, IAEA, Vienna (1962).

144. International Atomic Energy Agency, *Directory of Whole-Body Radioactivity Monitors*, IAEA, Vienna, 1970.

145. P. H. Abelson, *Science*, **193**, 4250, 279 (1976).

146. W. D. Metz, *Science*, **192**, 4246, 1320 (1976).

147. W. D. Metz, *Science*, **193**, 4247, 38 (1976).

148. W. D. Metz, *Science*, **193**, 4250, 307 (1976).

149. J. Walsh, *Science*, **193**, 4250, 305 (1976).

150. D. M. Smith, Respiratory-Tract Carcinogenesis Induced by Radionuclides in the Syrian Hamster, in *Proceedings of Symposium on Pulmonary Toxicology of Respirable Particles*, C. L. Sanders, F. T. Cross, G. E. Dagle, and J. A. Mahaffey, Eds., CONF-791002, Technical Information Center, U.S. Department of Energy, 1980, p. 575.

151. E. C. Anderson, L. M. Holland, J. R. Prine, and C. R. Richmond, Lung Irradiation with Static Plutonium Microspheres, in *Experimental Lung Cancer: Carcinogenesis and Bioassays*, E. Harbe and J. F. Park, Eds., Springer-Verlag, New York, 1974, p. 432.

152. L. M. Holland, J. R. Prine, D. M. Smith, and E. C. Anderson, Irradiation of the Lung with Static Plutonium Microemboli, in *The Health Effects of Plutonium and Radium*, W. S. S. Jee, Ed., J. W. Press, University of Utah, Salt Lake City, 1976, p. 127.

153. Chernobyl Doses across the Continent, *Nuclear News*, January 1987, p. 60.

154. *Health and Environmental Consequences of the Chernobyl Nuclear Power Plant Accident*, M. Goldman, R. Catlin, and L. Anspaugh, Eds., U.S. Department of Energy, Report DOE/ER-0332, June 1987.

155. D. A. Holaday, Uranium Mining Hazards, in *Uranium–Plutonium–Transplutonic Elements*, H. C. Hodge, J. N. Stannard, and J. B. Hursh, Eds., Springer-Verlag, New York, 1973, p. 300.

156. R. L. Gross, D. M. Smith, R. G. Thomas, G. Saccomanno, and R. Saunders, Immunological Assessment of Patients with Pulmonary Metaplasia and Neoplasia, in *Proceedings of Conference on Safe Handling of Chemical Carcinogens, Mutagens, Teratogens and Highly Toxic Substances*, Vol. 1, D. B. Walters, Ed., Ann Arbor Science Publishers, Ann Arbor, MI, 1980, p. 259.

157. G. Saccomanno, The Contribution of Uranium Miners to Lung Cancer Histogenesis, *Canc. Res.*, **82**, 43 (1982).

CHAPTER FOURTEEN

Nonionizing Electromagnetic Energies

William E. Murray, M.S., R. Timothy Hitchcock, M.S.P.H., CIH, Robert M. Patterson, Sc.D., CIH, and Sol M. Michaelson, D.V.M.*

1 INTRODUCTION

During the last 75 years, there has been marked development and increased utilization of equipment and devices for medical, industrial, telecommunications, consumer use, and military applications that emit one or more types of nonionizing radiant energy. These include ultraviolet (UV), visible (light), infrared (IR), microwave, radiofrequency (RF), and extremely low frequency (ELF) radiations and fields, all of which are components of the spectrum of electromagnetic waves.

Concomitant with this increased usage, there is growing concern in the military and civilian government agencies, industry, and professional societies as well as among the public regarding health hazards associated with the development, manufacture, and operation of devices that emit nonionizing radiant energies. To address these concerns, private organizations and government agencies have developed exposure guidance or standards to protect workers and the public against such hazards. For some nonionizing radiation sources, product standards have been established to limit the accessible radiation levels from these sources. The Radiation Control for Health and Safety Act of 1968 was enacted in the United States (Public Law 90-602) and gives the U.S. Food and Drug Administration (USFDA) the authority to develop

*Deceased.

Patty's Industrial Hygiene and Toxicology, Third Edition, Volume 3, Part B, Edited by Lewis J. Cralley, Lester V. Cralley, and James S. Bus.
ISBN 0-471-53065-4 © 1995 John Wiley & Sons, Inc.

standards for electronic products that emit particulate or electromagnetic ionizing radiation, electromagnetic nonionizing radiation, or infrasonic, sonic, or ultrasonic waves.

2 BIOPHYSICS

Electromagnetic radiation (EMR) can be defined as energy in motion that consists of oscillating electric (E) and magnetic (H) fields. The energies of the electromagnetic spectrum are propagated in the form of waves that act as small bundles of energy with many of the properties ordinarily ascribed to particles. These are called photons or quanta and have four important characteristics:

1. *Wavelength (λ).* The distance between two similar points on the wave is called the wavelength. [Unit: meter (m), km (10^3 m), mm (10^{-3} m), μm (10^{-6} m), nm (10^{-9} m).]

2. *Frequency (ν).* The frequency describes the number of wavelengths passing a given point per unit time. [Unit: hertz (Hz), kHz (10^3 Hz), MHz (10^6 Hz), GHz (10^9 Hz), THz (10^{12} Hz).]

3. *Velocity (c).* All EMR travels at the same speed. In a vacuum or air, the velocity is approximately 3×10^8 m/s. These three characteristics are related by the following equation:

$$c = 3 \times 10^8 \text{ m/s} = (\lambda)(\nu) \tag{1}$$

4. *Energy (Q).* The energy of a photon can be calculated as follows:

$$Q = h\nu = hc/\lambda \tag{2}$$

where h is Planck's constant ($=6.63 \times 10^{-34}$ joule second [J s])

Thus the photon energy (Q) is directly proportional to the frequency of the electromagnetic radiation and inversely proportional to the wavelength. Longer wavelength radiations have lower photon energy while shorter wavelength radiations have higher photon energy. Using the photon wavelength, the energy can be calculated in electron volts (eV) from the following equation:

$$Q = 1.26/\lambda \ (\mu\text{m}) \qquad \text{eV} \tag{3}$$
$$1 \text{ eV} = 1.6 \times 10^{-19} \text{ J}$$

To provide some perspective on the order of magnitude of the electron volt, it may be noted that 1 MeV (1,000,000 eV) is equivalent to the energy required to lift a 1-nanogram (ng) weight to a height of about 16 mm or $\frac{5}{8}$ of an inch. The thermal energy of motion of molecules at room temperature is about $\frac{1}{30}$ eV; the energy of bonds holding atoms together in molecules of chemical compounds ranges from fractions of 1 to approximately 4 eV, whereas the binding energy holding protons and neutrons together in the nucleus is in the millions of electron volts.

There are several types of EMR and they are commonly arranged in a spectrum by frequency, wavelength, and energy (see Table 14.1). As the frequency decreases,

Table 14.1 Energy, Wavelength and Frequency Ranges for Electromagnetic Radiation and Fields

Radiation Type	Energy per Photon	Wavelength Range	Frequency Range
Ionizing	>12.40 eV	<100 nm	$>3.00 \times 10^3$ THz
Ultraviolet (UV)	12.40–3.10 eV	100–400 nm	3.00–0.75×10^3 THz
Visible	3.10–1.63 eV	400–760 nm	7.50–3.95×10^2 THz
Infrared (IR)	1.63 eV–1.24 meV	760 nm–1 mm	395–0.30 THz
Microwave (MW)	1.24 meV–1.24 μeV	1 mm–1 m	300 GHz–300 MHz
Radio-frequency (RF)	1.24 μeV–1.24 peV	1 m–1 Mm	300 MHz–300 Hz
Extremely low frequency (ELF)	<1.24 peV	>1 Mm	<300 Hz

the energy of the emitted photons is insufficient, under normal circumstances, to dislodge orbital electrons and produce ion pairs. The minimum photon energies capable of producing ionization in water and atomic oxygen, hydrogen, nitrogen, and carbon are between 12 and 15 eV. Since these atoms constitute the basic elements of living tissue, about 12 eV may be considered the lower limit for ionization in biological systems. Thus 12.4 eV (corresponding to a wavelength of 100 nm in the UV region) is considered the dividing point between ionizing and nonionizing radiation (Faber, 1989). Although weak hydrogen bonds in macromolecules may involve ionization levels less than 12 eV, energies below this value can, on a biological basis, be considered as nonionizing.

Radiant energy can produce an effect only when it is absorbed by the body directly or as a surface stimulation that elicits a response. Interaction with matter follows the physical laws of nature. The primary modes of action of these radiations are either photochemical or thermal. According to the basic law of Grotthus and Draper, no photochemical reaction can occur unless radiant energy is absorbed. Such absorption requires that the energy of the photons be transferred to the absorbing molecules (Grossweiner and Smith, 1989).

The nonionizing radiant energies absorbed into the molecule either affect the electronic energy levels of its atoms or change the rotational, vibrational, and transitional energies of the molecules. In biological systems, energy transfer produces electron excitation, which can result in dissociation of the molecule if the bonding electrons are involved, dissipation of the excitation energy in the form of fluorescence or phosphorescence, the formation of free radicals, and degradation into heat. In the latter situation, absorption changes the vibrational or rotational energy and increases the kinetic energy of the molecule (Matelsky, 1968). The effects of interaction between radiant energy and biological systems are dependent on the energy of the radiated photons, the degree to which these photons are capable of penetrating into the system, and the ability of specific molecules to undergo chemical changes when these energies are absorbed.

It is also relevant to consider the time rate at which the energy is received (i.e., the power deposited). Even though high-power levels can be reached at longer wavelengths, ionization does not occur, but heating is possible. Thermal effects are produced upon absorption of energy in portions of the UV, visible, IR, microwave, and RF regions of the electromagnetic spectrum.

The relative effectiveness of different wavelengths in eliciting a specific photochemical response constitutes the *action spectrum* for that response. Photochemical reactions occur primarily upon absorption of radiant energy by specific target chromophores in the UV and visible portions of the spectrum.

3 CRITERIA FOR SETTING STANDARDS

Many of the electromagnetic energies at certain frequencies, power levels, and exposure durations can produce biological effects or injury depending on multiple physical and biological variables. Although systems or devices utilizing or emitting electromagnetic waves benefit mankind, the energies constitute acute and chronic exposure hazards to persons exposed to excessive levels of these energies.

Thus there is a need to set limits on the amount of exposure to nonionizing radiant energies individuals can accept with safety. Setting protection standards is a complicated process. The objectives of protection are to prevent acute effects and to limit the risks of late effects. Prevention of acute effects can be achieved quite readily when necessary. The second objective, to limit the risk of possible late effects, becomes more difficult.

It is important to consider some of the philosophic as well as practical aspects of standard development and promulgation. Protection standards should be based on scientific evidence but quite often are the result of empirical approaches to various problems reflecting the current status of our qualitative and quantitative knowledge. A numerical value for a standard implies a knowledge of the effect produced at a given level of stress and that both *effect* and *stress* are measurable. One problem is the definition of what an effect is and whether it can ultimately be shown to modify the life-style of the exposed individual or that individual's offspring (Taylor, 1970).

As in all biological processes, there is a certain range of levels between those that produce no observable effects and those that produce detectable effects. Since a detectable effect is not necessarily one that is irreparable or even a sign that the threshold for damage has been reached, the setting of permissible or allowable levels of exposure requires considerable caution and circumspection.

It must also be appreciated that in considering standards for different population groups, one has to use a certain amount of inference, calculation, and judgment. Before a precise evaluation of the risks can be made, the dose–effect relationship for such exposure must be considered, if it is known. It is important to know the rate and extent of recovery. It is also essential to know whether one can extrapolate from in vitro to in vivo studies or from lower animals to humans. Exposure risks must be assessed for appropriate standards to be set so that potential risk does not exceed a level judged to be acceptable.

There is no question that standards should be promulgated when need is demonstrated. They should be developed, however, by those with knowledge of the concepts of radiant energy interaction with the body as a whole or specific part (critical organs) of the body. In addition, a clear differentiation must be made between biologic effects per se that do not result in short-term or latent functional impairment against which the body cannot maintain homeostasis and effectiveness, and injury that may impair normal body activity by somatic change or induce genetic damage.

4 OPTICAL RADIATION

4.1 Background

4.1.1 Optical Spectrum

Ultraviolet (UV), visible, and infrared (IR) radiation compose the optical radiation region of the electromagnetic spectrum. The wavelength ranges from 100 nm in the UV to 1 mm in the IR, corresponding to energies of 12.4 eV to 1.24 meV, respectively. The 100-nm wavelength is generally considered to be the boundary between ionizing and nonionizing radiant energy.

Several schemes are used to divide the optical wavelengths into spectral bands;

but that of greatest importance to the photobiologist and health professional is that adopted by the Committee on Photobiology of the International Commission on Illumination [Commission International de l'Eclairage] (CIE, 1970). The UV region includes the wavelength range from 100 to 400 nm; it is further subdivided into the UV-A (near UV or black light) from 315 to 400 nm, UV-B (mid-UV) from 280 to 315 nm, and UV-C (far UV) from 100 to 280 nm. Wavelengths in the 100- to 180-nm range are known as the vacuum UV since they are readily absorbed in air and present no hazard. The UV-B and C together often are referred to as actinic UV. The visible region includes wavelengths from 400 to 760 nm. The IR region ranges from 760 nm to 1 mm. This region is subdivided into IR-A (near IR) from 760 to 1400 nm, IR-B (mid-IR) from 1.4 to 3.0 μm, and the IR-C (far IR) from 3.0 μm to 1 mm. Although sharp demarcation points have been given to the regions, there is no physical basis for these transition points. Rather they serve only as a convenient framework for discussing the biological effects and exposure hazards.

4.1.2 Interactions with Matter

When optical radiation impinges on matter, the possible modes of interaction include transmission, absorption, reflection, refraction, diffraction, and scattering. When striking an interface, optical radiation can be reflected. There are two types of reflection: specular (mirrorlike) and diffuse. The type of reflection depends on the relative smoothness of the surface struck by the incident radiation. If the surface irregularities are smaller than the wavelength of the incident radiation, the reflection will be specular. However, when the irregularities are much greater than the wavelength and oriented in a random fashion, diffuse reflection results and the incident radiation is reflected away for the interface in a random but predictable way.

Refraction or the bending of an optical radiation beam also occurs at an interface between two transmitting media that have different indexes of refraction. The degree of bending depends on the magnitude of the difference in the refractive index between the two materials. For a given material, the index also varies with the wavelength of the incident radiation. Thus, a prism can be used to separate a beam of white light into its component colors.

When interacting with a medium, a radiation beam can be reflected, absorbed, or transmitted. The fraction of the incident beam that is transmitted depends on the spectral content of the beam and the optical properties of the medium. The energy transmitted is governed by the following equation:

$$Q_\tau = Q_I - Q_A - Q_R \tag{4}$$

where Q_τ = energy transmitted
Q_I = energy incident on surface
Q_A = energy absorbed
Q_R = energy reflected

Other possible interaction modes that will only be mentioned here are interference, diffraction and (Rayleigh and Mie) scattering.

4.1.3 Quantities and Units

Unfortunately, a congruent system of quantities and units for use in describing or measuring electromagnetic energy across the whole spectrum does not exist although one has been proposed (Sliney and Wolbarsht, 1980). In the optical region, two systems of quantities and units are employed; radiometric and photometric. The radiometric is a physical system that can be used to describe the emission from, or exposure to, any source of optical radiation. The photometric system, however, is based on the response of the human eye to optical radiation and is applicable only in the visible portion of the spectrum. The relative luminous efficiency of the eye for photopic (day) vision peaks at about 555 nm and falls off rapidly to less than 10 percent of the maximum in the blue and orange wavelengths. Because of this variation in the luminous efficiency as a function of wavelength, any correspondence between the radiometric and photometric output of a visible radiation source is not predictable.

Some of the more commonly used quantities and units are shown in Table 14.2. Similar terms are shown in opposing columns, that is, irradiance vs. illuminance, radiance vs. luminance, and so forth. All of the quantities and units shown in Table 14.2 conform to the International System of Units [Système International (SI) d'Unités] (IRPA, 1991). Irradiance and radiant exposure are specified frequently in units of mW/cm^2 and mJ/cm^2, respectively ($1 \ W/m^2 = 0.1 \ mW/cm^2$; $1 \ J/m^2 = 0.1 \ mJ/cm^2$).

The term luminance, L, describes the brightness, or luminous intensity of an object per unit area of surface. There are several different measures of luminance; the basic unit is the candela per square meter (cd/m^2). In the English system, the unit is the footlambert (ftL). The conversion is $1 \ ftL = 3.426 \ cd/m^2$. Illuminance, E, is the luminous flux density impinging upon a surface. The common unit is the lumen per square meter (lm/m^2) or lux (lx). The English unit is the footcandle (ftcd), which is 10.76 lx. The candela is the unit of luminous intensity, I. The candela is equal to one lumen per unit solid angle [steradian (sr)]. One lumen is 0.00147 W at 555 nm. The relationship between the SI and English units is summarized below:

Illuminance: lm/m^2; lm/cm^2; lux (lx); footcandle (ftcd)

$$1 \ lm/m^2 = 10^{-4} \ lm/cm^2 = 1 \ lx = 0.929 \ ftcd$$

Luminance: cd/m^2; cd/m^2; footlambert (ftL)

$$1 \ cd/m^2 = 10^{-4} \ cd/cm^2 = 0.2919 \ ftL$$

4.2 Sources of Optical Radiation

Optical radiation in the wavelength range from 100 nm to 1 mm has energy of 12.4 eV to 1.24 meV, respectively. Photons having energies in the 1–100 eV range are usually associated with electron transitions in atoms and molecules whereas lower energies (1–100 meV) result from rotational and vibrational transitions in molecules.

Many materials at color temperatures above about 2500 kelvin (K), or otherwise excited to corresponding energies, may emit UV radiation as their excited atomic orbital electrons lose energy and return to ground state.

Table 14.2 Useful Radiometric and Photometric Quantities and Units[a,b]

Radiometric			Photometric		
Quantity	Symbol	SI Unit (Abbreviation)	Quantity	Symbol	SI Unit (Abbreviation)
Radiant energy	Q_e	joule (J)	Quantity of light	Q_v	lumen-second (lm s)
Radiant power or flux	Φ_e or P	watt (W)	Luminous flux	Φ_v	lumen (lm)
Irradiance (dose rate in photobiology)	E_e	watt per square meter (W/m²)	Illuminance (luminous density)	E_v	lumen per square meter (lm/m²)
Radiant intensity	I_e	watt per steradian (W/sr)	Luminous intensity (candlepower)	I_v	lumen per steradian (lm/sr) or candela (cd)
Radiance	L_e	watt per steradian and per square meter (W/sr m²)	Luminance	L_v	candela per square meter (cd/m²)
Radiant exposure (dose in photobiology)	H_e	joule per square meter (J/m²)	Light exposure	H_v	lux-second (lx s)
Optical density	D_e	unitless	Optical density	D_v	Unitless

[a] The quantities may be altered to refer to narrow *spectral* bands, in which case the term is preceded by the word *spectral*, and the unit is then per unit of wavelength and the symbol has a subscript λ. For example, spectral irradiance H_λ has units of W/(m² μm) or more often, W/(cm² nm).

[b] While the meter is the preferred unit of length, the centimeter is still the most commonly used unit of length for many of the above terms and the nm or μm are most commonly used to express wavelength.

Source: Adapted from Sliney and Wolbarsht (1980) and IRPA (1991).

At temperatures greater than 773 K, many solids and liquids emit radiation at wavelengths in the visible region. This thermoradiant emission is caused by translation of kinetic energy of the molecules of the substance to electron vibrational or rotational motion. Light is produced also by electric discharges, such as the spark discharges in gases under reduced pressure or by the bombardment of solids, liquids, or gases with ionizing radiation; by exothermic chemical reactions, such as the oxidation of various organic compounds (chemiluminescence); and, by photoluminescence when a dye or pigment, such as chlorophyll, absorbs UV radiation and emits light at longer wavelengths.

Infrared radiation is produced by the rotational and vibrational movements within molecules. All matter above absolute zero (0 K) generates kinetic energy and emits IR. All hot bodies emit IR and radiate IR to other objects with lower surface temperatures.

4.2.1 Characteristics of Optical Radiation Sources

Radiation sources are often classified as *point* or *extended sources*. Theoretically, no source is a point source because it has a finite size. Distant stars are treated as point sources, although the sun is certainly an extended source. The laser is the only man-made device that is treated as a point source. All other sources are extended sources.

The intensity of the radiation emitted by a point source varies with the distance from the source according to the inverse square law. The ratio of the irradiance (E_1) at distance r_1 to that (E_2) at distance r_2 is calculated as follows:

$$E_1/E_2 = (r_2)^2/(r_1)^2 \qquad (5)$$

This equation is a good approximation for extended sources at distances > 10 times the maximum dimension of the source. For lasers or other collimated sources, this law can be used only at very large distances from the source. The region in which the inverse square applies is called the *far field*. It does not hold in the *near-field* region at distances close to the source.

Another important property of optical sources is coherence. To be coherent, the emission must be at a single wavelength (temporal coherence) and be spatially in phase (spatial coherence). The laser is the only source commonly encountered that is classified as a coherent radiation source; all other sources emit incoherent radiation.

4.2.2 Sources of Exposure

There are a number of ways of classifying optical radiation sources. A simple scheme for categorizing these sources is the following (EHC, 1982):

(1) sunlight,
(2) lamps,
(3) incandescent (warm-body) sources, and
(4) lasers.

The sun provides the greatest component of human exposure to optical radiation for both the work force and the public; as much as one-third of the work force is exposed occupationally to sunlight. Some solar exposure is unavoidable in the course of one's lifetime. However, the amount of exposure varies tremendously with climate, geography, altitude, and occupational and recreational activities. The solar irradiance outside the atmosphere is about 135 mW/cm^2; atmospheric absorption reduces this level at sea level to a maximum of about 100 mW/cm^2 (Sliney and Wolbarsht, 1980). The solar spectrum peaks at about 450 nm; it drops off rapidly in the UV and more gradually in the IR region. Near-IR wavelengths are absorbed by CO_2, O_2, and H_2O molecules in the atmosphere. Virtually no UV below 290 nm reaches the earth's surface due to scattering in the ozone layer.

Many diverse lamps and lights are employed in industry (Sliney and Wolbarsht, 1980; Hitchcock, 1991). This classification includes the following types of sources: incandescent lamps, low-pressure discharge lamps, fluorescent lamps, high-intensity discharge (HID) lamps, short- (compact-) arc lamps, carbon arcs, cathode ray tubes (CRT), solid-state lamps, radioactive phosphor, gas lamps, and pulsed flash lamps. Several examples and the major emissions are shown in Table 14.3 (EHC, 1982).

Incandescent or warm-body sources are similar to the black-body radiator. The intensity and spectral distribution of the optical radiation emitted by such sources

Table 14.3 Common Exposures to Optical Radiation Sources

Sources	Wavelength Regions of Concern	Potential Exposures
Sunlight	Ultraviolet (UV), visible, near-infrared (IR)	Farming, construction, landscaping, other outdoor workers
Arc lamps	UV, visible, near-IR	Photoreproduction, optical laboratories, entertainment
Germicidal	Actinic, far UV	Hospitals, laboratories, medical clinics, maintenance
Hg-HID lamps	UV-A and blue light actinic UV (if broken envelope)	Maintenance, factories, warehouses, gymnasiums
Carbon arcs	UV, blue light	Laboratories, searchlight operations
Industrial infrared sources	IR	Steel mills, foundries, glassmaking, drying equipment
Metal halide UV-A lamps	Near UV, visible	Printing plant, maintenance, integrated circuit manufacturing
Sunlamps	UV, blue light	Tanning parlors, beauty salons, fitness parlors
Welding arcs	UV, blue light	Welders' helpers, welders, ancillary workers, bypassers

Source: Adapted from EHC (1982).

are a function of its absolute temperature. Although less intense than the theoretical black body, the radiance of sources, such as heated filaments, wires, or other objects, is proportional to the fourth power of its temperature (in K), and the wavelength of maximum emission is inversely proportional to that temperature. Thus, as its temperature increases, the peak emission occurs at a shorter wavelength. The spectrum of the sun outside the atmosphere is similar to that of a black-body radiator at a temperature of about 6000 K.

The acronym laser is derived from the process by which the coherent radiation is generated, that is, *l*ight *a*mplification by the *s*timulated *e*mission of *r*adiation. These devices emit coherent radiation with an output wavelength in the optical radiation region. Various materials have been found that lase at distinct wavelengths or are tunable over a specific wavelength range. The laser output beam ranges from single pulses shorter than 1 picosecond (ps) to continuous wave (CW). Generally speaking, four modes of pulsed operation are encountered with laser systems: (1) ultrashort (duration, 0.01–15 ps); (2) mode-locked (pulse envelope of ultrashort pulses; duration, 10–100 ns); (3) Q-switched (duration, 1–100 ns), (4) normal multiple-spike (pulse duration, 100 μs to 30 ms). In the CW mode, the exposure duration may last from a few milliseconds to seconds, minutes, or up to a whole day in some instances.

The extremely collimated character and monochromatic nature of the laser beam make this energy of great utility in industrial, scientific, medical, military, and communications applications. Examples of laser applications and specific lasers used in these applications are shown in Table 14.4.

Table 14.4 Common Laser Devices and Applications

Type	Wavelength(s)	Applications
Argon (Ar)	458–515 nm 350 nm	Instrumentation, holography, retinal photocoagulation, entertainment
Carbon dioxide (CO_2)	10.6 μm	Material processing, optical radar/ranging, surgery, instrumentation
Dye(s)	Variable 350 nm to 1 μm	Instrumentation
Excimer lasers	180–250 nm	Laser pumping, spectroscopy
Gallium arsenide (GaAs)	850–950 nm	Instrumentation, ranging, intrusion detection, communications, toys
Helium cadmium (HeCd)	325, 442 nm	Alignment, surveying
Helium neon (HeNe)	632.8 nm	Alignment, surveying, holography, ranging, intrusion detection, communications, entertainment
Neodymium glass (Nd-glass); neodymium yttrium-aluminum garnet (Nd-YAG)	1.06 μm	Material processing, instrumentation, optical radar/ranging, surgery, research
Ruby	694.3 nm	Material processing, holography, ranging, research

Source: Adapted from EHC (1982).

Lasers do not constitute an environmental hazard to the general public in the sense that air pollution, noise, radioactive fallout, and other contaminants do, except under rather special circumstances, such as surveying or range finding at a commercial airport, satellite tracking, air turbulence and pollution studies, laser illumination at art exhibits, and public displays or laser light shows.

4.3 Biological Effects

4.3.1 Introduction

Photobiology is the science that investigates the effects of optical radiation on living organisms. The study of photochemical reactions is important to our understanding of these effects because the optical radiation must be absorbed to initiate the chain of events leading to a biological effect (Grossweiner and Smith, 1989). If a molecule does not absorb the optical energy, then it cannot lead directly to an effect, although it may mediate the outcome. Since a given molecule absorbs only certain wavelengths, it exhibits a characteristic absorption spectrum. Knowledge of these absorption spectra is important in order to identify the molecules suspected to be associated with a certain biological effect. As mentioned previously, the energy absorbed by a molecule can be dissipated in heat or vibrational energy, charge transfer, free radical formation, fluorescence, or phosphorescence. Charge transfer and free radical formation can result in biochemical events in living organisms.

In the intact animal, incident optical radiation does not penetrate an appreciable distance into the body. Thus, the organs at risk are limited to the skin and the eye. The specific tissue at risk, the effects observed, and the threshold for injury vary considerably with the wavelengths to which the organism is exposed.

4.3.2 Action Spectra

The concept of the action spectrum was originated by nineteenth-century biologists after noting that the rate of plant growth varied with the wavelength of the ambient environment (Grossweiner, 1989). The action spectrum for a given biological effect can be derived by determining the threshold dose for that effect as a function of wavelength. A plot of the reciprocal of the threshold dose versus the incident wavelength yields that action spectrum. The relative spectral effectiveness is obtained by normalizing the curve to the wavelength of maximum sensitivity. Under certain conditions, the action spectrum will parallel closely the absorption spectrum of the chromophore thought to be associated with that effect (Faber, 1989). Examples of various normal and abnormal reactions to optical radiation for which action spectra have been established in humans include erythema and sunburn, melanogenesis, vitamin D production, solar urticaria, hyperbilirubinemia of prematurity, polymorphic photodermatitis, xeroderma pigmentosa, porphyria photosensitivity, lupus erythematosus (LE) and discoid LE, and photosentivity reactions to some drugs (Moss et al., 1977). Many of these reactions are induced by UV energy but some extend into the visible region.

4.3.3 Beneficial and Therapeutic Effects of Optical Radiation

Although most of the effects of exposure are harmful, especially at high intensities, some effects are beneficial or therapeutic. Vitamin D_3 is synthesized from 7-dehydrocholesterol in normal human skin by the UV-B in sunlight (285–310 nm). Exposure of the face, hands, and arms for short durations produces this vitamin in amounts similar to an oral dose of 200 international units (IU), or 5 μg, which is one-half of the recommended daily allowance (RDA) for vitamin D. Repetitive exposure of the whole body is equivalent to a chronic, daily ingestion of 10^4 IU, or 250 μg. Inadequate stores of vitamin D can result in reduced calcium absorption, decreased bone mineralization, fractures, and muscle weakness (Haddad, 1992).

Ultraviolet radiation (UVR) was used for therapeutic purposes first in the treatment of cutaneous tuberculosis in the early 1900s. Since then its therapeutic qualities have been directed at several conditions including vitiligo, psoriasis, mycosis fungoides, acne, eczema, and pityriasis rosea. In some treatments, photosensitizing agents enhance the effect of UVR, for example, 4,5′,8-trimethylpsoralen (TMP) for vitiligo, or produce an effect in combination with UVR when neither agent is effective separately, for example, 8-methoxypsoralen (8-MOP) with UV-A in treating psoriasis (Wulf and Urbach, 1988; Epstein, 1990). There have been suggestions that UV exposure may have beneficial effects on certain types of cancer including breast and colon cancer (Ainsleigh, 1993). Red laser light at 630 nm is used in photochemotherapy for certain tumors with hematoporphyrin. Blue light (450–500 nm) is employed to treat the hyperbilirubinemia of neonatal jaundice (Epstein, 1986).

4.4 Skin Effects

The skin is the largest of the body's organs. As the outer covering, the skin protects the body against optical radiation, heat, injury, infection, and many chemicals. Additionally, it regulates body temperature, stores water and fat, maintains the electrolyte balance, and assists in making vitamin D. As an environmental sensor, information about temperature, pressure, pain, vibration, and touch is transmitted to the brain.

The skin has two primary layers: the outer epidermis and the inner dermis, separated by the basal membrane. Below the dermis is the subcutis, composed of subcutaneous fat and other biological structures. The stratum corneum is the surface layer of the epidermis and consists mostly of flat, scalelike cells. The growth of new skin cells originates deep in the basal layer of the epidermis where cell division occurs in the germinative cells. The melanocytes, also located in this layer, manufacture the pigment melanin. The pigment granules (melanosomes) are passed upward through dendritic processes to keratinocytes where the pigment is distributed through the epidermis to the stratum corneum.

The thickness of the epidermis is fairly uniform throughout the body, about 50–150 μm; the stratum corneum is usually 10–20 μm thick, except on the palms of the hands and the soles of the feet where it is 500–600 μm thick.

The dermis or corium, is composed primarily of connective tissue, blood and lymph vessels, sweat and sebaceous glands, and hair follicles. The sweat is secreted via pores to the skin surface and helps to regulate body temperature through evaporative cooling. The connective tissue provides the skin with supportive strength and elasticity.

4.4.1 Optical Characteristics of the Skin

Actinic UVR (UV-B and C) is absorbed strongly by the skin. Wavelengths above 1.5 μm are completely absorbed in the surface layers of the skin because of the almost complete absorption of these radiations by water, and any heat produced is quickly dissipated. Less than 5 percent of the incident radiation is reflected at these wavelengths. Skin reflectance in the range 0.3–3 μm is highly variable with wavelength and skin pigmentation, with white skin having much greater reflectivity than black skin. Fair-skinned Caucasians reflect almost two-thirds of the incident light at 0.7 μm, about 50 percent more than moderately-pigmented Caucasians and twice that of black skin. Peak reflectance for black skin is about 45 percent in the near-IR region (Jacquez et al., 1955). The spectral reflectance curves are quite similar to the spectral irradiance of sunlight, thus reducing the thermal load from the sun.

The transmittance of human skin also is wavelength dependent. Less than one-fourth of the incident radiation in the UV-C range penetrates the stratum corneum. In the UV-B and A regions the photons reach the Malphigian cells. As the wavelength increases toward the visible region, the transmission increases from 10–34 percent, with the higher energy photons approaching the subcutaneous layer (Bruls et al., 1984). The skin transmittance continues to increase through the visible into the IR-A. Hardy et al. (1956) found that the maximum transmission occurred at 1.2 μm. The small (1 μm) melanin granules minimize the transmission of UVR by filtering the radiation through absorption and scattering. The granules scatter IR-A photons, and this partly explains the increased penetration of these wavelengths into the dermis and subcutis (EHC, 1982).

4.4.2 UVR Effects

4.4.2.1 UVR Mechanisms. The ultraviolet spectrum of concern includes the wavelengths from 180 to 400 nm. Wavelengths below 180 nm, referred to as vacuum ultraviolet, are readily absorbed in air and thus present no hazard.

Absorption of UVR by a chromophore results in photochemical reactions of great importance to photobiology. The absorption of this energy induces an excited state in the chromophore that can be relaxed by a number of means, for example, thermal decay and light emission. Proteins and nucleic acids are the most important chromophores in biological media. The pyrimidine structures in nucleic acids have their primary absorption peak at 265 nm; whereas absorption in protein is associated with the aromatic amino acids, tyrosine and tryptophan, with absorption peaks at 275 nm and 280 nm, respectively. The photon energy can also be absorbed by a photosensitizer that transfers the energy to a biochemical compound either directly or through an intermediate. Quenching agents, such as carotenes, vitamin E, and ascorbic acid, inhibit this process by bringing the excited molecule down to the ground state or neutralizing reactive intermediates (Faber, 1989).

The production of DNA strand breaks and covalent dimers of thymine and other pyrimidines accompanies UVR exposure (Peak et al., 1987). Such defects are repaired primarily by excising the dimers and rebuilding the intact DNA strand by replacing the normal bases in the proper sequence utilizing the dark repair enzyme system (Cleaver, 1974). However, photoreactivation also occurs with exposure to wavelengths in the UV-A and visible regions.

4.4.2.2 Acute Exposure. Common articles of clothing are at least somewhat opaque to UV, depending on the material and the density of the weave (Sliney and Wolbarsht, 1980). Whole-body exposure to UV radiation is possible but not usual. However, with the recent emphasis on suntanning both outdoors and in indoor tanning booths, the acceptability of briefer clothing styles and the transparency of newer fabrics and garments, the opportunity exists for more extensive, longer, and more intense exposure.

The most common adverse response of the skin is erythema or sunburn. The maximum erythemal effect is at 260 nm, with a secondary peak at 300 nm (Everett et al., 1969). The erythematous changes observed depend on the UVR dose and the spectral characteristics of the incident radiation, the pigmentation and exposure history of the exposed skin, and the thickness of the stratum corneum. The erythemal response (skin reddening) results from the vascular reaction of the skin to vasodilation and increased blood volume, and ranges from a slight reddening to severe blistering. The faint erythemal response seen during exposure, disappears quickly; a delayed response begins at 2–4 hr postexposure, peaks at 14–20 hr, and lasts for up to 48 hr.

More recently, Parrish et al. (1982) determined the minimum erythemal dose (MED) at 8 and 24 hr in previously untanned, very fair Caucasian volunteers at wavelengths from 250 to 435 nm. They also determined the minimal melanogenic dose (MMD) over the same wavelength range. As shown in Table 14.5, there are minima at 250 and 296 nm with a broad maximum near 280 nm. A small minimum was also observed at 365 nm. The threshold at wavelengths above 313 nm increases dramatically. No erythema or melanogenesis was observed at 435 nm. Overall, the curves for both effects are similar or indistinguishable; the MED for the 24-hr erythema was higher than that for the 8-hr MED at every wavelength. The authors conclude that the data support the premise that DNA is the primary chromophore responsible for both melanogenesis and erythema below 313 nm.

The erythema resulting from exposure to wavelengths shorter than 290 nm (below the range of the sun's rays at the earth's surface) appears to be a somewhat different response. The skin color is pink, not deep red, and the color intensity peaks at 8 hr rather than 14–20 hr, leading to speculation that they are slightly different syndromes.

The value of 15–20 mJ/cm^2 is variously referred to as the threshold dose for erythema (TDE), the dose for minimum perceptible erythema (MPE), or MED. These doses relate to lightly pigmented skin that has not been exposed to UVR (not conditioned) in the recent past. The defense measures taken by the skin upon exposure, melanogenesis and skin thickening subsequent to hyperplasia, will increase the threshold required to produce the same degree of reddening (Matelsky, 1968).

Almost all people will exhibit at least some degree of erythemal response to UVR.

Table 14.5 Minimal Erythemal Doses (MED) and Minimal Melanogenic Doses (MMD) in the Wavelength Range 250–405 nm for Fair-Skinned, Untanned Caucasians

Wavelength (nm)	8-h MED (mJ/cm^2)	7–14 day MMD (mJ/cm^2)
250	14	121
265	31	140
275	34	142
282	36	148
290	25	55
296	21	46
300	35	48
304	47	67
313	871	1,160
334	23,800	25,300
365	57,700	63,500
380	192,300	213,000
405	267,800	268,000

Source: Adapted from Parrish et al. (1982).

The skin types have been categorized into six classes based on the individual's history of sunburning or tanning. Types I and II always sunburn easily, with no or minimal tanning. Types III and IV experience moderate to minimal sunburn, while attaining a light to moderate tan. Types V and VI rarely sunburn and tan profusely (Council on Scientific Affairs, 1989).

The erythemal action spectrum has been studied by many investigators over the years, dating back to the original research by Hausser and Vahle in 1921 (reported in Urbach, 1987). The results have been similar but not identical. The MED for previously unexposed skin ranged from 20 to 50 mJ/cm^2. The standard erythemal curve shown in Figure 14.1 represents a compromise based on earlier data. This curve shows two maxima of erythemal effectiveness, one around 250 nm and another around 300 nm. The erythemal curve is based on very artificial measures of "redness" indicative of a lack of maturity in the experimental approach in this field. However, it may also reflect the difference between "sunburn" and the erythemal response to artificial UV sources, where there are differences in spectral content.

More recently, the CIE adopted a new reference spectrum for UVR-induced erythema based on a statistical analysis of MED studies published by several investigators between 1964–1982 (McKinlay and Diffey, 1987). The erythemal effectiveness (EE) can be calculated for a given wavelength range as follows:

$$EE(\lambda) = 1.0 \qquad (250 < \lambda < 298 \text{ nm})$$

$$EE(\lambda) = 10^{[0.094(298 - \lambda)]} \qquad (298 < \lambda < 328 \text{ nm})$$

$$EE(\lambda) = 10^{[0.015(139 - \lambda)]} \qquad (328 < \lambda < 400 \text{ nm})$$

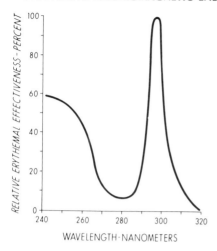

Figure 14.1. Standard erythemal curve.

This action spectrum is used only to determine the approximate erythemal efficacy of optical radiation sources between 250 and 400 nm.

Increased pigmentation, or suntanning, has an immediate and a delayed component. During and soon after exposure, immediate pigment darkening reaches a maximum at 1–2 hr and falls off during the following day. It is most noticeable in darkly pigmented individuals. Delayed pigment darkening, enhanced by the formation of new melanin granules, begins after 2–3 days, peaks at 2–3 weeks, and persists for several months.

4.4.2.3 Chronic Exposure. Frequent, prolonged exposure to sunlight especially of fair-skinned Caucasians results in marked morphologic changes often referred to as actinic skin. "Sailor's," "farmer's" and "fisherman's" skin are names given to actinic skin, which indicates their occupational origin (Knox et al., 1965). Oil field, pipeline, construction, and other workers who spend long periods of time outdoors without protecting their skin also develop this condition. The features of actinic skin include scaliness, coarse and fine wrinkling, telangiectasis, and irregular pigmentation. These are not harmful changes but may foreshadow the appearance of precancerous lesions such as actinic keratoses.

4.4.2.4 Photosensitivity. The initial studies of photosensitization were carried out in 1913 by a physician who injected himself with hematoporphyrin and then exposed his body to sunlight (Spikes, 1989). Since then it has been well established that certain chemicals encountered in topical or systemic medications, plants, or industrial operations are photosensitizing agents. Photosensitivity is a broad term that includes two types of reactions. Phototoxicity is the more common of the two types and can affect all individuals if the radiation dose or the dose of the photosensitizer is high enough. This phenomenon occurs with the initial insult and does not rely on an immunologic response. Photoallergy, on the other hand, is less common and is an acquired altered reactivity in the exposed skin resulting from an antigen–antibody or cell-mediated hypersensitivity to the agent (Epstein, 1989). Different mechanisms are involved but the basic principle is the same, that is, the photon energy is absorbed by the photosensitizer and transferred to the target molecule, resulting in an enhanced response or a lowered threshold for the response.

The USFDA published a listing of known medications that increase sensitivity to UV-A and light (Levine, 1990). The primary classes of photosensitizing medications include the following generic categories: antihistamines, coal tar and derivatives, oral contraceptives and estrogens, nonsteroidal anti-inflammatory drugs, phenothiazines, psoralens, sulfonamides, sulfonylureas, thiazide diuretics, tetracyclines, and tricyclic antidepressants. Hawk (1984) has compiled a similar list of potentially photosensitizing medications and other agents commonly encountered in the United Kingdom. Other photosensitizing agents found in consumer products including sunscreens were listed. Many plants such as figs, parsley, limes, parsnips, and pinkrot celery contain photosensitizing chemicals (Zaynoun et al., 1985). Industrial photosensitizers include coal tar, pitch, anthracene, naphthalene, phenanthrene, thiophene, and many phenolic agents.

4.4.2.5 Photocarcinogenesis. The carcinogenic effects of UV radiation have been studied extensively in both humans and animals. Excellent reviews of the literature are available (Blum, 1959, 1969; Epstein et al., 1971; Epstein, 1983; van der Leun, 1984). The three types of skin cancer of concern are squamous cell carcinoma (SCC) and basal cell carcinoma (BCC), referred to jointly as nonmelanoma skin cancer (NMSC), and cutaneous malignant melanoma (CMM). The National Cancer Institute estimates that there are about 600,000 new cases of NMSC in the United States each year; of these, about 80 percent are basal cell carcinomas (NCI, 1990). These cancers frequently are located on parts of the body that are exposed to the sun, that is, head, neck, arms, and hands, and occur most often in lightly pigmented Caucasians. The association between solar exposure and SCC is stronger than that for BCC (Faber, 1989). Four variables have been implicated in the development of NMSC: (1) lifetime sun exposure; (2) the intensity and duration of the UV-B component in sunlight; (3) genetic predisposition, as in individuals who have xeroderma pigmentosa, albinism, and so forth; and, (4) other factors unrelated to sunlight, for example, exposure to ionizing radiation, polycyclic aromatic hydrocarbons, and so forth (Epstein, 1989; NIH, 1989). In males treated with UV-A, UV-B, and 8-methoxypsoralen (a known photosensitizer) for psoriasis, the incidence of genital cancers (SCC) was as much as 300 times that in the general population, depending on the treatment regimen (Stern et al., 1990).

The incidence of CMM has increased rapidly not only in the United States but also worldwide in the past 30–40 years. In the United States alone, about 32,000 new cases of CMM were forecast for 1991, with 6000 deaths anticipated. From 1973 to 1988 the death rate for malignant melanoma among U.S. males increased 50 percent, a greater increase than that for any other cancer. For females, the death rate rose 21 percent. Although several factors, for example, the presence of moles, genetic predisposition, have been associated with this disease, the increased solar exposure, especially during recreational activities, has been implicated as a contributing factor (CDC, 1992), as well as intermittent exposure, that is "short bursts of intense exposure" (Nelemans et al., 1993).

Of the four types of melanoma, only lentigo maligna melanoma occurs on sun-exposed areas of the body (Koh, 1991). The occurrence of ocular melanoma is not thought to be related to sunlight or UVR exposure (Faber, 1989). The melanomas, unlike the NMSC, metastasize to other body tissues and organs and are often fatal.

In a case control study, patients with heavy freckling, a large number of moles, and an inability to tan had a significantly increased incidence of CMM (Dubin et al., 1986). Increased rates have also been associated with severe sunburns as children or adolescents in several epidemiologic studies (Marks and Whiteman, 1994). Koh (1991) hypothesized that melanoma risk appears to be more dependent on intermittent solar exposure, especially at an early age, than on cumulative exposure as NMSC does. He also pointed to the role of dysplastic nevi as indicators of an increased risk of CMM and as a potential precursor lesion from which the melanoma develops. In fact, about 50 percent of melanomas develop on a preexisting nevus (Cesarini, 1987).

The carcinogenic action spectrum for humans is known to be in the region between 280 and 320 nm, but may extend throughout the UV-A region. Cole et al. (1986) postulated that the most appropriate action spectrum for UVR-induced photocarcinogenesis was similar to that determined for acute skin edema in mice, with peak effectiveness at 260 and 297 nm. Diffey (1988), however, argues that the reference action spectrum for erythema in humans adopted by the CIE (McKinlay and Diffey, 1987) adequately approximates the photocarcinogenesis action spectrum. In both action spectra, wavelengths beyond 330 nm are more than 1000 times less effective in producing a carcinogenic response. An action spectrum for malignant melanoma was recently determined in hybrid platyfish and swordtails by irradiating the fish at 302, 313, 365, 405, and 436 nm. Surprisingly, melanoma induction at wavelengths greater than 313 nm was much greater than expected leading the researchers to conclude that the most effective wavelengths are not absorbed directly in DNA (Setlow et al., 1993).

Recent concern has been focused on the increased popularity of tanning booths and parlors, especially among young adults living in the United States, Canada, and some western European countries. Although most lamps now in use generate most of their radiation in the UV-A region, even the small percentage of UV-B emitted may present a long-term risk of cancer or skin damage.

Extensive data have been accumulated on UVR-induced NMSC in experimental animals. Repeated exposure to UVR results in the production primarily of SCC in the mouse and guinea pig, whereas both SCC and BCC are produced in rats. Epidermal DNA damage, inflammation, benign hyperplasia, and dysplasia precede the appearance of the malignancy (NIH, 1989). Although UV-B is a more effective carcinogen, nonetheless UV-A does damage DNA and can produce erythema and SCC in mice and guinea pigs. The action spectrum for photocarcinogenesis in mice is similar to that established for an acute response, with wavelengths above 330 nm contributing minimally to the process (Cole et al., 1986). Epstein (1989) has discussed the quantitative and qualitative aspects of experimental carcinogenesis in more detail.

4.4.2.6 Photoimmunology. Research in the past decade has done much to demonstrate that UVR exposure depresses both the systemic and local immunologic response. Morphologic, enzymatic, and antigenic changes have been observed in Langerhans cells and the density of these cells is reduced after UVR exposure (Council on Scientific Affairs, 1989). The loss of these cells results in a decreased antigen presentation function. Since Langerhans cells are located above the basal membrane of the epidermis, they are not protected by the melanocytes, and this effect may be independent of skin pigmentation (Morison, 1989).

In mice, UV-B exposure activates T-suppressor lymphocytes that interfere with the function of the immune surveillance system in monitoring tissues for the presence of malignant cells (Morison, 1989). This change is thought to be responsible for the animal's inability to reject skin cancers transplanted from genetically identical mice (Council on Scientific Affairs, 1989; Faber, 1989).

The evidence in humans is more indirect. Chronically immunosuppressed patients who have received organ transplants have an increased incidence of NMSC, especially SCC. The initial report by Walder et al. (1971) has since been confirmed by several investigators (Morison, 1989; Streilien, 1991).

UV-C, UV-B, sunlight, and UV-A-psoralen (PUVA) therapy have been shown to activate the human immunodeficiency virus (HIV) (Wallace and Lasker, 1992). This has been demonstrated in both in vitro and in vivo experiments (Valerie et al., 1988; Zmudzka and Beer, 1990; Yarosh et al., 1993) but not in a clinical evaluation (Horn et al., 1993). The mechanism is not understood but may be mediated through Langerhans cells, which have been found to harbor HIV in infected individuals (Wallace and Lasker, 1992), and are a target for UVR as discussed above.

4.4.2.7 Ozone Reduction and Disease. Data from both satellite measurements and ground-based observations support the conclusion that column ozone levels have been reduced. However, there are differences in the magnitude of the reductions between satellite and ground-based data. This may be due to the air pollutants, especially ozone, near the earth's surface, which is accounted for in ground-based measurements only (Cartalis et al., 1992; Frederick, 1993). The ozone reduction has not directly translated to increased levels of UVR at mid-latitudes (Frederick, 1993). Concern does exist at high southern latitudes that have reported a doubling of the UV-B levels during annual ozone holes in the springtime (Frederick and Alberts, 1991; Schein et al., 1993).

An evaluation for UV-induced acute skin and eye effects has been performed in Punta Arenas, Chile. Studies in humans focused on two groups of outdoor workers, shepherds and fishermen, with hospital workers serving as controls. No differences were seen between outdoor workers and controls (Schein et al., 1993).

4.4.2.8 Photoprotection. In the past decade, there has been considerable growth in the availability of, and emphasis on the need for, sunblocks and sunscreens that are applied to the skin to reduce cutaneous UVR exposure from solar radiation. A sunblock reduces exposure by reflecting and scattering the incident radiation. Zinc oxide and titanium dioxide are very effective sunblocks, reflecting up to 99 percent of the radiation both in the UV and visible, possibly into the infrared region also. Although very effective, the sunblocks are messy to use and some people think the sunblocks present an unacceptable appearance. Sunscreens, on the other hand, absorb the ultraviolet rays, usually over a limited wavelength range in the UV-B and A regions, and are considered to be drugs by the USFDA (Wuest and Gossel, 1992a). The agents include para-aminobenzoic acid (PABA) and its esters, benzophenones, salicylates, and anthranilates (Epstein, 1989). Usually in a gel or cream form in a lotion or other vehicle, the preparations have good substantivity allowing them to bind with the stratum corneum and resist removal by perspiration during heavy activity or by water when swimming. All provide protection in the UV-B range; only the benzophenones and anthranilates provide limited protection in the UV-A region. PABA preparations can discolor clothing and can cause contact-type eczematous dermatitis. The body's natural photoprotector, melanin, absorbs strongly in the UV-A (Kollias and Baqer, 1987).

Sunscreens are designed to protect against UVR-induced erythema and sunburn, and their efficacy has been demonstrated in humans. The sun protection factor (SPF = the ratio of the UV-B dose required to produce an erythema with protection to that required without protection) is based on their ability to do so. Although many commercial products claim very high SPFs, the increase in protection claimed above an SPF of 15 has not been documented (Wuest and Gossel, 1992b).

Sunscreens have also reduced the risk of skin cancer in animal studies (Wulf and Urbach, 1988) but did not protect against immunologic effects in in vitro experiments with human skin (Hersey et al., 1987). Quick-tanning preparations containing β-carotene or carthoxanthine will produce a "tan" by coloring the skin but offer no protection against the UVR (Epstein, 1989). Hawk et al. (1982) tested 55 commercial products available in the United Kingdom (45 sunscreens and 10 sunblocks) in 300 human volunteers. The sunscreens provided good protection against UV-B but only a few products were effective in the UV-A region, mostly at the short wavelengths. Sunblocks reflected the radiation throughout the UV region far into the visible (up to 650 nm). Stern et al. (1986) estimated that the lifetime incidence of NMSC could be reduced by 78 percent if a sunscreen with an SPF of 15 were used in the first 18 years of life. Since their effectiveness in preventing melanoma has not been established, other protective measures, for example, hats and protective clothing, avoidance of exposure during peak hours (10:00 A.M. to 2:00 P.M. sun time), should be employed (Koh and Lew, 1994).

As part of a sun awareness program, the Environmental Protection Agency (EPA) and the National Weather Service introduced the experimental UV index in the summer of 1994 (EPA, 1994a). The index is a "next-day forecast of the likely exposure to ultraviolet radiation for a particular location at noon" (EPA 1994b). The index is determined by estimating the noon-time value of the UVR radiant exposure, weighted by the effect of the cloud cover. These index values are assigned an exposure category (minimal, low, moderate, high, or very high) that indicates the potential for sunburn in fair-skinned individuals. The "low" category indicates the likelihood of an erythema resulting from a 20-min exposure while the erythemal exposure duration in "very high" category would be less than 5 min (EPA 1994a).

4.4.3 Visible Radiation

As mentioned previously, the skin reflects visible wavelengths strongly, as much as two-thirds of the incident radiation, depending on the wavelength. Since melanin absorbs strongly in this spectral region, lightly pigmented skin reflects more radiation. At wavelengths above 400 nm, high-intensity exposure to visible wavelengths can lead to thermocoagulation of skin similar to that produced by electrical or thermal burns. Several variables influence the threshold for and amount of damage done: the absorption and scattering by the skin at the wavelength of interest; the incident irradiance or radiant exposure; exposure duration; area of skin exposed; and the degree of vascularization of the irradiated tissue (Goldman et al., 1989).

Irradiation of a small (2 cm^2) area of skin will result in a first-degree burn at about 12 W/cm^2; a second-degree burn at 24 W/cm^2; and a third-degree burn at about 34 W/cm^2. However, most industrial sources do not emit sufficient intensities in areas where workers could potentially be exposed to produce thermal burns (Sliney and Wolbarsht, 1980). Extended exposure at lower intensities may result in a generalized heat stress response.

Using a tunable argon laser, thresholds for minimal skin reactions varied from 4.0 to 8.2 J/cm^2 in the wavelength range from 458 to 515 nm at an exposure duration of 1 s. At 694.3 nm, thresholds ranged from 0.25 to 0.34 J/cm^2 for a 75-ns pulse from a Q-switched ruby laser. With 2.5-ms pulses, the thresholds were 11–20 J/cm^2 on unpigmented skin and only 2.2–6.9 J/cm^2 on pigmented skin (Parrish et al., 1976). The variability in the thresholds was attributed to differences in skin absorption.

Individuals having certain diseases or genetic deficiencies are sensitive to light; but the mechanisms and action spectra are not well documented (Epstein, 1989). Some photosensitizing agents may have action spectra that reach into the visible region.

A rare reaction to fluorescent lighting has been observed in individuals who are sensitive to light. This sensitivity is manifested by urticaria (urticaria solaris) or by erythema and edema (erythema solare persistans) of exposed areas.

4.4.4 Infrared Radiation

There is little evidence that IR photons can enter into photochemical reactions in biological systems, probably because their low energy will not affect the electron energy levels of these atoms. Upon absorption the kinetic energy of the system is increased, producing a degradation of the radiant energy as heat. Thus, the primary response to IR exposure is thermal and the direct damage results from a temperature increase in the absorbing tissue. The increase depends on the wavelength, the parameters involved in heat conduction and dissipation, the intensity of the exposure, and the exposure duration.

Comprised mainly of water (60–70%), the skin absorbs IR radiation with an absorption pattern similar to water (Moss et al., 1989). Radiant energy in the near IR can be transmitted into the dermis with a maximum penetration at a wavelength of about 1.2 μm (Hardy et al., 1956). At wavelengths beyond 2 μm the skin is essentially opaque and exposure results in surface heating.

The most prominent effects of near IR include acute skin burn, increased vaso-dilation of the capillary beds, and an increased pigmentation that can persist for long periods of time. With continuous exposure to high-intensity IR, the erythematous appearance due to vasodilation may become permanent (Moss et al., 1989). Many factors mediate the ability to produce a skin burn. It is evident that the rate at which the temperature of the skin is permitted to increase is of prime importance. Necrosis of the eyelid has not been reported, primarily because thermal sensation or pain is encountered long before injury could occur. This provides a warning to the individual and thus a chance for self-protection. Repetitive exposure can, however, give rise to blepharitis, a chronic inflammation of the lids (Matelsky, 1968).

Hardy et al. (1951) observed that the initial skin temperature was important in thresholds obtained. The pain threshold varies linearly with skin temperature, and at a skin temperature of 45°C no further stimulus is required. Physiologically this means that the pain threshold depends only on skin temperature and not on the rate of skin heating nor on the rate of change of internal thermal gradients. The pain threshold among U.S. subjects was uniform and independent of gender or age, when measured under controlled conditions on the exposed skin of the forehead or back of the hand (Hardy, 1958).

It should be pointed out that the intensity of pain sensation does not depend solely on the peripheral excitation pattern sent from the pain receptors but also on many other factors influencing the central nervous system at that time, such as suggestions, attitude, and other psychological factors (Wolff and Hardy, 1947). In addition, subthreshold changes in skin temperature do occur without evoking temperature sensations. Whether these subthreshold changes play a role in thermoregulation is unknown. On the other hand, marked alteration in rate of change and magnitude of change in skin temperature can occur without evoking the temperature sensations usually reported. Thus, precooled skin can be rapidly heating without evoking sensations of warmth.

The stimulus giving rise to pain is the same as that giving rise to flexor reflex responses in vertebrates. When the skin temperature reaches 45°C, reflex responses are obtained. This type of stimulus, giving rise as it does to both pain and reflex activity, is termed noxious stimulation (Hardy, 1953). As the intensity of radiation above threshold is increased, increasing intensities of pain are perceived. There is a marked increase in pain between 45 and 52°C after which the increase in painfulness is abated. Maximal pain is perceived at skin temperature of approximately 65°C (Hardy et al., 1974).

Moritz and Henriques (1947) have shown that 45°C is also critical in producing cutaneous burns. Skin temperatures lower than 44–45°C rarely produce burns. It may be speculated that noxious stimulation results from chemical reactions in the skin probably involving inactivation of cellular protein (Henriques, 1947). The pain threshold is thereby determined as the lowest rate of inactivation of tissue proteins that, if sufficiently prolonged, will cause tissue damage. Pain is related to skin temperature only, whereas tissue damage is dependent on both the skin temperature and the duration of the hyperthermic episode (Hardy, 1958).

The tolerance limits of human skin for IR exposure have been determined by several workers. Lloyd-Smith and Mendelssohn (1948) found that an incident intensity of 0.17 W/cm^2 of near IR could just be tolerated by epigastric and interscapular skin areas of 144 cm^2. Approximately 25 percent of this energy would be reflected, so their result corresponds to a tolerated dose of 0.13 W/cm^2 (Whyte, 1951; Moss et al., 1989). On skin purposely blackened to obtain maximum absorption, the maximum temperature increase in situ that produced burns ranged from 56°C after a 0.5-s exposure to 5.6 J/cm^2 of filtered radiation to 15°C with a 100-s exposure to 13 J/cm^2 (Derksen et al., 1963). Later thresholds determined with a 75-ns pulse from a Q-switched Nd-glass laser operating at 1.06 μm ranged from 2.5 to 5.7 J/cm^2 while those from a 1-s exposure from a Nd-YAG at 1.064 μm were 46–78 J/cm^2. The threshold from a 1-s exposure from a CO_2 laser at 10.6 μm was 2.8 J/cm^2 (Parrish et al., 1976).

4.5   Ocular Effects

4.5.1   Anatomy

The adult human eye approximates a sphere with a diameter of about 2.5 cm. The outermost layer of the eyeball is the sclera that serves as the outer shell of the eye. The anterior portion of the sclera is the cornea, a transparent, highly refractive tissue. The corneal epithelium is the thin surface of the cornea, exposed to environmental insults and protected only by the tear layer. Surrounding the cornea in the anterior eye is the conjunctiva.

The aqueous humor fills the anterior chamber and provides nutrition to the lens. The highly vascular, pigmented iris controls the amount of light entering the eye by varying the pupil diameter from 2 to 8 mm, depending on the brightness of the environment or the object being viewed.

Directly behind the iris is the crystalline lens, an elastic, biconvex tissue suspended from the ciliary body by suspensory ligaments. The shape of the lens determines its refractive power and the effective focal length of the eye. This accommodation allows the eye to focus the image of objects viewed at varying distances onto the retina. The capsule is the outer covering of the lens; the inner portion is the cortex and the central area is the nucleus. Epithelial cells within the lens continue to grow throughout the person's lifetime forming the lens fibers. The lens is avascular and receives nutrition through the aqueous. The posterior portion behind the lens is mostly filled by the vitreous body, a colorless, gelatinous mass composed largely of water.

The retina is composed of the sensory elements and the neural cells of the eye organized into several complex layers. The rods and cones are the photoreceptors; there are approximately 125 million rods and 7 million cones. The rods are used primarily in night (scotopic) vision and have peak sensitivity at about 505 nm. The cones are used for daylight or color (photopic) vision with a maximal sensitivity at about 555 nm. The majority of the cones is located in the part of the retina called the macula. At its center is the fovea, containing only cones, that is responsible for peak visual acuity. The retinal pigment epithelium (RPE), located posterior to the rods and cones, performs diverse biological and physical functions and absorbs up to 50 percent of the photons striking the retina. Separating the RPE and choroid is Bruch's membrane, which functions as a blood-retinal barrier. The high vascular choroid supplies oxygen to the retina and maintains the eye at a uniformly warm temperature.

4.5.2 Optical Characteristics of the Eye

The cornea absorbs virtually all incident radiation at wavelengths below 290 nm and above 1400 nm. Between 290 and 380 nm, the lens absorbs much of the radiation transmitted through the cornea and aqueous. Transmittance through the ocular media is relatively high in the wavelength range from 500 to 900 nm. However, there are significant variations among the data obtained by several investigators using different techniques that remain unexplained (Moss et al., 1982). The ocular system is most sensitive to wavelengths from 400 to 760 nm. There is limited sensitivity to wavelengths outside this range both in the near-UV and near-IR regions. Since the lens absorbs UV-A radiation strongly, in the aphake (a person who has had the lens removed), these wavelengths are transmitted through the ocular media to the retina. Less than 100 photons are required to produce a visual response, and the optical gain of the human ocular system is about 100,000 times or greater. The normal human eye with an effective focal length of 0.017 m (17 mm) has a refractive power of about 59 diopters (D). About three-fourths of its refractive power is due to the cornea with the remainder from the lens (Sliney and Wolbarsht, 1980).

4.5.3 Ultraviolet Radiation

The cornea and conjunctiva are the primary absorbers of UVR in the UV-B and C ranges. Corneal transmission ranges from 60 to 83 percent in the UV-A band, with much of the energy absorbed by the lens.

4.5.3.1 Corneal Effects. Several corneal conditions are associated with UVR exposure: nuclear band keratopathies, pinguecula, pterygium, photokeratitis, photoconjunctivitis, and epidermoid carcinoma.

The keratopathies, that is, Labrador keratopathy and Bietti's corneal degradation, occur only in a few geographic regions and are prevalent in Labrador, Cameroon, and the Dahlak Islands. The strong association with UVR exposure results from the exact match of the distribution of nodular deposits in both eyes with the pattern of sunlight impinging upon the surface of the eyeball. The yellow deposits noted in the elderly and in outdoor workers are pinguecula. These deposits are found most often at the corneal-scleral junction on the nasal side of the eyeball. Pterygium is a benign hyperplasia characterized by deltoid-shaped growths of epithelial cells. The prevalence of this condition appears to have a latitudinal gradient peaking between 40°N and 40°S latitude. However, the relationship is not well documented and other factors, for example, physical trauma, may play an etiological role (Marshall, 1987).

Epidermoid carcinoma of the conjunctiva, a relatively rare neoplasm, occurs primarily in the inhabitants of tropical or subtropical climes. These moderately to poorly differentiated carcinomas are seen more frequently in darkly pigmented individuals (EHC, 1982; Faber, 1989).

Photokeratitis and photoconjunctivitis, also known as photoophthalmia, result from acute, high-intensity exposure to UV-B and C. Objective findings include lacrimation, irritation, and congestion of the conjunctiva; punctate staining of the cornea due to epithelial defects; and blepharospasm and photophobia accompanied by a sensation of sand in the eye and severe pain. These symptoms usually become apparent after a dose-dependent latency period of 2–12 hr, depending on the source irradiance and spectral distribution, and persist for up to 48 hr, usually without residual damage. Photochemical denaturation and coagulation of protein structures are the basic mechanisms of cell damage (Pitts et al., 1977). The thinness of the conjunctival epithelium and the almost complete absence of the stratum corneum and melanin granules deprive the conjunctiva of the protective action these structures afford the skin. There is, thus, no increasing tolerance developed to subsequent irradiation (Matelsky, 1968).

Threshold doses for humans range from 4 to 14 mJ/cm^2 between 220 and 310 nm (Pitts and Tredici, 1971; Pitts et al., 1977). An action spectrum has been established and the wavelength of minimum threshold occurs at 270 nm where nucleic acids and some proteins have absorption maxima. Reciprocity holds for exposure durations up to several hours; thus the tissue damage depends on the total dose (energy) received and not the dose rate.

Commonly referred to as "arc eye" or "welder's flash" by workers, this injury results from exposure of the unprotected eye to a welding arc or other artificial source rich in UV-B and C. Sunlight exposure produces these sequelae only in environments where highly (UVR) reflective materials are present, for example, snow (snow blindness) or sand.

4.5.3.2 Lenticular Effects. Virtually all UVR above 300 nm is transmitted by the cornea (Lerman, 1988). Before 10 years of age, the human lens transmits over 75 percent of the UVR in the 300–400 nm band whereas, in adults over 25 years, the transmittance is quite low, approaching zero. The nucleus of the lens becomes yellow in color during this time, protecting the retina against near UV- and blue-light-induced photochemical damage.

The only lenticular effect linked with UVR, primarily solar radiation, is the cataract. A cataract, which can be developmental or degenerative, is an opacity in the crystalline lens that interferes with the transmittance of visible wavelengths (light) through the normally transparent lens and may or may not result in visual impairment. The degenerative cataract is a manifestation of aging, systemic disease, trauma, or certain forms of radiant energies among others.

Precise definition and meticulous care are required in choosing proper criteria for designation of a cataract. Such criteria may include:

1. The most minute and subtle changes in the crystalline lens recognized by the examiner, regardless of whether they interfere with vision. The term cataract for such minute changes is not a very good one and may lead to misinterpretation if not more closely defined.

2. Lens changes that are obvious to any qualified ophthalmologist but do not severely interfere with visual acuity.

3. Lens changes that reduce visual acuity. These may not be in the category of mature cataracts depending on the clinical judgment of the examiner in interpreting changes observed in animal eyes and correlating them with possible deterioration of vision in humans.

4. Mature cataracts, where the lens is milky white and vision is reduced to light projection only (Geeraets, 1970).

Senile cataracts can be classified into three major groups: cortical cataract surrounding the cortex, nuclear cataract in the lens nucleus, and posterior subcapsular cataract that occurs beneath the posterior capsule of the lens. Cataracts are more prevalent in females and occur primarily in adults over 65 years of age. In the United States, estimates of the number of cataract operations range up to 1 million annually, and it is the third leading cause of blindness (Taylor et al., 1988). In addition to age and gender, other predisposing factors that have been suggested include family history, nutritional status, and certain medical conditions and medications.

An action spectrum has been established for transient and permanent lenticular opacities in rabbits. For exposures of less than one day, threshold radiant exposures for transient opacities in the 290–320 nm region ranged from 0.15 to 12.60 J/cm^2 and 0.5 to 15.5 J/cm^2 for permanent opacities. In both cases, the minimum threshold was at 300 nm. From 325 to 395 nm, cataracts could not be produced even at very high exposures. Primates were exposed only at the wavelength of maximum sensitivity in rabbits (300 nm), and the threshold radiant exposure was determined to be 0.12 J/cm^2 (Pitts et al., 1977). Thresholds for immediate cataracts have been established in primates for laser exposures at 325, 337, 351 and 363.8 nm (Zuclich and Connolly, 1976). The mechanism for UV-induced cataracts may be photochemical in the 295–320 nm range and thermal at longer wavelengths (Pitts et al., 1986).

Epidemiologic studies of human populations in six countries on four continents have attempted to discern if an association exists between sunlight exposure and the occurrence of cataracts, specifically senile cataracts. Despite differences in experimental design and confounding variables, generally the prevalence of cataracts is associated with some measure of solar exposure. Although individual solar or UV radiation exposures were not measured, diverse surrogates, for example, UV-B counts, hours of sunlight exposure, and geographic location, were used to characterize average daily exposures in areas under study. Relative risk or odds ratios calculated to determine the strength of the association ranged from 1.3 to 5.8, demonstrating a consistent association between UVR exposure and cataracts (Pitts et al., 1986; Zigman, 1986). In reviewing the literature, Waxler (1986) concluded that the threshold for cataract production was 2500 MEDs annually, corresponding to daily exposures of 7–90 mJ/cm^2 in the UV-B and 0.4–98 J/cm^2 in the UV-A range.

Chesapeake Bay watermen exposed to high cumulative UV-B levels experienced a significant increase in the occurrence of cortical cataracts. Watermen in the highest exposure quartile experienced a 3.3-fold increased risk compared to those in the lowest quartile. No association was found between UV-B exposure and nuclear cataracts or UV-A exposure and either type of cataract (Taylor et al., 1988).

From in vitro studies, several biochemical mechanisms have been postulated for

UV-induced damage to the lens. Photooxidative damage to the crystallins may affect proteins in the nucleus of the lens. Photooxidation of lipids in the lens membrane can inhibit crosslinking and insolubilization of proteins in the lens. Changes in DNA in the lens epithelium may suppress cellular function. However, these changes have not yet been shown to be related to cataractogenesis (Pitts et al., 1986). Also open to question is the role of UV-A as a cataractogenic agent either through thermal injury or by interacting with photosensitizing agents present or produced in the lens. Tryptophan absorbance extends into the UV-A, range and protein enzymes such as ATPase can be inactivated by UV-A exposure (Zigman, 1986). There is also some evidence that systemic photosensitizers, primarily drugs, may play a role in cataractogenesis (Dayhaw–Barker et al., 1986; Lerman 1986).

4.5.3.3 Retinal Effects. Exposure to UV-A and B, primarily from solar radiation, has been implicated as an etiologic agent in solar retinitis, cystoid macular edema (CME), and senile macular degeneration (SMD) (Marshall, 1987).

In adults transmission of these wavelengths through the ocular media is minimal except in aphakes. Although originally thought to be a thermal injury, solar retinitis (photoretinitis or eclipse blindness) is in fact a photochemical lesion resulting from absorption of UV-A and blue wavelengths. Experimental studies have determined that retinal damage occurs at lower exposures at 325 nm than at 441 nm (5.0 vs. 30 J/cm^2, respectively). The threshold for broad-band UV-A exposure is 0.09 J/cm^2 (Waxler, 1986).

4.5.4 Visible Radiation

Visible radiation or light is focused by the cornea and lens onto the fovea of the retina. Since the human eye is relatively transparent to light, it is transmitted through the ocular media without appreciable absorption. In this process the intensity of the light is concentrated by a factor of 10^4–10^6 over that incident on the pupil. Although these photons have relatively low energy (3.10–1.63 eV), they are unique in that they initiate a photochemical chain reaction in light-sensitive absorbers present in the cells of the retina, the end result of which is the sensation of vision (Matelsky, 1968). Vision involves the capacity of the eye to adapt to enormous ($\geq 10^6$) variations in ambient illumination. This adaptation is mediated by the two interspersed photoreceptive systems of the retina. The rods and cones of the retina transduce the light into neuroelectrical phenomena. The neural elements of the retina are gathered into the optic nerve at the blind spot and pass by discrete pathways to the highest visual center in the brain, the primary visual cortex located in the occipital lobe. Cone cells are responsible for the perception of color and good visual acuity at higher levels of ambient illumination. The rod cells function under reduced illumination and are used for night vision. Comprehensive information is available on vision as related to dark adaptation and night vision (Fisher et al., 1970). A detailed review of physiological optics has been prepared by Graham (1965).

The initiating event in the visual process is the absorption of light photons by pigments in photoreceptors in the retina. The rods contain rhodopsin that absorbs maximally at about 505 nm. The cones contain three visual pigments having absorption peaks at about 445 nm (blue cones), 540 nm (green cones), and 565 nm (red cones). It appears that the latter two cone pigments are involved in photopic (day) vision while scotopic (night) vision depends on rhodopsin. The mechanism of the photochemical reaction that occurs when these visible radiant energies are absorbed is not completely known. Several hypotheses have been proposed to explain the mechanism whereby neural stimulation results from absorption of one light quantum by one molecule of visual pigment (Wald, 1956; McConnell and Scarpelli, 1963; Rosenberg, 1966; Bonting and Bangham, 1967). It appears to be an initiating event that, once started, continues as an oxidation–reduction reaction in a series of stages involving intermediate degradation products of the pigment and 11-*cis*-retinal, a derivative of vitamin A (Adler, 1965; Dratz, 1989).

4.5.4.1 Retinal Effects.

An intensely brilliant light source such as the sun, carbon arc, or welder's arc may produce temporary or permanent retinal scotomas (photoretinitis) that occur when the retina is subjected to intense light without proper protection. Injuries to the eyes from observing eclipses of the sun have been known since earliest history and are known as eclipse blindness or solar retinitis. Reports of such injuries date back as far as Hippocrates (460–370 B.C.), and sporadic references to this have been made throughout the centuries (Geeraets, 1970).

Both exposure and biological factors affect the degree of hazard to the retina. Exposure factors include the size and type of optical source, its spectral intensity, and the exposure duration. Biological factors include the pupil size, retinal image quality, including the size and location of the image, and the spectral transmittance of the cornea, aqueous, lens, and vitreous, and the spectral absorption of the retina and choroid.

Geometrical optics can be employed to calculate the size of the image on the retina from most extended sources. Given the effective focal length (f_e) to be 1.7 cm, the retinal image size (d_r) can be calculated if the viewing distance (r) and the source size (D_L) are known:

$$d_r = D_L f_e / r \tag{6}$$

Using this, the relationship between retinal irradiance (E_r) and source luminance (L) can be derived for small angles:

$$E_r = (\pi d_e^2 L \tau)/4 \, f_e^2 \tag{7}$$

$$= 0.27 d_e^2 L \tau \tag{8}$$

where τ is the spectral transmission of the ocular media and d_e is the pupil diameter in cm. This relationship cannot be used for very small image sizes ($< 50 \, \mu$m). When viewing a point source, that is, a laser beam or extended source at large distances, diffraction theory can be used to estimate the small image size on the retina. However, corneal and lenticular aberrations and intraocular scattering of the light spread the image size to at least 10–20 μm. When the relaxed eye views a point source, the retinal irradiance is about a factor of 10^5 higher than the corneal irradiance (Sliney and Wolbarsht, 1980).

Retinal effects can occur through four mechanisms of interaction: (1) thermal, (2) photochemical, (3) elastic or thermoacoustic transient pressure waves, or (4) nonlinear phenomena. The latter two mechanisms are seen primarily with high-intensity laser pulses. With the elastic or acoustic transient pressure wave, a portion of the energy of the light pulse impinging on tissue is transduced to a mechanical compression wave (acoustic energy). A sonic transient pressure wave is built up that can rip and tear tissue. How this transient interacts with the thermal mechanism is unclear. Nonlinear phenomena occur only with ultrashort pulses and include direct electric-field effects, Raman and Brillouin scattering, and multiphoton absorption (EHC, 1982).

The absorption of visible radiation by tissue leads to the production of heat. At sufficient intensities, a rapid rise in temperature will result that can denature protein and inactivate enzymes. The greatest thermal stress is created around those portions of tissue that are the best radiation absorbers. If the absorption is rapid and localized, high temperatures will result. Steam production, evident only at high exposure levels, can be quite dangerous if it occurs in an enclosed and completely filled volume such as the eye. The thermal mechanism is the primary damage mechanism for 1-ms to 10-s exposure durations (EHC, 1982).

Long-term ($> 10–100$ s) exposure to light levels above those normally encountered in our environment may produce photochemical damage in the retina. Both the photoreceptors (rods and cones) and the RPE are subject to photochemical injury. In the photoreceptors, the action spectrum is very similar to the absorption spectrum of the photopigments, especially rhodopsin, whereas that for the RPE mimics the spectral absorption of melanin through the lower end of the visible (500–400 nm) region down into the UV-A region. This latter effect, commonly referred to as blue-light damage, may be related to the presence of free radicals produced during metabolic processes, perhaps abetted by photosensitizers normally present in the retina, but this has not been confirmed. Photochemical damage can be enhanced by the thermal mechanism that predominates in the 500–1400 nm range. Although there is no sharp cut-off point between the two types of damage, histological examination of the retinal lesions can be used to discriminate between them since the photochemical damage appears much more uniform over the lesion. It has been suggested that solar retinitis is similar to the blue-light damage and that chronic light exposure may be involved in retinal degeneration, senile macular degeneration, and cystoid macular edema (Ham et al., 1986).

Retinal damage occurs when excessive amounts of light (and near IR) energy penetrate the eye and are absorbed by the retinal tissues. It is estimated that approximately 5 percent of the incoming visible radiation is used for vision, with the remainder absorbed by pigment granules in the RPE and choroid. The type of damage resulting from excessive exposure to light ranges from a small, inconsequential scotoma in the peripheral retina to severe damage of the fovea, with consequent loss of visual acuity. The damage produced may be either temporary or permanent and recovery from minimal photochemical lesions has been observed over extended periods of time.

Thresholds for retinal damage have been determined for both incoherent and coherent optical radiation sources under a wide variety of exposure conditions. The body of data accumulated in these experiments is too large to document here. There are many variables that affect the magnitude of the threshold to at least some degree: wavelength, spectral intensity, pulse duration, exposure duration, retinal image size, and retinal radiant exposure. The threshold radiant exposure for retinal damage is 30 J/cm^2 for a 1000-s exposure at 441 nm (wavelength of maximum sensitivity for retinal photochemical damage), corresponding to a retinal irradiance of 30 mW/cm^2. At 633 nm, this threshold value has risen to 950 mW/cm^2, for the same exposure duration. The photochemical damage at 441 nm may be cumulative over 2 or more days. Threshold retinal irradiances for broad-band sources as low as 0.93 μW/cm^2 have been reported for all-day exposures. Visual impairment has been shown in monkeys exposed repeatedly to 6 and 100 mJ/cm^2 at 514 nm over many months (Waxler, 1986). Because of the lack of information on the human eye, threshold values are based on data obtained primarily on rabbits and monkeys. As in most other cases, extrapolation to the human eye must be made with caution.

4.5.4.2 Pulsating Light. Pulsating light sources, such as strobe lights operating at frequencies of a few hertz, do not constitute a hazard to most individuals and may be slightly irritating during the day or in high ambient lighting levels. Susceptible individuals, for example, persons with epilepsy, may experience seizures as a result of exposure to light sources flickering at a frequency of 8–16 Hz (flashes per second), a brightly lit pattern, or sudden appearance of a bright light. Approximately 1 percent of persons with epilepsy (< 1% of the general population) are thought to experience such seizures (Hardy, 1953; EHC, 1982). Various other activities that have triggered photic seizures include television viewing especially at close distances, reading, driving or riding in a car, waving the hands between the eyes and a light source, blinking, and observing objects such as rotating helicopter blades, home movies, red and white checked tablecloths, and patterns of sunlight and shadows. The effect is more common in young children, especially females.

4.5.4.3 Glare. Glare may produce a feeling of visual discomfort often due to the work of squinting in an effort to screen reflected nonparallel light rays. If the glare is substantial or frequently induced, it may result in tiredness, irritability, possibly headache, and a decrease in work efficiency (Geeraets, 1970). Glare can be differentiated into the following types:

1. Disability (veiling) glare. This is glare created by light uniformly superimposed on the retinal image that reduces contrast and, therefore, visibility.
2. Discomfort glare. This is adventitious light scattered in the ocular media so as not to form part of the retinal image.
3. Scotomatic or blinding glare (flash blindness). This is glare produced by light of sufficient intensity to alter the sensitivity of the retina.

Although all three types of glare are present in the case of high-intensity light, the effects of the first two are primarily evident only when the source is present. The third type, scotomatic or blinding glare, is especially significant in flash blindness where it produces symptoms (afterimages) that persist long after the light itself has vanished (see Flash Blindness).

Regardless of whether the glare source is direct or indirect (specularly or diffusely reflected), it can cause discomfort, can affect the visual performance, or both. The visual discomfort or annoyance from glare is a common well-understood experience and has been confirmed by many experiments. In connection with certain experimental studies, it has been found that people sometimes become more physically tense and restless under glare conditions.

The effect of glare on visual performance can be of serious consequence by itself; however, the visual discomfort brought about by glare can also be a matter of some concern. Although the cause is physical, the discomfort brought about by it is often of a subjective nature. The evaluation of discomfort, then, must make use of subjective responses as criteria. Involved in such procedures is the concept of borderline between comfort and discomfort (Luckiesh and Guth, 1949).

4.5.4.4 Flash Blindness. Flash blindness, a relatively new problem occurring only since the development of intense light sources brighter than the sun, is a temporary effect in which visual sensitivity and function is decreased severely in a very short time period. This normal visual response, whose complex mechanism is not well understood, is related to the eye's adaptive state and the size, spectral distribution and luminance of the light source (EHC, 1982). Ordinarily, the sequence of events following stimulation of the retina by a flash of light is the primary sensation of light followed by a series of positive and negative afterimages. The duration of visual loss depends on the degree of flash blindness and is proportional to the intensity and duration of the light exposure. Recovery of visual function does not entail a single, solitary mechanism, but rather it includes those visual components involved in dark adaptation as well as neurophysiologic mechanisms (Brown, 1973).

4.5.5 Infrared Radiation

The eye has mechanisms (the blink and pupil reflexes) to protect it from IR radiation, if it is accompanied by high-intensity visible light. Some industrial IR sources, however, are not found with intense light, so that these protective mechanisms are not always activated.

4.5.5.1 Corneal Effects. The cornea of the eye is highly transparent to IR-A wavelengths, has two water absorption bands at 1.43 and 1.96 μm, and becomes opaque to radiant energy above 2.5 μm (Moss et al., 1982). Thermal damage to the cornea probably occurs in the thin epithelium rather than in the deeper stroma. The threshold radiant exposure decreased from 5.5 to 1.25 kJ/cm^2 in the near IR, as the source irradiance was increased from 2.93 to 4.55 W/cm^2. The corneal damage, in rabbits, elicited ranges from epithelial haze to corneal erosion with a focused beam incident upon a miotic pupil. Complete recovery occurred, usually in 24 hr. The threshold in exposed primates was 8.0 kJ/cm^2 at an irradiance of 4.23 W/cm^2 (Pitts et al., 1980).

Excessive exposure may completely destroy the protective epithelium, with opacification of the stroma due to protein coagulation and would seriously interfere with vision. The probability of incurring such as insult is low except where highly collimated sources can irradiate the eye without producing the sensation of pain in the cornea or the surrounding skin tissue (Matelsky, 1968). However, chronic corneal exposure to subthreshold doses may lead to the occurrence of "dry-eye" characterized by conjunctivitis and decreased lacrimation (Moss et al., 1982). Davson (1963) has shown that the cornea and iris exhibit thermal sensitivity to IR. Humans experience a pain sensation with about a 10°C elevation in ocular temperature with a corneal irradiance of 8.4 W/cm^2 (Lele and Weddell, 1956).

Threshold data are available for infrared laser emissions at 1.54 μm (erbium), 2.02 μm (Tm: YAG), 2.7–3.0 μm (hydrogen fluoride {HF}), 5 μm (CO), and 10.6 μm (CO$_2$). The collective data indicate that thermal injury is the damage mechanism as expected (Sliney and Wolbarsht, 1980; McCalley et al., 1992).

4.5.5.2 Iritic Effects. Because the iris is a highly pigmented tissue, it is generally assumed that most of the IR radiation reaching it will be absorbed. However, this may not be the case since the melanin pigment in the iris is not a strong absorber of wavelengths beyond the visible region. The occurrence of thermal injury may result from the absorption of the visible radiation that usually accompanies the IR. Moderate doses can result in constriction of the pupil (miosis), hyperemia, and the formation of aqueous flares. More severe exposures may lead to muscle paralysis, congestion with hemorrhage, thrombosis, and stromal inflammation. Within a few days, necrosis of the iris may cause the formation of bleached atrophic areas. Pigmentation loss at the border of the iris follows in 2–4 days (Moss et al., 1982).

A stromal haze was observed in the rabbit iris at the threshold radiant exposure ranging from 4.0 to 1.25 kJ/cm^2 as the irradiance increased from 2.93 to 4.55 W/cm^2. All exposures were done with a focused beam incident upon a miotic pupil. In primates, the threshold radiant exposure was 8.0 kJ/cm^2 at an irradiance of 4.23 W/cm^2 (Pitts et al., 1980).

4.5.5.3 Lenticular Effects. The lens is avascular and thus unable to dissipate heat efficiently. It is relatively transparent to near-IR radiation but has two water absorption bands (0.98 and 1.2 μm). The lens grows throughout life and remains an actively metabolic tissue. This activity along with metabolic disorders, medications, and trauma can lead to the development of lenticular opacities that reduce the lens's optical clarity. These opacities, referred to as cataracts, may or may not interfere with visual acuity.

Damage to the lens of the eye from IR has been the subject of considerable investigation for many years. The term *glassworker's cataract* has become generic for lenticular opacities found in individuals exposed to processes hot enough to be luminous (Matelsky, 1968). The first report of lenticular damage associated with optical radiation (sunlight) was in 1739. Prior to 1900, several investigators observed that workers having chronic exposure to heat sources seemed to have an increased incidence of cataracts. The occupational groups mentioned were glassblowers, foundry and forge workers, cooks, and laundry workers, as well as those who worked in sunlight (Moss et al., 1982).

Robinson (1907) published the results of his investigations in England on the incidence of opacities on the posterior surface of the lens in the eyes of glassworkers that were different from senile cataracts in appearance. On his recommendation, the disease *radiation cataract* became scheduled as occupational in origin in England and by 1921 was copied into the U.S. Workmen's Compensation Act.

Dunn (1950) of Corning Glass Company was unable to find evidence of any ocular disturbance among glassworkers who had been exposed to 0.14 W/cm^2 for at least 20 years. Keating et al. (1955) found no posterior cortical changes in the eyes of iron rolling mill workers exposed to 0.084–0.418 W/cm^2 for an average of 17 years. However, the incidence of posterior capsular opacities was higher. The opacity originated in the capsular plane and extended to the cortex. This differs from what is classified as a cataract.

The mechanism for IR cataractogenesis has been the subject of considerable debate. The issue is whether the cataract results from: (1) a direct effect of the radiation on the lens fibers; (2) the temperature elevation resulting from IR absorbed directly by the lens; or (3) the temperature rise in the lens resulting from absorption of IR by the iris (Pitts et al., 1980). Further complications arise from differences in experimental techniques and because heat interferes with normal ciliary body function and lens metabolism (Moss et al., 1982). Vogt (1932) considered changes arising from the increase in lenticular temperature and direct effects on the lens fibers. Other evidence indicates that heat per se in the absence of IR can produce cataracts. Goldmann (1935; Goldmann et al., 1950) claimed that the cataracts were not due to the absorption of IR by the lens but rather from the temperature elevation in the lens resulting from the heating of the iris.

Pitts et al. (1980) could not produce lenticular damage in either rabbits or primates with exposure of the lens alone through the pupil even at corneal radiant exposures exceeding 15 kJ/cm^2. When the iris and the lens were irradiated, the threshold radiant exposure in the rabbit ranged from 4.0 to 2.25 kJ/cm^2 at irradiances of 2.93–4.44 W/cm^2. Minimal opacities were observed in primates exposed to 10 kJ/cm^2 at an irradiance of 4.23 W/cm^2. No permanent opacities could be produced in any animals. These results appear to support Goldmann's hypothesis that the iris must be involved in IR or heat-induced cataractogenesis at least with regard to acute exposure conditions.

4.5.5.4 *Retinal Effects.*

The retina is susceptible to near-IR wavelengths (760–1400 nm), since the ocular media are relatively transparent to these wavelengths. Absorption by the retina is strongest at the very shortest IR wavelengths, making it difficult to separate the thermal effects from thermally enhanced photochemical injury. For extended exposures, the corneal irradiance required to produce a minimal retinal lesion at 1064 nm is almost three orders of magnitude greater than that at 442 nm (Moss et al., 1982).

The mechanism of retinal damage involves the absorption of energy by the RPE and the rate of conduction of heat from this layer into the adjacent tissues, such as the choroid. If the heat is not dissipated rapidly enough and the temperature of the tissue rises about 20°C, irreversible thermal damage will result from the denaturation of protein and other macromolecules (Ham et al., 1986). The size or area of the image on the retinal choroid apparatus and the absorbed retinal irradiance (dose rate) are two of the most critical factors in the production of thermal injury.

4.6 Standards

Two general types of standards or guidelines can be employed to protect people from hazardous optical radiation; exposure and emission standards. Exposure standards are generic in nature and are used to limit the intensity of optical radiation to which a worker or the general public can be exposed, regardless of the source of that radiation exposure. Emission standards on the other hand apply only to a specific optical radiation source, such as high-intensity mercury vapor lamps, or a type of source such as the laser.

Such standards and guidelines are available from a number of sources including governmental agencies and private organizations. Examples of each type of standard/ guideline will be highlighted here. However, this overview is not intended to be a compendium of available standards and the reader is encouraged to seek out further sources of such information.

4.6.1 Exposure Standards

4.6.1.1 Ultraviolet Radiation. In 1948 the Council on Physical Medicine of the American Medical Association issued guidance for safe exposure to UVR emitted by germicidal lamps (Council on Physical Medicine, 1948). This group recommended that for the wavelength at which most of the energy is emitted, 253.7 nm, exposure should not exceed 0.5 μW/cm^2 for periods of 7 hr or less, nor 0.1 μW/cm in the case of continuous exposure. This standard applies only to 253.7-nm radiation, and the hazard from sources of ultraviolet other than germicidal lamps must take into account the relative effectiveness of the wavelengths produced.

Occupational exposure guidance has been developed by the International Non-Ionizing Radiation Committee [now the International Commission on Non-Ionizing Radiation Protection (ICNIRP)] of the International Radiation Protection Association (IRPA, 1985a, 1989). This guidance also applies to exposure of the general public.

The American Conference of Governmental Industrial Hygienists (ACGIH) has established Threshold Limit Values (TLVs) for occupational UVR exposure (ACGIH, 1993). The TLVs for UVR incident on skin or eye are similar to the Exposure Limits (ELs) established by IRPA.

The TLVs apply to UVR in the spectral band from 180 to 400 nm and represent conditions under which it is believed that nearly all workers may be exposed repeatedly day after day without adverse effect. These TLVs apply to exposure of eye and skin to solar radiation and all artificial UVR sources except lasers. These TLVs should not be used for photosensitive individuals, persons concomitantly exposed to systemic or topical photosensitizing agents, or to ocular exposure of aphakes. They should be used as guides in the control of UVR hazards for exposure durations from 0.1 s to the whole workday; they should not be used as fine lines between safe and dangerous levels of exposure.

As with all TLVs, these limits are intended for use in the practice of industrial hygiene and should be interpreted and applied only by a person trained in this or a related discipline. They are not intended for use, or for modification for use (1) in the evaluation or control of the levels of UVR or other physical agents in the community, (2) as proof or disproof of an existing physical disability, or (3) for adoption by countries whose working conditions differ from those in the United States. These values are reviewed annually by the Committee on Threshold Limits for Physical Agents for revisions or additions, as further information becomes available.

The UV radiant exposure incident on the unprotected skin or eye should not exceed the wavelength-dependent levels specified by either group (ACGIH, 1993; IRPA 1985a, 1989). The TLV is a minimum (3 mJ/cm^2) at 270 nm and increases on either side to 250 mJ/cm^2 at 180 nm and 100 J/cm^2 at 400 nm. In applying the TLV, the spectral irradiance of a broad-band source is measured over small, discrete band-widths, weighted using a spectral effectiveness factor, and summed over the UV range. The resulting effective irradiance is then compared to a monochromatic source emitting all of its energy at 270 nm. The effective irradiance of the source is calculated as follows:

$$E_{\text{eff}} = \sum E_\lambda S_\lambda \Delta\lambda \qquad (9)$$

where E_{eff} = effective irradiance (W/cm^2)
    E_λ = spectral irradiance (W/cm^2 nm)
    S_λ = spectral effectiveness factor (unitless)
    $\Delta\lambda$ = bandwidth (nm)

The permissible exposure time (in seconds) for exposure to actinic UVR (180–315 nm) is calculated as follows:

$$t = 0.003 \text{ J/cm}^2/E_{\text{eff}} \qquad (10)$$

For most white-light sources and all open arcs, calculating the effective irradiance in the actinic region should be sufficient. Only certain lamps designed to emit UV-A radiation would normally require spectral weighting from 315 to 400 nm.

For the UV-A spectral region, total irradiance incident on the unprotected eye should not exceed 1 mW/cm^2 for periods greater than 1000 s and a radiant exposure of 1 J/cm^2 for periods less than 1000 s. The IRPA EL specifies 1 J/cm^2 during an entire 8-hr day.

Conditioned (tanned) individuals can be exposed to levels exceeding these TLVs without an erythemal effect; however, this does not preclude the possibility of long-term skin damage, that is, skin aging or skin cancer.

4.6.1.2 Light and Near-Infrared Radiation.

The ACGIH has established a TLV for ocular exposure in any 8-hr workday in the spectral region from 400 to 1400 nm. The same qualifying statements made above for the UVR TLV also apply to this TLV. For white-light sources having a luminance exceeding 1 cd/cm^2 (2920 ftL), the spectral radiance and total irradiance of the source must be known.

The TLVs protect against the following hazards:

1. Retinal thermal injury in the 400–1400 nm range using a spectral burn hazard function (R_λ).

2. Retinal photochemical injury from chronic blue-light exposure in the 400–700 nm region using a blue-light hazard function (B_λ).

Workers who have had their lens(es) removed (aphakes) in cataract surgery have an increased risk of photochemical injury. Wavelengths in the 300–400 nm range can be transmitted to the retina if these workers do not have a UV-absorbing intraocular lens implant(s). The latter TLV is modified in this instance by substituting an aphake photic hazard function (A_λ) for the B_λ in the 300–700 nm region.

The TLV for near IR (770–1400 nm) protects against delayed effects on the lens (cataracts) by limiting exposure to 10 mW/cm^2. Sources used as heat lamps or lacking a strong visual stimulus require a more complex analysis using the spectral radiance (L_λ) and angular subtense of the source.

4.6.1.3 Lasers. The IRPA (1985b, 1988a) and ACGIH (1993) have established guidelines, ELs and TLVs, respectively, for ocular and skin exposure to laser radiation in the UV, visible, and IR regions (180 nm to 1 mm). The guidelines are comprehensive and complex, covering exposure durations from 1 ns to 30 ks. They are expressed in terms of radiant exposure or irradiance and are averaged over the limiting aperture. The present limiting apertures are 1 mm for both eye and skin with the following exceptions: (1) in the retinal hazard region (400–1400 nm), the aperture is 7 mm, and (2) between 0.1 and 1 mm, the aperture for eye and skin is 11 mm. Changes in these values have been proposed by ACGIH (1992), and the derivation of some aspects of the guideline have been criticized (Vos, 1993).

The guidelines are the same for ocular and skin exposure in the UV region (180–400 nm) and in the IR-B and C regions (1.401 μm to 1 mm) because of the comparability of the injury thresholds for cornea and skin at these wavelengths.

However, separate guidelines are provided in the retinal hazard region (400–1400 nm) because of the focusing effect by the ocular media on the retina. In this region the ocular exposure guidelines include both direct ocular exposure (intrabeam viewing and specular reflections) and extended source exposure (diffuse reflection).

Similar guidance is also available from the American National Standards Institute (ANSI, 1993). In addition, several states in the United States have laser exposure standards (Handren, 1991).

4.6.2 Emission (Product) Standards

4.6.2.1 Lasers. The USFDA, under the Radiation Control for Health and Safety Act of 1968 (Public Law 90-602), has authority to regulate electronic product radiation and has established performance standards for lasers operating in the wavelength range from 180 nm to 1 mm (USFDA, 1985a). The standard requires the manufacturer to classify laser products and incorporate certain control measures into the product to protect the user. A laser product is assigned to one of six classes based on the laser's capability to injure persons who are exposed to emissions from the device. The classes are defined as follows according to the accessible laser radiation levels emitted in any part of the optical radiation spectrum unless otherwise specified:

Class I levels are not considered to be hazardous under any viewing condition for any exposure duration. Laser devices in this class are often referred to as "eye-safe" lasers. However, many laser products are a higher class (more hazardous) laser embedded in a protective housing to limit the accessible levels to those required for Class I.

Class IIa levels are not considered to be a chronic viewing hazard for exposure durations ≤ 1000 s, but are for durations > 1000 s. This class applies only to visible lasers that have a radiant power ≤ 3.9 μW.

Class II levels are considered to be a chronic viewing (> 0.25 s) hazard. Only visible lasers (400–710 nm) with a radiant power of ≤ 1 mW are included in this class.

Class IIIa levels can be either an acute or chronic viewing hazard depending on the irradiance. The radiant power is limited to ≤ 5 mW in the visible region.

Class IIIb levels are considered to be an acute eye and skin hazard from the direct laser beam.

Class IV levels are considered to be an acute eye and skin hazard from direct or scattered laser radiation.

Both the American National Standards Institute (ANSI, 1993) and the ACGIH (1990a) have similar classification schemes for laser devices. There are some differences between the scheme used by the latter two organizations and the FDA product standard, for example, the FDA allows only visible lasers into Class IIIa whereas lasers having emissions outside the visible region are allowed under the ANSI and ACGIH classification scheme.

There is an international laser product standard published by the International Electrotechnical Committee (IEC) that uses a classification scheme similar to that of the FDA and has exposure guidance similar to the ANSI and ACGIH (Smith, 1991; Edmunds, 1991).

4.6.2.2 *Other Sources.* The USFDA also has two other optical radiation product standards:

1. High-intensity mercury vapor discharge lamps (21 CFR Part 1040.3; effective March, 1980). This standard requires lamp manufacturers to provide a label that will inform the user if the lamp will self-extinguish when the outer glass envelope is broken.
2. Sunlamps (21 CFR Part 1040.2; effective September, 1986). This standard requires sunlamp manufacturers to affix a warning (danger) label on each product, to provide the consumer with instructions for using the sunlamp, to incorporate a timer and a manual control to terminate the exposure, and to include protective eyewear with each sunlamp (USFDA, 1985b).

5 MICROWAVE AND RADIO-FREQUENCY ENERGIES

5.1 Introduction

For the purposes of this chapter, radio-frequency (RF) radiation is defined as electromagnetic radiation in the frequency range of 300 Hz to 300 GHz. Microwaves (MW) are a subset of the RF spectral region and include frequencies from 300 MHz to 300 GHz. Various order-of-magnitude band designations have been assigned to these lower-frequency electromagnetic energies, as shown in Table 14.6. Frequencies in the various bands are allocated for uses including aeronautical radio, navigation, citizens' radio, and broadcasting.

In addition to these, letter-band designations are used for microwave frequencies in industry (Table 14.7). According to Wilkening (1991), these designations are not universally accepted.

Specific frequencies are also designated for industrial, scientific, and medical (ISM) uses by the Federal Communications Commission (FCC). ISM frequencies

Table 14.6 Frequency Band Designations in the Radio-Frequency Spectrum

Frequency Range	Designation	Abbreviation
<300 Hz	Extremely low frequency	ELF
300 Hz to 3 kHz	Voice frequency	VF
3–30 kHz	Very low frequency	VLF
30–300 kHz	Low frequency	LF
300 kHz to 3 MHz	Medium frequency	MF
3–30 MHz	High frequency	HF
30–300 MHz	Very high frequency	VHF
300 MHz to 3 GHz	Ultra high frequency	UHF
3–30 GHz	Super high frequency	SHF
30–300 GHz	Extremely high frequency	EHF

Table 14.7 Microwave Letter Designations

Letter Designation	Frequency Range (GHz)
L	1.100–1.700
LS	1.700–2.600
S	2.600–3.950
C	3.950–5.850
XN	5.850–8.200
X	8.200–12.400
Ku	12.400–18.000
K	18.000–26.500
Ka	26.500–40.000

are reserved for uses other than communications. In addition to the frequencies in Table 14.8, other frequencies may also be used for ISM purposes internationally. For example, 433 MHz is used for medical hyperthermia.

When propagated, these electromagnetic energies are categorized into two discrete modes known as CW and pulsed. CW stands for continuous wave and is associated with sources that generate RF for relatively long periods of time such as communication transmitting devices and consumer products, whereas pulsed energies are associated with radar and industrial and medical equipment. A number of useful reviews of RF/MW energies are available (Cahill and Elder, 1984; Gandhi, 1990; Patterson and Hitchcock, 1990; Wilkening, 1991; EHC, 1993; NCRP, 1993; Hitchcock, 1994; Hitchcock and Patterson, 1994).

5.1.1 Quantities and Units

The quantities used to describe RF/MW energy are the electric-field strength (E), the magnetic-field strength (H), and the power density (W). Although these are vector quantities, E and H may be viewed as the strength or intensity of the two fields. The units of E and H are volt per meter (V/m) and ampere per meter (A/m), respectively. The squares of these units, V^2/m^2 and A^2/m^2 are also often used. Power density is the vector cross product of the electric and magnetic fields. The units are those of time-averaged energy flow, W/m^2, although mW/cm^2 is more commonly used. In measuring exposures, these units can be applied to either a peak value or to an average value. The latter is often called the root-mean-square (rms) value and is 0.707 times the peak value.

Other quantities used to describe the potential for biological effects are the specific absorption rate (SAR), the specific absorption (SA), and induced and contact currents. The SAR is the fundamental quantity of the exposure criteria and has units compatible with metabolic rate, watt per kilogram (W/kg). The SAR is proportional to the square of the electric-field strength within the body. The SA is the time integral of the SAR and may be viewed as the RF energy dose with units of joule per kilogram (J/kg). Currents may be induced in the body by an incident RF field or transferred from a conductive object to the body. Contact and induced currents are quantified using the unit of current, milliampere (mA).

Table 14.8 Industrial, Scientific, and Medical (ISM) Frequencies Assigned by the Federal Communications Commission

Frequency
13.56 MHz $\pm$ 6.78 kHz
27.12 MHz $\pm$ 160 kHz
40.68 MHz $\pm$ 20 kHz
915 MHz $\pm$ 25 MHz
2450 MHz $\pm$ 50 MHz
5800 MHz $\pm$ 75 MHz
24,125 MHz $\pm$ 125 MHz

5.1.2 Absorption

The absorption of RF/MW energy is directly dependent on the electrical properties of the absorbing medium; specifically, its relative permittivity (dielectric constant) and electrical conductivity. These properties change as the frequency of the applied electrical field changes. Values of relative permittivity and electrical conductivity and depth of penetration have been determined for many tissues (Schwan and Li, 1953, 1956; Kraszewski et al., 1982; Karolkar et al., 1985). Because of its tissue-penetrating capability, heat generated as a result of the transformation of RF/MW energy is directly proportional at any depth to the intensity of the energy present, neglecting heat transfer by other methods. The so-called volumetric heating that results therefrom is quite unlike that due to conductive heating.

When RF/MW energy is absorbed by any material, the energy is transformed into increased kinetic energy of the absorbing molecules, which, by increased collision with adjacent molecules, produces a general heating of the entire medium. The energy value of 1 quantum of RF/MW radiation is much too low (< 1 meV) to produce ionization, no matter how many quanta are absorbed.

Knowledge of the electrical properties of a tissue permits direct calculation of the absorption coefficient. Absorption coefficients of 2.5 and 0.6 for 10 and 30 GHz, respectively, have been obtained in skin. The reciprocals of these coefficients define the penetration depths, that is, the depth at which the incident intensity is reduced to $1/e$, or about 37 percent of its initial power. For 10 GHz this is about 4 mm, whereas for 3 GHz it is approximately 16 mm.

When considering the biological effects of RF/MW radiation, the wavelength and its relationship to the physical dimensions of objects exposed to radiation become important factors. Absorption of energy radiating from a source into space also depends on the relative absorption cross section of the irradiated object. Thus the size of the object with relation to the wavelength of the incident field plays a role. If a small part of the body were subjected to far-field exposure to radiant energy of relatively long wavelength, it is possible that this part could absorb more energy than that falling on its shadow cross section (Anne et al., 1961). The term *far field* refers to that region away from a source of electromagnetic energy where the power density decreases inversely with the square of distance, $1/r^2$, the so-called inverse square law.

One of the main difficulties in working with electromagnetic energy is the determination of how much of the incident energy is absorbed by the tissues. If a conductive object is introduced into the field, it perturbs the field, which means that it tends to distort the field pattern in a generally unpredictable manner. Therefore, the field strength at a particular location may be quite different when a conductive object (animal, cage, etc.) is introduced into the field.

The elucidation of the biological effects of exposure to RF/MW energies requires a careful review and critical analysis of available literature. This entails differentiating established effects and mechanisms from speculative and unsubstantiated reports. Most of the experimental data support the concept that the effects of exposure are primarily a response to heating or altered thermal gradients in the body, that is, most of the RF/MW-induced responses are explained by thermal energy conversion, almost exclusively as enthalpic energy phenomena. However, there are reports that indicate that observed effects are not associated with demonstrable increases in body temperature. These are called athermal and nonthermal effects. The former occurs when there are measurable physiological changes in which the body temperature is not measurably elevated, but sufficient energy is absorbed to activate thermoregulatory receptors and cause a physiological response. Nonthermal effects are responses due to low levels of exposure and cause no significant thermal input and, hence, no significant change in body temperature (Elder, 1987).

The organs and organ systems affected by exposure to RF/MW energies are susceptible in terms of functional disturbance, structural alterations, or both. Some reactions to MW/RF exposure may lead to measurable biological effects that remain within the range of homeostasis (normal physiological compensation) and are not necessarily hazardous, or may even improve the efficiency of certain physiological processes and can thus be used for therapeutic purposes. Some reactions, on the other hand, may lead to effects that may be potential or actual hazards to health.

The nonuniform, largely unpredictable distribution of energy absorption in tissues may give rise to increases in temperature and rates of heating that could result in unique biological effects. Nevertheless, it is important to recognize that the mammalian body normally is not a uniform incubator at 37°C but does contain significant temperature gradients in deep body organs that may act as stimuli to alter normal function both in the heated organ and in other organs or organ systems. Thus, indirect effects can be elicited in organs far removed from the site of the primary interaction.

Specific organ tissue systems may function at a significantly different rate if local thermal gradients are altered. Relatively large changes in circulation are provoked by quite small deviations from neutral temperature (Thauer, 1965). Body heat content is equilibrated by approach to equality of two overall processes, gain and loss.

Absorption of RF/MW energy leads to increased temperature when the rate of energy absorption exceeds the rate of energy dissipation. Whether the resultant increased temperature is diffuse or confined to specific anatomical sites depends on (a) the electromagnetic field characteristics and distributions within the body and (b) the passive and active thermoregulatory mechanisms available to the organism, such as heat radiation, conduction, convention, and evaporative cooling. The efficacy of heat convection between a body and its immediate environment is influenced by the environmental conditions.

5.2   Dosimetry

It has been shown that the absorption rate of RF energy in humans and test animals is frequency dependent. As noted, the RF absorption rate is the SAR, and attaining its maximum value depends upon the orientation of the body relative to the E- and H-field vectors. The maximum SAR occurs when the E-field vector is parallel with the long axis of the body. During whole-body irradiation of small animals, the maximum SAR apparently occurs at frequencies between about 0.5 and 3 GHz, depending on the species, and for humans at around 60–100 MHz with a peak at about 70 MHz for a standard man. At frequencies below 30 MHz, the SAR drops off rapidly and it also decreases at frequencies above 500 MHz.

As noted, whole-body SARs approach maximal values when the long axis of a body is parallel to the E-field vector and is about 0.4 times the wavelength of the incident field. At 2.45 GHz, (λ = 12.5 cm), for example, a standard man (long axis 175 cm) will absorb about half of the incident energy. Relative to standard man, a mouse would absorb about twice the electromagnetic energy, because its body more closely approaches the 0.4λ criterion for maximal absorbance. This requires that scaling be used in biological effects studies so that frequencies of interest in humans are appropriately studied in test animals. Detailed discussions that serve as bases for scaling have been presented by several authors (Guy et al., 1975b; Massoudi, 1976; Durney et al., 1978, 1986; Gandhi, 1980).

The SAR is now widely used by the experimentalist and has been integrated into exposure criteria. It is the time rate at which RF energy is imparted to an element of mass of a biological body (NCRP, 1981). Thus the SAR is applicable to any tissue or organ of interest or is expressed as a whole-body average and is specified in W/kg. The SAR depends on a finite period of exposure to yield the amount of energy absorbed by a given mass of material, which is termed the specific absorption (SA), that is, the time derivative of the SAR. The SA is the dose of absorbed RF/MW energy in J/kg.

The average, whole-body SAR as a function of frequency for different sizes of laboratory animals and humans has been calculated (Durney et al., 1978). This concept is useful in allowing limited extrapolation from one species to another, but it should be used cautiously in view of its limitations. Because different biological tissues have different dielectric properties and RF fields usually have an inhomogeneous distribution in space, energy deposition may be uneven in animals and people. Measures are put in place to control this as well as possible in the laboratory, but energy absorption and distribution may still be much different than the models predict.

5.3 Experimental Biological Effects

These include observations made in studies of test animals and cells. For the most part, human exposure criteria rely on observations from in vivo studies with rodents and monkeys.

5.3.1 Chromosomes and Cellular and Genetic Effects

Some investigators have reported chromosome changes in various plant and animal cells and tissue cultures (Janes et al., 1969; Heller, 1970). These studies have been criticized because the systems were subjected to a thermal stress; the chosen parameters of the applied field caused biologically significant, field-induced force effects in in vitro experiments, and many of these experiments have not yet been independently replicated.

Saffer and Profenno (1992) observed an increased activity of a marker gene, β-galactosidase in cultures of *Escherichia coli*, irradiated between 2 and 4 GHz to produce a 1°C temperature rise. They speculate that "small thermal gradients" may produce the effects. The magnitude of the increase in activity appears to be highly dependent on experimental methods and has not been replicated in an independent laboratory.

Murine L929 cells showed no differences in growth and survival at 2.45 GHz and SARs of 130 and 1300 W/kg. Gene expression of two interferon-regulated enzymes was monitored in a temperature-controlled experiment at 130 W/kg. Activity of 2-5A synthetase was unaffected, while changes were observed in RNase L, although this does not appear to be detrimental to the cell (Krause et al., 1991). MW-induced chromosomal aberrations that were observed in one study, were also seen when cells were heated by convection (Kerbacher et al., 1990). Czerska et al. (1992) found that temperature plays a significant role in lymphoblastoid transformation of human cells exposed to conventional heat or CW microwaves. Pulsed and CW microwaves acted differently in this experiment, although mechanisms supporting such an interaction are not known.

In vivo studies have evaluated somatic and germ cells. Some strains of mice have been found to be susceptible to MW-induced increases in complement receptor (CR^+) cells while other strains are not. This discovery led researchers to suggest that the MW-induced reversible increase in CR^+ was genetically controlled. In a study designed to examine this hypothesis, the results showed that there appears to be a genetic basis for increased CR-bearing B lymphocytes, and this control was affected by a single regulatory gene (Schlagel and Ahmed, 1982). Male rats were evaluated for mutagenesis and reproductive efficiency by exposure in utero and up to postpartum day 90 at 2.45 GHz. No differences were found between exposed and control animals in body and organ weights and sperm concentrations (Berman et al., 1980). EPA researchers also studied sperm mutagenesis, observing no differences in a dominant lethal assay in male rat pups exposed (100 MHz, 2.5–3 W/kg) 4 hr/day from day 6 of gestation up to day 97 of life (Smialowicz et al., 1981).

In this vein of research, there is a need for replication of published works and for experimental designs that evaluate potential mechanisms for these observations. Until that time, a reasonable conclusion is that there is no conclusive evidence for RF-induced genetic effects that are deleterious to test animals.

5.3.2 Development

In almost all instances where developmental effects have been observed, the reported effects may be ascribed to increased temperature caused by the exposure (O'Conner, 1980). Typically, whole-body SARs exceeding 10 W/kg are necessary to produce teratogenic effects. Studies in rats at 27.12 MHz have demonstrated that the intensity of observed teratogenic effects was associated not only with the dam's colonic temperature but also the time that the temperature remains in an elevated state (Lary et al., 1983a). A threshold for abnormalities in rats was found when the core-body temperature of the dam was raised above 41.5°C (Lary et al., 1986). No developmental abnormalities were found when rats were exposed at 100 MHz, 0.4 W/kg on days 6–11 of gestation for a total of 40 hr (Lary et al., 1983b). Increased teratogenicity has been reported from a combined effect of RF and the solvent 2-methoxyethanol (Nelson et al., 1991).

No abnormalities have been observed at 2.45 GHz with SARs from 3.6 to 38

W/kg in rats, mice, and hamsters (Chernovetz et al., 1975, 1977; Jensh et al., 1983; Berman et al., 1981, 1982a). Various abnormalities have been observed at 2.45 GHz, including cleft palate at 23–42 W/kg (Nawrot et al., 1981), immature skeletal development at 16.5 W/kg (Berman et al., 1982b), and exencephalies at ≤43 to ≤112 W/kg (Rugh et al., 1975).

Results of experiments at lower frequencies have been inconclusive. Female rats were exposed to a 20-kHz, 15-microtesla (μT) sawtooth magnetic field, 24 hr/day for 20 days. There were no significant differences in implantations, pre- and postimplantation losses, resorptions, malformed fetuses, minor malformations, living fetuses, and measures of dam and fetal body masses. The RF-treated group had more skeletal variants and minor skeletal anomalies than controls, which "are common in teratological studies." This finding was statistically significant when analyzed by the fetus but not when analyzed by the litter (Huuskonen et al., 1993).

Stuchly et al. (1988) also exposed female rats to magnetic fields in the form of sawtooth pulses (18 kHz; 0, 5.7, 23, or 66 μT). Two types of common skeletal variants were significantly increased at the two highest flux densities when analyzed by fetus or litter. This finding was characterized as typical teratologic "noise." This is supported by the average fetal weight data, which were not significantly different. Furthermore, fetal weights were higher and less variable for the exposed groups in comparison to the controls.

Mice were exposed to 20-kHz sawtooth pulses (3.6, 17, and 200 μT): 192 animals were used in 4 replicate experiments within each exposure group and the sham group. No statistically significant differences were observed in these major endpoints: embryo/fetal mortality, fetal malformations (external, visceral, and skeletal), and fetal growth. Although not published in a peer-reviewed journal, the study design included an audit committee that provided a scientific review (Wiley et al., 1990).

These exposures equate to rms values of 1.1, 5.1, and 60 μT and were designed to bracket potential video display terminal (VDT) operator exposures at 30 cm. The 3.6-μT level was selected "to correspond, after the application of current-induction-based scaling considerations, to actual exposure levels likely to be experienced by VDT operators" (Wiley et al., 1990).

5.3.3 The Gonads

The effect of MW radiation on the testes has been studied extensively, indicating that high power density exposure can affect spermatogenesis (Ely et al., 1964). Temporary sterility observed in male rats (SAR = 5.6 W/kg) is thought to be related to the heating of the organs (Berman et al., 1980). The thermal sensitivity of the testes is well established.

5.3.4 Neuroendocrine Effects

Some investigators believe that endocrine changes result from hypothalamic-hypophyseal stimulation due to thermal interactions at the hypothalamic or adjacent levels of organization, the hypophysis itself, or the particular endocrine gland or end organ under study (Lu et al., 1977, 1980a,b; Magin et al., 1977a,b; Lotz and Michaelson, 1978, 1979). According to other investigators, the observed changes are interpreted as resulting from direct MW interactions with the central nervous system (CNS) (Petrov and Syngayevskaya, 1970; Mikolajczyk, 1974, 1977; Novitskii et al., 1977). In either case neuroendocrine perturbations should not be considered as necessarily harmful because the function of the neuroendocrine system is to maintain homeostasis, and hormone levels will fluctuate normally to maintain organ stability (Lu et al., 1980a).

5.3.5 Immunology

Microwaves have been reported to induce an increase in the frequency of complement receptor-bearing lymphoid spleen cells in mice (Wiktor-Jedrezejczak et al., 1977). RF/MW-induced hyperthermia in mice has been associated with transient lymphopenia with a relative increase in splenic T and B lymphocytes (Liburdy, 1979) and decreased in vivo local delayed hypersensitivity. In one experiment, normal controls, sham-exposed, and MW-exposed (13 W/kg) hamsters were injected with a lethal dose of vesicular stomatitis virus 1 day after MW treatment. The mean survival time for MW-exposed animals was significantly longer. Furthermore, for MW treatment, 25 percent of the high-dose group and 33 percent of the low-dose group survived, while all of the control animals died. Survival was attributed to macrophage activation (Rama Rao et al., 1984). Using a similar protocol, an increase in antibody response in hamsters immunized with sheep red blood cells (a T-cell-dependent antigen) was noted (Rama Rao et al., 1985). The results of these studies support the hypothesis that the observed effects have a thermal basis. The influence of increased temperature on immune responses is well known (Roberts and Steigbigel, 1977).

5.3.6 Effects on the Nervous System

Nervous system effects include changes in biochemical markers of brain energy metabolism, although these were not observed consistently at the frequencies examined (Sanders and Joines, 1983; Sanders et al., 1984). It is reported that doubly ionized calcium may be influenced by RF fields that are amplitude-modulated by ELF signals. The efflux of calcium in these in vitro experiments is highly dependent on methodology (Michaelson, 1991); the results are not robust and cannot be easily replicated, which may be explained by methodological differences (Blackman et al., 1991). Changes have been observed in the EEG in some experiments (Shandala et al., 1981) but not in others (Kaplan et al., 1982).

A transient alteration was found in the permeability of small inert polar molecules across the blood–brain barrier (BBB) of MW-exposed rats (Oscar and Hawkins, 1977). Attempts to duplicate these findings have yielded equivocal results unless the brain is subjected to a large increase in temperature (Merritt et al., 1978; Preston et al., 1979). The BBB was not compromised when whole-body hyperthermia was produced in conscious, unrestrained rats by exposure at either 2.45 GHz or ambient heat of $42 \pm 2°C$. Also, MW exposure of animals suppresses the BBB permeability and this is temperature dependent (Williams et al., 1984). The findings are not consistent, and, furthermore, the design of some studies included the use of anesthetics, which may depress the core temperature and have an impact on the BBB.

Transient functional changes referable to the CNS have been reported after low-level (< 10 mW/cm^2) MW exposure of small laboratory animals. Although some reports describe the thermal nature of RF/MW energy absorption, others implicate nonthermal or "specific" effects at the molecular and cellular level. It should be noted, however, that specific RF/MW effects have not been experimentally verified. The first reports of a MW-induced effect on conditional response activity in experimental animals was made by investigators in Eastern Europe (Gordon et al., 1955). In subsequent years the study of nonthermal effects of MW effects gradually occupied the central role in electrophysiological studies in the Soviet Union.

5.3.7 Interaction with Psychoactive Drugs

Research has focused on combined interaction between psychoactive drugs and MW, primarily 2.45 GHz (Lai et al., 1987). Drugs were administered at doses that by themselves produce measurable effects in the test animals. Sodium pentobarbital prior to irradiation reduced sleeping times in unrestrained rabbits, but no significant differences were noted when MW exposure preceded the stimulant (Wangemann and Cleary, 1976).

Hypothermia was produced in rats by pentobarbital treatment. Animals were then exposed with their heads pointing toward (anterior) or away (posterior) from the MW source. The rate of recovery from hypothermia in animals receiving posterior exposure was significantly more rapid than either anteriorly exposed or sham-exposed animals (Lai et al., 1984a). Some studies have demonstrated that 2.45-GHz MW exposure can attenuate ethanol-induced hypothermia (Lai et al., 1984b; Hjeresen et al., 1988).

It is not known if subtle effects in humans are possible from low doses of drugs, solvents, or other neuroactive substances found in the workplace. Patients under anesthesia or medication in health care facilities who receive RF irradiation (e.g., from hyperthermia or electrosurgical units) are a more likely population.

5.3.8 Behavioral Effects

Studies have been conducted on the effects of RF/MW on the performance of learned tasks by rats, mice, chickens, dogs, and rhesus and squirrel monkeys. Many studies indicate that RF/MW exposure can be related to suppressed performance of a learned task and that a power density/dose rate and duration threshold for achieving the suppression exists. Studies demonstrate that disruption of learned behavior is closely related to elevated body temperature, typically by a $\geq 1\,°C$ increment (de Lorge, 1978, 1984). Some studies with pulsed MW have found behavioral effects; however, it has been noted that these effects observed may be due to MW hearing (Blackwell and Sanders, 1986; Raslear et al., 1993).

Animal behavior reflects adaptive brain–behavior patterns, and behavioral thermoregulation expresses an attempt to maintain a nearly constant internal thermal environmental. Changes in body temperature bring about not only autonomic drives but also behavioral drives (Stolwijk, 1977); thus MW can influence behavioral thermoregulation (D'Andrea et al., 1977; Stern et al., 1979; Adair and Adams, 1980).

5.3.9 Hematopoietic Effects

Although several investigators have found that the blood and blood-forming system are not consistently affected by acute or chronic MW exposure (Tyagin, 1957; Hyde and Friedman, 1968; Spalding et al., 1971; Smialowicz et al., 1981; Ragan et al., 1983; Wright et al., 1984; Toler et al., 1988), effects on hematopoiesis have been reported (Michaelson et al., 1964, 1967; Baranski, 1971; Czerski et al., 1974; Czerski, 1975; Stuchly et al., 1988). The degree of hematopoietic change is dependent on the field intensity, duration of exposure, and induced hyperthermia. In evaluating reports of hematologic changes, one must be aware of the relative distribution of blood cells in the population and the susceptibility to thermal influences. Early and sustained leukocytosis in animals exposed to thermogenic levels of MW levels may be related to stimulation of the hematopoietic system, leukocytic mobilization, or recirculation of sequestered cells.

5.3.10 Ocular Effects

Ocular studies have emphasized the lens and cornea. Other structures studied include the iris vasculature and the retina. Some researchers have observed no corneal effects (Williams and Finch, 1974; McAfee et al., 1979; Rotkovska et al., 1993), while others have reported injury to the endothelial (Kues et al., 1985; Kues and D'Anna, 1987) and epithelial (Trevithick et al., 1987) layers of the cornea.

Monkeys, exposed to pulsed MW at 2.45 GHz (local SAR = 2.65 W/kg) showed increased iris permeability (blood–aqueous barrier) to a tracer molecule and endothelial lesions. A correlation was noted between increased iris permeability and subsequent development and severity of endothelial effects (Kues and D'Anna, 1987). Pretreatment with two drugs used in the treatment of glaucoma, timolol maleate and pilocarpine, enhanced these effects (Kues and Monahan, 1992).

Two reports have claimed retinal changes in animals. Rabbits were exposed to 3.1-GHz pulsed MW at 57 mW/cm². Changes in retinal plexiform layers were observed in five animals (Paulsson et al., 1979). Restrained, unanesthetized primates were exposed to pulsed microwaves at 1.25 GHz and 4 W/kg for nine, 4-hr treatments. Changes in photoreceptor responses were reported when pretest and posttest values were compared. Histopathology indicated ''degenerative changes'' of the photoreceptors (Kues and Monahan, 1992).

During the past 25 years, numerous investigations with animals and several surveys among human populations have been devoted to assessing the relationship between MW exposure and cataractogenesis. It is significant that of the many experiments on rabbits by several investigators using various techniques, power density above 100–150 mW/cm² for 1 hr or longer appears to be the time–power threshold for cataractogenesis in the tested frequency range of 200–10,000 MHz (Michaelson, 1972, 1974; Guy et al., 1975c). The effective local SARs must exceed 150 W/kg for times greater than 20 min to produce cataracts (Elder, 1984). In other species, for example, dogs and nonhuman primates, the threshold appears to higher. Frequencies between 1 and 10 GHz appear to be the most biologically effective in test animals.

5.3.11 Auditory Phenomenon

The occurrence of an auditory phenomenon has been established in humans (Frey, 1961; Somer and von Gierke, 1964) and animals (Chou et al., 1982, 1985) at frequencies from 2.4 MHz to 8 GHz. The effects reported include audible clicking, hissing, buzzing, chirping, and popping sounds that seem to originate behind or inside the head. The mechanism is believed to be a thermoelastic expansion within the skull. The ensuing pressure wave "is detected by the hair cells in the cochlea via bone conduction" (Elder, 1984). Although not shown to be adverse, Lin (1989) cautions that the question of health risk from exposure to RF pulses at power levels well above threshold has not been answered.

In the MW range, exposures must be pulsed at a repetition rate of 0.5–1000 Hz, with pulse widths from 1–1000 s. Whole-body SA values in rats were 0.9–1.8 mJ/kg per pulse (Chou et al., 1985). At 2450 MHz the threshold for an audible response in humans is about 4 J/m^2 for pulse widths <30 μs (Guy et al., 1975; Lin, 1989). More recently this phenomenon has been associated with exposure to pulsed RF fields generated by coils in magnetic resonance imaging (MRI) systems, at frequencies from 2.4–170 MHz. With the head in the coil, the threshold energy was 16 ± 4 mJ/pulse for pulse widths of about 3–100 μs. With the coil located at the ear, the threshold was as low as 3 mJ/pulse (Roschmann, 1991).

5.3.12 Cancer

A small number of studies have examined the potential for tumor formation in animals and the potential for enhanced cell proliferation in in vitro studies. The animal studies are limited because they typically use one sex of one species at one dose rate. Just a few studies have actually been designed to assess the potential promotional effects of RF fields. One study used three bioassays (lung cancer colony, spontaneous breast tumors, and benzopyrene-induced skin cancer) to evaluate tumor promotion at two exposure levels: 2–3 W/kg and 6–8 W/kg. The study also used sham controls and a chronic confinement stress group that was maintained in smaller cages. For all measures, the high-SAR group had the greatest response, but this was consistently followed by the confinement stress group, then the low-SAR group (Szmigielski et al., 1982). In a long-term study benign tumors did not differ for MW-exposed and sham controls: 192 neoplastic lesions were identified, with the endocrine system having the "highest incidence of neoplasia in the aging rats, as is to be expected in this experimental animal" (Guy et al., 1985). No one type of primary malignant neoplasm was significantly increased, but when the data for primary malignant lesions at death were combined for all organs and tissues, the exposed animals had a significantly increased incidence (Johnson et al., 1984; Guy et al., 1985). In discussing this finding, the researchers state: "The collapsing of sparse data without regard for tissue origin is useful in detecting possible statistical trends, and the finding here of a relative increase in primary malignancies in the exposed animals is provocative; however, when this single finding is considered in the light of other parameters evaluated, it is questionable if the statistical difference reflects a true biological activity." (Guy et al., 1985; Kunz et al., 1985). Reviews of RF exposure and cancer are available for the interested reader (Kirk, 1984; Adey, 1988; Szmigielski and Gil, 1989).

5.4 Survey of Human Exposures

Studies evaluating potential health effects in humans include epidemiology, clinical investigations, and incident or case reports. Studies have focused on radar workers, VDT operators, physical therapists, heat sealer operators, and ham radio operators. A number of literature reviews that address human studies have been published and are available to the interested reader (Silverman, 1973, 1980, 1985; Albrecht and Landau, 1978; Michaelson, 1983; Hill, 1984; Roberts and Michaelson, 1985; Heynick, 1987; Dennis et al., 1991a,b).

5.4.1 Epidemiologic Studies

Most epidemiologic studies of RF/MW exposure have examined individuals exposed in the military or in industry. A few studies investigated populations living or working near generating sources (Lester and Moore, 1982a,b), although the results in at least one of these studies has been disputed (Lester and Moore, 1982a; Lester, 1985; Polson and Merritt, 1985).

Studies are primarily cross-sectional surveys or case reports. Some case-control designs have been utilized to evaluate general and specific causes of morbidity and mortality. Some studies provide insufficient detail to allow a clear understanding of the design strategy. Some clinical investigations contain elements of cross-sectional surveys, for example, both disease and exposure are ascertained simultaneously, but a control group is not included.

Information about health status has come from medical records, questionnaires, physical and laboratory examination, and vital statistics. Sources of exposure data include personnel records, questionnaires, environmental measurements, equipment emission measurements, and (assumed adherence to) established exposure limits. Accurate estimates of exposure or dose present formidable problems in most epidemiologic studies (Silverman, 1979, 1980, 1985).

An early clinical study of U.S. Navy personnel did not reveal any conditions that could be ascribed to radar exposure (Daily, 1943). Four years of surveillance of industrial radar workers did not reveal significant clinical or pathophysiologic differences between the exposed and control groups (Barron et al., 1955; Barron and Baraff, 1958). On the other hand, surveys of Eastern European workers revealed functional changes in the nervous and cardiovascular systems (Gordon, 1966, 1970; Sadcikova, 1974).

In an extensive survey of Polish workers exposed to microwaves for various periods (Siekierzynski et al., 1974; Czerski and Siekierzynski, 1975), no dependence of the incidence of disorders, such as organic lesions of the nervous system, changes in translucency of the ocular lens, primary disorders of the blood system, neoplastic diseases, or endocrine disorders, could be shown on exposure level, duration, or work history. The incidence of functional disturbances (neurasthenic syndrome, gastrointestinal tract disturbances, cardiovascular disturbances with abnormal ECG) reported in Eastern European publications (Gordon, 1966, 1970; Sadcikova, 1974) was found not to be related to the level or duration of occupational exposure. There were no instances of irreversible damage or disturbances caused by exposure to MW energy (Czerski and Piotrowski, 1972).

The Medical Follow-Up Agency of the U.S. National Academy of Sciences studied morbidity and mortality among 40,000 U.S. Navy personnel (Robinette et al., 1980) potentially exposed to radar. There was no indication of any adverse effect due to MW exposure. This study was preceded by a survey to investigate physiological and physical effects among U.S. Navy crewmen potentially exposed to 0.1–1 mW/cm^2 aboard an aircraft carrier (U.S. Senate, 1977). No significant differences were found with respect to task performance, physiological tests, or biological effects. Hematological findings were within the normal range.

A study of 4388 employees and 8283 dependents of the American Embassy in Moscow potentially exposed to 5–15 μW/cm^2 of MW for variable periods of time (9–18 hr/day) up to 8 months, showed no differences in mortality experience and various morbidity measures (Lilienfeld et al., 1978; U.S. Senate, 1979). Exhaustive comparative analyses were made of all symptoms, conditions, diseases, and causes of death among employees and dependent groups of adults and children. No differences in health status by any measure could be attributed to MW exposure. No genetic or other adverse biologic effects among employees and dependents attributable to microwave exposure could be established. An altered lymphocyte count was believed to be of microbial origin (U.S. Senate, 1979).

Hill (1988) performed a longitudinal study of the mortality experience of the technical and scientific staff who worked at the MIT Radiation Laboratory (Rad Lab) between October 1940 and January 1946. Technicians were not included in the study group, and they were likely to have received some of the highest MW exposures. Mortality data were coded for occupation and cause of death. Three control groups were used: U.S. white males, white male physician specialists, and internal comparisons within the Rad Lab cohort. No exposure or dose information was available, so exposures were estimated. Mortality analysis showed that approximately half the deaths were circulatory system diseases and a quarter due to neoplasms. When compared to U.S. white males, a nonsignificant increase in death due to mental, psychoneurotic, and personality disorders was observed. Standardized mortality ratios (SMRs) were significantly lower in the MW exposed group for other study measures compared to U.S. white males, which is attributed to the healthy worker effect. In comparison with physician specialists, SMRs were significantly increased for death due to digestive disease, Hodgkin's disease, malignant neoplasms of the gall bladder and bile ducts, and cirrhosis of the liver. Scrutiny of the 95 percent confidence interval (CI) shows that estimates associated with malignant neoplasms of the gall bladder and bile ducts are somewhat imprecise. Significantly reduced SMRs were observed for death from circulatory disease and external causes. No dose–response trend was found when the data were analyzed by exposure group. In summary, the study reported a generally positive mortality picture of a well-defined cohort. However, exposures were estimated and some misclassification of exposure potential may have occurred.

5.4.2 Ocular Effects

Numerous surveys of ocular effects of MW/RF energies in humans have been made. Most investigations have involved military personnel and civilian workers at military bases and in industrial settings. The principal factors of interest have been the significance of minor lens changes in the cataractogenic process and cataracts (opacities impairing vision). Several cases of cataract attributed to MW exposure have been reported but not substantiated. There is no clinical or experimental evidence that ocular lens damage allegedly due to MW exposure is morphologically different from lens abnormalities from other causes, including aging. It is also difficult to relate cause and effect because lens imperfections do occur in otherwise healthy individuals, especially with increasing age. Also drugs, industrial chemicals, other physical agents, and certain metabolic diseases are associated with cataracts.

Lenticular defects too minor to affect visual acuity have been studied as possible early markers of microwave exposure or precursors of cataracts. The studies have been mainly prevalence surveys, although the time periods are often variable or not specified; reexamination data rarely permit estimates of incidence. Some generalizations, however, can be made about observations of lens changes in microwave workers and comparison groups (Silverman, 1979, 1980).

1. Lens imperfections occur normally and increase considerably with age among employed males studied. There is evidence that lens changes increase with age even during childhood (Zydecki, 1974). By about age 50, lens defects have been reported in most comparison subjects, based on data from various studies.

2. Although a few reports suggested differences (Cleary and Pasternack, 1966; Majewska, 1968; Zydecki, 1974), there is no clear indication that minor lens defects are a marker for MW exposure in terms of type or frequency of changes, exposure factors, or occupation. The reported earlier appearance of lens defects in MW-exposed workers than in comparison groups is not convincing because there is considerable variation in the type, number, and size of defects recorded; in the scoring methods used by different observers; and in the numbers examined.

3. Clinically significant lens changes, which would permit selection of individuals to be followed up, have not been identified (Zaret et al., 1963).

4. There is no evidence from ophthalmological surveys to date that minor lens opacities are precursors of clinical cataracts; a case-control study of World War II and Koran War veterans was negative for cataracts (Cleary et al., 1965).

Neither definitions nor methods of detection of cataract are standardized (Silverman, 1979, 1980). The common meaning of cataract, a lens opacity that interferes with visual acuity, is open to many interpretations as to degree and nature of the opacity and loss of visual acuity (see section on Optical Radiation). Although many factors are associated with cataracts, many cataracts are loosely called "senile" when they occur after middle age. Alleged MW-induced cataracts are not distinguishable from other cataracts in the opinion of most ophthalmologists (Appleton, 1973; Shaklett et al., 1975; Hathaway et al., 1977; Hathaway, 1978).

The most prominent characteristic of cataract incidence is the age distribution. Estimates of incidence are not comparable because of differences in the population groups surveyed, in addition to the nonuniformities in detection and definition. However, all studies point to a low incidence until about the fifth decade of life, when a sharp increase occurs. Although not comparable with general population figures, recorded mean annual incidence rates are of the order of 2 per 100,000 (Silverman, 1979, 1980). In a preliminary U.S. estimate of the total prevalence of cataracts by age in the civilian, noninstitutionalized population aged 1–74 years, one or more cataracts was found in 9 percent of the population (Health Examination Statistics, 1979). For age groups under 45 years, the incidence increased gradually from 0.4 percent in ages 1–5 years to 4 percent in ages 35–44 years. The pronounced increase that occurs after age 45 reaches a maximum in the oldest group examined. Over one-half of the population in the 65–74 years age group had cataracts. Cataract incidence for active duty military personnel (who are mainly healthy, relatively young men) show a similar age dependence up to about age 55 (Odland, 1972).

5.4.3 Nervous System and Cardiovascular Effects

Clinical and laboratory studies of MW-exposed workers in Eastern European countries are reported to have shown CNS and cardiovascular reactions to MW/RF exposure (Gordon, 1970; Sadcikova, 1974; Baranski and Czerski, 1976). Functional disturbances of the CNS have been described as "radiowave sickness," also called the neurasthenic or asthenic syndrome. The symptoms and signs include headache, fatigability, irritability, loss of appetite, sleepiness, sweating, thyroid gland enlargement, difficulties in concentration of memory, depression, and emotional instability. The clinical syndrome is generally reversible if exposure is discontinued (Dodge, 1969; Healer, 1969; Dodge and Glaser, 1977). A form of neurocirculatory asthenia is also attributed to nervous system influence. Effects indicated by hypotonus, bradycardia, delayed auricular and ventricular conduction, decreased blood pressure, and ECG alterations in workers in RF/MW fields have been reported (Sadcikova and Orlova, 1958; Gordon, 1966, 1970). The identification and assessment of these poorly defined, nonspecific complaints and symptom complexes (syndromes) is extremely difficult (Silverman, 1968, 1973). These changes, however, do not diminish the capacity to work and are reversible (Peacock et al., 1971).

In a more recent study, 113 operators of plastic welding machines and 23 female controls were interviewed and their E- and H-field strength exposures were measured. Of the study subjects 38 were selected for a more extensive neurological examination of the hands, arms, shoulders, and neck. RF-exposed individuals reported more neurasthenia, fewer headaches, and about the same as controls for feeling tired. Forty percent of the total exposed group and 53 percent of the exposed women reported hand numbness (paresthesia) versus 22 percent of the controls. Use of a two-point discrimination test showed a significant decrease in response for RF exposure and a dose–response relationship between numbness and either machine type or E field. Around 32 percent of the subjects and 22 percent of the controls had disturbed sensory conduction velocities as measured by electroneurography, which also affects about 10 percent of the general population in Sweden. Of those receiving

RF exposure, 58 percent had signs of carpal tunnel syndrome and the remainder had "a more peripheral effect." The authors suggest that finger disturbance might be related to RF exposure (Kolmodin-Hedman et al., 1988). However, the significant findings in the two-point discrimination test were found by comparing men and women to women controls. Furthermore, it is possible that the effects preceded exposure, or could be associated with other workplace factors. For example, carpal tunnel syndrome could be associated with material handling. The reported ocular effects also could be associated with other workplace factors, since workers were also exposed to airborne contaminants associated with the heating process. Finally, the results of this study cannot be generalized because the number of subjects is small, and the control group is of one sex and much smaller than the exposed group.

In an evaluation undertaken by the U.S. Center for Devices and Radiological Health of the health status of male physical therapists, 3004 responses were received (58% response rate) from a self-administered questionnaire. For therapists older than 35 years, the prevalence of heart disease in the high- versus low-exposure group was significantly and consistently elevated. The prevalence of heart disease in the study group was less than that in a general population with similar sex, age, and race. Analysis by age group showed heart disease to be significantly elevated for shortwave diathermy and combined MW/shortwave diathermy, but not for MW diathermy (Hamburger et al., 1983). It is possible that the low response rate introduced a selection bias into this study.

Dose-related effects on blood pressure and unspecified blood disorders in a cohort of 3093 male physical therapists have been reported. No effects were found for heart disease. This study combined data on work with shortwave (RF energies, typically at 27.12 MHz) and MW diathermy instead of examining morbidity for either type. The results were not peer reviewed and were reported as an abstract from a society meeting (Stellman and Stellman, 1980).

5.4.4 Reproduction

In a study of 31 male MW-exposed workers, 22 men reported decreased sexual drive and disturbances with erection and orgasm and/or ejaculation. This was attributed to neurasthenic disturbances, which reportedly occurred in 25 of the men. Laboratory analysis found a number of significant decreases in semen quality (Lancranjan et al., 1975). These findings have not been reliably correlated with MW exposure. It is possible that some other environmental agent associated with MW work could be involved, a factor not considered in the design or analysis.

A nested case-control study was performed within a cohort of female physical therapists in Sweden. Cases included 36 women with infants who had died without a major malformation and all infants with major malformations; there were 67 controls: 33 percent of the cases and 14 percent of the controls had used shortwave (27.12 MHz) diathermy often or daily (Kallen et al., 1982).

A retrospective analysis of a cohort of about 5000 Finnish female physical therapists found a significant difference in spontaneous abortion when shortwave diathermy use was ≥ 5 hr/week and the length of pregnancy was > 10 weeks, but this result was not significant when adjusted for confounding. A significant difference in congenital malformations was observed when shortwave diathermy was used between 1 and 4 hr/week, but not when used for ≥ 5 hr/week, which may be due to recall bias. When adjusted for potential confounding, shortwave use > 1 hr/week was still significantly increased. The use of MW diathermy resulted in fewer cases of congenital malformation than observed for controls (Taskinen et al., 1990).

Concern about a cluster of 4 infants with malformations out of 25 pregnancies in a physical therapy staff initiated a Danish study. From a cohort of female physical therapists, congenital malformations were examined in a case-referent study of 54 cases and 248 controls. There were no significant findings by sex of the offspring for duration of exposure or for the level of peak exposure (Larsen, 1991). In a further examination, no differences were found in spontaneous abortion, stillbirth/death within 1 year, prematurity, and low birthweight. There were significant differences in the sex ratio, that is, fewer boys were born to mothers who had indicators of higher exposure. A significant difference in sex ratio was also found for women who use shortwave diathermy, especially units with plate electrodes. Fields around plate electrodes were the strongest of the three types of electrodes evaluated (Larsen and Skotte, 1991; Larsen et al., 1991).

An assessment of early fetal loss was made of female members of the American Physical Therapy Association who used MW or shortwave diathermy 6 months prior to conception or during the first trimester. The odds ratio for MW diathermy, adjusted for prior fetal loss, was marginally significant [odds ratio (OR) = 1.26, 95% CI = 1.00–1.59]. The OR increased with reported increases in monthly usage of MW diathermy, and this exposure–response effect was statistically significant. The opposite was observed for shortwave diathermy, where the OR decreased with increased use (Stewart and Ouellet-Hellstrom, 1991; Ouellet-Hellstrom and Stewart, 1993).

A number of studies have examined reproductive effects in VDT operators (Kurppa et al., 1985; Ericson and Kallen, 1986a,b; Nurminen and Kurppa, 1988; Bryant and Love, 1989; Brandt and Nielsen, 1990; Nielsen and Brandt, 1990, 1992; and Windham et al., 1990), finding no significant differences in study measures. In one study a slight increase in risk of spontaneous abortion was observed for all working women who used VDTs in current pregnancies (relative [RR] = 1.19, 90% CI = 1.09–1.30) but not in previous pregnancies. The authors attribute the result to recall bias (McDonald et al., 1988). Another study found that use of VDTs for more than 20 hr/week during the first trimester is associated with an increased incidence of spontaneous abortion (OR = 1.8, 95% CI = 1.2–2.8) (Goldhaber et al., 1988). However, it should be noted that these studies, and most others, did not measure exposures for any operators, but instead used "VDT use" as the chief criterion of exposure assessment. By so doing, these studies are evaluating the effects of the work environment on reproductive outcome and not specifically the effects associated with electromagnetic energies.

Only one study has included actual operator measurement data. Researchers evaluated reproductive outcomes in two groups of communications operators. The exposed cohort consisted of 323 directory-assistance operators who used cathode-ray-tube (CRT) VDTs at work. The control cohort was 407 general telephone operators who used workstations that had similar design. However, the visual displays used by controls did not emit very low frequency (VLF) fields unlike the CRT-type VDTs and were either neon glow tube or light-emitting diode (LED) displays. Information was collected by telephone interview and exposure verified by payroll records, along with extensive measurement data. Summary statistical values showed that average E and H fields incident at the operators were less than, respectively, 1.5 V/m and 0.417 A/m (Tell, 1990). Abdominal VLF exposure to VDT operators was significantly higher than in non-VDT workers. No significant differences were found between the exposed and control groups for live births, stillbirths, and spontaneous abortion during the first trimester of pregnancy. No dose–response effect was observed when data were analyzed by hours of VDT use/week. Spontaneous abortion was associated with a history of prior spontaneous abortion, alcohol consumption, cigarette smoking, and evidence of a thyroid disorder (Schnorr et al., 1991).

5.4.5 Growth and Development

A case-control study of Down's syndrome in relation to exposure to ionizing radiation yielded an unexpected finding regarding paternal exposure to radar (Sigler et al., 1965). Apparently, fathers of children with Down's syndrome gave more frequent histories of occupational exposure to radar during military service than did fathers of unaffected children, a difference that was of borderline statistical significance. Exposure during military service occurred before the birth of the affected child. After publication of the first report in 1965, expansion of the study group, follow-up of all fathers to obtain more detailed information about radar exposure, and a search of available military personnel records were all undertaken. The suggestive excess of radar exposure to fathers of babies with Down's syndrome was not confirmed on further study (Cohen et al., 1977).

A report of congenital anomalies at a U.S. Army base (Peacock et al., 1971) suggested that adjoining communities surrounding the base had reported cases of clubfoot that greatly exceeded the expected number. A more detailed investigation showed that there was a considerably higher rate of anomalies among births to military personnel than in the state as a whole. Analysis showed that apparently there were errors in the malformation data on the birth certificates and a probable overreporting from the base. Thus convincing evidence was lacking that radar exposure was related to congenital malformations (Burdeshaw and Schaffer, 1976). The higher malformation rate across a group of counties of the state was presumably environmentally induced, but no specific agent was suggested (Silverman, 1979, 1980).

A few human data are available from studies in which RF was used to therapeutically heat the pelvic region. Gellhorn (1928) reported the temperature in women was raised to 46°C. Although the author was concerned about possible harmful effects, he did not allude to specific complications. In another report four women were treated with MW diathermy (2.45 GHz) before or during pregnancy (Rubin and Erdman, 1959). Three women delivered normal infants; the fourth, who received eight treatments during the first 59 days of pregnancy, aborted on day 67 but delivered a normal baby after a subsequent pregnancy during which she again received MW treatment. The authors concluded that microwaves did not interfere with ovulation, conception, and pregnancy, but the study group is too small to generalize.

Microwave heating has been used to relieve the pain of uterine contractions during labor (Daels, 1973, 1976). The analgesic effect was found helpful in 2000 selected patients without obstetric pathology, and the babies were born healthy with good circulation. No evidence of injury was manifest in a 1-year follow-up of the children; there was no evidence of mental retardation. Four cases of chromosome anomalies in controls and two cases in the irradiated group were noted. It is important to note that the human fetus at parturition is almost fully developed; thus gross structural defects at this late stage of development would not be expected.

There are reported case studies of increases in congenital abnormalities in women working in RF fields in Eastern Europe (Marha et al., 1971), but there are no unequivocal reports of MW-induced human teratologies.

5.4.6 Cancer

Microwave-induced cancer has not been reported in medical surveillance examinations of microwave workers or military personnel (Silverman, 1979, 1980). Two cohort epidemiologic studies (Lilienfeld et al., 1978; Robinette et al., 1980) that investigated the question systematically did not show an excess of any form of cancer that could be interpreted as MW related (Silverman, 1979, 1980). A later study of naval personnel found an increased risk of leukemia in 1 of 95 occupations exposed to electromagnetic radiation. This was an electricians' mate, a job category with sub-RF exposure anticipated to be around 60 Hz (Garland et al., 1990). The cancer mortality experience of members of the MIT Rad Lab staff was generally favorable. Cancers accounted for about 25 percent of the deaths. Over 80 percent of the 99 neoplasms were from four major sites: digestive organs and peritoneum, respiratory systems, genitourinary organs, and lymphatic and hematopoietic tissue. A further examination of the 13 neoplasms of the lymphatic and hematopoietic tissues showed nothing remarkable (Hill, 1988).

Hamburger and colleagues (1983) found eight malignancies and three nonmalignant neoplasms in a cohort of 3004 male physical therapists. Three of these were malignant melanomas, which they conclude is "greater than might be expected for white males in the 35–39 age group." The neoplasms could not be associated with a specific type of diathermy (Hamburger et al., 1983).

In a population-based study, Milham determined proportionate mortality ratios (PMRs) for leukemia deaths for white males in Washington (WA) state. Occupational groups with potential RF exposure included radio and telegraph operators, television and radio repairmen, electrical engineers, and (arc) welders. Acute leukemia was significantly increased for TV and radio repairmen (Milham, 1982, 1985a).

A number of studies have evaluated cancer deaths among amateur radio operators. Although an avocation, these studies could have occupational implications. Cause of death was determined for males in the states of California (CA) and WA who have been members of the American Radio Relay League. Statistically increased causes of death were found for all leukemias, myeloid leukemia, acute myeloid leukemia, and chronic myeloid leukemia. From study of occupations of the Washington group, Milham concluded that work in occupations involving potential exposure to electromagnetic fields could probably not explain the observed excess (Milham, 1985b).

Milham extended this study through a search of FCC records for radio amateurs. The SMRs were significantly lower for death by all causes, all malignant neoplasms, all circulatory diseases, all respiratory diseases, and all accidents. Significantly increased SMRs were found for malignant neoplasms of other lymphatic tissue in CA and CA + WA but not for WA. Lymphatic and hematopoietic tissue malignancies were significantly elevated for CA, but not for either WA or the two states combined. Prostate cancer was significantly elevated in WA but not in CA or the combination. Leukemia was nonsignificantly increased, and acute myeloid leukemia was significantly increased for CA and WA. According to Milham, the risk estimates may be underestimates because females were excluded and U.S. death rates, not the death rates of CA and WA, were used. From occupational information on death certificates, Milham estimated that around 31 percent of the amateurs from WA had jobs that would have exposed them to electromagnetic fields at work. Comparable information from the CA group was not available (Milham, 1988a).

An evaluation was performed by amateur radio license class because this could be a crude measure of exposure duration. Technicians had elevated ratios for lymphatic and hematopoietic cancers, while those individuals with a general license had increased multiple myeloma and other lymphomas. Other SMRs were elevated, but none significantly. No general disease trend was observable in the data for death by all causes, all malignant neoplasms, brain cancer, lymphatic and hematopoietic cancer, all leukemias, and myeloid leukemia (Milham, 1988b).

The conclusions from Milham's studies are limited because exposure to electromagnetic fields is inferred. Two reports were hypothesis-generating studies (PMR) reported as letters to journal editors (Milham, 1985b, 1988c). The studies did not control for other potential exposures for these hobbyists, including power frequency electric and magnetic fields, electric shock, exposure to airborne contaminants from soldering and degreasing, and potential ingestion of lead and other metals from workbench contamination during soldering.

Anecdotal concerns have been raised in trade newspapers over the possibility that emissions from traffic-control radar units may be associated with cancer. Reported measurement data indicates that potential exposures are well below human exposure criteria (Pavlik, 1991; Whiteacre, 1991; Bitran et al., 1992). In the only epidemiological study, Davis and Mostofi (1993) reported a significant excess of testicular cancer in policemen in two counties in WA (observed/expected $[O/E] = 6.9$, $P < 0.001$). According to the authors these policemen had a practice in common of resting a radar gun in their lap while the gun was on. This hypothesis requires further study with a more rigorous design including a better defined exposed population, matched controls, confounder control, and exposure estimates including microwave levels and exposure duration.

5.4.7 Eastern European Reports

Nervous system perturbations and behavioral reactions in humans after exposure to MW energy have been reported mostly in Eastern European publications that describe subjective complaints consisting of fatigability, headache, sleepiness, irritability, loss of appetite, and memory difficulties (Gordon, 1960; Presman, 1968; Petrov, 1970;

Marha et al., 1971). Psychic changes that include unstable mood, hypochondriasis, and anxiety have also been reported. Most of the subjective symptoms are reversible (Dodge, 1969; Healer, 1969), and pathological damage to neural structures is insignificant (Orlova, 1971).

These reports of MW sickness were met with skepticism by Western scientists (Dodge, 1969; Healer, 1969). Western critiques noted that the findings were based on workplace surveys, and these surveys often lacked appropriate design elements, and relied too heavily on reported subjective complaints. Several reviewers (Michaelson, 1972, 1974, 1975; Guskova and Kochanova, 1975; Albrecht and Landau, 1979; Silverman, 1979, 1980) have noted the difficulties in establishing the presence, and quantifying the frequency and severity, of subjective complaints. Individuals suffering from a variety of chronic diseases may exhibit the same dysfunctions of the CNS and cardiovascular system as those reported to be a result of MW exposure (Guskova and Kochanova, 1975). Thus it is extremely difficult, if not impossible, to rule out other factors in attempting to relate MW exposure to clinical conditions. Although these reports are numerous, they lack the necessary level of detail that allows for a meaningful interpretation.

Also, the reports lack the necessary documentation of study methods, and when documented, some of the methods were unique and unfamiliar to Western scientists. Symptoms were reported at levels less than those viewed as safe in the West, around 10 mW/m² continued, and often less than 1 mW/m² (Dodge, 1969; Marha, 1971; Sadcikova, 1974; Baranski and Czerski, 1976).

5.4.8 Cutaneous Perception

Perception of microwave energy is a function of cutaneous thermal sensation or pain. The data on temperature sensation in Table 14.9 are for a forehead surface area of 37 cm². The threshold data for pain (Table 14.10) are for a 9.5-cm² area exposed at 3 GHz. Several studies suggest that a threshold sensation is obtained when the temperature of the warmth receptors in the skin is increased by a certain amount (ΔT) (Cook, 1952; Hendler et al., 1963; Hendler, 1968).

Table 14.9 Stimulus Intensity and Temperature Increase to Produce a Threshold Warmth Sensation

Exposure Time (s)	3 GHz Power Density (mW/cm²)	10 GHz Power Density (mW/cm²)	10 GHz Skin Temperature Increase (°C)	Far Infrared Power Density (mW/cm²)	Far Infrared Skin Temperature Increase (°C)
1	58.6	21.0	0.025	4.2–8.4	0.035
2	46.0	16.7	0.040	4.2	0.025
4	33.5	12.6	0.060	4.2	—

Source: Data from Hendler (1968) and Hendler et al. (1963).

Table 14.10 Threshold for Pain Sensation as a Function of Exposure Duration

Power Density (W/cm^2)	Exposure Duration (s)
3.1	20
2.5	30
1.8	60
1.0	120
0.83	>180

Source: Data from Cook (1952).

5.4.9 Critique of Human Studies

An important concept of disease causation is that, in general, disease is not caused by a single factor or agent but rather is influenced by multiple, interactive components, including subjects and their environment. Health effects or manifestations of disease have a spectrum of intensity ranging from the barely discernible and rapidly reversible symptomatic disorders, through an increasing gradient of severity to the point of irreversibility, and finally, to disease states of such gravity as ultimately to cause death. For electromagnetic fields, in common with most other agents, biological, chemical, or physical, the trivial end of this severity scale includes detectable effects, which are well within the range of physiological adaptation and do not constitute disease in any meaningful sense.

Eastern European reports describe such symptoms as listlessness, excitability, headache, drowsiness, fatigue, and cardiovascular deficits in persons occupationally exposed to electromagnetic fields. These symptoms are also caused by many other occupational factors, so it is not possible to define a cause–effect relationship. Many other factors in the industrial setting or home environment as well as psychosocial interactions can cause similar symptoms. In addition to smoking and obesity, genetic factors, and emotional stress, psychological personality factors that determine an individual's reactivity to environmental conditions are proven risk factors in the development of cardiac ischemia (Guskova and Kochanova, 1975). Thus these factors are important in assessing the effects of environmental insults.

Analysis of occupational exposure to MW/RF energies is fraught with many difficulties. Of utmost importance is the assessment of the relationship between exposure levels and the health status of the examined groups of workers. Quantitation of occupational exposure is extremely difficult and usually lacking in studies. This requires that exposure assessment utilize surrogate measures such as job classification, where exposure is inferred. An important limitation of epidemiologic studies of MW/RF exposure is the lack of recognized pathophysiologic manifestations at realistic levels of exposure as indicators for measuring the effects of the fields on humans. The problem of adequate control groups may be controversial, but in many studies appropriate control groups are lacking. The possible role of other environmental factors and of socioeconomic conditions must be taken into account. As often happens in such studies, it is difficult to show a causal relationship between a disease and the influence of environmental factors, at least in individual cases (Czerski and Siekierzynski, 1975).

5.5 Protection Guides and Standards

5.5.1 U.S. Exposure Standards

The first U.S. standards for limiting RF/MW exposure were introduced in the 1950s (Steneck et al., 1980; Wilkening, 1991). These were primarily for manufacturers and users of radar equipment. In 1966, the United States of America Standards Institute (USASI) [now the American National Standards Institute (ANSI)] published the first U.S. consensus standard, C95.1-1966. The exposure guideline was 10 mW/cm^2 averaged over any 6-min period, for frequencies from 10 MHz to 100 GHz (USASI, 1966). The ACGIH published a TLV in 1970 (ACGIH, 1970), and in 1971 the USASI standard was incorporated into the Occupational Safety and Health Standard for General Industry (OSHA, 1971). However, because of the use of advisory language, the exposure limit in the specific standard, 29 CFR 1910.97, has been ruled unenforceable.[1] A key legal case in 1987 was interpreted to allow OSHA inspectors to use state-of-the-art standards; this interpretation has been applied to RF/MW exposure criteria. The USASI standard was reissued with minor modifications in 1975 as an ANSI standard (ANSI, 1974), and then underwent a major revision in the 1982 (ANSI, 1982). The ACGIH made significant revisions to their guidelines in 1981 (ACGIH, 1981).

These standards published in the early 1980s incorporated information on frequency-dependent absorption, the concept of the SAR, and reversible behavior disruption in test animals as the primary biologic criterion. In 1985, the National Institute for Occupational Safety and Health (NIOSH) released a draft recommended exposure limit (REL) for RF/MW for external review, but this document was never officially issued (NIOSH, 1985). The Federal Communications Commission [FCC] adopted the 1982 ANSI standard for use in licensure activities of broadcast facilities (Cleveland, 1985; FCC, 1985), and the EPA proposed approaches for controlling public exposure to RF in 1986 (EPA, 1986). In the same year, the National Council on Radiation Protection and Measurements (NCRP) published their recommendations for acceptable exposures (NCRP, 1986). In 1990 the ACGIH recommended criteria for sub-RF fields and adjusted the lower bound of the RF TLV up to 30 kHz (ACGIH, 1990b). The C95.1 standard was revised in 1992 by a professional society, the Institute of Electrical and Electronics Engineers (IEEE), as IEEE C95.1-1991 (IEEE, 1992).

The IEEE standard applies to frequencies between 3 kHz and 300 GHz for individuals in controlled or uncontrolled environments. In the controlled environment, exposed individuals know that there is the potential for exposure. Individuals clas-

---

[1]Personal Communication from Jeffrey Snyder (1987), Health Standards Section, OSHA, Washington, D.C.

sified as being in the uncontrolled environment have "no knowledge or control of their exposure" (IEEE, 1992). Although the controlled environment is most generally applicable to the workplace, there are times when it could be applied in the home. For example, a radio amateur will be in the controlled environment at home because he/she has knowledge that his/her hobby will expose them to radio-frequency fields. The maximum permissible exposures (MPEs) for the two environments are in Tables 14.11 and 14.12. The MPEs are in terms of E- and H-field strength, and power density (S).

The IEEE standard, and most other standards that are currently in use, utilize five regions that define an envelope curve, as shown in Figure 14.2. In three regions the

Table 14.11 Maximum Permissible Exposures (MPE) for Controlled Environments

Frequency Range (MHz)	E-Field Strength (V/m)	H-Field Strength (A/m)	Power Density (mW/cm^2)	Averaging Time (min) E^2, H^2 or S
0.003–0.1	614	163	NA	6
0.1–3.0	614	16.3/f	NA	6
3–30	1842/f	16.3/f	NA	6
30–100	61.4	16.3/f	NA	6
100–300	61.4	0.163	1.0	6
300–3000	NA	NA	f/300	6
3000–15,000	NA	NA	10	6
15,000–300,000	NA	NA	10	616,000/f^2

[a]NA, not applicable; f, frequency in MHz.
Source: Data from IEEE (1992).

Table 14.12 Maximum Permissible Exposures (MPE) for Uncontrolled Environments

Frequency Range (MHz)	E-Field Strength (V/m)	H-Field Strength (A/m)	Power Density (mW/cm^2)	Averaging Time (min) E^2 or S	H^2
0.003–0.1	614	163	NA	6	6
0.1–1.34	614	16.3/f	NA	6	6
1.34–3.0	823.8/f	16.3/f	NA	f^2/0.3	6
3–30	823.8/f	16.3/f	NA	30	6
30–100	27.5	158.3/$f^{1.688}$	NA	30	0.636$f^{1.337}$
100–300	27.5	0.0729	0.2	30	30
300–3000	NA	NA	f/1500	30	NA
3000–15,000	NA	NA	f/1500	90,000/f	NA
15,000–300,000	NA	NA	10	616,000/$f^{1,2}$	NA

NA, not applicable; f, frequency in MHz.
Source: Data from IEEE (1992).

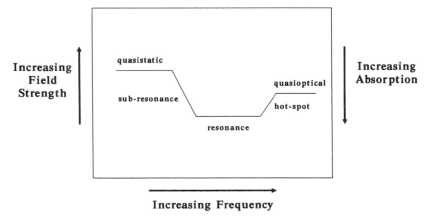

Increasing Frequency

Figure 14.2. Five spectral regions in the radio-frequency exposure limits shown graphically. A hypothetical human absorption spectrum is included to demonstrate the envelope nature of the curve.

MPE plateaus at invariant, but different, exposure values. In the other sloping, transition regions, the MPE varies with frequency and must be calculated. The envelope curve is necessary because of the frequency dependence of the SAR for the human body. Absorption is maximized in the whole-body resonance part of the spectrum from about 30–300 MHz and the MPE is at a minimum.

The derived exposure limits (workplace and controlled environment) are based on maintaining a whole-body average SAR of 0.4 W/kg. From studies of effects on animal behavior, the SAR threshold for harmful effects was established at 4 W/kg to which a safety factor of 10 was applied. Where whole-body exposure is concerned, the MPEs are believed to result in energy deposition averaged over the entire body for any 0.1-hr period of about 144 J/kg or less. This is equivalent to an SAR of about 0.40 W/kg or less, spatially and temporally averaged over the entire body mass.

The units of the derived limits include E-field strength (V/m) and H-field strength (A/m), or the square of these units (V^2/m^2 and A^2/m^2). The equivalent plane-wave power density (W/m^2 or mW/cm^2) may also be used at frequencies less than 300 MHz, but is usually restricted to use at MW frequencies, 300 MHz to 300 GHz.

For the IEEE standard, the spectral region where the SAR applies is 100 kHz to 6 GHz. In this region the derived limits, MPEs, must be frequency dependent to maintain the whole-body SAR at 0.4 W/kg. Above 6 GHz the penetration depth into tissues decreases, with penetration limited to the skin above 30 GHz where the "depth of penetration approaches a constant value of the same size as the penetration of infrared radiation" (Schwan and Piersol, 1954). In this region, quasioptical considerations apply. The power density is 10 mW/cm^2 at frequencies greater than 3 GHz for the controlled environment and 15 GHz for the uncontrolled environment. At the lower frequencies, the interaction with the body is defined in terms of static fields or slowly time-varying fields, and is called quasistatic.

The low-frequency H-field values were relaxed in the 1992 IEEE standard because they do not contribute as significantly to the whole-body SAR as do electric fields, nor are they as effective in inducing currents in the human body (Orcutt and Gandhi, 1988; Gandhi, 1988b). H-field levels are the same for the controlled and uncontrolled environments up to 30 MHz, then the levels for the uncontrolled environment decrease more rapidly from 0.54 A/m at 30 MHz to 0.0729 A/m at 100 MHz.

Averaging time for field-strength limits is 6 min for frequencies less than 15 GHz in the controlled environment. As shown in the last column of Tables 14.11 and 14.12, averaging time is based on the square of the E- and H-field strengths or on the power density (S). All averaging time calculations must use these quantities E^2, H^2, or S. Above 15 GHz averaging time, in minutes, decreases as $616,000/f^{1.2}$, where f is frequency in MHz. The averaging time decreases from 6 min at 15 GHz to 10 s at 300 GHz. This reduction was necessary to assure that skin burns did not occur in the quasioptical part of the spectrum (Petersen, 1991).

Low-frequency exposures may produce burns, shock, and high local SAR. At frequencies less than about 3–30 MHz, it is possible for ungrounded conductive objects that are influenced by the field to store RF energy as an electric charge. If a grounded person touches the object, RF currents may be discharged to the body. Objects that may satisfy the criteria of conductivity and isolation from earth include metallic roofs, metallic fences, vehicles, guy wires, scaffolding, rebar, aluminum studs, or part of a crane.

Modest values of RF current may cause perception. This manifests as a tingling or pricking sensation at frequencies less than about 100 kHz, and a sensation of warmth at higher frequencies (Dalziel, 1972; Gandhi, 1987). As the level of current increases, a sensation of pain may be elicited with the potential for RF-induced shock and burns becoming increasingly significant. Conductive objects can develop open-circuit voltages of hundreds of volts when exposed to the ANSI C95.1-1982 E-field limit between 0.3 and 3 MHz (Gandhi, 1990). When the circuit is closed by contact with a grounded human body, currents above the let-go current threshold can exist during finger contact.

Currents may also be induced within the human body when exposed to E fields where wavelengths are greater than about 2.5 times the length of the body (Guy, 1987). These currents flow through the body to ground, where they may be measured as the short-circuit current through the feet (Deno, 1977). These currents are associated with exposure to the E field, while the H field adds little to the current flowing to ground. The path to earth takes the current through the ankle, where the current flows in the high-conductivity muscle tissue (Dimbylow, 1988, 1991). This current flow may produce high currents and local SARs in the ankle (Gandhi et al., 1985; Gandhi, 1988a; Hoque and Gandhi, 1988).

For contact and induced currents, the IEEE has established the criteria shown in Tables 14.13 and 14.14, which apply at frequencies less than 100 MHz. The contact current limit is for grasping contact, while the induced current limit is for flow through one or both feet. Note that the allowable current through either foot is one-half that allowed through both feet. Averaging time for measurements of induced or contact currents is 1 s.

The need for both field strength and current limits at the lower frequencies reflects

Table 14.13 Maximum Permissible Induced and Contact Currents for Controlled Environments

Frequency Range (MHz)	Both Feet (mA)	Each Foot (mA)	Contact (mA)
0.003–0.1	$2000f$	$1000f$	$1000f$
0.1–100	200	100	100

f, frequency in MHz.
Source: Data from IEEE (1992).

Table 14.14 Maximum Permissible Induced and Contact Currents for Uncontrolled Environments

Frequency Range (MHz)	Both Feet (mA)	Each Foot (mA)	Contact (mA)
0.003–0.1	$900f$	$450f$	$450f$
0.1–100	90	45	45

f, frequency in MHz.
Source: Data from IEEE (1992).

the different exposure conditions. The possibility of contacting ungrounded conductive objects may not exist in all exposure environments. In this case field-strength measurements would suffice. If the potential for contact currents does exist, then the field-strength limits will not necessarily be sufficiently conservative, and the limits for current must also be used.

The ACGIH recommends TLVs for workers only as shown in Table 14.15. In comparison to pre-1989 recommendations, the ACGIH has adjusted the lower bound of its RF TLV from 10 to 30 kHz. Below 30 kHz, the so-called sub-RF limits are based on acceptable values of current density (ACGIH, 1992). In 1993 the ACGIH introduced a notice of intended change of the RF TLVs (ACGIH, 1993). Except for

Table 14.15 Threshold Limit Values (TLVs) for Radio Frequency and Microwave Radiation Exposure

Frequency Range (MHz)	E-Field Strength (V^2/m^2)	H-Field Strength (A^2/m^2)	Power Density (mW/cm^2)
0.03–3	377,000	2.65	100
3–30	$3770(900/f^2)$	$900/(37.7\,f^2)$	$900/f^2$
30–100	3770	0.027	1
100–1000	$3770(f/100)$	$f/3770$	$f/100$
1000–300,000	37,700	0.265	10

f, frequency in MHz.
Source: Data from ACGIH (1992).

the lower boundary, these values are identical to the IEEE limits for the controlled environment shown in Tables 14.11 and 14.13.

Other criteria include the recommendations by the NCRP and the International Radiation Protection Association (IRPA). The NCRP recommends limits for workers and members of the general public. The limits for the worker are identical to the ANSI C95.1-1982 criteria and are based upon an SAR of 0.4 W/kg averaged over a 6-min period (NCRP, 1986).

For exposures to members of the general public, the SAR is reduced by a factor of 5 to 0.08 W/kg, and the averaging time is increased by the same factor to 30 min. This approach varies the SAR between workers and members of the public but maintains the same total dose, 144 J/kg. If the exposure fields satisfy a given set of criteria for ELF amplitude-modulated fields, the exposure guidelines for the general public should be used for the occupational setting. In this case the carrier frequency must be modulated to a depth of ≥ 50 percent with a modulating signal having frequencies between 3 and 100 Hz. This is a unique provision among all present exposure standards, and the rationale is not clearly delineated. The recommendations are based on observations from primarily in vitro experiments on calcium efflux. These experiments are difficult to replicate, being highly dependent on methodology, and the effects are not robust. Furthermore, the NCRP offers no guidance to the health professional on where such fields might exist and how to evaluate them (NCRP 1986).

5.5.2 *International Exposure Standards*

A number of countries have published exposure guidelines. For comprehensive reviews and recommendations for specific countries that are not addressed below, the interested reader is directed to a number of references (Marha, 1971; Czerski, 1985; Repacholi, 1987, 1990; Stuchly, 1987, 1989; Bernhardt, 1988; Czerski and Bernhardt, 1989; Grandolfo and Hansson Mild, 1989; Szmigielski and Obara, 1989).

The International Commission on Nonionizing Radiation Protection (ICNIRP, once part of IRPA) has recommended the first international consensus standard on RF/MW. This organization makes recommendations for occupational and public exposures (Tables 14.16 and 14.17). The ICNIRP limits for the workplace are com-

Table 14.16 IRPA Limits for the Workplace

Frequency Range (MHz)	E-Field Strength (V/m)	H-Field Strength (A/m)	Power Density (mW/cm^2)
0.1–1	614	$1.6/f$	NA
>1–10	$614/f$	$1.6/f$	NA
>10–400	61	0.16	1.0
>400–2000	$3f^{1/2}$	$0.008f^{1/2}$	$f/400$
>2000–300,000	137	0.36	5.0

NA, not applicable; f, frequency in MHz.
Source: Data from IRPA (1988).

Table 14.17 IRPA Limits for the General Public

Frequency Range (MHz)	E-Field Strength (V/m)	H-Field Strength (A/m)	Power Density (mW/cm^2)
0.1–1	87	$0.23/f^{1/2}$	NA
>1–10	$87/f^{1/2}$	$0.23/f^{1/2}$	NA
>10–400	27.5	0.073	0.2
>400–2000	$1.375f^{1/2}$	$0.0037f^{1/2}$	$f/2000$
>2000–300,000	61	0.16	1.0

NA, not applicable; f, frequency in MHz.
Source: Data from IRPA (1988).

prised of both field-strength values and limits on contact and induced currents. These latter values are 200 mA for induced currents and 50 mA for contact currents between the frequencies of 0.1 and 10 MHz. The contact current value is also recommended for the general public (IRPA, 1984, 1988b).

In Canada the recommendations for federal workers made by the Bureau of Radiation and Medical Devices (BRMD) in Safety Code 6 include field-strength limits for workers and the general population. The squares of the field strengths for the working population are about five times those of the general public. Averaging time is 0.1 hr, although there are provisions for time periods less than this. The contact current limits for workers may cause perception in some individuals, while the values for the public will not (BRMD, 1991).

In the United Kingdom the National Radiological Protection Board (NRPB) recommends guidelines from 0 Hz to 300 GHz. The basic limit is called the "basic restriction" and includes the current density (A/m^2) from 1–100 kHz; the SAR from 100 kHz to 10 GHz; the power density from 10–300 GHz; and the SA for pulsed fields. From 0.1–10 MHz the basic restriction includes both the current density in the head, neck, and trunk, and the whole-body average SAR. The derived levels called "investigation levels" are specified in field strength or power density and are used "for investigating whether compliance with basic restrictions is achieved" (NRPB, 1993).

Recently revised limits are proposed in Germany. A proposal for physical agents in the European Community including RF and lower frequency fields was based on the limits recommended by the ICNIRP. Japan has also made a recommendation for a RF exposure limit.

5.5.3 Product Emission Standards

The Radiation Control for Health and Safety Act of 1968 (PL 90-602), administered by the USFDA provides authority for controlling radiation from electronic devices (USFDA, 1971). Their microwave oven standard requires that oven measurable leakage at 5 cm from any surface may not exceed 1 mW/cm^2 prior to purchase and 5 mW/cm^2 for the life of the product (USFDA, 1971).

The Canadian standard (IEC, 1976; Repacholi, 1978; Stuchly et al., 1979) restricts the maximum leakage to 1 mW/cm^2 at 5 cm from the oven (consumer, commercial, and industrial). If the cavity does not have a load, the acceptable leakage level is 5 mW/cm^2.

5.5.4 Problems and Recommendations

Elucidation of the biological effects of RF/MW exposure requires a careful review and critical analysis of the available literature. Such a review requires the differentiation of established effects and mechanisms from speculative and unsubstantiated reports. Most experimental studies support the concept that the biological effects of RF/MW exposure are primarily a response to hyperthermia or altered thermal gradients in the body. There are, nevertheless, areas of confusion, uncertainty, and disagreement.

There is a philosophical question about the definition of hazard. One objective definition of injury is an irreversible change in biological function as observed at the organ or system level. Thus it is possible to define a hazard as a probability of injury on a statistical basis. It is important to differentiate between the levels that are hazardous and those representing biological effects or perception. Obviously, this is because all effects are not necessarily hazards. In fact, some effects may be beneficial under appropriately controlled conditions, such as the medical use of diathermy and hyperthermia. RF/MW-induced effects must be understood sufficiently so that their clinical significance can be determined, their hazard potential assessed, and the appropriate benefit–risk analysis applied. It is important to determine whether an observed effect is irreparable, transient, or reversible. Of course, even reversible effects are unacceptable if they transiently impair the ability of the individual to function properly or to perform a required task.

A critical review of studies of the biological effects of RF/MW exposure indicates that many of the investigations suffer from inadequacies of either technical facilities and dosimetry determination or insufficient control of the biological specimens and criteria for biological change. More sophisticated conceptual approaches and rigorous experimental designs must be developed. There is a great need for systematic and quantitative comparative investigation of the biological effects, with the use of well-controlled experiments. This should be done by using sound biomedical and biophysical approaches at the various levels of biological organization from the subcellular to the whole animal on an integrated basis, with full recognition of the multiple associated and interdependent variables.

It is important that research be conducted in such a way that all aspects of the study are quantified, including the type and magnitude of the effect, and how the effect relates to the results obtained by other investigators. For RF/MW bioeffects, body size of the experimental animal must be taken into account. Absorption cross sections and internal heating patterns vary with body size and wavelength. Generally, a small animal exposed at high frequency at a given intensity will absorb more energy than a large animal exposed under the same conditions. The converse can hold at low frequencies. In the performance of experimental studies on animals, interspecies scaling factors must, therefore, be used for extrapolation to humans. Also the distribution of energy absorption may produce hot spots within the animal.

It is possible that this may lead an investigator to think that a low-level or a non-thermal effect is being manifested in an animal because the incident power is low, whereas in fact the absorbed energy may be localized in a specific region of the body and not contribute that greatly to a measurable temperature rise but still produce an effect. Studies require the selection of biomedical parameters that consider basic physiological functions and work capacity, identification of specific and nonspecific reactions, and differentiation of adaptational or compensatory changes from pathological manifestations.

Well-designed and appropriately controlled epidemiological and clinical investigations of groups of workers and others exposed to RF/MW should be fostered. Although a number of recent studies of radar workers and physical therapists have strong design elements, they share a major limitation with previous studies: no measured exposure or doses can be assigned to any individuals included in the studies. Studies of individuals exposed to MW/RF energies along with appropriate control groups should include a thorough analysis of the exposure environment, including cofactors as well as electromagnetic fields. There is always the danger that real factors may be overlooked, leading to false association with factors included in the study. Such interacting factors could be heat, cold, chemical agents, hypoxia, noise, other radiant energies such as x-rays, chronic disease, medication, stress, and ergonomic considerations.

Because of the difficulties in extrapolating from animal experiments to humans, epidemiological studies, including appropriate clinical and laboratory examinations, are essential to improve our understanding of possible health hazards from exposure to MW/RF energies. As noted by Silverman, it is difficult to identify exposed populations, select suitable controls, and obtain exposure data (Silverman, 1979, 1980). Some study groups already characterized can be improved by the acquisition of additional exposure data, some groups should be followed for longer periods of time, and some should be investigated for additional endpoints. Hypotheses that have been generated should be evaluated by independent investigators.

6 EXTREMELY LOW FREQUENCY FIELDS

6.1 Introduction

Early studies of possible health effects from occupational exposure to extremely low frequency (ELF, 30–300 Hz but usually applied to frequencies below 30 Hz also) electric and magnetic fields were published in the literature of the former USSR in the mid-1960s. Electric switchyard workers reported a variety of subjective complaints, including problems with their cardiovascular, digestive, and central nervous systems. A study of electric utility linemen in the United States at about the same time failed to find any adverse health effects. Similar studies of electric utility workers continued for the next decade, with similarly contradictory results. In the early 1980s, attention broadened to workers whose jobs fit the category "electrical occupation," and a number of analyses showed an increased risk, primarily of leukemia, for "electrical workers" in such varied occupations as electrical engineer, electronics engineer, television and radio repairman, electrician, motion picture projectionist, and telephone lineman. However, the actual exposures to electric and magnetic fields of these workers were not characterized, as is typical in these types of hypothesis-generating studies, and there were known confounding exposures. Nevertheless, the studies raised important questions, which are being pursued through well-designed epidemiologic studies. An important aspect of these latter studies is that exposure assessments are based on actual measurements.

To be able to evaluate health effects fully, the observational, epidemiologic studies must be accompanied by controlled laboratory experiments. When in vitro studies yield mechanisms for observed effects and in vivo studies confirm the possible occurrence of such effects, then there is a foundation for judging whether there is a cause-and-effect relationship. Such work is also being pursued in many laboratories throughout the world.

Industrial hygienists are increasingly asked to consider whether there are health effects from occupational exposure to electric and magnetic fields. The answer is not known at this time. This section provides background on the issue by explaining the nature of fields and their interaction with matter, sources of high strength fields, the status of extant biological and epidemiological research, and recommended exposure guidelines.

6.2 Extremely Low Frequency Fields

6.2.1 Background

When the distance from the source of electromagnetic radiation is large compared to the wavelength, the electric and magnetic fields are linked and must be considered together. This is referred to as the *far*, or *radiation*, zone. There is an electromagnetic field. With a relatively small 3-m wavelength, a television signal with a frequency on the order of 10^8 Hz, broadcasts, or radiates, as an electromagnetic quantity.

However, when the distance from the source is small with respect to the wavelength, the electric and magnetic fields are not linked. This distance is referred to as the *near*, or *static*, zone. The fields are independent and can be considered as separate entities. One is always in the near or static zone of ELF fields because of their long wavelength, on the order of 1000 km or longer. ELF fields behave and are treated as separate, independent, nonradiating electric and magnetic fields at any conceivable observation point. This is referred to as the *quasistatic approximation*. Also, in materials with the electric and magnetic properties of living tissues, ELF fields have long wavelengths and penetration ("skin") depths (the distance over which the field intensity decreases to 37%, or $1/e$ of its initial value). For example, the skin depth of a 60-Hz field in tissue is about 150 m. In interacting with humans and other living organisms, then, electric and magnetic fields behave independently. Hence, ELF fields are discussed here as though generated by orthogonal, quasistatic, electric, and magnetic field sources.

6.2.2 Field Definitions and Units

Electric fields are created by electric charges. The electric field, E, is defined by the magnitude and direction of the force it exerts on a static unit charge:

$$\mathbf{F} = q\mathbf{E} \tag{11}$$

where $\mathbf{F}$ = force [newton (N)]
$\quad q$ = charge [coulomb (C)]
$\quad \mathbf{E}$ = electric field (V/m)

(Vector quantities are in bold type.) The magnitude or intensity of $\mathbf{E}$ describes the voltage gradient, or the difference in voltage between two points in the field. E-field intensities near alternating current (ac) high-voltage transmission lines are in the range of kilovolt per meter (kV/m).

The magnetic flux density, $\mathbf{B}$, characterizes the magnetic field strength, $\mathbf{H}$, as $\mathbf{E}$ characterizes the electric field and is linearly related to the magnetic field through the following:

$$\mathbf{B} = \mu\mathbf{H} \tag{12}$$

where $\mathbf{B}$ = magnetic flux density [tesla (T)]
$\quad \mu$ = magnetic permeability
$\quad \mathbf{H}$ = magnetic field strength (A/m)

The magnetic permeability depends on the medium; the magnetic permeability of a vacuum is designated μ_0. Air and biological matter have permeabilities essentially equal to μ_0. This means that the magnetic flux density is unchanged by these materials.

B fields are created by moving charges, or currents. [In this section, the letter B will be used to denote the magnetic field unless the magnetic field strength (H) is intended.] This applies to all fields, whether they are from magnets, power lines, or the earth. Just as the E field is defined by the force on a unit charge, the B field is defined by the magnitude and direction of the force exerted on a moving charge or current:

$$\mathbf{F} = q(\mathbf{v} \times \mathbf{B}) \tag{13}$$

where $\mathbf{F}$ = force (N)
$\quad q$ = unit charge (C)
$\quad \mathbf{v}$ = velocity (m/s)

The magnetic flux density may be represented by lines of induction per unit area; the unit is the tesla, T. One tesla is 10,000 gauss (G), a frequently used engineering unit. The microtesla (μT) is more convenient for environmental levels. Convenient conversions are that 1 μT is 10 mG ($= 0.8$ A/m), and 1 mT is 10 G. The static magnetic flux density of the earth is about 50 μT.

A system of units that preceded the present SI system was the CGS system, which used units of centimeters, grams, and seconds. In the CGS system, the permeability, μ_0, is dimensionless and equal to one. As a result, $\mathbf{B}$ numerically approximates $\mathbf{H}$ in the CGS system, and the two came to be used interchangeably. The value of μ_0 in the SI system is $4\pi \times 10^{-7}$ henry per meter (H/m). This factor can be used to convert true magnetic field strength (H) (A/m) to magnetic flux density (B) (T) in air by the expression $\mathbf{B} = \mu_0\mathbf{H}$.

6.2.3 Interaction with Humans

E fields interact with humans through the outer surface of the body, inducing fields and currents within the body. Hair vibration or other sensory stimuli may occur in fields greater than 10 kV/m. A safety issue arises from currents induced in metal structures, which may produce shocks when humans contact the structure and provide a path to ground.

An E field will cause currents to flow in the body, as expressed by Ohm's law:

$$\mathbf{J} = \sigma\mathbf{E} \tag{14}$$

where $\mathbf{J}$ = induced current density (A/m^2)
σ = tissue conductivity [sievert per meter (S/m)]
$\mathbf{E}$ = electric field strength (V/m)

A grounded person in an electric field experiences a short-circuit current of approximately (Tenforde and Kaune, 1987):

$$I_{sc} = 15 \times 10^{-8}fW^{2/3}E_0 \tag{15}$$

where I_{sc} = short-circuit current (μA)
f = frequency (Hz)
W = weight (g)
E_0 = external electric field strength (V/m)

Thus, a person weighing 70 kg would have a total short-circuit current of about 153 μA in a 10-kV/m field. Deno (1977) and Kaune and Phillips (1980) investigated current flow in models of humans and laboratory animals exposed to 60-Hz E fields. Their data indicate that current densities induced in a grounded, erect person exposed to a 10-kV/m vertical E field are 0.55 μA/cm^2 through the neck and 2 μA/cm^2 through the ankles.

Time-varying B fields induce E fields, which in turn induce currents, in tissue in direct proportion to the magnetic flux density, the frequency of oscillation, and the radius of the current loop. The E field and current flow are perpendicular to the flux density. A vertically-directed flux density will cause current to flow in standing humans in loops whose plane is perpendicular to the vertical axis. For a sinusoidally varying flux density and a circular current flow, the current density can be expressed as:

$$J = \sigma\pi frB \tag{16}$$

where f = frequency (Hz)
r = loop radius (m)

With average tissue conductivity equal to 0.2 S/m, the current density at the perimeter of the torso of an adult can be approximated as:

$$J = 0.1fB \qquad (17)$$

The maximum current density induced in the normal residential environment is of the order of $\mu A/m^2$. For electric arc welders it may be of the order of mA/m^2.

With photon energy directly proportional to frequency, it is readily apparent that ELF fields, which have frequencies 13 orders of magnitude less than very weak ionizing radiation, and 5 orders of magnitude less than radiation associated with significant tissue heating, will cause neither ionization nor heating. The SAR may be calculated by the expression

$$SAR = \sigma E^2/2\rho \qquad (18)$$

where E = internal electric field strength (V/m)
   ρ = tissue density (g/cm^3)
   σ = tissue conductivity (S/m)

Because the body is a relatively good conductor, the highest internal field, that can be induced by an electric field strength in air, is about 1 V/m, which leads to an SAR of 10^{-4} W/kg, four orders of magnitude less than the resting metabolic rate (Tenforde, 1991). Pulsed B fields can produce higher internal E fields, but they are still too small to produce measurable tissue heating. Any interactions of ELF fields in air with humans are thus nonthermal. Biological effects have, however, been observed in laboratory studies, sparking a good deal of interest in developing theoretical descriptions of the phenomena. This is discussed later.

6.3 Sources and Exposures

6.3.1 Naturally Occurring Fields

The natural environmental static E field is about 130 V/m, formed by charge buildup from thunderstorms. The earth is negative relative to the upper atmosphere. Naturally occurring, oscillating ELF E fields exist primarily due to thunderstorms, and generally decrease in amplitude with increasing frequency. Field strengths range up to about 1 kV/m but may reach to 20 kV/m during thunderstorms. Frequencies may extend through the ELF range, to about 1 kHz. The typical atmospheric field strength at 50–60 Hz is only about 10^{-4} V/m. The static magnetic field of the earth is about 50 μT. Natural ELF B fields are quite low, about 10^{-6} μT at 60 Hz (EHC, 1984).

6.3.2 Man-Made Sources

E and B fields are produced by the generation, transmission, and use of electricity, and anything in this path is a potential exposure source, from the generator to the power lines to an electric drill or clock. Occupational sources and source characteristics are as varied as the occupational environments in which they are found, but two basic facts may be noted:

1. Sources of strong E fields will be associated with the presence of electrical charge, such as around high-voltage equipment.
2. Sources of high B fields will generally be characterized by high currents, such as around high-amperage equipment or locations of high current flow.

These need not be associated with heavy industry. Electrical transformers and wiring located in vaults in office buildings are nonindustrial sources of high fields.

E fields are typically about 10 kV/m directly under the conducting wires and 1 or 2 kV/m at the edge of the transmission line right-of-way. By contrast, 50- to 60-Hz electric fields from building wiring, power tools, and other electrical appliances typically range only up to 100 V/m (Miller, 1974).

Electrical appliances may have flux densities of about 10 μT in their very immediate vicinity. The home environment generally has levels of about 0.1 μT. Some occupational environments, such as those where induction heaters are in use, may have flux densities of 10–100 mT.

6.3.3 Occupational Exposure Levels

Data on field levels near specific sources are increasingly being generated and reported. Extant data have often been collected to characterize sources expected to produce high fields. For example, Stuchly and Lecuyer (1989) measured fields at the worker's position for 22 electric arc welding machines, which use high amounts of current. E fields were generally low, about 1 V/m, and the highest value was 300 V/m. The arithmetic mean was 47 V/m. Magnetic flux densities ranged from about 1 μT to a few hundred μT, and averaged 136 μT. The highest measurement was 1 mT at a worker's hand. The measurements were reported for the frequency of highest flux density. This was usually 60 Hz, although for some sources it was 120 or 180 Hz, due to the presence of harmonics. In addition, some machines had a ''highly complex frequency spectrum.''

As another example, Rosenthal and Abdollahzadeh (1991) measured flux densities in microelectronics fabrication rooms. In the aisles of the rooms, values ranged from 0.01 to 0.7 μT and averaged 0.07 μT. Higher levels were found near specific pieces of equipment. Levels ranged from 0.3 to 7 μT, with a mean of 3.1 μT at a distance of 2 ft from a furnace. Next highest was a sputterer, for which levels ranged from 0.2 to 1.5 μT and averaged 0.6 μT, also at 2 ft from the source. The authors estimated an 8-hr time-weighted-average exposure for a worker using the furnace to be 1.8 μT, and 0.6 μT for the worker using the sputterer.

Very high flux densities were measured at the worker's position at welding machines and steel furnaces by Lovsund et al. (1982) in Sweden. For welders, the highest level was 10 mT near a spot welder. An even higher value, 70 mT, was measured at an induction heater.

Although specific source contributions were not measured, the effects of large motors, compressors, and switchyard sources were captured by Cartwright et al. (1993) during 8-hr exposures of electrical workers in a petroleum refinery. Low-voltage electrical distribution workers had a higher geometric mean exposure than did high-voltage workers, 0.53 compared to 0.46 μT. Such exposures are not remarkable, but high exposures occurred in the latter group during one particular operation, with values ranging from 0.2 to 1.8 mT. The geometric mean exposure was 71 μT, and the mean peak was 870 μT. The results of these studies are summarized in Table 14.18.

Table 14.18 Exposure Data Measured for Occupations Having Relatively High Magnetic Flux Densities

Reference	Site, Location	Flux Density (μT)	
		Mean ± SD	Range
Stuchly and Lecuyer (1989)	Electric arc welders	136 ± 47	
Rosenthal and Abdollahzadeh (1991)	Microelectronics Fabrication rooms		
	Aisles	0.07	
	Aligner	0.26 ± 0.14	0.15–0.5
	Etcher	0.4 ± 0.25	0.1–0.7
	Sputterer	0.62 ± 0.77	0.15–1.5
	Furnace	3.1 ± 3.1	0.3–7.0
	Inspection station	0.27 ± 0.16	0.05–0.5
Lovsund et al. (1982)	Welding machines		
	Arc welding		100–6000
	Submerged melt		500–2000
	Flash weld		100–3000
	Spot weld		200–10,000
	Seam weld		4000–4500
	Electroslag refining		500–2000
	Steel furnaces		
	Arc furnace		100–1000
	Ladle furnace		200–8000
	Induction furnace		100–900
	Induction heater		1000–70,000
Cartwright et al. (1993)	Petroleum refinery (geometric mean values)		
	High-voltage workers	0.46 ± 0.41	
	Low-voltage workers	0.53 ± 0.34	
	Maintenance electricians	0.23 ± 0.20	
	Nonelectrical workers	0.17 ± 0.21	

Other studies have collected data for a variety of work environments. These have often focused on electric utility workers or on so-called electrical workers, whose work might be expected to expose them to high-intensity fields. The findings of many of these studies are summarized below.

Swedish workers in a 400-kV substation spent most of their time in fields below 5 kV/m, with brief exposures above 15 kV/m (Knave et al., 1979). A study of Canadian linemen and substation workers used measurements and task activity patterns to estimate daily exposures. Estimates ranged from 50 to 60 (kV/m) · hr for 500 kV linemen to 13 (kV/m) · hr for 115–230 kV lineman, and they were 12 (kV/m) · hr for substation workers (Stopps and Janischewskyj, 1979).

Using a personal exposure meter, which was worn on the arm, Male et al. (1984) measured exposures of electrical workers in the United Kingdom. The device had a threshold of 6.6 (kV/m) · hr. Among 166 transmission workers (equipment rated 132, 275, or 400 kV) and 121 distribution workers (equipment rated at 132 kV or below), only 26 transmission workers and two distribution workers had cumulative, 10-day exposures above the threshold. Among the 26 transmission workers, the median daily exposure value was 1.5 (kV/m) · hr per day, and the maximum value was 24.3 (kV/m) · hr per day.

Farmers whose land is crossed by high-voltage transmission lines represent another exposed population with higher-than-normal peak exposures. Using a combination of measured and modeled concentration data, it was determined that the annual exposure of this group might range from 10 to 120 (kV/m) · hr, with differences being attributed to the voltage of the lines (EPRI, 1985). Peak exposures ranged above 8 kV/m.

The Bonneville Power Administration (BPA, 1986) reported the use of personal electric field exposure monitors to measure cumulative exposures of utility employees. Highest exposures, 1.7 (kV/m) · hr, occurred for linemen. Exposures generally rose with the voltage of the equipment, and daily maxima ranged from 5.1 to 7.6 (kV/m) · hr.

More recently, Bowman et al. (1988) sampled 114 work sites. Electrical worker environments had geometric mean E fields of 4.6 V/m and flux densities of 0.5 μT. Secretaries had values of 2–5 V/m, 0.31 μT if they used a VDT, and 0.11 μT if they did not. For power line workers, the overhead line environment yielded geometric means of 160 V/m and 4.2 μT. A value of 5.7 μT was determined for underground lines. Other findings included 298 V/m and 3.9 μT at a transmission substation, and 72 V/m and 2.9 μT at a distribution substation. Radio and television repair shops yielded 45 V/m, while AC welding produced 4.1 μT. The data are summarized in Table 14.19.

Personal exposure data have been collected for work, nonwork, and sleep periods in a study of 36 Canadians—20 utility workers and 16 office workers (Deadman et al., 1988). The time-weighted average of one week's data yielded a geometric mean E field of 10 V/m. The utility workers' geometric mean B field exposure was 0.31 μT. It was 0.19 μT for the office workers. Both groups had a level of 0.15 μT while sleeping. While at work, the utility workers' exposures averaged 48.3 V/m and 1.66 μT. Office workers were exposed to a geometric mean level of 4.9 V/m and 0.16 μT. The data are summarized by occupation in Table 14.20.

Table 14.19 Summary of Spot Magnetic Flux Density Measurements by Bowman et al. (1988)

Occupation	Environment	Geometric Mean (μT)
Electricians	Industrial power supply	10.3
Power line workers	Underground lines	5.7
	Overhead lines	4.3
	Home hookups	0.1
Welders/flame cutters	TIG/AC	4.1
	TIG/DC	0.7
Power station operators	Transmission	3.9
	Distribution	2.9
	Generating	0.6
	Control room	0.2
Electronics assemblers	Sputtering	2.4
	Soldering	0.1
	Microelectronics	0.003
Projectionists	Xenon arc	1.4
Fork-lift operators	Battery powered	1.2
Electronics engineers/technicians	Laser laboratory	1.1
	Calibration laboratory	0.06
	Office	0.02
Radio/TV repairers	Repair shops	0.6
Radio operators	Dispatchers	0.03
Secretaries	VDT operators	0.3
	Other	0.1
Electrical workers		0.5
All occupations		0.5
Residences		0.06

Table 14.20 Summary Workplace Exposures Measured by Deadman et al. (1988)[a]

Occupation	E Field (V/m)	B Field (μT)	HFTE Field Geometric mean (ppm)
Apparatus electrician, transmission	181.7	3.44	0.862
Splicer, distribution	6.7	2.08	0.039
Lineman, distribution	62.5	1.45	0.286
Lineman, transmission	418.9	1.31	3.051
Apparatus mechanics	4.7	1.18	0.044
Generating station assistant operator	5.0	1.14	7.965
All occupations	48.3	1.66	0.331
Comparison group, office workers	4.9	0.16	0.002

[a]Values are expressed as geometric means of the average daily work exposures.

6.4 Biological and Health Effects

Biological effects from ELF E and B fields have been demonstrated in both in vitro and in vivo laboratory studies. A discussion of all the biological literature on ELF is beyond the scope of this section, and the reader is referred to the many excellent reviews (EHC, 1984, 1987; AIBS, 1985; NRPB, 1991a,b, 1992; ORAU, 1992). The more frequently cited studies and biological responses are described below.

6.4.1 In Vitro Studies

Two facts should be kept in mind regarding in vitro studies of cell cultures or tissue exposed to E and B fields. First, electric currents are often used rather than fields because the dosimetry is easier to determine. However, the electric currents applied in some studies may be unrealistically large or even impossible to achieve using externally applied fields in air. Second, when B fields are used, the induced circulating currents vary with radial distance from the center of the culture, which makes the dosimetry more difficult.

A study that fostered a great deal of follow-on laboratory and theoretical work was that of Bawin et al. (1975). Fresh chick brain tissue was spiked with the radioactive isotope, $^{45}Ca^{2+}$. Half of a brain was exposed to a 147-MHz field that was amplitude modulated at 16 Hz; the other half served as a control. Compared to the unexposed half, the exposed half had about a 20 percent increase in calcium exchange with the physiologic solution in which it was immersed. Other modulation frequencies were used, but none produced so large an effect. When plotted against modulation frequency, the relative amount of $^{45}Ca^{2+}$ exchanged followed a curve indicative of a resonant response. Without the 147-MHz carrier frequency, a 16-Hz E field had the opposite effect, namely, the exchange from the exposed half was decreased relative to the unexposed half.

In a follow-on study, Blackman et al. (1982) found an increase rather than a decrease in $^{45}Ca^{2+}$ exchange in chick brains exposed to combined 15-Hz electric and magnetic fields. Later work suggested that the effect depended on the relative orientation and strength of the earth's magnetic field (Blackman et al., 1985), and that it occurred in "windows" of field frequency and intensity (Blackman et al., 1989) and even temperature (Blackman et al., 1991).

Liboff (1985) suggested that the resonant response was due to a cyclotron resonance phenomenon operating on K^+ ions, which in turn were involved in transmembrane exchange with the $^{45}Ca^{2+}$ ions. Studies of rat operant behavior by Thomas et al. (1986), of diatom motility by Smith et al. (1987), and of lymphocyte incorporation of $^{45}Ca^{2+}$ by Liboff et al. (1987) supported the proposed resonance mechanism. However, later studies of turtle colon by Liboff et al. (1988), of cells by Parkinson and Hanks (1989), of lymphocytes by Prasad et al. (1991), and of rat operant behavior by Stern and Laties (1992) were all negative. The last two studies were noteworthy as they attempted to replicate the earlier work of Liboff et al. (1987) and of Thomas et al. (1986).

The suggestion that the movement of $^{45}Ca^{2+}$ in these studies is due to a cyclotron resonance mechanism driving the ions through cell membrane pores has been discounted on a number of grounds. First, the ions are hydrated, making their effective mass much greater than that which fits the requirements for cyclotron resonance. Second, the radius of the ion's orbit would be of the order of meters, rather than the submicrometer size of cell membrane pores. However, other theoretical descriptions have been proposed by Male and Edmonds (1990) and by Lednev (1991), using classical physics and quantum physics descriptions of the modification of energy levels of a charged oscillator in a magnetic field. The modification of energy levels changes the binding rate of ions to proteins, which in turn alters their biological activity. Theoretical problems also exist with these descriptions (Adair, 1992), and initial laboratory studies have been unable to confirm them (Bruckner-Lea et al., 1992).

Most evaluations of genetic toxicology have found no reliable effects, but there are scattered positive findings (Murphy et al., 1993; McCann et al., 1993). Reported observations with cells related to ELF and cancer include: increase in the activity of the growth-related enzyme ornithine decarboxylase and ELF exposure (Byus et al., 1987); inhibition of human breast cancer cells with melatonin (Hill and Blask, 1988); ELF-induced reduction of the inhibitory effect of melatonin on human breast cancer cell proliferation (Liburdy et al., 1993); and synergistic effects of 60-Hz magnetic fields and ionizing radiation in producing clastogenic changes in human lymphocytes (Hintenlang, 1993).

6.4.2 In Vivo Studies

Laboratory animal studies have investigated the possible effect of E and B fields on a broad range of systems and outcomes, including body weight, hematology and immunology, the endocrine, cardiovascular, and nervous systems, circadian rhythm, behavior, genetics, reproduction and development, and cancer. In general, the results have been negative, inconsistent, or of limited relevance to questions of adverse human health effects (NRPB, 1991b). Work that has received more attention because of its implications for gross human health effects is summarized below.

Graves and Reed (1985) exposed chicken embryos to 60-Hz, sinusoidal E fields ranging up to 100 kV/m. The chicken embryo provided a rapidly developing, and hence relatively susceptible, yet stationary organism, so that exposure levels were well documented. No effects were found on mortality, deformity, birth weight, or postnatal development.

A very large scale study of chicken embryos exposed to pulsed B fields was conducted in six laboratories in North America and Europe (Berman et al., 1990). The fields had a 1-μT flux density pulsed at 100 Hz, with a rise and fall time of 2 μs, and a 500-μs pulse duration. Two laboratories found a significant increase in abnormal embryos in the exposed groups. Pooled data from all six laboratories also were significant. However, the interaction between the incidence of abnormalities and laboratory was also significant, that is, the effect of exposure differed significantly among laboratories.

Studies of reproduction and development in Hanford miniature swine were reported by Sikov et al. (1987). In this multigenerational study, females were exposed to 60-Hz E fields of 35 kV/m for 20 hr/day. When these and an unexposed control group were bred with unexposed males, the control group had a greater rate of fetal malformations than the exposed group. This result reversed upon a second breeding. Breedings of the offspring produced similarly conflicting results.

Because of the ambiguities in the swine study, a study using a similar protocol was performed using rats (Rommereim et al., 1988, 1990). Exposure intensities included 0, or control, and 10, 65, and 130 kV/m. There were no significant increases in litters with malformations for exposed animals compared to controls.

A developmental toxicology study in rats was conducted in replicate using exposures to 60-Hz magnetic fields. Flux densities were 0.09 μT (sham exposure), 0.61 μT, and 1000 μT. There were no significant differences in fetal body weight or the incidence of malformations and variations among the exposure groups (Rommereim et al., 1991).

The possible effects of ELF fields on levels of the pineal hormone melatonin levels has been investigated. Night-time suppression of melatonin was found in rats exposed to 60-Hz E fields of about 2 and 60 kV/m (Wilson et al., 1981). Replication of the findings has been difficult. According to Wood (1993), the current density induced in some of the melatonin studies is of the order of 0.005 (in vitro) to 1 mA/m^2 (in vivo).

A possible role of ELF fields in carcinogenesis has been proposed by Stevens (1987; Stevens et al., 1992), who hypothesized that a reduction of melatonin, which has been suggested to have tumor suppression properties, may lead to greater tumor growth. Stevens suggested a testable hypothesis where test animals would be dosed with a tumor initiator and then exposed to ELF fields, to study the hypothesis that ELF fields are tumor promoters.

Using this or a similar protocol, the results of some experiments with test animals are supportive of a promotional effect. An increase in the incidence of mammary tumors was observed when rats were exposed to 50-Hz B fields subsequent to treatment with DMBA (100 μT) (Loscher et al., 1993) or nitrosomethyl urea (20 μT) (Beniashvili et al., 1991). Results of another study were equivocal as exposure at 50 Hz and 30 mT did not increase the incidence of mammary tumors in rats. However, the number of tumors/tumor-bearing animal was increased, but this was not observed in a replicate (Mevissen et al., 1993). In a study of mice with spontaneous mammary tumors, exposed animals (12, 100, or 460 Hz, 4.5 or 6 mT); survived longer than controls and had less metastasis of tumor cells (Bellossi et al., 1988).

McLean and colleagues (1991) studied the effects of magnetic fields (2 mT, 60 Hz) on skin cancer initiated with DMBA. No promotional effect was observed. To study co-promotional effects, following DMBA treatment, mice were treated with the tumor promoter 12-O-tetradecanoylphorbol-13-acetate (TPA, a phorbol ester). In one report, the time to appearance of skin tumors was shorter in the exposed animals, but this finding was not statistically significant (McLean et al., 1991). In another experiment, a higher percentage of the ELF-exposed animals had tumors, and the average tumors/mouse were higher at weeks 12 and 18 but not at the end of the experiment (week 23) (Stuchly et al., 1992).

Rannug and colleagues (1993) also examined the promotional effects of magnetic fields at 50 Hz (50 and 500 μT) on skin cancer in female mice treated with DMBA. No significant differences were observed in tumors or skin hyperplasia between exposed and control animals. In a follow-up study, the promotional effects on skin cancer of continuous vs. intermittent (15 s on/off) exposure to mice were investigated at the same frequency and magnetic flux density. Although no significant differences were observed between exposed and control animals, significant differences in the accumulated skin tumors were noted between continuously and intermittently exposed animals (Rannug et al. 1994).

Results from experiments by Leung et al. (1987, 1988) using DMBA followed by E-field exposure (60 Hz, 40 kV/m) showed no significant difference between exposed and sham-exposed groups of rats in mammary tumors. Exposures of leukemia-implanted mice to 60-Hz B fields of up to 500 μT did not affect survival (Thomson et al., 1988).

6.4.3 Human Studies

Studies of human volunteers exposed to combined E and B fields (9 kV/m and 20 μT) have found a statistically significant slower heart rate and a change in brain-evoked potential. No significant responses occurred at exposures of 12 kV/m and 30 μT. The observed responses appeared to be associated with changes in field conditions rather than exposure strength or duration (Graham et al., 1990; Cook et al., 1992).

Phosphenes are involved in an unusual visual phenomenon generated by stimuli other than light, producing visual sensations having the appearance of flickering white light. It is observed with the eyes open or closed and appears to be generated through stimulation of retinal tissue. Stimuli that have been shown to produce phosphenes include pressure, mechanical shock, chemical substances, sudden fright, and ELF E and B fields (Lovsund et al., 1980a). ELF E and B fields generate responses called electrophosphenes and magnetophosphenes, respectively.

Electro- and magnetophosphene sensitivity appears to be greatest around 20 Hz (Adrian, 1977; Lovsund et al., 1980a,b). The estimated values of current density necessary for electrophosphenes are 7–70 mA/m^2 (Carstensen et al., 1985). B levels of 10–12 mT have stimulated magnetophosphenes (Lovsund et al., 1980a). Studies of magnetophosphenes indicate that sensitivity changes with dark adaptation and is different for individuals with normal and deficient color vision (Lovsund et al., 1979).

6.4.4 Epidemiologic Studies

This review will address occupational epidemiologic studies of ELF exposures, which have focused on cancer. For a review of residential studies of cancer and ELF exposure, see the summary by Savitz (1993). None has employed a rigorous exposure assessment; rather, exposure has been inferred from job titles or some other surrogates (Patterson, 1992). Designs have ranged from analyses of cohort death records to case-control studies of occupational groups. These are outlined below; extensive discussion may be found in the reviews cited earlier, particularly that of the NRPB (1992).

An increased risk of various forms of cancer among those employed in electrical occupations, such as electricians, electronics engineers, or radio repairmen has been identified in statistical analyses of death records (Milham, 1982; Coleman et al., 1983; McDowell, 1983). These studies commonly employ a statistic known as the proportionate mortality ratio, or PMR (observed deaths/expected deaths, expressed as a percent), to identify increased risk. As discussed in the WHO report cited below (EHC, 1984), the associations were not consistent among the studies, and they often involved small numbers of cases. Deficiencies included a lack of consistency in designating occupational classification, no consideration of mobility between occupations, and lack of consideration of confounding exposures to other physical and chemical agents.

With respect to the studies of workers in the electric utility industry exposed to high-intensity electric fields, the WHO concluded (EHC, 1984, pp. 82, 84):

> Few physiological or psychological effects in human beings have been credibly related to electric field exposure. Such effects, when reported, have often been questionable for the following reasons:
>
> a. Monitoring of symptomatology was subjective and was frequently not well-defined.
>
> b. Quantitative evaluation of effects was either not performed or was not clearly described.
>
> c. Control populations were poorly matched with exposed groups or were absent.
>
> d. E fields had been confounded by secondary factors (e.g., microshocks).
>
> e. Observation periods were often short.
>
> f. Exposure levels varied widely or were not documented, making it difficult to estimate accurately the magnitude and duration of exposure.
>
> g. Numbers of subjects in many of the earlier studies were insufficient to establish the statistical significance of adverse effects.

A more recent review article (Coleman and Beral, 1988), which considered 11 separate studies of workers in electrical occupations, concluded that the most consistent finding was a small increase in the risk of leukemia. The relative risk (RR) exceeded a value of 1 in 9 of the studies; however, the risk was significantly different from 1 in only 2 of the 11. By combining the results from all studies, the authors calculated an RR = 1.18, with a 95% CI of 1.09–1.29. The elevated risk of leukemia was "partly or wholly due to a 46% increase in the risk of acute myeloid leukemia" [AML] (RR = 1.46, 95% CI 1.27–1.65). The authors warned that the results are equivocal with respect to cause, because electrical workers are also exposed to agents besides fields, some of which may be leukemogenic.

In contrast to the results cited above for leukemia, a case-control study of deaths due to primary brain cancer or leukemia found no increased risk of leukemia among those in electrical occupations [odds ratio (OR) = 1.0, 95% CI 0.8–1.2] (Loomis and Savitz, 1990). There was an increased risk of brain cancer (OR = 1.4, 95% CI 1.1–1.7), with electrical engineers and technicians, telephone workers, electric power workers, and electrical workers in manufacturing industries all showing elevated risk. Among these groups, the odds ratios for leukemia were from 1.1 to 1.5, but none was significantly different from an odds ratio of 1. The greatest risk was for acute lymphocytic leukemia (ALL).

For workers in electrical occupations, London et al. (1994) reported a modest dose–response relationship between the magnetic field exposure and all leukemia (OR per 1 μT increase in average magnetic field = 1.2, 95% CI = 1.0–1.5). For chronic myelogenous leukemia the OR = 1.6 (95% CI = 1.2–2.0) per 1 μT increase in average magnetic field.

Sahl et al. (1993) studied leukemia and brain cancer in electrical utility workers. He found no statistically significant excess of these diseases in the study population.

Welders are among the most highly exposed workers to power-frequency magnetic fields (see Table 14.18). Analyzing published data, Stern (1987) found the RR for leukemia among welders to be essentially unity.

Theriault and colleagues (1994) found that utility workers exposed above the median cumulative magnetic field exposure value (3.1 μT yr) had an elevated risk of acute non-lymphocytic leukemia (OR = 2.41, 95% CI = 1.07–5.44) and AML (OR = 3.15, 95% CI = 1.20–8.27). For mean exposure above 2 μT, the OR for acute non-lymphocytic leukemia was 2.36 (95% CI = 1.00–5.58), while for AML the OR = 2.25 (95% CI = 0.79–6.46). There was no clear dose–response relationship and no significant association was observed for the other 29 cancer types evaluated.

Tynes et al. (1992) have reported significantly increased standardized incidence ratios (SIRs) among a cohort of 37,945 Norwegian electrical workers for cancers of the colon, pancreas, larynx, breast, and bladder. The highest was for breast cancer (SIR 2.07, 95% CI 1.07–3.61). Among leukemias, AML and chronic myeloid leukemia (CML) were significantly elevated (SIR = 1.56 and 1.97; 95% CI 1.06–2.26 and 1.10–3.26, respectively), but not other types. Exposure was assessed by occupational title. When analyzed by occupation, only radio/television repairmen and power line workers had significantly elevated SIRs for leukemia. The study design did not allow for analysis of the effects of confounders such as solvent vapors and solder fumes.

A Danish study (Guenel et al., 1993), which also judged exposure by job title, found no increase in the risk of breast cancer or brain tumor. The ratio of observed to expected incidence of leukemia was significant among men but not women, having a value of 1.64 (95% CI 1.20–2.24).

In a cohort study of male telephone workers, Matanoski et al. (1991) reported an excess risk of male breast cancer among central office technicians. Two cases were observed in the cohort of 9561 workers while none were expected. However, a later investigation revealed that ionizing radiation sources were found in the workplace that had not been considered in the original study (NIOSH, 1993).

In contrast to the results of Tynes et al. (1992), Floderus et al. (1992) reported an increased risk of chronic lymphocytic leukemia (CLL) but not AML in a case-control study in Sweden.

The epidemiologic literature suffers from a lack of good exposure assessment, adequate consideration of confounders, and conflicting findings among studies. Many of the criticisms raised by the WHO (EHC, 1984) remain in later studies. Occupational epidemiological studies are continuing in many countries, with exposure assessment relying increasingly on extensive exposure data. Results from these studies will become available over the next few years and should shed further light on any potential adverse health effects of occupational exposure to electric and magnetic fields.

6.4.5 Effects on Implanted Cardiac Pacemakers

A number of studies have examined the effects of power-frequency E fields on implanted cardiac pacemakers (Butrous et al., 1982, 1983a,b; Moss and Carstensen, 1985). They have shown that under certain conditions, electromagnetic interference (EMI) can affect pacemaker function by causing (1) premature pacemaker impulse formation, (2) inhibition of pacemaker impulse formation, (3) reduction in pacemaker rate, or (4) reversion of pacemaker function to the asynchronous or fixed-rate mode of operation instead of demand operation. Moss and Carstensen (1985) observed effects in subjects having unipolar pacemakers who were exposed to power-frequency electric fields ranging from 2 to 9 kV/m with measured body currents of 47 to 175 μA. Implanted bipolar pacemakers were not affected, nor were all models of unipolar pacemakers affected under the test conditions. They commented that pacemakers are generally most sensitive to input signals of 1–2 mV in the 10–40 ms pulse width range, and that these can come from pectoral muscle activity as well as from EMI. They also commented that pacemaker manufacturers can incorporate appropriate circuits in their devices to eliminate the EMI problem.

A short review included an assessment of what is required for a pacemaker patient to be at severe risk from EMI (Griffin, 1986). The patient must (1) have a susceptible pacemaker, that is, one of the few unipolar models that have been found to be affected; (2) be completely dependent on the pacemaker at the time of interference; and (3) experience the effects of interference long enough to lose consciousness (5–10 s). The number of patients at such risk was estimated by using the following data: (1) unipolar design, 50 percent of pacemakers, and susceptible, 10–20 percent of unipolar models; and (2) completely dependent patient, 20–25 percent. Applying these data to 500,000 pacemaker patients gives 5000–12,500 at risk. It was suggested that standardization of pacemaker resistance to EMI would be the preferred mode of protection.

6.5 Exposure Guidelines

Ideally, exposure guidelines and standards are established on the basis of an accepted mechanism of interaction, dose–response studies in animals, and epidemiological evidence of similar effects in humans. None of this has occurred for ELF fields. However, because of concerns from workers and the general public, exposure guidance has been developed by a number of countries and organizations. The ACGIH, the National Radiological Protection Board (NRPB) in the United Kingdom, the German government, and the ICNIRP/IRPA have developed standards or guidelines for exposures in the ELF range. Only the ACGIH and IRPA guidance will be discussed here.

The exposure limit rationale is based on induced body currents. Both the ACGIH and IRPA developed guidance by limiting induced current densities in the body to those levels that occur normally, that is, up to about 10 mA/m^2 (higher current densities can also occur naturally in the heart). They acknowledged that biological effects have been demonstrated in laboratory studies at field strengths below those permitted by the exposure guidelines, but both concluded that there was no convincing evidence that occupational exposure to these field levels leads to adverse health effects (IRPA, 1990; ACGIH, 1990c,d). As expressed in the IRPA guidelines, the following biological effects are associated with this order of current density and above (Bernhardt et al., 1986; IRPA, 1990):

$1–10$ mA/m^2; minor biological effects have been reported

$10–100$ mA/m^2; well-established effects, including effects on the visual and nervous system

$100–1000$ mA/m^2; stimulation of excitable tissue; possible health hazards

>1000 mA/m^2; extrasystoles and ventricular fibrillation can occur

The IRPA guidelines recommend power-frequency electric field exposure limits of 10 kV/m for a whole working day, with a short-term limit of 30 kV/m. Interim times and field strengths are related by the formula;

$$t = 80/E \tag{19}$$

where t = allowed exposure duration (hr)
  E = electric field strength (kV/m)

The guidelines for magnetic flux density for occupational exposure are 0.5 mT for the entire workday, 5 mT for exposures of 2 hr or less, and 25 mT for exposure to limbs.

The ACGIH TLV for occupational exposure to ELF E fields states that exposure should not exceed 25 kV/m for frequencies from 0 to 100 Hz. For frequencies in the range of 100 Hz to 4 kHz, the TLV is given by

$$E_{TLV} = 2.5 \times 10^6/f \tag{20}$$

where E = electric field strength (V/m) (rms value)
  f = frequency (Hz)

A proviso is added for workers with cardiac pacemakers, limiting power-frequency exposures to 1 kV/m since EMI with pacemaker function may occur in some models at power-frequency electric fields as low as 2 kV/m.

The ACGIH TLV for magnetic fields limits routine occupational (rms) exposure to

$$B_{TLV} = 60/f \tag{21}$$

where B = magnetic field strength (mT)
  f = frequency (Hz)

At frequencies below 1 Hz, the TLV is 60 mT. At 60 Hz it is 1 mT. For workers with cardiac pacemakers, the limit is reduced to one-tenth of these values at frequencies equal to or above 6 Hz, and to 1.0 mT below 6 Hz.

For static fields, the 60 mT limit is a whole body TWA, and it is extended to 600 mT for the extremities. A flux density of 2 T is recommended as a ceiling value. It remains at 1 mT for cardiac pacemaker wearers.

REFERENCES

ACGIH (American Conference of Governmental Industrial Hygienists) (1970). *Threshold Limit Values of Physical Agents Adopted by ACGIH for 1970*, ACGIH, Cincinnati.

ACGIH (1981). *Threshold Limit Values for Chemical Substances and Physical Agents in the Workroom Environment with Intended Changes for 1982*, ACGIH, Cincinnati.

ACGIH (1990a). *A Guide for Control of Laser Hazards*, 4th ed., ACGIH, Cincinnati.

ACGIH (1990b). *1990–1991 Threshold Limit Values for Chemical Substances and Physical Agents and Biological Exposure Indices*, ACGIH, Cincinnati.

ACGIH (1990c). *Appl. Occup. Environ. Hyg.*, **5**, 734–737.

ACGIH (1990d). *Appl. Occup. Environ. Hyg.*, **5**, 884–892.

ACGIH (1992). *1992–1993 Threshold Limit Values for Chemical Substances and Physical Agents and Biological Exposure Indices*, ACGIH, Cincinnati.

ACGIH (1993). *1993–1994 Threshold Limit Values for Chemical Substances and Physical Agents and Biological Exposure Indices*, ACGIH, Cincinnati.

Adair, R. K. (1992). *Bioelectromagnetics*, **13**, 231–235.

Adair, E. R. and B. W. Adams (1980). *Bioelectromagnetics*, **1**, 1–20.

Adey, W. R. (1988). In *Nonionizing Electromagentic Radiations and Ultrasound* (National Council on Radiation Protection and Measurements [NCRP] Proceedings No. 8), NCRP, Bethesda, MD, pp. 88–110.

Adler, F. H. (1965). *Physiology of the Eye*, 4th ed., Mosby, St. Louis.

Adrian, D. J. (1977). *Radio Sci.*, **12**(6S), 243–250.

AIBS (American Institute of Biological Sciences) (1985). *Biological and Human Health Effects of Extremely Low Frequency Electromagnetic Fields* (Technical Report AD/A152 731), AIBS, Arlington, VA.

Ainsleigh, H. G. (1993). *Preventive Med.*, **22**, 132–140.

Albrecht, R. M. and E. Landau (1978). *Rev. Environ. Hlth.*, **III**, 43–57.

Anne, A., M. Saito, O. M. Salati, and H. P. Schwan (1961). In *Proceedings of 4th Annual Tri-Service Conference on Biologic Effects of Microwave Radiating Equipment: Biological Effects of Microwave Radiations*, M. F. Peyton, Ed., Plenum Press, New York, pp. 153–176.

ANSI (American National Standards Institute) (1974). *Safety Level of Electromagnetic Radiation with Respect to Personnel* (ANSI C95.1-1974), ANSI, New York.

ANSI (1982). *Safety Levels with Respect to Human Exposure to Radio Frequency Electromagnetic Fields 300 kHz to 100 GHz* (ANSI C95.1-1982), ANSI, New York.

ANSI (1993). *American National Standard for the Safe Use of Lasers* (ANSI Z136-1993), Laser Institute of America, Orlando, FL.

Appleton, B. (1973). *Results of Clinical Surveys for Microwave Ocular Effects*, DHEW (FDA) Pub. No. 73-803, U.S. Government Printing Office, Washington, D.C.

Baranski, S. (1971). *Aerospace Med.*, **42**, 1196–1199.

Baranski, S., and P. Czerski (1976). In *Biological Effects of Microwaves*, Dowden, Hutchison & Ross, Stroudsburg, PA, pp. 153–169.

Barron, C. I. and A. A. Baraff (1958). *J. Am. Med. Assoc.*, **168**, 1194–1199.

Barron, C. I., A. A. Love, and A. A. Baraff (1955). *J. Aviation Med.*, **26**, 442–452.

Bawin, S. M., L. Kaczmarek, and W. R. Adey (1975). *Ann. N. Y. Acad. Sci.*, **247**, 74–81.

Bellossi, A., A. Desplaces, and R. Morin (1988). *Can. Biochem. Biophys.*, **10**, 59–66.

Beniashvili, D. Sh., V. G. Bilanishvili, and M. Z. Menabde (1991). *Can. Lett.*, **61**, 75–79.

Berman, E., H. Carter, and D. House (1980). *Bioelectromagnetics*, **1**, 65–76.

Berman, E., H. Carter, and D. House (1981). *J. Microwave Power*, **16**, 9–13.

Berman, E., H. Carter, and D. House (1982a). *J. Microwave Power*, **17**, 107–112.

Berman, E., H. Carter, and D. House (1982b). *Bioelectromagnetics*, **3**, 285–291.

Berman, E., L. Chacon, D. House, B. A. Koch, W. E. Koch, J. Leal, S. Lovtrup, E. Mantiply, A. H. Martin, G. I. Martucci, K. H. Mild, J. C. Monahan, M. Sandstrom, K. Shamsaifar, R. Tell, M. A. Trillo, A. Ubeda, and P. Wagner (1990). *Bioelectromagnetics*, **11**, 169–187.

Bernhardt, J. H. (1988). *Radiat. Environ. Biophys.*, **27**, 1–27.

Bernhardt, J. H., H. J. Haubrich, G. Newi, N. Krause, and K. H. Schneider (1986). *Limits for Electric and Magnetic Fields in DIN VDE Standards: Considerations for the Range 0 to 10 kHz*, CIGRE, International Conference on Large High Voltage Electric Systems, 112 Boulevard Haussmann, 75008 Paris, August 27–September 4th.

Bitran, M. E., D. E. Charron, and J. M. Nishio (1992). *Microwave Emissions and Operator Exposures from Traffic Radars Used in Ontario*, Ministry of Labour, Weston, Ontario, Canada.

Blackman, C. F., S. G. Benane, L. S. Kinney, W. T. Joines, and D. E. House (1982). *Rad. Res.*, **92**, 510–520.

Blackman, C. F., S. G. Benane, J. R. Rabinowitz, D. E. House, and W. T. Joines (1985). *Bioelectromagnetics*, **6**, 327–338.

Blackman, C. F., L. S. Kinney, D. E. House, and W. T. Joines (1989). *Bioelectromagnetics*, **10**, 115–128.

Blackman, C. F., S. G. Benane, and D. E. House (1991). *Bioelectromagnetics*, **12**, 173–182.

Blackwell, R. P. and R. D. Saunders (1986). *Int. J. Radiat. Biol.*, **50**, 761–787.

Blum, H. F. (1959). *Carcinogenesis by Ultraviolet Light*, Princeton University Press, Princeton, NJ.

Blum, H. F. (1969). In *Biological Effects of Ultraviolet Radiation (With Emphasis on the Skin)*, F. Urbach, Ed., Pergamon Press, New York, pp. 543–549.

Bonneville Power Administration (BPA) (1986). *Analysis of BPA Occupational Electric Field Exposure Data*, U.S. Department of Energy, Bonneville Power Administration, Vancouver, WA.

Bonting, S. L. and A. D. Bangham (1967). *Exptl. Eye Res.*, **6**, 400–413.

Bowman, J. D., D. H. Garabrant, E. Sobel, and J. M. Peters (1988). *App. Ind. Hyg.*, **3**, 189–194.

Brandt, L. and C. V. Nielsen (1990). *Scand. J. Work Environ. Hlth.*, **16**, 329–333.

BRMD (Bureau of Radiation and Medical Devices) (1991). *Limits of Exposure to Radiofrequency Fields at Frequencies from 10 kHz–300 GHz* (Safety Code 6), Canada Communications Group, Ottawa, Ontario, Canada.

Brown, J. L. (1973). *Human Factors*, **6**, 503–516.

Bruckner-Lea, C., C. H. Durney, J. Janata, C. Rappaport, and M. Kaminski (1992). *Bioelectromagnetics*, **13**, 147–162.

Bruls, W. A. G., H. Slaper, J. C. van der Leun, and L. Berrens (1984). *Photochem. Photobiol.*, **40**, 485–494.

Bryant, H. and E. J. Love (1989). *Int. J. Epidemiol.*, **18**, 132–138.

Burdeshaw, J. A. and S. Schaffer (1976). *Factors Associated with the Incidence of Congenital Anomalies: A Localized Investigation*, Environmental Protection Agency, Final Report, Contract No. 68-02-0791, Washington, D.C.

Butrous, G. S., S. Meldrum, D. G. Barton, J. C. Male, J. A. Bonnell, and A. J. Camm (1982). *J. Royal Soc. Med.*, **75**, 327.

Butrous, G. S., R. S. Bexton, D. G. Barton, J. C. Male, and A. J. Camm (1983a). *Br. J. Ind. Med.*, **40**, 462.

Butrous, G. S., J. C. Male, R. S. Webber, D. G. Barton, S. J. Meldrum, J. A. Bonnell, and A. J. Camm (1983b). *PACE*, **6**, 1282.

Byus, C. V., S. E. Peiper, and W. R. Adey (1987). *Carcinogenesis*, **8**, 1385–1389.

Cahill, D. F. and J. A. Elder, Eds. (1984). *Biological Effects of Radiofrequency Radiation* (EPA-600/8-83-026F), National Technical Information Service (PB85-120848), Springfield, VA.

Carstensen, E. L., A. Buettner, V. L. Genberg, and M. W. Miller (1985). *IEEE Trans. Biomed. Eng.*, **BME-32**, 561–565.

Cartalis, C., C. Varotsos, and H. Feidas (1992). *Toxicol. Environ. Chem.*, **36**, 195–203.

Cartwright, C. E., P. N. Breysee, and L. Booher (1993). *App. Occup. Environ. Hyg.*, **8**(6), 587–592.

Centers for Disease Control (CDC) (1992). *MMWR*, **41**, 20–21, 27.

Cesarini, J. P. (1987). In *Human Exposure to Ultraviolet Radiation: Risks and Regulations*, W. F. Passchler and B. F. M. Bosnjakovic, Eds., Elsevier Science Publishers, Biomedical Division, Amsterdam, pp. 33–44.

Chernovetz, M. E., D. R. Justesen, N. W. King, and J. E. Wagner (1975). *J. Microwave Power*, **10**, 391–409.

Chernovetz, M. E., D. R. Justesen, and A. F. Oke (1977). *Radio Sci.*, **12**(6S), 191–197.

Chou, C. K., A. W. Guy, and R. Galambos (1982). *J. Acoust. Soc. Am.*, **71**, 1321–1334.

Chou, C. K., K. C. Yee, and A. W. Guy (1985). *Bioelectromagnetics*, **6**, 323–326.

Cleary, S. F. and B. S. Pasternack (1966). *Arch. Environ. Hlth.*, **12**, 23–29.

Cleary, S. F., B. S. Pasternack, and G. W. Beebe (1965). *Arch. Environ. Hlth.*, **11**, 179–182.

Cleaver, J. E. (1974). *Adv. Radiat. Biol.*, **4**, 1–75.

Cleveland, R. F. (1985). *IEEE Trans. Broadcast.*, **BC-31**, 81–87.

Cohen, B. H., A. M. Lilienfeld, S. Kramer, and L. C. Hyman (1977). In *Population Cytogenetics, Studies in Humans*, E. B. Hook and I. H. Porter, Eds., Academic Press, New York, pp. 301–352.

Cole, C. A., P. D. Forbes, and R. E. Davies (1986). *Photochem. Photobiol.*, **43**, 275–284.

Coleman, M. and V. Beral (1988). *Int. J. Epidemiol.*, **17**, 1–13.

Coleman, M., J. Bell, and R. Skeet (1983). *Lancet*, **i**, 982–983.

Commission International de l'Eclairage (CIE) (1970), *International Lighting Vocabulary*, Bureau Central de la Commission Internationale de l'Eclairage, Paris, p. 51.

Cook, H. F. (1952). *J. Physiol.*, **118**, 1–11.

Cook, M. R., C. Graham, C. D. Cohen, and M. M. Gerkovich (1992). *Bioelectromagnetics*, **13**, 261–285.

Council on Physical Medicine (1948). *J. Am. Med. Assoc.*, **137**, 1600–1603.

Council on Scientific Affairs (1989). *J. Am. Med. Assoc.*, **262**, 380–384.

Czerska, E. M., E. C. Elson, C. C. Davis, M. L. Swicord, and P. Czerski (1992). *Bioelectromagnetics*, **13**, 247–259.

Czerski, P. (1975). *Ann. N.Y. Acad. Sci.*, **247**, 232–242.

Czerski, P. (1985). *J. Microwave Power*, **20**, 233–239.

Czerski, P. and M. Piotrowski (1972). *Medycyna Lotnicza* (Polish), **39**, 127–139.

Czerski, P. and J. H. Bernhardt (1989). In *Electromagnetic Interaction with Biological Systems*, J. C. Lin, Ed., Plenum Press, New York, pp. 271–279.

Czerski, P. and M. Siekierzynski (1975). In *Fundamental and Applied Aspects of Non-Ionizing Radiations*, S. M. Michaelson, M. W. Miller, R. Magin, and E. L. Carstensen, Eds., Plenum Press, New York, pp. 367–375.

Czerski, P., E. Paprocka-Slonka, M. Siekierzynski, and A. Stolarska (1974). In *Biological Effects and Health Hazards of Microwave Radiation*, Polish Medical Publishers, Warsaw, pp. 67–74.

Daels, J. (1973). *Obstet. Gynecol.*, **42**, 76–79.

Daels, J. (1976). *J. Microwave Power*, **11**, 166–168.

Daily, L. (1943). *US Naval Med. Bull.*, **41**, 1052–1056.

Dalziel, C. F. (1972). *IEEE Spect.*, **9**, 41–50.

D'Andrea, J., O. P. Gandhi, and J. L. Lords (1977). *Radio Sci.*, **12**(6S), 251–256.

Davis, R. L. and F. K. Mostofi (1993). *Am. J. Ind. Med.*, **24**, 231–233.

Davson, H. (1963). *The Physiology of the Eye*, 2nd ed., Little, Brown, Boston.

Dayhaw-Barker, P., D. Forbes, D. Fox, S. Lerman, J. McGinness, M. Waxler, and R. Felten (1986). In *Optical Radiation and Visual Health*, M. Waxler and V. M. Hitchins, Eds., CRC Press, Boca Raton, FL, pp. 147–175.

Deadman, J. E., M. Camus, B. G. Armstrong, P. Heroux, D. Cyr, M. Plante, and G. Theriault (1988). *Am. Ind. Hyg. Assoc. J.*, **49**, 409–419.

de Lorge, J. O. (1978). In *Proceedings of the 1978 Symposium on Electromagnetic Fields in Biological Systems*, International Microwave Power Institute, Edmonton, Canada, pp. 215–228.

de Lorge, J. O. (1984). *Bioelectromagnetics*, **5**, 233–246.

Dennis, J. A., C. R. Muirhead, and J. R. Ennis (1991a). *J. Radiol. Prot.*, **11**, 3–12.

Dennis, J. A., C. R. Muirhead, and J. R. Ennis (1991b). *J. Radiol. Prot.*, **11**, 13–25.

Deno, D. W. (1977). *IEEE Trans. Power Apparat. Sys.*, **PAS-96**, 1517–1527.

Derksen, W. L., T. I. Monohan, and G. P. Delhery (1963). In *Temperature, Its Measurement and Control in Sciences and Industry, Part 3*, J. D. Hardy, Ed., Reinhold, New York, p. 171.

Diffey, B. L. (1988). *Phys. Med. Biol.*, **33**, 1187–1193.

Dimbylow, P. J. (1988). In *Proceedings of the Annual International Conference of the IEEE Engineering in Medicine and Biology Society* [88CH2566-8], G. Harris and C. Walker, Eds., IEEE, New York, pp. 894–895.

Dimbylow, P. J. (1991). *J. Radiol. Prot.*, **11**, 43–48.

Dodge, C. H. (1969). In *Biological Effects and Health Implications of Microwave Radiation*, S. F. Cleary, Ed., U.S. Government Printing Office, Washington, D.C., pp. 140–149.

Dodge, C. H. and Z. R. Glaser (1977). *J. Microwave Power*, **12**, 319–334.

Dratz, E. A. (1989). In *The Science of Photobiology*, 2nd ed., K. C. Smith, Ed., Plenum Press, New York and London, pp. 231–271.

Dubin, N., M. Moseson, and B. S. Pasternack (1986). In *Epidemiology of Malignant Melanoma, Recent Results in Cancer Research*, Vol. 2, R. P. Gallagher, Ed., Vol. 102, Springer-Verlag, Berlin, pp. 56–75.

Dunn, K. L. (1950). *Arch. Ind. Hyg. Occup. Med.*, **1**, 166–180.

Durney, C. H., C. C. Johnson, P. W. Barber, H. Massoudi, M. F. Iskander, J. L. Lords, D. K. Ryser, S. J. Allen, and J. C. Mitchell (1978). *Radiofrequency Radiation Dosimetry Handbook*, 2nd ed. (Report SAM-TR-78-22), USAF School of Aerospace Medicine, Brooks Air Force Base, TX.

Durney, C. H., H. Massoudi, and M. F. Iskander (1986). *Radiofrequency Radiation Dosimetry Handbook*, 4th ed. (USAFSAM-TR-85-73), USAF School of Aerospace Medicine, Brooks Air Force Base, TX.

Edmunds, H. D. (1991). In *Proceedings of the International Laser Safety Conference*, S. S. Charscan, Ed., Laser Institute of America, Orlando, FL, pp. 2-41–2-54.

EHC (Environmental Health Criteria) 23 (1982). *Lasers and Optical Radiation*, World Health Organization (WHO), Geneva.

EHC 35 (1984). *Extremely Low Frequency (ELF) Fields*, WHO, Geneva.

EHC 69 (1987). *Magnetic Fields*, WHO, Geneva.

EHC 137 (1993). *Electromagnetic Fields (300 Hz to 300 GHz)*, WHO, Geneva.

Elder, J. A. (1984). In *Biological Effects of Radiofrequency Radiation* (EPA-600/8-83-026F), D. F. Cahill and J. A. Elder, Eds., National Technical Information Service (PB85-120848), Springfield, VA, pp. 5-64–5-70.

Elder, J. A. (1987). *Hlth. Phys.*, **53**, 607–611.

Ely, T. S., D. E. Goldman, and J. Z. Hearon (1964). *IEEE Trans. Biomed. Eng.*, **BME-11**, 123–137.

EPA [Environmental Protection Agency] (1986). *Fed. Reg.*, **51**(146), 27313–27339.

EPA (1994a). *The Experimental Ultraviolet Index Factsheet: Explaining the Index to the Public* (EPA 430-F-94-017), EPA, Washington, D.C.

EPA (1994b), *Technical Appendices to the Experimental Ultraviolet Factsheet* (EPA 430-F-94-019), EPA, Washington, D.C.

EPRI [Electric Power Research Institute] (1985). *AC Field Exposure Study: Human Exposure to 60-Hz Electric Fields*, Technical Report EA-3993, EPRI, Palo Alto, CA.

Epstein, J. H. (1983). *J. Am. Acad. Dermatol.*, **9**, 487–502.

Epstein, J. H. (1989). In *The Science of Photobiology*, 2nd ed., K. C. Smith, Ed., Plenum Press, New York and London, pp. 155–192.

Epstein, J. H. (1990). *N. Engl. J. Med.*, **322**, 1149–1151.

Epstein, W. L., K. Fukuyama, and J. H. Epstein (1971). *Fed. Proc.*, **30**, 1766–1771.

Ericson, A. and B. Kallen (1986a). *Am. J. Ind. Med.*, **9**, 459–475.

Ericson, A. and B. Kallen (1986b). *Am. J. Ind. Med.*, **9**, 447–457.

Everett, M. A., R. M. Sayre, and R. L. Olsen (1969). In *The Biological Effects of Ultraviolet Radiation (With Emphasis on the Skin)*, F. Urbach, Ed., Pergamon Press, New York, pp. 181–186.

Faber, M. (Revised by J. C. van der Leun) (1989). In *Nonionizing Radiation Protection*, M. J. Seuss, Ed., World Health Organization, Regional Office for Europe, Copenhagen, pp. 13–48.

FCC [Federal Communications Commission] (1985). *Evaluating Compliance with FCC-Specified Guidelines for Human Exposure to Radiofrequency Radiation* (OST Bulletin No. 65), FCC, Washington, D.C.

Fisher, K. D., C. J. Carr, J. E. Huff, and T. E. Huber (1970). *Fed. Proc.*, **29**, 1605–1638.

Floderhus, B., T. Persson, C. Stenlund, G. Linder, C. Johansson, J. Kiviranta, H. Parsman, M. Lindblom, A. Wennberg, A. Ost, and B. Knave (1992). *Occupational Exposure to Electromagnetic Fields in Relation to Leukemia and Brain Tumors. A Case-Control Study*, Project Abstracts of The Annual Review of Research on Biological Effects of Electric and Magnetic Fields from the Generation, Delivery & Use of Electricity, p. A-52, San Diego, California, November 8–12. Available from W/L Associates, Ltd., 120 W. Church Street, Frederick, MD.

Frederick, J. E. (1993). *Photochem. Photobiol.*, **57**, 175–178.

Frederick, J. E. and A. D. Alberts (1991). *Geophys. Res. Lett.*, **18**, 1869–1871.

Frey, A. (1961). *Aerospace Med.* **32**, 1140–1142.

Gandhi, O. P. (1980). *Proc. IEEE*, **68**, 24–32.

Gandhi, O. P. (1987). *IEEE Eng. Med. Biol. Mag.*, **6**, 22–25.

Gandhi, O. P. (1988a). *Currents Induced in a Human Being for Electromagnetic Fields 10 kHz–50 MHz*, (Final Report on Project N 00014-86-K 0104), National Technical Information Service, (AD-A191 977) Springfield, VA.

Gandhi, O. P. (1988b). In *IEEE Instrumentation and Measurement Technology Conference* (IEEE Catalogue No. 88CH2569-2), IEEE, New York, pp. 109–113.

Gandhi, O. P., Ed. (1990). *Biological Effects and Medical Applications of Electromagnetic Energy*, Prentice-Hall, Englewood Cliffs, NJ.

Gandhi, O. P., I. Chatterjee, D. Wu, and Y.-G. Gu (1985). *Proc. IEEE*, **73**, 1145–1147.

Garland, F. C., E. Shaw, E. D. Gorham, C. F. Garland, M. R. White, and P. J. Sihsheimer (1990). *Am. J. Epidemiol.*, **132**, 293–303.

Geeraets, W. J. (1970). *Ind. Med.*, **39**, 441–450.

Gellhorn, S. (1928). *J. Am. Med. Assoc.*, **90**, 1005–1008.

Goldhaber, M. K., M. R. Polen, and R. A. Hiatt (1988). *Am. J. Ind. Med.*, **13**, 695–706.

Goldman, L., S. M. Michaelson, R. J. Rockwell, D. H. Sliney, B. M. Tengroth, and M. L. Wolbarsht (Revised by D. H. Sliney) (1989). In *Nonionizing Radiation Protection*, M. J. Seuss, Ed., World Health Organization, Regional Office for Europe, Copenhagen, pp. 49–84.

Goldmann, H. (1935). *Am. J. Ophthalmol.*, **18**, 590–591.

Goldmann, H., H. Koenig, and F. Maeder (1950). *Ophthalmologica*, **120**, 198–205.

Gordon, Z. V. (1960). *Gigyena Truda Akademiya Meditsina Nauk USSR*, **1**, 5–7.

Gordon, Z. V. (1966). *Biological Effect of Microwaves in Occupational Medicine*, Izvestiya Meditisina, Leningrad (TT 70-50087, NASA TT-633, 1970), p. 164.

Gordon, Z. V. (1970). In *Ergonomics and Physical Environmental Factors* (Occupational Safety and Health Series No. 21), International Labour Office, Geneva, pp. 159–172.

Gordon, Z. V., Y. A. Lobanova, and M. S. Tolgskaya (1955). *Gig. Sanit.*, **12,** 16–18.

Graham, C. H. (1965). *Vision and Visual Perception*, Wiley, New York.

Graham, C. M., M. R. Cook, and H. D. Cohen (1990). *Immunological and Biochemical Effects of 60-Hz Electric and Magnetic Fields in Humans*, Report No. DOE/CE/76246-T1, U.S. Department of Energy, P.O. Box 62, Oak Ridge, TN.

Grandolfo, M. and K. Hansson Mild (1989). In *Electromagnetic Biointeraction Mechanisms, Safety Standards, Protection Guides*, G. Franceschetti, O. P. Gandhi, and M. Grandolfo, Eds., Plenum Press, New York, pp. 99–134.

Graves, H. B. and T. J. Reed (1985). *Effects of 60-Hz Electric Fields on Embryo and Chick Development*, Technical Report EA 4161, Electric Power Research Institute, Palo Alto, CA.

Griffin, J. C. (1986). *Cardiac Pacemakers: Effects of Power Frequency Fields*, Proceedings of the International Utility Symposium, Health Effects of Electric and Magnetic Fields, Toronto.

Grossweiner, L. I. (1989). In *The Science of Photobiology*, 2nd ed., K. C. Smith, Ed., Plenum Press, New York and London, pp. 1–45.

Grossweiner, L. I. and K. C. Smith (1989). In *The Science of Photobiology*, 2nd ed., K. C. Smith, Ed., Plenum Press, New York and London, pp. 47–77.

Guenel, P., P. Rashmark, J. B. Anderson, and E. Lynge (1993). *Br. J. Ind. Med.*, **50,** 758–764.

Guskova, A. K. and Y. M. Kochanova (1975). *Gigyena Truda i Professional'nye Zabllolvaniya* (Moscow), **3,** 14–17.

Guy, A. W. (1987). *Hlth. Phys.*, **53,** 569–584.

Guy, A. W., C. K. Chou, J. C. Lin, and D. Christensen (1975a). *Ann N.Y. Acad. Sci.*, **247,** 194–218.

Guy, A. W., J. C. Lin, and C. K. Chou (1975b). In *Fundamental and Applied Aspects of Non-ionizing Radiation*, Plenum Press, New York, pp. 167–207.

Guy, A., J. C. Lin, P. O. Kramar, and A. F. Emery (1975c). *IEEE Trans. Microwave Theory Tech.*, **MTT-23,** 492–498.

Guy, A. W., C.-K. Chou, L. L. Kunz, J. Crowley, and J. Krupp (1985). *Effects of Long-Term Low-Level Radiofrequency Radiation Exposure on Rats, Vol. 9, Summary* (Report USAFSAM-TR-85-64), USAF School of Aerospace Medicine, Brooks Air Force Base, TX.

Haddad, J. G. (1992). *N. Engl. J. Med.*, **326,** 1213–1215.

Ham, Jr., W. T., R. G. Allen, L. Feeney-Burns, M. F. Marmor, L. M. Parver, P. H. Proctor, D. H. Sliney, and M. L. Wolbarsht (1986). In *Optical Radiation and Visual Health*, M. Waxler and V. M. Hitchins, Eds., CRC Press, Boca Raton, FL, pp. 43–67.

Hamburger, S., J. N. Logue, and P. M. Silverman (1983). *J. Chron. Dis.*, **36,** 791–802.

Handren, R. T. (1991). In *Proceedings of the International Laser Safety Conference*, S. S. Charscan Ed., Laser Institute of America, Orlando, FL, pp. 1-51–1-58.

Hardy, J. D. (1953). *J. Appl. Physiol.*, **5,** 725–739.

Hardy, J. D. (1958). In *Therapeutic Heat*, Vol. 2, S. Licht, Ed., Elisabeth Licht, New Haven, CT, pp. 157–178.

Hardy, J. D., H. G. Wolff, and H. Goodell (1947). *J. Clin. Invest.*, **26,** 1152–1158.

Hardy, J. D., H. Goodell, and H. G. Wolff (1951). *Science*, **115**, 149–150.

Hardy, J. D., H. T. Hammel, and D. Murgatroyd (1956). *J. Appl. Physiol.*, **9**, 257–264.

Hathaway, J. A. (1978). In *Currents Concepts in Ergophthalmology*, B. Tengroth and D. Epstein, Eds., Ergophthalmologica Internationalis, Stockholm, pp. 139–160.

Hathaway, J. A., H. Stern, E. M. Sales, and E. Leighton (1977). *J. Occup. Med.*, **19**, 683–688.

Hawk, J. L. M. (1984). *Clin. Exptl. Dermatol.*, **9**, 300–302.

Hawk, J. L. M., A. V. J. Challoner, and L. Chaddock (1982). *Clin. Exptl. Dermatol.*, **7**, 21–31.

Healer, J. (1969). In *Biological Effects and Health Implications of Microwave Radiation*, S. F. Cleary, Ed., U.S. Government Printing Office, Washington, D.C., pp. 90–97.

Health Examination Statistics (1979). U.S. Department of Health, Education and Welfare, National Center for Health Statistics, Hyattsville, MD.

Heller, J. H. (1970). In *Biological Effects and Health Implications of Microwave Radiation*, S. F. Cleary, Ed., Department of Health, Education and Welfare, Washington, D.C., pp. 116–121.

Hendler, E. (1968). In *Thermal Problems in Aerospace Medicine*, J. D. Hardy Ed., Unwin, Ltd., Old Woking, Surrey, pp. 149–161.

Hendler, E., J. E. Hardy, and D. Murgatroyd (1963). In *Temperature Measurement and Control in Science and Industry, Part 3, Biology and Medicine*, J. D. Hardy, Ed., Reinhold, New York, pp. 221–230.

Henriques, Jr., F. C. (1947). *Arch. Pathol.*, **43**, 489–502.

Hersey, P., M. MacDonald, C. Burns, S. Schibeci, H. Matthews, and F. J. Wilkinson (1987). *J. Invest. Dermatol.*, **88**, 271–276.

Heynick, L. (1987). *Critique of the Literature on Bioeffects of Radiofrequency Radiation: A Comprehensive Review Pertinent to Air Force Operations* (Report USAFSAM-TR-87-3), USAF School of Aerospace Medicine, Brooks Air Force Base, TX.

Hill, D. G. (1984). In *Biological Effects of Radiofrequency Radiation* (EPA-60018-83-026F), D. Cahill and J. Elder, Eds., National Technical Information Service (PB85-120848), Springfield, VA, pp. 5-112–5-122.

Hill, D. G. (1988). *A Longitudinal Study of a Cohort with Past Exposure to Radar: The MIT Radiation Laboratory Follow-up Study*, Ph.D. Dissertation, Johns Hopkins University, University Microfilms International, Ann Arbor, MI.

Hill, S. M. and D. E. Blask (1988). *Can. Res.*, **48**, 6121–6126.

Hintenlang, D. E. (1993). *Bioelectromagnetics*, **14**, 545–551.

Hitchcock, R. T. (1991). *Nonionizing Radiation Guide Series: Ultraviolet Radiation*, American Industrial Hygiene Association, Fairfax, VA.

Hitchcock, R. T. (1994). *Nonionizing Radiation Guide Series: Radio-Frequency and Microwave Radiation*, American Industrial Hygiene Association, Fairfax, VA, in press.

Hitchcock, R. T., and Patterson, R. M. (1994). *Radiofrequency and ELF Electromagnetic Energies: The Health Professional's Handbook*, Van Nostrand Reinhold, New York.

Hjeresen, D. L., A. Francendese, and J. M. O'Donnell (1988). *Bioelectromagnetics*, **9**, 63–78.

Hoque, M. and O. P. Gandhi (1988). *IEEE Trans. Biomed. Eng.*, **35**, 442–449.

Horn, T. D., W. L. Morison, H. Farzadegan, B. Z. Zmudzka, and J. Z. Beer (1993). *Clin. Res.*, **41**, 502A.

Huuskonen, H., J. Juutilainen, and H. Komulainen (1993). *Bioelectromagnetics*, **14**, 205–213.

Hyde, A. S. and J. J. Friedman (1968). In *Thermal Problems in Aerospace Medicine*, J. D. Hardy, Ed., Unwin, Ltd., Old Woking, Surrey, pp. 163–175.

IEC [International Electrotechnical Committee] (1976). *Particular Requirements for Microwave Cooking Appliances* (IEC EIC 335-25), Part, 2, IEC, Geneva.

IEEE [Institute of Electrical and Electronics Engineers] (1992). *Safety Levels with Respect to Human Exposure to Radio Frequency Electromagnetic Fields, 3 kHz to 300 GHz* (IEEE C95.1-1991), IEEE, New York.

IRPA [International Radiation Protection Association] (1984). *Hlth. Phys.*, **46**, 975–984.

IRPA (1985a). *Hlth. Phys.*, **49**, 331–340.

IRPA (1985b). *Hlth. Phys.*, **49**, 341–359.

IRPA (1988a). *Hlth. Phys.*, **54**, 573–574.

IRPA (1988b). *Hlth. Phys.*, **54**, 115–123.

IRPA (1989). *Hlth. Phys.*, **56**, 971–972.

IRPA (1990). *Hlth Phys.*, **58**, 113–122.

IRPA (1991). *IRPA Guidelines on Protection Against Nonionizing Radiation*, A. S. Duchêne, J. R. A. Lakey, and M. H. Repacholi, Eds., Pergamon Press, New York.

Jacquez, J. A., J. Huss, W. McKeehan, J. M. Dimitroff, and H. F. Kuppenheim (1955). *J. Appl. Physiol.*, **8**, 297–299.

Janes, D. E., W. M. Leach, W. A. Mills, R. T. Moore, and M. L. Shore (1969). *Nonionizing Radiat.*, **1**, 125–130.

Jensh, R. P., I. Weinberg, and R. L. Brent (1983). *J. Toxicol. Environ. Hlth.*, **11**, 23–35.

Johnson, R. B., L. L. Kunz, D. Thompson, J. Crowley, C. K. Chou, and A. W. Guy (1984). *Effects of Long-Term Low-Level Radiofrequency Radiation Exposure on Rats; Volume 7. Metabolism, Growth, and Development* (Report USAFSAM-TR-84-31), Brooks Air Force Base, TX.

Kallen, B., G. Malmquist, and U. Moritz (1982). *Arch. Environ. Hlth.*, **37**, 81–84.

Kaplan, J., P. Polson, C. Rebert, K. Lunan, and M. Gage (1982). *Radio Sci.*, **17**(5S), 135S–144S.

Karolkar, B. D., J. Behari, and A. Prim (1985). *IEEE Trans. Microwave Theory and Tech.*, **MTT-33**, 64–66.

Kaune, W. T., and R. D. Phillips (1980). *Bioelectromagnetics*, **1**, 117–130.

Keating, G. F., J. Pearson, J. P. Simons, and E. E. White (1955). *Arch. Ind. Hlth.*, **11**, 305–315.

Kerbacher, J. J., M. L. Meltz, and D. N. Erwin (1990). *Radiat. Res.*, **123**, 311–319.

Kirk, W. P. (1984). In *Biological Effects of Radiofrequency Radiation* (EPA-600/8-83-026F), D. Cahill and J. Elder, Eds., National Technical Information Service (PB85-12-0848), Springfield, VA, pp. 5-106–5-111.

Knave, B., F. Gamberale, S. Bergstrom, E. Birke, A. Iregren, B. Kolmodin-Hedman, and A. Wennberg (1979). *Scand. J. Work, Env. Hlth.*, **5**, 115–125.

Knox, J. M., R. G. Freeman, and R. Ogura (1965). *Dermatol. Invest.*, **4**, 205–212.

Koh, H. K. (1991). *N. Engl. J. Med.*, **325**, 171–182.

Koh, H. K., and R. A. Lew (1994). *J. Nat. Can. Inst.*, **86**, 78–79.

Kollias, N., and A. H. Baqer (1987). In *Human Exposure to Ultraviolet Radiation: Risks and Regulations*, W. F. Passchier and B. F. M. Bosnjakovic, Eds., Elsevier Publishers, Biomedical Division, Amsterdam, pp. 121–124.

Kolmodin-Hedman, B., K. Hansson Mild, M. Hagberg, E. Jonsson, M.-C. Andersson, and A. Eriksson (1988). *Int. Arch. Occup. Environ. Hlth.*, **60**, 243–247.

Kraszewski, A., M. A. Stuchly, S. S. Stuchly, and A. M. Smith (1982). *Bioelectromagnetics*, **3**, 421–432.

Krause, D., J. M. Mullins, L. M. Penafiel, R. Meister, and R. M. Nardone (1991). *Radiat. Res.*, **127**, 164–170.

Kues, H., and S. D'Anna (1987). In *Proceedings of the Ninth Annual Conference of the IEEE Engineering in Medicine and Biology Society*, Vol. 2, Institute of Electrical and Electronic Engineers, New York, pp. 698–700.

Kues, H. A., and J. C. Monahan (1992). *Johns Hopkins APL Tech. Dig.*, **13**, 244–254.

Kues, H., L. W. Hirst, G. A. Lutty, S. A. D'Anna, and G. R. Dunkelberger (1985). *Bioelectromagnetics*, **6**, 177–188.

Kunz, L. L., R. B. Johnson, D. Thompson, J. Crowley, C. K. Chou, and A. W. Guy (1985). *Effects of Long-Term Low-Level Radiofrequency Radiation Exposure on Rats; Volume 8. Evaluation of Longevity, Cause of Death, and Histopathological Findings* (USAFSAM-TR-85-11), USAF School of Aerospace Medicine, Brooks Air Force Base, TX.

Kurppa, K., P. C. Holmberg, K. Rantala, T. Nurminen, and L. Saxen (1985). *Scand. J. Work Environ. Hlth.*, **11**, 353–356.

Lai, H., A. Horita, C. K. Chou, and A. W. Guy (1984a). *Bioelectromagnetics*, **5**, 203–211.

Lai, H., A. Horita, C. K. Chou, and A. W. Guy (1984b). *Bioelectromagnetics*, **5**, 213–220.

Lai, H., A. Horita, C. K. Chou, and A. W. Guy (1987). *IEEE Eng. Med. Biol. Mag.*, **6**, 31–36.

Lancranjan, I., M. Maicanescu, E. Rafaila, I. Klepsch, and H. I. Popescu (1975). *Hlth. Phys.*, **29**, 381–383.

Larsen, A. I., (1991). *Scand. J. Work Environ. Hlth.*, **17**, 318–323.

Larsen, A. I., and J. Skotte (1991). *Am. J. Ind. Med.*, **19**, 51–57.

Larsen, A. I., J. Olsen, and O. Svane (1991). *Scand. J. Work Environ. Hlth.*, **17**, 324–329.

Lary, J. M., D. L. Conover, P. H. Johnson, and J. R. Burg (1983a). *Bioelectromagnetics*, **4**, 249–255.

Lary, J. M., D. L. Conover, and P. H. Johnson (1983b). *Scand. J. Work Environ. Hlth.*, **9**, 120–127.

Lary, J. M., D. L. Conover, P. H. Johnson, and R. W. Hornung (1986). *Bioelectromagnetics*, **7**, 141–149.

Lednev, V. V. (1991). *Bioelectromagnetics*, **12**, 71–76.

Lele, P. P., and G. Weddell (1956). *Brain*, **79**, 119–154.

Lerman, S. (1986). In *The Biological Effects of UVA Radiation*, Praeger, New York, pp. 231–251.

Lerman, S. (1988). *N. Engl. J. Med.*, **319**, 1475–1477.

Lester, J. R. (1985). *J. Bioelect.*, **4**, 129–131.

Lester, J. R. and D. F. Moore (1982a). *J. Bioelectricity*, **1**, 59–76.

Lester, J. R. and D. F. Moore (1982b). *J. Bioelectricity*, **1**, 77–82.

Leung, F. C., D. N. Rommereim, R. G. Stevens, and L. E. Anderson (1987). *Effects of Electric Field on Rat Mammary Tumor Development Induced by 7,12 Dimethyl-*

benz(a)anthracene (DMBA), in Abstracts, Ninth Annual Meeting of the Bioelectromagnetics Society, Portland, OR, p. 41.

Leung, F. C., D. N. Rommereim, R. G. Stevens, B. W. Wilson, R. L. Buschbom, and L. E. Anderson (1988). *Effects of Electric Fields on Rat Mammary Tumor Development Induced by 7,12 Dimethylbenz(a)anthracene (DMBA)*, in Abstracts, Tenth Annual Meeting of the Bioelectromagnetics Society, Stamford, CT, p. 2.

Levine, J. I. (1990). *Medications that Increase Sensitivity to Light: A 1990 Listing*, U.S. Department of Health and Human Services, HHS Pub. No. FDA 91-8280, Rockville, MD.

Liboff, A. R. (1985). In *Interactions Between Electromagnetic Fields and Cells*, A. Chiabrera, C. Nicolini, and H. P. Schwan, Eds., Plenum Publishing, New York, pp. 281–296.

Liboff, A. R., R. J. Rozek, M. L. Sherman, B. R. McLeod, and S. D. Smith (1987). *J. Bioelect.*, **6**, 13–22.

Liboff, A. R., W. C. Parkinson, and D. C. Dawson (1988). *Ion Cyclotron Resonance in Turtle Colon*, in Abstracts, Tenth Annual Meeting of the Bioelectromagnetics Society, Stamford, CT, p. 32.

Liburdy, R. P. (1979). *Radiat. Res.*, **77**, 34–36.

Liburdy, R. P., T. R. Sloma, R. Sokolic, and P. Yaswen (1993). *J. Pineal Res.*, **14**, 89–97.

Lilienfeld, A. M., J. Tonascia, S. Tonascia, C. H. Libauer, G. M. Cauthen, J. A. Markowitz, and S. Weida (1978). *Foreign Service Health Status Study: Evaluation of Health Status of Foreign Service and Other Employees from Selected Eastern European Posts*, National Technical Information Service (NTIS PB-288163), Springfield, VA.

Lin, J. C. (1989). In *Electromagnetic Interaction with Biological Systems*, J. C. Lin, Ed., Plenum Press, New York, pp. 165–177.

Lloyd-Smith, D. L., and K. Mendelssohn (1948). *Br. Med. J. No.*, **i**, 975–978.

Loomis, D. P., and D. A. Savitz (1990). *Br. J. Ind. Med.*, **47**, 633–638.

London, S. J., J. D. Bowman, E. Sobel, D. C. Thomas, D. H. Garabrant, N. Pearce, L. Bernstein, and J. Peters (1994). *Am J. Ind. Med.*, **26**, 47–60.

Loscher, W., M. Mevissen, W. Lehmacher, and A. Stamm (1993). *Can. Lett.*, **71**, 75–81.

Lotz, W. G., and S. M. Michaelson (1978). *J. Appl. Physiol. Respirat. Environ. Exercise Physiol.*, **44**, 438–445.

Lotz, W. G., and S. M. Michaelson (1979). *J. Appl. Physiol. Respirat. Environ. Exercise Physiol.*, **47**, 1284–1288.

Lovsund, P., P. A. Oberg, S. E. G. Nilsson, and T. Reuter (1980a). *Med. Biol. Eng. Comput.*, **18**, 326–334.

Lovsund, P., P. A. Oberg, and S. E. G. Nilsson (1979). *Radio Sci.*, **16**(65), 199–200.

Lovsund, P., P. A. Oberg, and S. E. G. Nilsson (1980b). *Med. Biol. Eng. Comput.*, **18**, 758–764.

Lovsund, P., P. A. Oberg, and S. E. G. Nilsson (1982). *Radio Sci.*, **17**(5S), 35S–38S.

Lu, S. T., N. Lebda, S. Pettit, and S. M. Michaelson (1977). *Radio Sci.*, **12(6S)**, 147–155.

Lu, S. T., N. Lebda, S. Pettit, and S. M. Michaelson (1980a). *J. Appl. Physiol. Respirat. Environ. Exercise Physiol.*, **48**, 927–932.

Lu, S. T., W. G. Lotz, and S. M. Michaelson (1980b). *Proc. IEEE*, **68**, 73–77.

Luckiesh, M., and S. K. Guth (1949). *Illum. Eng.*, **44**, 650–670.

Magin, R. L., S. T. Lu, and S. M. Michaelson (1977a). *IEEE Trans. Biomed. Eng.*, **24**, 522–529.

Magin, R. L., S. T. Lu, and S. M. Michaelson (1977b). *Am. J. Physiol.*, **233**, E363–E368.

Majewska, K. (1968). *Pol. Med. J.*, **38**, 989–994.

Male, J. C., and D. T. Edmonds (1990). *Ion Vibrational Precession: A Model for Resonant Biological Interactions with ELF*, in the Bioelectromagnetics Society Twelfth Annual Meeting Abstracts, San Antonio, TX, p. 98.

Male, J. C., W. T. Norris, and M. W. Watts (1984). *Exposure of People to Power-Frequency Electric and Magnetic Fields*, in *Proceedings of the 23rd Hanford Life Sciences Symposium*, Richland, WA, DOE 60 Symposium Series, CONF-841041, UC-97a, pp. 410–418.

Marha, K. (1971). *IEEE Trans. Microwave Theory Tech.*, **MTT-19**, 165–168.

Marha, K., J. Musil, and H. Tuha (1971). *Electromagnetic Fields and the Life Environment* (Trans.), San Francisco Press, San Francisco, p. 138.

Marks, R., and D. Whiteman (1994). *Br. Med. J.*, **308**, 75–76.

Marshall, J. (1987). In *Human Exposure to Ultraviolet Radiation: Risks and Regulations*, W. F. Passchler and B. F. M. Bosnjakovic, Eds., Elsevier Science Publishers, Biomedical Division, Amsterdam, pp. 125–142.

Massoudi, H. (1976). *Long Wavelength Analysis of Electromagnetic Power Absorption by Prolate Spheroidal and Ellipsoidal Models of Man*, Ph.D. Thesis, University of Utah, Salt Lake City, UT, p. 217.

Matanoski, G. M., E. A. Elliot, and P. N. Breysse (1991). *Lancet*, **337**, 737.

Matelsky, I. (1968). In *Industrial Hygiene Highlights, Vol. I*, L. V. Cralley, L. J. Cralley, and G. D. Clayton, Eds., Industrial Hygiene Foundation of America, Pittsburgh, pp. 140–179.

McAfee, R. D., A. Longacre, R. R. Bishop, S. T. Elder, J. G. May, M. G. Holland, and R. Gordon (1979). *J. Microwave Power*, **14**, 41–44.

McCally, R. L., R. A. Farrell, and C. B. Bargeron (1992). *Lasers Surg. Med.*, **12**, 598–603.

McCann, J., F. Dietrich, C. Rafferty, and A. O. Martin (1993). *Mutation Res.*, **297**, 61–95.

McConnell, D. G., and D. G. Scarpelli (1963). *Science*, **139**, 848.

McDonald, A. D., J. C. McDonald, B. Armstrong, N. Cherry, A. D. Nolin, and D. Robert (1988). *Br. J. Ind. Med.*, **45**, 509–515.

McDowall, M. E. (1983). *Lancet*, **8318**, 246.

McKinlay, A. F., and B. L. Diffey (1987). *CIE J.*, **6**, 17–22.

McLean, J. R. N., M. A. Stuchly, R. E. J. Mitchel, D. Wilkinson, H. Yang, M. Goddard, D. W. Lecuyer, M. Schunk, E. Callary, and D. Morrison (1991). *Bioelectromagnetics*, **12**, 273–287.

Merritt, J. H., A. F. Chamness, and S. J. Allen (1978). *Radiat. Environ. Biophys.*, **15**, 367–377.

Mevissen, M., A. Stamm, S. Buntenkotter, R. Zwingelberg, U. Wahnschaffe, and W. Loscher (1993). *Bioelectromagnetics*, **14**, 131–143.

Michaelson, S. M. (1972). *Proc. IEEE*, **60**, 389–421.

Michaelson, S. M. (1974). *Environ. Hlth. Perspect.*, **8**, 133–156.

Michaelson, S. M. (1975). In *The Foundations of Space Biology and Medicine*, Chapter 1, Vol. II, Book 2, M. Calvin and O. G. Gazenko, Eds., NASA, Washington, D.C., pp. 409–452.

Michaelson, S. (1983). In *Biological Effects and Dosimetry of Nonionizing Radiation Radiofrequency and Microwave Energies*, M. Grandolfo, S. Michaelson, and A. Rindi, Eds., Plenum Press, New York, pp. 589–609.

Michaelson, S. (1991). *Hlth. Phys.*, **61**, 3–14.

Michaelson, S. M., R. A. E. Thomson, M. Y. E. Tamami, H. S. Seth, and J. W. Howland (1964). *Aerospace Med.*, **35**, 824–829.

Michaelson, S. M., R. A. E. Thomson, and J. W. Howland (1967). *Biologic Effects of Microwave Exposure* (RADC: ASTIA Document No. AD 824-242), Griffis Air Force Base, New York.

Mikolajczyk, H. (1974). In *Biologic Effects and Health Hazards of Microwave Radiation*, Polish Medical Publishers, Warsaw, pp. 46–51.

Mikolajczyk, H. (1977). In *Biologic Effects of Electromagnetic Waves*, C. C. Johnson and M L. Shore, Eds., U.S. Department of Health, Education and Welfare, Rockville, MD, pp. 377–383.

Milham, S. (1982). *N. Eng. J. Med.*, **307**, 249.

Milham, S. (1985a). *Environ. Hlth. Perspect.*, **62**, 297–300.

Milham, S. (1985b). *Lancet*, **1**, 812.

Milham, S. (1988a). *Am. J. Epidemiol.*, **127**, 50–54.

Milham, S. (1988b). *Am. J. Epidemiol.*, **128**, 1175–1176.

Miller, D. A. (1974). In *Biological and Clinical Effects of Low-Frequency Magnetic and Electric Fields*, J. G. Llaurado et al., Eds., Charles C. Thomas, Springfield, IL, pp. 62–70.

Morison, W. L. (1989). *Photochem. Photobiol.*, **50**, 515–524.

Moritz, A. R., and F. C. Henriques, Jr. (1947). *Am. J. Pathol.*, **23**, 695–720.

Moss, A. J., and E. Carstensen (1985). *Evaluation of the Effects of Electric Fields on Implanted Cardiac Pacemakers*, Technical Report EA-3917, Electric Power Research Institute, Palo Alto, CA.

Moss, C. E., W. Murray, W. Parr, and D. Conover (1977). In *Occupational Diseases. . .A Guide to Their Recognition*, M. M. Key, A. F. Henschel, J. Butler, R. N. Ligo, and I. R. Tabershaw, Eds., U.S. Department of Health, Education and Welfare, Pub. No. DHEW (NIOSH) 77–181, Cincinnati, pp. 467–496.

Moss, C. E., R. J. Ellis, W. H. Parr, and W. E. Murray (1982). *Biological Effects of Infrared Radiation*, U.S. Department of Health and Human Services, Pub. No. DHHS (NIOSH) 82-109, Cincinnati.

Moss, C. E., R. J. Ellis, W. E. Murray, and W. H. Parr (Revised by B. M. Tengroth and M. L. Wolbarsht) (1989). In *Nonionizing Radiation Protection*, M. J. Seuss, Ed., World Health Organization, Regional Office for Europe, Copenhagen, pp. 85–115.

Murphy, J. C., D. A. Kaden, J. Warren, and A. Sivak (1993). *Mutation Res.*, **296**, 221–240.

Nawrot, P. S., D. I. McRee, and R. E. Staples (1981). *Teratology*, **24**, 303–314.

NCI [National Cancer Institute] (1990). *Skin Cancers: Basal Cell and Squamous Cell Carcinomas*, Research Report, U.S. Department of Health and Human Services, Pub. No. DHHS (NIH) 91-2977, Bethesda, MD.

NCRP [National Council on Radiation Protection and Measurements] (1981). *Radiofrequency Electromagnetic Fields* (NCRP Report No. 67), NCRP, Bethesda, MD, p. 134.

NCRP (1986). *Biological Effects and Exposure Criteria for Radiofrequency Electromagnetic Fields* (NCRP Report No. 86), NCRP, Bethesda, MD, pp. 271–289.

NCRP (1993). *A Practical Guide to the Determination of Human Exposure to Radiofrequency Fields* (NCRP Report No. 119), NCRP, Bethesda, MD.

Nelemans, P. J., H. Groenendal, L. A. L. M. Kiemeney, F. H. J. Rampen, D. J. Ruiter, and A. L. M. Verbeek (1993). *Environ. Hlth. Perspect.*, **101**, 252–255.

Nelson, B. K., D. L. Conover, W. S. Brightwell, P. B. Shaw, D. Werren, R. M. Edwards, and J. M. Lary (1991). *Teratology*, **43**, 621–634.

Nielsen, C. V., and L. Brandt (1990). *Scand. J. Work Environ. Hlth.*, **16**, 323–328.

Nielsen, C. V., and L. Brandt (1992). *Scand. J. Work Environ. Hlth.*, **18**, 346–350.

NIH [National Institutes of Health] Consensus Statement (1989). *Sunlight, Ultraviolet Radiation and the Skin*, **7**, NIH, Bethesda, MD.

NIOSH [National Institute for Occupational Safety and Health] (1985). *Recommended Occupational Exposure Standard for Radiofrequency Radiation* (External Review Draft—05/21/85), NIOSH, Cincinnati.

NIOSH (1993). *Health Hazard and Technical Assistance (HETA) Report: New York Telephone Company, White Plains, New York*, Report No. HETA 92-0009-2362, NIOSH, Cincinnati.

Novitskii, A. A., B. F. Murashov, P. E. Krasnobaev, and N. F. Markozova (1977). *Voen. Med. Zh.*, **8**, 53–56.

NRPB [National Radiological Protection Board] (1991a). *Biological Effects of Exposure to Non-Ionising Electromagnetic Fields and Radiation I. Static Electric and Magnetic Fields* (Report NRPB-R238), NRPB, Chilton, U.K., July 1991.

NRPB (1991b). *Biological Effects of Exposure to Non-Ionising Electromagnetic Fields and Radiation II. Extremely Low Frequency Electric and Magnetic Fields* (Report NRPB-R239), NRPB, Chilton, U.K., July 1991.

NRPB (1992). *Documents of the NRPB*, **3**(1), NRPB, Chilton, U.K., July 1992.

NRPB (1993). *Board Statement on Restrictions on Human Exposure to Static and Time Varying Electromagnetic Fields and Radiation* (Vol. 4, No. 5), NRPB, Chilton, UK.

Nurminen, T., and K. Kurppa (1988). *Scand. J. Work Environ. Hlth.*, **14**, 293–298.

O'Connor, M. E. (1980). *Proc. IEEE*, **68**, 56–60.

Odland, L. T. (1972). *J. Occup. Med.*, **14**, 544–547.

ORAU [Oak Ridge Associated Universities] (1992). *Health Effects of Low-Frequency Electric and Magnetic Fields*, ORAU, Report GPO#029-000-00443-9, U.S. Government Printing Office, Washington, D.C.

Orcutt, N., and O. P. Gandhi (1988). *IEEE Trans. Biomed. Eng.*, **35**, 577–583.

Orlova, T. N. (1971). In *Cerebral Mechanisms of Mental Illness*; *Kaanskiy Meditsinkiy Zhurnal*, 16–18.

Oscar, K. J., and T. D. Hawkins (1977). *Brain Res.*, **126**, 281–283.

OSHA [Occupational Safety and Health Administration] (1971). *Safety and Health Standards for General Industry* (29 CFR 1910.97), U.S. Government Printing Office, Washington, D.C.

Ouellet-Hellstrom, R., and W. F. Stewart (1993). *Am. J. Epidemiol.*, **138**, 775–786.

Parkinson, W. C., and C. T. Hanks (1989). *Bioelectromagnetics*, **10**, 129–145.

Parrish, J. A., R. R. Anderson, C. Y. Ying, and M. A. Pathak (1976). *J. Invest. Dermatol.*, **67**, 603–608.

Parrish, J. A., K. F. Jaenicke, and R. R. Anderson (1982). *Photochem. Photobiol.*, **36**, 187–191.

Patterson, R. M., and R. T. Hitchcock (1990). In *Health and Safety Beyond the Workplace*, L. V. Cralley, L. J. Cralley, and W. C. Cooper, Eds., Wiley & Sons, Inc., New York, pp. 143–176.

Patterson, R. M. (1992). *J. Exp. Anal. Env. Epidemiol.*, **2**, 159–176.

Paulsson, L.-E., Y. Hamnerius, H.-A. Hansson, and J. Sjostrand (1979). *Acta Ophthalmol.*, **57**, 183–197.

Pavlik, R. E. (1991). *Measurements for Sarasota County Sheriff's Office*, Division of Safety, Bureau of Consultation and Enforcement, Tampa, FL.

Peacock, P. B., J. W. Simpson, and C. A. Alford (1971). *J. Med. Assoc. State Ala.*, **41**, 42–50.

Peak, M. J., J. G. Peak, and B. A. Carnes (1987). *Photochem. Photobiol.*, **45**, 381–387.

Petersen, R. C. (1991). *Hlth. Phys.*, **61**, 59–67.

Petrov, I. R. (1970). *Influence of Microwave Radiation on the Organism of Man and Animals*, Meditsina Press (NASA TTF-708), Leningrad.

Petrov, I. R., and V. A. Syngayevskaya (1970). In *Influence of Microwave Radiation on the Organisms of Man and Animals*, Meditsina Press (NASA TTF-708), pp. 31–41.

Pitts, D. G., and T. J. Tredici (1971). *Am. Ind. Hyg. J.*, **32**, 235–246.

Pitts, D. G., A. P. Cullen, P. D. Hacker, and W. H. Parr (1977). *Ocular Ultraviolet Effects from 295 nm to 400 nm in the Rabbit Eye*, U.S. Department of Health, Education and Welfare, Pub. No. DHEW (NIOSH) 77-175, Cincinnati.

Pitts, D. G., A. P. Cullen, and P. Dayhaw-Barker (1980). *Determination of Ocular Threshold Levels for Infrared Radiation Cataractogenesis*, U.S. Department of Health and Human Services, Pub. No. DHHS (NIOSH) 80-121, Cincinnati.

Pitts, D. G., L. L. Cameron, J. G. Jule, S. Lerman, E. Moss, S. D. Varma, S. Zigler, S. Zigman, and J. Zuclich (1986). In *Optical Radiation and Visual Health*, M. Waxler and V. M. Hitchins, Eds., CRC Press, Boca Raton, FL, pp. 5–41.

Polson, P., and J. H. Merritt (1985). *J. Bioelect.*, **4**, 121–127.

Prasad, A. V., M. W. Miller, E. L. Carstensen, Ch. Cox, M. Azadniv, and A. A. Brayman (1991). *Rad. Environ. Biophys.*, **30**, 305–320.

Presman, A. S. (1968). *Electromagnetic Fields and Life* (Trans.), Plenum Press, New York, p. 332.

Preston, E., E. J. Vavasour, and H. M. Assenheim (1979). *Brain Res.*, **174**, 109–117.

Ragan, H., R. D. Phillips, R. L. Buschbom, R. H. Busch, and J. E. Morris (1983). *Bioelectromagnetics*, **4**, 383–396.

Rama Rao, G. R., C. A. Cain, and W. A. F. Tompkins (1984). *Bioelectromagnetics*, **5**, 377–388.

Rama Rao, G. R., C. A. Cain, and W. A. F. Tompkins (1985). *Bioelectromagnetics*, **6**, 41–52.

Rannug, A., T. Ekstrom, K. H. Mild, B. Holmberg, I. Gimenez-Conti, and T. J. Slaga (1993). *Carcinogenesis*, **14**, 573–578.

Rannug, A., B. Holmberg, T. Ekstrom, K. H. Mild, I. Gimenez-Conti, and T. J. Slaga (1994). *Carcinogenesis*, **15**, 153–157.

Raslear, T. G., Y. Akyel, F. Bates, M. Belt, and S. Lu (1993). *Bioelectromagnetics*, **14**, 459–478.

Repacholi, M. H. (1978). *J. Microwave Power*, **13**, 199–211.

Repacholi, M. H. (1987). *IEEE Eng. Med. Biol. Mag.*, **6**, 18–21.

Repacholi, M. H. (1990). In *Biological Effects and Medical Applications of Electromagnetic Energy*, O. P. Gandhi, Ed., Prentice-Hall, Englewood Cliffs, NJ, pp. 9–27.

Roberts, N. J., and S. M. Michaelson (1985). *Int. Arch. Occup. Environ. Hlth.*, **56**, 169–178.

Roberts, N. J., and R. T. Steigbigel (1977). *Infect. Immunol.*, **18**, 673–679.

Robinette, C. D., C. Silverman, and S. Jablon (1980). *Am. J. Epidemiol.*, **112**, 39–53.

Robinson, W. (1907). *Br. Med. J.*, **2**, 381–384.

Rommereim, D. N., R. L. Rommereim, L. E. Anderson, and M. R. Sikov (1988). *Reproductive and Teratologic Evaluation in Rats Chronically Exposed at Multiple Strengths of 60-Hz Electric Fields*, in Abstracts, Tenth Annual Meeting of the Bioelectromagnetics Society, Stamford, CT.

Rommereim, D. N., R. L. Rommereim, M. R. Sikov, R. L. Buschbom, and L. E. Anderson (1990). *Fund. Appl. Toxicol.*, **14**, 608–621.

Rommereim, D. N., R. L. Rommereim, D. L. Miller, R. L. Buschbom, and L. E. Anderson (1991). *Developmental Toxicology Evaluation of 60-Hz, Horizontal Magnetic Fields in Rats*, in *30th Hanford Symposium on Health and the Environment*, Richland, WA.

Roschmann, P. (1991). *Mag. Res. Med.*, **21**, 197–215.

Rosenberg, B. (1966). *Adv. Radiat. Biol.*, **2**, 193–241.

Rosenthal, F. S., and S. Abdollahzadeh (1991). *App. Occup. Env. Hyg.*, **6**, 777–784.

Rotkovska, D., J. Moc, J. Kautska, A. Bartonickova, J. Keptrova, and M. Hofer (1993). *Environ. Health Perspect.*, **101**, 134–136.

Rubin, A., and W. J. Erdman (1959). *Am. J. Phys. Med.*, **38**, 219–220.

Rugh, R., E. I. Ginns, H. S. Ho, and W. M. Leach (1975). *Radiat. Res.*, **62**, 225–241.

Sadcikova, M. N. (1974). In *Biologic Effects and Health Hazards of Microwave Radiation*, Polish Medical Publishers, Warsaw, pp. 261–267.

Sadcikova, M. N., and A. A. Orlova (1958). *Gigiena Truda i Professional'nye Zabolvaniya* (Moscow), **2**, 16–22.

Saffer, J. D., and L. A. Profenno (1992). *Bioelectromagnetics*, **13**, 75–78.

Sahl, J. D., M. A. Kelsh, and S. Greenland (1993). *Epidemiology*, **4**, 104–114.

Sanders, A. P., and W. T. Joines (1983). *Effects of 200, 591, and 2450 MHz Microwaves on Cerebral Energy Metabolism*, National Technical Information Service, Springfield, VA.

Sanders, A. P., W. T. Joines, and J. Allis (1984). *Bioelectromagnetics*, **5**, 419–433.

Savitz, D. (1993). *Am. Ind. Hyg. Assoc. J.*, **54**, 197–204.

Schein, O. D., B. Munoz, S. West, D. Duncan, J. Nethercott, K. Gelatt, C. Vicencio, and J. Honeyman (1993). *Ocular and Dermatologic Health Effects of Ultraviolet Exposure from the Ozone Hole in Southern Chile: A Pilot Project*, Wilmer Eye Institute, Johns Hopkins University, Baltimore, MD.

Schlagel, C. J., and A. Ahmed (1982). *J. Immunol.*, **129**, 1530–1533.

Schnorr, T. M., B. A. Grajewski, R. W. Hornung, M. J. Thun, G. M. Egeland, W. E. Murray, D. L. Conover, and W. E. Halperin (1991). *N. Engl. J. Med.*, **324**, 727–733.

Schwan, H. P., and K. Li (1953). *Proc. IRE*, **41**, 1735–1740.

Schwan, H. P., and K. Li (1956). *Proc. IRE*, **44**, 1572–1581.

Schwan, H. P., and G. M. Piersol (1954). *Am. J. Phys. Med.*, **33**, 371–404.

Setlow, R. B., E. Grist, K. Thompson, and A. D. Woodhead (1993). *Proc. Natl. Acad. Sci. USA*, **90**, 6666–6670.

Shaklett, D. E., T. J. Tredici, and D. L. Epstein (1975). *Aviat. Space Environ. Med.*, **46**, 1403–1406.

Shandala, M. G., M. I. Rudnev, and I. P. Los (1981). In *Proceedings of US-USSR Workshop on Physical Factors-Microwaves and Low Frequency Fields*, National Institute of Environmental Health Sciences, Research Triangle Park, NC, pp. 103–112.

Siekierzynski, M., P. Czerski, H. Milczarek, A. Gidynski, C. Czarnecki, E. Dzuik, and W. Jedrzejczak (1974). *Aerospace Med.*, **45**, 1143–1145.

Sigler, A. T., A. M. Lilienfeld, B. H. Cohen, and J. E. Westlake (1965). *Bull. J. Hopkins Hosp.*, **117**, 374–399.

Sikov, M. R., D. N. Rommereim, J. L. Beamer, R. L. Buschbom, W. T. Kaune, and R. D. Phillips (1987). *Bioelectromagnetics*, **8**, 229–242.

Silverman, C. (1968). *The Epidemiology of Depression*, Johns Hopkins University Press, Baltimore.

Silverman, C. (1973). *Am. J. Epidemiol.*, **97**, 219–224.

Silverman, C. (1979). *Bull. N.Y. Acad. Med.*, **55**, 1166–1181.

Silverman, C. (1980). *Proc. IEEE*, **68**, 78–84.

Silverman, C. (1985). In *Epidemiology and Quantitation of Environmental Risk in Humans from Radiation and Other Agents*, A. Castellani, Ed., Plenum Press, New York, pp. 433–458.

Sliney, D. H., and M. L. Wolbarsht (1980). *Safety with Lasers and Other Optical Sources*, Plenum Press, New York.

Smialowicz, R. J., J. S. Ali, E. Berman, S. J. Bursian, J. B. Kinn, C. G. Liddle, L. W. Reiter, and C. M. Weil (1981). *Radiat. Res.*, **86**, 488–505.

Smith, J. F. (1991). In *Proceedings of the International Laser Safety Conference*, S. S. Charscan, Ed., Laser Institute of American, Orlando, FL, pp. 2-1–2-12.

Smith, S. D., B. R. McLeod, and A. R. Liboff (1987). *Bioelectromagnetics*, **8**, 215–227.

Sommer, H., and H. von Gierke (1964). *Aerospace Med.*, **35**, 834–839.

Spalding, J. F., R. W. Freyman, and L. M. Holland (1971). *Hlth. Phys.*, **20**, 421–424.

Spikes, J. D. (1989). In *The Science of Photobiology*, 2nd ed., K. C. Smith, Ed., Plenum Press, New York and London, pp. 79–106.

Stellman, J., and S. Stellman (1980). *Am. J. Epidemiol.*, **112**, 442–443.

Steneck, N. H., H. J. Cook, A. J. Vander, and G. L. Kane (1980). *Science*, **171**, 1230–1237.

Stern, R. M. (1987). *Env. Hlth. Perspect.*, **76**, 221–229.

Stern, R. S., and Members of the Photochemotherapy Follow-up Study (1990). *N. Engl. J. Med.*, **322**, 1093–1097.

Stern, R. S., M. C. Weinstein, and S. G. Baker (1986). *Arch. Dermatol.*, **122**, 537–545.

Stern, S., and V. G. Laties (1992). Magnetic Fields and Behavior, in *Project Abstracts of The Annual Review of Research on Biological Effects of Electric and Magnetic Fields from the Generation, Delivery & Use of Electricity*, p. A-35, San Diego, CA. Available from W/L Associates, Ltd., 120 W. Church Street, Frederick, MD.

Stern, S., L. Margolin, B. Weiss, S.-T. Lu, and S. M. Michaelson (1979). *Science*, **206**, 1198–1201.

Stevens, R. G. (1987). *Am. J. Epidemiol.*, **125**, 556–561.

Stevens, R. G., S. Davis, D. B. Thomas, L. E. Anderson, and B. W. Wilson, (1992). *FASEB J.*, **6**, 853–860.

Stewart, W., and R. Ouellet-Hellstrom (1991). *Adverse Reproductive Events and Electromagnetic Radiation*, Johns Hopkins University (#1RO1 OH0 2373-01A1), National Technical Information Service (PB92-145796), Springfield, VA.

Stolwijk, J. A. J. (1977). *Fed. Proc.*, **36**, 1655–1658.

Stopps, G. J., and W. Janischewskyj (1979). *Epidemiological Study of Workers Maintaining HV Equipment and Transmission Lines in Ontario*, Canadian Electrical Association Research Report, Montreal, Quebec, Canada.

Streilein, J. W. (1991). *N. Engl. J. Med.*, **325**, 884–887.

Stuchly, M. A. (1987). *Hlth. Phys.*, **53**, 649–665.

Stuchly, M. A. (1989). In *Electromagnetic Interaction with Biological Systems*, J. C. Lin, Ed., Plenum Press, New York, pp. 257–269.

Stuchly, M. A., and D. W. Lecuyer (1989). *Hlth. Physics*, **56**, 297–302.

Stuchly, M. A., M. H. Repacholi, and D. Lecuyer (1979). *Hlth. Phys.*, **37**, 137–144.

Stuchly, M. A., J. Ruddick, D. Villeneuve, K. Robinson, B. Reed, D. W. Lecuyer, K. Tan, and J. Wong (1988). *Teratology*, **38**, 461–466.

Stuchly, M. A., J. R. N. McLean, R. Burnett, M. Goddard, D. W. Lecuyer, and R. E. J. Mitchel (1992). *Can. Lett.*, **65**, 1–7.

Szmigielski, S., and J. Gil (1989). In *Electromagnetic Biointeraction Mechanisms, Safety Standards, Protection Guides*, G. Franceschetti, O. P. Gandhi, and M. Grandolfo, Eds., Plenum Press, New York, pp. 81–98.

Szmigielski, S., and T. Obara (1989). In *Electromagnetic Biointeraction Mechanisms, Safety Standards, Protection Guides*, G. Franceschetti, O. P. Gandhi, and M. Grandolfo, Eds., Plenum Press, New York, pp. 135–151.

Szmigielski, S., A. Szudzinski, A. Pietraszek, M. Bielec, M. Janiak, and J. K. Wrembel (1982). *Bioelectromagnetics*, **3**, 179–191.

Taskinen, H., P. Kyyronen, and K. Hemminki (1990). *J. Epidemiol. Comm. Hlth.*, **44**, 196–201.

Taylor, H. R., S. K. West, F. S. Rosenthal, B. Munoz, H. S. Newland, H. Abbey, and E. A. Emmett (1988). *N. Engl. J. Med.*, **319**, 1429–1433.

Taylor, L. S. (1970). In *Proceedings of the Conference on Estimation of Low-Level Radiation Effects in Human Populations*, Argonne National Laboratory, Report No. ANL-7811, Argonne, IL, pp. 27–28.

Tell, R. (1990). *An Investigation of Electric and Magnetic Fields and Operator Exposure Produced by VDTs: NIOSH VDT Epidemiology Study*, Richard Tell Associates, Las Vegas, Nevada.

Tenforde, T. S. (1991). *Biochem. Bioenerg.*, **25**, 1–17.

Tenforde, T. S., and W. T. Kaune (1987). *Hlth. Phys.*, **53**, 585–606.

Thauer, R. (1965). In *Handbook of Physiology*, W. F. Hamilton, Ed., American Physiological Society, Washington, D.C., Section 2, Circulation III.

Theriault, G., M. Goldberg, A. B. Miller, B. Armstrong, P. Guenel, J. Deadman, E. Imberan, A. Chevalier, D. Cyr, and C. Wall (1994). *Am. J. Epidemiol.*, **139**, 550–572.

Thomas, J. R., J. Schrot, and A. R. Liboff (1986). *Bioelectromagnetics*, **7**, 349–357.

Thomson, R. A. E., S. M. Michaelson, and Q. A. Nguyen (1988). *Bioelectromagnetics*, **9**, 149–157.

Toler, J., V. Popovic, S. Bonasera, P. Popovic, C. Honeycutt, and D. Sgoutas (1988). *J. Microwave Power Electromag. Energy*, **23**, 105–136.

Trevithick, J., M. O. Creighton, M. Sanwal, and D. O. Brown (1989). In *Proceedings of the Ninth Annual Conference of the Engineering in Medicine and Biology Society* (87CH2513-0), IEEE, New York, pp. 695–697.

Tyagin, N. V. (1957). *Voyenno-Medit. Akad. Kirov.*, **73**, 116–126.

Tynes, T., A. Anderson, and F. Langmark (1992). *Am. J. Epidemiol.*, **136**, 81–88.

Urbach, F. (1987). In *Human Exposure to Ultraviolet Radiation: Risks and Regulations*, W. F. Passchier and B. F. M. Bosnjakovic, Eds., Elsevier Publishers, Biomedical Division, Amsterdam, pp. 3–16.

USASI [United States of America Standards Institute] (1966). *Safety Level of Electromagnetic Radiation with Respect to Personnel*, (USASI C95.1-1966) USASI, New York.

USFDA [U.S. Food and Drug Administration] (1971). In *Subchapter J—Radiological Health* (21 CFR Part 1030), U.S. Government Printing Office, Washington, D.C.

USFDA (1985a). *Performance Standards for Laser Products*, Title 21, Code of Federal Regulations, Part 1040.10, *Fed. Reg.*, **50**(161), 33688–33702.

USFDA (1985b). *Performance Standards for Sunlamp Products*, Title 21, Code of Federal Regulations, Part 1040.20, *Fed. Reg.*, **50**(173), 36550–36552.

U.S. Senate, Radiation Health and Safety (1977). 95th Congress, First Session on Oversight of Radiation Health and Safety, June 16, 17, 27, 28, and 29, 1977, Serial No. 95-49, pp. 284, 1195, 1196.

U.S. Senate, Committee on Commerce, Science, and Transportation (1979). *Microwave Irradiation of the U.S. Embassy in Moscow*, April 1979 (43-949), U.S. Government Printing Office, Washington, D.C.

Valerie, K., A. Delers, C. Bruck, C. Thiriart, H. Rosenberg, C. Debouck, and M. Rosenberg (1988). *Nature*, **333**(5), 78–81.

van der Leun, J. C. (1984). *Photochem. Photobiol., Yearly Rev.*, **39**, 861–868.

Vogt, A. (1932). *Klin. Monatsbl. Augenheilk.*, **89**, 256–260.

Vos, J. J. (1993). *Laser Anemometry Advances and Applications, SPIE*, **2052**, 467–474.

Wald, G. (1956). In *Enzymes: Units of Biological Structure and Function*, O. H. Gaebler, Ed., Academic Press, New York, p. 355.

Walder, B. K., M. R. Robertson, and D. Jeremy (1971). *Lancet*, **ii**, 1282–1283.

Wallace, B. M., and J. S. Lasker (1992). *Science*, **257**, 1211–1212.

Wangemann, R., and S. Cleary (1976). *Environm. Biophys.*, **13**, 89–103.

Waxler, M. (1986). In *Optical Radiation and Visual Health*, M. Waxler and V. M. Hitchins, Eds., CRC Press, Boca Raton, FL, pp. 183–204.

Whiteacre, R. T. (1991). *Electromagnetic Environment Measurements*, Battelle Memorial Institute, Columbus, OH.

Whyte, H. M. (1951). *Clin. Sci.*, **10**, 333–345.

Wiktor-Jedrezejczak, W., A. Ahmed, K. W. Sell, P. Czerski, and W. M. Leach (1977). *J. Immunol.*, **118**, 1499–1502.

Wiley, M., D. A. Agnew, J. M. Charry, M. Corey, T. Corey, S. Harvey, R. Kavet, and M. L. Walsh (1990). *Magnetic Field Rodent Reproductive Study* (Ontario Hydro Report # HSD-91-3), Ontario Hydro, Toronto, Ontario, Canada.

Wilkening, G. M. (1991). In *Patty's Industrial Hygiene and Toxicology*, 4th ed., Vol. 1, Part B, G. D. Clayton and F. E. Clayton, Eds., Wiley, New York, p. 700.

Williams, R. J. and E. D. Finch (1974). *Aerospace Med.*, **45**, 393–396.

Williams, W. M., S.-T. Lu, M. del Cerro, W. Hoss, and S. M. Michaelson (1984). *IEEE Trans. Microwave Theory Tech.*, **MTT-32**, 808–817.

Wilson, B. W., L. E. Anderson, D. I. Hilton, and R. D. Phillips (1981). *Bioelectromagnetics*, **2**, 371–380.

Windham, G. C., L. Fenster, S. Swan, and R. R. Neutra (1990). *Am. J. Ind. Med.*, **18**, 675–688.

Wolff, H. G. and J. D. Hardy (1947). *Physiol. Rev.*, **27**, 167–199.

Wood, A. W. (1993). *Australasian Phys. Eng.*, **16**, 1–21.

Wright, N. A., R. G. Borland, J. H. Cookson, R. F. Coward, J. A. Davies, and A. N. Nicholson (1984). *Radiat. Res.*, **97**, 468–477.

Wuest, J. R., and T. A. Gossel (1992a). *Kentucky Pharm.*, **55**(6), 185–188.

Wuest, J. R., and T. A. Gossel (1992b). *Kentucky Pharm.*, **55**(7), 213–216.

Wulf, H. C., and F. Urbach (1988). *Photodermatology*, **5**, 239–242.

Yarosh, D., L. Alas, J. Kibitel, A. O'Connor, J. Klein, and J. Morrey (1993). *J. Investig. Dermatol.*, **100**, 594.

Zaret, M. M., S. F. Cleary, B. Pasternack, M. Eisenbud, and H. Schmidt (1963). *A Study of Lenticular Imperfections in the Eyes of a Sample of Microwave Workers and a Control Population*, Rome Air Development Center, RADC-TDR-6310125, Rome, New York.

Zaynoun, S., L. A. Ali, K. Tenekjian, and A. Kurban (1985). *Clin. Exptl. Dermatol.*, **10**, 328–331.

Zigman, S. (1986). In *The Biological Effects of UVA Radiation*, Praeger, New York, pp. 252–262.

Zmudzka, B. Z., and J. Z. Beer (1990). *Photochem. Photobiol.*, **52**, 1153–1162.

Zuclich, J. A., and J. S. Connolly (1976). *Invest. Ophthalmol.*, **15**, 760–768.

Zydecki, S. (1974). In *Biologic Effects and Health Hazards of Microwave Radiation*, Polish Medical Publishers, Warsaw, pp. 306–308.

Index

729